**Progress in Systems and Control Theory**
Volume 4

M.A. Kaashoek
J.H. van Schuppen
A.C.M. Ran
Editors

# Robust Control of Linear Systems and Nonlinear Control

Proceedings of the International Symposium MTNS-89, Volume II

1990

Birkhäuser
Boston · Basel · Berlin

M.A. Kaashoek
Faculteit Wiskunde en Informatica
Vrije Universiteit
De Boelelaan 1081
1081 HV Amsterdam, The Netherlands

J.H. van Schuppen
Centre for Mathematics & Computer Science
P.O. Box 4079
1009 AB Amsterdam, The Netherlands

A.C.M. Ran
Faculteit Wiskunde en Informatica
Vrije Universiteit
De Boelelaan 1081
1081 HV Amsterdam, The Netherlands

Library of Congress Cataloging-in-Publication Data
Proceedings of the International Symposium MTNS-89 / M.A. Kaashoek, J.H. van Schuppen, A.C.N. Ran, editors.
p. cm. — (Perspectives in control theory ; v. 3–5)
"The 1989 International Symposium on the Mathematical Theory of Networks and Systems (MTNS) . . . held in Amsterdam, The Netherlands, June 19–23, 1989"—Pref.
Contents : v. 1. Realization and modelling in system theory — v. 2. Robust control of linear systems and nonlinear control — v. 3. Signal processing, scattering and operator theory, and numerical methods.
1. System analysis—Congresses. 2. Control theory—Congresses. I. Kaashoek, M. A. II. Schuppen, J. H. van. III. Ran, A. C. M. IV. International Symposium on the Mathematical Theory of Networks and Systems (9th : 1989 : Amsterdam, Netherlands) V. Series.
QA402.P766 1990
003—dc20 90-1003
CIP

Printed on acid-free paper.



ISBN-13:978-1-4612-8839-8 e-ISBN-13:978-1-4612-4484-4
DOI:10.1007/978-1-4612-4484-4

Camera-ready text provided by the individual authors.

9 8 7 6 5 4 3 2 1

# Preface

This volume is the second of the three volume publication containing the proceedings of the *1989 International Symposium on the Mathematical Theory of Networks and Systems (MTNS-89),* which was held in Amsterdam, The Netherlands, June 19-23, 1989

The International Symposia MTNS focus attention on problems from system and control theory, circuit theory and signal processing, which, in general, require application of sophisticated mathematical tools, such as from function and operator theory, linear algebra and matrix theory, differential and algebraic geometry. The interaction between advanced mathematical methods and practical engineering problems of circuits, systems and control, which is typical for MTNS, turns out to be most effective and is, as these proceedings show, a continuing source of exciting advances.

The second volume contains invited papers and a large selection of other symposium presentations in the vast area of robust and nonlinear control. Modern developments in robust control and H-infinity theory, for finite as well as for infinite dimensional systems, are presented. A large part of the volume is devoted to nonlinear control. Special attention is paid to problems in robotics. Also the general theory of nonlinear and infinite dimensional systems is discussed. A couple of papers deal with problems of stochastic control and filtering.

The titles of the two other volumes are: *Realization and Modelling in System Theory* (volume 1) and *Signal Processing, Scattering and Operator Theory, and Numerical Methods* (volume 3).

The Editors are most grateful to the about 300 reviewers for their help in the refereeing process. The Editors thank Ms. G. Bijleveld and Ms. L.M. Schultze for their professional secretarial assistance, and Mr. K. van 't Hoff for his programming support.

M.A. Kaashoek
J.H. van Schuppen
A.C.M. Ran

Amsterdam
February 1990

# Contents

## CONTRIBUTORS TO VOLUME 2

*E.H. Abed*
Dept. of Electrical Engineering, University of Maryland, College Park, MD 20742, U.S.A.

*G. Ammar*
Department of Mathematics, Northern Illinois University, DeKalb, IL 60115, U.S.A.

*B. d'Andréa-Novel*
CAI/Automatique, Ecole Nationale des Mines de Paris, 35 rue St. Honoré, F-77305 Fontainebleau, France

*J. Angeles*
Department of Mechanical Engineering, McGill University, 3840 University Street, Montreal, Quebec H3A 2A7, Canada

*K.J. Aström*
Department of Automatic Control, Lund Institute of Technology, Box 118, S-22100 Lund, Sweden

*J.A. Ball*
Virginia Polytechnic Institute Department of Mathematics, Virginia State University, Blacksburg VA-24861, U.S.A.

*G. Bastin*
Laboratoire d'Automatique, Universite Catholique de Louvain, Place Sainte-Barbe 2, B-1348 Louvain-la-Neuve, Belgium

*D. Biss*
Roche Products Ltd., Dalry, Ayrshire KA24 5JJ, United Kingdom

*S. Bittanti*
Dipartimento di Elettronica, Politecnico di Milano, Piazza Leonardi Da Vinci 32, I-20133 Milano, Italy

*J. Bontsema*
Mathematisch Instituut, RUG - Rijksuniversiteit Groningen, Postbus 800, 9700 AV Groningen, The Netherlands

*C.I. Byrnes*
Department of Systems Science and Mathematics, Washington University, Box 1040, St. Louis, MO 63130, U.S.A.

*G. Campion*
Laboratoire d'Automatique, Universite Catholique de Louvain, Place Sainte-Barbe 2, B-1348 Louvain-la-Neuve, Belgium

*C. Canudas de Wit*
LAG-ENSIEG-INPG, B.P. 46, 38402 Saint-Martin-d'Hères, France

*Daizhan Cheng*
Department of Mathematics, Texas Technical University, Lubbock, TX 79409, U.S.A.

*P. Colaneri*
Departimento di Elettronica, Politecnico di Milano, 20133 Milano, Italy

*R.F. Curtain*
Mathematisch Instituut, RUG - Rijksuniversiteit Groningen, Postbus 800, 9700 AV Groningen, The Netherlands

*S. Dancose*
Dept. of Mechanical Engineering, McGill University, 3840 University Street, Montreal, Quebec H3A 2A7, Canada

*W. Dayawansa*
Department of Mathematics, Texas Technical University, Lubbock, TX 79409, U.S.A.

*D.C. Deno*
Department EECS, University of California, Berkeley CA 94720, U.S.A.

*J. Doyle*
California University of Technology, Caltech 116-81, Pasadena, CA 91125, U.S.A.

*A. Elsayed*
Department of Electrical and Control Engineering, Industrial Control Unit, 204 George Street, Glasgow G1 1XW, United Kingdom

*J.C. Engwerda*
Faculteit der Economische Wetenschappen, Katholieke Universiteit Brabant, Postbus 90153, 5000 LE Tilburg, The Netherlands

*R.J.P. Figueiredo*
Dept. of Electrical and Computer Engineering, Rice University, Houston, TX 77251, U.S.A.

*N. Fixot*
Laboratoire d'Automatique, ENSIEG-INPG, B.P. 46, F-38402 Saint-Martin-d'Hères, France

*C. Foias*
Department of Mathematics, Indiana University, Bloomington, IN 47405, U.S.A.

*Y.K. Foo*
School of Electrical and Electronic Engineering, Nanyang Technological Institute, Nanyang Avenue, Singapore 2263, Singapore

*M.D. Fragoso*
Dept. of Research and Development, National Laboratory for Scientific Computation - LNCC/CNPq, Rua Lauro Muller, 455, 22.290, Rio de Janeiro, Brasil

*M. Fujita*
Dept. of Electrical and Computer Engineering, Kanazawa University, Kodatsuno 2-40-20, Kanazawa 920, Japan

*A.H.W. Geerts*
Faculteit der Wiskunde en Informatica, TUE - Technische Universiteit Eindhoven, Postbus 513, 5600 MB Eindhoven, The Netherlands

*D.S. Gilliam*
Department of Mathematics, Texas Technical University, Lubbock, TX 79409, U.S.A.

*K. Glover*
Department of Engineering, University of Cambridge, Trumpington Street, Cambridge CB2 1PZ, United Kingdom

*A. Gombani*
LADSEB - CNR, Corso Stati Uniti 4, I-35020 Padova, Italy

*M. Green*
Department Electrical Engineering, Imperial College of Science and Technology, London SW7 2BT, United Kingdom

*M.J. Grimble*
Department of Electrical and Control Engineering, Industrial Control Unit, 204 George Street, Glasgow G1 1XW, United Kingdom

*D.W. Gu*
Dept. of Engineering Science, University of Oxford, Oxford OX1 3PJ, United Kingdom

*Guanrong Chen*
Dept. of Electrical and Computer Engineering, Rice University, Houston, TX 77251, U.S.A.

*Han Zheng-zhi*
Department of Automatic Control, Shanghai Jiao-Tong University, Shanghai 200030, People's Republic of China

*S. Hara*
Dept. of Control Engineering, Tokyo Institute of Technology, Tokyo 152, Japan

*M.L.J. Hautus*
Faculteit der Wiskunde, TUE - Technische Universiteit Eindhoven, Postbus 513, 5600 MB Eindhoven, The Netherlands

*J.W. Helton*
Department of Mathematics, University of California, San Diego, CA 92110, U.S.A.

*D.J. Hill*
Dept. of Elect. Eng. & Computer Science, University of Newcastle, Rankin Drive, 2308 Newcastle, N.S.W., Australia

*D. Ho*
Department of Applied Mathematics, City Polytechnic of Hong Kong, Hong Kong, Hong Kong

*C.V. Hollot*
Dept. of Electrical and Computer Engineering, University of Massachusetts, Amherst, MA 01003, U.S.A.

*Hu Ting Shu*
Department of Automatic Control, Shanghai Jiao-Tong University, Shanghai, People's Republic of China

*H.J.C. Huijberts*
Faculteit der Toegepaste Wiskunde, UT - Universiteit Twente, Postbus 217, 7500 AE Enschede, The Netherlands

*M. Ikeda*
Department of Systems Engineering, Kobe University, Rokkodai Nada, Kobe 657, Japan

*H. Inaba*
Department of Information Sciences, Tokyo Denki University Hatoyama-Mach, Hiki Gun, Saitama 350-03, Japan

*B. Jakubczyk*
Institute of Mathematics, Polish Academy of Sciences, Sniadeckich 8, 00-950 Warsaw, Poland

*R. Kadiyala*
Department EECS, University of California, Berkeley CA 94720, U.S.A.

*N. Kalouptsidis*
Dept. of Physics, University of Athens, Panepiotimiopoli, 157 71 Athens, Greece

*E.M. Kasenally*
Department of Electrical Engineering, Imperial College of Science and Technology, Exhibition Road, London SW7 2BT, United Kingdom

*Kehui Wei*
DFVLR, Oberpfaffenhofen, D-8031 Wessling, B.R.D.

*H. Kimura*
Department of Mechanical Engineering, Osaka University, Suita 2-1 Yamada-oka, Osaka 565, Japan

*P.V. Kokotovic*
Coordinated Science Laboratory, University of Illinois. 1101 W. Springfield, Urbana, IL 61801, U.S.A.

*P. Kostelec*
Department of Mathematics, Darmouth College, Hannover, NH 03755, U.S.A.

*St. Kotsios*
Department of Physics, University of Athens, 157 71 Athens, Greece

*P.R. Kumar*
Coordinated Science Laboratory, University of Illinois, 1101 W. Springfield Avenue, Urbana, IL 61801, U.S.A.

*J. Lévine*
CAI/Automatique, Ecole Nationale des Mines de Paris, 35 Rue St. Honoré, F-77305 Fontainebleau, France

*D.J.N. Limebeer*
Department of Electrical Engineering, Imperial College of Science and Technology, Exhibition Road, London SW7 2BT, United Kingdom

*Lin Huang*
Dept. of Electrical and Computer Engineering, University of Massachusetts, MA 01003 Amherst, U.S.A.

*H. Logemann*
Institut fuer Dynamische Systeme, Universitaet Bremen, Postfach 330 440, D-2800 Bremen 33, B.R.D.

*J. Lund*
Dept. of Mathematics, Montana State University, Bozeman, MT 59717, U.S.A.

*A.M. Makowski*
Department of Electrical Engineering, University of Maryland, College Park, MD 20742, U.S.A.

*I.M.Y. Mareels*
Dept. of Elect. Eng. & Comp. Sience, University of Newcastle, Rankin Drive, Newcastle, N.S.W. 2308, Australia

*C. Martin*
Department of Mathematics, Texas Technical University, Lubbock, TX 79409, U.S.A.

*G.B. di Masi*
LADSEB - CNR, Corso Stati Uniti 4, I-35020 Padova, Italy

*R.M. Murray*
Department EECS, University of California, Berkeley CA 94720, U.S.A.

*D. Mustafa*
Department of Electrical Engineering L.I.D.S., M.I.T. - Massachusetts Institute of Technology, Cambridge MA 02139, U.S.A.

*H. Myung*
Department of Mathematics, University of Northern Iowa, Cedar Falls, IA 50614, U.S.A.

*R. Ortega*
DEPFI-UNAM, Ciudad Universitaria, P.O. Box 70-256, 04510 Mexico, D.F., Mexico

*N. Otsuka*
Department of Information Sciences, Hatoyama-Mach, Tokyo Denki University, Hiki-Gun, Saitama 350-03, Japan

*T.J. Owens*
Dept. of Electrical Eng., Brunel University Uxbridge, Middlesex UB8 3PH, United Kingdom

*J.R. Partington*
Systems Engineering Department, Cambridge University, Trumpington Street, CB2 1PZ Cambridge, United Kingdom

*I.R. Petersen*
Department of Electrical Engineering, Australian Defence Force Academy, University of New South Wales, Canberra ACT 2600, Australia

*K.S.J. Pister*
Department EECS, University of California, Berkeley CA 94720, U.S.A.

*S. Pohjolainen*
Dept. of Mathematics, Tampere University of Technology, P.O. Box 527, SF-33101 Tampere 10, Finland

*I. Postlethwaite*
Department of Engineering, University of Leicester, Leicester LE1 7RH, United Kingdom

*L. Praly*
Centre d'Automatique et Informatique, ENSMP, 35 Rue St. Honoré, F-77305 Fontainebleau, France

*A.J. Pritchard*
Control Theory Centre, Mathematics Institute, University of Warwick, CV4 7AL Coventry, United Kingdom

*A. Rantzer*
Optimization and Systems Theory, Royal Institute of Technology, S-100 44 Stockholm, Sweden

*W. Respondek*
Institute of Mathematics, Polish Academy of Sciences, Sniadeckich 8, 00-950 Warsaw, Poland

*A.A. Sagle*
Department of Mathematics, University of Hawaii-Hilo, Hilo, Hawaii 96720-4091, U.S.A.

*S.S. Sastry*
Dept. of Elect. Eng. & Computer Sci., University of California, Berkeley, CA 94720, U.S.A.

*L. Saydy*
Dept. of Electrical Engineering, University of Maryland, College Park, MD 20742, U.S.A.

*U. Shaked*
Dept. of Electronic Communications, Control and Computer Science, Tel Aviv University, Ramat-Aviv, tel-Aviv, Israel

*Shi Song-Jiao*
Dept. of Automatic Control, Shanghai Jiao-Tong University, Shanghai, People's Republic of China

*E. Shimemura*
Dept. of Electrical Engineering, Waseda University, Okubo 3-4-1, Shinjuku, Tokyo 160, Japan

*J.M. Soethoudt*
Faculty of Mathematics & Computer Science, Technical University Eindhoven, P.O. Box 513, 5600 MB Eindhoven, The Netherlands

*Y.C. Soh*
School of Electrical & Electronic Eng., Nanyang Technological Institute, Nanyang Avenue, 2263 Singapore, Singapore

*E.D. Sontag*
Department of Mathematics, Rutgers University, New Brunswick, NJ 08903, U.S.A.

*C.E. de Souza*
Dept. of Electrical and Computer Engineering, University of Newcastle, New South Wales, 2308, Australia

*R.B. Sowers*
Dept. of Electrical Engineering, University of Maryland, College Park, MD 20742, U.S.A.

*A. de Stefano*
Department of Mathematics, Darmouth College, NH 03755 Hannover, U.S.A.

*A.A. Stoorvogel*
Fac. Wiskunde en Informatica, TUE - Technische Universiteit Eindhoven, P.O.Box 513, 5600 MB Eindhoven, The Netherlands

*Hsiang-Ling Tan*
Dept. of Systems Engineering, Kobe University, Rokkodai, Nada, Kobe 657, Japan

*A. Tannenbaum*
Department of Electrical Engineering, Technion, Haifa 32000, Israel

*K. Tchoń*
Institute of Engineering Cybernetics, Technical University of Wroclaw, ul. Z. Janiszewskiego 11/17, 50-372 Wroclaw, Poland

*A. Teel*
Department EECS, University of California, Berkeley CA 94720, U.S.A.

*Tho Pham*
Dept. of ECE, University of Texas, Austin, TX 78712, U.S.A.

*A.L. Tits*
Dept. of Electrical Engineering, University of Maryland, College Park, MD 20742, U.S.A.

*P. Tomei*
Dipartimento di Ingegneria Elettronica, Seconda Universita di Roma 'Tor Vergata', Via O. Raimondo, 00173 Roma, Italy

*S. Townley*
Control Theory Centre Mathematics Institute, University of Warwick, CV4 7AL Coventry, United Kingdom

*H. Trentelman*
Faculteit Wiskunde en Informatica, TUE - Technische Universiteit Eindhoven, Postbus 513, 5600 MB Eindhoven, The Netherlands

*M.C. Tsai*
Dept. of Engineering, University of Leicester, Leicester LE1 7RH, United Kingdom

*J. Tsinias*
Dept. of Mathematics, National Technical University, Zografou Campus, 157 73 Athens, Greece

*K. Uchida*
Dept. of Electrical Engineering, Waseda University, Okubo 3-4-1, Shinjuku, Tokyo 160, Japan

*J.T. van der Vaart*
Faculteit Wiskunde, RUG - Rijksuniversiteit Groningen, Postbus 800, 9700 AV Groningen, The Netherlands

*M.S. Verma*
Dept. of ECE, University of Texas, Austin, TX 78712, U.S.A.

*M. Vidyasagar*
Dept. of Electrical Engineering, University of Waterloo, Waterlo, Ontario N2L 3H1, Canada

*D.I. Wallace*
Department of Mathematics, Darmouth College, Hanover NH 03755, U.S.A.

*L.L.M. van der Wegen*
Faculteit Toegepaste Wiskunde, UT - Universiteit Twente, Postbus 217, 7500 AE Enschede, The Netherlands

*J.A. Wolf*
Department of Mathematics, University of California, Berkeley, CA 94720, U.S.A.

*K. Woodgate*
Department of Electrical Engineering, University of Strathclyde, 204 George Street, Glasgow G1 1XW, United Kingdom

*Y. Yamamoto*
Dept. of Appl. Systems Science Faculty of Engineering, Kyoto University, Kyoto 606, Japan

*Yu Tang*
DEPFI-UNAM, Ciudad Universitaria, P.O. Box 70-256, 04510 Mexico, D.F., Mexico

*Zhang Zhong-jun*
Dept. of Automatic Control, Shanghai Jiao-Tong University, Shanghai 200030, People's Republic of China

*Zheng Yu-Fan*
Dept. of Mathematics, East-China Normal University, 3663 Northern Zhongshan Road, Shanghai 200062, People's Republic of China

*Zhong-Ling Xu*
Department of Electrical and Computer Engineering, University of Massachusetts, Amherst MA 01003, U.S.A.

*S.Q. Zhu*
Department of Mathematics and Computer Science, TUE - Technische Universiteit Eindhoven, P.O. Box 513, 5600 MB Eindhoven, The Netherlands

# NONLINEAR $H^\infty$ CONTROL THEORY:

# A LITERATURE SURVEY

JOSEPH A. BALL AND J. WILLIAM HELTON

Abstract

The central problem of $H^\infty$-control theory roughly is to optimize (by the choice of compensator in a standard feedback configuration) some worst case (i.e. infinity norm) measure of performance while maintaining stability. For the linear, time-invariant, finite-dimensional case, rather complete state space solutions are now available, and work has begun on understanding less restrictive settings. A recent new development has been the establishment of a connection with differential games and the perception of the $H^\infty$ -problem as formally the same as the earlier well established linear quadratic regulator problem, but with an indefinite performance objective. In this article we review the current state of the art for nonlinear systems. The main focus is on the approach through a global theory of nonlinear $J$-inner-outer factorization and nonlinear fractional transformations being developed by the authors. It turns out that the critical points arising naturally in this theory can also be interpreted as optimal strategies in a game-theoretic interpretation of the control problem.

## 1. INTRODUCTION.

Many control problems fit into the paradigm depicted in Figure 1.1.

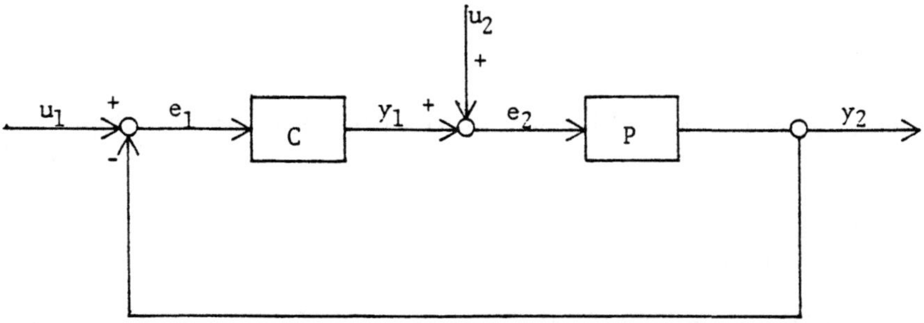


*Figure 1.1*

Here $P$ stands for the plant which one is stuck with and $C$ is the compensator to be designed. Usually the signals $u_1$, $u_2$, $y_1$, $y_2$, $e_1$, $e_2$ are taken to be vector valued functions of compatible sizes of the real variable $t$ representing time, with components in the extended $L^2$-space $L^2_e[0,\infty)$ of functions which are square integrable on any finite subinterval of the real line. Here $C$ and $P$ represent input-output maps with domain and range spaces equal to vector-valued $L^2_e[0,\infty)$ of appropriate sizes. Key properties for a general input-output (IO) map $H : u \to y$ are:

1) causality: $P_\tau H = P_\tau H P_\tau$ for all $\tau \geq 0$.

2) time-invariance: $[\mathrm{Hu}]\,(\cdot - \tau) = [Hu(\cdot - \tau)]$ for all $\tau \geq 0$.

3) stability: $\|P_\tau H(u)\|^2 \leq \delta(H)\|P_\tau u\|^2$ for some $\delta(H) < \infty$ for all $\tau \geq 0$.

Here $P_\tau : L^2_e[0,\infty) \to L^2_e[0,\infty)$ is defined by

$$[P_\tau u](t) = \begin{cases} u(t), & 0 \leq t \leq \tau \\ 0, & t > \tau \end{cases}$$

If $H$ is causal, linear, time-invariant and in addition *finite-dimensional* (LTIFD), then after Laplace transformation the input-output map $H$ can be viewed as multiplication by a proper rational matrix function $\hat{H}(s)$ (called the transfer function) on a space of functions analytic on some right half plane in the complex plane. In this case stability of $H$ corresponds to $\hat{H}$ having all poles in the left half plane.

The system $\Sigma$ in Figure 1.1 is said to be well-posed if it is possible to solve the system of equations on vector-valued $L^2_e[0,\infty)$ depicted by Figure 1.1

$$u_1 = e_1 + Pe_2$$

$$u_2 = -Ce_1 + e_2$$

for $e_1$, $e_2$ in terms of $u_1$, $u_2$. The resulting input-output map we then denote by $H_{\begin{bmatrix} e_1 \\ e_2 \end{bmatrix}, \begin{bmatrix} u_1 \\ u_2 \end{bmatrix}}$ : $\begin{bmatrix} u_1 \\ u_2 \end{bmatrix} \to \begin{bmatrix} e_1 \\ e_2 \end{bmatrix}$.

In the LTIFD case well-posedness means that the transfer function $(I + \hat{P}\hat{C})^{-1}$ is well-defined. Internal stability of the system $\Sigma$ amounts to the stability of the IO map $H_{\begin{bmatrix} e_1 \\ e_2 \end{bmatrix}, \begin{bmatrix} u_1 \\ u_2 \end{bmatrix}}$. In the LTIFD case, internal stability is equivalent to the four transfer functions $(I + \hat{P}\hat{C})^{-1}$, $-(I + \hat{P}\hat{C})^{-1}\hat{P}$, $(I + \hat{C}\hat{P})^{-1}\hat{C}$ and $(I + \hat{C}\hat{P})^{-1}$ being proper with all poles in the open left half plane.

The standard problem in $H^\infty$-control theory is to choose a compensator $C$ which optimizes some measure of performance subject to the side constraint that the associated closed loop system $\Sigma = \Sigma(P, C)$ be internally stable. In the linear $H^\infty$-theory the measure of performance $\mathcal{P}(\Sigma)$ is taken to be the induced operator norm of one of the IO maps $H$ associated with the system $\Sigma$ :

$$\mathcal{P}_H(\Sigma) = sup\{\|H(u)\|^2 : u \in L^2[0,\infty), \|u\|_2 \leq 1\}$$

For the LTIFD case, $\mathcal{P}_H(\Sigma)$ is simply the $H^\infty$-norm of the associated transfer function $\hat{H}(s)$.

Various examples for the choice of $H$ are:

*Weighted sensitivity:* $H = W \circ H_{e_1,u_1}$ (*$W$ = a weighting function*)

*Tracking:* $H = H_{y_2,u_1}$

*Robust stability with respect to additive plant perturbations:* $H = H_{y_1,u_1}$

In the definition of all these IO maps, the extraneous input $u_2$ is taken to be zero. All of these can be manipulated so as to fit the general paradigm depicted in Figure 1.2

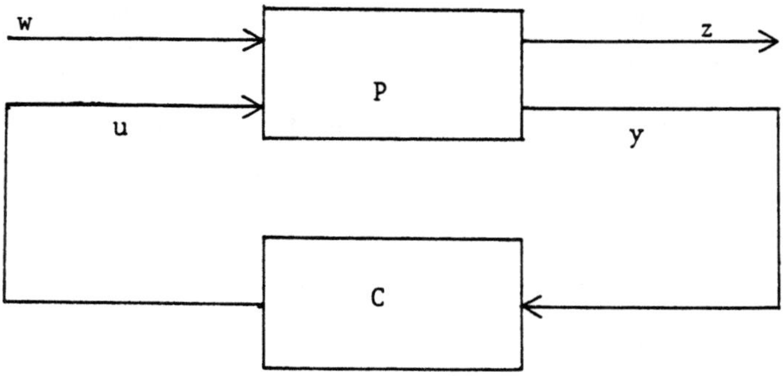


*Figure 1.2*

In this configuration, well-posedness means that the IO map $H_{\begin{bmatrix} z \\ y \\ v_2 \end{bmatrix}, \begin{bmatrix} w \\ u \\ v_1 \end{bmatrix}}$ in Figure 1.3 is well-defined, internal stability means that $H_{\begin{bmatrix} z \\ y \\ v_2 \end{bmatrix}, \begin{bmatrix} w \\ u \\ v_1 \end{bmatrix}}$ is stable, and the standard $H^\infty$-problem is to choose the compensator $C$ so as to minimize the induced operator norm of the IO map $H_{z,w}$ subject to the constraint of internal stability. All these ideas in greater detail and with more engineering motivation can be found in [D], [FD], [Fr].

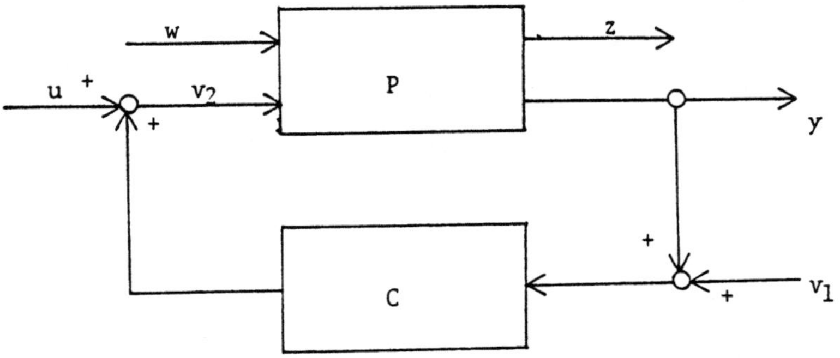


*Figure 1.3*

**2. THE BEGINNINGS OF A NONLINEAR THEORY.**

By a nonlinear $H^\infty$ control theory, we mean a theory for choosing a (possibly) nonlinear compensator $C$ which optimizes some worst case measure of performance $\mathcal{P}(\Sigma(P,C))$ for the system $\Sigma = \Sigma(P,C)$ as in Figure 1.1 or, more generally, Figure 1.2, subject to the side condition of internal stability, where now the plant $P$ may be nonlinear. Before such an ambitious theory was tackled, there appeared in the literature various preliminary probings. We mention three such thrusts.

a. NONLINEAR COPRIME FACTORIZATION

A well-known and well established approach for the linear $H^\infty$-problem is to use as a first step the Youla-Bongiorno-Lu parameterization of all stabilizing compensators. This is a powerful tool which has two advantages: (1) the free parameter sweeps through the linear space of all stable rational matrix functions, and (2) the performance transfer function becomes affine rather than linear fractional in the free parameter. The main ingredient for the Youla parameterization is a coprime factorization of the plant $P$ (see [Fr]). There has been a lot of work in recent years on extending this first step to nonlinear plants, i.e. on understanding coprime factorization for nonlinear plants and on using such a coprime factorization to parameterize all stabilizing (possibly nonlinear as well as time-varying) compensators for the nonlinear plant (see e.g. [DK, H, Kr, S, V] and other articles in these Proceedings). Of course this first step does not incorporate any measure of performance and hence is not yet from our point of view a nonlinear version of $H^\infty$ control theory.

b. NONLINEAR SOLUTIONS OF LINEAR PROBLEMS. An intermediate step in the understanding of performance for nonlinear systems is to understand the role of nonlinear and/or time varying solutions for linear time-invariant $H^\infty$ problems. This problem was analyzed by Khargonekar and his associates in a series of papers (see e.g. [KP]). A basic principle coming out of these studies was that for linear systems with no plant uncertainty one does not improve performance by choosing a nonlinear compensator, but in the presence of plant uncertainly it indeed may be possible to improve performance by using a nonlinear controller. In later work Ball-Helton-Sung [BHS] extracted some of the mathematical ideas of these studies so as to apply to interpolation problems independent of a control theory context, and made a few extensions.

c. MORE GENERAL PERFORMANCE MEASURES.

Still another direction is to consider other more realistic performance measure more complicated than the operator or infinity norms usually considered in $H^\infty$-control theory. This leads to interesting new problems and connections with other kinds of mathematics even for SISO linear systems, and is an ongoing project of the second author. For an overview of the current status of this area, see the survey article of the second author in these Proceedings.

## 3. PERFORMANCE OPTIMIZATION FOR NONLINEAR SYSTEMS: A GLOBAL APPROACH.

In this section we would like to present a guide to the reader of the approach to nonlinear $H^\infty$-control theory being pursued by us. Some work has already appeared in published or preprint form [BH2-BH8] and more is in preparation [BH9].

Consider the system $\Sigma$ as in Figure 1.2 where the compensator $C$ and the plant $P$ are assumed to be causal and time-invariant but may be nonlinear. Well-posedness and internal stability for the system $\Sigma$ are defined as in Section 1; for the nonlinear setting we have IO maps but never mention transfer functions; we work entirely in the time domain. For the measure of performance of the system $\Sigma$ we use the following nonlinear analogue of the induced operator norm of the IO map $H_{w,z}$ :

$$\mathcal{P}(\Sigma)^2 = sup\{\|H_{z,w}(w)\|_2^2/\|w\|_2^2 : w \in L^2[0,\infty)\} \tag{3.1}$$

Note that $\mathcal{P}(\Sigma) \leq \gamma$ if and only if

$$\hat{\mathcal{P}}_\gamma(\Sigma) = sup\{\|z\|_2^2 - \gamma^2\|w\|_2^2 : w \in L^2[0,\infty),\ z = H_{z,w}(w)\} \leq 0 \tag{3.2}$$

This formulation makes contact with game theory ideas.

Our approach to the nonlinear problem associated with the measure of performance (3.1) follows the approach for the linear case which uses $J$-inner-outer factorization. For purposes of the discussion here, we break this approach up into three steps, namely:

STEP 1. Reduction of the control problem to a $J$-inner-outer factorization problem. This involves justification of the use of the $J$-inner factor to parameterize all achievable performance transfer functions and of the outer factor to parameterize all stabilizing compensators for which the desired level $\gamma$ of performance is achieved.

STEP 2. Computation of the $J$-inner and outer factors.

We now discuss each step in more detail.

STEP 1. In the linear case, a standard procedure (see [Fr]) reduces the control problem to the model matching problem. For problems of the first kind (also known as the 1-block case), the model matching problem mathematically corresponds to classical Nevanlinna-Pick interpolation. The approach of [BH1] to interpolation then reduces the problem to one of $J$-inner-outer factorization, assuming that the classical Pick matrix positive definiteness test for existence of solutions is meant. For problems of the second kind (also known as the 2-block case), the same approach applies, but the $J$-inner-outer factorization problem is for a rectangular (rather than square) matrix function, and the associated $J$-inner factor is rectangular. For problems of the third kind (known as the 4-block case), the approach generalizes but is more complicated (see [BC] and [GGLD]). It has recently been observed by [HK], at least for the 2-block case, that $J$-inner-outer factorization can be applied at a much earlier stage; thereby they prove that one can bypass the Youla parameterization and obtain a parameterization of the performing, stabilizing compensators more directly.

In the nonlinear case a nonlinear analogue of the Youla parameterization exists [AD] which at least in special situations can be used to reduce the original control problem to a nonlinear model matching problem. For the 2-block case, the existence of a nonlinear $J$-inner-outer factorization leads to a parameterization of many solutions of the control problem (see [BH2], [BH4]); it is unknown at present if this parameterization gives all solutions. Understanding the action of the nonlinear fractional transformation involved requires a systematic analysis of the interconnection of nonlinear passive circuits and the use of degree theory to prove well-posedness. This is sketched in [BH3] with complete details in [BH7].

We expect that the control problem can be solved directly at an earlier stage via $J$-inner-outer factorization, just as in [HK] for the linear case. Understanding the full 4-block case for the nonlinear problem remains an open problem for future research.

STEP 2. For the linear case, very elegant state space solutions for $J$-inner and outer factors of a given rational matrix function and solutions of related interpolation and $H^\infty$-control problems have now appeared, e.g. [G, BR, BGR, Ki], even in the setting of the 4-block problem [BC, GD, GGLD]. For the nonlinear case, we follow the approach of [BR, BGR]. For mathematical convenience we consider the problem in discrete time; this amounts to assuming that the time variable $t$ assumes values in the nonnegative integers $\mathbb{Z}^+$ rather than the nonnegative real line $\mathbb{R}^+$. Thus $L_e^2[0,\infty)$ becomes the $l_e^{2+}$ of sequences indexed by $\mathbb{Z}^+$ and $L^2[0,\infty)$ becomes the space $\ell^{2+}$ of such sequences which are square-summable. We assume that we have an IO map $\mathcal{F}^\Sigma$ on $\ell^{2+}$ given by state space equations for the system

$$\Sigma : x_{n+1} = F(x_n,\ u_n),\ x_0 = 0$$

$$y_n = G(x_n,\ u_n)$$

and we wish to compute a state space representation for a system

$$\Theta : x_{n+1} = f(x_n,\ u_n),\ x_0 = 0$$

$$y_n = g(x_n,\ u_n)$$

which generates a $J$-inner factor IO map $\mathcal{F}^{\Theta}$. We assume here that $\Sigma$ is stable (in particular, $\mathcal{F}^{\Sigma}$ maps $\ell^{2+}$ into itself); a thorough understanding of the unstable case awaits further research. By a $J$-inner-outer factorization we mean a factorization of IO maps such that $\mathcal{F}^{\Sigma} = \mathcal{F}^{\Theta} \circ \mathcal{F}^{Q}$

(i) $\mathcal{F}^{\Theta}$ is $J$-inner

(ii) $\mathcal{F}^{Q}$ is outer i.e. both $\mathcal{F}^{Q}$ and $[\mathcal{F}^{Q}]^{-1}$ are stable.

Here $J$ inner corresponds to the physical notion of energy conserving and stable. The energy function $\rho_J$ is defined on $\vec{u} \in \ell^{2+}$ by

$$\rho_J(\vec{u}) = < J\vec{u}, \vec{u} >_{\ell^{2+}}$$

where $J$ is a constant signature matrix on vector valued $\ell^{2+}$. We say that $\mathcal{F}^{\Theta}$ is <u>$J$-lossless</u> provided that

$$\rho_J(\mathcal{F}^{\Theta}(\vec{u})) = \rho_J(\vec{u})$$

for $\vec{u} \in \ell^{2+}$ and that $\mathcal{F}^{\Theta}$ is <u>$J$-passive</u> if

$$\rho_J(\mathcal{F}^{\Theta}(\vec{u})) \le \rho_J(P_\tau(\vec{u}))$$

for all $\vec{u} = \ell^{2+}$ and each projection $P_\tau$ on to $\ell^2[0,\tau]$. Finally $\mathcal{F}^{\Theta}$ is $J$-inner if it is both $J$-lossless and $J$-passive. The construction of the system $\Theta$ (from which it is easy to find $Q$ via $Q = \Theta^{-1} * \Sigma$) breaks down into several steps.

(1)*Identification of the left null dynamics of* $\Theta$ *with the left null dynamics of* $\Sigma$. In the scalar linear case, this means that the transfer functions $\hat{\Theta}(z)$ and $\hat{\Sigma}(z)$ should have the same zeros in the unit disk.

(2)*Construction of the right pole dynamics of* $\Theta$. In the scalar linear case, this amounts to observing that the poles of $\hat{\Theta}(z)$ can be determined via Schwarz reflection from the zeros of $\hat{\Theta}(z)$ since $\hat{\Theta}(z)$ is to have modulus 1 values on the unit circle. A similar idea can be made precise for the linear MIMO case ([BR], [BGR]). In the nonlinear case, this step involves computation for each state $x$ of the critical point $u$ for the restriction of the quadratic form $\rho_J$ to the manifold of output strings $\{\mathcal{F}_x^\Sigma(\vec{u}) : \vec{u} \in \ell^{2+}\}$ associated with stable inputs $\vec{u} \in \ell^{2+}$ and initial state $x$.

(3)*Construction of the full state space representation* $(f,g)$. In the linear case, this amounts to a single $J$-Cholesky factorization. In the nonlinear case, one must perform a nonlinear Morse theoretic congruence with the state variable as parameter in a smooth way.

(4)*Check for passivity of* $\Theta$. In the linear case this amounts to checking that the solution to a certain Stein equation is positive definite. In the nonlinear case, the analogous object, an energy function on the state space which satisfies a nonlinear Stein equation, must be positive.

The formal recipe and flow chart together with general conditions for its validity is laid out in [BH5]; some of the ideas involved in the derivation as sketched in the above steps are given in [BH6]. Full details appear in [BH8].

Our constructions all assume that one has found critical points for a certain energy function associated with a control or factorization problem. The closest analogue of these critical points in engineering are saddle points or max.-min. points which occur in game theory (see [BH9]); note a max.-min. point of a function is always a critical point of that function. We thank D. Limebeer for introducing us to game theory in connection with linear control [LAKG, T1, T2].

In [BH9] we show that max.-min. points of the energy function lead directly to particular solutions of the equivalent control problem. In [BH8] we make more global assumptions and actually parameterize a large class of controllers in order to produce one controller.

## 4. PERFORMANCE OPTIMIZATION FOR NONLINEAR SYSTEMS: OTHER APPROACHES.

We mention that to our knowledge there are two other approaches to nonlinear $H^\infty$

control theory which have appeared in the literature.

The approach of Foias and Tannenbaum [BFHT1] [FT1], [FT2] is based on a power series representation for the plant. One can linearize the $n$-th $n$-homogeneous term of the power series by extending its action from the diagonal to the $n$-fold tensor product of the space of input signals. One can then do an iterative procedure of applying the linear commutant (more properly, intertwining) lifting theorem to each term of the power series. The resulting norm estimates for the nonlinear lift are not as sharp as in the linear case and the resulting power series representation for the lift may converge on a smaller ball than the power series for the original plant. The same iterative approach using linearization of the $n$-th term of the power series on the tensor space and iterating can also be used to yield a local nonlinear Beurling-Lax theorem [BFHT2]. In applications to the control problem [FT1, FT2], the measure of performance must be tailored to fit the iterated commutant lifting approach.

Chen and deFigueiredo [CdeF,deFC] localize the control problem to small balls and to plants having a simple parameterized form. Using a Lipschitz-norm measure of performance, they are able to reduce the control problem to a standard several-variable nonlinear optimization problem.

Research on this paper was supported in part by the National Science Foundation and the Air Force Office of Scientific Research.

## REFERENCES

[AD] V. Anantharam and C. A. Desoer, On the stabilization of nonlinear systems, IEEE Trans. Aut. Control AC-29 (1984), 569-573.

[BC] J. A. Ball and N. Cohen, Sensitivity minimization in $H^\infty$ norm: parameterization of all solutions, Int. J. Control 46 (1987), 785-816.

[BFHT1] J. A. Ball, C. Foias, J. W. Helton and A. Tannenbaum, On a local nonlinear commutant lifting theorem, Indiana Univ. Math. J. 36 (1987), 693-709.

[BFHT2] J. A. Ball, C. Foias, J. W. Helton and A. Tannenbaum, A Poincare-Dulac approach to a nonlinear Beurling-Lax-Halmos theorem, J. Math. Anal. Appl. 139 (1989), 496-514.

[BGR] J. A. Ball, I. Gohberg and L. Rodman, Two-sided Nudelman interpolation problem for rational matrix functions, to appear in Marcel-Dekkar volume dedicated to Mischa Cotlar.

[BH1] J. A. Ball and J. W. Helton, A Beurling-Lax theorem for the Lie group $U(m,n)$ which contains most classical interpolation, J. Operator Theory 9 (1983), 107-142.

[BH2] J. A. Ball and J. W. Helton, Sensitivity bandwidth optimization for nonlinear feedback systems, in Analysis and Control of Nonlinear Systems (ed. by C. I. Byrnes, C. F. Martin and R. E. Saeks), North Holland (1988), pp. 123-129.

[BH3] J. A. Ball and J. W. Helton, Well-posedness of nonlinear causal feedback systems, Proc. 26th IEEE Conf. on Decision and Control, Los Angeles (1987), pp. 152-154.

[BH4] J. A. Ball and J. W. Helton, Shift invariant manifolds and nonlinear analytic function theory, Integral Equations and Operator Theory 11 (1988), 615-725.

[BH5] J. A. Ball and J. W. Helton, Factorization of nonlinear systems: toward a theory for nonlinear $H^\infty$ control, Proc. 27th IEEE Conf. on Decision and Control, Austin (1988), 2376-2381.

[BH6] J. A. Ball and J. W. Helton, Interpolation problems for null and pole structure of nonlinear systems, Proc. 27th IEEE Conf. on Decision and Control, Austin (1988), 14-19.

[BH7] J. A. Ball and J. W. Helton, Interconnection of nonlinear causal systems, IEEE Trans. Aut. Control, to appear.

[BH8] J. A. Ball and J. W. Helton, Inner-outer factorization of nonlinear operators, preprint.

[BH9] J. A. Ball and J. W. Helton, $H^\infty$-control of nonlinear plants: connections with differential games, in preparation.

[BHS] J. A. Ball, J. W. Helton and C. H. Sung, Nonlinear solutions of Nevanlinna-Pick interpolation problems, Mich. Math. J. 34 (1987), 375-389.

[BR] J. A. Ball and A. C. M. Ran, Global inverse spectral problems for rational matrix functions, Linear Alg. Appl. 86 (1987), 237-282.

[CdeF] G. Chen and R. J. P. deFigueiredo, On robust stabilization of nonlinear control systems, Systems & Control Letters 12 (1989), 373-379.

[deFC] R. J. P. deFigueiredo and G. Chen, Optimal disturbance rejection for nonlinear control systems, IEEE Trans. Aut. Control, to appear.

[D] J. C. Doyle, Lecture Notes for ONR (Honeywell Workshop on Advances in Multivariable Control, Minneapolis, MN (1984).

[DK] C. A. Desoer and M. G. Kabuli, Right factorization of a class of time varying nonlinear systems, IEEE Trans. Aut. Control 33 (1988), 755-757.

[FT1] C. Foias and A. Tannenbaum, Weighted optimization theory for nonlinear systems, SIAM J. Control and Opt. 27 (1989), 842-860.

[FT2] C. Foias and A. Tannenbaum, Iterative commutant lifting for systems with rational symbol, in The Gohberg Anniversary Collection Vol. II: Topics in Analysis and Operator Theory, OT41 Birkhauser, Basel, (1989), pp. 255-277.

[Fr] B. Francis, A Course in $H_\infty$ Control Theory, Springer-Verlag, Berlin-New York, 1987.

[FD] B. Francis and J. Doyle, Linear control theory with an $H_\infty$ optimality criterion, SIAM J. Control Opt. 25 (1987), 815-844.

[GGLD] M. Green, K. Glover, D. Limebeer and J. Doyle, A $J$-spectral factorization approach to $H_\infty$ control, preprint.

[G] K. Glover, All optimal Hankel norm approximations of linear multivariable systems and their $L_\infty$ error bounds, Int. J. Control 39 (1984), 1115-1193.

[GD] K. Glover and J. C. Doyle, State-space formulas for stabilizing controllers, Systems and Control Letters 11 (1988), 167-172.

[HK] S. Hara and R. Kondo, Characterization and computation of $H_\infty$-optimal controllers in the state space, Proc. 27th IEEE Conf. on Decision and Control, Austin (1988), 20-25.

[H] J. Hammer, Fractional representations of nonlinear systems: a simplified approach, Int. J. Control 46 (1987), 455-472.

[Ki] H. Kimura, Conjugation, interpolation and model-matching in $H^\infty$, Int. J. Control (1989), 269-307.

[KP] P. R. Khargonekar and K. R. Poola, Uniformly optimal control of linear time-invariant plants: nonlinear time-varying controllers, Systems Control Letters 6 (1986), 303-308.

[Kr] A. J. Krener, Nonlinear controller design via approximate normal forms, in Proc. IMA Conf. on Signal Processing, Minneapolis (1988), to appear.

[LAKG] D. J. N. Limebeer, B. D. O. Anderson, P. P. Khargonekar and M. Green, A game theoretic approach to $H^\infty$ control for time varying systems, preprint.

[S] E. D. Sontag, Smooth stabilization implies coprime factorization, IEEE Trans. Aut. Control, 34 (1989), 435-443.

[T1] G. Tadmor, Worst case design in the time domain: the maximum principle and the standard $H_\infty$ problem, preprint.

[T2] G. Tadmor, The standard $H_\infty$ problem and the maximum principle: the general linear case, preprint.

[V] M. S. Verma, Coprime fractional representations and stability of nonlinear feedback systems, Int. J. Control, to appear.

Joseph A. Ball
Department of Mathematics
Virginia Tech
Blacksburg, VA 24061
U.S.A.

J. William Helton
Department of Mathematics
University of California at San Diego
La Jolla, CA 92093
U.S.A.

# Primitives for Robot Control

D. Curtis Deno  Richard M. Murray  Kristofer S. J. Pister
Shankar Sastry

**Abstract**

Inspired by the control system of the mammalian neuro-muscular system, we were motivated to develop a methodology for description of hierarchical control in a manner which is faithful to the underlying mechanics, structured enough to be used as an interpreted language, and sufficiently flexible to allow the description of a wide variety of systems. We present a consistent set of primitive operations which form the core of a robot system description and control language. This language is capable of describing a large class of robot systems under a variety of single level and distributed control schemes. We review a few pertinent results of classical mechanics, describe the functionality of our primitive operations, and present several different hierarchical strategies for the description and control of a two fingered hand holding a box.

## 1 Introduction

The complexity of compound, redundant robotic systems, both in specification and control, continues to present a challenge to engineers and biologists. Complex robot actions require coordinated motion of multiple robot arms or fingers to manipulate objects and respect physical constraints. As we seek to achieve more of the capability of biological robots, additional descriptive structures and control schemes are necessary. A major aim of this work is to propose such a specification and control scheme. The ultimate goal of our project is to build a high level task programming environment which is relatively robot independent.

In Section 2 we review the dynamics and control of coupled, constrained rigid robots in a Lagrangian framework. Section 3 contains definitions of the primitives of our robot control environment. Section 4 illustrates the application of our primitives to the description of a two fingered robot hand. We show that our environment can be used to specify a variety of control schemes for this hand, including a distributed controller which has a biological analog. In Section 5 we discuss future avenues of research. The remainder of this introduction presents motivation and background for our work, and an overview of the primitives we have chosen to use.

### 1.1 The Musculoskeletal System: Metaphor for a Robotic System

Motivation for a consistent specification and control scheme may be sought in our current knowledge of the hierarchical organization of mammalian motor systems. To some degree of accuracy, we may consider segments of limbs as rigid bodies connected by rotary joints. Muscles and tendons are actuators with sensory feedback which enter into low level feedback control at the spinal level [8]. Further up the nervous system, the brainstem, cerebellum, thalamus, and basal ganglia integrate ascending sensory information and produce coordinated motor commands. At the highest levels, sensory and motor cortex supply conscious goal-related information, trajectory specification, and monitoring.

The hierarchical structure of neuromuscular control is also evident from differences in time scale. The low-level spinal reflex control runs faster (loop delays of about 30 ms) than the high level feedback loops (100–200 ms delays). This distinction may be exploited by control schemes which hide information details from high level controllers by virtue of low level control enforcing individual details. These concepts are shown in Figure 1 where a drawing of neuromuscular control

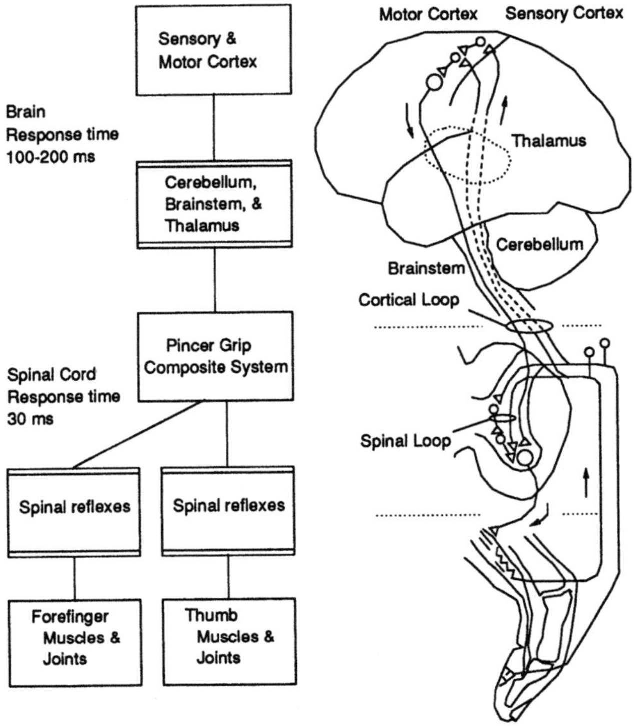


Figure 1: Hierarchical control scheme of a human finger. At the highest level, the brain is represented as sensory and motor cortex (where sensory information is perceived and conscious motor commands originate ) and brainstem and cerebellar structures (where motor commands are coordinated and sent down the spinal cord). A pair of fingers forms a composite system for grasping which is shown integrated at the level of the spinal cord. The muscles and sensory organs of each finger form low level spinal reflex loops. These low level loops respond more quickly to disturbances than sensory motor pathways which travel to the brain and back. Brain and spinal feedback controllers are represented by double lined boxes.

structures for a finger is juxtaposed with a block diagram to emphasize the hierarchical nature of the thumb-forefinger system for picking up objects.

Biological control systems commonly operate with constraints and redundancy. Kinematic constraints arise not only from joints which restrict the relative motion of adjacent limb segments, but also from contact with objects which leads to similar restrictions. Many musculo-skeletal subsystems possess kinematic and actuator redundancy which may be imagined to be resolved by effort and stability considerations. In any event, the neural controller directs a specific strategy and so expands a reduced set of control variables into the larger complete set.

In the sequel we shall see these concepts expressed in a notation which is faithful to the laws of mechanics and flexible enough to permit concise descriptions of robot motion control at various hierarchical levels.

## 1.2 Background

The robotics and control literature contains a number of topics which are related to the specification and control scheme of this paper.

**Robot programming languages**

Two directions of emphasis may be used to distinguish robot programming languages: traditional programming languages (perhaps including multitasking), and dynamical systems based descriptions of systems and control structures.

More traditional task specification languages include VAL II, AML, and Robot-BASIC [5, 19, 7, 18]. These languages are characterized by C, BASIC, or Lisp like data structures and syntax, coordinate frame specification and transformation primitives, sensor feedback conditionally controlling program flow, and motion between specified locations achieved through via points and interpolation. In a two stage hierarchy, low level controllers usually control joint angle trajectories which are specified by high level language statements and kinematics computations.

Brockett's Motion Description Language (MDL) is more closely aligned with dynamical systems theory. MDL employs sequences of triples $(u, k, T)$ to convey trajectory information, feedback control information, and time interval [6, 2] to an extensible Forth/PostScript like interpreter. The scheme described in this paper was inspired partly by descriptions of MDL. Our work explicitly utilizes geometric and inertial parameters together with the equations of motion to describe the organization and control of complex robots. MDL is less explicit on this matter but is more completely developed in the matter of sequences of motions.

**Distributed control, hierarchical control**

The nervous system controls biomechanical robots using both distributed controllers and hierarchical organization [8]. For example, spinal reflex centers can direct portions of gait in cats and the wiping motions of frog limbs without the brain. One reason for a hierarchical design is that high level feedback loops may respond too slowly for all of motor control to be localized there. Indeed the complexity and time delays inherent in biological motor control led the Russian psychologist Bernstein to conclude the brain could not achieve motor control by an internal model of body dynamics [10].

Centralized control has been defined as a case in which every sensor's output influences every actuator. Decentralized control was a popular topic in control theory in the late 1970's and led to a number of results concerning weakly coupled systems and multi-rate controllers [22]. Graph decomposition techniques, applied to the graph structures employed in a hierarchical scheme, permitted the isolation of sets of states, inputs, and outputs which were weakly coupled. This decomposition facilitated stability analyses and controller design. Robotic applications of hierarchical control are exemplified by HIC [3] which manages multiple low level servo loops with a robot programming language from the "traditional" category above. One emphasis of such control schemes concerns distributed processing and interprocess communication.

## 1.3 Overview of Robot Control Primitives

The fundamental objects in our robot specification environment are objects called robots. In a graph theoretic formalism they are nodes of a tree structure. At the lowest level of the tree are leaves which are instantiated by the `define` primitive. Robots are dynamical systems which are recursively defined in terms of the properties of their daughter robot nodes. Inputs to robots consist of desired positions and conjugate forces. The outputs of a robot consist of actual positions and forces. Robots also possess attributes such as inertial parameters and kinematics.

There are two other primitives which act on sets of robots to yield new robots, so that the set of robots is closed under these operations. These primitives (`attach` and `control`) may be considered as links between nodes and result in composite robot objects. Thus nodes closer to the root may possess fewer degrees of freedom, indicating a compression of information upon ascending the tree.

The `attach` primitive reflects geometrical constraints among variables and in the process of yielding another robot object, accomplishes coordinate transformations. `Attach` is also responsible for a bidirectional flow of information: expanding desired positions and forces to the robots below, and combining actual position and force information into an appropriate set for the higher level robot. In this sense the state of the root robot object is recursively defined in terms of the states of the daughter robots.

The `control` primitive seeks to direct a robot object to follow a specified "desired" position/force trajectory according to some control algorithm. The controller applies its control law (several different means of control are available such as PD and computed torque) to the desired and actual states to compute expected states for the daughter robot to follow. In turn, the daughter robot passes its actual states through the controller to robot objects further up the tree.

The block diagram portion of Figure 1 may be seen to be an example of a robot system comprised of these primitives. Starting from the bottom: two fingers are `defined`; each finger is `controlled` by muscle tension/stiffness and spinal reflexes; the fingers are `attached` to form a composite hand; the brainstem and cerebellum help `control` and coordinate motor commands and sensory information; and finally at the level of the cortex, the fingers are thought of as a pincer which engages in high level tasks such as picking.

# 2 Review of robot dynamics and control

In this section we review some basic results in dynamics and control of robot systems. Our goal is to give some insight into the mathematical framework underlying the primitives which we will be using. The basic result which we present is that even for relatively complicated robot systems, the equations of motion for the system can be written in a standard form. This point of view has been used by Khatib in his operational space formulation [12] and in some recent extensions [13]. The results presented in this section are direct extensions of those works, although the approach is different.

The dynamics for a robot manipulator with joint angles $\theta \in \mathbf{R}^n$ and actuator torques $\tau \in \mathbf{R}^n$ can be derived using Lagrange's equations and written in the form

$$M(\theta)\ddot{\theta} + C(\theta,\dot{\theta})\dot{\theta} = \tau \tag{1}$$

where $M(\theta) \in \mathbf{R}^{n\times n}$ is the inertia matrix for the manipulator and $C(\theta,\dot{\theta}) \in \mathbf{R}^{n\times n}$ is the Coriolis and centrifugal force matrix. For systems of this type, the inertia matrix is always symmetric and positive definite and it can be shown that $\dot{M} - 2C$ is skew symmetric (this requires some care in defining $C$). It is both the form and the structure of this equation that we will attempt to maintain in more complicated systems. For the moment we will ignore friction and gravity forces.

## 2.1 Change of coordinates

As our first exercise, we ask what effect a change of coordinates has on the form of the dynamics. Let $f : \mathbf{R}^n \to \mathbf{R}^n$ represent a locally invertible change of coordinates with $x = f(\theta)$ and $J = \frac{\partial f}{\partial \theta}$. Then $\dot{x} = J(\theta)\dot{\theta}$ and away from singularities we have $\dot{\theta} = J^{-1}(\theta)\dot{x}$. We can substitute this into the joint dynamics to obtain

$$(J^{-T}MJ^{-1})\ddot{x} + (J^{-T}M\dot{J}^{-1})\dot{x} + (J^{-T}CJ^{-1})\dot{x} = J^{-T}\tau \tag{2}$$

We see that this is close to our original form except for the second term. However, if we define

$$\begin{aligned}\tilde{M}(\theta) &= J^{-T}M(\theta)J^{-1}\\ \tilde{C}(\theta,\dot{\theta}) &= J^{-T}C(\theta,\dot{\theta})J^{-1} + J^{-T}M\dot{J}^{-1}\\ F &= J^{-T}\tau\end{aligned}$$

then we have

$$\tilde{M}(\theta)\ddot{x} + \tilde{C}(\theta,\dot{\theta})\dot{x} = F \tag{3}$$

It is easy to see that $\tilde{M}$ is symmetric and positive definite (away from critical points) and it can also be verified that $\dot{\tilde{M}} - 2\tilde{C}$ is skew-symmetric as before. Thus equation (3) has the same form and properties as the joint equations of motion and at least substituting for $\theta = f^{-1}(x)$, away from singularities, we can write

$$\tilde{M}(x)\ddot{x} + \tilde{C}(x,\dot{x})\dot{x} = F \tag{4}$$

which gives us an even closer correspondence. We also note that by definition $\tau = J^T F$ and so if $f$ is the forward kinematic function for the manipulator, $F$ corresponds to the Cartesian forces generated by the manipulator.

This simple result has some interesting consequences in control. Typically robot controllers are designed by placing a feedback loop around the joint positions (and velocities) of the robot. The controller generates torques which attempt to make the robot follow a prescribed joint trajectory. However, since the robot dynamics are of the same form in either joint or Cartesian space, we can

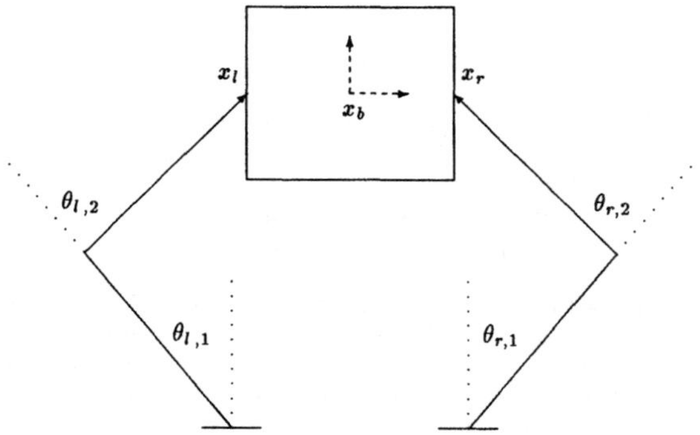


Figure 2: Planar two-fingered hand. Contacts are assumed to be maintained throughout the motion. Therefore the box position and orientation, $x$, form a generalized set of coordinates for the system.

just as easily write the control algorithm in Cartesian coordinates. In this case, we must take the output force from the controller and transform it back into joint torques, by premultiplying it by $J^T$. One advantage of this approach is that controller objectives are often specified in Cartesian space and hence it might be easier to perform the controller design and analysis in that space.

## 2.2 Constrained manipulators

We next demonstrate that more complicated robot systems can also be represented by dynamics in the same form as equation (3). For example, consider the control of a multi-fingered hand grasping a box (see Figure 2). If we let $\theta$ be the values of the joint variables for all of the joints and $x$ be the position and orientation of the box, then we can write $x = h(\theta)$ and we would like to use our previous analysis to claim that the dynamics of the system are given by

$$\tilde{M}(\theta)\ddot{x} + \tilde{C}(\theta, \dot{\theta})\dot{x} = F \tag{5}$$

where $F$ is the forces and torques exerted in the box's frame of reference. We must use a slightly different approach, however, since now we have not only a change of coordinates, but also a constraint and additional dynamics to include in our derivation. For now we make the assumption that our constraints are rigid (e.g., the fingers are connected to the box by ball and socket joints) in which case we can ignore all internal forces. This restriction can be lifted at the expense of additional complexity, as discussed in section 5.

As shown by Li, *et al.* [15], the constraint in this example can be written in the form

$$J(q)\dot{\theta} = G^T(q)\dot{x} \tag{6}$$

where $q = (\theta,\, x)$, $J$ is the Jacobian of the finger kinematic function and $G$ is the "grasp map" for the system. We will assume that $J$ is bijective in some neighborhood and that $G$ is surjective (this condition is necessary to insure *force closure* of a grasp, namely the ability to exert prescribed forces on an object). This form of constraint can also be used to describe a wide variety of other systems, including grasping with rolling contacts, surface following and coordinated lifting. For the primitives presented in the next section, we also assume that there exists a forward kinematic function between $\theta$ and $x$; that is, the constraint is holonomic. Non-holonomic constraints are a relatively straightforward extension but can cause difficulties in implementation. We shall discuss some of these details in section 5.

To include velocity constraints we must once again appeal to Lagrange's equations. A derivation of Lagrange's equations in the form we need can be found in Goldstein [9] or Rosenberg [20]. Using that derivation, the equations of motion for our constrained system can be written as

$$\tilde{M}(q)\ddot{x} + \tilde{C}(q, \dot{q})\dot{x} = F \tag{7}$$

where

$$\tilde{M} = M + GJ^{-T}M_\theta J^{-1}G^T \tag{8}$$

$$\tilde{C} = C + GJ^{-T}\left(C_\theta J^{-1}G^T + M_\theta \frac{d}{dt}\left(J^{-1}G^T\right)\right)$$

$$F = GJ^{-T}\tau$$

$M, M_\theta$ = inertia matrix for the box and fingers, respectively

$C, C_\theta$ = Coriolis and centrifugal terms

Thus we have an equation with similar form (and structure) to our "simple" robot. In the box frame of reference, $\tilde{M}$ is the mass of the effective mass of the box, and $\tilde{C}$ is the effective Coriolis and centrifugal matrix. These matrices include the dynamics of the fingers, which are being used to actually control the motion of the box. However the details of the finger kinematics and dynamics are effectively hidden in the definition of $\tilde{M}$ and $\tilde{C}$.

Again we note that even though we will write our controllers in terms of $F$, it is actually the joint torques which we are able to specify. Given the desired force in constrained coordinates, we can apply that force using an actuator force of $J^T G^+ \tau$, where $J$ and $G$ are as defined previously and $G^+$ is a pseudo inverse for $G$.

## 2.3 Control

To illustrate the control of robot systems, we look at two controllers which have appeared in the robotics literature. We start by considering systems of the form

$$M(q)\ddot{x} + C(q,\dot{q})\dot{x} + N(q,\dot{q}) = F \tag{9}$$

where $M(q)$ is a positive definite inertia matrix and $C(q,\dot{q})\dot{x}$ is the Coriolis and centrifugal force vector. The vector $N(q,\dot{q}) \in \mathbf{R}^n$ contains all friction and gravity terms and the vector $F \in \mathbf{R}^n$ represents generalized forces in the $x$ coordinate frame.

### Computed torque

Computed torque is an exactly linearizing control law that has been used extensively in robotics research. It has been used for joint level control [1], Cartesian control [16], and most recently, control of multi-fingered hands [15, 4]. Given a desired trajectory $x_d$ we use the control

$$F = M(q)\left(\ddot{x}_d + K_v\dot{e} + K_p e\right) + C(q,\dot{q})\dot{x} + N(q,\dot{q}) \tag{10}$$

where error $e = x_d - x$ and $K_v$ and $K_p$ are constant gain matrices. The resulting dynamics equations are linear with exponential rate of convergence determined by $K_v$ and $K_p$. Since the system is linear, we can use linear control theory to choose the gains ($K_v$ and $K_p$) such that they satisfy some set of design criteria.

The disadvantage of this control law is that it is not easy to specify the interaction with the environment. From the form of the error equation we might think that we could use $K_p$ to model the stiffness of the system and exert forces by commanding trajectories which result in fixed errors. Unfortunately this is not uniformly applicable as can be seen by examining the force due to a quasi-static displacement $\Delta x$:

$$\Delta F = M(q)K_p\Delta x \tag{11}$$

Since $K_p$ must be constant in order to prove stability, the resultant stiffness will vary with configuration. Additionally, given a desired stiffness matrix it may not be possible to find a positive definite $K_p$ that achieves that stiffness.

### PD + feedforward control

PD controllers differ from computed torque controllers in that the desired stiffness (and potentially damping) of the end effector is specified, rather than its position tracking characteristics. Typically, control laws of this form rely on the skew-symmetric property of robot dynamics, namely $\alpha^T\left(\dot{M} - 2C\right)\alpha = 0$ for all $\alpha \in \mathbf{R}^n$. Consider the control law

$$F = M(q)\ddot{x}_d + C(q,\dot{q})\dot{x}_d + N(q,\dot{q}) + K_v\dot{e} + K_p e \tag{12}$$

where $K_v$ and $K_p$ are symmetric positive definite. Using a Liapunov stability argument, it can be shown that the actual trajectory of the robot converges to the desired trajectory asymptotically [14]. Extensions to the control law result in exponential rate of convergence [23, 21].

This PD control law has the advantage that for a quasi-static change in position $\Delta x$ the resulting force is

$$\Delta F = K_p \Delta x \tag{13}$$

and thus we can achieve an arbitrary symmetric stiffness. Experimental results indicate that the trajectory tracking performance of this control law does not always compare favorably with the computed torque control law [17]. Additionally there is no simple design criteria for choosing $K_v$ and $K_p$ to achieve good tracking performance. While the stability results give necessary conditions for stability they do not provide a method for choosing the gains. Nonetheless, PD control has been used effectively in many robot controllers and has some computational features which make it an attractive alternative.

# 3 Primitives

In the previous section we saw that simple robots, Cartesian robots and constrained robots all have dynamics which can be written in the same form. Moreover, the rules for developing these dynamics are straightforward, at least in terms of representation. The problem of system and control specification is essentially that of solving the dynamics. The simple structure of this problem leads one to the natural conclusion that it should be automated. In this section we describe a set of primitives that give us the mathematical structure necessary to achieve this goal.

In the description that follows we will not use any particular programming environment or language. Ideas are freely borrowed from languages such as C, lisp and C++. As much as possible, we have tried to define the primitives so that they can be implemented in any of these languages. Whenever possible, we shall use mathematical symbols rather than functional notation. Details of implementation will be alluded to only in the interests of clarifying the presentation.

As our basic data structure, we will assume the existence of an object with an associated list of attributes. These attributes can be thought of as a list of name-value pairs which can be assigned and retrieved by name. A typical attribute which we will use is the inertia of a robot. The existence of such an attribute implies the existence of a function which is able to evaluate and return the inertia matrix of a robot given its configuration.

Attributes will be assigned values using the notation

$$\text{attribute} := \text{value}$$

Thus we might define our inertia attribute as

$$M(\theta) := \begin{bmatrix} m_1 l_1^2 + m_2 l_2^2 & m_2 l_1 l_2 \cos(\theta_1 - \theta_2) \\ m_2 l_1 l_2 \cos(\theta_1 - \theta_2) & m_2 l_2^2 \end{bmatrix} \tag{14}$$

In order to evaluate the inertia attribute, we would call $M$ with a vector $\theta \in \mathbf{R}^2$. This returns a $2 \times 2$ matrix which as defined above.

We will also assume that certain functions of attributes are readily available. Such functions might include evaluating the inverse or derivative of a function (when it exists); we will denote these simply as $M^{-1}$ or $\dot{M}$. In practice this could be implemented by defining a set of functions for each attribute which can evaluate the various forms of the attribute that will be used. These forms might be evaluated either numerically or symbolically, depending on the sophistication of the system.

Another frequently used function is *nil*() or just *nil*. This function, which does nothing, indicates an absence of data. It is used to avoid a situation where inappropriate data is returned when no data is available.

## 3.1 The robot object

The fundamental object used by all primitives is a *robot*. Associated with a robot are a set of attributes which are used to define it's behavior. Rather than fix the attributes associated with a robot, we wish to allow primitives to define new attributes as needed for their own use. All robots

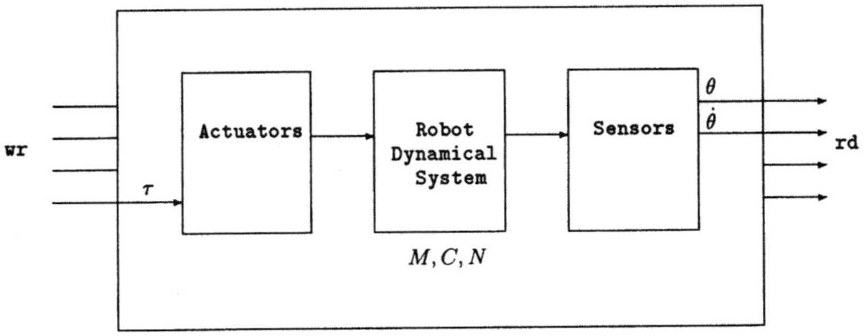


Figure 3: Example of the **define** primitive. The robot shown here corresponds to a robot with torque driven motors and only position and velocity sensing.

have a small set of attributes which must be defined. These attributes specify the basic properties which are used by all primitives:

| | |
|---|---|
| $M$ | inertia of the robot |
| $C$ | Coriolis/centrifugal vector |
| $N$ | friction and gravity vector |
| $rd()$ | function returning position and force information about the robot |
| $wr(x, \dot{x}, \ddot{x}, F)$ | function which sends information $(x, \dot{x}, \ddot{x}, F)$ to the robot |

The meaning of $M$, $C$, and $N$ should be clear from section 2. These are functions which are passed certain arguments (e.g. $x$, $\dot{x}$) and return a matrix, as described above for the inertia attribute.

The $rd$ function returns the current position, velocity, and acceleration of the robot, and the forces measured by the robot. Each of these will be vector quantity of dimension equal to the number of degrees of freedom of the robot. Typically a robot may only have access to its joint positions and velocities, in which case $\ddot{x}$ and $F$ will be *nil*.

The $wr$ function is used to specify an expected position and force trajectory that the robot is to follow. In the simplest case, a robot would ignore everything but $F$ and try to apply this force/torque at its actuators. As we shall see later, other robots may use this information in a more intelligent fashion. We will often refer to the arguments passed to write by using the subscript $e$. Thus $x_e$ is the desired position passed to the $wr$ function.

The task of describing a primitive is essentially the same as describing how it generates the attributes of the new robot. The following sections describe how each of the primitives generates these attributes. The new attributes created by a primitive are distinguished by a tilde over the name of the attribute.

### 3.2 DEFINE primitive

**Synopsis:**

`DEFINE(`$M$`, `$C$`, `$N$`, `$rd$`, `$wr$`)`

The **define** primitive is used to create a simple robot object. It defines the minimal set of attributes necessary for a robot. These attributes are passed as arguments to the **define** primitive and a new robot object possessing those attributes is created:

$$\begin{aligned}
\tilde{M}(\theta) &:= M(\theta) \\
\tilde{C}(\theta, \dot{\theta}) &:= C(\theta, \dot{\theta}) \\
\tilde{N}(\theta, \dot{\theta}) &:= N(\theta, \dot{\theta}) \\
\tilde{rd}() &:= rd() \\
\tilde{wr}(\theta_e, \dot{\theta}_e, \ddot{\theta}_e, \tau_e) &:= wr(\theta_e, \dot{\theta}_e, \ddot{\theta}_e, \tau_e)
\end{aligned}$$

Several different types of robots can be defined using this basic primitive. For example, a DC motor actuated robot would be implemented with a $wr$ function which converts the desired torque to a motor current and generates this current by communicating with some piece of hardware (such as a D/A converter). This type of robot system is shown in Figure 3. On the other hand, a stepper motor actuated robot might use a *wr* function which ignores the torque argument and uses the position argument to move the actuator. Both robots would use a *rd* function which return the current position, velocity, acceleration and actuator torque. If any of these pieces of information is missing, it is up to the user to insure that they are not needed at a higher level.

We may also define a ***payload*** as a degenerate robot by setting the *wr* argument to the *nil* function. Thus commanding a motion and/or force on a payload produces no effect. The only way to cause the object to move is to attach it to an actuated manipulator, the subject of the next section.

As an example of of the **define** primitive, consider the definition of a simple Cartesian manipulator for which the following functions have been defined:

$$M(\theta) := \begin{bmatrix} m_1 & 0 \\ 0 & m_2 \end{bmatrix}$$

$$\texttt{rd\_cartesian}() := \text{query the hardware and return current robot state}$$

$$\texttt{wr\_cartesian}(x_e, \dot{x}_e, \ddot{x}_e, F_e) := \text{output actuator torques to produce a Cartesian force } F$$

We can define a robot, **cartesian_robot**, using

**cartesian_robot = DEFINE($M$, 0, 0, rd_cartesian, wr_cartesian)**

A special case of this would be to define a planar payload (with the same mass distribution)

**payload = DEFINE($M$, 0, 0, *nil*, *nil*)**

## 3.3 ATTACH primitive

**Synopsis:**

**ATTACH(J, G, h, payload, robot-list)**

**Attach** is used to describe constrained motion involving a payload and one or more robots. **Attach** must create a new robot object from the attributes of the payload and of the robots being attached to it. The specification of the new robot requires a velocity relationship between coordinate systems ($J\dot{\theta} = G^T\dot{x}$), a kinematic function relating robot positions to payload position ($x = h(\theta)$), a payload object, and a list of robot objects involved in the contact.

The only difference between the operation of the **attach** primitive and the equations derived for constrained motion of a robot manipulator is that we now have a *list* of robots each of which is constrained to contact a payload. However, if we define $\theta_R$ to be the combined joint angles of the robots in **robot-list** and similarly define $M_R$ and $C_R$ as block diagonal matrices composed of the individual inertia and Coriolis matrices of the robots, we have a system which is identical to that presented previously. Namely, we have a "robot" with joint angles $\theta_R$ and inertia matrix $M_R$ connected to an object with a constraint of the form

(15) $$J\dot{\theta}_R = G^T\dot{x}$$

where once again $J$ is a block diagonal matrix composed of the Jacobians of the individual robots.To simplify notation, we will define $\mathcal{A} := J^{-1}G^T$ so that

(16) $$\dot{\theta}_R = \mathcal{A}\dot{x}$$

The attributes of the new robot can thus be defined as:

(17) $$\tilde{M} := M_p + \mathcal{A}^T M_R \mathcal{A}$$

(18) $$\tilde{C} := C_p + \mathcal{A}^T C_R \mathcal{A} + \mathcal{A}^T M_R \dot{\mathcal{A}}$$

(19) $$\tilde{N} := N_p + \mathcal{A}^T N_R$$

(20) $$\tilde{rd}() := (h(\theta_R),\ \mathcal{A}^+\dot{\theta}_R,\ \mathcal{A}^+\ddot{\theta}_R + \dot{\mathcal{A}}^+\dot{\theta}_R,\ \mathcal{A}^T\tau_R)$$

(21) $$\tilde{wr}(x_e, \dot{x}_e, \ddot{x}_e, F_e) := wr_R(h^{-1}(x_e),\ \mathcal{A}\dot{x}_e,\ \mathcal{A}\ddot{x}_e + \dot{\mathcal{A}}\dot{x},\ \mathcal{A}^{+T}F_e)$$

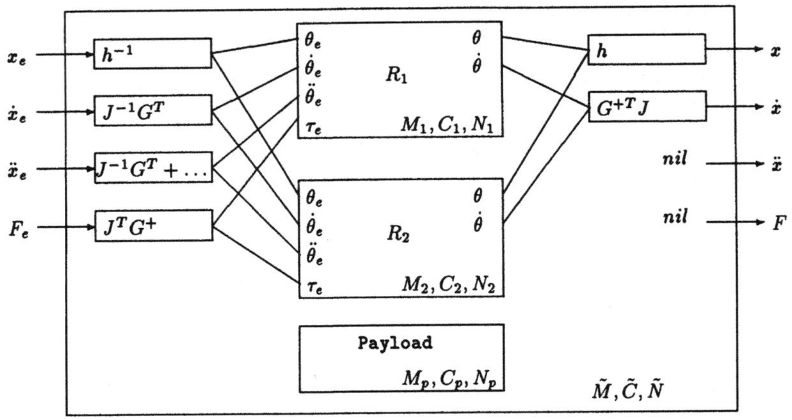


Figure 4: Data flow in a two robot **attach**. In this example we illustrate the structure generated by a call to **attach** with 2 robots and a payload (e.g. a system like Figure 2). The two large interior boxes represent the two robots, with their input and output functions and their inertia properties. The outer box (which has the same structure as the inner boxes) represents the new robot generated by the call to **attach**. In this example the robots do not have acceleration or force sensors, so these outputs are set to *nil*.

where $M_p, C_p, N_p$ are attributes of the payload, $M_R$ and $C_R$ are as described above and $N_R$ is a stacked vector of friction and gravity forces.

The $\tilde{rd}$ attribute for an attached robot is a function which queries the state of all the robots in **robot-list**. Thus $\theta_R$ in equation (20) is constructed by calling the individual *rd* functions for all of the robots in the list. The $\theta$ values for each of these robots are then combined to form $\theta_R$ and this is passed to the forward kinematic function. A similar computation occurs for $\dot{\theta}_R$, $\ddot{\theta}_R$ and $\tau_R$. Together, these four pieces of data form the return value for the $\tilde{rd}$ attribute.

In a dual manner, the $\tilde{wr}$ attribute is defined as a function which takes a desired trajectory (position and force), converts it to the proper coordinate frame and sends each robot the correct portion of the resultant trajectory.

The **attach** primitive also creates new attributes to store the constraint information (i.e. $J$ and $G$). These attributes are used by the internal functions which must evaluate the dynamics and input/output attributes of the robot.

A special case of the **attach** primitive is its use with a *nil* payload object and $G = I$. In this case, $M_p$, $C_p$, and $N_p$ are all zero and the equations above reduce to a simple change of coordinates, as shown in section 2.

An example of the operation of the **attach** primitive is summarized in Figure 4.

## 3.4 CONTROL primitive

**Synopsis:**

```
CONTROL(robot, controller)
```

The **control** primitive is responsible for assigning a controller to a robot. It is also responsible for creating a new robot with attributes that properly represents the controlled robot. The attributes of the created robot are completely determined by the individual controller. However, the *rd* and *wr* attributes will often be the same for different controllers. Typically the *rd* attribute for a controlled robot will be the same as the *rd* attribute for the underlying robot. That is, the current state of the controlled robot is equivalent to the current state of the uncontrolled robot. A common *wr* attribute for a controlled robot would be a function which saved the desired position, velocity, acceleration and force in a local buffer accessible to our controller. This configuration is shown in Figure 5.

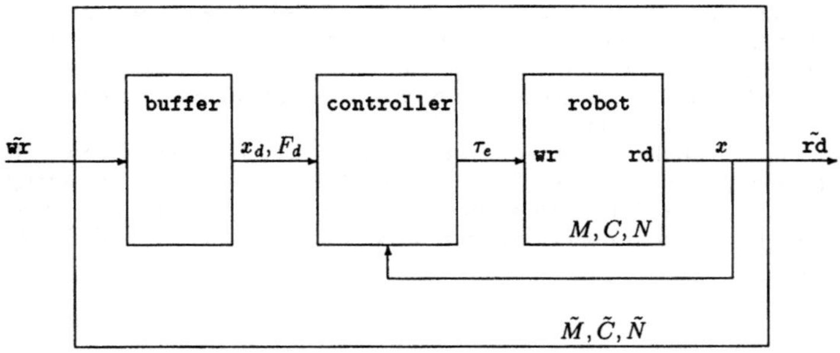


Figure 5: Data flow in a typical controlled robot. Information written to the robot is stored in an internal buffer where it can be accessed by the controller. The controller uses this information and the current state of the robot to generate forces which cause it to follow the desired trajectory.

The dynamic attributes $\tilde{M}$, $\tilde{C}$ and $\tilde{N}$ are determined by the controller. At one extreme, a controller which compensates for the inertia of the robot would set the dynamic attributes of the controlled robot to zero. This does not imply that the robot is no longer a dynamic object, but rather that controllers at higher levels can ignore the dynamic properties of the robot, since they are being compensated for at a lower level. At the other end of the spectrum, a controller may make no attempt to compensate for the inertia of a robot, in which case it should pass the dynamic attributes on to the next higher level. Controllers which lie in the middle of this range may partially decouple the dynamics of the manipulator without actually completely compensating for them. To illustrate these concepts, we give examples of two controllers and how they might be defined.

### Computed torque controller

As we mentioned in section 2, the computed torque controller is an exactly linearizing controller which inverts the nonlinearities of a robot to construct a linear system. This linear system has a transfer function equal to the identity map and as result has no uncompensated dynamics. The proper representation for such a system is

$$\begin{aligned}
\tilde{M} &:= 0\\
\tilde{C} &:= 0\\
\tilde{N} &:= 0\\
\tilde{rd}() &:= rd()\\
\tilde{wr}(x_e, \dot{x}_e, \ddot{x}_e, F_e) &:= (x_d = x_e,\ \dot{x}_d = \dot{x}_e,\ \ddot{x}_d = \ddot{x}_e,\ F_d = F_e)
\end{aligned}$$

The definition of the $\tilde{wr}$ attribute is intended to represent the buffering operation which we described above.

The control process portion of the controller is responsible for generating input robot forces which cause the robot to follow the desired trajectory (available in $x_d$). Additionally, the controller must determine the "expected" trajectory to be sent to lower level robots. For the computed torque controller we use the resolved acceleration [16] to generate this path. This allows computed torque controllers running at lower levels to properly compensate for nonlinearities and results in a linear error response. The methodology is similar to that used in determining that the dynamic attributes of the output robot should be zero. The control algorithm is implemented by the following equations:

$$\begin{aligned}
&(x, \dot{x}, \cdot, \cdot) = rd()\\
&F_e = M(\theta)(\ddot{x}_d + K_v(\dot{x}_d - \dot{x}) + K_p(x_d - x)) + C(x, \dot{x})\dot{x} + N(x, \dot{x}) + F_d\\
&\ddot{x}_e = \ddot{x}_d + K_v(\dot{x}_d - \dot{x}) + K_p(x_d - x)\\
&\dot{x}_e = \int_0^t \ddot{x}_e\\
&x_e = \int_0^t \dot{x}_e\\
&wr(x_e, \dot{x}_e, \ddot{x}_e, F_e)
\end{aligned}$$

where $rd$ and $wr$ are attributes of the robot which is being controlled. Note the existence of the $F_d$ term in the calculation of $F_e$. This is placed here to allow higher level controllers to specify not only a trajectory but also an force term to compensate for higher level payloads. In essence, a robot which is being controlled in this manner can be viewed as an ideal force generator which is capable of following an arbitrary path.

The computed torque controller defines two new attributes, $K_p$ and $K_v$, which determine the gains (and hence the convergence properties) of the controller. A variation of the computed torque controller is the feedforward controller, which is obtained by setting $K_p = K_v = 0$. This controller can be used to distribute nonlinear calculations in a hierarchical controller, as we shall see in section 4.

### PD controller

Unlike the computed torque controller, the PD controller does not compensate for the nonlinearity of the robot system. It simply uses a PD control law to improve tracking. Therefore, the dynamic properties of the output robot are identical to those of the input robot:

$$
\begin{aligned}
\tilde{M} &:= M \\
\tilde{C} &:= C \\
\tilde{N} &:= N \\
\tilde{rd} &:= rd \\
\tilde{wr}(x_e, \dot{x}_e, \ddot{x}_e, F_e) &:= (x_d = x_e,\ \dot{x}_d = \dot{x}_e,\ \ddot{x}_d = \ddot{x}_e,\ F_d = F_e)
\end{aligned}
$$

The control law is very simple

$$
\begin{aligned}
&(x, \dot{x}, \cdot, \cdot) = rd() \\
F_e &= K_v(\dot{x}_d - \dot{x}) + K_p(x_d - x) \\
\ddot{x}_e &= \ddot{x}_d \\
\dot{x}_e &= \dot{x}_d \\
x_e &= x_d \\
&wr(x_e, \dot{x}_e, \ddot{x}_e, F_e)
\end{aligned}
$$

Like the computed torque controller, the PD controller defines new attributes, $K_v$ and $K_p$, for use in setting the gains for the system. Setting $K_v = K_p = 0$ effectively disables the controller.

## 4 Examples

To make the use of the primitives more concrete we present some examples of a planar hand grasping a box (Figure 2) using various control structures. For all of the controllers, we will use the following functions

| | |
|---|---|
| $M_b$ | inertia matrix for the box in Cartesian coordinates |
| $M_l, M_r$ | inertia matrix for the left and right fingers |
| $C_b, C_l, C_r$ | Coriolis/centrifugal vector for box and fingers |
| $f$ | finger kinematics function, $f : (\theta_l, \theta_r) \mapsto (x_l, x_r)$ |
| $g$ | grasp kinematics function, $g : (x_l, x_r) \mapsto x_b$ |
| $J$ | finger Jacobian, $J = \frac{\partial f}{\partial \theta}$ |
| $G$ | grasp map, consistent with $g$ |
| `rd_left`, `rd_right` | read the current joint position and velocity |
| `wr_left`, `wr_right` | generate a desired torque on the joints |

where $\theta_l, \theta_r, x_l, x_r$, and $x_b$ are defined as in Figure 2.

### Example 1: High-level computed torque control

In this example we combine all systems to obtain a description of the dynamic properties of the overall system in box coordinates. Once this is done we can move the box by directly specifying the desired trajectory for the box. This structure is illustrated in Figure 6.

In terms of the primitives that we have defined, we build this structure from the bottom up

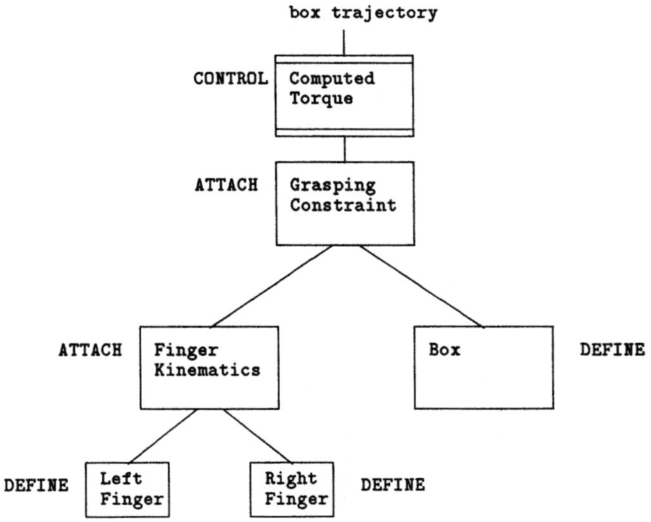


Figure 6: High level computed torque. The primitives listed next to the nodes in the graph indicate the primitive that was used to created the node. In this structure all dynamic compensation and error correction occurs at the top of the graph, using a complex dynamic model for the underlying system.

**left** = **DEFINE**($M_l$, $C_l$, **0**, **rd_left**, **wr_left**)
**right** = **DEFINE**($M_r$, $C_r$, **0**, **rd_right**, **wr_right**)
**fingers** = **ATTACH**($J$, **I**, $f$, **nil**, **left**, **right**)

**box** = **DEFINE**($M_b$, $C_b$, **0**, **nil**, **nil**)
**hand** = **ATTACH**(**I**, $G^T$, $g$, **box**, **fingers**)

**ct_hand** = **CONTROL**(**hand**, **computed-torque**)

It is useful to consider how the data flows to and from the control law running at the hand level. In the evaluation of $x_b$ and $\dot{x}_b$, the following sequence occurs (through calls to the *rd* attribute):

hand: asks for current state, $x_b$ and $\dot{x}_b$
    finger: ask for current state, $x_f$ and $\dot{x}_f$
        left: read current state, $\theta_l$ and $\dot{\theta}_l$
        right: read current state, $\theta_r$ and $\dot{\theta}_r$
    finger: $x_f, \dot{x}_f \leftarrow f(\theta_l, \theta_r), J(\dot{\theta}_l, \dot{\theta}_r)$
hand: $x_b, \dot{x}_b \leftarrow g(x_f), G^T \dot{x}_f$

Similarly, when we write a set of hand forces using the *wr* attribute, it causes another chain of events to occur: call sequence is generated

box: generate a box force $F_b$
    hand: generate finger force $GF_b$
        finger: generate joint torques $J^T GF_b$
            left: output torques conjugate to $\theta_l$
            right: output torques conjugate to $\theta_r$

Using the transformations given above it is straightforward to calculate the torque generated by the control law by expanding the structure of Figure 6 using the definition of the primitives.

$$\begin{pmatrix} \tau_l \\ \tau_r \end{pmatrix} = J^T F_{f,d}$$

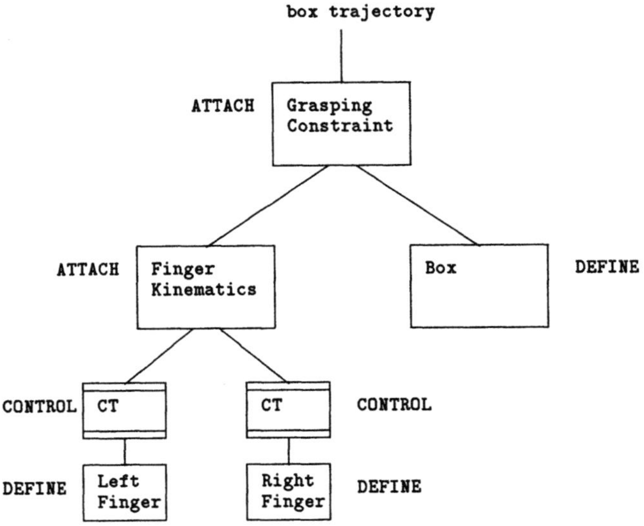


Figure 7: Low level computed torque. Computed torque controllers are used for the individual fingers to provide trajectory following capability in joint space. Since no controller is positioned above the box, the dynamics of the box are ignored even though the path is given in the box's frame of reference.

$$
\begin{aligned}
&= J^T G^+ F_{b,d} \\
&= J^T G^+ \left[M_h \left(\ddot{x}_{b,d} + K_v \dot{e} + K_p e\right) + C_h \dot{x}_b\right] \\
&\ \vdots \\
&= J^T G^+ \Big[ \left(M_b + G J^{-T} M_\theta J^{-1} G^T\right)\left(\ddot{x}_{b,d} + K_v \dot{e} + K_p e\right) + \\
&\qquad \left(C_b + G J^{-T} C_\theta J^{-1} G^T\right) \dot{x}_b + G J^{-T} M_\theta \frac{d}{dt}\left(J^{-1} G^T\right) \dot{x}_b \Big]
\end{aligned}
$$

This control law corresponds exactly to the generalized computed torque control algorithm presented by Li, *et al.* [15].

## Example 2: Low-level computed torque control

Another common structure for controlling robots is to convert all trajectories to joint coordinates and perform computed torque at that level. In a crude implementation one might assume the dynamic effects of the box were negligible and construct the following structure shown in Figure 7. The primitives used to define this structure are

```
left  = DEFINE(M_l, C_l, 0, rd_left, wr_left)
right = DEFINE(M_r, C_r, 0, rd_right, wr_right)
ct_left  = CONTROL(left, computed-torque)
ct_right = CONTROL(right, computed-torque)

fingers = ATTACH(J, I, f, nil, ct_left, ct_right)
box     = DEFINE(M_b, C_b, 0, nil, nil)

hand = ATTACH(I, G^T, g, box, fingers)
```

This controller is provably exponentially stable when the mass of the box is zero. However, this controller does not compensate for the mass of the box. As a result, we expect degraded performance if the mass of the box is large. Experimental results on a system of this form confirm our intuition [17].

## Example 3: Multi-level computed torque/stiffness control

As a final example, we consider a control structure obtained by analogy with biological systems in which controllers to run at several different levels simultaneously (see Figure 8). At the lowest level

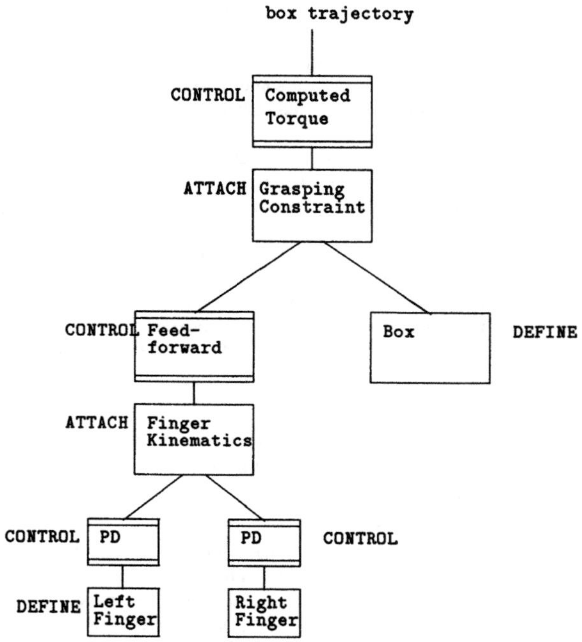


Figure 8: Multi level computed torque and stiffness (PD). Controllers are used at each level to provide a distributed control system with biological motivation, desirable control properties, and computational efficiency.

we use simple PD control laws attached directly to the individual fingers. These PD controllers mimic the stiffness provided by muscle coactivation in a biological system [11]. Additionally, controllers at this level might be used to represent spinal reflex actions. At a somewhat higher level, the fingers are attached and considered as a single unit with relatively complicated dynamic attributes and Cartesian configuration. At this point we employ a feedforward controller (computed torque with no error correction) to simplify these dynamic properties, as viewed by higher levels of the brain. With respect to these higher levels, the two fingers appear to be two Cartesian force generators represented as a single object.

Up to this point, the representation and control strategies do not explicitly involve the box, a payload object. These force generators are next attached to the box, yielding a robot with the dynamic properties of the box but capable of motion due to the actuation in the fingers. Finally, we use a computed torque controller at the very highest level to allow us to command motions of the box without worrying about the details of muscle actuation. By this controller we simulate the actions of the cerebellum and brainstem to coordinate motion and correct for errors.

The structure in Figure 8 also has interesting properties from a more traditional control viewpoint. The low level PD controllers can be run at high servo rates (due to their simplicity) and allow us to tune the response of the system to high frequency perturbations. The Cartesian feedforward controller permits a distribution of the calculation of nonlinear compensation terms at various levels, lending itself to multiprocessor implementation. Finally, using a computed torque controller at the highest level gives the flexibility of performing the controller design in the task space and results in a system with linear error dynamics. Another feature is that if we ignore the additional torques due to the PD terms, the joint torques generated due to an error in the box position are the same as those of the high level computed torque scheme presented earlier.

# 5 Discussion

The work presented in the previous sections is a starting point for the development of what we hope will be a high level robot programming environment. The primitives of the previous sections will represent the lowest levels of this system. However, these primitives need further refinement since even fairly simple robot systems violate some of the assumptions we have made. In this section we discuss some of the limitations of the current system and offer possible approaches for their solution.

Our implementation goals include a real time digital control system driven by our description language. A symbolic processor (such as Mathematica or MACSYMA) will interpret this language and generate the specified control structure. This control will then be executed by a multiprocessor real time control system. This allows for the exciting possiblility of specifying a control hierarchy and obtaining experimental results from that controller in a matter of minutes. Unfortunately, the output from a symbolic processor, even after application of its simplification routines, is far from terse. However, the structure of our primitives allows for a natural division of computation, and an initial investigation into the feasibility of the system indicated the computational burden should be manageable using moderate resources such as a VME based 68020 multiprocessor system.

## 5.1 Geometric issues

The manipulator Jacobian matrix $J$ and grasp matrix $G$ play key roles in this work and we next consider more carefully the consequences of violating our assumptions.

**Redundancy: $J$ or $G$ nonsquare**

There are two flavors of redundancy which we have ignored in the primitives presented so far. The first of these, which we shall call kinematic redundancy, occurs when we have a robot with more degrees of freedom than needed to position and orient the end effector. The result of this redundancy is that the forward kinematic function, $f$ is no longer locally bijective—for every end effector position there exists a continuum of joint angles which maintain that position. The other kind of redundancy which is common to biological systems is actuator redundancy, namely, the presence of multiple muscles which can exert torques about a single joint. This form of redundancy is more subtle and we defer its discussion to future work.

Problems with kinematic redundancy appear mainly in the `attach` primitive. There we assumed the existence of the an inverse function for $h$, the map between joint variables and object coordinates. If any of the robots being attached to the object is redundant such an inverse function is not uniquely defined. Furthermore, the Jacobian of the robot forward kinematics will no longer be square, resulting in a $J$ which is not square and hence not invertible. Currently we assume that a redundancy resolving controller is used between a redundant robot definition and a subsequent attach.

The major consideration with these resolutions of redundancy is to use redundancy to improve performance. Studies of human motion control indicate that redundancy plays roles other than obstacle avoidance and flexibility, e.g. achieving desirable stiffness properties [11]. A classical use of redundant motion in robotics is to specify a cost function and use the redundancy of the manipulator to attempt to minimize this cost function. This method is equivalent to specifying a velocity in the tangent space to the internal motion manifold. If we extend our definition of the *wr* function so that it takes not only an "external" trajectory, but also an internal trajectory (which might be represented as a cost function or directly as a desired velocity in the internal motion directions) then this internal motion can be propagated down the graph structure.

A similar situation occurs with internal or constraint forces. As noted in section 2, forces which lie in the null space of the grasp map cause no net motion of the system. These forces can be useful in the case of grasping. Here sufficient internal forces are applied to insure that the forces exerted by the fingers on the object lie in the friction cone defined by the contact. Thus we might also extend our primitives to propagate an internal force through the graph structure.

In summary, degrees of freedom are not completely lost due to a constraint or resolved redundancy. Instead, we have shifted motion or force from external to internal. A complete approach to control must also provide for feedback control of these internal motions and forces.

#### Singularities: $J, G$ singular

Throughout the definitions of the primitives (and even the underlying dynamics) we assumed that $J^{-1}$ was well defined. This is not the case if one of the manipulators in the system goes through a kinematic singularity. In this case the calculations performed in the primitives will become unstable, demanding large joint velocities and torques to achieve a specific trajectory. Similarly, when the grasp matrix $G$ approaches a singularity, force closure is lost and large forces may be required. If it is not possible to avoid singularities at the path planning stage then it may be acceptable to tolerate some trajectory error.

#### Nonholonomic constraints

We have assumed so far the the constraints applied to the system are holonomic. For some common systems, such as grasping with rolling in three dimensions, it can turn out that $J$ and $G$ are well defined and full rank but there exists no forward kinematic function such that $x = h(\theta)$. Such systems can still be controlled, however, with some restrictions.

Control laws commonly use the position of the object as part of the feedback term. If this position cannot be calculated from $\theta$ then we must retrieve it from some other source. One possibility is to use an external sensor which senses $x$ directly, such as a camera or tactile array. The function to" read the sensor" could be assigned to the payload *rd* function and `attach` could use this information to return the payload position when queried. Another possible approach is to integrate the object velocity (which is well defined) to bookkeep the payload position.

### 5.2 Implementation issues

One of the major future goals of this research is to implement the primitives presented here on a real system. This requires that efforts be made toward implementing primitives in as efficient fashion as possible.

#### On-line vs off-line

The first choice to be made in implementing the primitives is deciding where computation should occur. It is possible that the entire set of primitives could be implemented off-line. In this case, a controller-generator would read the primitives and construct suitable code to control the system. Such an approach is only possible if the basic contact structure is specified ahead of time.

A more realistic approach is to split the computation burden more judiciously between on-line and off-line resources. Symbolically calculating the attributes of the low-level robots and storing these as precompiled functions might enable a large number of systems to be constructed without too much prior knowledge about the structure of the graph describing the system.

Off line computation of controller outputs may be done if an open loop control strategy is acceptable. Such an approach, which resembles simulation of the composite robot system, will require numeric integration of the equations of motion in place of the physical system. If the models are faithful, such off-line techniques also may be used to compensate for low bandwidth communication channels and time delays. On-line implementations are expected to be more robust to sensor and modeling error, at the cost of real-time computation and data flow requirements.

#### Multirate controllers

Although the expressions employed are continuous time, in practice digital computers will be relied upon for discrete time implementations. This raises the issue of computation rates and whether lower rates may be practical for higher level robots/controllers. In the case of mammalian motor control, higher level feedback appears to occur at a slower rate—due in part to transmission delays. However, humans are able to perform tasks accurately in spite of this slow (and hence low gain) high-level feedback. Both hardware and wetware implementations may benefit from distributed gain and multiple rate controllers.

# 6 Conclusion

Working from a physiological motivation we have developed a set of robot description and control primitives consistent with Lagrangian dynamics. Starting from a description of the inertia, sensor, and actuator properties of individual robots, these primitives allow for the construction of a composite constrained motion system with control distributed at all levels. The resulting hierarchical system can be represented as a tree structure in a graph theoretic formalism, with sensory data fusion occurring as information flows from the leaves of the tree (individual robots and sensors) toward the root, and data expansion as relatively simple motion commands at the root of the tree flow down through contact constraints and kinematics to the individual robot actuators.

In future, we hope to extend these basic description and control primitives into a more complete task level programming environment which is device independent to some extent. The output of such a system may be employed to generate a multiprocessor based control structure. Further work will include extending the primitives to allow specification of internal constraint forces and specification of internal motion of redundant manipulators in a more general way.

The primitives that we have defined are intended to be useful on a theoretical level as well as in a real time control system. This structure for describing hierarchical robot control systems may also assist the study of biological motion control. These primitives provide a conceptual framework in which to develop hypotheses and integrate experimental data.

**Acknowledgements**

D.C. Deno was supported by NEI EY05913, the Smith-Kettlewell Eye Research Foundation, and the Rachael C. Atkinson Fellowship Award. K.S.J. Pister, R.M. Murray, and S.S. Sastry were supported in part by the NSF under grant numbers DMC 84-51129, ECS 87-19298 and the Air Force Office of Scientific Research (AFSC) under grant number F49620-87-C0041.

# References

[1] A. K. Bejczy. Robot arm dynamics and control. Technical Report 33-699, Jet Propulsion Laboratory, 1974.

[2] Roger W. Brockett. On the computer control of movement. Technical Report CICS-P-31, Center for Intelligent Control Systems, Harvard Univ., November 1987.

[3] Dayton Clark. HIC: An operating system for hierarchies of servo loops. In *Proceedings 1989 IEEE International Conference on Robotics and Automation*, pages 1004–1009. IEEE Computer Society Press, 1989.

[4] A. Cole, J. Hauser, and S. Sastry. Kinematics and control of multifingered hands with rolling contact. In *IEEE International Conference on Robotics and Automation*, pages 228–233, 1988.

[5] John J. Craig. *Introduction to Robotics: Mechanics and Control. 2nd Ed.*, chapter 9–13. Addison-Wesley, 1989.

[6] Victor Eng. *The MDL Programmer's Reference Manual.* Harvard Robotics Laboratory, 1988. First Revision.

[7] K.S. Fu, Raphael C. Gonzalez, and C. S. George Lee. *Robotics: control, sensing, vision, and intelligence*, chapter 9. McGraw-Hill, Inc., 1987.

[8] Claude Ghez. Introduction to the motor system. In Eric R. Kandel and James H. Schwartz, editors, *Principles of Neural Science, 2nd Ed.*, chapter 33. Elsevier, 1985.

[9] H. Goldstein. *Classical Mechanics.* Addison Wesley, Reading, Mass., 1981.

[10] G. Hinton. Some computational solutions to Bernstein's problems. In H. T. A. Whiting, editor, *Human Motor Actions — Bernstein Reassessed*, chapter 4b. Elsevier Science Publishers B.V., 1984.

[11] Neville Hogan, Emilio Bizzi, F. A. Mussa-Ivaldi, and Tamar Flash. Controlling multijoint behavior. *Exercise and Sport Sciences Reviews*, 15:153–190, 1987.

[12] O. Khatib. A unified approach for motion and force control of robot manipulators: The opertional space formulation. *IEEE Journal of Robotics and Automation*, RA-3(1):43–53, February 1987.

[13] O. Khatib. Augmented object and reduced effective inertia in robot systems. *Automatic Control Conference*, pages 2140–2147, 1988.

[14] D. Koditschek. Natural motion for robot arms. In *IEEE Conference on Decision and Control*, pages 733–735, 1984.

[15] Z. Li, P. Hsu, and S. Sastry. On kinematics and control of multifingered hands. In *IEEE International Conference on Robotics and Automation*, pages 384–389, 1988.

[16] J. Y. S. Luh, M. W. Walker, and R. P. Paul. Resolved acceleration control of mechanical manipulators. *IEEE Transactions on Automatic Control*, AC-25, 1980.

[17] R. Murray and S. Sastry. Control experiments in planar manipulation and grasping. In *IEEE International Conference on Robotics and Automation*, 1989.

[18] R. P. Paul and V. Hayward. Robot control and computer languages. In *Theory and Practice of Robots and Manipulators*, 1984. Proceedings of RoManSy '84: The Fifth CISM-IFTOMM Symposium.

[19] Richard P. Paul. *Robot manipulators: mathematics, programming, and control*, chapter 10. MIT Press, 1981.

[20] R. M. Rosenberg. *Analytical Dynamics of Discrete Systems*. Plenum Press, New York, 1977.

[21] N. Sadegh. *Adaptive Control of Mechanical Manipulators: Stability and Robustness Analysis*. PhD thesis, Department of Mechanical Engineering, University of California, Berkeley, California, 1987.

[22] N. R. Sandell, Jr., P. Varaiya, M. Athans, and M. G. Safonov. Survey of decentralized control methods for large scale systems. *IEEE Transactions on Automatic Control*, AC-23:108–128, 1978.

[23] J. E. Slotine and W. Li. On the adaptive control of robot manipulators. *International Journal of Robotics Research*, 6:49–59, 1987.

# OPTIMAL FREQUENCY DOMAIN DESIGN VS. AN AREA OF SEVERAL COMPLEX VARIABLES

J. William Helton

**Abstract.** $H^\infty$ design theory has developed rapidly in the last decade, both mathematically and from the standpoint of applications. The talk addresses one very natural direction one can take in going beyond this. The goal is to develop a systematic mathematical theory of worst case frequency domain design where stability is the key constraint. This leads immediately to a collection of mathematics problems which have strong connections with ongoing developments in several complex variables. The main point of this article is to point out these connections.

## INTRODUCTION

The development of a systematic theory of worst case design in the frequency domain where stability is the key consideration is developing rapidly and moving in several directions. An example of a major success in the subject is on the paradigm mixed sensitivity problem for multiport control systems. Here a beautiful and powerful theory has emerged. When one looks beyond the paradigm for new directions in the subject several areas are quickly suggested by physical considerations.

a. time varying systems,

b. non-interacting systems: sparsity patterns,

c. non-linear systems,

d. realistic performance specs.

This talk focuses on the last topic, since the first is already well developed, a systematic theory of the second appears to be exceptionally hard, and Joe Ball will lecture on the third topic.

For completeness we note that there are several other directions. One is infinite dimensional $H^\infty$ control. It was not listed since philosophically it is the same as finite dimensional $H^\infty$ control, however, mathematically it is much more difficult and so is something of a subject unto itself. Another very different direction lies in adapting $H^\infty$ control to radically different uses, such as adaptive control.

As most MTNSer's know there are various areas of several complex variables which connect seriously with electrical engineering and these are admirably described in Bose's book **[Bose]**. In the talk we show how unflinching pursuit of an $H^\infty$ design theory leads to a serious connection between engineering and yet another area of SCV.

It is interesting to compare this with the early days of $H^\infty$ design. There a key element was the discovery that a well developed area of mathematics, interpolation and commutant lifting theory solved paridigm amplifier and control problems. This led to extremely fast engineering progress. I doubt that the connections pointed out in this article will have an instantaneous effect. The reason is that the mathematical area involved is not yet so well developed; much remains poorly understood and it has no tradition of numerical work. Hopefully, though over the long haul progress by each community will substantially benefit both.

## I. PROBLEM STATEMENT

A basic problem in designing stable systems is this:

*At each $e^{i\theta} \in \Pi$ we are given a region $S_\theta \subset \mathbb{C}^N$ (which represents specs to be met at 'frequency' $e^{i\theta}$).*

(OPT′) ***Find a vector valued function $f$ analytic in the unit disk $\Delta$ and continuous on its closure $\bar{\Delta}$, such that***

$$f(e^{i\theta}) \in S_\theta \quad \forall\, \theta.$$

Here $\Pi$ stands for the unit circle $\{|z| = 1\}$ in $\mathbb{C}$ and henceforth $A_N$ denotes the set of functions $f$ on $\Pi$ which are analytic on $\Delta$ and continuous in $\bar{\Delta}$. It is easy to put many design problems into this form, (e.g. the Horowitz templates of control) so the issue quickly becomes mathematical: computing solutions and developing a useful qualitative theory.

I always think of a picture in connection with (OPT′).

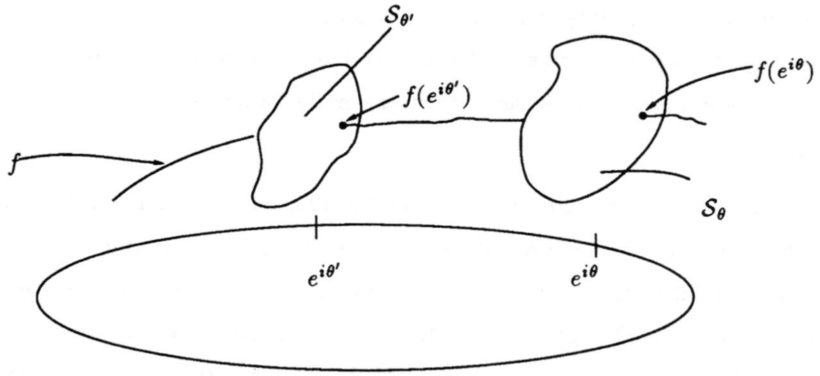


Figure I.1.

The problem (OPT′) is very closely related to

(OPT) *Given* $\Gamma \geq 0$ *a map from* $\Pi \times \mathbb{C}^N$ *to* $R$ *(which is a performance measure). Find* $\gamma^* \geq 0$ *and* $f^* \in A_N$ *which solve*

$$\gamma^* = \inf_{f \in A_N} \sup_\theta \Gamma(e^{i\theta}, f(e^{i\theta})) = \sup_\theta \Gamma(e^{i\theta}, f^*(e^{i\theta})).$$

Indeed (OPT′) is a graphical version of

($\text{OPT}_c$) *For a fixed (performance level)* $c$ *find a function* $f \in A_N$ *with*

$$\Gamma(e^{i\theta}, f(e^{i\theta})) \leq c, \ \forall \theta.$$

(That is, $f$ produces performance better than $c$). To see that (OPT′) solves ($\text{OPT}_c$) we start with $\Gamma$ and denote its sublevel sets corresponding to $c$ by

$$(I.1) \qquad \mathcal{S}_\theta(c) \triangleq \{z \in \mathbb{C}^N;\ \Gamma(e^{i\theta}, z) \leq c\}.$$

Now we take $\mathcal{S}_\theta = \mathcal{S}_\theta(c)$ in (OPT′) and see that

$$f(e^{i\theta}) \in \mathcal{S}_\theta \quad \text{if and only if} \quad \Gamma(e^{i\theta}, f(e^{i\theta})) \leq c.$$

Thus $f \in A_N$ solves (OPT′) if and only if $f$ solves ($\text{OPT}_c$). Conversely, to go from ($\text{OPT}_c$) to (OPT′) one merely builds a *defining function* $\Gamma$ for the given set $\mathcal{S}_\theta$, that is, build a $\Gamma$ which satisfies

$$\Gamma(e^{i\theta}, \ ) = 1 \quad \text{on} \quad \partial\mathcal{S}_\theta$$

$$\Gamma(e^{i\theta}, \ ) < 1 \quad \text{inside} \quad \partial\mathcal{S}_\theta$$

$$\Gamma(e^{i\theta}, \ ) > 1 \quad \text{outside} \quad \partial\mathcal{S}_\theta.$$

Then one solves ($OPT_1$) for that $\Gamma$. The problem is formidable, but what is surprising and makes a good story for a general audience is how this problem meshes with a branch of intense research within the mathematical several complex variables community.

The fact that pursuits and results on this problem now connect solidly with established SCV theory is a recent development (the last three to four years). Indeed I am grateful to R. Rochberg for first suspecting work of my colleagues and I on (OPT) was related to his work on "harmonic continuation" of sets. Then Wermer patiently explained basics of polynomial hulls (§IIIb) when we met at several conferences. Z. Slodkowski also helped by sending valuable reprints and providing more explanations. Connections with Kobayashi distance problems, §IIIc were developed jointly in discussions with my colleague Jim Agler. As a consequence of (OPT) developing independently from these other lines of SCV we are now at a place where someone who studies the SCV literature carefully might make considerable progress on it.

We give a warning: Descriptions here are for functions analytic in the disk which may or may not be real on the real axis. The effect of this important restriction is not traced throughout this article and in several key cases it has not been analyzed. For $N = 1$, Theorem IV.1 leads us to believe that many difficulties imposed by this restriction are surmountable. Also convex problems will be well behaved.

## II. ENGINEERING OCCURENCES OF (OPT)

First we give generic engineering motivation for (OPT). Then we give specific examples.

### (a) **Motivation**

The (OPT) problem is central to the design of a system where specifications are given in the frequency domain and stability is a key issue. Suppose our objective is to design a system part of which we are forced to use (in control it is called the plant) and part of which is designable

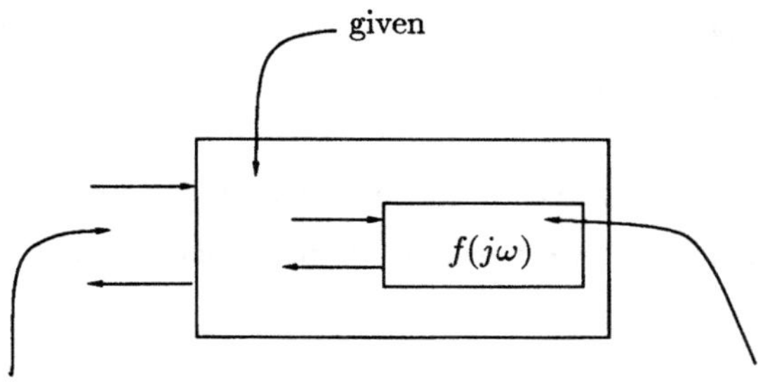


Figure II.1.

The objective of the design is to find the admissible $f$ which gives the best performance. The "worst case" is the frequency $\omega$ at which

$$\sup_{\omega} \Gamma(\omega, f(j\omega))$$

occurs. One wants to minimize this over all admissible $f$. The stipulation that the designable part of the circuit be stable amounts to requiring that $f$ has no poles in the R.H.P. In other words $f \in A_N$ (R.H.P.). This is exactly the (OPT) problem for the R.H.P and, of course, conformally transforming R.H.P. to $\Delta$ transforms this problem to precisely the (OPT) problem we stated in §1. However, in this section we shall stay with the R.H.P. and $jw$-axis formulation of problems as is conventional in engineering. . Even when parts of the system other than the designable part are in $H^\infty$ one can frequently reparametrize to get (OPT). Consequently (OPT) arises in a large class of problems.

Indeed the (OPT) problem is so basic that I am fond of calling it *the fundamental $H^\infty$ problem of control.* This sits in distinction to the fundamental problem of $H^\infty$-control. I don't know what that problem is.

(b) **Examples**

**Example 1.** The famous "mixed sensitivity" performance measure of control is

(IIb.1) $$\tilde{\Gamma}(\omega, T) \triangleq W_1(j\omega)\,|T-1|^2 + W_2(j\omega)\,|T|^2$$

where $W_1$ weights low frequencies and $W_2$ weights high frequencies. The *basic $H^\infty$ control problem* is

*Find a compensator which produces an internally stable system with acceptable performance (mixed sensitivity) over all frequencies.*

This converts directly to an (OPT) problem over functions $T$ analytic in the R.H.P. (denoted $A(R.H.P.)$) which meet interpolation conditions

$$\text{(INT)} \qquad T(\xi_k) = r_k \quad k = 1, 2, \ldots, m$$

imposed by the engineering system one wants to control. Then we get the mathematical statement:

**Basic $H^\infty$ Control Problem:** Find such a $T$ which gives a certain performance

$$\Gamma(\omega, T(j\omega)) \leq c\,.$$

This is the ($\text{OPT}_c$) problem except for the interpolation constraints.

The interpolation constraints are easily dealt with by a reparametrization in function space. For example, if $m = 2$, then express $T$ as

$$\text{(IIb.2)} \qquad T(\xi) = \frac{1}{(1+\xi)^2}\left[r_1 \frac{(\xi - \xi_2)}{\xi_1 - \xi_2}(1+\xi_1)^2 + r_2 \frac{(\xi - \xi_1)}{\xi_2 - \xi_1}(1+\xi_2)^2\right] + \frac{(\xi - \xi_1)(\xi - \xi_2)}{(1+\xi)^2} H(\xi)$$

where $H$ is in $A$(R.H.P.). Clearly $T$ sweeps through the desired class as $H$ sweeps through $A$(R.H.P.). Abbreviate (IIb.2) to $T = a + bH$ and substitute into $\tilde{\Gamma}$ to define

$$\Gamma(\omega, H) \triangleq \tilde{\Gamma}(\omega, a + bH)\,.$$

Then ($\text{OPT}_c$) for $\Gamma$ is equivalent to the basic control problem.

The graphical version of this problem goes as follows. First note via simple algebra that the sublevel sets $\mathcal{S}_\omega(c)$ of $\tilde{\Gamma}$ are always disks and that they vary in a pattern like this:

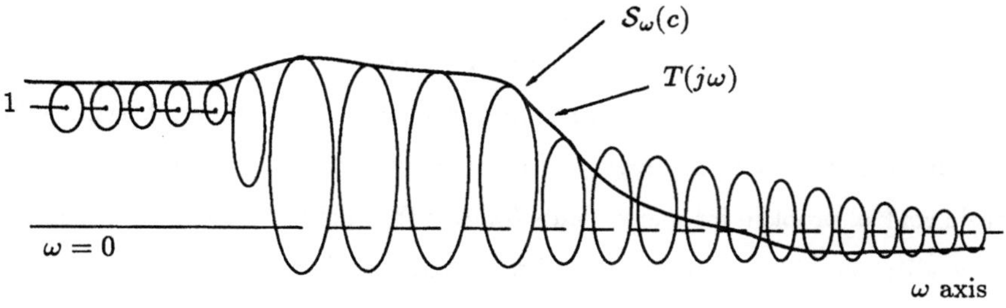


Figure II.2.

Thus finding $T$ in $A$ meeting (INT) whose values on the $j\omega$ axis lie in $\mathcal{S}_\omega(c)$ is equivalent to the basic problem.

Historically, the algebraic formulation (IIb.1) and graphical formulation Fig. II.2 together with effective proposals for solution were done independently in 1983 by Kwaakernak **[Kw]** and Helton **[H1]**, respectively. Also Doyle **[Dy1]** simultaneously gave a different approach which also was both physically correct and an effective computationally. The original paper of Zames and Francis **[ZF]** took $W_2 = 0$.

I might insert a caveat here to any practically oriented individual. Physically one is not given $W_1$ and $W_2$. The basic $H^\infty$ control problem is a great abbreviation of the design process. In my opinion very little intelligent discussion about design can be carried out at this level of abbreviation.

Indeed a salty comment of mine along these lines pertains to a debate which persists in the $H^\infty$ control community. The issue is whether or not a control theory will ever exist which in practice sets the weights $W_1$, $W_2$ once and for all. Opponents contend that in an industrial setting enormous tuning of the weights $W_1$, $W_2$ must occur. I don't wish to take sides here on the outcome. What impresses me about this debate is less the arguments of one side or the other, but the fact that it has not evolved significantly in the last five years. I (somewhat tongue in cheek) attribute this lack of progress to the physical imprecision of talking primiarly in terms of mixed sensitivity (IIb.1) and $W_1$, $W_2$. The $W_1$, $W_2$ are in fact derivable (with explicit formulas) from more primitive specifications such as tracking error, gain-phase margins, bandwidth constraints, etc. (c.f. **[H1]**, **[H2]**, **[BHMer]**). If the $H^\infty$ culture commonly used this or an even more precise language possibly the debate on tuning could advance to a higher plane.

It seems to me that another practical issue might be worth emphasizing (since it is frequently treated incorrectly [**A**]). It is an outgrowth of the fact that the basic control problem for many plants (ones with a pole or zero on the $jw$ axis) is ill conditioned. This is because the basic control problem has not been correctly posed; the difficulty lies in the fact that the usual notion of internal stability:

(IIb.3) $$T,\ (I+PC)^{-1},\ (I+PC)^{-1}P,\ C(I+PC)^{-1} \quad \text{are in} \quad H^\infty$$

while philosophically correct is too loose practically speaking. In particular a designer must have specified initially enough constraints to have produced numbers (or bounded functions)

$$M_1,\ M_2,\ M_3,\ M_4$$

so that

(IIb.4)

$$|T| \le M_1,\ \ |(I+PC)^{-1}| \le M_2,\ \ |C(I+PC)^{-1}| \le M_3,\ \ |(I+PC)^{-1}P| \le M_4\,,$$

through the entire R.H.P. or even on a slightly larger region. Note that (IIb.4) is just a strengthened form of (IIb.3), in that (IIb.3) says that these functions must be bounded in the R.H.P., but does not say by how much. The point is that we must a priori say what these bounds are.

As mentioned before, forgetting constraints (IIb.4) becomes deadly at a $jw$-axis pole or zero of $\mathbb{P}$. For example, when $P(jw_0) = 0$ a naive $H^\infty$ solution produces compensators with $|C(jw_0)|$ of arbitrarily large size. Fortunately, adding constraints (IIb.4) to the standard $H^\infty$ control solution is easy to do using a function space reparameterization like (IIb.2). The interested reader is referred to Part I [**BHMer**] which does an example thoroughly; also [**H2**] mentions this tersely.

**Example 2.** Power mismatch (cf. [**H4,5**], [**Y**], [**YS**]).

**Example 3.** Two competing constraints typically yield $\mathcal{S}_\theta$ which are intersections of two disks, etc.

**Example 4.** Plant uncertainty naturally leads to (OPT) problems with very complicated $\Gamma$. Our formulation is to start with a performance measure $\tilde{\Gamma}$ which depends on what one believes the plant $P$ to be at frequency $\omega$ and the choice $T$ of the designable parameter at $\omega$. The basic design optimization problem is:

$$\text{(UNCOPT}'\text{)} \qquad \inf_{T\in A_N^\infty} \sup_{\omega} \sup_{p\in R_\omega} \tilde{\Gamma}(\omega,p,T(j\omega)).$$

Here $R_\omega$ denotes the range of values $p$ at frequency $\omega$ which you believe the plant $P(j\omega)$ might actually take. For this problem "tightening the specs" amounts to calculating the "tightened" performance measure

$$\text{(UNC)} \qquad \Gamma(\omega,T) = \sup_{p\in R_\omega} \tilde{\Gamma}(\omega,p,T).$$

After this is done solving the full (UNCOPT′) problem is equivalent to (OPT).

Plant uncertainty when treated in this way simply amounts to a mathematization of the age old engineering adage:

**In the presence of uncertainty tighten the specs.**

## III. CONNECTIONS WITH TRADITIONAL MATHEMATICS

In this section we work on the unit disk $\Delta$, the unit circle $\Pi$ rather than on the R.H.P. and the $j\omega$-axis. Thus, $\mathcal{S}_\omega$ becomes $\mathcal{S}_\theta$, etc.

(a) **Quasicircular (OPT) problems**

Call a $\Gamma$ with sublevel sets which are disks (even higher dimensional ones) *quasicircular*. A marvelous piece of luck is that when the $\mathcal{S}_\theta$ are disks (even in higher dimensions), the (OPT′) problem was solved by pure mathematicians starting at the turn of the century with $N = 1$ and moving to higher $N$ in the 1950's and 60's with operator theorists. This was first used effectively by Saito and Youla (when $N = 1$) and by Helton (when $N \geq 1$) on amplifier problems. The techniques were later taken up by Zames and Francis who learned them from Helton and introduced them to control. In an independent movement Tannenbaum solved a paradigm ($N = 1$) control problem with this mathematics. Subsequent to this exciting insight there has been a rush of activity called $H^\infty$ control and most mathematical development in the West since the mid seventies has either been done by engineers or by associated mathematicians. This engineering oriented community has made major contributions to the mathematics of the problem particular in the context of a state space theory. There is so much good work along these lines that we neither survey nor list references here, but refer the reader to the books [**BGR**], [**Fr**], [**H3**], [**Her**].

When we turn to $\mathcal{S}_\theta$ which are not disks all of the mathematics of the previous paragraph fails. I know of no (OPT′) problem with an "explicit solution" for cases where the $\mathcal{S}_\theta$ are not "generalized disks." One must either develop completely new techniques or find them in a different mathematics literature. That is the subject of this talk.

(b) **Polynomial hulls**

This is the main mathematical connection we describe in this talk. Define the *envelope of solutions* to (OPT′) to be

$$E(\mathcal{S}) \triangleq \{(\xi, f(\xi)) : \text{ all } f \in A_N, \quad f(e^{i\theta}) \in \mathcal{S}_\theta \text{ all } \theta, \ |\xi| \leq 1\}.$$

This is a natural object physically in that it is the envelope of all design possibilities. Its *cross sections* $E(\mathcal{S})|_\xi$ are defined to be $E(\mathcal{S})|_\xi \triangleq \{z : (\xi, z) \in E(\mathcal{S})\} = \{f(\xi) : f \in A_N, \ f(e^{i\theta}) \in \mathcal{S}_\theta \ \forall \theta\}$.

A surprising thing is that a classical mathematical object of a very different character, the polynomial hull $\mathbf{P}(\mathcal{S})$ of the set

$$\mathcal{S} = \{(e^{i\theta}, \mathcal{S}_\theta) : 0 \leq \theta \leq 2\pi\},$$

is informative in understanding the envelopes. The reason put loosely is that in many (and possibly all nice) cases *it is the "envelope" of all solutions to* (OPT′). Recall that the *polynomial hull* $\mathbf{P}(S)$ of a set $S \subset \mathbb{C}^{1+N}$ is

$$\mathbf{P}(S) \triangleq \{w \in \mathbb{C}^{1+N}: |p(w)| \leq \max_{v \in S} |p(v)| \quad \text{for all} \quad p \in \mathcal{P}^{1+N}\}.$$

Here $\mathcal{P}^{1+N}$ denotes the set of all polynomials on $\mathbb{C}^{1+N}$. The set $S$ is called *polynomially convex* provided $S$ equals its convex hull. An excellent basic reference on this topic is [**W1**].

It is well known that

**Theorem IIIb.1.**

$$\mathcal{S} \cup E(\mathcal{S}) \subset \mathbf{P}(\mathcal{S}).$$

**Proof.** Of course $\mathcal{S} \subset \mathbb{P}(\mathcal{S})$ so we must show that $E(\mathcal{S}) \subset \mathbb{P}(\mathcal{S})$. Now each point $(\xi_0, z_0)$ in $E(\mathcal{S})$ lies on the graph $\{(\xi, f(\xi));\ \xi \in \text{ disk}\}$ of some solution $f$ to (OPT′). To show that $(\xi_0, z_0)$ also lies in $\mathbb{P}(\mathcal{S})$ select any polynomial $p$ in $\mathcal{P}^{1+N}$ and note by the standard one complex variable maximum principle that

$$|p(\xi_0, z_0)| = |p(\xi_0, f(\xi_0))| \leq \sup_{\theta} |p(e^{i\theta}, f(e^{i\theta}))| .$$

Since $f$ solves (OPT′) we have $f(e^{i\theta})$ is in $\mathcal{S}_\theta$ and conclude that

$$|p(\xi_0, z_o)| \leq \sup_{(\xi,z)\in\mathcal{S}} |p(\xi, z)| .$$

The fact that this holds for all $p \in \mathcal{P}^{1+N}$ is precisely the statement that $(\xi_0, z_0) \in \mathbb{P}(\mathcal{S})$. ∎

An open question which is the subject of substantial mathematical research is

$(\mathbb{P} = E)$ For nice $\mathcal{S}$ does

$$\mathcal{S} \cup E(\mathcal{S}) = \mathbb{P}(\mathcal{S})?$$

The implication of this (when it is true) is that computing the envelope of solutions to (OPT′) is equivalent to computing $\mathbb{P}(\mathcal{S})$, since after all $\mathcal{S}$ is already known. Thus we have a radically different way of characterizing $E(\mathcal{S})$ which might ultimately be used for computation or to gain qualitative insight. While for pathological $\mathcal{S}$ the answer to $(\mathbb{P} = E)$ is no [**W2**], the answer is yes in many cases and we now turn to that issue.

The condition $(\mathbb{P} = E)$ is known to be true in cases which include:

**Theorem IIIb.2.** (Alexander-Wermer [**AW1**], Slodkowski [**Sl1**]). *If the sets $\mathcal{S}_\theta$ are all convex and uniformly (in $\theta$) bounded, then $\mathcal{S} \cup E(\mathcal{S}) = \mathbb{P}(\mathcal{S})$.*

For $N = 1$ Slodkowski [**Sl2**] obtained:

**Theorem IIIb.3.** *For $N = 1$ if the sets $\mathcal{S}_\theta$ are connected, simply connected, smoothly varying in $\theta$ with boundaries $\partial\mathcal{S}_\theta$ which are analytic arcs, then $\mathcal{S} \cup E(\mathcal{S}) = \mathbb{P}(\mathcal{S})$.*

Also in 1988 Marshall and Helton (see [**MH**]) independently discovered a proof which we sketch in §IV, since it is in the spirit of the (OPT) problem.

Now we backtrack. To this point we have the formal definition of polynomial hull and motivation (via envelopes) for studying it. Let us give some vague intuition through pictures and examples of what the polynomial hull is. We begin by defining the *cross-sections* or the *fibers* of the polynomial hull $\mathbf{P}(\mathcal{S})$ over $\xi$ to be

$$\mathbf{P}(\mathcal{S})_\xi = \{z \in \mathbf{C}^N : (\xi, z) \in \mathbf{P}(\mathcal{S})\}.$$

One can abbreviate this to $\mathcal{S}_\xi$, but it is an abuse of notation, since $\mathcal{S}_{e^{i\theta}}$ in this notation corresponds to the $\mathcal{S}_\theta$ we have been using. The following figure illustrates the gross nature of $\mathbf{P}(\mathcal{S})$

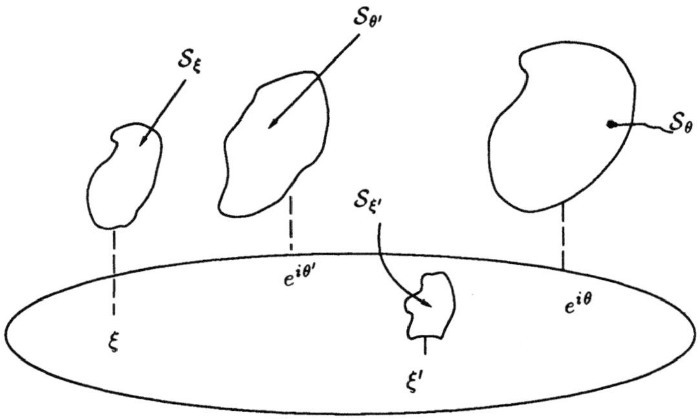


Figure III.1.

in that the $\mathcal{S}_\theta$ are prescribed on the unit circle and the $\mathcal{S}_\xi$ are a canonical uniquely determined extension of these sets into the unit disk. A basic fact in the subject (cf. Lemma 11.1 [**W1**]) is

**Theorem IIIb.4.** *If for any $\xi$ inside the open unit disk the fiber $\mathcal{S}_\xi$ is nonempty, then the fiber of the polynomial hull over every point in the unit disk is nonempty.*

That is, the polynomial hull either extends the "specs" $\{\mathcal{S}_\theta\}$ over the entire unit disk or over none of it. Clearly in the second case no solutions to $(\mathrm{OPT}_c)$ for $\mathcal{S}$ exist, so the envelope of solutions is empty.

A simple explicit example also helps in understanding polynomial hulls.

*Example.* Each $\mathcal{S}_\theta$ is a disk in $\mathbb{C}$ with center at the origin and radius $R(e^{i\theta})$. What is $\mathbf{P}(\mathcal{S})$? Answer: Its fibers are disks centered at 0, thus we need only compute the radius of the disk $\mathcal{S}_\xi$ at each $\xi$. It turns out to be $|\alpha(\xi)|$ where $\alpha(\xi)$ is the outer Wiener-Hopf (spectral) factor $\alpha(e^{i\theta})\overline{\alpha(e^{i\theta})} = R(e^{i\theta})^2$ of $R^2$.

From this example we see that computing polynomial hulls is bound up with and in general more complicated than computing Wiener-Hopf factorizations.

**Results about polynomial hulls.** A brief bibliography is included at the end of this paper. For $N = 1$ there is a strong

**Theorem IIIb.5.** (Forstneric [**F**]) *If* (OPT′) *has a solution* $f^*$ *whose values lie in the interior of* $\mathcal{S}_\theta$ *uniformly in* $\theta$, *then there is a function* $\varphi\colon \bar{\Delta} \times \Pi \to \mathbb{C}$ *which satisfies*

(i) *Fix* $\tau$, $\varphi(\ ,\tau)\colon \Delta \to \mathbb{C}$ *is analytic with no zeroes in* $\bar{\Delta}$.

(ii) *Fix* $\xi \in \bar{\Delta}$, *then* $\varphi(\xi,\ )\colon \Pi \to \mathbb{C}$ *is a smooth* $1-1$ *map onto the curve which is its range.*

*The boundary* $\partial E(\mathcal{S})$ *of* $E(\mathcal{S})$ *satisfies*

$$\text{(T)} \qquad \mathcal{S} \cup \partial E(\mathcal{S}) = \{(\xi, \varphi(\xi,\tau) + f^*(z))\colon \xi \in \Delta \quad \tau \in \Pi\}\,.$$

*This completely characterizes* $E(\mathcal{S})$ *provided that one can obtain* $\varphi$. *Forstneric's proof is not constructive.*

Note that when the $\mathcal{S}_\theta$ are all disks there is a linear fractional parametrization of all solutions of (OPT′) due to Nevanlinna for $N = 1$ and for $N \geq 1$ to Adamajan-Arov-Krein, see also [**ACF**]. The formula in this parametrization easily parametrizes $\partial E(\mathcal{S})$ and produces the $\varphi$ satisfying (i), (ii) and (iii). Thus Theorem IIIb.5 might be seen as a weaker Nevanlinna parametrization which holds in very general problems.

Another completely different finding about polynomial hulls for $N = 1$ actually seems like another type of duality. It says that each polynomial hull is the set of all singularities of certain classes functions on $\mathbb{C}^2$ analytic near $\infty$, see [**AW2**] and [**W3**]. How this pertains to $N \geq 2$ is unclear.

In another direction mathematicians try to provide explicit formulas and a concrete analysis of particular model cases. In particular Alexander and Wermer

[**AW3**] just did a successful study of the situation which each $\mathcal{S}_\theta$ is an interval $[a(e^{i\theta}), b(e^{i\theta})]$ in $\mathbb{C}$ with $a \in \mathcal{A}$. They analyzed the polynomial hull of $\mathcal{S} = \{(e^{i\theta}, \mathcal{S}_\theta)\}$ for various classes of functions $b$.

**A computational possibility.** We conclude this subsection by mentioning a radical possibility for performing $H^\infty$ (OPT) calculations. Let $\gamma^*$ and $f^*$ denote the solution to (OPT) and suppose that we know ($\mathbb{P} = E$) is true. Then we know that

$$c < \gamma^* \quad \text{if and only if} \quad E(\mathcal{S}(c)) \quad \text{is empty}$$

$$\text{if and only if} \quad \mathbb{P}(\mathcal{S}(c)) = \mathcal{S}\,.$$

This suggests that to compute whether ($\text{OPT}_c$) has a solution for a particular $c$ we could do a computation with $\mathbb{P}(\mathcal{S}(c))$. It turns out that it suffices to compute whether $\mathbb{P}(\mathcal{S}(c))$ has any point in it of the form $(0, a)$, that is a point for which

$$|p(0,a)| \le \sup_{z \in \mathcal{S}_\theta(c)} |p(e^{i\theta}, z)| \tag{IIIb.1}$$

for all $p(\xi, z)$ polynomials in $\xi \in \mathbb{C}$ and $z \in \mathbb{C}^N$. Recall that $\mathcal{S}_\theta(c)$ is defined in (I.1).

Theorem IIIb.4 says that $\mathbb{P}(\mathcal{S})$ contains such a point $(0, a)$ if and only if the envelope of solutions to (OPT) is nontrivial. Therefore $c \ge \gamma$ if and only if a point $(0, a)$ satisfying (IIIb.1) exists; provided ($\mathbb{P} = E$) is true.

Now inequalities like (IIIb.1) don't seem easy to manipulate with package software. However, one could work with an analog concocted by Merino and me. Define a hull $R\mathbb{P}(\mathcal{S})$ of $\mathcal{S} = \{(e^{i\theta}, \mathcal{S}_\theta)\colon -\pi \le \theta \le \pi\}$ to be

$$R\mathbb{P}(\mathcal{S}) \triangleq \{(\xi, z)\colon \operatorname{Re} p(\xi, z) \ge 0 \text{ for all } p \in \mathcal{P}^{1+N} \text{ satisfying } \operatorname{Re} p(s) \ge 0 \text{ on all } s \in \mathcal{S}\} \tag{IIIb.2}$$

One expects that in nice cases $R\mathbb{P}(\mathcal{S}) = \mathbb{P}(\mathcal{S})$. Now inequalities of the form $\operatorname{Re} p(\xi, z) \ge 0$ are convex and consequently (once discretized can be treated with a linear programming package).

Now we are a little more specific about how one converts this to a linear program:

To test if $(0, a) \in R\mathbb{P}(\mathcal{S})$,

$$\inf_{p \in \mathcal{P}^{1+N}} \operatorname{Re} p(0, a) \triangleq \eta(a) \tag{IIIb.3}$$

subject to

$$\text{Re}\,p(s) \geq 0 \quad \text{for all} \quad s \in \mathcal{S}\,.$$

Clearly $(0, a) \in \mathbb{R}\mathbb{P}(\mathcal{S})$ if and only if the answer is $\geq 0$. We are interested in vectors $a$ with real entries and in $\max_{a \in R^N} \eta(a)$. It is $\geq 0$ if and only if $\mathbb{R}\mathbb{P}(\mathcal{S})$ has nontrivial fiber over $s = 0$. Possibly a Broyden, etc. method would work reasonably for this stage of the computation.

Finally we add one more layer of detail to our description of the computation of (IIIb.3). This is the crudest possible method and surely one could improve it easily. For pedagogical simplicity take $N = 2$. Discretize $\partial\mathcal{S}$ by selecting many points $s^{k,\ell} \triangleq (e^{i\theta_k}, (z_1^{k\ell}, z_2^{k,\ell}))$ in $\partial\mathcal{S}$. The variables in the linear program are numbers $x_{m,n,k}$ which are the coefficients of the polynomials $p$. Then (IIIB.3) becomes, for fixed $a = (a_1, a_2)$

(IIIb.4) Minimize Re $\sum_{m,n=0}^{M} x_{m,n,0}(a_1)^m(a_2)^n$ subject to ($x$ under "Minimize")

$$\text{Re} \sum_{r=0}^{K} \sum_{m,n=0}^{M} x_{m,n,r}(e^{i\theta_k})^r(z_1^{k\ell})^m(z_2^{k\ell})^n \geq 0$$

for all $s^{k,\ell}$.

This is a standard linear program (once each complex number $x_{m,n,r}$ is expressed as a sum of 4 nonnegative numbers). Here we would hope that small $M$ say 2 or 3 would suffice to solve many interesting problems. In this case running time often would be proportional to

$$[(K+1)(M+1)^2]^3 C$$

with $C$ depending in a gentle way on the number of constraint points $s^{k,\ell}$.

**(c) The Kobayashi distance between two points $q$ and $r$ in a domain $\mathcal{D}$ in $\mathbb{C}^N$.**

The basic issue is to study metrics on $\mathcal{D}$ which are invariant under biholomorphic maps $b: \mathcal{D} \leftrightarrow \mathcal{D}$. If $\mathcal{D}$ is the unit disk in $\mathbb{C}$ there is essentially one, the Poincaré metric $\rho(z_1, z_2) = \text{arctanh}\left|\frac{z_1 - z_2}{1 - \bar{z}_1 z_2}\right|$. In higher dimensions there are many inequivalent invariant metrics on $\mathcal{D}$. One is the Caratheodory metric defined by

(CAR) $$c(q, r) = \sup_F \rho(F(q), F(r))$$

over $F: \mathcal{D} \to \Delta$ which is analytic on $\mathcal{D}$. Another is the true Kobayashi metric similar to the "one disk" Kobayashi distance between $q$ and $r$ defined by

$$\text{(KOB)} \qquad \kappa(q,r) = \inf_{\xi_1,\xi_2 \in \text{disk}} \rho(\xi_1,\xi_2)$$

where $\xi_1, \xi_2$ must have the property that there exists an analytic function $f$: disk $\to$ $\mathcal{D}$ such that $f(\xi_1) = q$ and $f(\xi_2) = r$. For given $q, r$ and $\xi_1, \xi_2$ which achieve $\inf \rho$ in (KOB) an analytic function $f$: disk $\to \mathcal{D}$ satisfying $f(\xi_1) = q$ and $f(\xi_2) = r$ is called a *Kobayashi extremal* for the given data $q, r$. Also an $F$ on which $\sup \rho$ in (CAR) is attained is called a *Caratheodory extremal.*

**Properties**:

1. $c$ and $\kappa$ are biholomorphic invariants.

**Proof**. Trivial.

2. $c(q,r) \leq \kappa(q,r)$.

**Proof**. Let $f^*(\xi_1) = q$, $f^*(\xi_2) = r$ and $F^*(q) = \xi_1'$, $F^*(r) = \xi_2'$ denote (KOB) and (CAR) extremals respectively. Then $g \triangleq F^* \circ f^*$: disk $\to$ disk and $g(\xi_1) = \xi_1'$, $g(\xi_2) = \xi_2'$. Schwartz's lemma says $\rho(g(\xi_1), g_2(\xi_2)) \leq \rho(\xi_1, \xi_2)$. This implies $\rho(F^*(q), F^*(r)) \leq \rho(\xi_1, \xi_2)$ for all $(\xi_1, \xi_2)$ admissible in (KOB), so $\rho(F^*(q), F^*(r)) \leq \inf_{\xi_1,\xi_2} \rho(\xi_1, \xi_2)$. ∎

The observation we make here is that computation of Kobayashi extremals is a very special case of (OPT′); at least if we are willing to iterate through a sequence of (OPT′) problems. To wit: Given $q, r$ select two (candidate) points $0, \xi_2 \in$ disk $\subset \mathbb{C}$. We want to know if $\exists f$: disk $\to \mathcal{D}$ satisfying $f(0) = q$ and $f(\xi_2) = r$. If such $f$ exists $\kappa(q,r) \leq \rho(0, \xi_2)$. By repeatedly guessing $\xi_2 \geq 0$ one can actually compute $\kappa(q,r)$.

To convert this to an (OPT′) problem we use the same type of reparametrization in function space which you saw in Example 1 on $H^\infty$ control. Let $a$: disk $\to$ $\mathbb{C}_N$ be analytic and satisfy the interpolating condition $a(0) = q$ and $a(\xi_2) = r$. Let $b$: disk $\to \mathbb{C}$ be analytic and satisfy $b(0) = 0 = b(\xi_2)$ and write $f$ as

$$f = a + bH$$

$H \in A_N$. Now such $f$ meets interpolating constraints but may not satisfy $f(\xi) \in \mathcal{D}$ for all $|\xi| \leq 1$. To invoke this condition we transform $\mathcal{D}$ to

$$\text{(IIIc.1)} \qquad \mathcal{S}_\theta \triangleq \frac{\mathcal{D} - a(e^{i\theta})}{b(e^{i\theta})}$$

because $\mathcal{S}_\theta$ has the property

$$H(e^{i\theta}) \in \mathcal{S}_\theta \quad \text{if and only if} \quad f(e^{i\theta}) \in \mathcal{D}.$$

There is one more thing to check: We require $f(\xi) \in \mathcal{D}$ for $\forall \xi \in$ disk, but we only have $f(e^{i\theta}) \in \mathcal{D}$ for all $e^{i\theta}$. If $\mathcal{D}$ is polynomially convex then the two properties are equivalent ([**K**, §3.1]). So we obtain

**Theorem IIIc.1.** *For a polynomially convex domain $\mathcal{D} \subset \mathbb{C}^N$, the od-Kobayashi distance $\kappa(q,r) \leq \rho(0,\xi_2)$ if and only if there is a solution $H^*$ to* (OPT′) *for the sets $\mathcal{S}_\theta$ given by* (IIc.1). *If $\xi_2$ produces equality, then $f^* = a + bH^*$ is a Kobayashi extremal.*

This is the main conclusion of the section: *Computing od-Kobayashi distances is a special case of* (OPT′).

What are the consequences of this? Clearly, computational tools provided by the engineering culture could be used to do experiments with the od-Kobayashi distance. In a more theoretical vein while theorems on $\kappa$ are probably too special for direct engineering use, they suggest generalizations which might be useful.

For example, the deepest result in the subject (see [**L**] Proc. ICM'86) is

**Theorem.** (Lempert). *If $\mathcal{D}$ is convex then the Caratheodory metric equals the od-Kobayashi distance equals the true Kobayashi metric.*

One major component of the proof is a characterization of $f^*$ the Kobayashi extremal. Note that Theorem IV.5 on our list of results also characterizes $f^*$. Indeed roughly these two theorems were discovered independently by Lempert (somewhat earlier) and Helton (more generally). Possibly other techniques of Lempert's will prove valuable. For example, he proves that $f^*$ is continuous. For (OPT) this is not known for $N > 1$ even when $\Gamma$ is a convex function. Maybe Lempert's techniques apply here.

The treatment of Kobayashi distance given here while capturing much of its content was done in an overly simple setting (focusing on the somewhat unusual "one disk" Kobayashi distance). This was done in order to avoid requiring the reader to recall the set up of differential geometry. Now we assume that the reader is familiar with this, in particular with the fact that metrics can typically be defined infinitesimally. The infinitesmal form of the Kobayashi metric is defined on the tangent space $\mathcal{D} \times \mathbf{C}^N$ of $\mathcal{D}$ to be

$$\text{(IIIc.3)} \qquad F_k(q,\eta) = \inf\{\alpha > 0 \colon\ \exists f \in A_N \quad f\colon \Delta \longrightarrow \mathcal{D}, \\ f(0) = q, \quad f_1'(0) = (\eta/\alpha, 0, \ldots, 0)^\dagger\}$$

Here $f_1$ is the first component of the vector valued function $f$. One might intuitively think of $F(q,\eta)$ as a directional derivative of $\kappa(q,r)$ discussed above.

The only point we wish to make here is that $F_k(q,\eta)$ can be computed by solving a succession of (OPT′) problems, in very much the same was that $\kappa$ was related to (OPT′) by (IIIc.1). To write down this relationship take $a_\alpha(\xi) = q + \left(\frac{\eta\xi}{\alpha}, 0, \ldots, 0\right)^\dagger$ and $b(\xi) = \xi_1^2$. Then as $H$ sweeps $A_N$, the function $f = a_\alpha + bH$ sweeps those functions $f \in A_N$ which satisfy $f(0) = a_\alpha(0) = q$ and $f'(0) = a_\alpha'(0) = (\eta, \alpha, 0, \ldots, 0)^\dagger$.

Suppose that we are given a defining function $\rho$ for the domain $D$. That is, $\rho$ is smooth and

$$\rho(z) < 1 \quad \text{for} \quad z \in D,$$

$$\rho(z) \geq 1 \quad \text{for} \quad z \notin D.$$

Then to produce an (OPT) problem define $\Gamma_\alpha$ by

$$\text{(IIIc.4)} \qquad \Gamma_\alpha(e^{i\theta}, z) = \rho(a_\alpha + bz) \quad \forall z \in \mathbf{C}^N$$

and denote by $\gamma_\alpha^*$ its solution

$$\gamma_\alpha^* \triangleq \inf_{f \in A_N} \sup_\theta \Gamma(e^{i\theta}, f(e^{i\theta})).$$

**Corollary IIIc.1.** *We have*

$$F_k(q,\eta) = \inf_\alpha \{\gamma_\alpha^* < 1\}.$$

**Proof.** Any $h \in A_N$ for which $\Gamma_\alpha(e^{i\theta}, h(e^{i\theta})) < 1$ produces $f = a_\alpha + bh \in A_N$ satisfying $\rho(f(e^{i\theta})) < 1$. That is $f: \bar{\Delta} \to \mathcal{D}$, and by construction $f$ satisfies the requirements to be in the set (IIIc.3). Thus $F_k(q, \eta) \leq \alpha$ if $\gamma_\alpha < 1$. This argument is reversable.

A solution $f^*$ to (OPT) for optimal $\alpha$ is called a Kobayashi extremal. Properties proceed much as before with $\kappa(q, r)$.

We conclude by mentioning that the Caratheodory distance is in some sense a dual to the Kobayashi extremal problem. The polynomial hull $\mathbf{P}(\mathcal{S})$ of $\{(e^{i\theta}, \mathcal{S}_\theta)\}$ with $\mathcal{S}_\theta$ given by (IIIc.1) is in another sense a dual construction to the Kobayashi extremal. Are these two seemingly different types of duality related? For example, can one use Theorem IIIb.2 and IIIc.1 below to prove Lempert's theorem?

### (d) Miscellaneous

It might help some engineers interested in reading SCV literature to remark on some standard terminology. The subject we have been discussing is very much bound up with what are called ***analytic disks***. The idea is that we have a domain $\mathcal{D}$ and are interested in ways in which the unit disk in the complex plane can fit in $\mathcal{D}$ (usually while meeting other constraints). That is, we are interested in non-constant analytic maps $f$: disk $\to \mathcal{D}$ and their image which is a set called an analytic disk. Properties and uses of analytic disks are discussed to some extent in [**K**, Ch. 3].

(OPT) is clearly a matter of finding analytic disks meeting certain constraints.

Another construct closely related to (OPT) is that of the ***analytic multifunction***. A multifunction on the disk is a function whose value at a point $\xi \in$ disk is a set in $\mathbf{C}^N$. There is a notion of a set valued function being ***analytic***. These are studied heavily by Slodkowski, see [**Sl3**] for a survey of results and definitions.

The value of all of this to us is that a polynomial hull $\mathbf{P}(\mathcal{S})$ introduced in section IIIb is the graph of an analytic multifunction, that is, the function $\xi \to \mathcal{S}_\xi$ is an analytic multifunction. Thus theory developed for analytic multifunctions applies to our problems.

## IV. QUALITATIVE RESULTS ON (OPT)

One of the most useful things to a person who is using a computer program to solve problems like (OPT) is a knowledge of the fundamental qualitative properties of solutions. Then when the program produces odd answers, or as will sometimes happen, fails to converge the user can have an idea of what is happening or at least eliminate some possibilities. Our standard assumption on the (OPT) problem is:

(SA) $\Gamma$ depends smoothly on $\theta$, is real analytic in $z$ (and in $\bar{z}$) and has gradient $\overline{\frac{\partial\Gamma}{\partial z}(e^{i\theta}, z)}$ which never vanishes when $\Gamma(e^{i\theta}, z) = \gamma^*$. The sets $\mathcal{S}_\theta(\gamma^*)$ are connected, simply connected, have nonempty interior, and are uniformly bounded in $\theta$.

While $\gamma^*$ may not be known in advance in a particular situation, one might verify that all $\mathcal{S}_\theta(\gamma)$ for a wide range of $\gamma$ satisfy these conditions; this is because the conditions are not very restrictive.

We now give a list of results. It basically follows the lines of **[BHM]** and updates that list somewhat in keeping with **[MH]**.

### Existence, Smoothness, and Uniqueness of Solutions

**Theorem IV.1.** **[MH]**. *Suppose $N = 1$ and* (SA) *holds. An $H^\infty$ solution $f^*$ to* (OPT) *exists. That solution $f^*$ is smooth*[1] *(i.e., $f^* \in C^\infty$). If $f_0 \in A_1$ is a local optimum to* (OPT) *and if $\frac{\partial\Gamma}{\partial z}(e^{i\theta}, f_0^*(e^{i\theta}))$ never vanishes on $\Pi$, then $f_0 = f^*$. If $\Gamma(e^{i\theta}, z) = \Gamma(e^{-i\theta}, \bar{z})$ for all $\theta$ and $z$, then $f^*$ is real on the real axis.*

When $N \geq 1$ we have

**Theorem IV.2.** **[HH]**. *If $\mathcal{S}_\theta(\gamma^*)$ is strictly convex (uniformly in $\theta$), then an $H_N^\infty$ solution $f^*$ to* (OPT) *exists and it is unique. Also $f^*(e^{i\theta}) \in \partial\mathcal{S}_\theta$ for a.e. $\theta$.*

**Theorem IV.3.** **[MH]**. *Suppose* (SA) *holds and that each $\mathcal{S}_\theta$ is polynomially convex. Then an $H_N^\infty$ solution $f^*$ to* (OPT) *exists. Moreover, if a sequence $f_k \in H_N^\infty$ approximately solves* (OPT) *(in the sense* $\operatorname*{ess\,sup}_\theta \Gamma(e^{i\theta}, f_k(e^{i\theta})) = \gamma^k$ *with $\gamma^k \searrow \gamma^*$). Then a subsequence which converges in normal family sense has as its limit a function $f_\infty$ in $H_N^\infty$ which satisfies $\Gamma(e^{i\theta}, f_\infty(e^{i\theta})) \leq \gamma^*$ almost everywhere.*

**Example 1.** (Helton, Merino **[HMer2]**).

[1] Originally S. Hui **[Hu]** proved Theorem IV.1 under the additional assumption of convexity. Also he showed that $f^*$ extends analytically across the circle.

$$\Gamma(e^{i\theta}, z_1, z_2) = |100 + e^{i\theta} z_1 + 0.1(z_1 z_2 + z_1 + z_2)|^2$$
$$+ |100 + e^{i\theta} z_2 + 0.1(z_1 z_2 + z_1 + z_2)|^2$$
$$+ \epsilon(|z_1|^2 + |z_2|^2)$$

for $\epsilon$ a real parameter, $0 < \epsilon < 19$ is strictly plurisubharmonic in $z$, but (OPT) has two local solutions:

$$f_1^* = (c, -c) \quad \text{and} \quad f_2^* = (-c, c)$$

with $c = 5\sqrt{2(19 - \epsilon)}$. For $\epsilon$ near 19 both $f_1$ and $f_2$ belong to the same connected component of the $\gamma^*$ sublevel set of $\Gamma$.

**Stopping Criteria**

**Theorem IV.4.** [**H3**]. *Suppose $N = 1$ and* (SA) *holds. Suppose $f^*$ is a smooth function in $H^\infty$ for which the function $a(e^{i\theta}) \triangleq \frac{\partial \Gamma}{\partial z}(e^{i\theta}, f^*(e^{i\theta}))$ never vanishes. Then $f^* \in A$ is a solution to* (OPT) *if and only if*

*(i)* $\Gamma(e^{i\theta}, f^*(e^{i\theta})) =$ *constant in* $\theta$.

*(ii)* wno $a > 0$.

Here wno $a$ means winding number of the function $a$ about 0.

**Theorem IV.5.** [**Mer**]. *For generic $\Gamma$ the solution $f^*$ to* (OPT) *produces* $a(e^{i\theta}) = \frac{\partial \Gamma}{\partial z}(e^{i\theta}, f^*(e^{i\theta}))$ *with* wno $a = 1$.

For $N \geq 1$ this generalizes to a reasonable extent. Now we have functions $a_j(e^{i\theta}) \triangleq \frac{\partial \Gamma}{\partial z_j}(e^{i\theta}, f^*(e^{i\theta}))$ for $j = 1, \ldots, N$. If these functions are continuous and extend meromorphically onto the disk, then define a generalized winding number by

wno $(a_1, \ldots, a_N) \triangleq$ number of common zeroes of $a_1, \ldots, a_N$ inside the disk minus their total number of poles inside the disk.

Here multiplicity must be counted.

**Theorem IV.6.** [**H6**]. *Suppose $\Gamma$ is smooth and that $f^*$ is a function for which the functions $a_j$ have no common zero on $\Pi$. If $f^*$ is a strict local solution to* (OPT), *then*

(i) $\Gamma(e^{i\theta}, f^*(e^{i\theta})) = constant\ in\ \theta.$

(ii) wno $(a_1, \ldots, a_N) > 0$.

*Conversely, if the sublevel sets $S_\theta(\gamma^*)$ of $\Gamma$ are strictly convex in $z$, then an $f^*$ satisfying* (i), (ii) *is a solution to* (OPT).

A forthcoming paper of Merino and Helton **[HMer2]** treats non-convex problems (very successfully for $N \leq 2$). Also we give alternatives to wno in the spirit of **[L]**) and discuss some implications for computation. See also **[Sl4]** for some related results.

This theorem provides useful diagnostics for a computer program. As the program progresses and generates approximate optima $f_k$, we expect the function $\Gamma(e^{i\theta}, f_k(e^{i\theta}))$ to become increasingly flat. Also behavior of the winding number $\frac{\partial\Gamma}{\partial z}(e^{i\theta}, f_k(e^{i\theta}))$ diagnostic can be monitored and seems to indicate how close one is getting to a local solution $f^*$.

**(OPT) vs. Forstneric**

We conclude with a few words about the Helton-Marshall proof of Theorem IIIb. This gives an interesting perspective on how the theory of (OPT) described in §IV meshes with the theory of polynomial hulls in §IIIb.3. As we shall see Forstnerics theorem and Theorems IV.1, IV.3 have a somewhat complementary relationship and together they combine to give Theorem IIIb.3 which says that

$$E(\mathcal{S}) \cup \mathcal{S} = \mathsf{P}(\mathcal{S})$$

for nice $\mathcal{S}$ when $N = 1$. We say that the graph $\{(\xi, f(\xi)) : |\xi| \leq 1\}$ for $f: \bar{\Delta} \to \mathbf{C}_N$ which is contained in $\mathsf{P}(\mathcal{S})$ is a *selection* (or an *analytic selection*) of $\mathsf{P}(\mathcal{S})$.

**Proof of Theorem IIIb.3.** The primary ingredients are Forstneric's theorem on a polynomial hull $\mathsf{P}(\mathcal{S})$: If $\mathsf{P}$ contains one selection $\{\xi, f^*(\xi)\}$, then $\mathsf{P}$ is swept out by sections, and our Theorem saying (OPT) has a unique solution when $N = 1$.

Let $\mathcal{S}$ be the given set. Let $\mathcal{S}^r$ denote a family of nice sets which expand $\mathcal{S}$; here $r \geq 1$, and $\mathcal{S}^1 = \mathcal{S}$ (see Fig. IV.1):

Choose the expansion to contain large enough sets so that for some $r$ the set $\mathsf{P}(\mathcal{S}^r)$ has an analytic selection. The family $\mathcal{S}^r$ of sets could be regarded as the

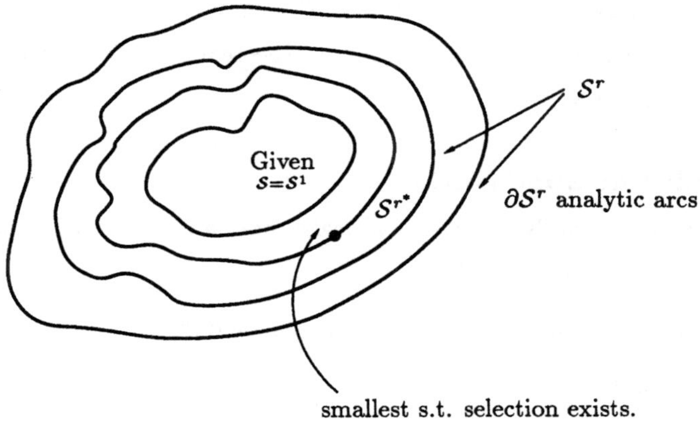


Figure IV.1.

sublevel sets for an (OPT) problem. Namely, define $\Gamma$ by

$$\Gamma(e^{i\theta}, z) \triangleq \min_r \{r \colon z \in \mathcal{S}_\theta^r\},$$

then its sublevel sets are indeed the sets $\mathcal{S}_\theta^r$. The statement, $\mathbf{P}(\mathcal{S}^1)$ contains no analytic selection is equivalent to:

The solution $r^0$ to this (OPT) problem is greater than one.

Indeed $r^0$ is the smallest $r$ so that $\mathbf{P}(\mathcal{S}^r)$ has an analytic selection.

**Lemma.** *If $r \downarrow r^0$ and all $\mathbf{P}(\mathcal{S}^r)$ have smooth $H^\infty$ selections in their interior, then $\mathbf{P}(\mathcal{S}^{r^0})$ is swept out by $H^\infty$ selections.*

**Proof.** Normal families plus Forstneric. That is: pick $(\xi_0, z_0) \in \mathbf{P}(\mathcal{S}^{r^0})$ with $|\xi_0| < 1$. We know by Forstneric that $\exists f^r \in A$ such that $f^r(\xi_0) = z_0$ and $f^r(e^{i\theta}) \in \mathcal{S}_\theta^r$ for all $\theta$. Normal families tells us $\exists$ subsequences $f^{r_n} \to f^*$ on compacta. Now $f^*(e^{i\theta}) \in \mathcal{S}_\theta^{r^0}$ a.e. (by Theorem IV.3) and $f^*(\xi_0) = z_0$, so $(\xi_0, z_0) \in$ selection. ∎

By the lemma $\mathbf{P}(\mathcal{S}^{r^*})$ is swept out by solutions $f^*$ to (OPT) in $H^\infty$. By Theorem IV.1 there is only one such $f^*$ and it is smooth. Thus $\mathbf{P}(\mathcal{S}^{r^*}) \equiv \{(\xi, z) \colon f^*(\xi) = z\}$.

This immediately implies $\mathbf{P}(\mathcal{S}^{r^*}) = \mathbf{P}(\mathcal{S}^1)$, since if $\mathbf{P}(\mathcal{S}^{r^*}) \supset\supset \mathbf{P}(\mathcal{S}^1)$, then $\mathbf{P}(\mathcal{S}^1)$ has a *trivial*[2] fiber over some $|\xi| < 1$. Theorem IIIb.4 implies $\mathbf{P}(\mathcal{S}^1) = \mathcal{S}$, contrary to assumption.

## V. NUMERICAL SOLUTIONS

This section is hardly complete and makes only a few remarks. The subject naturally divides into numerics for

(1) $\Gamma$ which are quasi-circular (sublevel sets are disks).

(2) General $\Gamma$.

The first subject contains the numerics of $H^\infty$ control and so is a huge field within the engineering (not the numerical analysis) community. I shall not discuss it.

Numerical efforts on (OPT) for general $\Gamma$ and related problems are carried out by various groups using very different methods.

(1) Peak point methods—Mayne-Polak-Salucidean, Fan-Tits
(2) Linear programming—Streit, Boyd, Daleh, Pearson.
(3) Convex programming—Boyd.
(4) $\mu$-synthesis—Doyle, the Honeywell group, Chu, Lenz, etc. (solves UNCOPT)
(5) Quasi-circular gradient Newton—Merino-Helton.
(6) Frequency dependent, conformal mapping ($N = 1$)—Sideris.

Codes are available from several of these groups, including Fan-Tits, Streit, Boyd, Helton-Merino. Efforts (1), (2) and (3) are carried out independently of numerical efforts in classical $H^\infty$ control while (4), (5), and (6) iterate classical $H^\infty$ control solutions.

Now we turn to numerical theory on (OPT). Theorems in section IV give qualitative perspective to users or developers of programs. For example, Theorem IV.6 gives diagnostics which should be helpful to many computer programs. More details can be found in **[BHMer]**, **[H-Mer]**.

Of all this we emphasize one simple property of the (OPT) problem which we suspect has broad numerical implications. Certainly it has a profound effect on

---

[2] observed by a student, M. Lawrence, at University of Washington.

the methods of Merino and myself. The issue is one of ***strong directional uniquess.*** Suppose we are given $\Gamma$ and a solution $f^*$ to (OPT) for $\Gamma$. We say that $f^*$ has order $p$ directionality provided that for each $h \in A_N$, there exists a constant $c_h > 0$, so that

$$\text{(SU}^p\text{)} \qquad \sup_\theta \Gamma(e^{i\theta}, f^*(e^{i\theta}) + th(e^{i\theta})) \geq \gamma^* + |t|^p c_h$$

for all small real numbers $t$. Moreover, if $p = 1$, then $f^*$ is called a directionally strongly unique solution to $\Gamma$.

**Theorem V.1.** [**HMer**] *Suppose* $\Gamma$ *is nice. If* $N = 1$, *then all solutions to* (OPT) *are directionally strongly unique. If* $N \geq 1$, *then no solution is directionally strongly unique.*

The proof is easy (see §2.d [**HMer**]).

We believe that when directional strong uniqueness (DSU) holds (OPT) is much better behaved numerically then when it fails. This is based on

(1) Extensive computer experiments using the Helton-Merino "gradient Newton" descent methods, see [**HMer**]. We are certainly curious to know if other methods are sensitive to the DSU distinction.

(2) Theoretical estimates [**HMer**] on our gradient-Newton methods suggest strongly that DSU is important.

As far as our computational efforts are concerned we consider the main open question to be that of adapting our algorithms to improve their behavior when DSU fails as well as determining its effect on other algorithms.

Research was supported in part by the Air Force Office of Scientific Research and the National Science Foundation.

## Bibliography

[ACF] Gr. Arsene, Z. Ceauşescu, and C. Foiaş, On intertwining dilations. VII, Complex Analysis Joensuu 1978, Proc., Colloq., Univ. Joensuu, Joensuu, 1978, Lecture Notes in Math., vol. 747, Springer-Verlag, 1979, pp. 24–45.
Gr. Arsene, Z. Ceauşescu, and C. Foiaş, On intertwining dilations. VIII, J. Operator Theory **4** (1980), 55–91.

[A] Anonymous.

[AW1] H. Alexander and J. Wermer, Polynomial hulls with convex fibers, Math. Ann. **271** (1985), 99–109.

[AW2] H. Alexander and J. Wermer, Envelopes of holomorphy and polynomial hulls, Math. Am. **281** (1988), 13–22.

[AW3] H. Alexander and J. Wermer, Polynomial hulls of sets with intervals as fibers, J. Complex Variables.

[BGR] J. Ball, I. Gohberg and L. Rodman. Book, to appear.

[BHM] J. Bence, J. W. Helton and D. Marshall, $H^\infty$ optimization, Proc., Conference on Decision and Control, Athens, 1986.

[BHMer] J. Bence, J. W. Helton and O. Merino, A primer in $H^\infty$ control, documentation for software: UCSD, 1985.

[Bose] N. K. Bose, Multidimensional systems: theory and applications, IEEE, J. Wiley, 1979.

[Dy] J. C. Doyle, Synthesis of robust controllers, Proc., IEEE Conference on Decision and Control, December, 1983, pp. 109–114.

[F] F. Forstneric, Polynomial hulls of sets fibered over the circle, Indiana J. of Math., **37** (1988), 869–889.

[Fr] B. A. Francis, *A course in $H^\infty$ control theory*, Lecture Notes in Control and Information Sci., vol. 88, Springer-Verlag, 1986.

[H1] J. W. Helton,An $H^\infty$ approach to control, IEEE Conference on Decision and Control, San Antonio, December 1983

[H2] J. W. Helton, Worst case analysis in the frequency domain: an $H^\infty$ approach to control, IEEE Trans. Auto. Control, AC-30 (1985), 1154–1170.

[H3] J. W. Helton, et.al, *Operator Theory, Analytic Functions, Matrices, and Electrical Engineering*, in Regional Conference Series in Mathematics, No. 68, Amer. Math. Soc., 1987.

[H4] J. W. Helton, Non-Euclidean functional analysis and electronics, Bull. Amer. Math. Soc., **7** (1982), 1–64.

[H5] J. W. Helton, Broadband gain equalization directly from data, IEEE Trans. Circuits and Systems, CAS-28 (1981), 1125–1137.

[H6] J. W. Helton, Optimization over spaces of analytic functions and the Corona problem, **13** (1986), 359–375.

[HMer1] J. W. Helton and O. Merino, A novel approach to accelerating Newton's method for sup-norm optimization arising in $H^\infty$-control, preprint.

[HMer2] J. W. Helton and O. Merino, Conditions for Optimality over $H^\infty$, in preparation.

[K] S. Kranz, *Function Theory of Several Complex Variables*, Wiley, New York, 1982.

[Her] R. Herman, book, to appear.

[KW] Kwaakernak, Robustness optimization of linear feedback systems, Proc., Conference on Decision and Control, December 1983.

[MH] D. Marshall and J. W. Helton, Frequency domain design and analytic selections, Indiana Journal of Math, 1990.

[Mer] O. Merino, Optimization over spaces of analytic functions, UCSD Thesis, 1988.

[L] L. Lempert, Complex geometry in convex domains, Proceedings Internation Congress of Mathematics, Berkeley 1986.

[Sl1] Z. Slodkowski, Polynomial hulls with convex sections and interpolating spaces, Proc. Ann. Math. Soc., **96** (1986), 25–260.

[Sl2] Z. Slodkowski, Polynomial hulls in $\mathbb{C}^2$ and quasicircles, preprint.

[Sl3a] Z. Slodkowski, An analytic set valued selection and its application to the Corona theorem, to polynomial hulls, and joint spectra Trans. Amer. Math. Soc., **294** (1986), 367–377.

[Sl3b] Z. Slodkowski, Analytic set-valued functions and spectra, Math. Ann., **256** (1981), 363–386.

[Sl4] Z. Slodkowski, Polynomial hulls with convex fibers and complex geodesics, preprint.

[W1] J. Wermer, *Banach Algebras and Several Complex Variables*, Springer-Verlag Grad Texts #35 (Second Edition), 1976.

[W2] J. Wermer, Polynomially convex hulls and analyticity, Arkiv für Matematik, **20** (1982), 129–135.

[W3] J. Wermer, Balayage and polynomial hulls, preprint.

[Y] D. C. Youla, A new theory of broadband impedance matching, IEEE CAS, 1964.

[YS] D. C. Youla and M. Saito, Interpolation with positive real functions, J. Franklin Inst., **284** (1967), 77–108.

[ZF] G. Zames and B. A. Francis, Feedback and minimax sensitivity, Advanced Group for Aerospace Research and Development NATO Lectures Notes No. 117, Multivariable Analysis and Design Techniques.

# FEEDBACK STABILIZATION OF NONLINEAR SYSTEMS

Eduardo D. Sontag

*Abstract*

This paper surveys some well-known facts as well as some recent developments on the topic of stabilization of nonlinear systems.

## 1 Introduction

In this paper we consider problems of local and global stabilization of control systems

$$\dot{x} = f(x,u)\,, \quad f(0,0) = 0 \tag{1}$$

whose states $x(t)$ evolve on $\mathbb{R}^n$ and with controls taking values on $\mathbb{R}^m$, for some integers $n$ and $m$. The interest is in finding feedback laws

$$u = k(x)\,, \quad k(0) = 0$$

which make the closed-loop system

$$\dot{x} = F(x) = f(x, k(x)) \tag{2}$$

asymptotically stable about $x = 0$. Associated problems, such as those dealing with the response to possible input perturbations $u = k(x) + v$ of the feedback law, will be touched upon briefly.

We assume that $f$ is smooth (infinitely differentiable) on $(x,u)$, though much less, –for instance a Lipschitz condition,– is needed for many results.

The discussion will emphasize intuitive aspects, but we shall state the main results as clearly as possible. The references cited should be consulted, however, for all technical details. Some comments on the contents of this paper:

- We do not consider control objectives different from stabilization, such as decoupling or disturbance rejection.
- Except for some remarks, we consider only state (rather than output) feedback.
- The survey talk centers on questions of possible regularity (continuity, smoothness) of $k$. This focus leads to natural mathematical questions, and it may be argued that that regular feedback is more "robust" in various senses. But –and to some extent this is emphasized by those negative results that are presented– it is often the case that discontinuous control laws must be considered (sliding mode controllers, or piecewise smooth feedback, for instance). In addition, non-continuous-time feedback (sampled control, pulse-width modulation), is often used in practice and is also not covered.

- The assumption that $k(0) = 0$ is quite natural; it says that no energy should need to be pumped into the system when it is at rest. The theory that results when this requirement is not imposed is also of great interest, however.
- Another related interesting set of problems ("practical" stabilization) deals with bringing states close to certain sets rather than to the particular state $x = 0$.

Space constraints force us to be selective in our coverage. Such selectivity will imply, as is often the case with surveys, some emphasis towards the speaker's favorite topics. Hopefully the inclusion of an additional bibliography –see the end of the paper– makes up for some of the omitted material.

### 1.1 *What regularity will be imposed on k ?*

The main questions that we want to address involve, as pointed out above, regularity of $k$. The requirements away from 0, whether $k$ should be, say, $C^0$, $C^1$, or $C^\infty$, appear to be not very critical; as we see later, it is often possible to "smooth out" a feedback law that is merely continuous. (Of course, if $k$ is not smooth enough, questions arise regarding uniqueness of trajectories for the closed-loop system (2).) Much more critical is the behavior of $k$ at the origin. Because of these facts, and in order to simplify the presentation, we shall consider just two types of feedback; the issues arising for these are quite typical of the general problems. We shall say that $k : \mathbb{R}^n \to \mathbb{R}^m$, $k(0) = 0$, is:

- **smooth:** if $k \in C^\infty(\mathbb{R}^n)$ .
- **almost smooth:** if $k \in C^\infty(\mathbb{R}^n \setminus \{0\})$ and $k \in C^0(\mathbb{R}^n)$ .

The problems of finding stabilizing feedback laws of these two types are *very different*: consider for instance the system

$$\dot{x} = x + u^3$$

which can be globally stabilized by the *almost* smooth law

$$u := -\sqrt[3]{2x}$$

resulting in

$$\dot{x} = -x$$

but cannot even be locally stabilized by a smooth $u = k(x)$, since for any such $k$ one would necessarily have $k(x) = O(x)$ so that the closed-loop system

$$\dot{x} = x + O(x^3)$$

is unstable.

It is probably fair to say that until now the most elegant local theory has been developed for the smooth case, while the most elegant global results are those that have been obtained for almost smooth stabilization.

## 2 Asymptotic Stability

As with regularity, there are also many possible notions of stabilization. These can be classified under two broad categories:

- **State-Space:** There is a map $k$ such that the system

$$\dot{x} = f(x, k(x))$$

has $x = 0$ as a locally or globally asymptotically stable point. We call this *local* or *global*, *smooth* or *almost-smooth*, stabilization, depending on the regularity required of $k$.

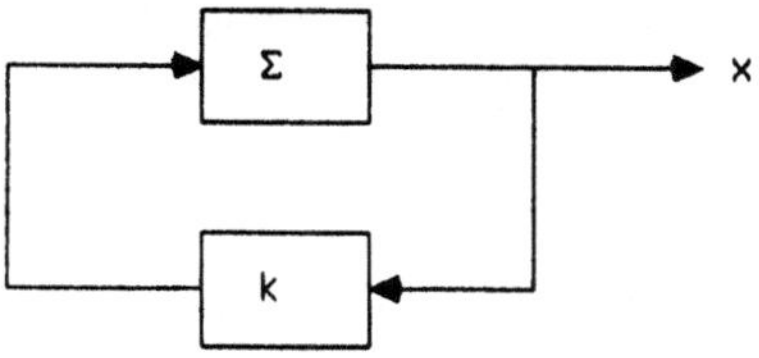


FIGURE 1: Pure state-feedback configuration

- **Operator-Theoretic:** There is a $k$ so that the initialized system

$$\dot{x} = f(x, k(x) + u) \ , \ \ x(0) = 0$$

induces a stable operator $u \mapsto x$. There are many possible, nonequivalent, definitions of stability for operators; this point will be discussed again later. This notion is of interest when studying stability under persistent or decaying input perturbations, and when trying to obtain *Bezout* factorizations for nonlinear systems.

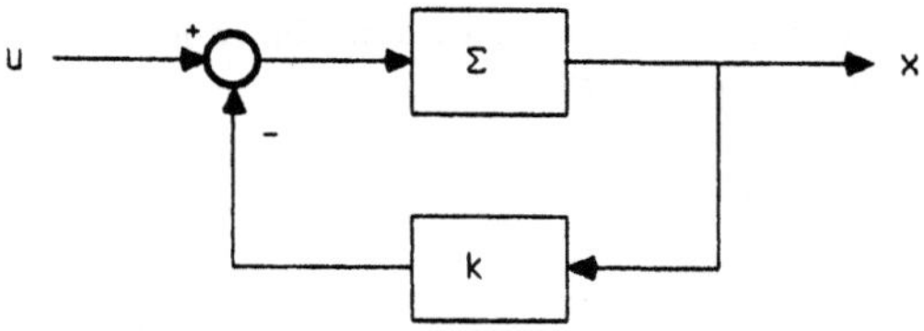


FIGURE 2: Additive state-feedback configuration

An alternative is to allow for an additional feedforward term, say with the same regularity as $k$. Such a variation appears when studying coprime, not necessarily Bezout, factorizations.

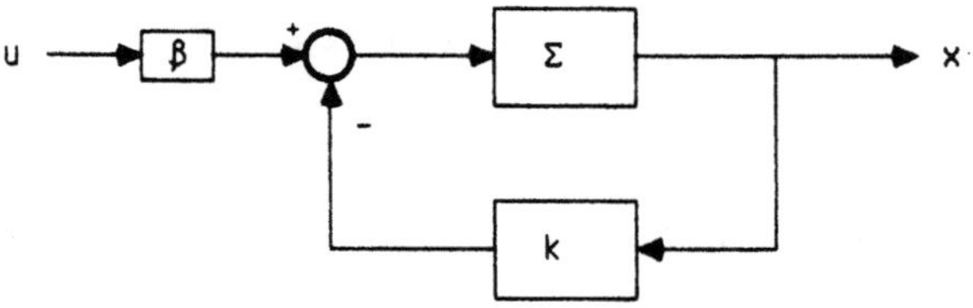


FIGURE 3: State-feedback with input weighing

We shall concentrate on pure state-feedback problems, but will also explain how some operator-type results can be obtained as a consequence of these.

### 2.1 *Asymptotic controllability*

An obvious necessary condition for state-space stabilizability is the corresponding open-loop property of (null-) **asymptotic controllability**: for each small $x_0$ there must exist some measurable, locally essentially bounded control $u(\cdot)$ defined on $[0,+\infty)$ such that, in terms of the trajectory $x(\cdot)$ resulting from initial $x_0$ and input $u$, (a) $x(t)$ is defined for all $t$ and $x(t) \to 0$ as $t \to \infty$, (b) this happens with no large excursions (stability), and (c) since $k$ is continuous at the origin, $u(t) \to 0$. This property can be summarized by the statement: *for each $\varepsilon > 0$ there is some $\delta > 0$ such that, for each $|x_0| < \delta$ there is some $u(\cdot)$ so that*

$$x(t), u(t) \to 0$$

*and also*

$$|x(t)| + |u(t)| < \varepsilon \;\; \forall t$$

where $x(\cdot)$ is the trajectory starting at $x_0$ and applying $u$.

(We use bars $|\xi|$ to denote any fixed norms in $\mathbb{R}^n$ and $\mathbb{R}^m$.)

For global stabilization, one has the additional property that *for every* $x_0$ there must exist a control $u$ so that $x(t) \to 0$; we call this **global asycontrollability.**

Observe that, for systems with no controls, classical asymptotic stability is the same as asycontrollability.

For operator-theoretic stabilizability, one has necessary **bounded-input bounded-output** or "input to state stability" necessary properties. These will be mentioned later.

The main basic questions are, for the various variants of the above concepts:

| *To what extent does asycontrollability imply stabilizability?* |
|---|

Such converse statements hold true for linear finite dimensional time-invariant systems, but are in general false, as we discuss next.

# 3 Case $n = m = 1$

To develop some intuition, it is useful to start with the relatively trivial case of single-input one-dimensional systems. Many of the remarks to follow are taken from [28].

For the system (1), asycontrollability means that for each $x$, or at least for small $x$ in the local case, there must exist some $u$ so that

$$xf(x,u) < 0$$

(see [28] for a detailed proof). Consider the set

$$\mathcal{O} := \{(x,u) \,|\, xf(x,u) < 0\}$$

and let

$$\pi : (x,u) \mapsto x$$

be the projection on the first coordinate. Then, global asycontrollability implies that

$$\pi\mathcal{O} = \mathbb{R}^n \backslash \{0\}$$

while local asycontrollability says that this projection contains a neighborhood of zero; in addition, a local property about $(0,0)$ also holds, since $u$ must be small if $x$ is small.

On the other hand, if $k$ is any feedback law giving asycontrollability, it must hold that $k$ provides a section over $\mathbb{R}^n \backslash \{0\}$ of the projection $\pi$, i.e.

$$(x, k(x)) \in \mathcal{O} \;\; \forall\, x \neq 0$$

Thus the main problem is essentially that of finding regular sections of $\pi$.

Using this geometric intuition, it is easy to construct examples of systems which are asycontrollable but for which there is no possible almost-smooth –or for that matter, not even just $C^0$ away from zero– feedback stabilizer. For instance

$$\dot{x} = x\left[(u-1)^2 - (x-1)\right]\left[x - 2 + (u+1)^2\right]$$

is so that $\mathcal{O}$ consists of the two components

$$\mathcal{O}_1 = \{(u-1)^2 < x - 1\}$$

and

$$\mathcal{O}_2 = \{(u+1)^2 < 2 - x,\ x \neq 0\}$$

and hence admits no continuous stabilizer, even though it is clearly asycontrollable. (See Figure 4: darkened area is the complement of $\mathcal{O}$; note that no continuous curve is contained entirely in $\mathcal{O}$ and projects onto the $x$-axis.) On the other hand, in this example it is easy to construct a controller –a section of the projection with $k(0) = 0$– that is everywhere smooth except for a single discontinuity.

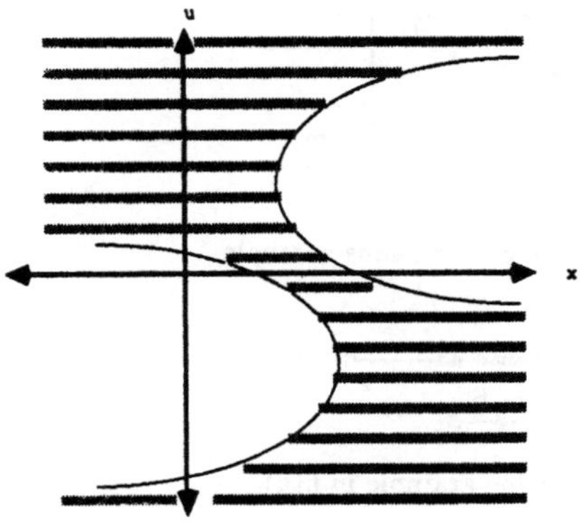


FIGURE 4: No continuous sections

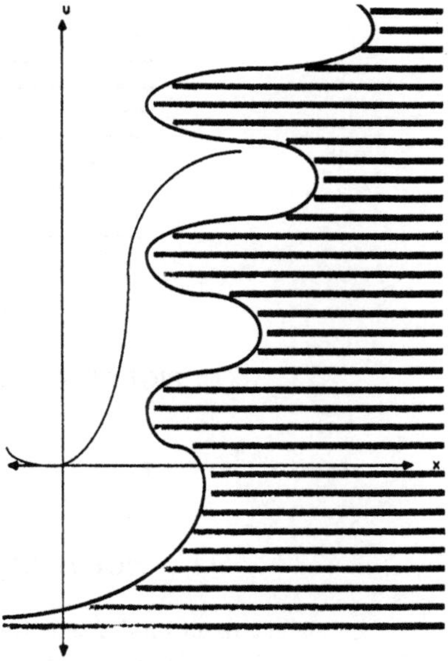


FIGURE 5: Semiglobal vs. global

This counterexample is based on the impossibility of choosing controls; the paper [30] provides examples where not even a continuous choice of state *trajectories* is possible.

The graphical technique allows answering other questions, such as those in [28] regarding the possibility of non-Lipschitz stabilizers even when there are none that are Lipschitz. In [31], the authors discuss "semiglobal" stabilizers in comparison with global ones: The question is whether it may be the case that for each compact subset of the state space there is a feedback stabilizer, but that there is none that works globally. They provide a counterexample analogous to the one illustrated graphically in Figure 5, the darkened area corresponding to the complement of $\mathcal{O}$. Note that for each fixed interval on the $x$-axis there are obviously smooth sections of the projection –as indicated by a curve–, but there can be no global sections.

An interesting fact for one-dimensional systems is that there are always rather regular *time-varying* feedback stabilizers. For the precise definition of smooth time-varying and more generally dynamic stabilizers see the reference [28]; essentially one obtains a smooth stabilizer for the system obtained by adding a parallel integrator. The idea of the proof in [28] is easier to understand with an example. In Figure 6a, again with the darkened area corresponding to the complement of $\mathcal{O}$, we consider two possible feedback laws, illustrated by their graphs. There is no way to obtain a continuous stabilizing feedback law, i.e. one whose graph stays entirely in $\mathcal{O}$. But the idea is to oscillate very fast between the two indicated (non-stabilizing) laws. Let $B = B_t$ denote the set of $x$'s where at any given time $t$ the feedback law satisfies $xf(x, k(t,x)) < 0$ (Figure 6b). This set oscillates, and we design the time variation so that it moves to the left slowly but it moves to the right fast (for $x > 0$, and the converse for $x < 0$). A state $x > 0$ to the right of $B$ will continue moving to the left, towards the origin, until it hits the set $B$. At that point, it will move in an undesired direction, but will do so only for a very small time duration, with a net effect of a leftward move. The above reference provides a complete proof.

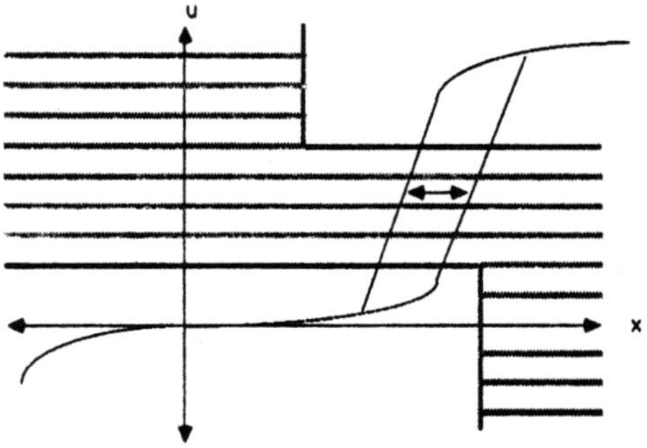


FIGURE 6(a): Time-varying continuous example

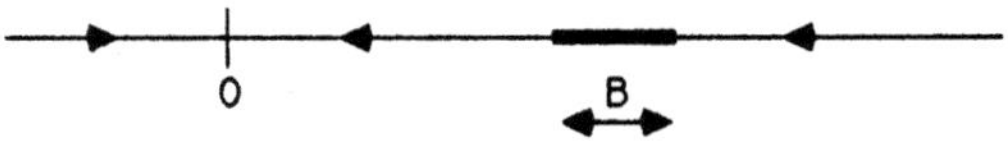


FIGURE 6(b): Bad set for example in 6(a)

A different result on dynamic feedback stabilization of one-dimensional systems holds for analytic $f$, and is given in the work [8]. It is shown there that asycontrollability is equivalent to almost-smooth stabilization of the enlarged system

$$\dot{x} = f(x,y)\,,\quad \dot{y} = u\ .$$

Later we shall see examples of systems (in higher dimensions) for which not even dynamic stabilization can be done continuously.

# 4 General $n, m$ – Main Techniques

The one-dimensional case illustrated that smooth or almost-smooth stabilizers may fail to exist even if the system is asycontrollable. We now survey the more general case, concentrating on the following techniques:

1. First order methods (linearization)
2. Topological techniques
3. Lyapunov functions
4. Relation to operator-theoretic stability
5. Decomposition approaches

We will not cover, due to time and space limitations, the very interesting work being done on special cases such as two-dimensional systems ([5], [4]) and in particular the use of center manifold techniques and perturbation analysis (see e.g. [1],[2]).

# 5 First-Order Techniques

We review here some facts that apply to the problem of **local, smooth** stabilizability. [The example $\dot{x} = x + (-\sqrt[3]{2x})^3$, discussed earlier, shows that these techniques do not say anything interesting regarding *almost* smooth feedback.] Write

$$\dot{x} = Ax + Bu + o(x,u) \qquad (\Sigma)$$

and call $\Sigma$ *first-order* (or "hyperbolically") *stabilizable* if the linearized system $\dot{x} = Ax + Bu$ is asycontrollable, or equivalently, if there exists a matrix $F$ so that

$$A + BF$$

is a Hurwitz matrix. This property is also equivalent to the requirement that

$$\operatorname{rank}[sI - A, B] = n \quad \text{whenever} \quad \operatorname{Re} s \geq 0$$

(PBH condition).

For each stabilizing feedback matrix $F$ for the linear part, the linear law $u = Fx$ is also a local stabilizer for the nonlinear system, and the following classical result is obtained:

**Theorem 1.** $\Sigma$ first-order stabilizable $\Rightarrow$ $\Sigma$ locally smoothly stabilizable.

Recall that this is proved by showing that a quadratic Lyapunov function for $\dot{x} = (A + BF)x$ is also a local Lyapunov function for the closed loop system

$$\dot{x} = (A + BF)x + o(x)$$

–see e.g. [36].

The converse of Theorem 1 is obviously false; for instance the system

$$\dot{x} = u^3$$

has a non-asycontrollable first-order part $\dot{x} = 0$ but the smooth (even linear) feedback law $u = -x$ results in $\dot{x} = -x^3$ which is asymptotically stable. However, this example illustrates what can be said about the converse. Note that even though the linearized system is not asymptotically stabilizable, its only uncontrollable eigenvalue has zero real part. In addition, the stability that can be achieved is not *exponential*, but is "slower" than exponential.

One says that the origin is exponentially stable for $\dot{x} = f(x)$ if there exist positive constants $\lambda$ and $M$ so that

$$|x(t)| \leq Me^{-\lambda t}|x(0)|$$

for all small enough initial states and all $t \geq 0$. By *smooth exponential stabilizability* we mean that there is a smooth $k$ so that the closed loop system (2) is locally exponentially stable. The next two results then hold:

**Theorem 2.** $\Sigma$ is locally smoothly stabilizable $\Rightarrow$ $\operatorname{rank}[sI - A, B] = n \ \forall \operatorname{Re} s > 0$.

**Theorem 3.** $\Sigma$ first-order stabilizable $\Longleftrightarrow$ $\Sigma$ exponentially stabilizable.

The first of these is proved by appealing to the standard controllability decomposition: If the rank condition fails, under the variables in this decomposition the closed-loop system corresponding to any smooth feedback law must result in block equations

$$\begin{aligned}\dot{x}_1 &= (A_1 + B_1 F)x_1 + A_2 x_2 + o(x)\\ \dot{x}_2 &= A_3 x_2 + o(x)\end{aligned}$$

where $A_3$ has some eigenvalue with strictly positive real part . But then Lyapunov's Second Theorem on Stability, or one of its variants such as Cheataev's Theorem, –applied to the $x_2$-equation,– implies that the closed-loop system is unstable, contradicting the assumption.

The second result is "folk" knowledge, and an analogous result for arbitrary-rate stabilization was given in [12]. A sketch of its proof is as follows. Sufficiency is proved as with Theorem 1. Conversely, assume that $k$ is a smooth feedback stabilizer, and look at the closed-loop system. Again via the controllability decomposition, the problem reduces to showing that the eigenvalues of the linearization of an exponentially stable equation must have negative real part. Let $\lambda$ be as in the definition of exponential stability, and consider the change of variables $z(t) := e^{\frac{\lambda}{2}t}x(t)$ which results in an equation

$$\dot{z}(t) = (\frac{\lambda}{2}I + A)z + g(z,t)$$

where $g(z,t)$ is $o(z)$ uniformly on $t$. Since $x(t)$ decays at rate $\lambda$, it follows that $z$ decays at rate $\lambda/2$, and hence the $z$ equation is asymptotically stable. From Cheataev's Theorem, one concludes that $\frac{\lambda}{2}I + A$ has all eigenvalues with real part $\leq 0$, from which it follows that all eigenvalues of $A$ have strictly negative real part, as wanted.

The gap in the characterization of local smooth stabilizability is due to the possible modes corresponding to $\operatorname{Re} s = 0$, i.e. the "critical" cases where $\operatorname{rank}[sI - A, B] < n$ for some purely imaginary $s$. This is precisely the point at which Center Manifold Techniques become important.

# 6 Topological Techniques

In this section we review some topological considerations that establish limitations on what almost smooth feedback can achieve. (In fact, the limitations will apply also to even weaker types of feedback.)

To motivate, let's start with an example due to Brockett. Consider the 3-dimensional 2-control system

$$\begin{aligned}\dot{x}_1 &= u_1\\ \dot{x}_2 &= u_2\\ \dot{x}_3 &= u_2 x_1 - u_1 x_2\end{aligned}$$

for which

$$[sI - A, B] = \begin{pmatrix} s & 0 & 0 & 1 & 0\\ 0 & s & 0 & 0 & 1\\ 0 & 0 & s & 0 & 0\end{pmatrix}$$

looses rank at $s = 0$. First-order tests for smooth stabilization are thus inconclusive, except for the fact that *exponential* stability can't be achieved. On the other hand, this system is completely controllable, since it is a system of the type

$$\dot{x} = u_1 g_1(x) + u_2 g_2(x)$$

("symmetric" system with "no drift term") and

$$\det(g_1, g_2, [g_1, g_2]) = 2 \neq 0$$

everywhere, where $[g_1, g_2]$ denotes the Lie bracket. The system is in particular asycontrollable, since controllability is preserved using arbitrarily small $u_1, u_2$. This suggests that the system might be smoothly stabilizable. But in fact it isn't. Consider the mapping

$$(x, u) \mapsto f(x, u) \tag{3}$$

which here is

$$\mathbb{R}^5 \to \mathbb{R}^3 : (x_1, x_2, x_3, u_1, u_2)' \mapsto (u_1, u_2, u_2 x_1 - u_1 x_2)' .$$

No points of the form

$$\begin{pmatrix} 0 \\ 0 \\ \varepsilon \end{pmatrix}, \quad \varepsilon \neq 0$$

are in its image, so the system can't be smoothly stabilizable, by *Brockett's necessary condition*:

**Theorem 4.** If $\Sigma$ is almost smoothly stabilizable then the image of (3) contains some neighborhood of zero.

For linear systems, Brockett's condition is that

$$\operatorname{rank}[A, B] = n$$

which is the case $s = 0$ of the PBH criterion.

Theorem 4 was given in [6]. It reduces to the purely differential-equation result that the image of $F(x) = f(x, k(x))$ must contain a neighborhood of zero if the closed-loop vector field $F$ is asymptotically stable. The following elementary proof was suggested to us by Roger Nussbaum (ca. 1982), and is analogous to those proofs given in [37] and [15].

Consider the closed-loop system $\dot{x} = F(x(t))$ and let $\Phi$ denote the flow associated to this. Then

$$H(x, t) := \frac{1}{t}\left[\Phi\left(\frac{t}{1-t}, x\right) - x\right] \quad , \quad t \in [0, 1]$$

is a homotopy between $F(x)$ and $-x$. (As $t \to 1^-$, the flow converges uniformly to zero by asymptotic stability, while as $t \to 0^+$ this is $F(x)$ by the definition of flow.) From this and the fact that $F$ can have no zeroes –equilibria of the ode– outside $x = 0$, one concludes that $F$ must have topological degree $(-1)^n$ with respect to all points $p$ near 0, and so $F(x) = p$ is solvable for all such $p$.

The above proof can be extended to show that not even "practical stability" can be achieved, in the sense that one looks for stabilizers defined away from 0 and with the property that closed-loop trajectories converge to a neighborhood of the origin. Moreover, even arbitrary continuous feedbacks (satisfying conditions of existence and uniqueness of trajectories) are ruled out by the theorem. In [37], it is shown that global attractivity is also ruled out, even if local asymptotic stability is not required to hold.

Note that when a system fails Brockett's test, it cannot be stabilized by almost smooth *dynamic* feedback either, in the sense that any extended system

$$\begin{aligned} \dot{x} &= f(x, u) \\ \dot{z} &= v \end{aligned}$$

where $v$ is a new control, and $z$ is a new state of state variables, will still fail the test.

## 6.1 Other Topological Techniques

Consider now the following two-dimensional, single-control system ([3])

$$\dot{\xi} = g(\xi)u$$

where

$$\xi = \begin{pmatrix} x \\ y \end{pmatrix}, \quad g(\xi) := \begin{pmatrix} x^2 - y^2 \\ 2xy \end{pmatrix} .$$

(This system represents the real and imaginary parts of the one-dimensional complex system $\dot{z} = uz^2$.) For each control, one can move at different velocities along the integral curves of $\dot{\xi} = g(\xi)$. These curves are the circles centered on the $y$-axis and passing through zero, plus the positive and negative $x$-axis; see Figure 7. Thus the system is asycontrollable, and in fact every state can be controlled in finite time to the origin. As opposed to the previous example, however, this one does pass Brockett's test, and linear tests are also inconclusive. We now show that this system is not almost-smoothly stabilizable, even locally, and use this to illustrate another technique.

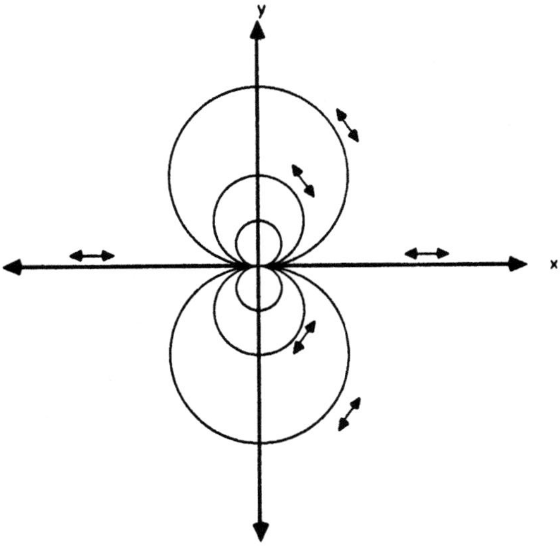


FIGURE 7: Orbits of $g$

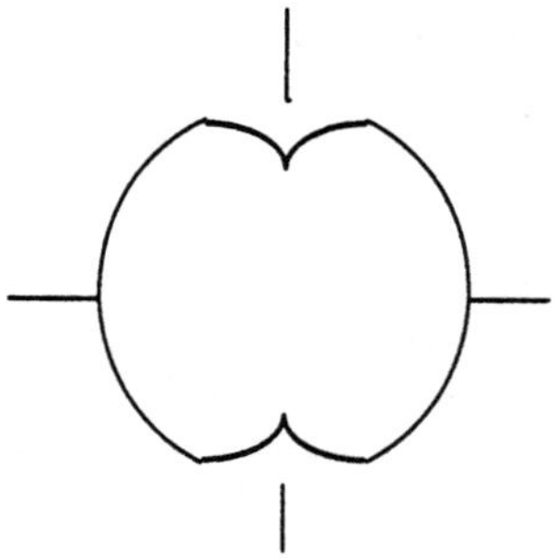

FIGURE 8: Clf level sets

Assume that there is a feedback law stabilizing this system on some open set $U$ containing the origin. Consider the closed-loop system

$$\dot{\xi} = g(\xi)k(\xi)$$

that results; by assumption the left-hand side is at least Lipschitz away from the origin, so this is a well-posed differential equation.

Choose any circular orbit of $g$ which is entirely contained in $U$. Then the restriction of the closed-loop equation to this circle provides a differential equation which is globally asymptotically stable on the circle. But this is impossible, because of the following fact:

**Theorem 5.** If a differential equation on a manifold $M$, $\dot{x} = F(x), F(x_0) = 0$, has $x_0$ as a globally asymptotically stable state, then $M$ must be contractible.

The only property needed for this result is that solutions exist and be unique, plus continuous dependence. The proof is almost trivial; see below. A somewhat stronger statement, often refered to as "Milnor's Theorem", asserts that $M$ must in fact be diffeomorphic to an Euclidean space, but the above version seems to be enough for most applications.

To prove the Theorem, just note that the map

$$H(x,t) := \Phi\left(\frac{t}{1-t}, x\right) \quad , \quad t \in [0,1]$$

provides a homotopy between the identity and the constant map $H(x,1^-) \equiv x_0$; here $\Phi$ is the flow induced by $F$ as before.

For the particular example that we had above, this is all very intuitive: for $y > 0$ and $x > 0$ near the origin, we must move to the left –stability part of "asymptotic stability"–, and for $x < 0$ to the right. Continuing back along any fixed circle, we reach a point where we must move both to the left and right, which would create a discontinuity of the feedback law, unless we passed first through zero which would create a nonzero equilibrium.

In this example, in fact, not even attractivity (all trajectories converging to zero) can hold with continuous feedback. This is because such a feedback law must satisfy

$$k\begin{pmatrix}-1\\0\end{pmatrix} > 0 \quad \text{and} \quad k\begin{pmatrix}1\\0\end{pmatrix} < 0$$

(since on the $x$-axis the equation is $\dot{x} = x^2u, \dot{y} = 0$). Thus any curve between $(-1,0)'$ and $(1,0)'$ will be so that $k$ has some zero somewhere on it, giving a new equilibrium point of the closed-loop system.

Theorem 5 implies that mechanical models with a noncontractible phase space –rigid body orientations, for example– give rise to systems that cannot be smoothly, or in any reasonable sense continuously, globally stabilizable.

# 7 Lyapunov Functions

Assume that $\Sigma$ is globally almost-smoothly stabilizable. The closed-loop system (2) being globally asymptotically stable, standard inverse Lyapunov theorems (see for instance Theorem 14 in [18],) can be used to conclude that there exists a proper ($V(x) \to \infty$ as $x \to \infty$), positive definite ($V(x) > 0$ for $x > 0$, $V(0) = 0$) function $V$ so that

$$L_F V(x) = \nabla V(x) F(x) < 0 \quad \forall x \neq 0$$

which implies in open-loop terms that

$$(\forall x \neq 0)(\exists u) \quad \nabla V(x) f(x,u) < 0$$

and in addition, by continuity of $k$ at 0, the property

$$(\forall \varepsilon > 0)(\exists \delta > 0) \quad \left[0 < |x| < \delta \;\Rightarrow\; \min_{|u| \leq \varepsilon} \nabla V(x) f(x,u) \; < 0\right] .$$

We call such a function $V$ a **control-Lyapunov function** ("clf"). (In the terminology of [26], this would be a clf which satisfies the *small control property.*) The above-mentioned theorems show that there always exists a *smooth* clf if $\Sigma$ is almost-smoothly stabilizable.

Intuitively, a clf is an "energy" function which at each nonzero $x$ can be decreased by applying a suitable open-loop control, and this control can be picked small if $x$ is small.

It is not hard to show that the existence of a clf implies asycontrollability. In fact, this implication holds even if we ask only that $V$ be continuous. In that case the gradient may be meaningless, so we replace the defining condition by

$$(\forall \varepsilon > 0)(\exists \delta > 0) \quad \left[ 0 < |x| < \delta \Rightarrow \min_{\|\omega\| \leq \varepsilon} D^+ V_\omega(x) \; < 0 \right]$$

where $D^+$ indicates, as usual in the literature on nonsmooth Lyapunov functions, the Dini derivative

$$D^+ V_\omega(x_0) := \limsup_{t \to 0+} \frac{V(x(t)) - V(x_0)}{t}$$

and $x(t)$ is the trajectory corresponding to the measurable control $\omega$ (the norm is the sup norm).

To state the next two results, we assume for simplicity that the system (1) is affine in controls, a class that includes most examples of interest and which allows us to avoid "relaxed" controls. For

$$\dot{x} = f_0(x) + G(x)u \,, \quad f_0(0) = 0 \,, \quad G(x) \in \mathbb{R}^{n \times m} \;\; \forall x$$

we have:

**Theorem 6.** $\Sigma$ is asycontrollable $\iff$ it admits a $C^0$ clf.

**Theorem 7.** $\Sigma$ is almost smoothly stabilizable $\iff$ it admits a $C^\infty$ clf.

Thus we know that there is no possible *smooth* clf for the example seen before whose orbits are circles (Figure 7), since there are no almost-smooth stabilizers. But this system is asycontrollable, so we know that there do exist *continuous* clf's. Figure 8 illustrates what a typical level set for one such clf may look like; note the singularity due to lack of smoothness.

Theorem 6 was proved in [24], and is based on the solution of an appropriate optimal control problem. "Relaxed" controls are used there, because the more general case of systems not affine in controls is treated, but the proof here is exactly the same. Also, the "small-control" property didn't play a role in that reference, but as remarked there –top of page 464–, the proof can be easily adapted.

Theorem 7, which we will refer to as *Artstein's Theorem*, was originally given in [3], which also discussed the example in Figure 7. It has since been rediscovered by others, most notably in [32] and other work by that same author. In every case, the proof is based on some sort of partition of unity argument, but we sketch below a simple and direct proof. This result is very powerful; for instance, it implies:

**Corollary.** If there is a continuous function $k : \mathbb{R}^n \to \mathbb{R}^m$ with $k(0) = 0$ and such that

$$\dot{x} = f_0(x) + G(x)k(x)$$

has the origin as a globally asymptotically stable point then there is also an almost-smooth global stabilizer.

Since solutions may not be unique, the assumption is that for every trajectory the asymptotic stability definition holds. By Kurzweil's Theorem, –see the discussion in [3]– there is a smooth clf, and hence by Theorem 7 there is an almost smooth feedback as desired. This explains our earlier remarks to the effect that the precise degree of regularity away from zero seems to be not very critical, so long as at least continuity holds.

A proof of Artstein's Theorem is as follows. For simplicity, we consider just the case $m = 1$ and a system $\dot{x} = f_0(x) + ug(x)$, but for $m > 1$ the proof is entirely analogous.

As explained earlier, one implication is immediate from the converse Lyapunov theorems. Assume then that $V$ is a smooth clf, and let

$$a(x) := \nabla V(x).f_0(x) \ , \quad b(x) := \nabla V(x).g(x)$$

Then

$$\text{the pair } (a(x), b(x)) \text{ is stabilizable } \forall x \neq 0$$

(for each fixed $x$ as an $n = m = 1$ LTI system). On the other hand, an almost-smooth feedback law that stabilizes and so that the same $V$ is a Lyapunov function for the closed-loop system is a $k(\cdot)$ so that

$$a(x) + k(x)b(x) < 0 \ \ \forall x \neq 0$$

and is smooth for $x \neq 0$ and satisfies $k(x) \to 0$ as $x \to 0$. This is basically a problem on "Families of Systems", if we think of $(a(x), b(x))$ as a parameterized set of one-dimensional LTI systems. We use a technique due to Delchamps ([9]) in order to construct $k$. Consider the LQ Problem

$$\min_u \int_0^\infty u^2(t) + b^2 y^2(t)\, dt$$

for each fixed $x$, where the "$y$" appearing in the integral is a state variable for the system

$$\dot{y} = ay + bu \ .$$

This results in a feedback law $u = ky$ parameterized by $x$. Moreover, note that when $x$ is near zero also $b = b(x)$ is small, by continuity and the fact that, because $V$ has a local minimum at the origin, $\nabla V(0) = 0$. Therefore one may expect that when $x$ is near zero the $b^2$ term gives more relative weighting to the $u^2$ term, forcing small controls and thus continuity of the feedback at the origin.

Explicitely solving the corresponding algebraic Riccati equation results in the feedback law

$$k \ := \ -\frac{a + \sqrt{a^2 + b^4}}{b}$$

which is analytic in $a, b$; the apparent singularity at $b = 0$ is "removable", and the feedback is 0 at those points with $b(x) = 0$. Further, as proved in [26], this is $C^0$ at the origin, as desired.

The same formula shows how to obtain a feedback law *analytic* on $x \neq 0$ provided that $f_0, g, V$ be analytic. A different construction can be used to prove that there is a rational feedback stabilizer if $f_0, g, V$ are given by rational functions, but it is not yet clear if this rational stabilizer can be made continuous at the origin.

The above formula for a stabilizing feedback law can be compared to the alternative proposed in [32], which is

$$k(x) \ = \ -\chi\frac{a}{b} - b$$

where $\chi : \mathbb{R}^n \to [0,1]$ is any function such that $\chi \equiv 1$ where $a \geq 0$ and $\chi \equiv 0$ about $b = 0$. (Such functions exist, but are hard to construct explicitely.)

Note that when it is known that $a \leq 0$ for all $x$, one may try the feedback law $k(x) := -b$. If there is sufficient "transversality" between $f_0$ and $g$ a LaSalle invariance argument establishes stability. The assumption that for some $V$ there holds $a \leq 0$ everywhere is valid for instance if one knows that $a \equiv 0$ for such a $V$, which in turn happens with conservative systems. This idea, apparently first studied in [11], gave rise to a large literature on feedback stabilization; see for instance [21], [10], [16], and references there. For example, consider the system ([11])

$$\begin{aligned} \dot{x}_1 &= x_2 \\ \dot{x}_2 &= -x_1 - x_1 u \end{aligned}$$

for which $V := (1/2)(x_1^2 + x_2^2)$ satisfies $a \equiv 0$. The feedback law $k(x) := -b(x) = -x_1x_2$ leads to a Liénards-type closed-loop equation, which can be proved asymptotically stable using the invariance principle. This function $V$ is not a clf in our (strict) sense, since one can not guarantee

$$(\forall x \neq 0)(\exists u) \quad \nabla V(x) f(x, u) < 0$$

but just the corresponding weak inequality. However, one can still try to apply the above control law, and the formula gives in this case precisely the same feedback, $-x_1x_2$ (we thank Andrea Bacciotti for pointing this out to us).

# 8 Input-to-State Stability

The paper [25] studied relations between state-space and operator notions of stabilization. One such notion is that of *input to state stabilization*, which deals with finding a feedback law $k$ so that, for the system

$$\dot{x} = f(x, k(x) + u) \tag{4}$$

in Figure 2, a strong type of bounded-input bounded-output behavior is achieved. We do not give here the precise definition of input-to-state stable system (ISS), save to point out that such stability implies asymptotic stability when $u \equiv 0$ as well as bounded trajectories for bounded controls; see also [27] for related properties. The main Theorem from [25] is:

**Theorem 8(a).** If the system $\dot{x} = f_0(x) + G(x)u$ is globally smoothly (respectively, almost smoothly) stabilizable then there exists a smooth (respectively almost smooth) $k$ so that (4) is ISS.

Note that, in general, a different $k$ is needed than the one that stabilizes; for instance

$$\dot{x} = -x + (x^2 + 1)u$$

is already asymptotically stable, i.e. $k \equiv 0$ can be used, but the constant input $u \equiv 1$ produces an unbounded trajectory –and a finite escape time from every initial state. On the other hand, $k(x) = -x$ gives an ISS closed-loop system.

The result holds also locally, of course. Further, there is a generalization to systems which are not necessarily linear in controls:

**Theorem 8(b).** If the system $\dot{x} = f(x, u)$ is smoothly (respectively, almost smoothly) stabilizable then there exists a smooth (respectively almost smooth) $k$ and an everywhere nonzero smooth scalar function $\beta$ so that the system

$$\dot{x} = f(x, k(x) + \beta(x)u)$$

in Figure 3 is ISS.

# 9 Decomposition Methods

Consider a cascade of systems as in Figure 9,

$$\begin{aligned} \dot{z} &= f(z, x) \\ \dot{x} &= g(x, u) \end{aligned}$$

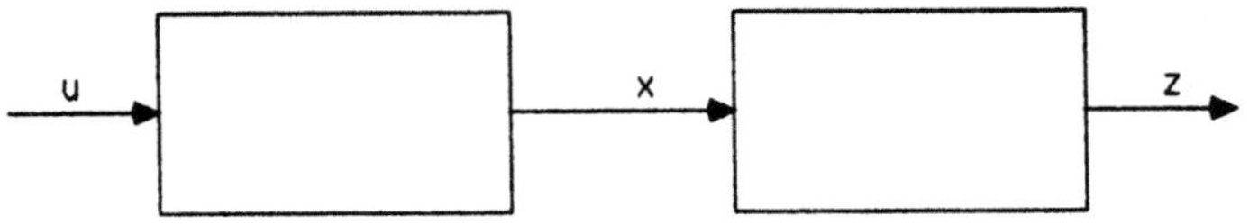


FIGURE 9: Cascade of systems

Many authors have studied the following question: If the system $\dot{z} = f(z,x)$ is stabilizable (with $x$ thought of as a control) and the same is true for $\dot{x} = g(x,u)$, what can one conclude about the cascade? More particularly, what if the "zero-input" system $\dot{z} = f(z,0)$ is already known to be asymptotically stable?

There are many reasons for studying these problems ([34], [17], [7]):

- They are mathematically natural;
- In "large scale" systems one can often easily stabilize subsystems;
- Many systems, e.g. "minimum phase" ones, are naturally decomposable in this form;
- In "partial linearization" work, one has canonical forms like this;
- Sometimes two-time scale design results in such systems.

The first result along these lines is local, and it states that a cascade of locally asymptotically stable systems is again asystable. One can also say this in terms of stabilizability of the $x$-system, since any stabilizing feedback law $u = k(x)$ can be also thought of as a feedback $u = k(x,z)$:

**Theorem 9.** If $\dot{z} = f(z,0)$ has 0 as an asymptotically stable state and if $\dot{x} = g(x,u)$ is locally smoothly stabilizable then the cascade is also locally smoothly stabilizable.

This follows from classical "total stability" theorems, and was proved for instance in [34] and in a somewhat different manner in [27] using Lyapunov techniques. The same result holds for almost-smooth stabilizability.

There is also a global version of the above:

**Theorem 10.** If $\dot{z} = f(z,0)$ has 0 as a globally asymptotically stable state and if $\dot{x} = g(x,u)$ is globally smoothly stabilizable then the cascade is also smoothly globally stabilizable, *provided* that the system $\dot{z} = f(z,x)$ be ISS.

The last condition can be weakened considerably, to the statement: If $x(t) \to 0$ as an input to the $z$-subsystem, then for every initial condition $z(0)$, the trajectory $z(\cdot)$ is defined globally and it remains bounded. (The theorem shows that in fact it must then also go to zero.)

For a proof, see [27]. Under extra hypotheses on the system, such as that $f$ be globally Liptschitz, the ISS (or the BIBS) conditions can be relaxed –the paper [31] provides a detailed discussion of this issue, which was previously considered in [37] and [19].

Consider now the more general case in which the ISS condition fails. The last statement in Theorem 10 suggests first making the $z$-system ISS, using Theorem 8(b), and thus proving stabilizability of the composition. The problem with this idea is that the feedback law cannot always be implemented through the first system. One case when this idea works is what is called the "relative degree one" situation in zero-dynamics studies. Given is a system

$$\begin{aligned} \dot{z} &= f(z,x) \\ \dot{x} &= u \end{aligned}$$

where $x$ and $u$ now have the same dimensions. Assume that $k$ and $\beta$ have been found making the system

$$\dot{z} = f(z, k(z) + \beta(z)x)$$

ISS with $x$ as input. Then, with the change of variables

$$x = k(z) + \beta(z)y$$

(recall that $\beta(z)$ is always nonzero), there results a system of the form

$$\begin{aligned} \dot{z} &= f(z, k(z) + \beta(z)y) \\ \dot{y} &= \frac{1}{\beta(x)}[h(z, y) + u] \end{aligned}$$

with $h$ a smooth function. Then $u := -\beta(x)y + h(z, y)$ stabilizes the $y$-subsystem, and hence also the cascade by Theorem 10.

Other, previous, proofs of this "relative degree one" result were due to [14], in the context of "PD control" of mechanical systems, as well as [32] and [7]. In [29], an application to rigid body control is given, in which the equations naturally decompose as above. Another such example is the following one. Assume that we wish to stabilize

$$\begin{aligned} \dot{z} &= x^3 \\ \dot{x} &= u \end{aligned}$$

and note that $u := K(z) = -z$ stabilizes the first system. Since

$$\dot{z} = (u - z)^3$$

is ISS –because $z(u - z)^3 < 0$ for large $z$ and bounded $u$,– one can chose $\beta = 1$ in the above construction. There results the smooth feedback law

$$u = -z - x - x^3$$

stabilizing the system.

## 10 Why Continuous Feedback?

Since smooth or even continuous feedback may be unachievable, one should also study various techniques of discontinuous stabilization, and this is in our view the most important direction for further work. Here we limit ourselves to a few references:

- Techniques from *optimal control* theory typically result in such stabilizing feedbacks;
- There are many classical techniques for discontinuous control, such as *sliding mode systems* (see e.g. [33]);
- A *piecewise-analytic* synthesis of controllers was shown to be possible under controllability and analyticity assumptions on the original system ([30]);
- If constant-rate sampling is allowed, *piecewise-linear* feedback can often be implemented ([22]);
- Pulse Width Modulated control is related to sampling and becoming popular (see e.g. [20]).

## 11 Output Feedback

Typically only partial measurements are available for control. Some authors have looked at output stabilization problems, and in particular the separation principle for observer/controller configurations; see e.g. [35].

For linear systems, one knows that output (dynamic) stabilizability is equivalent to stabilizability and detectability. A generalization of this theorem, when discontinuous control is allowed, was obtained in [23], based on the stability of the subsystem that produces zero output when the zero input is applied, a notion of detectability for nonlinear systems. Very little is still known in this area, however.

## 12 Additional Bibliography

There follow here some additional references related to the general topic of this paper. We'd like to acknowledge Andre Bacciotti's help in compiling this list.


- Abed E.H., and Fu J.H., "Local Feedback Stabilization and Bifurcation Control, II. Stationary Bifurcation," *Systems and Control Letters* **8**(1987): 467-473
- Aeyels D., "Stabilization by Smooth Feedback of the Angular Velocity of a Rigid Body," *Systems and Control Letters* **5**(1985): 59-63.
- Aeyels D., "Local and Global Stabilizability for Nonlinear Systems," in *Theory and Applications of Nonlinear Control Systems*, C.I. Byrnes and A. Lindquist, Eds, North Holland 1986, pp. 93-106
- Aeyels D., and M. Szafranski, "Comments on the Stabilizability of the Angular Velocity of a Rigid Body," *Systems and Control Letters*, **10**(1988): 35-39.
- Andreini, A., A. Bacciotti, and G. Stephani, "Global stabilizability of homogeneous vector fields of odd degree," *Systems and Control Letters*, **10** (1989): 251-256.
- Andreini, A., A. Bacciotti, and G. Stefani, "On the Stabilizability of Homogeneous Control Systems," Proc.8th Int. Conf. Analysis and Optim. of Systems, Antibes 1988, INRIA, *Lect. Notes in Control and Inf. No. 111*, Springer Verlag, 1988, pp. 239-248.
- Andreini, A., A. Bacciotti, P. Boieri, and G. Stefani, "Stabilization of Nonlinear Systems by means of Linear Feedbacks," Preprints of Nonlinear Control Conference, Nantes 1988.
- Bacciotti A., "Remarks on the Stabilizability Problem for Nonlinear Systems," in *Proc.25th CDC, Athens*, IEEE, 1986, pp. 2054-2055.
- Bacciotti A., "The Local Stabilizability Problem for Nonlinear Systems," *IMA J.Math. Control and Information* **5**(1988): 27-39.
- Bacciotti, A., and P. Boieri, "An Investigation of Linear Stabilizability of Planar Bilinear Systems," Proc. IFAC Symposium on Nonlinear Control Systems Design, Capri, 1989.
- Banks, S.P., "Stabilizability of finite- and infinite-dimensional bilinear systems, *IMA J. of Math Control and Inf* **3**(1986): 255-271.
- Barmish, B.R., M.J. Corless, and G. Leitmann, "A new class of stabilizing controllers for uncertain dynamical systems," *SIAM J. Control and Opt.* **21**(1983): 246-255.

- Bethash S., and S. Sastry, "Stabilization of Nonlinear Systems with Uncontrollable Linearization," *IEEE Trans. Autom. Control* **AC-33**(1988): 585-591.
- Bhatia, N.P., and G.P. Szegö, *Stability Theory if Dynamical Systems*, Springer, Berlin, 1970.
- Byrnes C.I., and A. Isidori, "Global Feedback Stabilization. of Nonlinear Systems," *Proc. 24th IEEE CDC, Ft.Lauderdale*, IEEE Publications, 1985, pp. 1031-1037
- Byrnes, C.I., and A. Isidori, "Local Stabilization of Minimum-Phase Nonlinear Systems," *Systems and Control Letters* **11**(1988): 9-17.
- Chow, J.H., and P.V. Kokotovic, "A two-stage Lyapunov-Bellman design of a class of nonlinear systems," *IEEE Trans. Autom. Cntr.* **TAC-26** (1981): 656-663.
- Corless, M.J., and G. Leitmann, "Continuous state feedback guaranteeing uniform boundedness for uncertain synamical systems," *IEEE Trans. Autom. Control*, **26**(1981): 1139-1144.
- Crouch, P.E., "Spacecraft attitude control and stabilization: applications of geometric control theory," *IEEE Trans. Autom. Control*, **AC-29**(1984): 321-333.
- Dayawansa W.P., C.F. Martin, and G. Knowles, "Asymptotic Stabilization of a Class of Smooth Two Dimensional Systems," to appear in *SIAM J.Cntr and Optimization.*
- Gauthier J.P., and G. Bornard, "Stabilization des systemes non lineaires," in *Outils Mathematiques pour l'Automatique, l'Analyse des Systemes et le Traitement du Signal*, Ed. I.D. Landau, vol.1, Publ.CNRS, 1981, Paris, pp. 307-324.
- Gauthier J.P., and G. Bornard, "Stabilization of Bilinear Systems, Performance Specification and Optimality," in Analysis ans Optimization of Systems (INRIA), Ed. Bensoussan and Lions, *Lect. Notes Contr. Inf. Sc. No.28*, Springer Verlag, Berlin, 1980, pp. 125-140
- Glass, M., "Exponential stability revisited," *Int.J. Control* **46**(1987): 1505-1510.
- Hermes, H., "On the synthesis of a stabilizing feedback control via Lie algebraic methods," *SIAM J.Cntr. and Opt.* **18**(1980): 352-361.
- Hermes, H., "On a Stabilizing Feedback Attitude Control," *J. Optim. Theory and Appl.*, 31 (1980): 373-384.
- Huijberts, H.J.C., and A.J.van der Schaft, "Input-output decoupling with stability for Hamiltonian systems," to appear in *Math. of Control, Signals, and Systems.*
- Irving, M., and P.E. Crouch, "On Sufficient Conditions for Local Asymptotic Stability of Nonlinear Systems whose Linearization is Uncontrollable," Control Theory Centre Report No. 114, University of Warwick
- Kawski, M., "Stabilizability and Nilpotent Approximations," *Proc. 1988 IEEE CDC*, IEEE Publications, 1988.
- Koditschek, D.E., and K.S. Narendra, "Stabilizability of Second-Order Bilinear Systems," *IEEE Trans. Autom. Ctr.* **AC-28**(1983): 987-989.
- Kokotovic, P.V., and H.J. Sussmann, H.J., "A positive real condition for global stabilization of nonlinear systems," Report 89-01, *SYCON - Rutgers Center for Systems and Control*, Jan. 1989.

- Korobov, V.I., "Controllability and Stability of Certain Nonlinear Systems," *Differentsial'nye Uranveniya* Vol. 9 (1973): 614-619 (in Russian, translation on *Differential Equations*, pp. 466-469).
- Longchamp, R., "State-feedback control of bilinear systems," *IEEE Trans. Autom. Cntr.* **25**(1980): 302-306.
- Longchamp, R., "Controller Design for Bilinear Systems," *IEEE Trans. Autom. Ctr.* **AC-25**(1980): 547-548.
- Luesink, R., and H. Nijmeijer, "On the Stabilization of Bilinear Systems via Constant Feedback," to appear in *Linear Algebra and its Applications.*
- Marino R., "Feedback Stabilization of Single-input Nonlinear Systems," *Systems and Control Letters* **10**(1988): 201-206.
- Pedrycz, W., "Stabilization of Bilinear Systems by a Linear Feedback Control," *Kybernetika* **16** (1980): 48-53.
- Quinn J.P., "On Russell's Method of Controllability via Stabilizability," *J. Opt. Theory and Appl.* **52**(1987): 279-291.
- Quinn J.P., "Stabilization of Bilinear Systems by Quadratic Feedback Control," *J. Math. Anal. Appl.* **75**(1980): 66-80.
- Roxin, E., "Stability in general control systems," *J.of Diff. Eqs.* **19**(1965): 115-150.
- Ryan, E.P., and N.J. Buckingham, "On asymptotically stabilizing feedback control of bilinear systems," *IEEE Trans. Autom. Cntr.* **28**(1983): 863-864.
- Saskena, V.R., J.O'Relly, and P.V. Kokotovic, "Singular perturbations and two-scale methods in control theory: survey 1976-1983," *Automatica* **20**(1984): 273-293.
- Singh S.N., "Stabilizing Feedback Control for Nonlinear Hamiltonian Systems and Nonconservative Bilinear Systems in Elasticity," *J. Dyn. Syst., Meas. and Control, Transac. of ASME* **104**(1982): 27-32.
- Tsinias J., "Stabilization of Affine in Control Nonlinear Systems," *Nonlinear Analysis TMA* **12**(1988): 1283-1296.
- Tsinias, J., and N. Kalouptsidis, "Prolongations and stability analysis via Lyapunov functions of dynamical polysystems," *Math. Systems Theory* **20**(1987): 215-233.
- Van der Schaft, A.J., "Stabilization of Hamiltonian Systems," *J. Nonlinear Analysis, MTA* **10**(1986): 1021-1035.


*This Research was supported in part by US Air Force Grant AFOSR-88-0235.*

## 13 References


1. Abed, E.H., and J-H. Fu, "Local stabilization and bifurcation control, I. Hopf bifurcation," *Systems and Control Letters* **7** (1986): 11-17.

2. Aeyels, D., "Stabilization of a class of nonlinear systems by a smooth feedback control," *Systems and Control Letters*, **5** (1985): 289-294.

3. Artstein, Z., "Stabilization with relaxed controls," *Nonl. Anal., TMA* **7** (1983): 1163-1173.

4. Baccioti, A., and P. Boieri, "Linear stabilizability of planar nonlinear systems," to appear in *Math. of Control, Signals, and Systems.*

5. Boothby, W.M., and R. Marino, "Feedback stabilization of planar nonlinear systems," *Systems and Control Letters* **12**(1989): 87-92.

6. Brockett, R.W., "Asymptotic stability and feedback stabilization," in *Differential Geometric Control theory* (R.W. Brockett, R.S. Millman, and H.J. Sussmann, eds.), Birkhauser, Boston, 1983, pp. 181-191.

7. Byrnes, C.I. and A. Isidori, "New results and counterexamples in nonlinear feedback stabilization," *Systems and Control Letters*, **12**(1989), Number 5.

8. Dayawansa, W.P., and C.F. Martin, "Asymptotic stabilization of two dimensional real-analytic systems," *Systems Control Lett.* **12**(1989): 205-211.

9. Delchamps, D.F., "Analytic stabilization and the algebraic Riccati equation," *Proc. IEEE Conf. Dec. and Control* (1983) :1396-1401.

10. Gutman, P.O., "Stabilizing controllers for bilinear systems," *IEEE Trans. Autom. Control*, **26**(1981): 917-922.

11. Jurdjevic, V. and J.P. Quinn, "Controllability and stability," *J.of Diff.Eqs.* **28** (1978): 381-389.

12. Kalouptsidis, N., and J. Tsinias, "Stability improvement of nonlinear systems by feedback," *IEEE Trans. Autom. Crt.* **29**(1984): 346-367.

13. Kawski, M., "Stabilization of nonlinear systems in the plane," *Systems and Control Letters*, **12**(1989): 169-176.

14. Koditschek, D.E., "Adaptive techniques for mechanical systems," *Proc.5th. Yale Workshop on Adaptive Systems*, pp. 259-265, Yale University, New Haven, 1987.

15. Krasnoselskii, M.A., and P.P. Zabreiko, *Geometric Methods of Nonlinear Analysis*, Springer, Berlin, 1983.

16. Lee, K.K., and Arapostathis, A., "Remarks on smooth feedback stabilization of nonlinear systems," *Systems and Control Letters*, **10** (1988): 41-44.

17. Marino R., "High-Gain Stabilization and Partial Feedback Linearization," *Proceedings of 25th CDC, Athens*, IEEE Publ., 1986, pp.209-213

18. Massera, J.L., "Contributions to stability theory," *Annals of Math*, **64**(1956): 182-206. Erratum in *Annals of Math*, **68**(1958): 202.

19. Sastry, S.I., and A. Isidori, "Adaptive control of linearizable systems," Memo UCB/ERL M87/53, March 1988, U.California, Berkeley.

20. Sira-Ramirez, H. "A geometric approach to pulse-width-modulated control in dynamical systems," *IEEE Trans. Automatic Control*, **34**(1989): 184-187.

21. Slemrod, M., "Stabilization of bilinear control systems with applications to nonconservative problems in elasticity," *SIAM J.Control and Opt.* **16**(1978): 131-141.

22. Sontag, E.D., "Nonlinear regulation: The piecewise linear approach," *IEEE Trans. Autom. Control* **AC-26**(1981): 346-358.

23. Sontag, E.D., "Conditions for abstract nonlinear regulation," *Information and Control*, **51**(1981): 105-127.

24. Sontag, E.D., "A Lyapunov-like characterization of asymptotic controllability," *SIAM J. Control and Opt.*, **21**(1983): 462-471.

25. Sontag, E.D., "Smooth stabilization implies coprime factorization," *IEEE Trans. Automatic Control*, **34**(1989): 435-443.

26. Sontag, E.D., "A 'universal' construction of Artstein's theorem on nonlinear stabilization," *Systems and Control Letters*, **13** (1989): No.2.

27. Sontag, E.D., "Further facts about input to state stabilization", to appear in *IEEE Trans. Automatic Control*, 1990.

28. Sontag, E.D., and H.J. Sussmann, "Remarks on continuous feedback," *Proc. IEEE Conf. Dec. and Control, Albuquerque, Dec.1980.*

29. Sontag, E.D., and H.J. Sussmann, "Further comments on the stabilizability of the angular velocity of a rigid body," *Systems and Control Letters*, **12** (1989): 213-217.

30. Sussmann, H.J., "Subanalytic sets and feedback control," *J.Diff.Eqs.* **31**(1979):31-52.

31. Sussmann, H.J., and P.V. Kokotovic, "The peaking phenomenon and the global stabilization of nonlinear systems," Report 89-07, *SYCON - Rutgers Center for Systems and Control*, March 1989.

32. Tsinias, J., "Sufficient Lyapunovlike conditions for stabilization," *Math. of Control, Signals, and Systems* **2**(1989): 343-357.

33. Utkin, V.I., *Sliding Modes and Their Application in Variable Structure Systems*, Mir, Moscow, 1978.

34. Vidyasagar, M., "Decomposition techniques for large-scale systems with nonadditive interactions: stability and stabilizability," *IEEE Trans. Autom. Ctr.* **25**(1980): 773-779.

35. Vidyasagar, M., "On the stabilization of nonlinear systems using state detection," *IEEE Trans. Autom. Ctr.* **25**(1980): 504-509.

36. Vidyasagar, M., *Nonlinear Systems Analysis*, Prentice-Hall, 1978.

37. Varaiya, P.P. and R. Liu, "Bounded-input bounded-output stability of nonlinear time-varying differential systems," *SIAM J.Control* **4**(1966): 698-704.

38. Zabczyk J., "Some Comments on Stabilizability," *Applied Math Optimiz.* **19**(1989): 1-9.

# A MONOTONICITY RESULT FOR THE PERIODIC RICCATI EQUATION

Sergio Bittanti, Patrizio Colaneri

Abstract

The differential Riccati equation with periodic coefficients is considered in this paper. Attention is focused on the symmetric and periodic solutions, in particular on the strong solution. It is proven that the strong solution is greater than or equal to any solution of another periodic Riccati equation with coefficients which are suitably related to those of the original equation.

Key words : periodic linear systems, periodic Riccati equation, strong solution, stabilizability of periodic systems

## 1. INTRODUCTION

In [6], H.K.Wimmer proved an interesting monotonicity result for algebraic Riccati equations. This result points out the monotonic behaviour of the solutions when the equation coefficients are suitably modified.

The purpose of this paper is to supply the (nontrivial) generalization of Wimmer's theorem to differential Riccati equations with periodic coefficients. As a byproduct, a number of properties concerning the periodic Riccati equation (in standard and nonstandard form) are also obtained.

In the sequel, some basic notions of PSICO (Periodic Systems, Identification, Control and Optimization) will be used, see [1] for a survey. In particular, given a T-periodic matrix $A(\cdot)$, i.e. $A(t+T)=A(t)$, $\forall$ t, the associated transition matrix will be denoted by $\Phi_A(t,\tau)$. The matrix $\Phi_A(T,0)$ is called monodromy matrix of $A(\cdot)$. Its eigenvalues, which determine the system stability, are called characteristic multipliers. By Floquet theory, see e.g. [7], $\Phi_A(T,0) = e^{AT}$, where the eigenvalues of A are named characteristic exponents.

The structural property of interest herein is stabilizability. A number of different yet equivalent characterizations of stabilizability have been recently obtained [2]. In particular, it can be shown that a periodic pair $(A(\cdot),B(\cdot))$, where $A(t)\in R^{nxn}$ and $B(t)\in R^{nxm}$, is stabilizable if and only if, for each characteristic multipliers $\lambda$ of $A(\cdot)$

$$\left.\begin{cases} \Phi_A(T,0)'\eta = \lambda\eta \\ \Phi_A(0,t)'B(t)'\eta=0, \ \forall t\in[0,T] \end{cases}\right\} \Rightarrow \quad \eta = 0 .$$

## 2. THE PERIODIC RICCATI EQUATION

Consider the periodic Riccati equation

$$- \dot{P}(t) = A(t)'P(t) + P(t)A(t) - P(t)B(t)B(t)'P(t) + Q(t) \tag{1}$$

where, as already stated, $A(t)\in R^{nxn}$, $B(t)\in R^{mxn}$; moreover, $Q(t)\in R^{nxn}$. Matrices $A(\cdot)$, $B(\cdot)$ and $Q(\cdot)$ are periodic of period T. Note that *no definiteness* assumption on Q(t) is made.

Letting

$$M(t) = \begin{bmatrix} Q(t) & A(t)' \\ A(t) & -B(t)B(t)' \end{bmatrix}, \tag{2}$$

equation (1) can be also written as

$$-\dot{P}(t) = [I_n \ P(t)]' \ M(t) \begin{bmatrix} I_n \\ P(t) \end{bmatrix} . \tag{3}$$

Amongst the Symmetric T-Periodic (SP) solutions of (1), the attention will be focused on the so called <u>strong</u> <u>solution</u>. AnSP solution $P_s(\cdot)$ is said to be strong if the characteristic multipliers of $\hat{A}(\cdot) = A(\cdot) - B(\cdot)B(\cdot)'P_s(\cdot)$ belong to the <u>closed</u> unit disk. This notion can be seen as a generalization of the well known concept of stabilizing solution (characteristic multipliers of $\hat{A}(\cdot)$ belonging to the <u>open</u> unit disk).

The SP solutions of equation (3) will be compared with the SP solutions of the following Riccati equation

$$-\dot{\tilde{P}}(t) = [I_n \ \tilde{P}(t)]' \ \tilde{M}(t) \begin{bmatrix} I_n \\ \tilde{P}(t) \end{bmatrix} \qquad (4)$$

where

$$\tilde{M}(t) = \begin{bmatrix} \tilde{Q}(t) & \tilde{A}(t)' \\ \tilde{A}(t) & -\tilde{B}(t)\tilde{B}(t)' \end{bmatrix} . \qquad (5)$$

Here, $\tilde{A}(\cdot)$, $\tilde{B}(\cdot)$, $\tilde{Q}(\cdot)$ are T-periodic matrices of the same dimensions of $A(\cdot)$, $B(\cdot)$ and $Q(\cdot)$ respectively.

The comparison will be made under the basic *monotonicity assumption*

$$M(t) \geq \tilde{M}(t), \ \forall \ t \in [0,T].$$

This assumption reflects into a monotonicity property of the SP solutions of the two periodic Riccati equations (3) and (4). Precisely, under suitable hypothesis, any SP strong solution of (3) turns out to be greater then any SP solution of (4). In fact, under the same hypothesis, there is at most one SP strong solution of (3), which is also maximal.

3. MAIN RESULT

The proof of the main result relies on the following basic Lemma concerning the periodic differential Riccati equation (of special type)

$$-\dot{W}(t) = F(t)'W(t) + W(t)F(t) + W(t)G(t)G(t)'W(t) + H(t)'H(t) \qquad (6)$$

$F(t)$; $G(t)$ and $H(t)'H(t)$ are T-periodic matrices of the same dimension of $A(t)$, $B(t)$ and $Q(t)$, respectively.

We begin with by stating a preliminary result, the proof of which is omitted for the sake of conciseness.

Microtheorem

Consider equation (6) and suppose that $F(\cdot)$ has $n_1$ characteristic multipliers belonging to the open unit disk and $n_2=n-n_1$ characteristic multipliers belonging to the unit circle.
Then, there exists a differentiable T-periodic nonsingular transformation $S(\cdot)$: $R \Rightarrow R^{nxn}$ such that

$$\hat{F}(t) = S(t)F(t)S(t)^{-1} + S(t)S(t)^{-1}$$

is constant and given by

$$\hat{F}(t) = \bar{F} = \begin{bmatrix} F_1 & 0 \\ 0 & F_2 \end{bmatrix} \quad \forall\ t\ ,$$

where the eigenvalues of $F_1$ anf $F_2$ coincides with the characteristic exponents of $\hat{F}(\cdot)$ with negative or null real part, respectively.

Moreover, letting

$$\hat{G}(t) = S(t)G(t)\ ,\ \forall\ t\ ,\quad \hat{H}(t) = H(t)S(t)^{-1},\ \forall\ t$$

and defining

$$\dot{\hat{W}}(t) = \bar{F}'\hat{W}(t) + \hat{W}(t)\bar{F} + \hat{W}(t)\hat{G}(t)\hat{G}(t)'\hat{W}(t) + \hat{H}(t)'\hat{H}(t) \tag{7}$$

the SP solutions of equation (6) are in one-to-one correspondence with the SP solution of equation (7), with

$$\hat{W}(t) = (S(t)^{-1})'W(t)S(t)^{-1}. \blacksquare$$

Notice that, from this result, it follows that

$$W(t) \geq 0 \ , \ \forall \ t \in [0,T] \quad \Leftrightarrow \quad \hat{W}(t) \geq 0 \ , \ \forall \ t \in [0,T].$$

Lemma

Suppose that $(F(\cdot), G(\cdot))$ is stabilizable and that the characteristic multipliers of $F(\cdot)$ belong to the closed unit disk.
Then an SP solution of (6), if any, is positive semidefinite at each time point:

$$\hat{W}(t) \geq 0 \ , \ \forall \ t.$$

Proof

Let $n_2$ be the number of characteristic multipliers of $F(\cdot)$ with null real part.

- Thanks to the previous Microtheorem, it can be assumed without any loss of generality that

$$F(t) = \text{const.} = \begin{bmatrix} F_1 & 0 \\ 0 & F_2 \end{bmatrix} \tag{8}$$

where $F_1$ and $F_2$ are square matrices the eigenvalues of which are the characteristic exponents with negative or null real part, respectively.

- We first prove that

$$W(t)\begin{bmatrix} 0 \\ \xi_2 \end{bmatrix} = 0 \ , \ \forall \ t \ , \ \forall \ \xi_2 \in R^{n_2} \tag{9}$$

and

$$H(t)'H(t)\begin{bmatrix} 0 \\ \xi_2 \end{bmatrix} = 0 \ , \ \forall \ t, \ \forall \ \xi_2 \in R^{n_2} \ . \tag{10}$$

To this purpose, consider a particular vector $x_2$ such that

$$e^{F_2 T} x_2 = \lambda \ x_2.$$

Obviously, $|\lambda|$-land

$$e^{FT} y = \lambda \ y \tag{11}$$

with $y' = [0 \ \ x_2']$ . Consider then

$$\dot{z}(t) = e^{Ft} y \ . \tag{12}$$

From (1), premultiplying by $z(t)^*$ and postmultiplying by $z(t)$, it follows that

$$-\frac{d}{dt}(z(t)^* W(t)z(t)) = z(t)^*( W(t)G(t)G(t)' + H(t)'H(t) )z(t). \tag{13}$$

By integrating (13) over $[0,T]$, the periodicity condition $W(T)=W(0)$ and (11) and (12) entail that

$$G(t)'W(t)z(t) = 0, \; \forall \; t \in [0,T] \tag{14}$$

$$H(t)'H(t)z(t) = 0, \; \forall \; t \in [0,T] \; . \tag{15}$$

Now, postmultiplying (6) by $z(t)$ and using (14) and (15) one obtains

$$\frac{d}{dt}\,(W(t)z(t)) = -\,F\,'W(t)z(t) \; .$$

Therefore

$$W(t)z(t) = e^{-F\,'T}W(0)y \; . \tag{16}$$

Substituting (16) into (14), it follows that

$$G(t)'\,e^{-F't}W(0)y = 0, \; \forall \; t \in [0,T]. \tag{17}$$

From (11) and (12)

$$z(T) = e^{FT}y = \lambda\, y.$$

Hence, taking into account the periodicity of $W(\cdot)$ and equation (16),

$$e^{F'T}W(0)y = e^{F'T}W(T)z(T)\lambda^{-1} = \lambda^{-1}W(0)y \tag{18}$$

i.e $W(0)y$ is an eigenvector of $e^{FT}$ associated with the modulus one

eigenvalue $\lambda^{-1}$ .
In conclusion, a vector $\eta$ has been found such that

$$\Phi_F(T,0)'\eta = \lambda\eta ,$$

$$G(t)'\Phi_F(0,t)'\eta = 0, \ \forall \ t \in [0,T] ,$$

see (17) and (18). The stabilizability of (F(·), G(·)) implies that $\eta$ = 0, namely W(0)y = 0, and, from (16),

$$W(t)z(t) = 0 \ , \ \forall \ t \in [0, T] \ . \tag{19}$$

It is possible to generalize the previous arguments to show that (15) and (19) hold true even if, in (12), y is replaced by a generalized eigenvector associated with $\lambda$. Then, by considering all eigenvalues of $F_2$, the following identities are obtained:

$$H(t)'H(t)e^{Ft}\begin{bmatrix} 0 \\ \xi_2 \end{bmatrix} = 0 \ , \ \forall \ t \in [0,T], \ \forall \ \xi_2 \in R^{n_2} \tag{20}$$

$$W(t)e^{Ft}\begin{bmatrix} 0 \\ \xi_2 \end{bmatrix} = 0 \ , \ \forall \ t \in [0,T], \ \forall \ \xi_2 \in R^{n_2} \ . \tag{21}$$

Since $e^{Ft}$ is nonsingular and block diagonal, according to partition (8), the conclusion given by (9) and (10) can be easily deduced from (20) and (21).

- It is important to observe that (9) and (19) entail that W(t) and H(t)'H(t) have the following structure

$$W(t) = \begin{bmatrix} W_1(t) & 0 \\ 0 & W_2(t) \end{bmatrix} \tag{22}$$

$$H(t)'H(t) = \begin{bmatrix} H_1(t)'H_1(t) & 0 \\ 0 & 0 \end{bmatrix} \tag{23}$$

where $W_1(t)$ and $H_1(t)'H_1(t)$ have the same dimensions of matrix $F_1$.
By means of partitions (22) and (23), the Riccati equation (6) can be reduced to the following one:

$$-\dot{W}_1(t) = F_1'W_1(t) + W_1(t)F_1 + W_1(t)G_1(t)G_1(t)'W_1(t) + H_1(t)'H_1(t) \tag{24}$$

where $G_1(t)G_1(t)'$ is given by the first $n-n_2$ rows and $n-n_2$ columns of $G(t)G(t)'$.

- Suppose that $\bar{W}_1(t)$ is an SP solution of (24). Replacing $W_1(t)G_1(t)G_1(t)'W_1(t)$ by $W_1(t)\bar{G}_1(t)G_1(t)'W_1(t)$, equation (24) becomes a Lyapunov equation with periodic coefficients. Since $F_1$ is asymptotically stable, this Lyapunov equation admits a unique SP solution $W_1(t) = \bar{W}_1(t)$ which is positive semidefinite $\forall$ t, see e.g. [3] and this completes the proof. ∎

Theorem

Suppose that $(A(\cdot),B(\cdot))$ is stabilizable and that equation (1) has at least an SP solution. Then

- The Riccati equation (1) has a strong SP solution $P_s(\cdot)$.

- Suppose also that $M(t) \geq \tilde{M}(t)$ , $\forall\ t \in [0,T]$, where $M(t)$ and $\tilde{M}(t)$ are given by (2) and (5) respectively. If $\tilde{P}(t)$ is any SP solution of equation (6), then

$$P_s(t) \geq \tilde{P}(t) \ , \ \forall\ t \in [0,T].$$

Proof

For the existence of $P_s(\cdot)$ see [6]. Let

$$\bar{M}(t) = M(t) - \tilde{M}(t) \quad \text{and} \quad W(t) = P_s(t) - \tilde{P}(t) \quad .$$

Note that, by assumption, $\bar{M}(t) \geq 0$, $\forall\ t \in [0,T]$. From (3) and (4), it can be shown that $W(\cdot)$ satisfies equation (6) with

$$F(t) = A(t) - B(t)B(t)'P_s(t),$$

$$G(t) = B(t) \quad \text{and}$$

$$H(t)'H(t) = [I_n \quad \tilde{P}(t)]\ \bar{M}(t) \begin{bmatrix} I_n \\ \tilde{P}(t) \end{bmatrix} .$$

Since $P_s(\cdot)$ is a strong solution of equation (1), the characteristic multipliers of $F(\cdot)$ belong to the closed unit disk. Then, the statement is a straighforward consequence of the previous Lemma.

Corollary

Suppose that $((A(\cdot), B(\cdot))$ is stabilizable and that equation (1) has at least an SP solution. Then, the strong solution $P_s(\cdot)$ is unique. Moreover, $P_s(\cdot)$ is maximal, i.e. $P_s(t) \geq \tilde{P}(t)$, $\forall\ t$, for any SP solution of equation (1).

Proof

The strong solution is maximal, as it can be easily concluded from the theorem above by setting $\bar{M}(t) = 0$, $\forall\ t$. Uniqueness of the maximal solution is obvious.■

Note that the above conclusions hold with no assumption on the definiteness of $Q(t)$. If, besides the stabilizability hyphothesis, the assumption is made that $Q(t)$ is positive semidefinite for each t, then the strong solution does exist, is unique and maximal, and turns out to be positive semidefinite as well, see [4] and [5].

REFERENCES

[1] S.Bittanti, "Deterministic and stochastic linear periodic systems", in Time Series and Linear Systems (S.Bittanti Ed.), Springer Verlag, Berlin, pp. 141-182, 1986.

[2] S.Bittanti and P.Bolzern, "Four equivalent notions of stabilizability of periodic systems", Proc. of 3rd American Control Conference, San Diego, CA, pp. 1321-1323, 1984.

[3] S.Bittanti, P.Bolzern and P.Colaneri, "The extended Lyapunov Lemma", Automatica, Vol. 21, N. 5, pp. 603-605, 1985.

[4] S.Bittanti and P.Colaneri, "Periodic solutions of the nonstandard periodic Riccati equation", 26th Conference on Mathematical Theory of Network and Systems, Phoenix, 1987.

[5] S.Bittanti, P.Colaneri, G. De Nicolao, "On an iterative procedure for periodic Riccati equation computation", Centro Teoria dei Sitemi C.N.R.,Milano, Int. Rep. 86-009.

[6] H.K.Wimmer, " Monotonicity of maximal solutions of algebraic Riccati equations", Systems and Control Letters, 5, pp. 317-319, 1985.

[7] V.A.Yakubovich and V.M.Starzhinskii, "Linear differential equations with periodic coefficients", Wiley, New York (translated from Russian, 1972).

# RESULTS ON GENERALISED RICCATI EQUATIONS ARISING IN STOCHASTIC CONTROL

Carlos E. de Souza and Marcelo D. Fragoso

**ABSTRACT**: This deals with a **generalized** version of the standard matrix Riccati equations which arises in certain stochastic optimal control problems. A novelty here, regarding previous works, is that it is assumed that the systems are not necessarily detectable, including those having nonobservable modes on the imaginary axis. The collection of results which are derived in this paper includes, *inter alia*, the following: a) existence and uniqueness of nonnegative definite solutions of ,the generalized algebraic Riccati equations which give rise to stable closed loop systems in the case of non-detectable systems; b) new convergence results for the solution of the generalized Riccati differential equation under relatively weaker assumptions.

## 1. INTRODUCTION

This paper considers a variant of well-known standard matrix Riccati equations which arises in certain stochastic optimal control problems. What differentiates this class of matrix Riccati equations from the standard one is an additional linear positive operator in the unknown matrix, as in equations (2.1a) and (2.2). As this class encompasses the standard one, and for the sake of nomenclature easiness, we append the name generalized here to distinguish the later from the former whenever the additional non-trivial linear positive operator appears, i.e. hence-forth (2.1a) shall be called **generalized Riccati differential equation** (**GRDE**) and (2.2) **generalized algebraic Riccati equation (GARE)**. A typical situation where these equations arise is, for example, in the stochastic optimal control problem of linear systems with markovian jumping parameters and quadratic cost, as, for instance, in [4], [6] and [8]. In this case, the additional linear operator appears in connection with the coupling of a set of Riccati equations.

Although generalized Riccati equations have only recently appeared more frequency in the literature (see [5] for details), a study of this class of Riccati equations can be traced back to [7] in which a number of results have been established subject to both stabilizability and observability assumptions.

However, the results referred to above are not exhaustive, as they cannot handle non-detectable systems. Results in this direction for the standard Riccati equations have recently been carried out in [1] and [3].

The prime concern of this paper then, is to establish results for the generalized Riccati equations to handle systems not necessarily detectable, along the same lines as in [3]. This is accomplished here by establishing conditions for existence and uniqueness of nonnegative definite solution of the GARE, which gives rise to a stable closed-loop system. Furthermore, the convergence of the solution of the GRDE is investigated under weaker assumptions.

## 2. PROBLEM STATEMENT

In this paper we shall be interested in the asymptotic properties of the solution of the following GRDE

$$\dot{P}(t) + P(t)A + A^TP(t) - P(t)BN^{-1}B^TP(t) + \Pi[P(t)] + C^TC = 0; \quad t \in [t_0,T] \tag{2.1a}$$

$$P(T) = P_T \geq 0 \tag{2.1b}$$

where $A,B,C$ are matrices of dimension respectively $n\times n$, $n\times m$, $p\times n$; $N,P$ denote symmetric matrices of dimension respectively $m\times m$, $n\times n$ with $N$ assumed positive definite and $\Pi(\cdot)$ denotes a positive linear map of the class of symmetric $n\times n$ matrices into itself, i.e. $P\geq 0$ implies $\Pi(P)\geq 0$, where $P\geq 0$ (respectively, $P>0$) stands for nonnegative (respectively, positive) definite matrix and $A^T$ denotes the transpose of $A$.

We will also investigate the solutions of the following GARE

$$PA + A^TP - PBN^{-1}B^TP + \Pi(P) + C^TC = 0 \tag{2.2}$$

These solutions play an important role because, as we shall see later, under certain conditions the solution of the GRDE will converge to a nonnegative definite solution, $P_s$, of the GARE which gives rise to a stable matrix

$$\bar{A}_s = A - BN^{-1}B^TP_s \tag{2.3}$$

Throughout the paper, the notation $\lambda(A)$ denotes the eigenvalues of the matrix A, Re(•) stands for the real part of a complex number, I denotes the identity matrix and $\|A\|$ will be used to denote the spectral norm of the matrix A defined as its maximum singular value.

An equation of type (2.1) arises, for example, in the finite-time horizon $[t_0,T]$, optimal control problem of linear time-invariant systems with state-dependent noise and quadratic cost [8]. In this situation the state feedback control gain matrix is given by $K(t) = N^{-1}B^TP(t)$, $t \in [t_0, T]$. Furthermore, under certain conditions $K_s = N^{-1}B^TP_s$, with $P_s$ as before, corresponds to the control gain matrix for the infinite-time horizon optimal control. Therefore, the convergence of P(t) to $P_s$ as $t \to -\infty$ is fundamental as it ensures that the finite-time horizon optimal control policy will tend to the stationary infinite-time horizon optimal control as the time interval $(T-t_0) \to \infty$.

<u>Definition 2.1</u> (**Strong Solution**): A real symmetric nonnegative definite solution, $P_s$, of the GARE is called a **strong solution** if the corresponding system matrix $\bar{A}_s$ has all its eigenvalues in the closed left half plane.

<u>Defintion 2.2</u> (**Stabilizing Solution**): A real symmetric nonnegative definite solution, $P_s$, of the GARE is called a **stabilizing solution** if the corresponding system matrix $\bar{A}_s$ is stable.

In order to ensure existence of a solution to (2.2) and uniqueness of the strong solution, we will require:

<u>Assumption A.1</u>: The operator $\Pi(\cdot)$ is such that

(i) $$\left\| \int_0^\infty \exp(A^Tt)\, \Pi(I)\exp(At)dt \right\| < 1,$$

in the case where $Re[\lambda(A)] \leq 0$ and all the modes of (C,A) are unobservable, or

(ii) $$\inf_K \left\| \int_0^\infty \exp[(A-BK)^T t]\Pi(I)\exp(A-BK)t]dt \right\| < 1,$$

otherwise.

Remark 2.1: Assumption A.1(ii) was introduced in [7] and expresses the fact that $\Pi$ is not too large. Note that for the case where $Re[\lambda(A)] \leq 0$ and all the modes of $(C,A)$ are unobservable, Assumption A.1(i) is a natural extension of Assumption A.1(ii) as in this situation $K=0$ is the optimal feedback gain matrix, corresponding to the strong solution. Assumption A.1(i) is fundamental for establishing uniqueness of the strong and stabilizing solution of the GARE in the case of non-observable systems (in the control context).

## 3. THE GENERALIZED ALGEBRAIC RICCATI EQUATION

In this section we discuss properties of solutions of the GARE. Conditions for existence and uniqueness of **strong solutions** to the GARE for systems not necessarily detectable will be investigated. The existence of the **strong solution** is established subject only to a stabilizability assumption.

The proof of the main theorems in this section require the following intermediate results. For detail of the proofs see [2].

Lemma 3.1: Let $A$ be stable matrix and $Q$ be symmetric nonnegative definite. Furthermore, assume that

$$\left\| \int_0^\infty \exp(A^T t)\Pi(I)\exp(At)dt \right\| < 1. \tag{3.1}$$

If $P$ is symmetric and a solution of the equation

$$PA + A^T P + \Pi(P) = -PBN^{-1}B^T P - Q \tag{3.2}$$

then $P$ is nonnegative definite.

Lemma 3.2: Let $P_s \geq 0$ be a strong solution of the GARE. If $\Pi$ satisfies Assumption A.1 then, we have

$$\left\| \int_0^\infty \exp[(A-BK_s)^T t]\Pi(I)\exp[(A-BK_s)t]dt \right\| < 1 \tag{3.3}$$

where $K_s=N^{-1}B^TP_s$ is the state feedback gain.

<u>Lemma 3.3</u>: Let $\Pi$ satisfy Assumption A.1 and assume that the GARE

$$PA + A^TP - PBN^{-1}B^TP + \Pi(P) + C^TC = 0 \tag{3.4}$$

has a stabilizing solution $P_s$. Furthermore, let $P_1$ be symmetric and satisfy

$$P_1A + A^TP_1 - P_1BN_1^{-1}B^TP_1 + \Pi(P_1) + C_1^TC_1 = 0 \tag{3.5}$$

If $C^TC \geq C_1^TC_1$ and $N \geq N_1$ then $P_s \geq P_1$.

We consider in the following, conditions for existence and uniqueness of stabilizing solution of the GARE. The lemma discussed below is an extension of Theorem 4.1 of [7] to handle systems not necessarily observable.

<u>Lemma 3.4</u>: If $(A,B)$ is stabilizable, $(C,A)$ is detectable and $\Pi$ satisfies Assumption A.1 then, the GARE has a unique stabilizing solution.

The next theorems discuss the questions of the existence and uniqueness of strong solution of the GARE.

**THEOREM 3.1**: If $(A,B)$ is stabilizable and $\Pi$ satisfies Assumption A.1, then a strong solution of the GARE exists and is unique.

**THEOREM 3.2**: If $(A,B)$ is stabilizable, $(C,A)$ has no unobservable modes on the imaginary axis and $\Pi$ satisfies Assumption A.1 then, the strong solution of the GARE is also the stabilizing solution.

The next theorem establishes sufficient conditions for the strong solution to be the unique solution of the GARE in the class of nonnegative definite matrices.

**THEOREM 3.3**: If $(A,B)$ is stabilizable, $(C,A)$ has no unobservable mode in the open right half plane and $\Pi$ satisfies Assumption A.1 then, the strong solution is the only nonnegative definite solution of the GARE.

The last theorem in this section deals with the existence of positive definite stabilizing solution of the GARE.

**THEOREM 3.4**: If $(A,B)$ is stabilizable, $(C,A)$ has no unobservable mode in the closed left half plane and $\Pi$ satisfies Assumption A.1 then, the stabilizing solution of the GARE exists and is positive definite.

## 4. THE GENERALIZED RICCATI DIFFERENTIAL EQUATION

In the sequel we shall investigate the asymptotic behaviour of the solution of the GRDE. The results are new and apply to systems not necessarily detectable including those having unobservable modes on the imaginary axis.

Initially we shall present some monotonicity properties of solutions of the GRDE.

Lemma 4.1: Let $P_1(t)$ and $P_2(t)$ be the solutions of two GRDE's (2.1), with the same matrices $A$ and $B$ but possibly different $C$ matrices, $C_1$ and $C_2$, possibly different $N$ matrices, $N_1$ and $N_2$ and possibly different terminal conditions, $P_{1T}$ and $P_{2T}$ respectively. If $C_1^TC_1 \geq C_2^TC_2$, $N_1 \geq N_2$ and $P_{1T} \geq P_{2T}$ then, $P_1(t) \geq P_2(t)$, $t_0 \leq t \leq T$.

Corollary 4.1: Let $P(t)$ be the solution of the GRDE, and suppose there exists a nonnegative definite solution, $P_1$, of the GARE. If $P_T \geq P_1$ then $P(t) \geq P_1$ on $[t_0,T]$.

Lemma 4.2: Let $P(t)$ be the solution of the GRDE. If for some $t_1$, $\dot{P}(t_1) \geq 0$ (respectively, $\dot{P}(t) \leq 0$) then, $\dot{P}(t) \geq 0$ (respectively, $\dot{P}(t) \leq 0$) for all $t \leq t_1$.

The lemma above shows that monotonicity of $P(t)$ in a neighbourhood of $t=0$ will imply the monotonicity of $P(t)$ for all $t \leq 0$. This fact will allow us to establish the convergence of $P(t)$ under relatively weaker assumptions.

In the sequel we will investigate the convergence of the solution of the GRDE to the strong solution of the GARE. We will consider (2.1) with $t_0 = -\infty$ and $T=0$. We present three theorems, the first one is an extension of Theorem 2.1(iv) of [7] in which the observability requirement is weakened to detectability while the second and third apply to non-detectable systems.

**THEOREM 4.1**: If $(A,B)$ is stabilizable, $(C,A)$ is detectable and $\Pi(\cdot)$ satisfies Assumption A.1 then

$$\lim_{t \to -\infty} P(t) = P_s$$

where $P(t)$ is the solution to the GRDE with terminal condition $P_0 \geq 0$ and $P_s$ is the unique stabilizing solution of the GARE.

**THEOREM 4.2**: Let $P(t)$, $t \leq 0$, be the solution of the GRDE with terminal condition $P_0$. Suppose $(A,B)$ is stabilizable, $(C,A)$ has no unobservable modes on the imaginary axis and $\Pi(\cdot)$ satisfies Assumption A.1. Then, subject to either $P_0 > 0$ or $P_0 - P_s \geq 0$

$$\lim_{t \to -\infty} P(t) = P_s$$

where $P_s$ is the unique stabilizing solution of the GARE.

**Theorem 4.3**: Let $P(t)$, $t \leq 0$, be the solution of the GRDE with terminal condition $P_0$. If $(A,B)$ is stabilizable, $\Pi(\cdot)$ satisfies Assumption A.1 and $P_0 - P_s \geq 0$ then

$$\lim_{t \to -\infty} P(t) = P_s$$

where $P_s$ is the unique strong solution of the GARE.

ACKNOWLEDGEMENT

This research was partly supported by CNPq, Brazil, while the first author was visiting the National Laboratory for Scientific Computation - LNCC/CNPq, Rio de Janeiro, Brazil.

REFERENCES

[1] S.W. Chan, G.C. Goodwin and K.S. Sin, *Convergence properties of the Riccati difference equation in optimal filtering of nonstabilizable systems*, IEEE Trans. Auto. Control, AC-29 (1984), 110-118.

[2] C.E. de Souza and M.D. Fragoso, *Results on generalized Riccati equations arising in stochastic control*, Technical Report EE8829, Department of Electrical and Computer Engineering, University of Newcastle, N.S.W., Australia, 1988.

[3] C.E. de Souza, M.R. Gevers and G.C. Goodwin, *Riccati equations in optimal filtering of nonstabilizable systems having singular state transition matrices*, IEEE Trans.Auto.Control, AC-31 (1986),831-838.

[4] M.D. Fragoso, *On a partially observable LQG problem for systems with Markovian jumping parameters*, Syst.Control.Lett.,10 (1988) 349-356.

[5] M.D. Fragoso, *Optimal control for a class of noisy linear systems with Markovian jumping parameters and quadratic cost*, to appear.

[6] D.D. Sworder, *Feedback control of a class of linear systems with jump parameters*, IEEE Trans. Auto. Control, AC-18 (1969), 9-14.

[7] W.M. Wonham, *On a matrix Riccati equation of stochastic control*, SIAM J. Control, 4 (1968), 681-697.

[8] W.M. Wonham, *Random differential equations in control theory*, in Probabilistic Methods in Applied Mathematics, A.T. Bharucha Reid, ed., Vol.2, Academic Press, New York, 1970, 131-212.

# LQ-PROBLEM: THE DISCRETE-TIME TIME-VARYING CASE

J.C. Engwerda

Abstract

In this paper we solve the Linear Quadratic (LQ) regulator problem for discrete-time time-varying systems. By making an appropriate state-space decomposition of the system, sufficient conditions are derived under which this LQ-problem is solvable and, moreover, the closed-loop system becomes exponentially stable.

These conditions are extensions of the time-invariant notions of stabilizability and detectability. Unfortunately, in general these conditions are not necessary. The approach we take provides, however, also a good insight into the difficulties that occur if one looks for both necessary and sufficient conditions solving the problem.

Keywords

Linear discrete-time time-varying systems, stabilizability, detectability, state-space decomposition.

## 1. Introduction

In the past much research has been done on the subject under which conditions the linear quadratic regulator problem has a solution if the considered system is time-varying, see e.g. Kwakernaak et al [8], Hager et al [7] and Anderson et al [1,2].

In Anderson et al [2] it was claimed that under a uniform stabilizability and uniform detectability condition the Kalman filter, the dual of the LQ problem, is exponentially stable (under the usual system noise assumptions). Engwerda showed by means of a counterexample in [5], however, that this claim is not correct. He shows that the definitions given by Anderson et al of uniform stabilizability and uniform detectability do not imply that the system is stabilizable and detectable, respectively. More in particular the example shows that the intuïtion and definition of uniform detectability (i.e. lemma 2.2 in Anderson et al [2]) do not correspond.

For that reason Engwerda formulated in the same paper new conditions which imply (exponential) stabilizability and detectability of the system.

These new conditions are formulated in terms of a transformed system that is obtained by applying an appropriate state-space decomposition. In this paper we use and extend this analysis in order to obtain sufficient conditions under which the LQ-problem has a solution with the property that it makes the closed-loop system exponentially stable. The paper is organized as follows.

First, we introduce in section 2 the notions of uniform periodic smooth exponential stabilizability and detectability respectively. Then, we show in section 3 that under these conditions the LQ problem has a solution. Consecutively, we show that when the resulting optimal state-feedback control is applied the system becomes exponentially stable.

The paper ends with some concluding remarks.

## 2. Preliminaries

In this paper we will be dealing with the linear time-varying discrete-time system:

$$\Sigma_y: \begin{array}{lll} x(k+1) & = & A(k)x(k)+B(k)u(k);\; x(k_0)=\bar{x} \\ y(k) & = & C(k)x(k), \end{array}$$

where $x(k)\in\mathbb{R}^n$ is the state of the system, $u(k)\in\mathbb{R}^m$ the applied control and $y(k)\in\mathbb{R}^r$ the output at time k. Here, we assume that all matrices A(.), B(.) and C(.) are bounded.

Since the system is time-varying it is convenient to have the notation:

Let N be any positive number, then

$A(k+N,k) := A(k+N-1) \ldots\ldots A(k)$ if $N>1$

$:= I$ if $N=1$

$S[k,k-N] := [B(k)|A(k+1,k)B(k-1)|\ldots|A(k+1,k-N+1)B(k-N)]$

$W[k,k-N] := [C(k)^T|\ldots\ldots|\{C(k+N)A(k+N,k)\}^T]^T$

$v[k,\ell] := (v^T(k),\ldots\ldots,v^T(\ell))^T$

$v[k,.] := (v^T(k),v^T(k+1),\ldots\ldots)^T$

$x(k,k_0,\bar{x},u)$ is the state of the system at time k resulting from the initial state $\bar{x}$ at time $k_0$ if the input $u[k_0,k-1]$ is applied

$y(k,k_0,\bar{x},u) := C(k)x(k,k_0,\bar{x},u)$. □

Using this notation we can give now easily formal definitions of several notions that are used later on in this paper.

Definition 1

The initial state $\bar{x}$ of the system $\Sigma_y$ is said to be

* exponentially stable at $k_0$ if there exist positive constants $\alpha$ and M such that $\|x(k,k_0,\bar{x},0)\| \leq Me^{-\alpha(k-k_0)}\|\bar{x}\|$ for any $k>k_0$
* exponentially stabilizable at $k_0$ if there exists a control sequence $u[k_0,.]$, with the property that u(.) converges exponentially fast to zero, and positive constants $\alpha$ and M such that $\|x(k,k_0,\bar{x},u)\| \leq Me^{-\alpha(k-k_0)}\|\bar{x}\|$ for any $k>k_0$.
* unobservable at $k_0$ if $y(k,k_0,\bar{x},0)=0$ for any $k\geq k_0$.
* exponentially detectable at $k_0$ if there exists a finite integer $N>0$ such that $\bar{x}$ modulo $X_e^-(A(.,k_0))$ is determined from any $y[k_0,k_0+N-1]$ and $u[k_0,k_0+N-2]$. Here $X_e^-(A(.,k_0))$ is the linear subspace consisting of all exponentially stable states at time $k_0$.

Like all exponentially stable states, the set of all unobservable states at $k_0$ constitute a linear subspace. We denote it by $U_{k_0}$. Now, $\Sigma_y$ is called observable at $k_0$ if x=0 is the only unobservable state at $k_0$. Moreover, we say that $\Sigma_y$ is exponentially stable (respectively stabilizable, exponentially detectable) at $k_0$ if any initial state of $\Sigma_y$ possesses the corresponding property at $k_0$.
Using these concepts the notion of uniform exponential stabilizability and detectability are defined as follows.

Definition 1 (continued):

$\Sigma_y$ is called uniformly exponentially stabilizable (respectively detectable) if $\Sigma_y$ has the corresponding property at any time $k\geq k_0$ and, moreover, the constants $\alpha(k)$ and $M(k)$ appearing in the definitions satisfy the inequalities $\alpha(k)\geq\bar{\alpha}>0$ and $M(k)\leq\bar{M}<\infty$ for some $\bar{\alpha}$ and $\bar{M}$. □

Another notion that plays an important role in our analysis is the concept of reachability. Formal, we call a state $\bar{x}$ reachable (from zero) if there exists a control sequence $u[N,k-1]$ with $-\infty<N<k$ such that $x(k,N,0,u)=\bar{x}$. The linear subspace of all reachable states at time k is denoted by $R_k$.
Now, Engwerda showed in [4,6] that both $R_k$ and $U_k$ are $A(k)$-invariant. These properties are used in lemma 2, where we give an equivalent system representation of $\Sigma_y$. To that extent we introduce the state-space decomposition

$$X_1(k) := R_k \cap U_k;$$
$$X_1(k) \oplus X_2(k) := R_k;$$
$$X_1(k) \oplus X_2(k) \oplus X_3(k) := \mathbb{R}^n,$$

where $X_1$, $X_2$ and $X_3$ are chosen orthogonal.
Then, we have

Lemma 2
There exists an orthogonal state-space transformation $x(.)=T'(.)x'(.)$, which does not effect the boundedness property of the system parameters, such that $\Sigma_y$ is described by the recurrence equation.

$$\Sigma_y': \begin{bmatrix} x_1'(k+1) \\ x_2'(k+1) \\ x_3'(k+1) \end{bmatrix} = \begin{bmatrix} A_{11}'(k) & A_{12}'(k) & A_{13}'(k) \\ 0 & A_{22}(k) & A_{23}'(k) \\ 0 & 0 & A_{33}'(k) \end{bmatrix} \begin{bmatrix} x_1'(k) \\ x_2'(k) \\ x_3'(k) \end{bmatrix} + \begin{bmatrix} B_1'(k) \\ B_2'(k) \\ B_3'(k) \end{bmatrix} u'(k)$$

$$y(k) = (0 \quad C_1'(k) \quad C_2'(k))x'(k) ,$$

where

$\Sigma_1'$: $x_1'(k+1) = A_{11}'(k)x_1'(k) + B_1'(k)u'(k)$ is reachable at any time $k \geq k_0$.

$\Sigma_2'$: $x_2'(k+1) = A_{22}'(k)x_2'(k) + B_2'(k)u'(k) + A_{23}'(k)A_{33}'(k+1,k_0)x_3'(k_0)$
$y(k) = C_2'(k)x_2'(k)$ is both reachable and observable at any time $k \geq k_0$;

$\Sigma_3'$: $x_3'(k+1) = A_{33}'(k)x_3'(k)$. □

In order to obtain sufficient conditions for exponential stabilizability and detectability of $\Sigma_y$ at $k_0$ we introduce the notions of periodic smooth controllability and observability. Roughly spoken, we say that a system is periodically smoothly controllable if there exists a finite time period such that whenever such a time period has passed, the system has been at least once controllable during that period.

Definition 3.
$\Sigma_y$ is called periodically smoothly controllable at $k_0$ if there exist positive constants $\varepsilon$ and $k_1$ such that for all k>0 there exists an integer $k_2(k)$ in the interval $[k_0+(k-1)\ k_1,\ k_0+k\ k_1]$ for which $S[k_2-2\ k_1,k_2]S^T[k_2-2\ k_1,k_2] \geq \varepsilon\ I$.
Similarly we say that $\Sigma_y$ is periodically smoothly observable at $k_0$ if there exist positive constants b and $k_1$ such that for all k>0 there exists an integer $k_2(k)$ in the interval $[k_0+(k-1)\ k_0,\ k_0+k\ k_0]$ for which $W[k_2,k_2+2\ k_1]\ W^T[k_2,k_2+2\ k_1] \geq b\ I$.

Instead of periodic smooth controllability (respectively observability) of $\Sigma_y$ we often use the phraseology periodic smooth controllability of the pair (A(.), B(.)) and observability of the pair (C(.), A(.)), respectively. □

With the notation of lemma 2, we then have as a special case from theorem 20 of Engwerda [5]:

Theorem 4:
$\Sigma_y$ is both exponentially stabilizable and exponentially detectable at $k_0$ if the following three conditions are satisfied:

i) $\Sigma_1'$ is uniformly exponentially stable;

ii) $\Sigma_2'$ is both periodically smoothly controllable and observable at $k_0$;

iii) $\Sigma_3'$ is exponentially stable at $k_0$.

From this theorem we immediately have

Corollary 5:
$\Sigma_y$ is both uniformly exponentially stabilizable and uniformly exponentially detectable if

i) $\Sigma_1'$ is uniformly exponentially stable

ii) $\Sigma_2'$ is both periodically smoothly controllable and observable at $k_0$

iii) $\Sigma_3'$ is uniformly exponentially stable.

In the sequel conditions i) upto iii) in corollary 5 are called the exponential stabilizability and detectability (E.S.D.) conditions. Note that for time-invariant systems these three conditions are necessary too.

## 3. The solution of the LQ control problem.

In this section we consider the LQ optimal control problem:

(1) $$\min_{u[k_0,..]} \lim_{N\to\infty} J_N, \text{ subject to } \Sigma_y$$

where $$J_N = \sum_{k=k_0}^{k_0+N-1} \{\|y(k)\|^2 + \|u(k)\|^2_{R(k)}\} + \|y(k_0+N)\|^2.$$

Here $\|u(k)\|^2_{R(k)}$ equals $u^T(k)\ R(k)\ u(k)$. In the sequel we take without loss of generality $k_0 = 0$, and we denote $C^TC$ be denoted by Q. Furthermore, we assume that the following, the so-called Sufficient Control Existence (S.C.E.), conditions are satisfied.

i) The E.S.D. conditions of corollary 5

(S.C.E.) ii) a) $R(k) \geq \beta I$ for some $\beta > 0$, for all $k \geq 0$

or b) $B^T(k)Q(k+1)B(k) \geq \beta_1 I$, $Q(k) \geq \beta_2 I$ and $R(k) \geq 0$ for some $\beta_i > 0$, $i = 1,2$ for all k.

We will show that under these conditions an optimal control for the LQ problem exists and is given by:

(2) $u(k) = - F(k)\ x(k)$

where $F(k) = (R(k) + B^T(k)K(k+1)B(k))^{-1}\ B^T(k)K(k+1)A(k)$,
and $K(k)$ is given by $K(k) = \lim_{N\to\infty} K_N(k)$
where $K_N(k)$ is obtained from the recursive equation:

(RRE): $K_N(k) = A^T(k)\{K_N(k+1) - K_N(k+1)B(k)(R(k) + B^T(k)K_N(k+1)B(k))^{-1}.$
$.\ B^T(k)K_N(k+1)\}A(k) + Q(k)$, $K_N(N) = Q(N)$.

Moreover, we show that if this optimal feedback controller (2) is used to regulate the system, the closed-loop system becomes exponentially stable.

<u>Theorem 6:</u>
Let $\Sigma_y$ satisfy the S.C.E. conditions.
Then, controller (2) minimizes $\lim_{N\to\infty} J_N$.

<u>Proof:</u>
First, consider the optimal control problem $\min_{u[0,N-1]} J_N$ subject to $\Sigma_y$.
The optimal control for this problem is:

(3) $u_N(k) = - F_N(k)\ x(k)$

where $F_N(k) = (R(k) + B^T(k)K_N(k+1)B(k))^{-1}\ B^T(k)K_N(k+1)A(k)$, and $K_N(k)$ is given by the resursive equation (RRE).
Moreover, we have that the corresponding minimal control cost equals $\bar{J}_N := x^T(0)K_N(0)x(0)$ (see e.g. Bertsekas [3]).

Since, due to our assumptions, $\Sigma_y$ is exponentially stabilizable we have that there exists a control sequence such that $\lim_{N\to\infty} J_N$ remains finite. Now, $\bar{J}_N$ is a monotonically increasing sequence.
Consequently, $\lim_{N\to\infty} K_N(0)$ exists. Moreover, since $\Sigma_y$ is uniformly exponentially stabilizable, a similar reasoning shows that $\lim_{N\to\infty} K_N(k)$ exists for any k.
So, we have shown now that $\lim_{N\to\infty} u_N(k)$ exists. Denote this limit by $\bar{u}(k)$.
Due to the monotonicity property of $\bar{J}_N$ we can apply Bellman's principle to conclude that

$$\lim_{N\to\infty} \bar{J}_N \leq \min_{u[0,.]} \lim_{N\to\infty} J_N.$$

So, the only thing left to be proved is that

$$\lim_{N\to\infty} \bar{J}_N \geq \min_{u[0,.]} \lim_{N\to\infty} J_N.$$

This can be done by using some elementary analysis. Since $J_N$ consists of the sum of positive functions, Fatou's lemma (see Rudin [9]), Theorem 11.31) can namely be applied to conclude that the order of taking limits and summations can be interchanged (for a more detailed proof see Engwerda [6]). This completes the proof. □

To prove exponential stability of $x(k+1) = (A-BF)(k)x(k)$ we need an extended result of Lyapunov's lemma. This result can be proved along the lines the proof of the corresponding property for uniformly stabilizable and detectable systems in Anderson et al [2].

<u>Lemma 7:</u> (Extended lemma of Lyapunov).
Let A(.) and H(.) be bounded.
Suppose that (A(.), H(.)) is periodically smoothly observable and that there is a bounded positive semi-definite symmetric matrix sequence P(k) satisfying $A^T(k)P(k+1)A(k) - P(k) = -H^T(k)H(k)$ on $[0,\infty)$.
Then $x(k+1) = A(k)\, x(k)$ is exponentially stable. □

We are now able to prove the main result of this paper.

Theorem 8:

Let the S.C.E. conditions be satisfied.

Then there exist constants $M<\infty$ and $\alpha > 0$ such that $\| (A-BF)(k,0) \| \leq Me^{-\alpha k}$.

Proof:

We know from theorem 6 that we can associate the following control problem with $(A-BF_N)(.)$:

$$\min_{u[0,\, N-1]} J_N, \text{ subject to } \Sigma_y.$$

We reconsider this minimization problem.

From lemma 2 we have that this problem can be rewritten as:

$$\min_{u'[0,N-1]} \sum_{k=0}^{N-1} \{\| x_2'(k) \|^2_{C_2'^T(k)\, C_2'(k)} + \| x_3'(k) \|^2_{C_3'^T(k) C_3'(k)} +$$

$$\| u'(k) \|^2_{R(k)}\} + \| x_2'(N) \|_{C_2'^T(N) C_2'(N)} + \| x_3'(N) \|^2_{C_3'^T(N) C_3'(N)},$$

subject to $\Sigma_y$.

According to the proof of theorem 6 the optimal control is given by (3).

Substitution of the system parameters yields (by induction on k) that $K_N'(k)$ has the following structure:

$$K_N'(k) = \begin{bmatrix} 0 & 0 & 0 \\ 0 & K_{22} & K_{23} \\ 0 & K_{23} & K_{33} \end{bmatrix}_N' (k) \qquad \text{(i)}$$

and consequently $F_N'(k) = (0 \mid F_2 \mid F_3)_N'(k)$ (ii)

Since $K_N'(.)$ converges to $K'(.)$ and $F_N'(.)$ to $F'(.)$ it is clear that $K'(.)$ and $F'(.)$ have the structure of (i) and (ii), respectively. Since $K'(.)$ converges for any k we have from (RRE), moreover, that $K'(.)$ satisfies the recurrence equation

$$K'(k) = A'^T(k)\{ K'(k+1) - K'(k+1)B'(k)(R(k) + B'^T(k)K'(k+1)\, B'(k))^{-1} B'^T(k)K(k+1)\}A'(k) + Q'(k),$$

which can be rewritten as:

$$K'(k) = (A-BF)'^T(k)\ K'(k+1)\ (A-BF)'(k) + (Q+F^TRF)'(k).$$

In particular it follows now, by substitution of all the system parameters, that

$$K_{22}'(k) = (A_{22}-B_2F_2)'^T(k)\ K_{22}'(k+1)\ (A_{22}-B_2F_2)'(k) + (C_2^TC_2 + F_2^TRF_2)'(k)$$

with $F_2'(k) = (R(k) + B_2'^T(k)K_{22}'(k+1)B_2'(k))^{-1}B_2'^T(k)K_{22}'(k+1)A_{22}'(k)$.

From the S.C.E. conditions it follows that $K_{22}'(.)$ and $F_2'(.)$ are bounded.
Now, let $D' := (C_2'^T \mid F_2'^T R^{\frac{1}{2}})^T$

then, $A_{22}' - B_2'F_2' = A_{22}' - [0 \mid B_2']\ R^{-\frac{1}{2}}\ D'$ and $C_2'^TC_2' + F_2'^TRF_2' = D'^TD'$.

Since the observability property of $(A_{22}', C_2')$ implies that $(A_{22}', D')$ has the same property, it is easily shown that $(A_{22}'-B_2'F_2', D')$ is periodically smoothly observable too (see e.g. Anderson [2]).
Application of lemma 7 yields now that $(A_{22}-B_2F_2)'(.)$ is exponentially stable.
Since the feedback gain F' does not influence the exponential stability of $\Sigma_1'$ and $\Sigma_3'$, we conclude that the overall CL-system, $(A-BF)'(.)$, is exponentially stable.
Finally, we note that since the transformation matrix $T'(k)$ is bounded, exponential convergence of $x'(k)$ implies that the same property holds for the original state of the system $x(k)$. Which completes the proof. □

## Concluding Remarks

In this paper we solved the discrete-time time-varying LQ optimal control problem under some weak conditions on the system. These conditions were formulated in terms of a transformed system that was obtained by making use of several invariance properties of the system.
A major problem occurring was to find a suitable state-space representation. This, since the prerequisite that the convergence properties of the transformed and original system must coincide, reduces the class of admissible transformations.

Fortunately, we succeeded in finding such a transformation which, moreover, was useful when we had to prove that the closed-loop system is exponentially stable if the optimal LQ control is applied.
An interesting question which remains to be solved is whether the LQ-problem has also an exponentially stabilizing solution when the system is both uniformly exponentially stabilizable and detectable (in the sense defined here). A subsequent question, which immediately arises if the answer to the previous one is affirmative, is then to give both necessary and sufficient conditions (that can a priori be verified) that guarantee this uniform exponential stabilizability and detectability.

Last, but not least, we note that the obtained results can be used in a straightforward manner to solve e.g. the LQG-problem, the EQL-problem (see Engwerda [6]) and the Kalman filter problem for discrete-time time-varying systems.

## References

[1] Anderson, B.D.O. and Moore, J.B., 1980, Coping with singular transition matrices in estimation and control stability theory; International Journal of Control, pp. 571 - 586.

[2] Anderson, B.D.O. and Moore, J.B., 1981, Detectability and Stabilizability of time-varying discrete-time linear systems; Siam Journal Control and Optimization, Vol. 19, no. 1, pp. 20 - 32.

[3] Bertsekas, D.P., 1976, Dynamic Programming and Stochastic Control (New York, Academic Press).

[4] Engwerda, J.C., 1988, Control aspects of linear discrete time-varying systems; International Journal of Control, Vol. 48, no.4, pp. 1631 - 1658.

[5] Engwerda, J.C., 1988-a, Stabilizability and detectability of discrete-time time-varying systems; Memorandum Cosor 88-16, Technical University Eindhoven, to appear in I.E.E.E. transactions on Automatic Control.

[6] Engwerda, J.C., 1988-b, Regulation of Linear Discrete Time-Varying Systems; Ph.D. Thesis, Technical University Eindhoven.

[7] Hager, W.W. and Horowitz, L.I., 1976, Convergence and stability properties of the discrete Riccati operator equation and the associated optimal control and filtering problems; Siam Journal Control and Optimization, Vol. 14, no. 2, pp. 295 - 312.

[8] Kwakernaak, H. and Sivan, R., 1971, Linear Optimal Control Systems (New York, John Wiley & Sons).

[9] Rudin W., 1976, Principles of Mathematical Analysis, third edition (Tokyo, McGraw-Hill).

# The output-stabilizable subspace and linear optimal control

A.H.W. Geerts & M.L.J. Hautus

**Abstract**

Properties of a certain subspace are linked to well-known problems in system theory.

**Keywords**

Output stabilizability, linear-quadratic problem, singular controls, structure algorithm, dissipation inequality.

## 1. Introduction

Consider the following finite-dimensional linear time-invariant system $\Sigma$:

$$\dot{x}(t) = Ax(t) + Bu(t),\ x(0) = x_0 \ , \tag{1.1a}$$

$$y(t) = Cx(t) + Du(t) \ , \tag{1.1b}$$

where for all $t \geq 0$, $x(t) \in \mathbb{R}^n$, $u(t) \in \mathbb{R}^m$ and $y(t) \in \mathbb{R}^r$, and the input $u(\cdot)$ is required to be an element of

$$C^m_{sm} := \{u : [0, \infty) \mapsto \mathbb{R}^m \mid \exists_{\varepsilon>0} \exists_{v \in C^\infty((-\varepsilon,\infty) \mapsto \mathbb{R}^m)} \forall_{t \geq 0} : u(t) = v(t)\} \ ,$$

the class of *smooth* controls. Moreover, without loss of generality, we may assume that $[B'\ D']'$, $[C\ D]$ is injective and surjective, respectively.

For the case $D = 0$, we now recall Wonham's *Output Stabilization Problem* ([11, Section 4.4]):

(OSP): Given the system $\Sigma$ with $D = 0$. Find a feedback map $F : \mathbb{R}^m \mapsto \mathbb{R}^n$ such that with the input $u = Fx$, we have $y(t) \to 0$ for any initial value $x_0$.

If this problem has a solution, then $\Sigma$ (with $D = 0$) is called *output stabilizable*. A necessary and sufficient condition for the output stabilizability was provided by [11, Theorem 4.4]. A slightly different formulation of the condition was given in [5, Theorem 4.10], where it was shown that OSP has a solution if and only if $\mathbb{R}^n = \mathbf{S}^-$, where the subspace $\mathbf{S}^-$ was defined in terms of $(\xi, \omega)$-representations. More generally, this subspace also plays a role in the output-stabilization problem under disturbances, i.e., the problem of achieving BIBO stability in the presence of a disturbance input term $Eq$. Then, it turns out, the condition is: $\mathrm{im}(E) \subset \mathbf{S}^-$.

Next, let

$$J(x_0, u) := \int_0^\infty y'(x_0, u)\, y(x_0, u)\, dt \ , \tag{1.2}$$

with $y(x_0, u) = Cx(x_0, u) + Du$ (compare (1.1b)), and $x(x_0, u)$ denotes the solution of (1.1a) for given $x_0$ and $u \in C^m_{sm}$. We introduce the Linear-Quadratic optimal Control Problem:

(LQCP): for all $x_0$, determine $J^-(x_0) := \inf\{J(x_0, u) \mid u \in C^m_{sm}\}$ and, if for all $x_0 \in \mathbb{R}^n$, $J^-(x_0) < \infty$, then compute, if one exists, all *optimal* controls (i.e. all controls $u^* \in C^m_{sm}$ such that $J^-(x_0) = J(x_0, u^*)$).

We will call LQCP *solvable* if for all $x_0, J^-(x_0) < \infty$ and if for every $x_0$ there exists an *optimal* input $u^*$ (i.e. an input $u^*$ such that $J^-(x_0) = J(x_0, u^*)$). In this paper we shall see that the subspace $\mathbf{S}^-$ is relevant for the issue of LQCP-solvability.

The above-mentioned problem is called *regular* if $\ker(D) = 0$ and *singular* if $\ker(D) \neq 0$. The regular case is well established and considered classical. Curiously, the problem of finding necessary and sufficient conditions for solvability of the problem has found little attention, even in the regular case. Usually, one is satisfied with the statement that the problem is solvable if $(A, B)$ is stabilizable (see e.g. [10, Propositions 9-10]). Of course, this condition is not necessary (if $C = 0$, then $u \equiv 0$ is optimal for all $x_0$). Now recently ([1]), a necessary and sufficient condition of solvability was given for the regular case in terms of the stabilizability of a suitable defined quotient system.

If the problem is *singular*, then it is known that optimal inputs need not exist within the class $C^m_{sm}$ ([7, Example 2.11]). With a reformulation in the style of [7] incorporating *distributions* as possible inputs, this extra difficulty can be dealt with and it is proven in [2] that the input class $C^m_{\mathrm{imp}}$ of *impulsive-smooth* distribution on $\mathbb{R}$ with support on $[0, \infty)$ ([7, Definition 3.1]) is large enough to be representative for the system's behaviour under general distributions as inputs. A distribution $u \in C^m_{\mathrm{imp}}$ can be written as a sum of a function $u_2 \in C^m_{sm}$ and an impulsive distribution $u_1$ with support in $\{0\}$. Obviously, we require $u \in C^m_{\mathrm{imp}}$ to be such that for every $x_0$ the resulting output $y(x_0, u)$ has no impulsive component, and the (system dependent) space of these inputs is denoted $U_\Sigma$. In [2, Proposition 4.5] an explicit description for this input class is given by means of a dual version of Silverman's structure algorithm. With the help of this *generalized dual structure algorithm* ([2, Section 4]), the necessary and sufficient condition for solvability of LQCP given in [1] can be generalized to singular problems ([1, Remark 5]).

In the present paper, it will be shown that the latter condition is equivalent to the condition $\mathbf{S}^- = \mathbb{R}^n$. In other words, output stabilizability is necessary and sufficient for solvability of LQCP. This intuitively rather obvious condition turns out to be relatively difficult to prove.

In the sequel we will need the following well-known concepts. Let $\mathbf{V} = \mathbf{V}(\Sigma) = \{x_0 \in \mathbb{R}^n \mid \exists_{u \in C_m^n} : y(x_0, u) \equiv 0\}$ (the *weakly unobservable* subspace), then ([7, Theorem 3.10]) $\mathbf{V}$ is the largest subspace $\mathbf{L}$ for which there exists a feedback $F$ such that $(A + BF)\,\mathbf{L} \subset \mathbf{L}$, $(C + DF)\,\mathbf{L} = 0$. Dually, $\mathbf{W} = \mathbf{W}(\Sigma)$ (the *strongly reachable* subspace) is the smallest subspace $\mathbf{K}$ for which there exists an "output injection" $G$ such that $(A + GC)\,\mathbf{K} \subset \mathbf{K}$, $\mathrm{im}(B + GD) \subset \mathbf{K}$ ([7, Theorem 3.15]) and $\mathbf{W} \subset \langle A \mid \mathrm{im}(B)\rangle$ (the reachable subspace). It is easily established that $\mathbf{W} = 0$ if and only if $\ker(D) = 0$.

Next, if $K \in \mathbb{R}^{n \times n}$ and $F(K) := \begin{bmatrix} C'C + A'K + KA & KB + C'D \\ B'K + D'C & D'D \end{bmatrix}$ (the dissipation matrix), then $K$ is said to satisfy the *dissipation inequality* if $K \in \Gamma := \{K \in \mathbb{R}^{n \times n} \mid K = K',\ F(K) \geq 0\}$ ([9]). Note that $\Gamma \neq \varnothing$ $(0 \in \Gamma)$. If $T(s) := D + C(sI - A)^{-1}B$ $(s \in \mathbb{C})$ (the *transfer function*), and $\rho :=$ normal rank $(T(s))$, then it is proven in [8] that

**Lemma 1.1**

If $K \in \Gamma$, then rank $(F(K)) \geq \rho$.

Set $\Gamma_{\min} := \{K \in \Gamma \mid \mathrm{rank}\,(F(K)) = \rho\}$. This subset of $\Gamma$ is of importance because of the next result from [2].

**Proposition 1.2**

If $(A, B)$ is stabilizable, then there exists an element $K^- \in \Gamma_{\min} \cap \{K \in \Gamma \mid K \geq 0\}$ such that, for all $x_0$, $J^-(x_0) = x_0' K^- x_0$.

If $\ker(D) = 0$ and

$$\Phi(K) := C'C + A'K + KA - (KB + C'D)(D'D)^{-1}(B'K + D'C)\ , \tag{1.3}$$

then it is easily seen ([9]) that $\Gamma_{\min} = \{K \in \mathbb{R}^{n \times n} \mid K = K',\ \Phi(K) = 0\}$, the set of solutions of the algebraic *Riccati* equation. Now a second major observation of this paper is, that $\Gamma_{\min} \cap \{K \in \Gamma \mid K \geq 0\} \neq \varnothing$ if and only if $\mathbf{S}^- = \mathbb{R}^n$. Hence, in the regular case, there exists a positive semi-definite solution of the algebraic Riccati equation if and only if $\Sigma$ is output stabilizable.

## 2. The dual structure algorithm and the output-stabilizable subspace

If $q_0 := \text{rank}(D)$, then there exists a regular transformation $S_0$ such that $DS_0 = [D_0, 0]$ with $D_0$ left invertible and we will take $S_0 = I_m$ if $q_0 = m$ (note that $S_0$ can be chosen such that $S_0^{-1} = S_0'$). Set $BS_0 =: [\bar{B}_0, \tilde{B}_0]$, then substitution of $u = S_0[\bar{w}_0', \tilde{w}_0']'$ into (1.1) yields

$$\dot{x} = Ax + \bar{B}_0\bar{w}_0 + \tilde{B}_0\tilde{w}_0, x_0, y = Cx + D_0\bar{w}_0 \ , \tag{2.1}$$

and $\tilde{B}_0$ is left invertible, $\text{im}(\tilde{B}_0) \subset \mathbf{W}$. This input transformation corresponds to the first part of step 0 of the generalized dual structure algorithm ([2, Section 4]). Notice that $\tilde{B}_0$ is not appearing if $q_0 = m$. In fact, the dual algorithm is a void concept if $\ker(D) = 0$. If $\ker(D) \neq 0$, then this algorithm transforms the given system $\Sigma$ into a system $\Sigma_\alpha$ ($\alpha$ an integer, not less than 1) of the form

$$\dot{x}_\alpha = Ax_\alpha + \underline{\bar{B}}\bar{w}_\alpha + \hat{B}\hat{w}, x_0 \ , \tag{2.2a}$$

$$y = Cx_\alpha + \underline{D}\bar{w}_\alpha \ , \tag{2.2b}$$

where $\underline{\bar{B}} = [\bar{B}_0, \bar{B}_{\text{add}}]$, $\underline{D} = [D_0, D_{\text{add}}]$, $\bar{B}_{\text{add}}$ is an $n \times (\rho - q_0)$ real matrix which is such that $\text{im}(\bar{B}_{\text{add}}) \subset A(\mathbf{W})$, $D_{\text{add}}$ is a $r \times (\rho - q_0)$ real left invertible matrix, and $\text{rank}(\underline{D}) = \rho$, $C(\mathbf{W}) \subset \text{im}(\underline{D})$ and $\text{im}(\hat{B}) \subset \mathbf{W}$. Moreover, the control $u \in C^m_{\text{imp}}$ and the input $[\bar{w}_\alpha' \ \hat{w}']'$ are linked by $u = H(p)[\bar{w}_\alpha' \ \hat{w}']'$, where $H(s)$ is an invertible polynomial matrix, $p$ stands for the derivative of Diracs $\delta$ distribution and $H(p)$ thus is the matrix-valued distribution obtained by substituting $s = p$ into $H(s)$. Finally, for all $t > 0$, we have that $(x(x_0, u)(t) - x_\alpha(x_0, [\bar{w}_\alpha', \hat{w}']')(t)) \in \mathbf{W}$. Now, let us apply to (2.2) the preliminary state feedback law

$$\bar{w}_\alpha = -(\underline{D}'\underline{D})^{-1}\underline{D}'Cx_\alpha + \hat{w}_\alpha \ . \tag{2.3}$$

Then we get

$$\dot{x}_\alpha = \underline{A}x_\alpha + \underline{\bar{B}}\hat{w}_\alpha + \hat{B}\hat{w}, \ x_0, y = \underline{C}x_\alpha + \underline{D}\hat{w}_\alpha \tag{2.4a}$$

$$\text{with} \ \ \underline{A} := A - \underline{\bar{B}}(\underline{D}'\underline{D})^{-1}\underline{D}'C, \ \underline{C} := (I_r - \underline{D}(\underline{D}'\underline{D})^{-1}\underline{D}')C \ . \tag{2.4b}$$

From [2, Lemmas 4.2 - 4.4] and the above we then have the following.

**Proposition 2.1**

a) $\underline{A}\mathbf{W} \subset \mathbf{W}$.

b) $\mathbf{V}(\Sigma_\alpha) = \mathbf{V}(\Sigma) + \mathbf{W}(\Sigma) = \langle \ker(\underline{C}) \mid \underline{A} \rangle$.

c) $\langle A \mid \text{im}(\underline{\bar{B}}, \hat{B}) \rangle + \mathbf{W} = \langle A \mid \text{im}(B) \rangle$.

One consequence of Proposition 2.1 is, that $y$ is *independent* of $\hat{w}$; we may just as well take $\hat{w} = 0$. Now let us define (where $y(\infty)$ denotes $\lim_{t\to\infty} y(x_0, u)(t)$)

$$\mathbf{T}_1 := \{x_0 \in I\!R^n \mid \exists_{u \in U_\Sigma} : y(\infty) = 0\} \tag{2.5a}$$

and $$\mathbf{T}_2 := \{x_0 \in I\!R^n \mid \exists_{u \in U_\Sigma} : J(x_0, u) < \infty\} \ , \tag{2.5b}$$

then we establish that $\mathbf{T}_1 = \{x_0 \mid \exists_{\hat{w}_\alpha,\, \text{smooth}} : (\underline{C}x_\alpha + \underline{\hat{D}}\hat{w}_\alpha)(\infty) = 0\}$ and $\mathbf{T}_2 = \{x_0 \mid \exists_{\hat{w}_\alpha,\, \text{smooth}} : \int_0^\infty [\underline{C}x_\alpha + \underline{D}\hat{w}_\alpha]'[\underline{C}x_\alpha + \underline{D}\hat{w}_\alpha]\, dt < \infty\}$ with $x_\alpha(t) = \exp(At)\, x_0 + \int_0^t \exp(A\,(t-\tau))\, \underline{\bar{B}}\hat{w}_\alpha(\tau)\, d\tau$ and hence $\mathbf{T}_{1,2}$ are $\Sigma_\alpha$-invariant ([5, Def. 2.2]).

Next, let

$$\mathbf{S}^-(\Sigma) := X^-(A) + <A \mid \text{im}(B)> + \mathbf{V}(\Sigma) \tag{2.6}$$

(where $X^-(A)$ denotes the stable subspace of $A$). Then it is rather obvious that $\mathbf{S}^-(\Sigma) \subset \mathbf{T}_i (i = 1, 2)$ and that (Proposition 2.1) $\mathbf{S}^-(\Sigma_\alpha) = \mathbf{S}^-(\Sigma) =: \mathbf{S}^-$. Therefore ([5, Remark 2.26]) $\mathbf{T}_{1,2}$ are *strongly* $\Sigma_\alpha$-invariant and we thus have found that $\mathbf{V}(\Sigma_\alpha) \subset \mathbf{S}^- \subset \mathbf{T}_i$ and $\underline{A}\mathbf{V}(\Sigma_\alpha) \subset \mathbf{V}(\Sigma_\alpha)$, $\underline{A}\mathbf{S}^- \subset \mathbf{S}^-$, $\underline{A}\mathbf{T}_i \subset \mathbf{T}_i (i = 1, 2)$. Let $X_2, X_3, X_4$ be such that $\mathbf{V}(\Sigma_\alpha) \oplus X_2 = \mathbf{S}^-$, $\mathbf{S}^- \oplus X_3 = \mathbf{T}_1$, $\mathbf{T}_1 \oplus X_4 = I\!R^n$. By choosing appropriate basis matrices for these subspaces, (2.4a) (with $\hat{w} = 0$) transforms into

$$\begin{bmatrix} \dot{x}_1 \\ \dot{x}_2 \\ \dot{x}_3 \\ \dot{x}_4 \end{bmatrix} = \begin{bmatrix} A_{11} & A_{12} & A_{13} & A_{14} \\ 0 & A_{22} & A_{23} & A_{24} \\ 0 & 0 & A_{33} & A_{34} \\ 0 & 0 & 0 & A_{44} \end{bmatrix} \begin{bmatrix} x_1 \\ x_2 \\ x_3 \\ x_4 \end{bmatrix} + \begin{bmatrix} B_1 \\ B_2 \\ 0 \\ 0 \end{bmatrix} \hat{w}_\alpha \, , \begin{bmatrix} x_{01} \\ x_{02} \\ x_{03} \\ x_{04} \end{bmatrix} ,$$

$$y = [0\ C_2\ C_3\ C_4] \begin{bmatrix} x_1 \\ x_2 \\ x_3 \\ x_4 \end{bmatrix} + \underline{D}\hat{w}_\alpha, \quad \left( [C_2\ C_3\ C_4], \begin{bmatrix} A_{22} & A_{23} & A_{24} \\ 0 & A_{33} & A_{34} \\ 0 & 0 & A_{44} \end{bmatrix} \right) \text{ is observable} . \tag{2.7}$$

Note that $\sigma(A_{33}) \cup \sigma(A_{44}) \subset \bar{\mathbb{C}}^+$ (since $X^-(A) \subset \mathbf{S}^-(\Sigma_\alpha)$). Now take a point $x_0 \in \mathbf{T}_1$, i.e. $x_{04} = 0$ in (2.7) (and thus $x_4 \equiv 0$). Since $\underline{D}'\underline{C} = 0$ and $\underline{D}$ is left invertible, it follows that $(C_2x_2 + C_3x_3)(\infty) = 0$, $\hat{w}_\alpha(\infty) = 0$, and thus ([3, Chapter 3]) that $x_2(\infty) = 0, x_3(\infty) = 0$ (i.e., $x(x_0, u)(t)$ converges to $\mathbf{V} + \mathbf{W}(t \to \infty)$). Hence, necessarily, $x_{03} = 0$ and we establish that $\mathbf{T}_1 = \mathbf{S}^-$. In the same way we find that $\mathbf{T}_2 = \mathbf{S}^-$. If for every $F \in I\!R^{m \times n}$, we define the spaces

$$\mathbf{T}_1^F := \{x_0 \in I\!R^n \mid \text{if } u = Fx, \text{ then } y(x_0, u)(\infty) = 0\} \ , \tag{2.8a}$$

$$\mathbf{T}_2^F := \{x_0 \in I\!R^n \mid \text{if } u = Fx, \text{ then } \int_0^\infty y'(x_0, u) y(x_0, u) dt < \infty\} \ , \tag{2.8b}$$

we thus have arrived at our first main result.

**Theorem 2.2**

Consider the system $\Sigma$ and the corresponding subspaces defined above. Then $\mathbf{T}_i = \mathbf{S}^-$ and $\mathbf{T}_i^F \subset \mathbf{S}^-$ for every $F \in \mathbb{R}^{m \times n}$. In addition, there exists an $F \in \mathbb{R}^{m \times n}$ such that $\mathbf{T}_i^F = \mathbf{S}^-$ $(i = 1, 2)$.

Proof. Let $F \in \mathbb{R}^{m \times n}$ be given. If we use the feedback $u = Fx$, then the resulting output $y$ will tend to zero exponentially fast when either $x_0 \in \mathbf{T}_1^F$ or $x_0 \in \mathbf{T}_2^F$ and thus $\mathbf{T}_1^F = \mathbf{T}_2^F$. In addition, it is trivial that $\mathbf{T}_1^F = \mathbf{T}_2^F \subset \mathbf{T}_i$ $(i = 1, 2)$. The fact that there exists an $F$ such that $\mathbf{T}_1^F = \mathbf{S}^-$ is known (compare [5]). The rest follows from the above.

Because of the relation $\mathbf{T}_1^F = \mathbf{S}^-$ for some $F$, we will refer to $\mathbf{S}^-$ as the **output-stabilizable** subspace.

## 3. The dual structure algorithm and optimal control

Let us reconsider the LQCP and assume that $\mathbf{S}^- = \mathbb{R}^n$. According to Theorem 2.2, we can reformulate this as: For every $x_0$ there exists an input $u \in U_\Sigma$ such that $J(x_0, u) < \infty$. Clearly this is a *necessary* condition for the solvability of LQCP. Since $y = \underline{C}x_\alpha + \underline{D}\hat{w}_\alpha$ with $\underline{D}$ left invertible, we are left with a *regular* LQCP by taking $\hat{w} = 0$ in (2.4a). Hence we may apply the second part of the proof of the main Theorem in [1] and state that the algebraic Riccati equation associated with (2.4a), $\tilde{\Phi}(K) = 0$ with

$$\tilde{\Phi}(K) := \underline{C}'\underline{C} + \underline{A}'K + K\underline{A} - K\underline{\bar{B}}(\underline{D}'\underline{D})^{-1}\underline{\bar{B}}'K \ , \tag{3.1}$$

has a solution $K^- \geq 0$ and that every other solution $K \geq 0$ of $\tilde{\Phi}(K) = 0$ satisfies $K \geq K^-$. The optimal cost for LQCP, $J^-(x_0)$, equals $x_0'K^-x_0$ for all $x_0$, $\ker(K^-) = \mathbf{V} + \mathbf{W}$ and, in addition, for every $x_0$ an optimal control for LQCP exists (see for details [2, Theorem 4.5]) and thus the condition $\mathbf{S}^- = \mathbb{R}^n$ is also *sufficient* for solvability of LQCP. Now in [2, Section 6] the next result is proven.

**Proposition 3.1**

$$\Gamma = \{K \in K', \mathbf{W} \subset \ker(K), \tilde{\Phi}(K) \geq 0\} \ ,$$

$$\Gamma_{\min} = \{K \in \mathbb{R}^{n \times n} \mid K = K', \mathbf{W} \subset \ker(K), \tilde{\Phi}(K) = 0\} \ .$$

Consequently, we observe that $K^- \in \Gamma_{\min} \cap \{K \in \Gamma \mid K \geq 0\}$ and every other $K \in \Gamma_{\min} \cap \{K \in \Gamma \mid K \geq 0\}$ satisfies $K \geq K^-$ (compare Proposition 1.2). Note that $\tilde{\Phi}(K) = \Phi(K)$ if $\ker(D) = 0$. Therefore, in the regular case, $K^-$ represents the smallest positive semi-definite solution of (1.3). On the other hand, if $\Gamma_{\min} \cap \{K \in \Gamma \mid K \geq 0\} \neq \emptyset$, then ([3, Chapter 3]) $\mathbf{S}^- = \mathbb{R}^n$. Hence

**Theorem 3.2**

$\mathbf{S}^- = \mathbb{R}^n$ if and only if $\Gamma_{\min} \cap \{K \in \Gamma \mid K \geq 0\} \neq \emptyset$. In addition, if the latter set is nonempty, then the smallest element of this set, $K^-$, represents the optimal cost for the LQCP.

Note that the characterization of $K^-$ as given above is formulated directly in terms ot the *original* system data $(A, B, C, D)$. Moreover, this representation of the optimal cost includes the singular as well as the regular case. Finally, we mention that a condition for output stabilizability can be given in the spirit of [4]. In fact, a more general formulation is

**Proposition 3.3** ([3, Chapter 3])

Let $\mathbf{T}$ be a $\Sigma$-invariant subspace. Then $X^-(A) + \langle A \mid \mathrm{im}(B)\rangle + \mathbf{T} = \mathbb{R}^n$ if and only if $\forall_{\lambda \in \bar{\mathbb{C}}^+}$ $\forall_{\eta \in \mathbb{C}^n} : [\eta(A - \lambda I_n, B) = 0 \text{ and } \eta\mathbf{T} = 0] \Rightarrow \eta = 0$.

The condition for output stabilizability is obtained by taking $\mathbf{T} = \mathbf{V}$.

**Remarks**

1. While proving out main Theorem 2.2, we established that if $u \in U_\Sigma$ is such that $y(x_0, u)(\infty) = 0$ or $J(x_0, u) < \infty$, then $x(x_0, u)(t)$ converges to $\mathbf{V} + \mathbf{W}$ $(t \to \infty)$, but *not* necessarily to $\mathbf{V}$ (for a counterexample, see [6]), unless (of course) $\mathbf{W} = 0$, i.e. $\ker(D) = 0$.
2. Since $\mathbf{S}^- \subset \tilde{\mathbf{T}}_1 := \{x_0 \mid \exists_{u \in \mathbb{C}^m_-} : y(\infty) = 0\} \subset \mathbf{T}_1 = \mathbf{S}^-$, we find that $\tilde{\mathbf{T}}_1 = \mathbf{T}_1$, and, analogously, that $\tilde{\mathbf{T}}_2 := \{x_0 \mid \exists_{u \in \mathbb{C}^m_-} : J(x_0, u) < \infty\} = \mathbf{T}_2$. In fact, this can be seen directly as $\mathbf{T}_i = \mathbf{W} + \tilde{\mathbf{T}}_i = \tilde{\mathbf{T}}_i$ because $\mathbf{W} \subset \langle A \mid \mathrm{im}(B)\rangle \subset \tilde{\mathbf{T}}_i$ $(i = 1, 2)$.
3. If $\mathbb{R}^n := \mathbb{R}^n/_{(\mathbf{V}+\mathbf{W})}$, $\bar{\underline{A}} : \overline{\mathbb{R}}^n \to \overline{\mathbb{R}}^n$ denotes the induced map of $\underline{A}$ defined by $\bar{\underline{A}}\bar{x} := \overline{(\underline{A}x)}$ $(\bar{x} = x + (\mathbf{V} + \mathbf{W}))$ and $\bar{\bar{\underline{B}}} : \mathbb{R}^m \to \overline{\mathbb{R}}^n$ is defined by $\bar{\bar{\underline{B}}}\, u := \overline{(\underline{B}u)}$, then it can be seen (e.g. compare [2, Lemma 5.6]) that the condition in Proposition 3.3 with $\mathbf{T} = \mathbf{V}$ is equivalent to: $(\bar{\underline{A}}, \bar{\bar{\underline{B}}})$ is stabilizable. Hence, in accordance with [1, Remark 5], the latter condition is necessary and sufficient for the solvability of LQCP.

**Acknowledgement**

This research was supported by the Netherlands organization for scientific research (N.W.O.).

**References**


[1] Ton Geerts, "A necessary and sufficient condition for solvability of the linear-quadratic control problem without stability", *Syst. & Contr. Lett.*, vol. 11, pp. 47-51, 1988.

[2] Ton Geerts, "All optimal controls for the singular linear-quadratic problem without stability; a new interpretation of the optimal cost", *Lin. Alg. & Appl.*, vol. 116, pp. 135-181, 1989.

[3] Ton Geerts, PhD Thesis, forthcoming.

[4] M.L.J. Hautus, "Stabilization, controllability and observability of linear autonomous systems", *Nederl. Akad. Wetensch. Proc.* Ser. A, 73, pp. 448-455, 1970.

[5] M.L.J. Hautus, "$(A, B)$-invariant and stabilizability subspaces, a frequency domain description", *Automatica*, vol. 16, pp. 703-707, 1980.

[6] M.L.J. Hautus, "Strong detectability and observers", *Lin. Alg. & Appl.*, vol. 50, pp. 353-368, 1983.

[7] M.L.J. Hautus & L.M. Silverman, "System structure and singular control", *Lin. Alg. & Appl.*, vol. 50, pp. 369-402, 1983.

[8] J.M. Schumacher, "The role of the dissipation matrix in singular optimal control", *Syst. & Contr. Lett.*, vol. 2, pp. 262-266, 1983.

[9] J.C. Willems, "Least squares stationary optimal control and the algebraic Riccati equation", *IEEE Trans. Automat. Contr.*, vol. AC-16, pp. 621-634, 1971.

[10] J.C. Willems, A. Kitapçi & L.M. Silverman, "Singular optimal control: A geometric approach", *SIAM J. Contr. & Opt.*, vol. 24, pp. 323-337, 1986.

[11] W.M. Wonham, *Linear Multivariable Control: A Geometric Approach*, Springer, New York, 1979.

# THE DECOMPOSITION OF (A,B)-INVARIANT SUBSPACES AND ITS APPLICATIONS

HAN,ZHENG-ZHI AND ZHANG,ZHONG-JUN
ZHENG,YU-FAN

ABSTRACT: In this paper, (A,B)-radical subspaces are defined and the (A,B)-invariant subspaces are decomposed into a direct sum of radical subspaces. With such a decomposition, it is found that the (A,B)-invariant subspaces have the similar geometric structure to A-invariant subspaces. the uniqueness of this decomposition is investigated. These results are used to describe the characteristic of assignable eigenstructure.

## 1. INTRODUCTION

In the geomatric approach of linear system theory, the concept of (A,B)-invariant subspace is of vital importance. Although this concept was defined about twenty years ago, it is still a subject under current discussion. Many contributors continue investigating the structure of (A,B)-invariant subspaces via different approachs, such as: polynomail models [3], state space [9], matrix fraction[8] and so on. It is well known that an A-invariant subspace can be decomposed into a direct sum of some cyclic subspaces, called radical subspaces [4]. In this paper a similar concept, named (A,B)-radical subspace, is defined and it is used in the decomposition of the (A,B)-invariant subspaces. It will be shown in this paper that there is a similarity of the geometric structures betweem (A,B)-invariant subspaces and A-invariant subspaces. The difference between them is that the decomposition of the (A,B)-invariant subspaces may be not unique in general. The condition of the uniqueness is derived in this paper.

The text is organized as follows: In section 2 some preliminaries of (A,B)-characteristic subspace are given. Section 3 investigates the decomposition of (A,B)-invariant subspaces. In section 4 the characterization of assignable eigenstructure is obtained by using the decomposition.

## 2. PRELIMINARIES

Consider the following linear control system

$$\dot{x} = Ax + Bu \tag{2.1}$$

where $A \in R^{n\times n}$ and $B \in R^{n\times m}$. It is assumed that $rank\ B = m$ and the system (2.1) is controllable. Let $\mathcal{X}$ and $\mathcal{U}$ denote the state space and input space, respectively, with the state space being complex, i.e., $\mathcal{X} \cong C^n$ .

The class of (A,B)-invariant subspaces is denoted by In(A,B). If $\mathcal{V} \in$ In(A,B), then $\underline{F}(\mathcal{V})$ denotes the set of the feedbacks F:$\mathcal{X} \longrightarrow \mathcal{U}$, such that (A+BF)$\mathcal{V} \subset \mathcal{V}$. Let $\varphi(\lambda)$ be the characteristic subspace relative to $\lambda$,[11]. the following lemma plays a fundamental role for $\varphi(\lambda)$.

Lemma 2.1 [11]: Let $x \in \mathcal{X}$ and $x \neq 0$. There exists F: $\mathcal{X} \longrightarrow \mathcal{U}$ such that $x$ is an eigenvector with eigenvalue $\lambda$ of A+BF if and only if $x \in \varphi(\lambda)$.

If k is an integer, denote $\underline{k}$ the set {1,2,...,k}. When S is a set, denote card{S} the number of elements in S. Under compatible coordinate transformation T:$\mathcal{X} \longrightarrow \mathcal{X}$, input transformation G:$\mathcal{U} \longrightarrow \mathcal{U}$, the closed-loop system will be

$$\dot{x} = T^{-1}(A+BF)Tx + T^{-1}BGv$$

where,

$$T^{-1}(A+BF)T = block\ diag\{\begin{pmatrix} 0 & 1 & 0 & \cdots & 0 & 0 \\ 0 & 0 & 1 & \cdots & 0 & 0 \\ \cdot & \cdot & \cdot & \cdots & \cdot & \cdot \\ 0 & 0 & 0 & \cdots & 0 & 1 \\ 0 & 0 & 0 & \cdots & 0 & 0 \end{pmatrix} \in R^{v_i \times v_i},\ i \in \underline{m}\} \quad (2.2.1)$$

$$T^{-1}BG = block\ diag\{(0\ 0\ \cdots\ 0\ 1)^t \in R^{v_i \times 1}\} \quad (2.2.2)$$

Lemma 2.2 [11]: In a suitable basis of the state space, $\varphi(\lambda)$ may be written as

$$\varphi(\lambda) = span\{block\ diag\ \{(1\ \lambda\ \cdots \lambda^{v_i-1})^t, i \in \underline{m}\}\} \quad (2.3)$$

where, the symbol "t" stands for transpose.

$\{\varphi(\lambda),\ \lambda \in C\}$ is a subset of the family of all subspaces in $\mathcal{X}$. It is known that both $\varphi(\lambda)+\varphi(\mu)$ and $\varphi(\lambda)\cap\varphi(\mu)$ are subspace in $\mathcal{X}$ for any $\lambda, \mu \in C$. With respect to the two operations of subspaces, the (A,B)-characteristic subspaces have following important properties.

Lemma 2.3: Let $\lambda_j, j \in \underline{t}$, be different complex numbers, then

$$dim\{\varphi(\lambda_1)+\cdots+\varphi(\lambda_t)\} = \sum_{j=1}^{t} card\{v_i \geq j,\ i \in \underline{m}\}.$$

Lemma 2.3 implies that $dim\ \varphi(\lambda)$=m for every $\lambda \in C$.
Define a subspace $\mathcal{B}_0$ in $\mathcal{B}$(:= Im B) as follows:

$$\mathcal{B}_0 = \mathcal{B} \cap A^{-1}\mathcal{B} = \{x;\ x \in \mathcal{B}, Ax \in \mathcal{B}\} = \varphi(0).$$

The following lemma describes the meaning of $\mathcal{B}_0$.

Lemma 2.4:
1) If $\lambda, \mu \in C$ and $\lambda \neq \mu$, then $\varphi(\lambda)\cap\varphi(\mu) = \mathcal{B}_0$;
2) $\mathcal{B}_0 = \mathcal{B}$ if and only if $\mathcal{B} = \mathcal{X}$.

## 3. STRUCTURE OF (A,B)-INVARIANT SUBSPACE

We will decompose (A,B)-invariant subspaces by (A,B)-radical subspace to be defined in this section. By this decomposition it is shown that the geometric structure of the (A,B)-invariant subspaces is similar to that of the A-invariant subspaces.

Let $a(\lambda)$ be a polynomial vector such that for any $\lambda_0 \in C$, $a(\lambda_0) \in \varphi(\lambda_0)$, then the $a(\lambda)$ is celled a (A,B)-characteristic vector-polynomial. In this section and the next one, $a(\lambda)$, $b(\lambda), \cdots$ represent the (A,B)-characteristic vector-polynomials, in general. When the $\lambda$ is fixed, the $a(\lambda)$ is a vector. Let $a(\lambda) \in \varphi(\lambda)$ and denote $a_1(\lambda) = a(\lambda)$. If $\frac{d}{d\lambda}a(\lambda) \neq 0$, then it is possible to define that $a_2(\lambda) = \frac{d}{d\lambda}a(\lambda) + b_1(\lambda)$ for some $b_1(\lambda) \in \varphi(\lambda)$. In general, if $a_{p-1}(\lambda)$ is defined, and $\frac{d^{p-1}}{d\lambda^{p-1}}a(\lambda) \neq 0$, then $a_p(\lambda)$ may be defined as following:

$$a_p(\lambda) = \frac{1}{(p-1)!}\frac{d^{p-1}}{d\lambda^{p-1}}a(\lambda) + \frac{1}{(p-2)!}\frac{d^{p-2}}{d\lambda^{p-2}}b_1(\lambda) + \cdots + b_{p-1}(\lambda) \tag{3.1}$$

for some $b_{p-1}(\lambda) \in \varphi(\lambda)$.

Definition 3.1: Let $a_1(\lambda), \cdots, a_p(\lambda)$ be defined as above, and denote

$$\mathcal{D}_p\{a(\lambda); b_1(\lambda), \cdots, b_{p-1}(\lambda)\} = span\{a_1(\lambda), a_2(\lambda), \cdots, a_p(\lambda)\}$$

The $\mathcal{D}_p\{a(\lambda); b_1(\lambda), \cdots, b_{p-1}(\lambda)\}$ is called an (A,B)-radical subspace with eigenvector $a(\lambda)$.

By definition the (A,B)-radical subspace seems to depend on the choice of the $b_j(\lambda), j \in \underline{p-1}$, in fact, it is not so. The following proposition shows that. So, at moment, it is denoted as $\mathcal{D}_p\{a(\lambda)\}$.

Proposition 3.1: Let $\mathcal{D}_p\{a(\lambda)\}$ be defined as above, then $dim\ \mathcal{D}_p\{a(\lambda)\} = p$

Proof: By lemma 2.2, $a(\lambda)$ may be written as

$$a(\lambda) = (block\ diag\{(1\ \lambda\ \cdots\ \lambda^{v_i-1})^t, i \in \underline{m}\}) \cdot (u_1\ \cdots\ u_m)^t. \tag{3.2}$$

where,$(u_1\ \cdots\ u_m)^t$ is a constant vector. Let $b_j(\lambda), j \in \underline{m}$, be written as

$$b_j(\lambda) = (block\ diag\{(1\ \lambda\ \cdots\ \lambda^{v_i-1})^t, i \in \underline{m}\}) \cdot (b_1^j\ \cdots\ b_m^j)^t$$

Let $u_i$ be the first nonzero component of vector $(u_1\ \cdots\ u_m)$. Denote the subvactor consisting of the $(\sum_{j=1}^{i-1} v_j + 1)$-th up to the $(\sum_{j=1}^{i} v_j)$-th component of $a_j(\lambda)$ as $a_j^i(\lambda)$. Then, for every $j \in \underline{p}$

$$a_j^i(\lambda) = \frac{u_i}{(j-1)!}\frac{d^{j-1}}{d\lambda^{j-1}}\begin{pmatrix}1\\ \lambda\\ \vdots\\ \lambda^{v_i-1}\end{pmatrix} + \frac{b_i^1}{(j-2)!}\frac{d^{j-2}}{d\lambda^{j-2}}\begin{pmatrix}1\\ \lambda\\ \vdots\\ \lambda^{v_i-1}\end{pmatrix} + \cdots + b_i^{j-1}\begin{pmatrix}1\\ \lambda\\ \vdots\\ \lambda^{v_i-1}\end{pmatrix} \tag{3.3}$$

where $u_i \neq 0$. Denote $\varphi^i(\lambda) = span\{(1\ \lambda\ \cdots\ \lambda^{v_i-1})^t\}$. It is clear that $\varphi^i(\lambda) = span\{a_1^i(\lambda)\}$. Thus, we can see that $\frac{d^{p-1}}{d\lambda^{p-1}}a(\lambda) \neq 0$ implies $v_i \geq p$. By (3.3) it is true that

$$span\{a_1^i(\lambda), \cdots, a_p^i(\lambda)\} = span\{a_1(\lambda), \frac{d}{d\lambda}a_1(\lambda), \cdots, \frac{d^{p-1}}{d\lambda^{p-1}}a_1(\lambda)\}$$

$$= span\begin{pmatrix} 1 & 0 & \cdots & 0 \\ \lambda & 1 & \cdots & 0 \\ \vdots & \vdots & \cdots & \vdots \\ \lambda^{p-1} & (p-1)\lambda^{p-2} & \cdots & (p-1)! \\ \vdots & \vdots & \cdots & \vdots \\ \lambda^{v_i-1} & (v_i-1)\lambda^{v_i-2} & \cdots & \frac{(v_i-1)!}{(v_i-p)!}\lambda^{v_i-p} \end{pmatrix}$$

Therefore, that $dim\ span\{a_1^i(\lambda), \cdots, a_p^i(\lambda)\} = p$ implies that $dim\mathcal{D}_p\{a(\lambda)\} = p$.

The following lemma is used to prove that $\mathcal{D}_p\{a(\lambda)\} \in In(A, B)$, the proof of this lemma is trivial and omitted.

Lemma 3.1: Let $a(\lambda)$ be a vector polynomial and $\mathcal{V}$ a given subspace, if for every $\lambda \in C$, $a(\lambda) \in \mathcal{V}$, then $\frac{d}{d\lambda}a(\lambda) \in \mathcal{V}$, too.

Theorem 3.1: For every integer $p$, $\mathcal{D}_p\{a(\lambda)\} \in In(A, B)$.
Proof: When $p = 1$, $\mathcal{D}_p\{a(\lambda)\} = span\{a_1(\lambda)\}$. $a_1(\lambda) = a(\lambda) \in \varphi(\lambda)$, i.e.,

$$(A - \lambda I)a_i(\lambda) \in \mathcal{B}. \tag{3.4}$$

(3.4) shows that $Aa(\lambda) \in \mathcal{B} + span\{a(\lambda)\}$.

For the case of $p \geq 2$ we prove that the validity of the following relationship:

$$(A - \lambda I)a_i(\lambda) - a_{i-1}(\lambda) \in \mathcal{B} \tag{3.5}$$

By induction, let $a_0(\lambda) = 0$, (3.4) shows that when $i = 1$, (3.5) is valid. We assume that (3.5) hold for all $i \leq j - 1$. Now we consider the case of $i = j$. Differentiating the equation (3.4) (j-1) times, by Leibnitz Formula, along with Lemma 3.1, it yields that

$$(A - \lambda I)\frac{d^{j-1}}{d\lambda^{j-1}}a_1(\lambda) - (j-1)\frac{d^{j-2}}{d\lambda^{j-2}}a_1(\lambda) \in \mathcal{B}. \tag{3.6}$$

(3.6) is equivalent to

$$(A - \lambda I)\frac{1}{(j-1)!}\frac{d^{j-1}}{d\lambda^{j-1}}a_1(\lambda) - \frac{1}{(j-2)!}\frac{d^{j-2}}{d\lambda^{j-2}}a_1(\lambda) \in \mathcal{B}. \tag{3.7}$$

As $b_1(\lambda), \cdots, b_{j-1}(\lambda) \in \varphi(\lambda)$, by the assumption of induction, the following relationship is obtained

$$(A - \lambda I)\{\frac{1}{(j-2)!}\frac{d^{j-2}}{d\lambda^{j-2}}b_1(\lambda) + \cdots + \frac{d}{d\lambda}b_{j-2}(\lambda) + b_{j-1}(\lambda)\} -$$

$$\{\frac{1}{(j-3)!}\frac{d^{j-3}}{d\lambda^{j-3}}b_1(\lambda) + \cdots + b_{j-2}(\lambda)\} \in \mathcal{B}. \tag{3.8}$$

(3.7) and (3.8) imply that (3.5) holds for $i = j$.

In linear algebra it is known that every radical subspace contains only one-dimensional eigensubspace. Therefore, from the generator point of view, it is the minimal subspace. The following theorem shows that the (A,B)-radical subspace has the same property.

Theorm 3.2: Let $F \in \underline{F}(\mathcal{D}_p\{a(\lambda)\})$, then restricting A+BF to $\mathcal{D}_p\{a(\lambda)\}$, $a(\lambda)$ is the unique eigenvector. Furthermore, when $p \geq 2$, the eigenvalue of $a(\lambda)$ is $\lambda$.

Proof: First, we prove that $\mathcal{D}_p\{a(\lambda)\} \cap \varphi(\lambda) = span\{a(\lambda)\}$ by taking the following steps:

1) For every j, with $1 < j \le p$, $a_j(\lambda) \notin \varphi(\lambda)$. Let $u_i$, $a_j^i(\lambda)$ and $\varphi^i(\lambda)$ be the notations used in the proof of proposition 3.1, thus, $j > 1$ implies $v_i \ge j > 1$. If $a_j(\lambda) \in \varphi(\lambda)$, then $a_j^i(\lambda) \in \varphi^i(\lambda)$. Since $u_i \ne 0$ and $v_i \ge j$, the vectors in (3.3),i.e., $\frac{u_j}{(j-1)!}\frac{d^{j-1}}{d\lambda^{j-1}}(1\ \lambda\ \cdots\ \lambda^{v_i-1})^t, \cdots, (1\ \lambda\ \cdots\ \lambda^{v_i-1})^t$, are independent. It follows that $a_j^i(\lambda) \notin \varphi^i(\lambda)$ contradicting the assumption.

2) For every j, $1 < j \le p$, if $c(\lambda) \in \mathcal{D}_{j-1}\{a(\lambda)\}$, then $a_j(\lambda) + c(\lambda) \notin \varphi(\lambda)$.

Since $c(\lambda) \in \mathcal{D}_{j-1}\{a(\lambda)\}$, there exist $v_1(\lambda), \cdots, v_{j-1}(\lambda) \in \varphi(\lambda)$ (some of them may be zero vectors) such that

$$c(\lambda) = \frac{1}{(j-2)!}\frac{d^{j-2}}{d\lambda^{j-2}}v_1(\lambda) + \cdots + v_{j-1}(\lambda))$$

Consequently,

$$a_j(\lambda) + c(\lambda) = \frac{1}{(j-1)!}\frac{d^{j-1}}{d\lambda^{j-1}}a(\lambda) +$$

$$\frac{1}{(j-2)!}\frac{d^{j-2}}{d\lambda^{j-2}}(b_1(\lambda) + v_1(\lambda)) + \cdots + (b_{j-1}(\lambda) + v_{j-1}(\lambda)) \tag{3.9}$$

(3.9) shows that $a_j(\lambda) + c(\lambda)$ has the same form as $a_j(\lambda)$. Therefore, by using the technique developed in the proof of step 1), it can be verified that $a_j(\lambda) + c(\lambda) \notin \varphi(\lambda)$.

Let $\lambda$ be replaced by another comlex number $\mu$, it is enough to verify that $a_j(\lambda) \notin \varphi(\mu)$, moreover, it is only necessary to prove that the following matrix M has rank j+1.

$$M = \begin{pmatrix} 1 & 1 & 0 & \cdots & 0 \\ \mu & \lambda & 1 & \cdots & 0 \\ \vdots & \vdots & \vdots & \cdots & \vdots \\ \mu^{j-1} & \lambda^{j-1} & (j-1)\lambda^{j-2} & \cdots & 1 \\ \vdots & \vdots & \vdots & \cdots & \vdots \\ \mu^{v_i-1} & \lambda^{v_i-1} & (v_i-1)\lambda^{v_i-2} & \cdots & \frac{(v_i-1)!}{(v_i-p)!(j-1)!}\lambda^{v_i-j} \end{pmatrix}$$

Using induction, we can show that the determinant of the first (j+1) rows of M is equal to $(\lambda - \mu)^j$. The details are omitted.

Theorem 3.2 illustrates that for every $F \in \underline{F}(\mathcal{D}_p\{a(\lambda)\}$, with respect to A+BF, $\mathcal{D}_p\{a(\lambda)\}$ is a radical subspace with eigenvector $a(\lambda)$.

Theorem 3.3: If $\mathcal{V} \in In(A, B)$, then there exist $a(\lambda_j), i \in \underline{p};\ b(\mu_j), j \in \underline{q};\ c(\eta_k), k \in \underline{r}; \cdots$, such that

$$\mathcal{V} = [\oplus_{i=1}^{p}\mathcal{D}_i\{a(\lambda_i)\}] \oplus [\oplus_{j=1}^{q}\mathcal{D}_2\{b(\mu_j)\}] \oplus [\oplus_{k=1}^{r}\mathcal{D}_3\{c(\eta_k)\}] \oplus \cdots \tag{3.10}$$

Proof: Take $F \in \underline{F}(\mathcal{V})$, then $\mathcal{V}$ is an (A+BF)-invariant subspace. Thus, $\mathcal{V}$ can be decomposed into a direct sum of (A+BF)-radical subspaces [4]. i.e.,

$$\mathcal{V} = [\oplus_{i=1}^{p}\mathcal{V}_i^1(x_{1i}] \oplus [\oplus_{j=1}^{q}\mathcal{V}_j^2(x_{2j}] \oplus [\oplus_{k=1}^{r}\mathcal{V}_k(x_{3k}] \oplus \cdots$$

where $\mathcal{V}_i^1(x_{1j})$ indicates an 1-dimensional cyclic subspace with eigenvector $x_{1j}$. Exactly, $\mathcal{V}_i^1(x_{1j})$ is just the same as the eigensubspace of A+BF, let $\lambda_i$ be eigenvalue associated to $x_{1j}$, then $\mathcal{V}_i^1(x_{1j}) = span\{x_{1i}\} = \mathcal{D}_1\{x_{1i}(\lambda)\}$. $\mathcal{V}_j^2(x_{2j})$ is a

2-dimensional cyclic subspace with eigenvetor $x_{2j}$ whose eigenvalue is denoted by $\mu_j$. Hence, there exists $x_j$ such that $(A+BF-\mu_j I)x_j = x_{2j}$, and $\mathcal{V}_j^2(x_{2j}) = span\{x_{2j}, x_j\}$. Since $x_{2j}$ is the eigenvector of A+BF with eigenvalue $\mu_j$, it follows that $x_{2j} \in \varphi(\mu_j)$ by Lemma 2.1. Therefore, it is possible to write $x_{2j} = b(\mu_j)$, and

$$(A+BF-\mu_j I)b(\mu_j) = 0 \tag{3.11}$$

Differentiating (3.11) with respect to $\mu_j$, (3.11) yields

$$(A+BF-\mu_j I)\frac{d}{d\mu_j}b(\mu_j) = b(\mu_j) = x_{2j} \tag{3.12}$$

As a consequence of (3.12), we have

$$x_j = \frac{d}{d\mu_j}b(\mu_j) + b_1(\mu_j) \tag{3.13}$$

for some $b_1(\mu_j) \in \varphi(\mu_j)$.

(3.11) and (3.13) lead to the assertion that $\mathcal{V}_j^2(x_{2j}) = \mathcal{D}_2\{b(\mu_j)\}$. In a similar way, it can be verified that $\mathcal{V}_k^3(x_{3k}) = \mathcal{D}_3\{c(\eta_k)\}$ for some $c(\eta_k) \in \varphi(\eta_k)$. The details are also omitted.

For a given $\mathcal{V} \in In(A,B)$ in the decomposion (3.10), the eigenvalues $\lambda_i$, $i \in \underline{p}$; $\mu_j$, $j \in \underline{q}$; $\eta_k$, $k \in \underline{r}; \cdots$, as well as the integers $p, q, r, \cdots$, may be changed along with the selection of feedback $F \in \underline{F}(\mathcal{V})$. The following theorem gives the condition of uniqueness for such decomposition.

Theorem 3.4: Suppose $\mathcal{V} \in In(A,B)$, then decompsition (3.10) is unique if and only if $\mathcal{V} \cap \mathcal{B} = 0$.

Proof: Necessity: Let $\mathcal{R}^*$ denote the maximal (A,B)-controllability subspace contained in $\mathcal{V}$, then $\mathcal{R}^* = < A+BF/\mathcal{V}\cap\mathcal{B} >$ for every $F \in \underline{F}(\mathcal{V})$ [10]. If $\mathcal{V}\cap\mathcal{B} \neq 0$, then $\mathcal{R}^* \neq 0$. The eigenvalues of $A+BF \mid \mathcal{R}^*$ may be variable along with the selection of F. It yield that the decomposition (3.10) is also variable.

Sufficiency: If there exist different decompositions of $\mathcal{V}$, then the Jordan form of $A+B\mathcal{V} \mid \mathcal{V}$ is variable along with the selectoin of F. Take a feedback $F_1 \in \underline{F}(\mathcal{V})$, under a compatible basis of state space, $\mathcal{V}$ and $(A+BF_1\ B)$ have the following forms

$$\mathcal{V} = span\begin{pmatrix} I_t \\ 0 \end{pmatrix}, \quad (A+BF_1\ B) = \begin{pmatrix} A_{11} & A_{12} & B_1 \\ 0 & A_{22} & B_2 \end{pmatrix}$$

where $t = dim\ \mathcal{V}$. $F_2$ is another feedback with $F_2 \in \underline{F}(\mathcal{V})$ such that the Jordan form of $A+BF_2 \mid \mathcal{V}$ is different from that of $A+BF_1 \mid \mathcal{V}$. Setting $F = F_2 - F_1$, $F$ is partitioned as $F = (F^1\ F^2)$, then

$$A+BF_2 = A+BF_1+BF = \begin{pmatrix} A_{11}+B_1F^1 & A_{12}+B_1F^2 \\ B_2F^1 & A_{22}+B_2F^2 \end{pmatrix}$$

As $F_2 \in \underline{F}(\mathcal{V})$, it is necessary that $B_2F^1$=0. Moreover, the Jordan form of $A_{11} + B_1F^1$ is different form that of $A_{11}$. Hence, $B_1F^1 \neq 0$. Select the independent columns from $F^1$, and then extend them to be an $m \times m$ nonsigular matrix G. We get

$$\begin{pmatrix} B_1 \\ B_2 \end{pmatrix} G = \begin{pmatrix} B_{11} & B_{22} \\ 0 & B_{22} \end{pmatrix}$$

with $B_{11} \neq 0$. Clearly $\mathcal{V} \supset Im \begin{pmatrix} B_{11} \\ 0 \end{pmatrix} \neq 0$ contradicting $\mathcal{V} \cap \mathcal{B} = 0$.

To conclude this section, we give a corollary of Theorem 3.4 for the case of m=1.

Corollary 3.1: When m=1, the decomposition (3.10) is unique if and only if $n > dim\ \mathcal{V}$.

Proof: When m=1, the matrix B reduces to a vector b. We consider the intersection $span\{b\} \cap \mathcal{V}$. If $b \in span\{Im\ b \cap \mathcal{V}\}$, then for every $f^t \in \underline{F}(\mathcal{V})$, $\mathcal{V} \supset \mathcal{R}^* =< A + bf^t / Im\ b >=< A/Im\ b >= \mathcal{X}$. It leads to $\mathcal{V} = \mathcal{X}$ contradicting $n > dim\ \mathcal{V}$. By Theorem 3.4, the conclusion comes immediately.

The authors were told by the referees that for the case m=1 the decomposition (3.10) has been treated by Gohberg *et al*[5].

## 4. THE ASSIGNABLE EIGENSTRUCTURE

In the last decade, the problem of eigenstructure assignment has been investigated. This investigation follows two directions. One is to seek the characters of assignable eigenstructure [6,7]. The other is to treat the design methods [1,2]. Most of the contributed papers dealt with the second direction. From Lemma 2.1 and Lemma 2.3 almost all vectors in the state space can not meet the requirement of assignable eigenvectors. Therefore, it is very necessary to find a characterization of eigenvector assignabilty. Any method becomes ineffective if we can not give the assignable vectors.

In this section, this characterization will be described by many of the (A,B)-radical subspaces and the decomposition of (A,B) invariant subspaces.

Definition 4.1: [6] The ordered set $\{x_i; i \in \underline{p}\}$ is called an assignable radical system with eigenvalue $\lambda$ if there exists a feedback $F{:}\mathcal{X} \longrightarrow \mathcal{U}$ such that

$$(A + BF - \lambda I)x_i = x_{i-1}\ for\ i \in \underline{p} \tag{4.1}$$

where $x_0 = 0$. This radical system is denoted by $\{\lambda; x^i, i \in \underline{p}\}$. The set of several systems $\{\lambda_j;\ x_i^j, j \in \underline{n}_j\}$, $j \in \underline{s}$, is often called eigenstructure. If $\sum_{j=1}^{s} n_j = n$, then the eigenstructure is complete.

Lemma 4.1 [6]: The set of $\{\lambda_j;\ x_i^j, j \in \underline{n}_j\}$, $j \in \underline{s}$, is an assignable eigenstructure if and only if

1) $x_i^j, i \in \underline{n}_j, j \in \underline{s}$, are independent vectors;

2) $(A - \lambda_j I)x_i^j - x_{i-1}^j \in \mathcal{B}$ for every $i \in \underline{n}_j$ and $j \in \underline{s}$.

The second condition of Lemma 4.1 implies $x_i^j \in \varphi(\lambda_j)$ for every $j \in \underline{s}$. From the proof of Theorem 3.1, there certainly exist $b_k(\lambda_j) \in \varphi(\lambda_j)$, $k \in \{1, 2, \cdots, n_{j-1}\}$, such that

$$x_j(\lambda) = \frac{1}{(j-1)!}\frac{d^{j-1}}{d\lambda^{j-1}}x(\lambda)(\lambda) + \frac{1}{(j-2)!}\frac{d^{j-2}}{d\lambda^{j-2}}b_1(\lambda) + \cdots + b_{j-1}(\lambda).$$

Therefore, using the concept of (A,B)-radical subspace, Lemma 4.1 is written as follws:

Theorem 4.1: The set of $\{\lambda_j;\ x_i^j,\ i \in \underline{n}_j\},\ j \in \underline{s}$, is an assignable eigenstructure if and only if

1) $x^j = a_i(\lambda_j)$, for every $i \in \underline{n}_j$ and $j \in \underline{s}$.

2) $dim\ span\{x_i^j, i \in \underline{n}_j, j \in \underline{s}\} = \sum_{j=1}^{s} n_j$.

5. CONCLUSION

In this paper, a new concept of (A,B)-radical subspace is defined and its properties are treated. Using this concept, an (A,B)-invariant subspace is decomposed into a direct sum of (A,B)-radical subspaces. This decomposition shows that the geometric structure of the (A,B)-invariant subspaces is completely similar to that of A-invariant subspaces. From this decomposition, we get a deeper understanding of the linear systems. The notion of (A,B)-radical subspaces and the decomposition of (A,B)-invariant subspaces are useful for the synthesis of linear systems.

AKNOWLEGMENT: This work was supported by Natual Science Fund of China.

REFERENCES

[1] Chang M-I.J.: Eigenstructure assignment by state feedback. Proc. 1984 America Contr. Conf. pp.372-377.
[2] Fahmy M., J.O'Reilly: On the eigenstructure assignment in linear multivariable systems. IEEE Trans. AC-27, 690-693, 1982.
[3] Fuhrman P.A., J.C.William : A study of (A,B) invariant subspaces via polynomial moedls. Int. J. Contr Vol.30 467-497,1980.
[4] Gantmacher F.R.: Theory of Matrices, Chelsea, 1959.
[5] Gohberg, Lancaster, Rodman: Invariant subspaces of matrices with applications.
[6] Han Zheng-zhi: Eigenstructure assignment using dynamical compensator, Int. J. Contr. Vol.49, 233-245, 1989.
[7] Moore B.C.: On the flexibility offered by state feedback in multivariable systems beyond closed loop eigenvalue assignment, IEEE Trans. AC-21,689-691,1976.
[8] Shamir T.: Matrix continued fraction are directly related to the maximal (A,B) invariant subspaces contained in Ker C. IEEE Trans. AC-32, 632-635,1987.
[9] Solok M.K.: A direct computational method for determining the maximal (A,B) invariant subspace contained in Ker C, IEEE Trans. AC-31, 349-352,1986.
[10] Wonham W.M. Linear Multivariable Control, Springer-verlag,1979.
[11] Zheng Yu-fan Han Zheng-zhi: What can be done for linear systems by feedback Proc. 9th IFAC World Cong. Vol.3, 1109-1414, 1984, Hungary.

# The Set of Feedback Matrices that Assign the Poles of a System

Hu Ting-Shu, Shi Song-Jiao

The set of all the feedback matrices that assign the poles of a system is determined through a function mapping $R^{m\times n}$ into $R^{m\times n}$. It is proved in the paper that the domain of the function is a dense open set in $R^{m\times n}$, and the range of it is exactly the set of all the $F$'s that make $A+BF$ similar to $A^*$. The set of all the $F$'s that assign the eigenvalues of $A+BF$ is the union of a finite number of such ranges. This function can be used to optimize other performance indexes of a system under the constraint of pole assignment.

NOTATION

| | |
|---|---|
| $R$ | : field of real numbers |
| $R^{m\times n}$ | : set of $m\times n$ matrices with elements in $R$ |
| $R_f$ | : range of function $f$ |
| $D_f$ | : domain of function $f$ |
| $CsA$ | : column vector form of matrix $A$ |
| $RsA$ | : row vector form of matrix $A$ |
| $A\otimes B$ | : Kronecker product of matrices $A$ and $B$ |

## 1. Introduction

Consider the system given by the following state-space representation:

$$\dot{x}(t) = Ax(t)+Bu(t) \tag{1}$$

where $A\in R^{n\times n}$, $B\in R^{n\times m}$, $x(t)$ is an $n$-dimensional state vector and $u(t)$ is an $m$-dimensional input vector. The feedback control law is given by:

$$u(t) = Fx(t). \tag{2}$$

The pole assignment problem is to find some $F$ so that the eigenvalues of $A+BF$ is a given set $\lambda_1,\lambda_2, \ldots ,\lambda_n$.

If system (1) is controllable, any pole assignment requirement can be satisfied. Furthermore, if $m>1$, the number of feedback matrices $F$'s that satisfy the same pole placement requirement is infinite. The extra freedom of $F$ is discussed by O'REILLY and FAHMY [4], and is utilized to satisfy other criteria by RAMAR and GOURISHANKER [2,3].

In this article, we will give a new approach to determine all the $F$'s that satisfy the pole placement requirement.

## 2. FUNCTION $F$, ITS DOMAIN AND RANGE

At first we introduce a function which maps an almost free space onto the set of all the $F$'s that make $A+BF$ similar to a certain $A^*$.

DEFINITION. Suppose $(A,B)$ is controllable. $A^*$ is an $n\times n$ matrix with the desired eigenvalues $\lambda_1,\lambda_2, \ldots ,\lambda_n$. $A$ and $A^*$ have no common eigenvalues. Then a function $f$ mapping $R^{m\times n}$ into $R^{m\times n}$ is defined as follows:

Let $U \in R^{m\times n}$, if the solution $V$ of

$$AV-VA^* = -BU \tag{3}$$

is nonsingular, then

$$F = UV^{-1} \tag{4}$$

is the image of $U$ under $f$, and $V$ is the image of $U$ under $f_0$. We denote the functions by $F=f(U)$ and $V=f_0(U)$. The domain of $f$, denoted by $D_f$, is defined as all the $U$'s that make $V$ nonsingular, and the range of $f$ is the image of $D_f$ under $f$, which is denoted by $R_f$.

REMARK.

a. The condition that $A$ and $A^*$ have no common eigenvalues is very important. Because it guarantees that (3) has a unique solution. Since (3) is equivalent to

$$(I\otimes A-(A^*)'\otimes I)CsV = -(I\otimes B)CsU. \tag{3'}$$

If $A$ and $A^*$ have common eigenvalues, $I\otimes A-(A^*)'\otimes I$ will be singular and (3) will have no solution or have infinite solutions, hence the definition is invalid.

b. The condition that $(A,B)$ is controllable is also necessary. If not so, $V=f_0(U)$ will be singular for all $U\in R^{m\times n}$, so the domain of $f$ is empty. This argument can be proved in the following way: If there is a $U$ so that $V=f_0(U)$ is nonsingular, then $A+BUV^{-1}=VA^*V^{-1}$, which means that $(A,B)$ is controllable, because $A$ and $A^*$ have no common eigenvalues.

c. Even if the above conditions are satisfied, the domain of $f$ may also be empty. This is possible only if the controllability indexes of $(A,B)$ and $(A^*,B)$ are different. If the eigenvalues of $A^*$ are distinct, and the above conditions are met, $D$ will not be empty.

In the following we will show some important properties of the range and domain of $f$.

THEOREM 1. *The range of $f$ is the set of all the $F$'s that make $A+BF$ similar to $A^*$.*

PROOF.

i) If $F \in R_f$, then there is a $U \in D_f$ so that $V = f_0(U)$ is nonsingular and $F = UV^{-1}$, from

$$A - VA^*V^{-1} = -BUV^{-1}$$

we have $A+BUV^{-1} = VA^*V^{-1}$ since $F = UV^{-1}$, so $A+BF$ is similar to $A^*$.

ii) If $A+BF$ is similar to $A^*$, then there exists a transformation matrix $V$ so that

$$A+BF = VA^*V^{-1}, \quad AV - VA^* = -BFV^{-1}.$$

Let $U = FV^{-1}$, then $F$ is the image of $U$ under $f$, therefore $F \in R_f$. □

THEOREM 2. *If the domain of $f$ is not empty, it's a dense open set in $R^{m \times n}$. (Here, the norm of an $R^{m \times n}$ element is defined as $\|U\| := \Sigma_i \Sigma_j |U_{ij}|$, and the concept of distance and open set is naturally derived).*

PROOF. At first, let's denote the largest element of a matrix $X$ by $\max[X] = \max\{|X_{ij}|\}$.

i) Suppose $U_0 \in D_f$, then $V_0 = f_0(U_0)$ is nonsingular, hence there exists a $\epsilon$, if $\max[\Delta V] < \epsilon$, $V_0 + \Delta V$ is also nonsingular. For any $\Delta U$: $\|\Delta U\| < \epsilon / \max[(I \otimes A - (A^*)' \otimes I)^{-1}(I \otimes B)]$, $\max[f_0(U_0)] < \epsilon$, therefore,

$$f_0(U_0 + \Delta U) = f_0(U_0) + f_0(\Delta U) = V_0 + f_0(\Delta U) \quad \text{is nonsingular,}$$

which implies that $U_0 + \Delta U \in D_f$, thus $D_f$ is open.

ii) $D_f$ is dense. It only needs to be shown that for any $U_0 \notin D_f$, every neighborhood of $U_0$ contains at least one $U \in D_f$.

Since $D_f$ is not empty, there exists a $U$, such that $f_0(U_1) = V_1$ is nonsingular. Suppose $V_0 = f_0(U_0)$ is singular, then

$$f_0(U_0 + \epsilon U_1) = V_0 + \epsilon V_1 = [V_0 V^{-1} + \epsilon I]V_1.$$

Let $\beta$ be the eigenvalue of $V_0 V^{-1}$ which is the closest to the origin except for zero eigenvalues, and the distance is $\delta$, then

$$f_0(U_0+\epsilon U_1) \quad \text{is nonsingular whenever} \quad 0<\epsilon<\delta.$$

Therefore, every neighborhood of $U_0$ contains a $U \in D_f$. This completes the proof. □

In the definition of $f$, there is a condition that $A$ and $A^*$ have no common eigenvalues. If the eigenvalues of $A$ contain some of the desired eigenvalues $\lambda_i$, we can choose a $F_0$ so that $A+BF_0$ has no common eigenvalues with $A^*$. Let $A_0=A+BF_0$, then the range of the map:

$$A_0V-VA^* = -BU, \quad F = UV^{-1}$$

equals the set of $F$'s that make $A_0+BF$ similar to $A^*$. Therefore, $\{F_1\colon F_1=F+F_0,\ F\in R_f\}$ is the set of all $F$'s that make $A+BF$ similar to $A^*$.

It is easy to see that if $A^*$ is replaced by any $n\times n$ matrix that is similar to $A^*$, the range and domain of $f$ will be the same. So, in the pole placement problem, $A^*$ is usually chosen to be a Jordan canonical form for simplicity. In such a case, solving the Sylvester equation (3) is equivalent to solving $n$ $n$-ordered linear algebraic equations.

If the desired eigenvalues of $A+BF$ are distinct, it has only one Jordan canonical form. In general, we can list all the possible Jordan forms of $A+BF$: $A_1^*, A_2^*, \ldots, A_N^*$. For each $A_i^*$,

$$AV-VA_i^* = BU \quad \text{and} \quad F = UV^{-1}$$

define a function $f_i$. Let the range of $f_i$ be $R_{fi}$, then $\bigcup_{i=1}^{N} R_{fi}$ equals the set of $F$'s which assign the poles of (1) to given values. Since some Jordan form cannot be actually reached, $R_{fi}$ may be empty for some $i$.

EXAMPLE. Suppose $m=1$, $A^*=\text{diag}[\lambda_1,\lambda_2,\ldots,\lambda_N]$, $[A,b]$ be controllable. Let $U=[u_1,u_2,\ldots,u_n]$, where $u_i$, $(i=1,\ldots,n)$ are scalars, then the solution of (3) is,

$$V = [(\lambda_1I-A)^{-1}b,\ldots,(\lambda_nI-A)^{-1}b]\ \text{diag}[u_1,u_2,\ldots,u_n].$$

Since $[A,b]$ is controllable, there must exist a $U$ which makes $V$ nonsingular, so $[(\lambda_1I-A)^{-1}b,\ldots,(\lambda_nI-A)^{-1}b]$ is nonsingular. Moreover, whenever $U$ has no zero element, $V$ is nonsingular. Such $U$'s form a dense open set in $R^n$. But the range of $f$ contains only one element:

$$\begin{aligned} F &= UV^{-1}=[u_1,u_2,\ldots,u_n]\text{diag}[u_1^{-1},u_2^{-1},\ldots,u_n^{-1}]\star \\ &\quad \star[(\lambda_1I-A)^{-1}b,\ldots,(\lambda_nI-A)^{-1}b]^{-1} \\ &= [1\ 1\cdots 1][(\lambda_1I-A)^{-1}b,\ldots,(\lambda_nI-A)^{-1}b]^{-1}. \end{aligned}$$

## 3. Application

When a practical system is to be designed, not only its eigenvalues are required to lie at some points, but also other performance indexes must be satisfied. For example, the eigenvectors are required to be orthogonal, and the sensitivity of the eigenvalues to be low, etc. Knowing the set of all the $F$'s which assign the poles of a system, we can choose $F$ in this set to make other performance index optimal.

**Example.** To make the eigenvectors near orthogonal under the constraint of pole placement, we may choose the performance index as:

$$J = 1/2\, tr[V'V + TT'] \qquad (T = V^{-1})$$

where $V$ is the eigenvector matrix of $A + BF$. It is easy to see that the set of all the possible $V$ equals the range of $f_0$. Therefore $J$ is a functional of $U \in R^{m \times n}$.

It can be proved that:

$$\partial J / \partial U = B'X' \tag{5}$$

where $X$ satisfies: $A^* X - XA = V' - TT'T$.

**Proof.**

$$\begin{aligned} \partial J / \partial U_{ij} &= \partial[1/2 tr(V'V + TT')]/\partial U_{ij} = tr[V'(\partial V/\partial U_{ij}) + T'(\partial T/\partial U_{ij})] \\ &= tr[V'(\partial V/\partial U_{ij}) - T'T(\partial V/\partial U_{ij})T] = tr[V' - TT'T](\partial V/\partial U_{ij}) \\ &= -Rs[V' - TT'T]\{(I \otimes A - (A^*)' \otimes I)^{-1}(I \otimes B)\}\partial CsU/\partial U_{ij} \end{aligned}$$

so,

$$[Cs(\partial J/\partial U)]' = -Rs[V' - TT'T]\{(I \otimes A - (A^*)' \otimes I)^{-1}(I \otimes B)\}$$

let,

$$Rs(X) = -Rs[V' - TT'T](I \otimes A - (A^*)' \otimes I)^{-1} \tag{6}$$

since $Rs(X)(I \otimes B) = Rs(XB)$ then, $\partial J/\partial U = (XB)'$ from (6), we have (5)□

From this result, we can optimize $J$ through gradient method. In the process of optimization, if $V$ is singular, we can continue the process by making a small change on $U$. After the optimal $U$ is evaluated, compute $F^* = U^*(V^*)^{-1}$.

## 4. Conclusion

For a multi-input linear multivariable system, the number of $F$'s which assign the poles of the system is infinite. Using the classical method, we can get only finite number of $F$'s. This paper presents a way through which all such $F$'s are represented as the range of a function. In applications, this function can be used to optimize other performance indexes of a system.

REFERENCES
[1]. C.T. CHEN (1984). *Linear System Theory and Design,* New York, Holt, Rinehart and Winston.
[2] K. RAMAR, V. GOURISHANKER (1976). Utilization of the design freedom of pole assignment feedback controllers of unrestricted rank. *Int. J. Contr. 24,* 423-430.
[3] V. GOURISHANKER, K. RAMAR (1976). Pole assignment with minimum eigenvalue sensitivity to plant parameter variation. *Int. J. Contr. 24,* 493-540.
[4] J. O'REILLY, M.M. FAHMY (1985). The minimum number of degrees of freedom in state feedback control. *Int. J. Contr. 41,* 749.

# Model matching for linear Hamiltonian systems

H.J.C. Huijberts

**Abstract**

We solve the Hamiltonian model matching problem by formulating it as a Hamiltonian disturbance decoupling problem with observation feedback and disturbance measurements. It turns out that the conditions for solvability of the Hamiltonian model matching problem are the same as those for solvability of the "normal" model matching problem. A procedure for reducing the compensator order is given.

## 1 Introduction

The Hamiltonian model matching problem (HMMP) consists of designing a compensating feedback for a given Hamiltonian system in such a way that the resulting input-output behavior matches that of a prespecified Hamiltonian model.

The "normal" model matching problem (MMP) for linear systems has been solved in different set ups by several authors (see e.g. [6,7,8,10]). In [4,9] it is shown that we can formulate MMP as a disturbance decoupling problem with observation feedback and disturbance measurements (DDOFM).

In this paper we solve HMMP by formulating it as a Hamiltonian disturbance decoupling problem with observation feedback and disturbance measurements (HDDOFM) and solving this associated problem using techniques also employed in e.g. [12,13].

The organization of the paper is as follows. In Section 2 some preliminary definitions and new and already existing invariance results are given. In Section 3 we solve HDDOFM. It turns out that the conditions for solvability of HDDOFM are the same as those for solvability of DDOFM. Furthermore, we can solve HDDOFM by means of a compensator with the same order as the dimension of the original system. In Section 4 we give a procedure for reducing the compensator order. In Section 5 finally, we formulate HMMP and, using the results of the foregoing sections, we give conditions for the solvability of HMMP. It turns out that these conditions are the same as the conditions for solvability of MMP.

## 2 Preliminaries

We consider a minimal linear time-invariant system:

$$(1) \quad \begin{cases} \dot{x} &= Ax + Bu \\ y &= Cx \end{cases}$$

where $x \in \mathbb{R}^q, u \in \mathbb{R}^m, y \in \mathbb{R}^p$ and $A, B, C$ are constant matrices of appropriate dimensions.

We recall briefly from [3,13]: a subspace $\mathcal{V} \subset \mathbb{R}^q$ is called *(A,B)-invariant* if $A\mathcal{V} \subset \mathcal{V} + Im\, B$, it is called *(C,A)-invariant* if $A(\mathcal{V} \cap Ker\, C) \subset \mathcal{V}$ and it is called *(C,A,B)-invariant* if it is $(C, A)$-invariant as well as $(A, B)$-invariant. $(A, B)$-invariance $((C, A)$-invariance) $((C, A, B)$-invariance) is equivalent to the existence of a matrix $F$ $(G)$ $(K)$ such that $(A + BF)\mathcal{V} \subset \mathcal{V}$ $((A + GC)\mathcal{V} \subset \mathcal{V})$ $((A + BKC)\mathcal{V} \subset \mathcal{V})$. For an $(A, B)$-invariant subspace $\mathcal{V}$ the set of *friends of* $\mathcal{V}$, denoted $\mathcal{F}(\mathcal{V})$, consists of all matrices $F$ satisfying $(A + BF)\mathcal{V} \subset \mathcal{V}$.

We call two $(C, A, B)$-invariant subspaces $\mathcal{V}_1, \mathcal{V}_2$ *compatibly (C,A,B)-invariant* if there is a matrix K such that $(A + BKC)\mathcal{V}_i \subset \mathcal{V}_i$ $(i = 1,2)$. In [14] it is proved that $\mathcal{V}_1$ and $\mathcal{V}_2$ are compatibly $(C, A, B)$-invariant if and only if

$$(2) \quad \begin{bmatrix} A & 0 \\ 0 & -A \end{bmatrix} \left( \mathcal{V}_1 \tilde{\oplus} \mathcal{V}_2 \cap Ker[C \;\; C] \right) \subset \left( \mathcal{V}_1 \tilde{\oplus} \mathcal{V}_2 + Im \begin{bmatrix} B \\ B \end{bmatrix} \right)$$

Here $\mathcal{V}_1 \tilde{\oplus} \mathcal{V}_2 := \left\{ \begin{bmatrix} v_1 \\ v_2 \end{bmatrix} \mid v_1 \in \mathcal{V}_1, \; v_2 \in \mathcal{V}_2 \right\}$.

One can easily verify that as a consequence of (2) we have:

**Corollary 2.1** *Two $(C, A, B)$-invariant subspaces $\mathcal{V}_1$ and $\mathcal{V}_2$ are compatibly $(C, A, B)$-invariant if $\mathcal{V}_1 \subset \mathcal{V}_2$.*

□

Now assume that there exists a non-singular map $J : \mathbb{R}^q \mapsto \mathbb{R}^q$ satisfying $J = -J^T$. From the non-singularity and the skew-symmetry of $J$ it follows that $q$ is necessarily even, say $q = 2n$. It can be proven (cf. [1]) that there exist bases for $\mathbb{R}^{2n}$ in which $J$ has the form $\begin{bmatrix} 0 & -I_n \\ I_n & 0 \end{bmatrix}$. Such a $J$ is called a *symplectic form* on $\mathbb{R}^{2n}$, and $(\mathbb{R}^{2n}, J)$ is called a *symplectic space*. We call a linear system (1) on $(\mathbb{R}^{2n}, J)$ a *Hamiltonian system* if $A^T J + JA = 0$ and $B^T J = C$ (cf. [11]).

Prototypic examples of Hamiltonian systems are conservative mechanical systems (in the linear case: systems consisting of masses and springs). For these systems we can interpret $\frac{1}{2}x^T JAx$ as the internal energy of the system, $u$ as the (generalized) forces applied to the system and $y$ as the (generalized) displacements along the line of action of the exerted forces (see e.g. [11]).

A subspace $\mathcal{V}$ will be called a *symplectic subspace* of $(\mathbb{R}^{2n}, J)$ if $J \mid_{\mathcal{V}}$ is non-singular (or equivalently: $J \mid_{\mathcal{V}}$ is a symplectic form on $\mathcal{V}$). By $\mathcal{V}^{\perp}$ we denote the orthogonal complement of $\mathcal{V}$ w.r.t. $J$, i.e. $\mathcal{V}^{\perp} := \{x \in \mathbb{R}^{2n} \mid x^T Jv = 0 \;\; \forall v \in \mathcal{V}\}$. We can prove quite easily that $(\mathcal{V}^{\perp})^{\perp} = \mathcal{V}$ and that $\mathcal{V}$ is a symplectic subspace of $(\mathbb{R}^{2n}, J)$ if and only if $\mathcal{V} \cap \mathcal{V}^{\perp} = \{0\}$. A subspace $\mathcal{V}$ will be called *isotropic* if $\mathcal{V} \subset \mathcal{V}^{\perp}$, *Lagrangian* if $\mathcal{V} = \mathcal{V}^{\perp}$ and *co-isotropic* if $\mathcal{V} \supset \mathcal{V}^{\perp}$.

We will call a feedback $u = Fx$ for a Hamiltonian system (1) a *Hamiltonian feedback* if the system after feedback is still a Hamiltonian system on $(\mathbb{R}^{2n}, J)$. Assuming that $B$ is injective (or equivalently that $C$ is surjective), which we can assume without loss of generality, a feedback $u = Fx$ is a Hamiltonian feedback if and only if $F = KC$, where

$K = K^T$ (cf. [12]). Hence Hamiltonian feedback is necessarily observation feedback. It is not difficult to prove that a subspace $\mathcal{V}$ is $(C,A,B)$-invariant if and only if $\mathcal{V}^\perp$ is $(C,A,B)$-invariant and that it can be made invariant by Hamiltonian feedback if and only if $\mathcal{V}^\perp$ can be made invariant by the same feedback. Hence $\mathcal{V}$ can be made invariant by Hamiltonian feedback if and only if $\mathcal{V}$ and $\mathcal{V}^\perp$ are compatibly $(C,A,B)$-invariant. Thus, using Corollary 2.1 we have:

**Corollary 2.2** *$\mathcal{V}$ can be made invariant by Hamiltonian feedback if $\mathcal{V}$ is $(C,A,B)$-invariant and isotropic (Lagrangian) (co-isotropic).*

□

**Remark 2.3** *For Lagrangian subspaces this was already shown in a different fashion in [12].*

□

## 3 The disturbance decoupling problem with observation feedback and disturbance measurements for Hamiltonian systems

We will first formulate and solve the disturbance decoupling problem with observation feedback and disturbance measurements (DDOFM), compare [12,13]. Let $(A,B,C)$ be a linear time-invariant system on $\mathbb{R}^q$. Suppose that there are disturbances that can be measured influencing the system and that we are particularly interested in regulating a part of the state space. In formulas:

$$(3)\quad \begin{cases} \dot{x} &= Ax + Bu + Ed,\ d \in \mathbb{R}^r \\ y &= Cx \\ z &= Dx,\ z \in \mathbb{R}^s \end{cases}$$

with $d$ the disturbances and $z$ the to-be-regulated variables. Then DDOFM consists of finding an integer $\mu$ and constant matrices $K$ and $G$ of appropriate dimensions such that after application of the compensator:

$$(4)\quad \begin{cases} \dot{x}_c &= B_c u_c \\ y_c &= C_c x_c \end{cases}$$

where $x_c, u_c, y_c \in \mathbb{R}^\mu$ and $B_c, C_c$ are invertible constant matrices, and the feedback:

$$(5)\quad \begin{bmatrix} u \\ u_c \end{bmatrix} = K \begin{bmatrix} y \\ y_c \end{bmatrix} + Gd$$

the transfer matrix from $d(\cdot)$ to $z(\cdot)$ equals zero.

Now define:

$$x_E := \begin{bmatrix} x \\ x_c \end{bmatrix},\ u_E := \begin{bmatrix} u \\ u_c \end{bmatrix},\ y_E := \begin{bmatrix} y \\ y_c \end{bmatrix},\ A_E := \begin{bmatrix} A & 0 \\ 0 & 0 \end{bmatrix},\ B_E := \begin{bmatrix} B & 0 \\ 0 & B_c \end{bmatrix},$$

$$E_E := \begin{bmatrix} E \\ 0 \end{bmatrix},\ C_E := \begin{bmatrix} C & 0 \\ 0 & C_c \end{bmatrix},\ D_E := \begin{bmatrix} D & 0 \end{bmatrix}.$$

Then (3) and (4) yield the following extended system on $\mathbb{R}^q \times \mathbb{R}^\mu$:

$$(6) \quad \begin{cases} \dot{x}_E &= A_E x_E + B_E u_E + E_E d \\ y_E &= C_E x_E \\ z_E &= D_E x_E \end{cases}$$

For a given subspace $\mathcal{V}_E \subset \mathbb{R}^q \times \mathbb{R}^\mu$ we define a subspace $p(\mathcal{V}_E) \subset \mathbb{R}^q$ by:

$$(7) \quad p(\mathcal{V}_E) := \left\{ x \in \mathbb{R}^q \mid \exists x_c \in \mathbb{R}^\mu : \begin{bmatrix} x \\ x_c \end{bmatrix} \in \mathcal{V}_E \right\}$$

**Proposition 3.1** *DDOFM can be solved for (6) by a static observation feedback (5) if and only if there is a $(C_E, A_E, B_E)$-invariant subspace $\mathcal{V}_E$ contained in $Ker\, D_E$ satisfying:*

$$(8) \quad Im\, E_E \subset \mathcal{V}_E + Im\, B_E$$

**Proof**
Follows by combining the results of [5] and [2]. See also [13].

□

**Proposition 3.2** *There is a $(C_E, A_E, B_E)$-invariant subspace $\mathcal{V}_E$ contained in $Ker\, D_E$ satisfying (8) if and only if there is an $(A, B)$-invariant subspace $\mathcal{V}$ contained in $Ker\, D$ satisfying:*

$$(9) \quad Im\, E \subset \mathcal{V} + Im\, B$$

**Proof**
*(necessity)*
Assume that there is a $(C_E, A_E, B_E)$-invariant subspace $\mathcal{V}_E$ contained in $Ker\, D_E$ satisfying (8). Now (8) implies: $Im\, E = p(Im\, E_E) \subset p(\mathcal{V}_E + Im\, B_E) = p(\mathcal{V}_E) + Im\, B$, because $B_c$ is invertible. Moreover, $p(\mathcal{V}_E)$ is $(A, B)$-invariant and contained in $Ker\, D$ (cf. [14]). Hence there is an $(A, B)$-invariant subspace contained in $Ker\, D$ satisfying (9).
*(sufficiency)*
Let $\mathcal{V}$ be an $(A, B)$-invariant subspace contained in $Ker\, D$ satisfying (9). Let $\mu := \dim(\mathcal{V})$, $T : \mathcal{V} \mapsto \mathbb{R}^\mu$ an isomorphism and $\mathcal{V}_E := \left\{ \begin{bmatrix} v \\ Tv \end{bmatrix} \mid v \in \mathcal{V} \right\}$. Then it is easy to see that $\mathcal{V}_E$ is a $(C_E, A_E, B_E)$-invariant subspace contained in $Ker\, D_E$ satisfying (8).

□

Propositions (3.1) and (3.2) immediately result in:

**Theorem 3.3** *DDOFM is solvable for (3) if and only if $Im\, E \subset \mathcal{V}^* + Im\, B$, where $\mathcal{V}^*$ is the maximal $(A, B)$-invariant subspace contained in $Ker\, D$.*

□

Note that in fact Proposition (3.2) and Theorem (3.3) give a procedure for constructing a compensator of order $\dim(\mathcal{V}^*)$ (see also [13]).

Now assume that $q = 2n$ and let $J$ be a symplectic form on $\mathbb{R}^{2n}$ such that $A^TJ + JA = 0$, $B^TJ = C$. Then the Hamiltonian DDOFM (HDDOFM) consists of finding an integer $\nu$, a symplectic form $J_c$ on $\mathbb{R}^{2\nu}$ and matrices $K = K^T, G$ such that after application of the Hamiltonian compensator:

$$(10)\quad \left\{ \begin{array}{rcl} \dot{x}_c & = & J_c^{-1}u_c \\ y_c & = & -I_{2\nu}x_c \end{array} \right.$$

and the feedback (5) the transfer matrix from $d(\cdot)$ to $z(\cdot)$ equals zero. The special choice of the compensator and of $K$ implies that the state matrix of the closed loop system is Hamiltonian w.r.t. the symplectic form $J_E = \begin{bmatrix} J & 0 \\ 0 & J_c \end{bmatrix}$.

It turns out that the conditions for solvability of DDOFM are also necessary and sufficient conditions for solvability of HDDOFM:

**Theorem 3.4** *HDDOFM is solvable for (3) if and only if $Im\,E \subset \mathcal{V}^* + Im\,B$, where $\mathcal{V}^*$ is the maximal $(A, B)$-invariant subspace contained in $Ker\,D$.*

**Proof**
*(necessity)*
Follows from necessity for solvability of DDOFM.
*(sufficiency)*
Let $\nu = n$ and $J_c = J$. Without loss of generality we can assume that $J = \begin{bmatrix} 0 & -I_n \\ I_n & 0 \end{bmatrix}$.

Let $R = \begin{bmatrix} I_n & 0 \\ 0 & -I_n \end{bmatrix}$. Then $R$ satisfies $R = R^{-1} = R^T$ and $RJ + JR = 0$. Now define a subspace $\mathcal{V}_E$ of $\mathbb{R}^{2n} \times \mathbb{R}^{2n}$ by: $\mathcal{V}_E := \left\{ \begin{bmatrix} v \\ Rv \end{bmatrix} \mid v \in \mathcal{V}^* \right\}$. Then it is straightforward to check that $\mathcal{V}_E$ is $(C_E, A_E, B_E)$-invariant and contained in $Ker\,D_E$. Furthermore, since $R^TJR = -J$, we have that $\mathcal{V}_E$ is an isotropic subspace. Hence there is a $K = K^T$ that makes $\mathcal{V}_E$ invariant. Noting that $Im\,E_E \subset \mathcal{V}_E + Im\,B_E$, we can find $G$ such that $Im\,(E_E + B_EG) \subset \mathcal{V}_E$. Hence we have solved HDDOFM.

□

**Remark 3.5** *Note that, similar to DDOF, the sufficiency part of the proof of Theorem 3.4 is constructive: it gives a procedure for constructing a compensator of order 2n.*

□

## 4 Reduction of the compensator order for HDDOFM

In section 3 we gave procedures for constructing compensating feedbacks that solve DDOFM and HDDOFM. For DDOFM the order of the compensator was equal to $\dim(\mathcal{V}^*)$, whereas for HDDOFM the order of the compensator was equal to the dimension of the state space of (3). In this section we will try to reduce the order of the compensator that solves HDDOFM. For this, we would like to use (a modified version of) the procedure for DDOFM. However, there are two problems to this. Firstly, we want the state space of the compensator to be a symplectic space. Secondly, it is not certain beforehand that we can find a symmetric $K$. In the following procedure we will give a clue to the solution of these problems:

**Procedure**

1. Calculate $\mathcal{V}^*$.
2. Let $\tilde{\mathcal{V}}$ be a symplectic subspace of $(\mathbb{R}^{2n}, J)$ such that $\mathcal{V}^* \subset \tilde{\mathcal{V}}$. Assume that $\dim(\tilde{\mathcal{V}}) = 2k$.
3. Choose a basis $(s_1, \cdots, s_{2n})$ for $\mathbb{R}^{2n}$ such that:
   $\tilde{\mathcal{V}} = span\{s_1, \cdots, s_{2k}\}$, $\tilde{\mathcal{V}}^{\perp} = span\{s_{2k+1}, \cdots, s_{2n}\}$ and $S^T J S = \begin{bmatrix} \tilde{J} & 0 \\ 0 & \bar{J} \end{bmatrix}$, where $S = col(s_1, \cdots, s_{2n})$ and $\tilde{J} = \begin{bmatrix} 0 & -I_k \\ I_k & 0 \end{bmatrix}$.
4. Let $J_c := \tilde{J}$, $\bar{R} := \begin{bmatrix} I_k & 0 \\ 0 & -I_k \end{bmatrix}$, $R := [\bar{R}\ 0]S^{-1}$. Define a subspace $\mathcal{V}_E^* \subset \mathbb{R}^{2n} \times \mathbb{R}^{2k}$ by:
   $\mathcal{V}_E^* := \left\{ \begin{bmatrix} v \\ Rv \end{bmatrix} \mid v \in \mathcal{V}^* \right\}$. Then $\mathcal{V}_E^*$ is an isotropic $(C, A, B)$-invariant subspace.
5. Determine $K = K^T$,$G$ satisfying $(A_E + B_E K C_E)\mathcal{V}_E^* \subset \mathcal{V}_E^*$ and $Im(E_E + B_E G) \subset \mathcal{V}_E^*$.

Then a compensator (10) on $(\mathbb{R}^{2k}, J_c)$ together with a feedback (5) solves HDDOFM.

□

**Comment**
We will give a brief comment on some of the steps of the above procedure.

2. Given $\mathcal{V}^*$, we can always find a $\tilde{\mathcal{V}} \supset \mathcal{V}^*$ that is a symplectic subspace of $(\mathbb{R}^{2n}, J)$ in the following way: There is a basis $(w_1, \cdots, w_r)$ for $\mathcal{V}^*$ such that $\mathcal{V}^* \cap \mathcal{V}^{*\perp} = span\{w_1, \cdots, w_s\}$. Then we can take $\tilde{\mathcal{V}} = span\{w_1, \cdots, w_r, Jw_1, \cdots, Jw_s\}$. Note however that $\tilde{\mathcal{V}}$ is in general not uniquely determined and that $\tilde{\mathcal{V}}$ obtained in this way is of maximal dimension.
3. See [1] for the basis choice.
4. $\mathcal{V}_E^*$ is indeed an isotropic subspace of $(\mathbb{R}^{2n} \times \mathbb{R}^{2k}, J_E)$. This can easily be checked by using the fact that $\bar{R}^T J_c \bar{R} = -J_c$.
5. Because $\mathcal{V}_E^*$ is an isotropic subspace, we can always find $K = K^T$ that satisfies $(A_E + B_E K C_E)\mathcal{V}_E^* \subset \mathcal{V}_E^*$ (see Corollary (2.2)). Letting $F \in F(\mathcal{V}^*)$, $\hat{R} := S \begin{bmatrix} \bar{R} \\ 0 \end{bmatrix}$, it is easy to check that we can take $K = \begin{bmatrix} 0 & -F\hat{R} \\ -\hat{R}^T F^T & -\hat{R}^T F^T C \hat{R} - J_c R(A + BF)\hat{R} \end{bmatrix}$.

□

Hence by the above procedure we can construct a compensating feedback with a compensator of order $\dim(\tilde{\mathcal{V}})$, where $\tilde{\mathcal{V}}$ is some symplectic subspace containing $\mathcal{V}^*$. Thus, if $\mathcal{V}^*$ is a symplectic subspace, a compensator of order $\dim(\mathcal{V}^*)$ does the job.

## 5 The Hamiltonian model matching problem

We will use the results of the foregoing sections to give a solution to the Hamiltonian model matching problem (HMMP). Given a Hamiltonian system $(A_p, B_p, C_p)$ on the symplectic space $(\mathbb{R}^{2n_p}, J_p)$ called the *plant*, and a Hamiltonian system $(A_m, B_m, C_m)$ on the symplectic space $(\mathbb{R}^{2n_m}, J_m)$ called the *model*, HMMP consists of finding a compensating feedback for the plant in such a way that the model and the plant after feedback constitute the same input-output behavior. Furthermore we demand the compensator and the feedback to be Hamiltonian. Analogous to [4,9] we can define HMMP as a HDDOFM with:

$$A = \begin{bmatrix} A_p & 0 \\ 0 & A_m \end{bmatrix},\ B = \begin{bmatrix} B_p \\ 0 \end{bmatrix},\ E = \begin{bmatrix} 0 \\ B_m \end{bmatrix},\ C = \begin{bmatrix} C_p & 0 \end{bmatrix},\ D = \begin{bmatrix} C_p & -C_m \end{bmatrix},$$

$$J = \begin{bmatrix} J_p & 0 \\ 0 & J_m \end{bmatrix}$$

Then we have as an immediate consequence of Theorem (3.4):

**Theorem 5.1** *HMMP is solvable if and only if* $Im \begin{bmatrix} 0 \\ B_m \end{bmatrix} \subset \mathcal{V}^* + Im \begin{bmatrix} B_p \\ 0 \end{bmatrix}$, *where* $\mathcal{V}^*$ *is the maximal* $(A, B)$*-invariant subspace contained in* $Ker\, D$.

□

**Remark 5.2**

1. *Note that the condition for solvability of HMMP given in the above theorem is equivalent to the one given in e.g. [7] for solvability of the "normal" model matching problem.*

2. *We can construct a compensating feedback for the solution of HMMP by using the procedure given in section 4. After we have applied the compensating feedback, the input-output behavior of the compensated plant will be that of a Hamiltonian system. However, it should be noted that by introducing the new input matrix* $(E + BG)$ *in general there will not be duality between inputs and outputs w.r.t. the symplectic form* $J_E$ *any more. This can have two reasons. Firstly, existence of a symplectic form for a system with Hamiltonian input-output behavior is only guaranteed if the system is minimal (cf. [11]), whereas it might very well be that the compensated plant is not minimal. Secondly, if the system is indeed minimal and thus existence of a symplectic form is guaranteed, the introduction of the new input matrix may imply that the symplectic form w.r.t. which the system is Hamiltonian changes.*

□

### Acknowledgments

I would like to thank A.J. van der Schaft and H. Nijmeijer for some useful discussions.

### References

[1] R. Abraham, J.E. Marsden, **Foundations of mechanics** (2nd. edition), Benjamin/Cummings, London, 1978.

[2] S.P. Bhattacharyya, *Disturbance rejection in linear systems*, **Intern. J. Systems Sci.** 5 (1974), 633-637.

[3] G. Basile, G. Marro, *Controlled and conditioned invariant subspaces in linear system theory*, **J. Optimiz. Theory Appl.** 3 (1969), 306-315.

[4] E. Emre, M.L.J. Hautus, *A polynomial characterization of (A,B)-invariant and reachability subspaces*, **SIAM J. Control Optimiz.** 18 (1980), 420-436.

[5] F. Hamano, K. Furuta, *Localization of disturbances and output decomposition in decentralized linear multivariable systems*, **Int. J. Control** 22 (1975), 551-562.

[6] J.W. Howze, C. Thisayakorn, R.K. Cavin III, *Model following using partial state feedback*, **IEEE Trans. Aut. Control** 21 (1976), 844-846.

[7] M. Malabre, *Structure à l'infini des triplets invariants. Application à la poursuite parfaite de modèle*, pp. 43-53 in **Analysis and optimization of systems** (A. Bensoussan and J.L. Lions, eds.), Lecture notes in control and information sciences 44, Springer, Berlin, 1982.

[8] A.S. Morse, *Structure and design of linear model following systems*, **IEEE Trans. Aut. Control** 18 (1973), 346-354.

[9] A.S. Morse, *Minimal solutions to transfer matrix equations*, **IEEE Trans. Aut. Control** 21 (1976), 131-133.

[10] B.C. Moore, L.M. Silverman, *Model matching by state feedback and dynamic compensation*, **IEEE Trans. Aut. Control** 17 (1972), 491-497.

[11] A.J. van der Schaft, **System theoretic descriptions of physical systems**, CWI Tracts 3, CWI, Amsterdam, 1984.

[12] A.J. van der Schaft, *Disturbance decoupling by observation feedback for Hamiltonian systems*, **Systems Control Lett.** 2 (1983), 286-291.

[13] J.M. Schumacher, *Compensator synthesis using (C,A,B)-pairs*, **IEEE Trans. Aut. Control** 25 (1980), 1133-1138.

[14] J.W. van der Woude, **Feedback decoupling and stabilization for linear systems with multiple exogenous variables**, Ph.D. Thesis, Technical University of Eindhoven, 1987.

CONNECTIVE STABILIZATION OF LARGE-SCALE SYSTEMS:
A STABLE FACTORIZATION APPROACH

Masao Ikeda and Hsiang-Ling Tan

**Abstract:** This paper considers a decentralized stabilization problem for large-scale linear systems composed of interconnected subsystems. The proper stable factorization approach is taken to design local dynamic controllers for the subsystems. Conditions are presented under which we can stabilize a given large-scale system so that stability of the resultant closed-loop system is robust to bounded perturbations of the interconnections among subsystems.

# 1 Introduction

Decentralized stabilization problems have been considered extensively for large-scale systems composed of a number of interconnected subsystems [2]. The information structure constraint imposed commonly is conformable to the subsystems, which is because autonomy of each subsystem is desired in large-scale systems explicitly or implicitly. Then, the overall closed-loop system is also a collection of closed-loop subsystems with the same interconnection structure as in the open-loop system. It is usually required that stability of the overall closed-loop system is preserved under perturbations in the interconnections such that the strength of each coupling between any two subsystems is bounded. This kind of stability property is referred to as connective stability [6]. It is obviously necessary for connective stability that each closed-loop subsystem is stable.

The methods of constructing connectively stable large-scale control systems, reported so far [2], are mostly those by state feedback or observer-based feedback. The main objective of this paper is to present a new method for such stabilization using the proper stable factorization approach [7]. The most fundamental and significant result of the factorization approach is the parametrization of all centralized stabilizing controllers. We apply the result to each subsystem to define a local stabilizing controller with an unspecified parameter. Then, we tune the parameter to make the overall closed-loop system connectively stable. This is possible under an M-matrix condition [1] described by bounds of the interconnections and minimized norms of transfer matrices of the closed-loop subsystems.

A graph-theoretic stabilizability condition is presented, where a directed graph is defined by the structure of the interconnections and solvability of two-sided linear equations of rational matrices associated with open-loop subsystems.

**Notations:** A rational matrix in $s$ with real coefficients is said to be stable if it is analytic in the closed right half complex plane $\mathbf{C}_+$ (excluding $s=\infty$). We use $\mathbf{R_s}$ and $\mathbf{R_{ps}}$ to denote respectively the sets of stable and proper stable rational matrices. If a matrix belonging to $\mathbf{R_{ps}}$ has an inverse in $\mathbf{R_{ps}}$ , we say that it is $\mathbf{R_{ps}}$-unimodular. The norm of a rational matrix $F \in \mathbf{R_{ps}}$ is defined by

$$\|F\| = \sup_{\omega} | (F(i\omega)) |,$$

where $\omega$ is a real number and $|\cdot|$ denotes a norm of the indicated complex matrix, which is induced by an $l_p$ vector norm, $p = 1, 2, \infty$.

# 2 Problem Formulaton

The large-scale interconnected system we deal with is a so-called input-output decentralized system [6] described by

$$\begin{aligned} \mathsf{S}: \quad \dot{x}_i &= A_i x_i + B_i u_i + \sum_{j=1}^{n} G_i E_{ij} H_j x_j \\ y_i &= C_i x_i, \qquad i = 1, 2, \ldots, n, \end{aligned} \tag{2.1}$$

which is composed of $n$ subsystems

$$\begin{aligned} \mathsf{S_i}: \quad \dot{x}_i &= A_i x_i + B_i u_i + G_i v_i \\ y_i &= C_i x_i \\ w_i &= H_i x_i, \qquad i = 1, 2, \ldots, n \end{aligned} \tag{2.2}$$

connecting with each other through the static interconnections

$$v_i = \sum_{j=1}^{n} E_{ij} w_j, \qquad i = 1, 2, \ldots, n. \tag{2.3}$$

In (2.1), (2.2) and (2.3), $x_i$ is the state, $u_i$ is the control input, $y_i$ is the measured output, $v_i$ is the interconnection input, and $w_i$ is the interconnection output of the subsystem $\mathsf{S_i}$. The matrices $A_i, B_i, C_i, G_i, H_i, E_{ij}$ are constant and of appropriate dimensions. We assume that the pair $(A_i, B_i)$ is stabilizable and $(C_i, A_i)$ is detectable.

For stabilization of the system $\mathsf{S}$ of (2.1), we apply local controllers

$$\begin{aligned} \mathsf{LC_i}: \quad \dot{z}_i &= F_i z_i + M_i y_i \\ u_i &= J_i z_i + K_i y_i, \qquad i = 1, 2, \ldots, n, \end{aligned} \tag{2.4}$$

to the individual subsystems $\mathsf{S_i}$. In (2.4), $z_i$ is the state of the local controller $\mathsf{LC_i}$, and $F_i, M_i, J_i, K_i$ are constant matrices of appropriate dimensions. The set of $\mathsf{LC_i}$ $(i = 1, 2, ..., n)$ constitutes a decentralized controller

$$\mathsf{DC}: \quad \{\mathsf{LC_1},\ \mathsf{LC_2},\ \ldots,\ \mathsf{LC_n}\} \tag{2.5}$$

for the overall system S. Then, the overall closed-loop system is written as

$$\mathsf{S^c}: \quad \begin{bmatrix} \dot{x}_i \\ \dot{z}_i \end{bmatrix} = \begin{bmatrix} A_i + B_i K_i C_i & B i J_i \\ M_i C_i & F_i \end{bmatrix} \begin{bmatrix} x_i \\ z_i \end{bmatrix} + \sum_{j=1}^{n} \begin{bmatrix} G_i \\ 0 \end{bmatrix} E_{ij} [H_j \ \ 0] \begin{bmatrix} x_j \\ z_j \end{bmatrix}, \quad i = 1, 2, ..., n. \tag{2.6}$$

In this paper, we consider decentralized stabilization under the existence of perturbations in interconnection matrices $E_{ij}$. We assume that there exist nonnegative numbers $\xi_{ij}$ such that $\|E_{ij}\| \leq \xi_{ij}$ for the original and any purturbed $E_{ij}$, where we note the norm $\|\cdot\|$ is equal to $|\cdot|$ for constant matrices. Then, we introduce:

**Definition 1** *We say that the system* S *of* (2.1) *is connectively stabilizable if there is a decentralized controller* DC *of* (2.5) *such that the overall closed-loop system* $\mathsf{S^c}$ *of* (2.6) *is stable for any interconnection matrices* $E_{ij}$ *satisfying specified upper bounds.*

It is obvious that when the system S is connectively stabilized, the subsystems

$$\mathsf{S_i^c}: \quad \begin{bmatrix} \dot{x}_i \\ \dot{z}_i \end{bmatrix} = \begin{bmatrix} A_i + B_i K_i C_i & B i J_i \\ M_i C_i & F_i \end{bmatrix} \begin{bmatrix} x_i \\ z_i \end{bmatrix}, \quad i = 1, 2, ..., n. \tag{2.7}$$

of the closed-loop system $\mathsf{S^c}$ are stable. Therefore, each local controller $\mathsf{LC_i}$ of (2.4) has to be at least a stabilizing controller for the subsystem $\mathsf{S_i}$ of (2.2). This implies that our task of connective stabilization is to select appropriate local controllers each from the set of stabilizing controllers for the corresponding subsystems. For this reason, we employ the proper stable factorization approach [7] to design the local controllers. The most fundamental result of the approach is the parametrization of all stabilizing controllers for a given system.

# 3 Preliminaries

To apply the factorization approach, we represent the subsystem $\mathsf{S_i}$ of (2.2) by its transfer matrix

$$\mathsf{S_i}: \quad \begin{bmatrix} w_i \\ y_i \end{bmatrix} = \begin{bmatrix} Z_{11}^i & Z_{12}^i \\ Z_{21}^i & Z_{22}^i \end{bmatrix} \begin{bmatrix} v_i \\ u_i \end{bmatrix}, \tag{3.1}$$

where $Z^i_{pq}(p,q=1,2)$ are defined as

$$Z^i_{11} = H_i(sI-A_i)^{-1}G_i \qquad Z^i_{12} = H_i(sI-A_i)^{-1}B_i$$
$$Z^i_{21} = C_i(sI-A_i)^{-1}G_i \qquad Z^i_{22} = C_i(sI-A_i)^{-1}B_i$$

and we use the same notations $u_i, y_i, v_i, w_i$ in the $s$ domain as in the time domain. We note that the transfer matrix from the control input $u_i$ to the measured output $y_i$ is $Z^i_{22}$, which is strictly proper. For stabilization of $\mathsf{S_i}$, we factorize $Z^i_{22}$ as

$$Z^i_{22} = N_iD_i^{-1} = \widetilde{D}_i^{-1}\widetilde{N}_i, \tag{3.2}$$

where $N_i, D_i \in \mathbf{R_{ps}}$ and $\widetilde{D}_i, \widetilde{N}_i \in \mathbf{R_{ps}}$ satisfy

$$\begin{bmatrix} Q_i & P_i \\ -\widetilde{N}_i & \widetilde{D}_i \end{bmatrix} \begin{bmatrix} D_i & -\widetilde{P}_i \\ N_i & \widetilde{Q}_i \end{bmatrix} = \begin{bmatrix} I & 0 \\ 0 & I \end{bmatrix} \tag{3.3}$$

for some $P_i, Q_i, \widetilde{P}_i, \widetilde{Q}_i \in \mathbf{R_{ps}}$. Then, the set of all stabilizing controllers for the subsystem $\mathsf{S_i}$ of (3.1) is given by

$$\mathsf{LC_i} : u_i = K_i(R_i)y_i \tag{3.4}$$

where

$$K_i(R_i) = -(\widetilde{P}_i + D_iR_i)(\widetilde{Q}_i - N_iR_i)^{-1} \tag{3.5}$$

and $R_i \in \mathbf{R_{ps}}$ is an unspecified parameter [7]. The time-domain realization of this controller is $\mathsf{LC_i}$ of (2.4).

The overall system $\mathsf{S}$ is now described in terms of transfer matrices. We define

$$\begin{aligned} w &= [w_1^t \; w_2^t \; \dots \; w_n^t]^t \qquad v = [v_1^t \; v_2^t \; \dots \; v_n^t]^t \\ y &= [y_1^t \; y_2^t \; \dots \; y_n^t]^t \qquad u = [u_1^t \; u_2^t \; \dots \; u_n^t]^t \\ Z_{pq} &= diag[Z^1_{pq},\; Z^2_{pq},\; \dots,\; Z^n_{pq}],\;\; (p,q=1,2) \\ E &= [E_{ij}]_{i,j=1,2,\dots,n}, \end{aligned}$$

and write

$$\begin{bmatrix} w \\ y \end{bmatrix} = \begin{bmatrix} Z_{11} & Z_{12} \\ Z_{21} & Z_{22} \end{bmatrix} \begin{bmatrix} v \\ u \end{bmatrix} \tag{3.6}$$

$$v = Ew. \tag{3.7}$$

We also collect local controllers $\mathsf{LC_i}$ of (3.4) to form a decentralized control law

$$\mathsf{DC} : u = K(R)y \tag{3.8}$$

where

$$\begin{aligned} K(R) &= diag[K_1(R_1),\; K_2(R_2),\; \dots,\; K_n(R_n)] \\ R &= diag[R_1,\; R_2,\; \dots,\; R_n]. \end{aligned}$$

When we apply the local controllers $\mathsf{LC_i}$ of (3.4) to the disconnected subsystems $\mathsf{S_i}$ of (3.1), the closed-loop subsystems are stable for any $R_i \in \mathbf{R_{ps}}$. The transfer matrix $T^i_{wv}(R_i)$ from the interconnection input $v_i$ to the interconnection output $w_i$ is computed as

$$T^i_{wv}(R_i) = T^i_1 - T^i_2 R_i T^i_3 \tag{3.9}$$

which is an affine function of the parameter $R_i$, where

$$\begin{aligned} T^i_1 &= Z^i_{11} - Z^i_{12}\widetilde{P}_i\widetilde{D}_i Z^i_{21} \\ T^i_2 &= Z^i_{12} D_i \\ T^i_3 &= \widetilde{D}_i Z^i_{21}. \end{aligned} \tag{3.10}$$

The matrix $T^i_1$ is equivalent to the transfer matrix from $v_i$ to $w_i$ of the subsystem $\mathsf{S_i}$ stabilized by the local controller $K_i(0) = -\widetilde{P}_i\widetilde{Q}_i^{-1}$, and hence belongs to $\mathbf{R_{ps}}$. $T^i_2$ and $T^i_3$ also belong to $\mathbf{R_{ps}}$ because $T^i_{wv}(R_i)$ belongs to $\mathbf{R_{ps}}$ for any $R_i \in \mathbf{R_{ps}}$.

Now, we consider the overall closed-loop system, which is described by the equations (3.6), (3.7), and (3.8). The system can be viewed as a feedback system composed of stable blocks

$$T_{wv}(R) = diag[T^1_{wv}(R_1),\ T^2_{wv}(R_2),\ \ldots,\ T^n_{wv}(R_n)] \tag{3.11}$$

and $E$. Then, the overall closed-loop system is stable if and only if

$$W = I - ET_{wv}(R) \tag{3.12}$$

is $\mathbf{R_{ps}}$-unimodular [7].

# 4 Connective Stabilization

We saw that our decentralized stabilization problem is to determine a block-diagonal $R \in \mathbf{R_{ps}}$ so that the matrix $W$ of (3.12) is $\mathbf{R_{ps}}$-unimodular for any interconnection matrix $E = [E_{ij}]$ such that $\|E_{ij}\| \le \xi_{ij},\ i,j = 1,2,...,n$, where $\xi_{ij}$ are specified upper bounds of individual $E_{ij}$. For this purpose, we use the idea of decomposition-aggregation method [6], and define an aggregated matrix

$$\overline{W} = I - \overline{E}\,\overline{T}_{wv}(R), \tag{4.1}$$

where

$$\overline{E} = [\|E_{ij}\|] \tag{4.2}$$

$$\overline{T}_{wv}(R) = diag[\|T^1_{wv}(R_1)\|,\ \|T^2_{wv}(R_2)\|,\ \ldots,\ \|T^n_{wv}(R_n)\|] \tag{4.3}$$

are constituted of the norms of submatrices in $E$ and $T_{wv}(R)$. Then, $W$ of (3.12) is $\mathbf{R_{ps}}$-unimodular if the matrix $\overline{W}$ of (4.1) is an M-matrix [3].

For $\overline{W}$ of (4.1) to be an M-matrix, we need to choose $R_i \in \mathbf{R_{ps}}$ so that $\|T^i_{wv}(R_i)\|$ is sufficiently small. If $\overline{W}$ is not an M-matrix even for the infimum of $\|T^i_{wv}(R_i)\|$ with respect to $R_i$, it can never be made so by changing $R_i \in \mathbf{R_{ps}}$. Since decreasing $\|E_{ij}\|$ does not violate the M-matrix property of $\overline{W}$, testing the property is needed only for $\|E_{ij}\| = \xi_{ij}$. From these discussions, we define the matrices

$$\Xi = [\xi_{ij}] \tag{4.4}$$

$$\Pi = diag[\pi_1,\ \pi_2,\ \ldots,\ \pi_n] \tag{4.5}$$

where

$$\pi_i = \inf_{R_i \in \mathbf{R_{ps}}} \|T^i_{wv}(R_i)\|. \tag{4.6}$$

We use

$$\widehat{W} = I - \Xi\Pi \tag{4.7}$$

to state the following:

**Lemma 1** *The system* S *of* (2.1) *is connectivly stabilizable if the matrix* $\widehat{W}$ *of* (4.7) *is an* M*-matrix.*

To present a connective stabilizability condition on the subsystems, we note that

$$\pi_i = 0 \tag{4.8}$$

if the equation

$$T^i_2 X^i T^i_3 = T^i_1 \tag{4.9}$$

has a solution $X^i$ in $\mathbf{R_s}$, where the matrices $T^i_1, T^i_2, T^i_3$ are those in the definition (3.9) of $T^i_{wv}(R_i)$. This fact is obvious by setting $R_i = X^i$ in $T^i_{wv}(R_i)$ when $X^i$ is proper, and can readily be shown using a proper approximation of $X^i$ when $X^i$ is improper [5]. Although $T^i_1, T^i_2$ and $T^i_3$ are not unique, which are defined by the coprime factorization of $Z^i_{22}$, it can be shown [3] that solvability of the equation (4.9) does not depend on the choice of $T^i_1, T^i_2$ and $T^i_3$. Therefore, we use a particular coprime factorization of $Z^i_{22}$ [4] here to test the solvability.

$$\begin{aligned} N_i &= C_i(sI - A^i_K)^{-1}B_i & D_i &= K_i(sI - A^i_K)^{-1}B_i + I \\ \widetilde{N}_i &= C_i(sI - A^i_L)^{-1}B_i & \widetilde{D}_i &= C_i(sI - A^i_L)^{-1}L_i + I \\ P_i &= K_i(sI - A^i_L)^{-1}L_i & Q_i &= -K_i(sI - A^i_L)^{-1}B_i + I \\ \tilde{P}_i &= K_i(sI - A^i_K)^{-1}L_i & \tilde{Q}_i &= -C_i(sI - A^i_K)^{-1}L_i + I \end{aligned} \tag{4.10}$$

where $A^i_K = A_i + B_iK_i$, $A^i_L = A_i + L_iC_i$, and $K_i$, $L_i$ are matrices such that $A^i_K$, $A^i_L$ are stable. This factorization yields the following lemma [3], which implies that we do not need to factorize $Z^i_{22}$ to see whether there holds $\pi_i = 0$.

**Lemma 2** *If the equation*

$$\begin{bmatrix} sI - A_i & B_i \\ H_i & 0 \end{bmatrix} \begin{bmatrix} V_{11}^i & V_{12}^i \\ V_{21}^i & V_{22}^i \end{bmatrix} \begin{bmatrix} sI - A_i & G_i \\ C_i & 0 \end{bmatrix} = \begin{bmatrix} sI - A_i & G_i \\ H_i & 0 \end{bmatrix} \tag{4.11}$$

*has a solution* $(V_{11}^i, V_{12}^i, V_{21}^i, V_{22}^i)$ *in* $\mathbf{R_s}$, *then* $\pi_i = 0$.

Now, we associate a directed graph with the system S of (2.1). We first consider a graph describing the interconnection pattern, in which node $i$ represents the subsystem $\mathsf{S_i}$ and the branch from node $j$ to node $i$ corresponds to a nonzero $E_{ij}$. We refer to this graph as $\Gamma$. If the equation (4.11) for $\mathsf{S_i}$ is solvable in $\mathbf{R_s}$, we then remove all the branches which go into or go out of node $i$. We denote this graph by $\tilde{\Gamma}$, and present the main result of this paper.

**Theorem 1** *If there is no directed loop in the graph* $\tilde{\Gamma}$, *then the system* S *of* (2.1) *is connectively stabilizable.*

Proof: A necessary and sufficient condition for the matrix $\widehat{W}$ of (4.7) to be an M-matrix is that its leading principal minors are all positive [1]. We note that the $k$-th leading principal minor of $\widehat{W} = I - \Xi\Pi$ can be expressed as $1 - *$, where $*$ is composed of products of $\pi_i$ and $\xi_{ij}$ along the directed loops in the subgraph of $\Gamma$ containing nodes $1, 2, \ldots, k$ and branches among them. We see from Lemma 2 and the definition of the graph $\tilde{\Gamma}$ that such products are 0 under the condition of the present theorem. Then, all the leading principal minors of $\widehat{W}$ are 1, and $\widehat{W}$ is an M-matrix. Lemma 1 completes the proof.

**Remark 1** *A way of investigating solvability of the equation* (4.11) *is reduction of the matrix equation to a set of scalar equations. We can do this by transforming*

$$\begin{bmatrix} sI - A_i & B_i \\ H_i & 0 \end{bmatrix} \quad \textit{and} \quad \begin{bmatrix} sI - A_i & G_i \\ C_i & 0 \end{bmatrix}$$

*into the Smith forms using elementary row and column operations over the polynomial ring, which do not violate the* $\mathbf{R_s}$ *property of the solution* $(V_{11}^i, V_{12}^i, V_{21}^i, V_{22}^i)$.

**Remark 2** *Sufficient conditions for* (4.11) *to be solvable, which can be tested readily, are:*

1. $\begin{bmatrix} sI - A_i & G_i \\ C_i & 0 \end{bmatrix}$ *has full column rank for* $s \in \mathbf{C}_+$;

   *Range* $G_i \subset$ *Range* $B_i$

*2.* $\begin{bmatrix} sI - A_i & B_i \\ H_i & 0 \end{bmatrix}$ *has full row rank for* $s \in \mathbf{C}_+$;

$Null\ H_i \supset Null\ C_i$

*3.* $\begin{bmatrix} sI - A_i & G_i \\ C_i & 0 \end{bmatrix}$ *has full column rank for* $s \in \mathbf{C}_+$;

$\begin{bmatrix} sI - A_i & B_i \\ H_i & 0 \end{bmatrix}$ *has full row rank for* $s \in \mathbf{C}_+$.

# 5 Concluding Remarks

We have applied the factorization approach to a decentralized stabilization problem for large-scale interconnected systems. The factorization adopted in this paper was that on the subsystem level, but not on the overall system level. This is reasonable when connective stabilization is considered, where perturbations of the interconnections among subsystems are supposed.

# References

[1] M. Araki and B. Kondo, Stability and transient behavior of composite nonlinear systems, *IEEE Trans. Automatic Control*, Vol.AC-17, pp.537-541, 1972.

[2] M. Ikeda, Decentralized control of large scale systems, *Three Decades of Mathematical System Theory,* H. Nijmeijer and J. M. Schumacher, eds., Lecture Notes in Control and Information Sciences, Springer Verlag, Heidelberg, 1989 (to appear).

[3] M. Ikeda and H. L. Tan, Decentralized control for connective stability: A factorization approach, *Proc. China-Japan Joint Symposium on System Control Theory and Its Applications*, pp.15-20, 1989.

[4] C. N. Nett, C. A. Jacobson and M. J. Balas, A connection between state-space and doubly coprime fractional representations, *IEEE Trans. Automatic Control*, Vol.AC-29, pp.831-832, 1984.

[5] H. L. Tan and M. Ikeda, Decentralized stabilization for expanding construction of large-scale systems, *IEEE Trans. Automatic Control* (submitted).

[6] D. D. Siljak, *Large-Scale Dynamic Systems: Stability and Structure,* North-Holland, New York, 1978.

[7] M. Vidyasagar, *Control System Synthesis: A Factorization Approach,* MIT Press, Cambridge, MA, 1985.

# RICCATI EQUATIONS, ALGEBRAS, AND INVARIANT SYSTEMS

Hyo Chul Myung and Arthur A. Sagle

Riccati or quadratic differential equations are constructed in terms of algebras. The idea is to use the structure of algebras as semisimplicity, radical, automorphisms to help determine the behavior of solutions, decoupling, equilibrium, stability before doing detailed calculations. Examples concerning geodesics, mechanics, predator-prey model, and the general solution are given by algebras. Relationships are given concerning the radical of an algebra and bifurcations, the stability of equilibrium and root space decompositions, the domains of attraction and automorphism groups.

## 1. Riccati equations and algebras.

A nonassociative algebra [7] is a vector space A over a field (usually real numbers R) with a bilinear multiplication $\beta: A\times A\to A$; denote this structure by $(A,\beta)$ or just A when $\beta$ is understood. For example, if A is an associative algebra, let $A^+=(A,\beta)$ be the commutative Jordan algebra with multiplication $\beta(X,Y)=XY+YX$. Thus if A is the nxn matrix algebra, the Jordan algebra $A^+$ has the symmetric matrices as a subalgebra. Similarly one may form the anticommutative Lie algebra $A^-=(A,\beta)$ with $\beta(X,Y)=[X,Y]=XY-YX$. Generalizations of these algebras have appeared in many applications as noted below.

An automorphism of an algebra A is a nonsingular linear transformation $\phi\in GL(A)$, the general linear group of A, such that $\phi\beta(X,Y)=\beta(\phi X,\phi Y)$ for all $X,Y\in A$. The set of all automorphisms, Aut(A), is a closed (Lie) subgroup of GL(A). A derivation D of A is a linear transformation $D:A\to A$ such that $D\beta(X,Y)=\beta(DX,Y)+\beta(X,DY)$ for all $X,Y\in A$. The set of all derivations, DerA, forms a Lie subalgebra of gl(A) which is the Lie algebra of GL(A). Furthermore for $D\in \mathrm{DerA}$ the exponential series $\exp D=I+D+D^2/2!+\dots$ is in AutA; i.e., DerA is the Lie algebra of the Lie group AutA; see [4].

An _ideal_ I of an algebra $(A,\beta)$ is a subspace of A such that $\beta(I,A)\subset I$ and $\beta(A,I)\subset I$. As in associative algebras, the quotient algebra A/I can be formed and the map $A\to A/I:x\to x+I$ is an algebra homomorphism. A is a _simple_ algebra if $\beta(A,A)\neq 0$ and A has no proper ideals; i.e., no proper homomorphisms. An algebra is _semisimple_ if it is the direct sum of ideals which are simple algebras. The _radical_, RadA, of an algebra is the smallest ideal of A such that A/RadA is semi-simple or the zero algebra. The radical is usually related to nilpotent elements in the algebra and RadA=(RadM)A where M is the associative algebra generated by the right and left multiplication functions $R(Z):X\to\beta(X,Z)$ and $L(W):X\to\beta(W,X)$; see [1,7].

_Definition_. Let $A=(R^n,\beta)$ be an algebra over R. A _Riccati_ or _quadratic_ differential equation is of the form

$$\dot{X} = C + TX + \beta(X,X) \equiv E(X)$$

where $C\in A$, $T:A\to A$ is linear, and $\dot{X}=dX/dt$; see [2,9].

_Remark_. Let N be an equilibrium point of the above Riccati vector field E; that is, $E(N)=0$. Then the translation $Y=X-N$ gives a Riccati equation with zero constant term which we henceforth assume.

By Taylor's theorem, Riccati equations occur as the quadratic approximation to the vector field equation $\dot{X}=F(X)\sim F(0)+F^1(0)X+F^2(0)X^2/2!$ where the algebra multiplication on $R^n$ satisfies $\beta(X,X)=F^2(0)X^2/2!$. Also Riccati equations occur in linear systems with quadratic cost, and in the differential geometry of invariant systems [5]. Thus let G be a connected Lie group and let H be a closed (Lie) subgroup with Lie algebras g and h respectively. The homogeneous space G/H is _reductive_ if there is a subspace m of g such that g=m+h (direct sum) and (AdH)(m)⊂m; i.e., [h,m]⊂m. For example, let g and h be semisimple and $m=h^{\perp}$ relative to the Killing form of g. For a reductive space there is a bijective correspondence between the set of G-invariant connections ∇ on G/H and the set of algebras $(m,\alpha)$ with $AdH\subset Aut(m,\alpha)$; i.e., $ad\, h\subset Der(m,\alpha)$. In particular a curve $\sigma(t)$ in G/H is a geodesic if its tangent field $X(t)=\dot{\sigma}(t)$ satisfies the Riccati equation

$$\dot{X} + \alpha(X,X) = 0.$$

Next let G/H be a configuration space for an invariant system with nondegenerate Lagrangian. Then a solution $\sigma(t)$ to the corresponding Euler-Lagrange equation satisfies an extended Euler field equation which reduces to the above geodesic equation when the Lagrangian is given by kinetic energy. More general quadratic equations occur when the Lagrangian is not given by kinetic energy [5].

Thus quadratic equations

$$\begin{aligned} \dot{x}_1 &= \Sigma a_{1j}x_j + \Sigma b^1{}_{ij}x_ix_j \\ &\vdots \\ \dot{x}_n &= \Sigma a_{nj}x_j + \Sigma b^n{}_{ij}x_ix_j \end{aligned}$$

may be written in the form $\dot{X}=TX+\beta(X,X)$ with the quadratic part an algebra multiplication $\beta$. Further examples in this context include the Volterra-Lotka predator-prey model, interaction equations, Lorenz system, Rossler equations, etc.

The Van der Pol and Duffing equations also yield quadratic equations since the following can be shown.

Proposition. Let $x^{(n)}=P(x,x^{(1)}, \ldots x^{(n-1)})$ be a differential equation where $P(z_1, \ldots z_n)$ is a polynomial in the z's. Then there exists a quadratic system $\dot{X}=\beta(X,X)$ whose solution gives the solution to the polynomial differential equation.

2. Structure properties.

The structure properties of an algebra are related to the behavior of solutions; for example, the radical and bifurcations, and identities and periodic points. Using the notation $\beta(X,Y)=XY$ we first consider the case $T=0$. Thus the Riccati equation becomes $\dot{X}=X^2$ and using the product rule, the series solution $X(t)$ with $X(0)=X$ is

$$X(t) = X^{[1]} + tX^{[2]} + t^2X^{[3]}/2! + \cdots$$

where $X^{[1]}=X, X^{[2]}=X^2, X^{[3]}=XX^2+X^2X, \ldots, X^{[k+1]}=\Sigma\binom{k-1}{j}X^{[j+1]}X^{[k-j]}$ is homogeneous of degree k+1 and in the subalgebra $R[X]$ generated by X. Thus the solution is in this subalgebra which is also invariant under the flow $F_t$ of the Riccati equation [2]. In case the solution is periodic with period $\tau$ and the algebra A has a right identity element e, then $X(t+\tau)=X(t)$ implies the identity

$$\exp(t+\tau) - \exp tX = \exp \tau X - e, \text{ for } X = X(0)$$

in the subalgebra $R[X]$ where $\exp sX=\Sigma s^n X^{[n]}/n!$ with $X(0)=e$. Furthermore X can not be nilpotent.

Remarks. (i) If A is power-associative (so that $R[X]$ is associative for each $X\varepsilon A$), then $X^{[n]}/(n-1)!=X^n$, which is the usual power in A.

(ii) If the algebra $A=A_1+\ldots+A_k$ is semisimple, then $A_iA_j=0$ if $i\neq j$. For $X(t)=\Sigma X_i(t)$, the Riccati equation decouples as $\Sigma\dot{X}_i=\dot{X}=(\Sigma X_i)^2=\Sigma X_i^{\,2}$ so that $\dot{X}_i=X_i^{\,2}$ in the simple algebra $A_i$.

(iii) In case $T\neq0$, a homogonization process in [3,9] allows $\dot{X}=TX+\beta(X,X)$ to be imbedded into an equation $\dot{\tilde{X}}=\tilde{\beta}(\tilde{X},\tilde{X})$ given by an algebra $\tilde{A} = (R^{n+1},\tilde{\beta})$. The original solution $X(t)$ is easily obtained from the solution $\tilde{X}(t)$ by setting $x_{n+1}=1$.

Next consider the bifurcation of a 1-parameter family of algebras $A_\lambda$ and the corresponding Riccati equation. As an example, with $\lambda\varepsilon R$ let $A_\lambda=(R^2,\beta_\lambda)$ have basis $\{e_1,e_2\}$ and let $\beta_\lambda$ be given by

$$\beta_\lambda(e_1,e_1) = 0,\ \beta_\lambda(e_1,e_2) = \beta_\lambda(e_2,e_1) = e_2,\ \beta_\lambda(e_2,e_2) = \lambda e_2$$

Then for $\lambda<0$, the algebras $A_\lambda$ are simple but for $\lambda=0$, $A_\lambda$ is nilpotent and equals its radical; that is, the family of algebras $A_\lambda$ changes its structure from semisimple to having a radical at the bifurcation point $\lambda=0$. For the corresponding Riccati equation let $X=x_1e_1+x_2e_2$, then $\dot{x}_1e_1+\dot{x}_2e_2=\dot{X}=\beta_\lambda(X,X)=\lambda x_2^2e_1+2x_1x_2e_2$ so that $\dot{x}_1=\lambda x_2^2$ and $\dot{x}_2=2x_1x_2$. For $\lambda<0$ the solution is a bounded ellipse and for $\lambda=0$ the solution is an unbounded ray; that is, $\lambda=0$ is a bifurcation point for the corresponding system.

Usually radicals consist of nilpotent elements so that $X^{[N+1]}=0$. Thus the solution $X(t)= X^{[1]}+..+t^{N-1}X^{[N]}/(N-1)!$ is an unbounded polynomial in $R[X]$. Whereas with semisimplicity, there is usually associated a nondegenerate bilinear form $C(X,Y)$ which is often positive definite and conserved by the solution, and consequently gives a bounded solution; the following is an application.

<u>Proposition</u>. Let $G/H$ be a reductive space with decomposition $g=m+h$ and let $\{(m,\alpha_\lambda,C_\lambda):\lambda\varepsilon R\}$ be a family of algebras which give metric connections on $G/H$. Let the kinetic energy given by $K_\lambda(X)=\frac{1}{2}C_\lambda(X,Y)$ be conserved on the trajectory of the Riccati (geodesic or Euler) equation $\dot{X}+\alpha_\lambda(X,X)=0$. If $\lambda=0$ is a bifurcation point for the family of algebras where $(m,\alpha_0,C_0)$ has nilpotent radical $R$ of index $\geq 3$, then $\lambda=0$ is a bifurcation point of the Riccati system with $X(0)\varepsilon R$.

## 3. <u>Critical elements and automorphisms</u>.

Let $\dot{X}=TX+X^2\equiv E(X)$ and let $P\varepsilon A$, then the <u>linearization of E</u> at P is it's derivative, $E^1(P)$, at P which acts on $Y\varepsilon A$:

$$E^1(P)Y = \lim_{h\to 0}\big(E(P + hY) - E(P)\big)/h = [T + L(P) + R(P)]Y.$$

Thus $E^1(P)=T+L(P)+R(P)$ and $L(P)+R(P)=L^+(P)$, the left

multiplication in the commutative algebra $A^+$: $(L(P)+R(P))X = PX+XP = L^+(P)X$ in $A^+$. It appears possible to work with a commutative algebra, however for the Volterra-Lotka model

$$E^1(N) = L(N)$$

the left multiplication in A. Next decompose the algebra into its stable, center and unstable components $A=A_s(N)+A_0(N)+A_u(N)$ where $A_s(N), A_0(N), A_u(N)$ correspond to the real parts of the eigenvalues of $E^1(N)$ being $<0$, $=0$, $>0$ respectively. For the Volterra-Lotka model, $E^1(N)=L(N)$ puts us into the familiar situation of decomposing a nonassociative algebra (or its complexification) into its root spaces $A=\Sigma A(N,\lambda)$ relative to $L(N)$.

For the Riccati equation $\dot{X}=E(X)$, let $\mathrm{Aut}E=\{\phi\varepsilon GL(A): E\phi=\phi E\}$. This is the solution preserving linear group which is a Lie group. A straight forward calculation shows

$$\mathrm{Aut}\ E = \{\phi\varepsilon \mathrm{Aut}A^+ : T\phi=\phi T\}$$

with Lie algebra $\mathrm{Der}E=\{D\varepsilon \mathrm{Der}A^+ : [D,T]=0\}$. If $A^+$ is a semisimple algebra with a right or left identity, then AutE is determined by $A^+$ as follows. Let $\phi=\exp D\varepsilon \mathrm{Aut}E$ where $D\varepsilon \mathrm{Der}A^+$, then from [4,7] D is contained in the Lie algebra $M^-$ where M is the associative algebra generated by left and right multiplication functions; i.e., the identity component of AutE is determined by $A^+$. If AutE is connected, then the identity components generates AutE and therefore is determined by $A^+$. The above derivation D is called "inner" and often has explicit formulas.

Let $\Gamma$ denote the set of <u>critical</u> <u>elements</u> for the quadratic equation $\dot{X}=E(X)$, i.e. $\Gamma$ is the set of equilibrium points and periodic solutions. Automorphisms help locate $\Gamma$ and describe the symmetry observed in domains of attraction.

<u>Theorem</u>. Let $(\mathrm{Aut}E)_0$ denote the connected component of AutE and let $\Gamma$ be as above. Then

(i) $(\mathrm{Aut}E)\cdot\Gamma=\Gamma$.

(ii) If $\Gamma$ is finite, then $(\mathrm{DerE})\cdot\Gamma=0$; that is, $\Gamma$ is in the set of fixed points of $(\mathrm{AutE})_0$.

(iii) Let $\mathrm{Att}(\gamma)$ be the domain of attraction of $\gamma\varepsilon\Gamma$. If $\Gamma$ is finite, then

a) $(\mathrm{AutE})_0\cdot\mathrm{Att}(\gamma)=\mathrm{Att}(\gamma)$.

b) There exist $X_a\varepsilon\mathrm{Att}(\gamma)$ such that $\bigcup_a(\mathrm{AutE})_0\cdot X_a=\mathrm{Att}(\gamma)$.

Remarks. (i) Proofs use $\exp sD \in \mathrm{Aut}\,E$ for all $D\varepsilon\mathrm{DerE}$ and $s\varepsilon R$, and $\phi F_t(X)=F_t(\phi X)$ for $\phi\varepsilon\mathrm{AutE}$ where $F_t(X)$ is the flow of E with $F_0(X)=X$.

(ii) For example, let G/H be a reductive space with G and H semi-simple, and g-m+h where $m=h^{\perp}$ as before. Let $\dot{X}=E(X)$ be a quadratic equation with $\mathrm{ad}\,h \subset \mathrm{Der}\,E$ and let $K_0=\{X\varepsilon m:(\mathrm{adU})X=0 \text{ all } U\varepsilon h\}$. If $\Gamma$ is finite then $\Gamma\subset K_0$; see [6] for specific matrix examples involving $K_0$. It has been conjectured [8] that the set of periodic solutions for a quadratic equation is finite.

The second author thanks the Mathematics Department at the University of Utah for their kind hospitality while doing this work.

References

[1] A. A. Albert, The radical of a nonassociative algebra, Bull. A.M.S. 48 (1942), 891-897.

[2] L. Markus, Quadratic differential equations and nonassociative algebras, Ann. Math. Studies 45 (1960), Princeton Univ. Press, 185-213.

[3] M. Peschel and W. Mende, The Predator-Prey Model, Springer-Verlag (1986).

[4] A. Sagle and R. Walde, Introduction to Lie Groups and Lie Algebras, Academic Press, 1973.

[5] A. Sagle, Invariant Lagrangian mechanics, connections and nonassociative algebras, Algebras, Groups and Geo. 3 (1986), 199-263.

[6] A. Sagle, Jordan algebras and connections on homogeneous spaces, Trans. A.M.S. 187 (1974).

[7] R. Schafer, Introduction to Nonassociative Algebras, Academic Press, 1966.

[8] J. Sotomayor and R. Paterlini, Quadratic vector fields with finitely many periodic orbits, Geometric Dynamics, Springer Lecture Notes #1007 (1983).

[9] S. Walcher, Algebras and Differential Equations, Hadronic Press, 1989.

# Maximal order reduction of proper transfer function matrices.

**Thomas John Owens**

**Abstract**

A parameterization of the class of linear state-feedback controllers that assign a set of desired self-conjugate eigenvalues to the closed-loop system is applied to give a method for maximal order reduction of proper transfer function matrices. By making the maximum number of closed-loop modes unobservable, while retaining arbitrary assignment of the remaining modes, a lower-order transfer function matrix is obtained. The method establishes that results concerning the existence, number, and cancellation of zeros of proper transfer-function matrices may be applied in the response insensitivity problem. The main result is a class of maximal order reducing fixed-gain state-feedback controllers explicitly specified by a set of free parameters which may be chosen to satisfy additional design requirements.

## 1. Introduction

In Owens [11, 12] a method of maximal order reduction for square multivariable strictly proper invertible transfer function matrices was presented. Maximal order reduction is carried out by applying state feedback to the original system such that the number of observable modes is reduced and a lower minimal realization can be found. This maximal order reduction problem is of interest, for example, because of its application to first-order multivariable design.

In this paper the above mentioned method of maximal order reduction is extended to multivariable systems with proper transfer function matrices. This problem has previously been considered by Antsaklis [3]. The extension is of interest, for example, because of its application to the problem of exact model matching for systems with proper transfer function matrices. Furthermore, it will enable results concerning the existence, number and cancellation, of zeros of such systems to be applied in response insensitivity problem (Owens and O'Reilly [14]. Owens [13]).

For ease of exposition we begin by considering regular proper transfer function matrices. A constant polynomial matrix is called regular if it is square and has a nonzero determinant. The related problem of finding input vectors which generate zero output vectors for commonly used systems of the form

$$\dot{x} = Ax + Bu, \; y = Cx + Du \qquad (1.1)$$

was considered by Al-Nasr [1]. This problem, for general state-space models of the form

(1.1) has been addressed by Amin and Hassan [2].

The eigenvector shifting problem plays a central role in the suggested method. The problem is tackled in this paper using a parameterization of the class of linear state feedback controllers which assign a set of desired self-conjugate eigenvalues to the closed-loop system (Fahmy and O'Reilly [6, 7], Roppenecker [15, 16]). For ease of exposition we make four simplifying though inessential assumptions, that the algebraic multiplicity of the left-half plane zeros is equal to their geometric multiplicity, that the observable closed-loop eigenvalues are assigned distinct values not equal to the corresponding open-loop eigenvalues or to the left-half plane zeros.

The problem is stated in section 2 and analysed in section 3. It is established that there exists a parametric class of maximal order reducing controllers. An algorithm for identifying the class is given in section 4. In section 5, the extension of the results obtained in preceeding sections to nonregular transfer function matrices is discussed.

## 2. Statement of the problem

Let the system with 'proper' regular transfer function matrix R(s) be described by the following irreducible state space system

$$\dot{x} = Ax + Bu \tag{2.1a}$$

$$y = Cx + Du \tag{2.1b}$$

where $x \in \mathbb{R}^n$, $y, u \in \mathbb{R}^p$, and A, B, C, D are constant matrices of appropriate dimensions. (ie R ($\infty$) = D is bounded).

The system transfer function matrix is given by

$$R(s) = C(sI - A)^{-1}B + D \tag{2.2}$$

Denoting by $\tilde{T} \simeq (\tilde{t}_1^T \quad ,\ldots.,\tilde{t}_n^T)$ and $\tilde{V} \simeq (\tilde{v}_1, ,\ldots.,\tilde{v}_n)$, the eigenrow and eigenvector frames, and by $\Lambda$ the spectral matrix of A, then

$$A = \tilde{V}\Lambda\tilde{T} \tag{2.3}$$

In the case where A has simple spectral structure, $\Lambda$ = diag($\lambda_i$) the transfer function matrix can be expressed in the dyadic form

$$R(s) = \sum_{i=1}^{n} \frac{C\tilde{v}_i\tilde{t}_i^T B}{(s - \lambda_i)} + D \tag{2.4}$$

If a state feedback control law of the form

$$u = Kx + r, \ r \in \mathbb{R}^p, K \in \mathbb{R}^{p \times n} \tag{2.5}$$

is applied to (2.1a, b), the nominal closed-loop system obtained is of the form

$$x = A_c x + Br \qquad A_c \simeq A + BK \quad (2.6a)$$
$$y = (C + DK)x + Dr \quad (2.6b)$$

The complete class of state feedback controllers for the system model (2a.1a, b) that assign a prescribed spectrum of distinct self-conjugate eigenvalues $(\lambda_1, \ldots., \lambda_n)$ to the closed-loop system is parameterized (Fahmy and O'Reilly [6, 7]) by the real feedback gain matrix

$$K(F) = FV^{-1}(F) \quad (2.7)$$

where

$$F = [f_1, \ldots., f_n] \quad (2.8)$$

is a matrix of free parameter vectors $f_i$, i = 1,....,n. To guarantee that K is real, we choose $f_i \in \mathbb{R}^p$ for a real eigenvalue, whereas for a complex-conjugate pair of eigenvalues $\lambda_i, \lambda_j = \lambda_i^*$, $f_i, f_j = f_i^* \in \mathbb{C}^p$ and $V = [v_1, \ldots., v_n]$ (2.9) is a matrix of correspondingly linearly independent eigenvectors $v_i = (\lambda_i I - A)^{-1} Bf_i$, i = 1,....,n (2.10)

Under a control law of the form (2.5) the closed-loop transfer-function matrix can be expressed in the dyadic form

$$R_c(s) = \sum_{i=1}^{n} \frac{(C + DK)v_i t_i^T B + D}{(s - \lambda_i)} \quad (2.11)$$

substituting (2.7) and (2.10) into (2.11) gives (2.12)

$$\sum_{i=1}^{n} \frac{(Cv_i + Df_i)t_i^T B + D}{(s - \lambda_i)} \quad (2.12)$$

Thus, if there exists j, such that

$$Cv_j + Df_j = 0 \qquad (2.13)$$

the mode $\lambda_j$ disappears from the transfer-function matrix and consequently does not affect the input-output relations. This implies that a lower-order model which excludes this unobservable mode can be used for a new minimal realization of the system. Obviously, such a procedure is practically acceptable as long as the unobservable mode is stable

$$(Re(\lambda_j) < 0).$$

Since one has no control on C or D it is reasonable to alter the system in such a way that the maximum number of stable modes will become unobservable. It is well known that the observability properties of a system are not invariant under state feedback. Thus it is of interest to examine the possibility of achieving the above requirements by means of state feedback. The problem is stated as follows.

Given the linear time-invariant system $S_o(A, B, C, D)$ of equations (2.1a,b) find a stable feedback matrix K such that the closed-loop system $S_c(A + BK, B, C + DK, D)$ has the maximum possible number of stable unobservable modes.

## 3. Analysis

Substituting (2.10) into (2.13) a necessary

and sufficient condition for the closed-loop characteristic frequency $z_i$ and its corresponding eigenvector $v_i^u$ to be unobservable is

$$[C(z_iI - A)^{-1}B + D]f_i^u = 0 \qquad (3.1)$$

A solution to (3.1) can only exist if $C(sI - A)^{-1}B + D$ looses rank at $z_i$. That is, if

$z_i$ is an invariant zero (MacFarlane and Karcanias [8]) of the system $S_o$ (A, B, C, D). $f_i^u$ is then referred to as an <u>input zero direction</u> and $v_i^u = (z_i I - A)^{-1} Bf_i$ as the <u>state zero direction</u>. It is clear from (2.10), (3.1) that the invariant zeros, state zero directions and input zero directions are invariant under state feedback.

In a previous paper (Mita [9]) the invariant zeros have been defined as the poles of the maximal unobservable subspace (MUS), which are invariant under state feedback.

Amin and Hassan [2] have developed algorithms for the determination of the invariant zeros and zero directions of general state space models of the form (1.1).

<u>Definition 3.1</u> (MacFarlane and Karcanias [8] The geometric multiplicity of an invariant zero is defined as the rank deficiency of $R(z_i) = C(z_i I - A)^{-1} B + D$.

<u>Lemma 3.1</u> (van Dooren [18]) Let $S_o$(A, B, C, D) be an irreducible state-space system of a regular transfer function matrix R(s) and

assume for the moment that $D = R(\infty)$ is regular.
The zeros of R(s) are then finite and n in number, multiplicity counted, and are the eigenvalues of $\hat{A} \simeq A - BD^{-1}C$. One can associate to $\hat{A}$ so-called invariant zero directions which are the eigenvectors, or in the defective case the principal vectors of $\hat{A}$.
When D is singular, but the system matrix

$$P(s) \simeq \left[\begin{array}{c|c} sI_n - A & B \\ \hline -C & D \end{array}\right] \qquad (3.2)$$

is still irreducible and invertible then R(s) has some infinite zeros. The generalized eigenvectors corresponding to these infinite zeros could be defined as 'infinite zero directions'.

Remark 3.1 The finite zeros of R(s) in Lemma 3.1 are the invariant zeros of R(s). This means that each left-half plane invariant zero has corresponding input zero direction and state zero direction unique up to a scalar multiple.

Remark 3.2 If $p_1$ is the rank deficiency of D, then R(s) has h infinite zeros where $h = \sum_{i=1}^{p_1} h_i$, $i = 1,\ldots.,p_1$ the order (multiplicity) of the infinite zeros. For a regular transfer function matrix the number of invariant zeros is given by n - h (Amin and Hassan [2]).

Remark 3.3 The state-space model of (2.1a, b) is assumed to be minimal. When a system is minimal its invariant zeros are the transmission zeros of the system (Emami-Naeini and van Dooren [5]). Therefore, in the following we will refer to the zeros of the system.

By Lemma 3.1 and the parameterization of (2.7-2.10) it is possible to state the following.

Theorem 3.1 Given the linear system $S_o(A, B, C, D)$ with left-half plane zeros $z_i$, i = 1,....,q, there always exists a class of (r x n) state feedback matrices K that assign the pairs ($z_i$ ,$v_i^u$ ), i = 1,......,q as the closed-loop eigenvalues and their corresponding eigenvectors, such that the n - q observable eigenvalues of $S_o$ may be arbitrarily assigned.

## 4. Algorithm for maximal order reduction

Given $S_o(A, B, C, D)$ of equations (2.1a, b) the following steps are pursued:

Step 1 The zeros $z_1,....,z_w$ of $S_o$ are computed. It is assumed, without loss of generality, that

$Re(z_i) < 0$, i = 1,....,q, $Re(z_i) \geqslant 0$, i = q+1,....,w.

Step 2 The vectors $f_1,....,f_q$, that satisfy $[C(z_iI - A)^{-1}B + D]f_i = 0$, i = 1,....,q, are computed. Having determined $f_1,....,f_q$, the corresponding linearly independent vectors $v_1$, ,....,$v_q$ are computed as $v_i = (z_iI - A)^{-1}Bf_i$, i = 1,....,q.

Step 3 The remaining observable modes have eigenvalues which may be arbitrarily assigned to $\lambda_i$ by taking $v_i = (\lambda_i I - A)^{-1} Bf_i$ $i = q+1,\ldots,n$, subject to the restriction that $f_{q+1},\ldots,f_n$, be selected such that $v_1,\ldots,v_n$, are linearly independent.

Step 4 Find K using equation (2.7).

Remark 4.1 The minimum number of free parameters (degrees of freedom after eigenvalue assignment) in the pxn parameter matrix F for distinct eigenvalues $\lambda_i$ $(i = 1,\ldots,n)$ is (O'Reilly and Fahmy [10]) $nx(p-1)$. Thus, step 2 utilises all the design freedom available in $f_i$, $i = 1,\ldots,q$.

Remark 4.2 No proper transfer function matrix may be reduced to first-order type. This is not so for strictly proper transfer function matrices (Owens [11]).

Remark 4.3 If no invariant zero has $\mathrm{Re}(z_i) = 0$, $\mathcal{V} = \{v_1,\ldots,v_q\}$ is what in Antsaklis [3] has been defined as the supremal output-nulling (A, B)-invariant subspace.

Remark 4.4 The major limitation of this approach to order reduction is the requirement that left half-plane zeros exist. The extent of the order reduction is dependent on the number of such zeros.

Example

Consider the system (example 2 of Amin and

Hassan [2]) given by

$$A = \begin{bmatrix} -1 & 0 & 1 \\ 0 & 0 & 0 \\ 1 & 0 & -1 \end{bmatrix}, \quad B = \begin{bmatrix} 0 & 1 \\ 1 & 0 \\ 0 & 0 \end{bmatrix} \qquad (4.1a)$$

$$C = \begin{bmatrix} 0 & 1 & 0 \\ 1 & 0 & 0 \end{bmatrix} \quad D = \begin{bmatrix} 0 & 0 \\ 1 & 0 \end{bmatrix} \qquad (4.1b)$$

Step 1 The system (4.1a, b) has one invariant zero at $z = -1$. Hence, $q = 1$. (van Dooren [18] also gives a numerically stable algorithm for computing the zero of $S_o(A, B, C, D)$).

Step 2

$$R(-1) = \begin{bmatrix} -1 & 0 \\ 1 & 0 \end{bmatrix}$$

Therefore, $R(-1)f_1 = \begin{bmatrix} 0 \\ b \end{bmatrix}$

$b \in \mathbb{R}$. Take $f_1 = \begin{bmatrix} 0 \\ -1 \end{bmatrix}$

without loss of generality since the actual number of degrees of freedom in $f_i$ is (p-1) (O'Reilly and Fahmy [10]). For this choice of $f_1$, $v_1 = \begin{bmatrix} 0 \\ 0 \\ 1 \end{bmatrix}$. $f_1$ and $v_1$

agree with the zero directions computed by Amin and Hassan [2].

Step 3 Suppose, arbitrarily, that it is desired to shift the other two modes to -0.5, -3

$$(-0.5I - A)^{-1} B = \begin{bmatrix} 0 & -2/3 \\ -2 & 0 \\ 0 & -4/3 \end{bmatrix}$$

$$(-3I - A)^{-1} B = \begin{bmatrix} 0 & -2/3 \\ -1/3 & 0 \\ 0 & 1/3 \end{bmatrix}$$

If we denote $f_2 = \begin{bmatrix} a \\ b \end{bmatrix}$, $f_3 = \begin{bmatrix} c \\ d \end{bmatrix}$, $a, b, c, d \in \mathbb{R}$.
Then we have that

<u>Step 4</u> The <u>class</u> of feedback matrices K which solve the maximal order reduction problem is parameterized by

$$K = \begin{bmatrix} 0 & a & c \\ -1 & b & d \end{bmatrix} \begin{bmatrix} 0 & -2/3b & -2/3d \\ 0 & -2a & -1/3c \\ 1 & -4/3b & 1/3d \end{bmatrix}^{-1}$$

Subject to the inverse in (4.2) existing.

## <u>5. Model reduction of nonregular proper transfer function matrices.</u>

Consider now systems represented by state space models of the form (2.1a, b) where $x \in \mathbb{R}^n$, $y \in \mathbb{R}^p$, and $u \in \mathbb{R}^m$.

For <u>all</u> invertible systems Amin and Hassan [2] have shown that the zeros and zero directions can be determined through the calculation of the eigenvalues and eigenvectors of a matrix having the dimension of the MUS. A system is invertible if Rank R(s) = r = min(m, p). This means that under the four simplifying through inessential assumptions stated in the introduction the design procedure of section 4 may be applied directly with the following

qualifications: (i)
Davidson and Wang [4] established that if $p \neq m$ then for almost all systems (2.1a, b) there exist no zeros for the system; If $p > m$ this can be seen, intuitively, to be so from (3.1).
(ii) If $p < m$ the zero directions corresponding to an invariant zero will be nonunique. The nonuniqueness for a given $z_i$ being parameterized by the design freedom remaining in $f_i$ after it has been restricted such that equation (3.1) is satisfied.
(iii) When $p < m$, order reduction may be carried out without resort to pole/zero cancellation (Antsaklis [3]). In fact, arbitrarily assignable eigenvalues may be made unobservable. For a particular choice of unobservable eigenvalues attainment of the maximal order reduction may require the assignment of specific zero directions to the zeros cancelled by poles.
For the case $p < m$, the question of how many of the n system modes can be made unobservable by state feedback is addressed.
Theorem 5.1 An upper bound on the number of modes that can be made unobservable is $n - \bar{q}$, where

$$\bar{q} = \min_{\lambda_i \in \mathbb{C}^-} \{\text{rank } [C(\lambda_i I - A)^{-1}B - X]\}$$

and X corresponds to those rows of $C(\lambda_i I - A)^{-1}B$ for which the corresponding row of D has one or more nonzero elements.
proof Analogous to that of theorem 4.2 of

Owens and O'Reilly [14].

For the case $p < m$, the algorithm for maximal order reduction becomes steps (a), (b), (c), and (e) of the design procedure of Owens and O'Reilly [14], substituting C for DA, and D for DB.

Remark 5.1 The relationship of the maximal order reduction problem to the insensitivity problem of Owens and O'Reilly [14] highlights the nature of ill-conditioning effect of making closed-loop modes unobserverable (Owens [13]).

Remark 5.2 Antsaklis [3] Lemma 10 establishes that if maximal order reduction is achieved all stable zeros of the system are cancelled by closed-loop poles. This result is of significance in the response insensitivity problem.

Finally it is noted that, a system is noninvertible if R(s) looses rank independently of s; it follows from (3.1) that the system has an infinite number of zeros. Such a system is defined as a degenerate system. This is a case of extreme control difficulty (Rosenbrock [17]). In view of our objectives in studying the problem considered, consequently, is not discussed further here.

## 6. Conclusions

A simple method for identifying the class of state-feedback controllers for exact cancellation of modes from a proper transfer-function matrix of a system by making them unobservable with simultaneous allocaion of the rest has been presented. It has been assumed that all the parameters in A, B, C, and D are given and exact. A slight ignorance of the model or change in the system may lead to the reappearance of the cancelled dipoles in the expression for the closed-loop transfer-function matrix of the original system. However, as the maximal order reduction does not exploit all the available degrees of freedom exploitation of the remaining degrees of freedom may lead to a reduced order model which is less sensitive to variations in A, B, C, and D (Owens and O'Reilly [14]). The method establishes that results concerning the existence, number, and cancellation of zeros of proper systems may be applied in the response insensitivity problem (Owens and O'Reilly [14]), Owens [13]). The relationship to the response insensitivity problem highlights the nature of the ill-conditioning that results from order reduction.

The method is constructive and does not require subspace terminology.

**Acknowledgements**

The support and encouragement of Professor G.Musgrave is gratefully acknowledged.

**References**

[1] Al-Nasr, N., 1982, On zeros of time-invariant multivariable linear systems containing input derivatives, International Journal of Control 35: 749-753.

[2] Amin, M.H., and Hassan, M.M., 1988, Determination of invariant zeros, International Journal of Control 47: 1011-1041.

[3] Antsaklis, P., 1980, Maximal order reduction and supremal (A, B)-invariant and controllability subspaces, IEEE Transactions on Automatic Control 25: 44-49.

[4] Davidson, E.J., and Wang, S.H., 1974, An algorithm for the calculation of transmission zeros of linear multivariable systems, Automatica, 10: 643-658.

[5] Emami-Naeini, A., and van Dooren, P., 1982, Computation of zeros of linear multivariable systems, Automatica, 18: 415-430.

[6] Fahmy, M.M., and O'Reilly, J., 1982, On eigenstructure assignment in linear multivariable systems, IEEE Transactions on Automatic Control 27: 690-693.

[7] Fahmy, M.M., and O'Reilly, J., 1983, Eigenstructure assignment in linear multivariable systems- A parametric solution, IEEE Transactions on Automatic Control 28: 990-994.

[8] MacFarlane, A.G.J., and Karcanias, N., 1976, Poles and zeros of linear multivariable systems: a survey of the algebraic, geometric and complex variable theory, International Journal of Control 24: 33-74.

[9] Mita, T., 1977, On maximal unobservable subspace, zeros and their applications, International Journal of Control 25: 885-899.

[10] O'Reilly, J., and Fahmy, M.M., 1985, The minimum number of degrees of freedom in state feedback control, International Journal of Control 41: 749-768.

[11] Owens, T.J., 1988, Parametric state feedback control with model reduction, International Journal of Control 47: 1299-1306.

[12] Owens, T.J., 1988, Parametric state feedback control with exact model matching, in CONTROL 88, IEE Conference Publication No. 285.

[13] Owens, T.J., 1988, Parametric output feedback control with response insensitivity, International Journal of Control 48: 1213-1239.

[14] Owens, T.J., and O'Reilly, J., 1987, Parametric state feedback control with response insensitivity, International Journal of Control 45: 791-809.

[15] Roppenecker, G., 1983, Minimum norm output feedback design under specified eigenvalue areas, Systems and Control Letters

3: 101-103.

[16] Roppenecker, G., 1986, On parametric state feedback design, International Journal of Control 43: 793-804.

[17] Rosenbrock, H.H., 1974, Computer-aided-design of control systems, Academic Press.

[18] van Dooren, P.M., 1981, The generalized eigenstructure problem in linear systems theory, IEEE Transactions on Automatic Control 26: 111-129.

# THE MATCHING CONDITION AND FEEDBACK CONTROLLABILITY OF UNCERTAIN LINEAR SYSTEMS

Ian R. Petersen

**Keywords:** Robust control, uncertain systems, controllability, feedback controllability

**Abstract**
This paper considers a problem of controllability for a class of linear uncertain systems. The uncertain systems under consideration contain norm bounded time-varying uncertainty. The paper introduces a new notion of controllability referred to as feedback controllability. Roughly speaking, an uncertain system is feedback controllable if there exists a time varying linear state feedback control such that with any initial condition, the closed loop system state converges to zero in a finite time. The main result of the paper shows that if the uncertain system satisfies a certain matching condition then the system will be feedback controllable. This matching condition is also known to be a sufficient condition for the stabilizability of the uncertain system.

## I. INTRODUCTION

In order to develop a theory of robust control system design, it is natural to consider linear dynamical systems containing time-varying unknown-but-bounded uncertain parameters. This leads to the notion of uncertain linear systems; e.g., see [1]. Given that the notion of controllability plays an important role in the theory of linear time-invariant systems, one might expect that some notion of controllability will play an important role in the theory of uncertain linear systems. This paper introduces a new notion of controllability for uncertain linear systems. This notion of controllability is referred to as Feedback Controllability. A system is feedback controllable if there exists a linear time-varying state feedback control law such that the following condition holds: Given any admissible uncertainty and any initial state, the state of the closed loop system converges to zero in a finite time.

For linear time-invariant systems, the usual definition of controllability is in terms of open loop control. However, one could equivalently define controllability in terms of closed loop control. At this point, it should be noted that when one introduces uncertainty into the system, the equivalence between open loop and closed loop control no longer applies. Indeed, one would in general expect closed loop control to be better able to cope with uncertainty than open loop control. Thus, it is somewhat surprising that most of the existing papers on the controllability of uncertain systems have dealt with open loop control; e.g., see [2]-[5]. However, reference [6] deals with a notion of modal controllability for uncertain systems which is related to our notion of closed loop controllability.

The main result of this paper shows that if the uncertain systems under consideration satisfy a certain matching condition (e.g., see [1]) then they will be feedback controllable. The method used in proving feedback controllability involves the use of a time-varying quadratic 'Lyapunov' function. This Lyapunov function is constructed by solving a Riccati differential equation. This Riccati equation is of the type which arises in linear optimal control. In fact, our approach to feedback controllability of uncertain linear systems is closely related to the Riccati equation approach to the stabilization of uncertain linear systems; e.g. see [7]-[10].

## II. FEEDBACK CONTROLLABILITY

The linear time invariant system

$$\dot{x}(t) = Ax(t) + Bu(t) \tag{2.1}$$

is said to be controllable if given any $T > 0$ and any initial condition $x(0) = x_0$, there exists a control function $u(t)$ such that $x(T) = 0$. A standard result in linear systems theory relates the controllability of the system (2.1) to the <u>controllability gramian</u>

$$W(t,T) \overset{\Delta}{=} \int_t^T e^{A(t-\tau)} BB' e^{A'(t-\tau)} d\tau.$$

Indeed, the system (2.1) will be controllable if and only if W(t,T) is positive-definite for all t < T. Furthermore, a suitable control function is given by

$$u(t) = -B' e^{-A't} W(0,T)^{-1} x_0; \tag{2.2}$$

e.g. see [11]. The control law given in (2.2) is an open loop control. However, it is straightforward to verify that this control law is equivalent to the following

$$u(t) = -B' W(t,T)^{-1} x(t); \quad t \in (0,T). \tag{2.3}$$

This control law is of the form u(t) = K(t)x(t) and will ensure $x(t) \to 0$ as $t \to T$. However when t = T, W(t,T) = 0 and hence this control law is not defined for t = T.

The above discussion of feedback controllability for linear time invariant systems provides the motivation for our definition of feedback controllability for uncertain linear systems. We consider uncertain systems of the form

$$\dot{x}(t) = Ax(t) + Bu(t) + DF(t)\left[E_1 x(t) + E_2 u(t)\right];$$
$$\|F(t)\| \le 1 \tag{$\Sigma$}$$

where $x(t) \in \Re^n$ is the <u>state</u>, $u(t) \in \Re^m$ is the <u>control input</u> and $F(t) \in \Re^{p \times q}$ is a norm bounded <u>matrix of uncertain parameters</u>. That is, F(t) is a matrix of measurable functions such that $||F(t)|| \le 1$ for all t. ($\|\cdot\|$ denotes the induced matrix norm) It is assumed that the uncertain system ($\Sigma$) satisfies the following assumptions:

A1. $E_2'E_2 = I$;

A2. $E_2'E_1 = 0$.

<u>Notation</u>: For any matrix R, the notation $\mathcal{N}(R)$ denotes the null space $\mathcal{N}(R) \overset{\Delta}{=} \{x: Rx = 0\}$. For any symmetric matrix M, $\lambda_{max}(M)$ denotes the maximum eigenvalue of the matrix M.

<u>Definition 2.1</u>: The system ($\Sigma$) is said to be <u>feedback controllable</u> if given any $T > 0$ there exists a continuous time-varying feedback gain matrix $K(t)$ defined on $(0,T)$ such that the following condition holds: If we apply the state feedback control $u(t) = K(t)x(t)$, then given any initial condition $x(0) = x_0$ and any admissible uncertainty $F(t)$, the solution to ($\Sigma$) will satisfy $x(t) \to 0$ as $t \to T$.

<u>Definition 2.2</u>: The <u>controllability gramian</u> associated with the system ($\Sigma$) is defined by

$$W_c(t,T) \stackrel{\Delta}{=} \int_t^T e^{A(t-\tau)}\left[BB' - DD'\right]e^{A'(t-\tau)}\, d\tau.$$

By analogy with equation (2.3), one might expect that a suitable feedback control law is given by

$u(t) = -B'W_c(t,T)^{-1}x(t)$. However, a slight modification to this control law is required. Indeed as in Theorem 13.2 of [11], it is straight forward to verify that $W_c(t,T)$ satisfies the differential equation

$$\frac{d}{dt}W_c(t,T) = AW_c(t,T) + W_c(t,T)A' - BB' + DD';\ W_c(T,T) = 0. \tag{2.4}$$

Our required feedback gain matrix will be obtained by solving the Riccati differential equation

$$\dot{S}(t) = AS(t) + S(t)A' - BB' + DD' + S(t)E_1'E_1S(t);\ S(T) = 0. \tag{2.5}$$

<u>Lemma 2.1</u>: <u>Suppose Riccati equation</u> (2.5) <u>has a solution on the interval</u> $(M, T]$. <u>Then</u>

$$S(t) \le W_c(t,T) \quad \underline{\text{for all}}\ t \in (M,T).$$

<u>Proof</u>: It follows from the optimal control interpretation of (2.5) (see Theorem 21-1 of [11]) that given any $t_0 \in (M, T)$

$$x'S(t_0)x = \min_{u(\cdot)} \int_{t_0}^T \left\{u(\tau)'u(\tau) + x(\tau)'(BB' - DD')x(\tau)\right\}d\tau:$$

$$\dot{x}(t) = -A'x(t) + E_1'u(t); \quad x(t_0) = x.$$

However, if we let $u(t) = 0$ then the resulting value of the cost functional is

$$J(u) = \int_{t_0}^{T} x' e^{A'(\tau - t_0)} (BB' - DD') e^{-A(\tau - t_0)} x \, d\tau = x' W_c(t_0, T) x.$$

Thus, we must have $x' S(t_0) x \leq x' W_c(t_0, T) x$. Since $x \in \Re^n$ was arbitrary, we conclude that $S(t) \leq W_c(t, T)$ for all $t \in (M, T)$. []

In the discussion of controllability of linear time invariant systems, we saw that the condition $W(t,T) > 0$ for all $t < T$ ensured the controllability of the system. We now give a condition on the uncertain system $(\Sigma)$ which will ensure that $W_c(t, T) > 0$ for all $t < T$. This condition is referred to as the matching condition and similar conditions arises frequently in the stabilization of uncertain systems; see [1], [12] and [13]. In the sequel, we will show that this condition ensures that Riccati equation (2.5) has a positive-definite solution on the interval $(0,T)$. Furthermore, we will show that the matching condition is a sufficient condition for the feedback controllability.

Definition 2.3: The uncertain system $(\Sigma)$ is said to satisfy the matching condition if:

M1. $BB' - DD' = GG' \geq 0$;

M2. The pair $(A,G)$ is controllable.

Remarks: Our matching condition is a generalization of the matching conditions given in references [1], [12] and [13]. Indeed, the matching condition given in [12] would require that then exists a matrix $M$ such that $D = BM$, where $I - MM' > 0$ and $(A, B)$ is controllable. If these conditions hold then it is straightforward to verify that conditions M1 and M2 will hold with $G = B(I-MM')^{1/2}$. We also observe that if condition M1 holds then we must have $\mathcal{N}(B') \subset \mathcal{N}(D')$. Thus, there exists a matrix $M$ such that $D = BM$.

Proposition 2.1: *If the system $(\Sigma)$ satisfies the matching condition then the controllability gramian $W_c(t,T)$ will be positive definite for all $t < T$.*

Proof: If the system $(\Sigma)$ satisfies the matching condition then the controllability gramian is

$$W_c(t,T)=\int_t^T e^{A(t-\tau)}GG'e^{A'(t-\tau)}\,d\tau \tag{2.6}$$

Furthermore, the pair $(A,G)$ is controllable. Using this fact, it now follows that $W_C(t,T)$ will be positive-definite for all $t < T$; e.g., see Theorem 13.3 in [11]. []

Remark: The condition $W_C(t,T) > 0$ for all $t < T$ is referred to as the 'relative controllability condition' in the literature on linear quadratic differential games; see [14]. Given the connection between linear quadratic differential games and the stabilization of uncertain systems (pointed out in [10]), it might be expected that this condition would be important in the study of feedback controllability of uncertain systems.

Theorem 2.1: *The matching condition is a sufficient condition for the feedback controllability of the uncertain system* $(\Sigma)$.

In order to prove this theorem, we must first establish a number of preliminary results.

Lemma 2.2: *Suppose the system* $(\Sigma)$ *satisfies the matching condition. Then the Riccati equation* (2.5) *will have a positive definite solution for all* $t \in (0,T)$.

Proof: If the system $(\Sigma)$ satisfies the matching condition then Riccati equation (2.5) becomes

$$\dot{S}(T)=AS(t)+S(t)A'-GG'+S(t)E_1'E_1S(t);S(T)=0. \tag{2.7}$$

It follows from Theorem 24.1 of [11] that there exists an $\varepsilon > 0$ such that (2.7) has a solution on the interval $(T-\varepsilon,T]$. Let the interval on which a solution to (2.7) exists be $(M,T]$. Thus $-\infty \le M < T$. Now let $\Phi(t,T)$ be the state transition matrix of the system $\dot{x}(t)=\left[-A'-\frac{1}{2}E_1'E_1S(t)\right]x(t)$. It follows from (2.7) that

$$\dot{S}(T)=-S(t)[-A'-\tfrac{1}{2}E_1'E_1S(t)]'-S(t)[-A'-\tfrac{1}{2}E_1'E_1S(t)]-GG';\ S(T)=0.$$

This leads to

$$S(t) = \int_t^T \Phi(\tau,t)' G G' \Phi(\tau,t) d\tau \tag{2.8}$$

for $t \in (M,T]$. It follows immediately that $S(t) \geq 0$ for all $t \in (M,T]$. Also, Lemma 2.1 implies that $S(t) \leq W_C(t,T)$ for all $t \in (M,T]$. Hence, $S(t)$ cannot have a finite escape time and thus (2.7) has a solution for all $t < T$.

In order to show that $S(t)$ is positive-definite for all $t < T$, we return to equation (2.8). Using this equation, it follows that given any $t_1 < t_2 \leq T$ then

$$S(t_1) \geq S(t_2) \tag{2.9}$$

Now suppose that there exists a $t_0 < T$ such that $S(t_0)$ is singular and let $x_0 \in \mathcal{N}(S(t_0))$ be given. It follows from (2.9) that $x_0' S(t) x_0 = 0$ for all $t \in (t_0,T]$. Thus, we must have

$$\frac{d}{dt} x_0' S(t) x_0 = x_0' \dot{S}(t) x_0 = 0$$

for all $t \in (t_0,T]$. Furthermore, using the continuity of $\dot{S}(t)$, it follows that $x_0' \dot{S}(t_0) x_0 = 0$. However, (2.9) implies that $\dot{S}(t) \leq 0$ for all $t < T$. Hence, we must have $\dot{S}(t_0) x_0 = 0$ and $S(t_0) x_0 = 0$. Returning to Riccati equation (2.7), it follows that

$$\begin{aligned} 0 = x_0' \dot{S}(t_0) x_0 &= x_0' A S(t_0) x_0 + x_0' S(t_0) A' x_0 - x_0' G G' x_0 + x_0' S(t_0) E_1' E_1 S(t_0) x_0 \\ &= x_0' G G' x_0 \end{aligned}$$

and thus $G' x_0 = 0$. Also,

$$\begin{aligned} 0 = \dot{S}(t_0) x_0 &= A S(t_0) x_0 + S(t_0) A' x_0 - G G' x_0 + S(t_0) E_1' E_1 S(t_0) x_0 \\ &= S(t_0) A' x_0. \end{aligned}$$

Thus, $A' x_0 \in \mathcal{N}(S(t_0))$. However, since $x_0 \in \mathcal{N}(S(t_0))$ was arbitrary, we conclude that $A' \mathcal{N}(S(t_0)) \subset \mathcal{N}(S(t_0))$. Hence, $\mathcal{N}(S(t_0))$ is a non-trivial $A'$ - invariant subspace contained in $\mathcal{N}(G')$. This contradicts the controllability of $(A,G)$. Therefore, we must have $S(t_0) > 0$. []

<u>Lemma 2.3</u>: <u>Given any</u> $x \in \Re^n$ <u>and any admissible uncertainty</u> $F(t)$ <u>for the system</u> $(\Sigma)$, <u>then</u>

$$2x'S(t)^{-1}DF(t)\left[E_1 - E_2BS(t)^{-1}\right]x \le x'S(t)^{-1}\left[BB' + DD'\right]S(t)^{-1}x + x'E_1'E_1x.$$

<u>Proof</u>: Given any $x \in \Re^n$ and any admissible uncertainty $F(t)$, we have $F(t)'F(t) \le I$ and

$$0 \le \left\| D'S(t)^{-1}x - F(t)\left[E_1 - E_2B'S(t)^{-1}\right]x \right\|^2$$

$$= x'S(t)^{-1}DD'S(t)^{-1}x - 2x'S(t)^{-1}DF(t)\left[E_1 - E_2B'S(t)^{-1}\right]x$$

$$+ x'[E_1 - E_2B'S(t)^{-1}]'F(t)'F(t)[E_1 - E_2B'S(t)^{-1}]x$$

$$\le x'S(t)^{-1}DD'S(t)^{-1}x - 2x'S(t)^{-1}DF(t)\left[E_1 - E_2B'S(t)^{-1}\right]x$$

$$+ x'[E_1 - E_2B'S(t)^{-1}]'[E_1 - E_2B'S(t)^{-1}]x$$

$$= x'S(t)^{-1}DD'S(t)^{-1}x - 2x'S(t)^{-1}DF(t)\left[E_1 - E_2B'S(t)^{-1}\right]x$$

$$+ x'E_1'E_1x + x'S(t)^{-1}BB'S(t)^{-1}x$$

using assumptions A1 and A2. The required inequality now follows immediately. []

<u>Proof of Theorem 2.1</u>: Suppose that the system $(\Sigma)$ satisfies the matching condition and let $S(t)$ be defined by (2.5). Using Lemma 2.2, it follows that $S(t) > 0$ for all $t \in (0,T)$. Hence, we can define $P(t) \triangleq S(t)^{-1} > 0$. The derivative of $P(t)$ is given by

$$\frac{d}{dt}P(t) = -P(t)\dot{S}(t)P(t).$$

Hence using (2.7), we conclude that $P(t)$ satisfies the Riccati equation

$$\dot{P}(t) = -A'P(t) - P(t)A + P(t)GG'P(t) - E_1'E_1. \tag{2.10}$$

We now construct a lower bound on $P(t)$. Indeed, let

$$\rho \triangleq \max_{\tau \in [0,T]} \left\{\lambda_{max}(e^{-A\tau}GG'e^{-A'\tau})\right\}.$$

It follows from the definition of $W_C(t,T)$ that

$$W_c(t,T) \le \int_t^T \rho I d\tau = \rho(T-t)I$$

for all $t \in [0,T]$. Hence, using Lemma 2.1 we conclude that $S(t) \le r(T-t)I$ for all $t \in [0,T]$. Therefore

$$P(t) \ge \frac{I}{\rho(T-t)} \tag{2.11}$$

for all $t \in [0, T)$. We now let $K(t)$ be defined by

$$K(t) = -B'P(t) \tag{2.12}$$

for $t \in [0, T)$. This results in the closed loop system

$$\dot{x}(t) = \left(A - BB'P(t) + DF(t)\left[E_1 - E_2B'P(t)\right]\right)x(t); \quad \|F(t)\| \le 1. \tag{2.13}$$

In order to show that all solutions to (2.13) satisfy $x(t) \to 0$ as $t \to T$, we propose to use the 'Lyapunov function'

$$V(x,t) = x'P(t)x. \tag{2.14}$$

The Lyapunov derivative corresponding to (2.13) and Lyapunov function (2.14) is

$$\begin{aligned}
\dot{V}(x,t) &= x'\dot{P}(t)x + 2x'P(t)\dot{x} \\
&= x'\left\{-A'P(t) - P(t)A - P(t)GG'P(t) - E_1'E_1\right\}x \\
&\qquad + 2x'P(t)\left\{A - BB'P(t) + DF(t)\left[E_1 - E_2B'P(t)\right]\right\}x \\
&= x'\left\{P(t)BB'P(t) - P(t)DD'P(t) - E_1'E_1 - 2P(t)BB'P(t))\right\}x \\
&\qquad + 2x'P(t)DF(t)\left[E_1 - E_2B'P(t)\right]x \\
&\le -x'\left\{P(t)(BB' + DD')P(t) + E_1'E_1\right\}x \\
&\qquad + x'\left\{P(t)(BB' + DD')P(t)\right\}x + x'E_1'E_1x \\
&= 0
\end{aligned}$$

using (2.10), (2.13), matching condition M1 and Lemma 2.3. Using the fact that $\dot{V}(x,t) \le 0$, it now follows that if $x(t)$ is a solution of (2.13) with $x(0) = x_0$ then $x(t)'P(t)x(t) \le x_0'P(0)x_0$ for all $t \in (0,T)$. However, using (2.11) this implies that

$$\frac{\|x(t)\|^2}{\rho(T-t)} \le x'_0 P(0) x_0 \quad \text{for all } t \in (0,T).$$

Hence $\|x(t)\|^2 \le \rho(T-t) x'_0 P(0) x_0$ for all $t \in (0,T)$. Therefore, $x(t) \to 0$ as $t \to T$. []

Remark: The above theorem has shown that the matching condition is a sufficient condition for feedback controllability. Furthermore, it is well known that in general the matching condition is a sufficient condition for the stabilizability of an uncertain system; e.g., see [1] or [13]. In particular, for the uncertain systems considered in this paper, it is straightforward to show that our matching condition is a sufficient condition for stabilizability; e.g., see [10].

**Acknowledgement**: This work was supported by the Australian Research Council.

REFERENCES

[1] G. Leitmann, "Guaranteed asymptotic stability for some linear systems with bounded uncertainties," J. Dynam. Syst., Meas. and Contr., vol. 101, no. 3, pp. 212-216, 1979.

[2] B.R. Barmish, "Controllability of linear systems in the presence of nonseparable uncertainty," Proc. 19th Allerton Conf. on Comm, Contr. and Comp., Monticello, Illinois, 1981.

[3] B.R. Barmish, "Global and point controllability of uncertain dynamical systems," IEEE Trans. on Aut. Contr., vol. AC-27, no. 2, pp. 399-408, 1982.

[4] B.R. Barmish and W.E. Schmitendorf, "The associated disturbance-free system: a means for investigating the controllability of a disturbed system," J. Optim. Theory and Applic., vol. 38, no. 4, pp. 525-540, 1982

[5] I.R. Petersen and B.R. Barmish, "A conjecture concerning certain notions of controllability for a class of uncertain systems," Syst. Contr. Lett., vol. 2, no. 6, pp. 359-262, 1983.

[6] I.R. Petersen, "Notions of stabilizability and controllability for a class of uncertain linear systems," Int. J. Contr. vol. 46, no. 2, pp. 409-422, 1987.

[7] I.R. Petersen and C.V. Hollot, "A Riccati equation approach to the stabilization of uncertain linear systems," Automatica, vol. 22, no. 4, pp. 397-411, 1986.

[8] I.R. Petersen, "A stabilization algorithm for a class of uncertain linear systems," Syst. Contr. Lett., vol. 8, pp.351-357, 1987.

[9] I.R. Petersen, "Stabilization of an uncertain linear system in which uncertain parameters enter into the input matrix," to appear in SIAM J. Contr. Optimiz., Nov, 1988.

[10] P.P. Khargonekar, I.R.Petersen and K. Zhou, "Robust stabilization of uncertain systems and $H^{\infty}$ optimal control," submitted to IEEE Trans on Aut. Contr..

[11] R.W. Brockett, Finite Dimensional Linear Systems, Wiley, New York, 1970.

[12] B.R. Barmish, I.R. Petersen and A. Feuer, "Linear ultimate boundedness control of uncertain dynamical systems," Automatica, vol. 19, no. 5, pp. 523-532, 1983.

[13] I.R. Petersen, "Structural stabilization of uncertain systems necessity of the matching condition," SIAM J. Contr. Optimiz., vol. 23, no. 2, pp. 286-296, 1985.

[14] Y.C. Ho, A.E. Bryson and S. Baron, "Differential games and optimal pursuit-evasion strategies," IEEE Trans. on Aut. Contr., vol. AC-10, no. 4, pp. 395-389, 1965.

# CONVERGENCE PROPERTIES OF INDEFINITE LINEAR QUADRATIC PROBLEMS WITH RECEDING HORIZON

H.L. Trentelman & J.M. Soethoudt

Abstract: In this paper we study the following question: given a finite dimensional linear system together with a finite horizon (possibly indefinite) quadratic cost functional, when does the corresponding optimal cost converge to the optimal cost of the corresponding infinite horizon problem, as the length of the horizon tends to infinity? For the case that the linear quadratic problems are regular we establish necessary and sufficient conditions for this convergence to hold.

## 1. INTRODUCTION

The finite horizon linear quadratic control problem for the linear time-invariant system

(1.1) $\dot{x}(t) = Ax(t) + Bu(t), \quad x(0) = x_0 \in \mathbb{R}^n$

is concerned with choosing a control function u such that the cost functional

(1.2) $J_T(x_0,u) := \int_0^T \omega(x(t),u(t))dt$

is minimized. Here, $\omega$ is a real quadratic form on $\mathbb{R}^n \times \mathbb{R}^m$ given by

(1.3) $\omega(x,u) = x^TQ\,x + 2u^TS\,x + u^TR\,u$ .

In the above expressions, $A \in \mathbb{R}^{n\times n}$, $B \in \mathbb{R}^{n\times m}$, $Q \in \mathbb{R}^{n\times n}$, $S \in \mathbb{R}^{m\times n}$ and $R \in \mathbb{R}^{m\times m}$. It is assumed that Q is symmetric and that R is positive definite. Apart from this definiteness assumption on R, *we allow $\omega$ to be indefinite.*

In order for the integral in (1.2) to be well-defined, we restrict the control functions to be elements of the class $L_2[0,T]$ of all $\mathbb{R}^m$-valued functions that are square integrable over [0,T]. The optimal cost for the above problem is then defined as

(1.4) $V_T^+(x_0) := \inf \{ J_T(x_0,u) \mid u \in L_2[0,T] \}$.

In addition to the above, the *infinite* horizon linear quadratic problem deals with minimizing the indefinite integral

(1.5) $J(x_0,u) := \lim_{T\to\infty} J_T(x_0,u)$.

Let $L_{2,loc}(\mathbb{R}^+) := \bigcap_{T\geq 0} L_2[0,T]$ denote the space of all $\mathbb{R}^m$-valued locally square integrable functions on $\mathbb{R}^+$. The functional $J(x_0,u)$ is defined for all functions $u \in L_{2,loc}(\mathbb{R}^+)$ for which the limit in (1.5) exists in the sense that it is finite or infinite. This class of functions is denoted by

$$(1.6) \qquad U(x_0) := \{ u \in L_{2,loc}(\mathbb{R}^+) \mid \lim_{T\to\infty} J_T(x_0,u) \text{ exists in } \mathbb{R}^e := \mathbb{R} \cup \{-\infty,+\infty\}\}$$

The optimal cost for the above infinite horizon linear quadratic problem is

$$(1.7) \qquad V_f^+(x_0) := \inf \{ J(x_0,u) \mid u \in U(x_0) \}.$$

In this paper we are interested in the question whether the optimal cost for the finite horizon problem (1.4) converges to the optimal cost for the infinite horizon problem (1.7) as T tends to infinity. Of course, if the quadratic form $\omega$ is positive semi-definite then it is well-known ( see [6] ) that indeed $V_T(x_0) \to V_f^+(x_0)$ $(T\to\infty)$ for all $x_0$ (provided that (A,B) is stabilizable). It turns out however that if $\omega$ is indefinite then this convergence no longer holds in general. In this paper we establish necessary and sufficient conditions for convergence to hold.

Actually, we shall treat the above question of convergence in the following, more general context. Let $N \in \mathbb{R}^{n\times n}$ be a symmetric matrix, let $T > 0$ and consider the finite horizon problem with cost functional

$$(1.8) \qquad J_{T,N}(x_0,u) := J_T(x_0,u) + x^T(T)Nx(T) .$$

The second term in the above represents a penalty on the terminal state. The matrix N is allowed to be indefinite. The optimal cost associated with the latter problem is given by

$$(1.9) \qquad V_{T,N}^+(x_0) := \inf\{ J_{T,N}(x_0,u) \mid u \in L_2[0,T] \} .$$

In addition to this finite horizon problem we consider the infinite horizon problem of infimizing (1.5) under the constraint that $Nx(t)$ converges to zero as t tends to infinity. More specifically, let

$$U_N(x_0) := \{ u \in U(x_0) \mid \lim_{t\to\infty} Nx(t) = 0 \}$$

and consider the problem of infimizing (1.5) over the class $U_N(x_0)$. The optimal cost for this problem is given by

$$(1.10) \qquad V_N^+(x_0) := \inf \{ J(x_0,u) \mid u \in U_N(x_0) \}.$$

The latter optimization problem was studied in detail in [8]. Of course, the problem (1.7) can be reobtained from this formulation as a special case by taking $N = 0$ (see also [9]). Another special case of (1.10) is obtained by taking $N = I$ , the identity matrix. This special case was treated in [10].

Now, in this paper we shall ask ourselves the question: when does the optimal cost for the finite horizon problem (1.9) converge to the optimal cost for the infinite horizon problem (1.10) as $T \to \infty$ ?

We conclude this introduction by noting that the questions to be studied here have been studied before in [2] and [11]. However, in these references only the case that both $\omega \geq 0$ and $N \geq 0$ was considered, while we intend to treat the most general case that $\omega$ and N are allowed to be indefinite.

## 2 The algebraic Riccati equation

The characterization of the optimal costs for the infinite horizon problems (1.7) and (1.10) centers around the algebraic Riccati equation (ARE):

$$(2.1) \qquad A^TK + KA + Q - (KB + S^T)R^{-1}(B^TK + S) = 0.$$

Let $\Gamma$ be the set of all real symmetric solutions of (2.1). According to [10] , if (A,B) is controllable and $\Gamma \neq \emptyset$ then there is exactly one $K \in \Gamma$ such that the matrix $A_K := A - BR^{-1}(B^TK + S)$ has all its eigenvalues in $\mathbb{C}^- \cup \mathbb{C}^0$ and exactly one $K \in \Gamma$ such that $A_K$ has all its eigenvalues in $\mathbb{C}^+ \cup \mathbb{C}^0$. Here, we define $\mathbb{C}^-(\mathbb{C}^0,\mathbb{C}^+) := \{ s \in \mathbb{C} \mid \text{Re } s < 0 \}$ ( Re s = 0, Re s > 0 ). These elements of $\Gamma$ are denoted by $K^+$ and $K^-$, respectively. It can be shown that if $K \in \Gamma$ then $K^- \leq K \leq K^+$. We denote $A_{K^-}$ by $A^-$ and $A_{K^+}$ by $A^+$.

If $M \in \mathbb{R}^{nxn}$ then $\mathcal{X}^-(M)$ ( $\mathcal{X}^0(M)$, $\mathcal{X}^+(M)$ ) denotes the span of all generalized eigenvectors of M corresponding to its eigenvalues in $\mathbb{C}^-$ ( $\mathbb{C}^0$, $\mathbb{C}^+$ ).

Let $\Omega$ denote the set of all $A^-$–invariant subspaces of $\mathcal{X}^+(A^-)$. Let $\Delta := K^+ - K^-$ ( the 'gap' of the ARE ). The following result states that there exists a bijection between $\Omega$ and $\Gamma$ :

<u>Theorem 2.1</u> ( [10],[3],[7] ). Let (A,B) be controllable and assume that $\Gamma \neq \emptyset$. If $\mathcal{V} \in \Omega$ then $\mathbb{R}^n = \mathcal{V} \oplus \Delta^{-1}\mathcal{V}^\perp$. There exists a bijection $\gamma : \Omega \to \Gamma$ defined by

$$\gamma(\mathcal{V}) := K^-P_{\mathcal{V}} + K^+(I - P_{\mathcal{V}}),$$

where $P_{\mathcal{V}}$ is the projector along $\Delta^{-1}\mathcal{V}^\perp := \{ x \in \mathbb{R}^n \mid \Delta x \in \mathcal{V}^\perp \}$. If $K = \gamma(\mathcal{V})$ then $\mathcal{X}^+(A_K) = \mathcal{V}$, $\mathcal{X}^0(A_K) = \ker \Delta$ and $\mathcal{X}^-(A_K) = \mathcal{X}^-(A^+) \cap \Delta^{-1}\mathcal{V}^\perp$. □

If $K = \gamma(\mathcal{V})$ then K is said to be supported by $\mathcal{V}$.

## 3 The infinite horizon problem with asymptotic constraints

In this section we briefly recall the results from [8] on the infimization problem (1.10). Let $\mathcal{L}$ be a subspace of $\mathbb{R}^n$. A symmetric matrix $K \in \mathbb{R}^{nxn}$ is called negative semi-definite on $\mathcal{L}$ if the following two conditions hold: (i) $\forall\, x \in \mathcal{L} : x^TK\,x \leq 0$ (ii) $\forall\, x \in \mathcal{L} : x^TK\,x = 0 \Leftrightarrow Kx = 0$.

If $\mathcal{V} \subseteq \mathbb{R}^n$ and $M \in \mathbb{R}^{nxn}$ then $<\mathcal{V}|M>$ will denote the largest M–invariant subspace of $\mathcal{V}$. A key role in the characterization of the optimal cost $V_N^+(x_o)$ is played by the subspace

$$(3.1) \qquad \mathcal{V}_N := <\ker N \cap \ker K^- |\, A^-> \cap\, \mathcal{X}^+(A^-).$$

Observe that $\mathcal{V}_N \in \Omega$. Thus, with $\mathcal{V}_N$ there corresponds exactly one solution of the ARE. This solution $\gamma(\mathcal{V}_N)$ is denoted by $K_N^+$. The following theorem is the main result of [8]:

Theorem 3.1 Let (A,B) be controllable. Assume that $\Gamma \neq \emptyset$ and that $K^-$ is negative semi-definite on ker N. Then we have

(i) $V_N^+(x_0) = x_0^T K_N^+ x_0$ for all $x_0 \in \mathbb{R}^n$.

(ii) For all $x_0 \in \mathbb{R}^n$ there exists an optimal $u^*$ if and only if $\ker \Delta \subseteq \ker N \cap \ker K^-$.

(iii) If $\ker \Delta \subset \ker N \cap \ker K^-$ then there exists exactly one optimal input and, moreover, this input is given by the feedback control law $u = -R^{-1}(B^T K_N^+ + S)x$. □

## 4 THE FINITE HORIZON PROBLEM WITH ENDPOINT PENALTY

In this section we consider the finite horizon problem of infimizing the cost functional (1.8). We note that for the case that both $\omega \geq 0$ and $N \geq 0$ this problem is quite standard and is treated, for example, in [5]. The general case however is slightly more complicated. The following result can be found in [1,p.131]:

Lemma 4.1 Suppose there exists on the interval $[0,\infty)$ a solution $K(t) = K_N(t)$ of the Riccati differential equation (RDE):

$$\begin{aligned} \dot{K}(t) &= A^T K(t) + K(t)A + Q - (K(t)B + S^T)R^{-1}(B^T K(t) + S), \\ K(0) &= N. \end{aligned} \tag{4.1}$$

Then for all $T > 0$ and for all $x_0 \in \mathbb{R}^n$ we have

$$V_{T,N}^+(x_0) = x_0^T K_N^+(T) x_0.$$

For all $T > 0$ and $x_0 \in \mathbb{R}^n$ there is exactly one $u^* \in L[0,T]$ such that $J_{T,N}(x_0,u^*) = V_{T,N}^+(x_0)$. This input is given by the feedback law $u = -R^{-1}(B^T K(T-t)+S)x$, $t \in [0,T]$. □

It is well-known that if $\omega \geq 0$ and $N \geq 0$ then (4.1) indeed has a unique solution on $[0,\infty)$. In the general case, the RDE need not have a solution on a given interval. We do have the following:

Lemma 4.2 Suppose that (A,B) is controllable and that $\Gamma \neq \emptyset$. If $N - K^- \geq 0$ then (4.1) has a unique solution $K_N(t)$ on $[0,\infty)$. In fact, $K_N(t) = K^- + D(t)$, where D(t) is the unique solution on $[0,\infty)$ of

$$\begin{aligned} \dot{D}(t) &= A^{-T} D(t) + D(t)A^- - D(t)BR^{-1}B^T D(t), \\ D(0) &= N - K^- . \end{aligned} \tag{4.2}$$

(with $A^- = A_{K^-}$, see section 2).

Proof The fact that (4.2) has a unique solution D(t) on $[0,\infty)$ is standard (see for example [4,cor. 2.4.4]). It is then a matter of straightforward calculation to show that

$K^- + D(t)$ satisfies (4.1). Uniqueness of $K^- + D(t)$ then follows from the uniqueness of $D(t)$. □

## 5 CONVERGENCE OF THE OPTIMAL COST

In this section we shall give a formulation of our main result. Before doing this , we state the following lemma:

Lemma 5.1 Let $\mathcal{L}$ be a subspace of $\mathbb{R}^n$ and let $K \in \mathbb{R}^{n \times n}$ be a symmetric matrix. Then K is negative semi-definite on $\mathcal{L}$ if and only if there exists a symmetric matrix N such that $\ker N = \mathcal{L}$ and $N - K \geq 0$.

Proof A proof of this can be given similar to the proof of [8,lemma 3.3]. □

Consider the problems (1.9) and (1.10). In the remainder of this section we assume that (A,B) is controllable and that $\Gamma \neq \emptyset$. According to the previous lemma, if $N - K^- \geq 0$ then $K^-$ is negative semi-definite on ker N. Conversely, if $K^-$ is negative semi-definite on ker N then one can always find a symmetric matrix $N_1$ such that $\ker N_1 = \ker N$ and $N_1 - K^- \geq 0$.

We now formulate our main result:

Theorem 5.2 Assume that (A,B) is controllable, $\Gamma \neq \emptyset$, $N - K^- \geq 0$ and $\ker \Delta \subset \ker N \cap \ker K^-$. Then $V^+_{T,N}(x_0) \to V^+_N(x_0)$ $(T \to \infty)$ for all $x_0 \in \mathbb{R}^n$ if and only if $\ker N \cap \ker K^-$ is $A^-$-invariant.

Our proof of theorem 5.2 runs along a series of lemmas that we consider to be interesting in their own right. Due to lack of space the proofs of these lemmas are deferred to a future paper. Our first lemma deals with an arbitrary system (A,B) and an arbitrary matrix $R > 0$, independent of the previous context. Consider the standard Riccati differential equation

$$\dot{P}(t) = A^T P(t) + P(t)A - P(t)BR^{-1}B^T P(t), \quad P(0) = P_0, \tag{5.1}$$

together with the standard algebraic Riccati equation

$$A^T P + PA - PBR^{-1}B^T P = 0. \tag{5.2}$$

Recall that if $P_0 \geq 0$ and (A,B) is controllable, then (5.1) has a unique solution $P(t) \geq 0$ on $[0,\infty)$ (see [4]). Also, (5.2) has at least one solution ( $P=0$ ). Let $P^+$ be the largest real symmetric solution of (5.2).

Lemma 5.3 Assume that (A,B) is controllable and $\sigma(A) \subset \mathbb{C}^+$. Then $P^+ > 0$. For any $P_0 > 0$ we have $\lim_{t \to \infty} P(t) = P^+$. □

Our following result again deals with the Riccati differential equation (6.1). The result is, in a sense, the converse of the previous lemma:

Lemma 5.4 Assume $P_0 \geq 0$ and assume that $\lim_{t\to\infty} P(t) =: P \in \mathbb{R}^{n\times n}$ exists. If $P > 0$ then $P_o > 0$. □

We now return to the original context of this paper. Consider the system (1.1), together with the quadratic form $\omega$ given by (1.3). Recall that we denote $A_{K^-}$ by $A^-$ (see section 2). An important role is played by the following algebraic Riccati equation in the unknown D:

$$(5.3) \qquad A^{-T}D + DA^{-} - DBR^{-1}B^{T}D = 0$$

We make the following observation:

Lemma 5.5 Assume (A,B) is controllable and $\Gamma \neq \emptyset$. Let $K \in \mathbb{R}^{n\times n}$ be symmetric. Then K is a solution of (2.1) if and only if $D = K - K^-$ is a solution of (5.3). In particular $D = 0$ and $D = \Delta$ ($= K^+ - K^-$) are solutions of (5.3). In fact, 0 and $\Delta$ are the extremal solutions of (5.3) in the sense that any solution of (5.3) satisfies $0 \leq D \leq \Delta$. □

Finally, we shall need the following result:

Lemma 5.6 Assume that (A,B) is controllable and $\Gamma \neq \emptyset$. Assume that $N - K^- \geq 0$. Let D(t) be the solution of (4.3).Then for all $t \geq 0$ we have:

$$\langle \ker N \cap \ker K^- | A^- \rangle \subset \ker D(t). \ □$$

We have now collected the most important ingredients that will be used in our proof of theorem 5.2. In order to give this proof we shall make a suitable decomposition of the state space. Let $\mathcal{V}_N$ be the subspace defined in (3.1). Define

$$\mathcal{X}_1 := \mathcal{V}_N$$
$$\mathcal{X}_2 := \ker \Delta ,$$
$$\mathcal{X}_3 := \mathcal{X}^-(A^+) \cap \Delta^{-1}\mathcal{V}_N^{\perp} .$$

Denote $A_N := A_{K_N^+}$. According to theorem 2.1 we have $\mathcal{X}_1 = \mathcal{X}^+(A_N)$, $\mathcal{X}_2 = \mathcal{X}^0(A_N)$ and $\mathcal{X}_3 = \mathcal{X}^-(A_N)$. Hence $\mathbb{R}^n = \mathcal{X}_1 \oplus \mathcal{X}_2 \oplus \mathcal{X}_3$ . With respect to this decomposition we have

$$(5.4) \qquad A^- = \begin{pmatrix} A_{11} & 0 & A_{13} \\ 0 & A_{22} & A_{23} \\ 0 & 0 & A_{33} \end{pmatrix}$$

for given matrices $A_{ij}$ This follows from the fact that both $\mathcal{X}_1$ and $\mathcal{X}_2$ are $A^-$-invariant. Note that $\sigma(A_{33}) \subset \mathbb{C}^+$. According to theorem 2.1, $\mathcal{X}_2 \oplus \mathcal{X}_3 = \Delta^{-1}\mathcal{X}_1^{\perp}$. Since also $\mathcal{X}_2 = \ker \Delta$, we have

$$(5.5) \qquad \Delta = \begin{pmatrix} \Delta_{11} & 0 & 0 \\ 0 & 0 & 0 \\ 0 & 0 & \Delta_{33} \end{pmatrix}$$

with $\Delta_{11} > 0$ and $\Delta_{33} > 0$. Finally, we partition

$$B = \begin{bmatrix} B_1 \\ B_2 \\ B_3 \end{bmatrix}.$$

Proof of theorem 5.2 From the assumption $\ker \Delta \subset \ker N \cap \ker K^-$ it follows that $\mathcal{X}_1 \oplus \mathcal{X}_2 \subset \ker N \cap \ker K^-$. Hence N and $K^-$ have the form

$$N = \begin{bmatrix} 0 & 0 & 0 \\ 0 & 0 & 0 \\ 0 & 0 & N_{33} \end{bmatrix}, \quad K^- = \begin{bmatrix} 0 & 0 & 0 \\ 0 & 0 & 0 \\ 0 & 0 & K^-_{33} \end{bmatrix},$$

with $N_{33} - K^-_{33} \geq 0$. Since $K^+ = K^- + \Delta$, we have

$$K^+_N = \begin{bmatrix} 0 & 0 & 0 \\ 0 & 0 & 0 \\ 0 & 0 & K^-_{33} + \Delta_{33} \end{bmatrix}.$$

(Recall that $K^+_N = K^-P + K^+(I - P)$, where P is the projector onto $\mathcal{X}_1 = \mathcal{V}_N$ along $\Delta^{-1}\mathcal{V}_N^{\perp} = \mathcal{X}_2 \oplus \mathcal{X}_3$ ). By combining the above we see that

$$K^+_N - K^- = \begin{bmatrix} 0 & 0 & 0 \\ 0 & 0 & 0 \\ 0 & 0 & \Delta_{33} \end{bmatrix}.$$

Let D(t) be the solution of the Riccati differential equation (4.2). By lemma 5.6 and the fact that $\mathcal{X}_2 = \ker \Delta$ is $A^-$-invariant, we have

$$D(t) = \begin{bmatrix} 0 & 0 & 0 \\ 0 & 0 & 0 \\ 0 & 0 & D_{33}(t) \end{bmatrix}$$

for some $D_{33}(t)$. By writing out (4.2) in the decomposition employed, we see that $D_{33}(t)$ is the unique solution of

$$(5.6) \qquad \begin{aligned} &D_{33}(t) = A^T_{33} D_{33}(t) + D_{33}(t) A_{33} - D_{33}(t) B_3 R^{-1} B_3^T D_{33}(t) \\ &D_{33}(0) = N_{33} - K^+_{33}. \end{aligned}$$

also, $K^+_N - K^-$ is a solution of the algebraic Riccati equation (5.3) (see lemma 5.5 ). This implies that $\Delta_{33}$ is a solution to

$$(5.7) \qquad A^T_{33} D_{33} + D_{33} A_{33} - D_{33} B_3 R^{-1} B_3^T D_{33} = 0.$$

In fact, since $\Delta$ is the largest solution of (5.3), $\Delta_{33}$ is the largest solution of (5.7). ('$\Rightarrow$ ') Assume that $V^+_{T,N}(x_0) \to V^+_N(x_0)$ for all $x_0$ or, equivalently, $K_N(t) \to K^+_N$ $(t \to \infty)$. Here, $K_N(t)$ is the unique solution of (4.1). Then we have $K_N(t) - K^- \to K^+_N - K^-$ $(t \to \infty)$.

Now, the point is that $K_N(t) - K^- = D(t)$. Hence we find that $D_{33}(t) \to \Delta_{33} (t \to \infty)$. Since $\Delta_{33} > 0$ it follows from lemma 5.4 that $N_{33} - K^-_{33} > 0$. We now prove the $A^-$-invariance of $\ker N \cap \ker K^-$. Let $x \in \ker N \cap \ker K^-$, $x = (x_1, x_2, x_3)$. Then $(N - K^-)x = 0$ whence $(N_{33} - K^-_{33})x_3 = 0$ so $x_3 = 0$. Thus $A^-x = (A_{11}x_1, A_{22}x_2, 0)$. Since $\mathcal{X}_1 \oplus \mathcal{X}_2 \subset \ker N \cap \ker K^-$, the claim follows.

('$\Leftarrow$') Assume $\ker N \cap \ker K^-$ invariant under $A^-$. We then claim that $\mathcal{X}_1 \oplus \mathcal{X}_2 = \ker N \cap \ker K^-$. Indeed,

$$\begin{aligned}\mathcal{X}_1 \oplus \mathcal{X}_2 &= \{\langle \ker N \cap \ker K^- | A^- \rangle \cap \mathcal{X}^+(A^-)\} \oplus \ker \Delta \\ &= \{(\ker N \cap \ker K^-) \cap \mathcal{X}^+(A^-)\} \oplus \ker \Delta \\ &= (\ker N \cap \ker K^-) \cap (\ker \Delta \oplus \mathcal{X}^+(A^-)),\end{aligned}$$

where the last equality again uses the assumption $\ker \Delta \subset \ker K^- \cap \ker N$. Now, $\ker \Delta \oplus \mathcal{X}^+(A^-) = \mathcal{X}^0(A^-) \oplus \mathcal{X}^+(A^-) = \mathbb{R}^n$. This proves the claim. It follows from this that $N_{33} - K_{33} > 0$. Hence, since $\sigma(A_{33}) \subset \mathbb{C}^+$, the solution $D_{33}(t)$ of (5.6) converges to $\Delta_{33}$, the largest solution of (5.7) (see lemma 5.3). In turn this implies that $D(t) \to K^+_N - K^-$ or, equivalently, that $K^+_N(t) \to K^+_N$. Thus $V^+_{T,N}(x_0) \to V^+_N(x_0)$ $(T \to \infty)$ for all $x_0$. This completes our proof of theorem 5.2. □

## REFERENCES


[1] R.W. Brockett, *Finite Dimensional Linear Systems*, New York, Wiley, 1970.

[2] F.M. Callier & J.L. Willems, 'Criterion for the convergence of the solution of the Riccati differential equation' *IEEE Trans. Aut. Control*, AC–26, p. 1232–1242, 1981.

[3] W.A. Coppel, 'Matrix quadratic equation', *Bull. Austral. Math Soc.*, Vol. 10, p. 377–401, 1974.

[4] O. Hijab, *Stabilization of Control Systems*, New York, Springer Verlag, 1987.

[5] H. Kwakernaak & R. Sivan, *Linear Optimal Control Systems*, Wiley, New York, 1972.

[6] K. Mårtensson, 'On the matrix Riccati equation', *Information Sciences*, Vol. 3, p. 17–49, 1971

[7] M Shayman, 'Geometry of the algebraic Riccati equation – part 1', *SIAM Journal of. Contr. & Opt.*, Vol. 21, no 3, p. 375–393, 1983.

[8] J.M. Soethoudt & H.L. Trentelman, 'The regular indefinite linear quadratic problem with linear endpoint constraints', *Systems & Contr. Letters*, Vol 12, No. 1, p. 23–31, 1989.

[9] H.L. Trentelman 'The regular free–endpoint linear quadratic problem with indefinite cost', *SIAM Journal of Cont. & Opt.*, Vol 27, No. 1, p. 27–42, 1989.

[10] J.C. Willems, 'Least squares stationary optimal control and the algebraic Riccati equation', *IEEE Trans. Aut. Control*, AC 16, no.6, p. 621–634, 1971.

[11] J.L. Willems & F.M. Callier, 'Large finite horizon and infinite horizon LQ–optimal control problems', *Optimal Contr. Appl. & Meth.*, Vol. 4, p 31–45, 1983.

# Generalized Stability of Linear Singularly Perturbed Systems Including Calculation of Maximal Parameter Range

*E.H. Abed, L. Saydy and A.L. Tits*

**Abstract**

The guardian map theory of generalized stability of parametrized linear time-invariant systems is used to prove new results on stability of linear time-invariant singularly perturbed systems. The results give necessary and sufficient conditions for generalized stability of the perturbed system for all sufficiently small values of the singular perturbation parameter, and, moreover, yield the exact parameter range for stability. Thus, the results generalize significantly the classical Klimushev-Krasovskii Theorem, while at the same time providing closed-form expressions for the maximal parameter range for stability.

## 1. Introduction

In this paper, the "guardian map" approach to the study of generalized stability of parametrized families of linear time-invariant systems recently reported in [11–13] is utilized to obtain several new conclusions regarding generalized stability of linear singularly perturbed systems. Here, generalized stability refers to stability with respect to a given domain in the complex plane. Consider the system

$$\dot{x} = Ax + By \tag{1a}$$

$$\epsilon\dot{y} = Cx + Dy \tag{1b}$$

in which $\epsilon > 0$ is a small real parameter, $x$ and $y$ are vectors in $\mathbb{R}^n$ and $\mathbb{R}^m$ respectively, and $A, B, C, D$ are matrices of appropriate dimensions. System (1) is referred to as a singularly perturbed system since its dimension drops from $n+m$ to $n$ when the parameter $\epsilon$ is formally set to 0. Theorem 1 below is a classical result giving *sufficient* conditions for the asymptotic stability of (1) for all sufficiently small values of the singular perturbation parameter $\epsilon$ and has been derived by several authors.

**Theorem 1.** [8], [4] Let $D$ be nonsingular. If the matrices $A_0 := A - BD^{-1}C$ and $D$ are Hurwitz stable, then there exists an $\bar{\epsilon} > 0$ such that the null solution of system (1) is asymptotically stable for all $\epsilon \in (0, \bar{\epsilon})$.

Since the parameter $\epsilon$ typically represents a small physical quantity over which one has little or no control, it is of practical significance to find explicit upper bounds $\epsilon_0$ on $\epsilon$ ensuring that the conclusion of the theorem above is valid for all $\epsilon \in (0, \epsilon_0)$. The aim of this paper is three-fold: (i) We extend the analysis from the case of Hurwitz stability to that of generalized stability relative to many domains of practical interest; (ii) We obtain *necessary and sufficient* conditions for stability; and (iii) The parameter range for generalized stability is obtained *exactly*.

Previous results on upper bounds on the singular perturbation parameter for stability have been obtained by several authors, including Zien [14], Javid [6], Sandell [10] , Chow [3], Khalil [7], Balas [2] and Abed [1]. Recently, Feng [5] used frequency domain stability analysis to characterize the maximal parameter range on $\epsilon$ for *Hurwitz* stability of (1), *under the hypotheses of Theorem 1.* In the present paper, however, no hypotheses are employed regarding the system (1).

The paper is organized as follows. In Section 2, the concept of guardian map is recalled and a result of interest to this paper is given. In Section 3, we obtain necessary and sufficient conditions for generalized stability of the singularly perturbed system (1) and give an explicit expression for the largest upper bound on $\epsilon$ for which Theorem 1 still holds. In Section 4, we provide a one-shot test for stability of System (1). More precisely, it is shown that System (1) is stable for all sufficiently small values of $\epsilon$ if and only if it is stable for one specially constructed value of $\epsilon$. In Section 5, we present an example. Finally, a brief discussion is given in Section 6.

## 2. Guardian maps

The guardian map approach was introduced in [12], [11] as a unifying tool for the study of generalized stability of parametrized families of matrices or polynomials. A basic review of the essentials now follows.

**Definition 1.** Let $\mathcal{S}$ be an open subset of $\mathbb{R}^{n\times n}$ and let $\nu$ map $\mathbb{R}^{n\times n}$ into $\mathbb{C}$. We say that $\nu$ *guards* $\mathcal{S}$ if for all $x \in \overline{\mathcal{S}}$, the equivalence

$$x \in \partial\mathcal{S} \quad \Longleftrightarrow \quad \nu(x) = 0 \tag{2}$$

holds.

The map $\nu$ is said to be *polynomic* if it is a polynomial function of the entries of its argument.

For the purposes of this paper, the set $\mathcal{S}$ will be a (*generalized*) *stability set*, i.e., a set of the form

$$\mathcal{S}(\Omega) := \{A \in \mathbb{R}^{n\times n} : \sigma(A) \subset \Omega\}, \tag{3}$$

where $\Omega$ is an open subset of the complex plane which is symmetric with respect to the real axis.

The next two examples provide the simplest useful illustrations of the concept of guardian map: both the set of Hurwitz stable matrices (or polynomials) and the set of Schur stable matrices (or polynomials) are guarded.

*Example 2.1.* The map $\nu: A \mapsto \det(A \oplus A)$ guards the set of $n \times n$ Hurwitz stable matrices $\mathcal{S}(\overset{\circ}{\mathbb{C}}_-)$. This follows from the property that the spectrum of $A \oplus A$ consists of all pairwise sums of eigenvalues of $A$.

*Example 2.2.* The map $\nu: A \mapsto \det(A \otimes A - I \otimes I)$ guards the set of Schur stable matrices, i.e., of matrices with eigenvalues in in the open unit disk. This follows from the property that the spectrum of $A \otimes A$ consists of all the pairwise products of eigenvalues of $A$.

For more examples of guarded sets $\mathcal{S}(\Omega)$, the reader is referred to [11] where it is shown that in fact, many stability sets of practical interest enjoy the guardedness property with polynomic corresponding guardian maps.

Theorem 2 below gives a necessary and sufficient condition for stability of parametrized families of matrices relative to domains of the complex plane corresponding to guarded stability sets. Let $r = (r_1, \ldots, r_k) \in U$, where $U$ is a pathwise connected subset of $\mathbb{R}^k$, and let $A(r)$ be a matrix in $\mathbb{R}^{n\times n}$ which depends continuously on the parameter vector $r$. Given an open subset $\Omega$ with guarded stability set $\mathcal{S}(\Omega)$, we seek basic conditions for $A(r)$ to lie within $\mathcal{S}(\Omega)$ for all values of $r$ in $U$.

**Theorem 2.** Let $\mathcal{S}(\Omega)$ be guarded by the map $\nu$. Then the family $\{A(r) : r \in U\}$ is stable relative to $\Omega$ if and only if (i) it is nominally stable, i.e., $A(r^0) \in \mathcal{S}(\Omega)$ for some $r^0 \in U$, and (ii) $\nu(A(r)) \neq 0$ for all $r \in U$.

## 3. Main result

Define the matrix

$$J(r) = \begin{bmatrix} A & B \\ rC & rD \end{bmatrix} \tag{4}$$

where $r := \epsilon^{-1}$ is large when $\epsilon$ is small. Stability of the null solution of (1) is identical to stability of the matrix $J(r)$.

We now proceed to study stability of $J(r)$ relative to an open subset $\Omega$ of the complex plane for which $\mathcal{S}(\Omega)$ is endowed with a polynomic guardian map $\nu$. Since $\nu$ is polynomial in its argument, and $J(r)$ depends linearly on $r$, we can write $\nu(J(r))$ as a polynomial in $r$:

$$\nu(J(r)) = \nu_0 + \nu_1 r + \ldots + \nu_{s-1} r^{s-1} + \nu_s r^s =: \nu(r). \tag{5}$$

Here, $s$ is the degree of the polynomial. The following cases present themselves.

*Case 1*: *$\nu$ identically zero.* In this case, the matrix $J(r)$ is unstable relative to $\Omega$ for each $r > 0$. This follows immediately since $\nu$ guards $\mathcal{S}(\Omega)$.

*Case 2*: *$\nu$ not identically zero.* It follows that $\nu(r)$ has finitely many zeros. If the polynomial $\nu(r)$ has no positive real zeros, then $J(r)$ does not cross $\partial\mathcal{S}(\Omega)$ as $r$ varies in $(0, +\infty)$. Thus the family $\{J(r) : r \in (0, +\infty)\}$ lies entirely within either $\mathcal{S}(\Omega)$ or $\mathrm{int}((\mathcal{S}^c(\Omega))$, the interior of the complement of $\mathcal{S}(\Omega)$. To determine which situation prevails, it suffices to test whether $J(r) \in \mathcal{S}(\Omega)$ or $J(r) \in \mathrm{int}\,(\mathcal{S}^c(\Omega))$ for an arbitrarily chosen $r$ in $(0, +\infty)$. If on the other hand $\nu(r)$ has $\ell \geq 1$ real positive zeros $0 < \alpha_1 < \cdots < \alpha_\ell$, then Theorem 2 implies that $J(r) \in \mathcal{S}(\Omega)$ for *all* $r > \alpha_\ell$ if and only if $J(r) \in \mathcal{S}(\Omega)$ for an arbitrarily chosen $r > \alpha_\ell$. It is also clear that in this case, the largest neighborhood of $+\infty$ in which $J(r) \in \mathcal{S}(\Omega)$ is $(\alpha_\ell, +\infty)$.

These remarks are now summarized in the theorem below. In the sequel, if System (1) is stable relative to $\Omega$ for all sufficiently small values of $\epsilon$, then the smallest value of $\epsilon$ for which (1) is unstable will be denoted by $\epsilon^*$.

**Theorem 3.** Let the domain $\Omega$ be guarded by a polynomic map $\nu$, with $\nu(r)$ as in Eq. (5).

(a) If $\nu(r)$ vanishes identically, then the singularly perturbed system (1) is unstable relative to $\Omega$ for all $\epsilon > 0$.

(b) If $\nu(r)$ does not vanish identically and has no positive real zeros, then the singularly perturbed system (1) is stable relative to $\Omega$ for all sufficiently small $\epsilon$ if and only if it is stable relative to $\Omega$ for an arbitrarily chosen $\epsilon > 0$. In the latter case, $\epsilon^* = +\infty$.

(c) Finally, let $\nu(r)$ have a largest positive real zero $\alpha_\ell$. Then System (1) is stable relative to $\Omega$ for all sufficiently small $\epsilon$ if and only if it is stable relative to $\Omega$ for an arbitrarily chosen $\epsilon < 1/\alpha_\ell$. We will then have

$$\epsilon^* = \frac{1}{\alpha_\ell}$$

In fact, we can state the following extension of the preceding result, which has no counterpart in the literature. It addresses the possibility of marginal stability in the singularly perturbed system (1) for finitely many values of $\epsilon$ in a maximal interval of stability, showing how calculations similar to the above can be performed even for this case.

**Theorem 4.** In the setting of Theorem 3(c), and using notation defined above, System (1) is stable relative to $\Omega$ for all but finitely many values of $\epsilon$ in the interval $(0,\ \epsilon^{**})$, where

$$\epsilon^{**} := \frac{1}{\alpha^{**}}$$

with $\alpha^{**} := \min_i \alpha_i$ such that for arbitrarily chosen $r_i \in (\alpha_i,\ \alpha_{i+1}),\quad i = 1, \ldots, \ell,$

($\alpha_{\ell+1} := +\infty$), $J(r_i)$ is stable relative to $\Omega$. Moreover, $\epsilon^{**}$ provides the largest parameter range for which this conclusion holds.

## 4. A one-shot test

The results of Section 2 are conceptually simple and can be implemented easily. The question arises as to whether or not a simple one-shot test exists by which one can immediately ascertain stability of (1) for all sufficiently small $\epsilon$ or the lack thereof. Such a test is given next.

By a well-known theorem (e.g. [9]), all the zeros of the polynomial (5) lie within the disc in the complex plane centered at the origin and of radius

$$R := 1 + \max_{i<s} \frac{|\nu_i|}{|\nu_s|} \tag{6}$$

Thus, generalized stability of the matrix $J(r)$ at an arbitrary $r > R$ is *equivalent* to its generalized stability *for all* sufficiently large $r$. This result is now recorded as Theorem 5, a significant generalization of the classical Theorem 1 of Klimushev and Krasovskii.

**Theorem 5.** Let $\mathcal{S}(\Omega)$ be guarded be a polynomic map $\nu$ of the form (5), and assume that $\nu(J(r))$ is not identically zero. Then there exists an $\bar{\epsilon} > 0$ such that System (1) is stable relative to $\Omega$ for all $\epsilon \in (0, \bar{\epsilon})$ if and only if (1) is stable relative to $\Omega$ for the value

$$\epsilon := \frac{1}{1 + \max_{i<s} \frac{|\nu_i|}{|\nu_s|}}. \tag{7}$$

## 5. Example

In this example, considered in [14], $\Omega$ is the open left-half plane and System (1) is specified by

$$A = \begin{bmatrix} -0.2 & 0.5 \\ 0 & -0.5 \end{bmatrix}, \quad B = \begin{bmatrix} 0 & 0 & 0 \\ 1.6 & 0 & 0 \end{bmatrix},$$

$$C = \begin{bmatrix} 0 & 0 \\ 0 & 0 \\ -275 & -56.9 \end{bmatrix} \quad \text{and} \quad D = \begin{bmatrix} -14.3 & 85.8 & 0 \\ 0 & -25 & 75 \\ -333 & -115 & -186 \end{bmatrix}.$$

In [14], it was shown that the singularly perturbed system under consideration is Hurwitz stable for all $\epsilon \in (0, 5.27)$. Following the method of this paper, we find that the largest interval of stability is in fact $(0, 67.26)$.

## 6. Discussion

The classical Klimushev-Krasovskii Theorem has been generalized in several directions, one of which is the formulation of explicit necessary and sufficient conditions for the stability conclusion to hold. The maximal parameter interval (and therefore the *best possible* upper bound on $\epsilon$) for generalized stability has been obtained. This computation does not require any assumption on the system. Besides this, the new results treat the broader issue of generalized stability with respect to a large class of domains in the complex plane which are of practical and theoretical significance. These domains are those endowed with polynomic guardian maps [11], [12]. Finally, we note that the results of this paper may be extended to the more general situation in which the generalized stability set is endowed with a polynomic *semiguardian map*, as defined in [11]. Detailed statements of these results will be presented elsewhere.

## Acknowledgments

The authors thank Professor G.L. Blankenship for raising a question which led to their considering this application of guardian maps. This work was supported in part by the National Science Foundation's Engineering Research Centers Program: NSFD CDR 8803012, as well as by the NSF under Grants ECS-86-57561, DMC-84-51515 and by the Air Force Office of Scientific Research under Grant AFOSR-87-0073.

## References

[1] E.H. Abed, "A new parameter estimate in singular perturbations," *Systems and Control Letters* 6 (1985), 193–198.

[2] M.J. Balas, "Stability of distributed parameter systems with finite-dimensional controller-compensators using singular perturbations," *J. Math. Anal. Appl.* 99 (1984), 80–108.

[3] J.H. Chow, "Asymptotic stability of a class of non-linear singularly perturbed systems," *J. Franklin Institute* 305 (1978), 275–281.

[4] C.A. Desoer & M.J. Shensa, "Networks with very small and very large parasitics: Natural frequencies and stability," *Proc. IEEE* 58 (1970), 1933–1938.

[5] W. Feng, "Characterization and computation for the bound $\epsilon^*$ in linear time-invariant singularly perturbed systems," *Systems and Control Letters* 11 (1988), 195–202.

[6] S.H. Javid, "Uniform asymptotic stability of linear time-varying singularly perturbed systems," *J. Franklin Institute* 305 (1978), 27–37.

[7] H.K. Khalil, "Asymptotic stability of nonlinear multiparameter singularly perturbed systems," *Automatica* 17 (1981), 797–804.

[8] A.I. Klimushev & N.N. Krasovskii, "Uniform asymptotic stability of systems of differential equations with a small parameter in the derivative terms," *Appl. Math. Mech.* 25 (1962), 1011–1025.

[9] M. Marden, *The geometry of the zeros of a polynomial in a complex variable*, American Mathematical Society, 1949.

[10] N.R. Sandell, Jr., "Robust stability of systems with application to singular perturbations," *Automatica* 15 (1979), 467–470.

[11] L. Saydy, A.L. Tits & E.H. Abed, "Guardian maps and the generalized stability of parametrized families of matrices and polynomials," *Mathematics of Control, Signals, and Systems* (to appear).

[12] L. Saydy, A.L. Tits & E.H. Abed, "Robust stability of linear systems relative to guarded domains," *Proc. 27th IEEE Conf. on Decision and Control, Austin, Texas* (1988).

[13] L. Saydy, *Studies in robust stability*, Ph.D. Dissertation, Dept. of Electrical Engineering, University of Maryland, College Park, 1988.

[14] L. Zien, "An upper bound for the singular parameter in a stable, singularly perturbed system," *J. Franklin Institute* 295 (1973), 373–381.

# CONVEX COMBINATIONS OF HURWITZ FUNCTIONS AND ITS APPLICATIONS TO ROBUSTNESS ANALYSIS

Y. K. Foo and Y. C. Soh

**Abstract**
In this paper, we shall show that a family of analytic functions, constructed from the convex hull of a finite number of vertex functions, will have no zero within a simply-connected region in the complex plane if and only if all its edge functions have no zero within the simply-connected region. The result is in fact a generalization of the Edge Theorem which has been derived for polynomial functions [1,2,3]. We then proceed to show how the result can be used to analyse the stability of uncertain systems.

## 1. Introduction

Over the recent years, there has been a considerable amount of literature dealing with the stability of a family of polynomial functions [1-12]. The main emphasis of these works is to find some simple necessary and sufficient conditions for checking the stability of a whole family of polynomials. One of the most celebrated work in this respect is the work of Kharitonov [4] where it is shown that a family of interval polynomials is stable if and only if <u>four</u> specially constructed extreme polynomials are stable. But the extension of Kharitonov's result to the discrete-time systems is limited [6-8,12]. Furthermore, for more general stability regions, one will have to check all extreme polynomials of the interval polynomials [9-11].

If we are given a more general family of polynomials other than interval polynomials, then it is not sufficient to check all the vertex polynomials. In fact, for a polytope of polynomials, we have to check all the edge polynomials [1-3]. This result is now known as the Edge Theorem. Basically, the Edge Theorem states that the polytope of polynomials is stable if and only if all the exposed edges of the polytope of polynomials are stable.

In this paper we shall generalize the result by showing that the edge theorem can also be applied to a more general class of analytic functions, and not restricted to polytopes of polynomials. We shall present the result by means of a simple graphical proof. We shall also discuss the application of the result to robust stability analysis of systems under structured perturbations.

## 2. Problem Formulation

Consider a family of functions (real or complex), continuous with respect to a complex variable s on the boundary of a simply-connected region D, which is described by

$$S_\phi \triangleq \text{conv}\ \{\phi_i(s)\ ,\quad i = 1, 2, \ldots m\} \tag{2.1}$$

where $\phi_i(s)$ are the <u>vertex functions</u> of $S_\phi$. Our problem is to derive a result which states that $S_\phi$ will have no zero within D if and only if a finite subset of $S_\phi$ has no zero in D. Towards this end, we define the set of <u>edge functions</u> as follows :

$$S_{\phi e} \triangleq \{\phi(s)\ :\ \phi(s) = \alpha\phi_i(s) + (1-\alpha)\phi_j(s)\ ;\ \alpha \in [0, 1],\ \ i,j = 1, 2, \ldots m\} \tag{2.2}$$

In the following section, we shall present a result which states that $S_\phi$ has no zero within D if and only if $S_{\phi e}$ has no zero in D.

## 3. The Main Result

<u>Definition 3.1</u> : Let G denotes a strongly connected graph in the complex plane. Then a point $\omega$ in the plane is said to be enclosed in G if and only if $\omega$ is enclosed by some simple closed curve G' and $G' \subseteq G$. A region R in the complex plane is said to be enclosed in G if every point in R is enclosed in G.

<u>Theorem 3.1</u> : Consider a family of functions described by

$$S_\phi \triangleq \text{conv}\ \{\phi_i(s),\ i = 1, 2, \ldots m\} \tag{3.1}$$

which is continuous on the boundary of a simply-connected region D in the complex plane. Suppose that $\phi_i(s)$ are analytic in D. Then $S_\phi$ will have exactly n zeros in D if and only if $S_{\phi e}$ has exactly n zeros in D, where

$$S_{\phi e} = \{\phi(s) : \phi(s) = \alpha\phi_i(s) + (1-\alpha)\phi_j(s) \\ ; \alpha \in [0, 1], \quad i,j = 1, 2, \dots m\} \tag{3.2}$$

Proof : Necessity is obvious. To prove the sufficiency part, we first note that if $\phi_i(s)$ are analytic in D, then any $\phi(s) \in S_\phi$ is also analytic in D. Next we note that for any fixed s, say $s = s_o$, on $C_D$ where $C_D$ is the contour of D, $\phi_i(s_o)$ is a point in the complex plane. Furthermore,

$$\alpha\phi_i(s_o) + (1-\alpha)\phi_j(s_o) \quad , \quad \alpha \in [0, 1] \tag{3.3}$$

will sweep out a straight line from $\phi_j(s_o)$ to $\phi_i(s_o)$ as $\alpha$ increases from 0 to 1. Hence at $s = s_o$, all the edge functions will form a strongly connected graph (see for example figure 3.1 where four vertex functions are considered), and for any $\phi(s) \in S_\phi$, $\phi(s_o)$ is enclosed by the strongly connected graph.

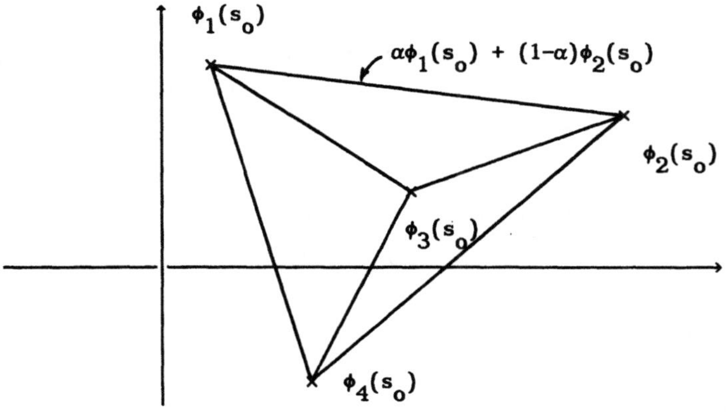


Figure 3.1

By the Argument principle, the image of the edges (i.e. the strongly connected graph) will encircle the origin of the complex plane exactly n times since each edge function has exactly n zeros in D. This implies that the boundary of the image of $S_\phi$ will encircle the origin exactly n times since the boundary is simply a subset of the image of the edges. Now, for any $\phi(s) \in S_\phi$, $\phi(s)$ will lie within the region bounded by the boundary. Thus $\phi(s)$ will encircle the origin of the complex plane exactly n times. Then the Argument Principle will imply that $\phi(s)$ has exactly n zeros in D. This completes the proof. ΔΔΔ

Corollary 3.1 : Let $S_f$ be a family of polynomials (real or complex) described by $S_f \triangleq \text{conv}\ \{f_i(s);\ i = 1, 2, \ldots m\}$ where $f_i(s)$ are the vertex polynomials. Then $S_f$ has no zero in a simply-connected region D in the complex plane if and only if $S_{fe}$ has no zero in D, where

$$S_{fe} \triangleq \{f(s) \ :\ f(s) = \alpha f_i(s) + (1-\alpha)f_j(s) \\ ;\ \alpha \in [0, 1],\ \ i,j = 1, 2, \ldots m\} \qquad (3.4)$$

Proof : Follows from Theorem 3.1 (with n = 0) since polynomials functions are entire functions. ΔΔΔ

Corollary 3.2 : Let $S_H$ be a family of rational functions described by $S_H \triangleq \text{conv}\ \{H_i(s);\ i = 1, 2, \ldots m\}$ where $H_i(s)$ are the vertex rational functions which have no pole in a simply-connected region in the complex plane D. Then $S_H$ has no zero in D if and only if all its edge rational functions have no zero in D. The edge rational functions are defined by

$$S_{He} \triangleq \{H(s) \ :\ H(s) = \alpha H_i(s) + (1-\alpha)H_j(s) \\ ;\ \alpha \in [0, 1],\ \ i,j = 1, 2, \ldots m\} \qquad (3.5)$$

Proof : Follows from Theorem 3.1. ΔΔΔ

Before we end this section, we shall discuss the application of the results to robustness analysis of unceratin systems. It is well known that the characteristic equations of a set of interval matrices are contained in the polytope of polynomials constructed from the vertex matrices. This result can be generalized. For example, in recent studies on the robust stability of linear time-invariant MIMO systems [14-18], it has been shown that robustness problem is equivalent to the problem of determining the non-singularity of $\det[I + M(s)\Delta(s)]$ for all s on the boundary of D. If at each frequency each element of $\Delta(s)$ is contained in some polygons, then $\det[I + M(j\omega_o)\Delta(j\omega_o)]$ $(s = j\omega_o)$ is non-singular if the convex closure of the image of the Cartesian product of these polygons under the mapping $\phi = \det[I + M(j\omega_o)V_i(j\omega_o)]$,

$i = 1, 2, \ldots \prod_{k=1}^{n^2} v_k$, does not contain the origin; $v_k$ denotes the number of vertices of the polygon which contain the frequency response map of the $k^{th}$ element of $\Delta(j\omega_o)$, and $V_i(j\omega_o)$'s denote the $\prod_{k=1}^{n^2} v_k$ vertex matrices constructed from the vertices of the polygon [17]. This problem can be transformed into one of determining the stability of a convex combinations of analytic functions as stated below:

<u>Propositon 3.1</u> : Let $\phi(s,\Delta) = \det[I + M(s)\Delta(s)]$ where $M(s)$ and $\Delta(s)$ are D-stable. Let the elements of $\Delta(s)$ be independent of each other and are each contained by some polygon. Let $\{V_i(s),\ i = 1, 2, \ldots \prod_{k=1}^{n^2} v_k\}$ be the set of vertex matrices constructed from the vertices of the polygons. Assume that $V_i(s)$'s are also analytic in D. Under these conditions, $\phi(s,\Delta) \neq 0$ for all $s \in C_D$ and $\Delta(s) \in \text{conv}\{V_i(s),\ i = 1, 2, \ldots \prod_{k=1}^{n^2} v_k\}$ (this defines a hyper-rectangle) if

$$S_\phi \triangleq \text{conv}\{\phi_i(s) : \phi_i(s) = \det[I + M(s)V_i(s)]\} \tag{3.6}$$

has no zero in D.

<u>Proof</u> : It is easy to prove that the image of $S_\phi$ in the complex plane coincides with the convex closure of the images of $\phi_i(s)$. Hence, the convex closure of the images of $\phi_i(s)$ will not contain the origin if and only if the image of $S_\phi$ in the complex plane does not contain the origin. If $S_\phi$ is analytic in D, then the image of $S_\phi$ will neither pass through nor encircle the origin. This completes the proof. ΔΔΔ

**4. Checking the Stability of a Convex Combinations of Functions**

Let us now consider a family of functions described by

$$S_f \triangleq \text{conv}\{f_i(s),\ i = 1, 2, \ldots m\}$$

where $f_i(s)$ are the vertex functions analytic in the CRHP. Let $\hat{f}(s)$ be an arbitrary function which is Hurwitz and analytic in the CRHP. Define $F_i(s) \triangleq f_i(s)/\hat{f}(s)$. It is then clear that $f_i(s)$ will have no zero in the CRHP if and only if $F_i(s)$ have none, and $\text{conv}\{f_i(s),\ i = 1, 2, \ldots$

m} neither encircles nor encloses the origin if and only if neither does conv$\{F_i(s),\ i = 1, 2, \dots m\}$, as s traverses the contour of D, $C_D$.

Lemma 4.1 : Conv $\{f_i(s),\ i = 1, 2, \dots m\}$ neither encircles nor encloses the origin as s traverses $C_D$ if and only if there exist a function $\hat{f}(s)$ which has no zero and is anlytic in the CRHP, and a function $h(s) = e^{i\theta(s)}$ continuous in $s \in C_D$ such that

$$\mathrm{Re}\{\mathrm{conv}\{h(s)f_i(s)/\hat{f}(s),\ i = 1, 2, \dots m\}\} > 0 \quad ; \text{all } s \in C_D \tag{4.1}$$

Proof : With no loss of generality, we let $\hat{f}(s)$ be any function which is Hurwitz and analytic in the CRHP such that $F_i(s)$ is proper, where $F_i(s) = f_i(s)/\hat{f}(s),\ i = 1, 2, \dots m$. Let $\theta_{max}(s)$ and $\theta_{min}(s) : C_D \to \mathbb{R}$ be two continuous (except possibly at the origin) functions defined by

$$\theta_{max}(s) \triangleq \max_i \{\underline{/F_i(s)}\} \ ; \quad 0 \le \theta_{max}(0) \le \pi \tag{4.2}$$

and

$$\theta_{min}(s) \triangleq \min_i \{\underline{/F_i(s)}\} \ ; \quad 0 \le \theta_{min}(0) \le \pi \tag{4.3}$$

Note that $\theta_{max}(s)$ and $\theta_{min}(s)$ are well defined on every $s \in C_D$ if and only if none of the Nyquist plots of $F_i(s),\ i = 1, 2, \dots m$, passes through the origin. Define

$$\theta_d(s) = \theta_{max}(s) - \theta_{min}(s) \tag{4.4}$$

and

$$\theta_m(s) = \frac{1}{2}[\theta_{max}(s) + \theta_{min}(s)] \tag{4.5}$$

Obviously, if $\theta_m(s)$ is not continuous at the origin, then there exist $F_i(s),\ i = 1, 2, \dots m$, that encircle the origin as s traverses $C_D$. This implies that there exists no continuous $\theta(s)$ which make

$$\mathrm{Re}\{\mathrm{conv}\{h(s)f_i(s)/\hat{f}(s),\ i = 1, 2, \dots m\}\} > 0 \quad ; \text{all } s \in C_D \tag{4.6}$$

Thus the existence of such continuous $\theta(s)$ implies that $\theta_m(s)$ is continuous at the origin and thus we may choose $\theta(s) = \theta_m(s)$. Conversely, continuity of $\theta_m(s)$ at the origin implies that we may choose $\theta(s) = \theta_m(s)$ as well.

If $\theta(s) = \theta_m(s)$, then (4.1) can be satisfied if and only if $\theta_d(s) < \pi$. If $\theta(s) < \pi$, then it is obvious that $\text{conv}\{F_i(s),\ i = 1, 2, \ldots m\}$ neither encircles nor encloses the origin and hence neither does $\text{conv}\{f_i(s),\ i = 1, 2, \ldots m\}$. Conversely, if $\theta_d(s) \geq \pi$, we may then find two members $f_i(s), f_j(s) \in S_f$ such that $\angle f_i(s) - \angle f_j(s) = \pi$ at some $s = s_o \in C_D$. Note that the edge joining these two vertices is a straight line in the complex plane. It thus follows that there exists an $f(s)$ where

$$f(s) = \alpha f_i(s) + (1-\alpha)f_j(s) \qquad \alpha \in [0, 1]$$

such that $f(s) = 0$ at $s = s_o$. Thus $\text{conv}\{f_i(s),\ i = 1, 2, \ldots m\}$ must enclose the origin at $s_o$ and this completes the proof. $\Delta\Delta\Delta$

We are now in a position to state the following main theorem.

<u>Theorem 4.1</u> : The family of functions analytic in the CRHP

$$S_f \triangleq \text{conv}\{f_i(s),\ i = 1, 2, \ldots m\} \qquad (4.7)$$

is Hurwitz if and only if

(i) $f_i(s)$ is Hurwitz, $i = 1, 2, \ldots m$, and

(ii) $\theta_d(s)$ as defined by (4.4) satisfies

$$\theta_d(s) < \pi \quad , \quad \text{all } s \in C_D \qquad (4.8)$$

<u>Proof</u> : Obvious, since Lemma 4.1 implies that the image of $S_f$ in the complex plane, as s traverses $C_D$, neither encloses nor encircles the origin if and only if (i) and (ii) are satisfied. Apply the Principle of Argument completes the proof. $\Delta\Delta\Delta$

<u>Comment 4.1</u> : Theorem 4.1 indicates that a necessary condition for the Hurwitz property of $S_f$ is that $f_i(s)$ is Hurwitz for all i. It will be nice in practice if it is not necessary to explicitly check the Hurwitz property of each of $f_i(s)$. The following corollary to Theorem 4.1 will therefore be useful computationally. □□□

Corollary 4.1 : Let $\theta_{max}(s)$ and $\theta_{min}(s)$ be defined as in (4.2)-(4.3). Then a necessary and sufficient condition for $S_f$ to be Hurwitz is that all the following conditions are satisfied:

(i) $\theta_{max}(s)$ and $\theta_{min}(s)$ are well-defined for all $s \in C_D$, that is, there does not exist $s_o \in C_D$ such that $f_i(s_o) = 0$ for some i.

(ii) $\theta_{max}(s)$ is continuous at the origin, that is

$$\lim_{\epsilon \to 0} [\theta_{max}(j\epsilon) - \theta_{max}(-j\epsilon)] = 0 \tag{4.9}$$

(iii) $\theta_d(s) < \pi$, all $s \in C_D$.

Proof : Conditions (i) and (iii) are obviously necessary in view of Lemma 4.1. To prove that (ii) is also necessary, we note that a necessary condition for (iii) to be satisfied is that $\lim_{\pm\epsilon \to 0}[\theta_d(j\epsilon)] = 0$. Thus the discontinuity of $\theta_{max}(s)$ at the origin and the satisfaction of (iii) implies $\theta_m(s)$ of equation (4.5) is discontinuous at the origin. Thus the proof of Lemma 4.1 indicates that conv$\{f_i(s), i = 1, 2, \ldots m\}$ cannot be possibly be Hurwitz. To prove sufficiency, assume (i), (ii) and (iii) are satisfied, then it is clear that there exists arbitrarily small positive $\epsilon$ such that (4.1) is satisfied with $\theta(s) = [\frac{\pi}{2} - \epsilon] - \theta_{max}(s)$. This completes the proof. ΔΔΔ

## 5. Conclusions

The main purpose of this paper is to present a generalized edge theorem which is applicable to a wide class of analytic functions. We have shown that a family of functions defined by the convex hull of a finite number of vertex analytic functions has no zero within a simply-connected region if and only if all its edge functions have no zero in the region. Extensions of the result to multiply-connected regions are obvious by considering some "ugly" Nyquist contour [13].

It is shown that the Edge Theorem can be very useful in studying the robust stability of feedback systems under highly structured perturbations. Furthermore, a simple method for checking the zero clustering property of convex combinations of two analytic functions is presented. In particular, corollary 4.1 should be useful in the stability analysis of interval matrices as well as robustness analysis of feedback systems.

**References**

[1] A.C. Bartlett, C.V. Hollot and H. Lin, "Root Locations of an Entire Polytope of Polynomials : It Suffices to Check the Edges," in *Proc. Amer. Contr. Conf.*, Minneapolis, MN, 1987. Also in *Mathematics of Control, Signals, and Systems*, 1 (1988), 61-71.

[2] M. Fu and B.R. Barmish, "Polytopes of Polynomials with Zeros in a Prescribed Region," submitted to *IEEE Trans. Auto. Contr.* for publication, (1988).

[3] A.C. Bartlett, C.V. Hollot, "A Necessary and Sufficient Condition for Schur Invariance and Generalised Stability of Polytope of Polynomials," *IEEE Trans. Auto. Contr.*, 33 (1988), 575-578.

[4] V.L. Kharitonov, "Asymptotic Stability of an Equilibrium Position of a Family of Systems of Linear Differential Equations," *Differential'nye Uravneniya*, 14, No. 11 (1978), 1483-1485.

[5] N.K. Bose, "A System-Theoretic Approach to Stability of Sets of Polynomials," *Contemp. Math.*, 47 (1985), 25-34.

[6] C.V. Hollot and A.C. Bartlett, "Some Discrete-Time Counterparts to Kharitonov's Stability Criterion for Uncertain Systems," *IEEE Trans. Auto. Contr.*, 31 (1986), 355-356.

[7] J. Cieslik, "On Possibilities of the Extension of Kharitonov's Stability Test for Interval Polynomials to the Discrete-Time Case," *IEEE Trans. Auto. Contr.*, 32 (1987), 237-238.

[8] F. Kraus, B.D.O. Anderson and M. Mansour, "Robust Schur Polynomial Stability and Kharitonov's Theorem," *Int. J. Contr.*, 47 (1988), 1213-1225.

[9] C.B. Soh and Y.C. Soh, "On the Existence of Kharitonov-Like Theorems," *Technical Report*, School of Electrical and Electronic Engineering, NTI, Singapore 2263, (1988).

[10] I.R. Petersen, "A Class of Stability Regions for which a Kharitonov-Like Theorem Holds," *Proc. of IEEE Conf. Decision and Control*, Los Angeles, (1987).

[11] C. B. Soh and C.S. Berger, "Damping Margin of Polynomials with Perturbed Coefficients," *IEEE Trans. Auto. Contr.*, 32 (1988), 509-511.

[12] N.K. Bose and E. Zeheb, "Kharitonov's Theorem and a Stability Test of Multidimensional Digital Filters," *IEE Proc.*, 133, Pt.6 (1986), 187-190.

[13] R. DeCarlo and R. Saek, "The Encirclement Condition : An Approach Using Algebraic Topology," *Int. J. Contr.*, 26 (1977), 279-287.

[14] J. C. Doyle and G. Stein, "Multivariable Feedback Design : Concepts for a Classical/Modern Synthesis," *IEEE Trans. Auto. Contr.*, 26 (1981), 14-16.

[15] N. A. Lehtomaki, N. R. Sandell, Jr. and M. Athens, "Robustness Results in Linear-Quadratic Gaussian Based Multivariable Control Design," *IEEE Trans. Auto. Contr.*.26 (1981), 75-93.

[16] I. Postlethwaite and Y. K. Foo, "Robustness with Simultaneous Pole and Zero Movement Across the $j\omega$-axis," *Automatica*, 21 (1985), 433-443.

[17] M. Saeki, "A Method of Robust Stability Analysis with Highly Structured Uncertainties," *IEEE Trans. Auto. Contr.*, 31 (1986), 935-940.

[18] J. C. Doyle, "Analysis of Feed-back Systems with Structured Uncertainties," *IEE Proc., Pt. D*, 129 (1982), 242-250.

# DESIGNING STRICTLY POSITIVE REAL TRANSFER FUNCTION FAMILIES: A NECESSARY AND SUFFICIENT CONDITION FOR LOW DEGREE AND STRUCTURED FAMILIES

C.V. Hollot, Lin Huang and Zhong-Ling Xu

ABSTRACT

Consider a transfer function family n(s)/d(s) where n(s) $\varepsilon$ N with N being an interval of polynomials. In this paper we study the problem of designing a d(s) such that n(s)/d(s) is strictly positive real for all choices n(s) from N. A necessary condition for the existence of such a d(s) is that N be stable. We show that this condition is also sufficient for low degree systems (degree $\leq$ 3) and when N has some added structure.

## 1. INTRODUCTION

An important problem in both parameter identification (using output-error schemes) and in adaptive control is to design a transfer function which is strictly positve-real invariant (SPR-invariant); e.g., see [1] and [5]. This means the following: Given a family N of stable (numerator) polynomials n(s), find a (denominator) polynomial d(s) such that deg [d(s)] = deg [n(s)] and the transfer function n(s)/d(s) satisfies

$$\text{Real } [n(j\omega)/d(j\omega)] > 0, \quad \text{for all } \omega \geq 0 \qquad \text{(SPR)}$$

and for all selections n(s) from N. If such a d(s) exists we say that the family of transfer functions

$$\Gamma(N,d) \stackrel{\Delta}{=} \{n(s)/d(s): n(s) \ \varepsilon \ N\}$$

is SPR-invariant.

In this paper we show that if N is a stable interval polynomial family and of low degree (n $\leq$ 3) or of a particular "structure" (the particular structures will be introduced in Section 2), then there always exists a d(s) such that $\Gamma$(N,d) is SPR-invariant. An interval of polynomials N is given by

$$N: \quad a_0 s^n + a_1 s^{n-1} + a_2 s^{n-2} + \ldots + a_n; \quad a_i \ \varepsilon \ [\underline{a}_i, \bar{a}_i] \qquad (1.1)$$

and is stable if and only if the four so-called Kharitonov (corner) polynomials

$$K_1(s) = \underline{a}_0 s^n + \underline{a}_1 s^{n-1} + \bar{a}_2 s^{n-2} + \bar{a}_3 s^{n-3} + \dots + \quad (1.2a)$$

$$K_2(s) = \bar{a}_0 s^n + \bar{a}_1 s^{n-1} + \underline{a}_2 s^{n-2} + \underline{a}_3 s^{n-3} + \dots + \quad (1.2b)$$

$$K_3(s) = \underline{a}_0 s^n + \bar{a}_1 s^{n-1} + \bar{a}_2 s^{n-2} + \underline{a}_3 s^{n-3} + \dots + \quad (1.2c)$$

$$K_4(s) = \bar{a}_0 s^n + \underline{a}_1 s^{n-1} + \underline{a}_2 s^{n-2} + \bar{a}_3 s^{n-3} + \dots + \quad (1.2d)$$

are stable; e.g., see [3]. Thus, the conditions for our results are easily tested.

## 2. THE MAIN RESULT

To begin, assume N is an interval of polynomials and for $n(s) \varepsilon N$ and $\omega \geq 0$ write

$$n(j\omega) = h_n(-\omega^2) + j\omega g_n(-\omega^2). \quad (2.1)$$

The polynomials $h_n$ and $g_n$ are the even and odd portions of $n(s)$. For $\omega \geq 0$ define

$$h_+(-\omega^2) \triangleq \max_{n \varepsilon N} h_n(-\omega^2); \quad h_-(-\omega^2) \triangleq \min_{n \varepsilon N} h_n(-\omega^2) \quad (2.2a)$$

$$g_+(-\omega^2) \triangleq \max_{n \varepsilon N} g_n(-\omega^2); \quad g_-(-\omega^2) \triangleq \min_{n \varepsilon N} g_n(-\omega^2). \quad (2.2b)$$

The Kharitonov polynomials can be expressed in terms of $h_+$, $h_-$, $g_+$ and $g_-$. If n = even, then

$$K_1(s) = h_-(s^2) + sg_+(s^2), \quad (2.3a)$$
$$K_2(s) = h_+(s^2) + sg_-(s^2), \quad (2.3b)$$
$$K_3(s) = h_-(s^2) + sg_-(s^2), \quad (2.3c)$$
$$K_4(s) = h_+(s^2) + sg_+(s^2). \quad (2.3d)$$

For n = odd

$$K_1(s) = h_+(s^2) + sg_-(s^2), \quad (2.4a)$$
$$K_2(s) = h_-(s^2) + sg_+(s^2), \quad (2.4b)$$
$$K_3(s) = h_+(s^2) + sg_+(s^2), \quad (2.4c)$$
$$K_4(s) = h_-(s^2) + sg_-(s^2). \quad (2.4d)$$

The even and odd components of the Kharitionov polynomials play a crucial role in the statement and proof of our main result. Before stating this theorem we'll need some lemmas. The first lemma states that SPR-invariance of $\Gamma(N,d)$ is equivalent to $K_i(s)/d(s)$, $i = 1,2,3,4$, satisfying Condition (SPR). This result first appeared in [2] and amounts to a Kharitonov-like result for SPR-invariance.

**Lemma 2.1 (See [2] for proof): Consider N an interval polynomial as in (1.1). If there exists a d(s) such that $K_i(s)/d(s)$, i=1,2,3,4, satisfies Condition (SPR), then $\Gamma(N,d)$ is SPR-invariant.** ∇∇∇

This lemma requires the ratio Real $[K_i(j\omega)/d(j\omega)] > 0$ for i = 1,2,3,4 and for all $\omega > 0$. A weaker equivalent condition for SPR-invariance is available and is needed in proving our main results. In the next lemma we'll relax the conditions in Lemma 2.1 and show that Real $[K_i(j\omega)/d(j\omega)] > 0$, i = 1,2,3,4 holding over portions of $[0, +\infty)$ is sufficient for SPR-invariance. To state this result we'll need to study the special zero patterns of $h_+(u)$, $h_-(u)$, $g_+(u)$ and $g_-(u)$ which arise in the problem formulation. To describe these patterns let

$$u \stackrel{\Delta}{=} -\omega^2 \tag{2.5}$$

and assume that the Kharitonov polynomials $K_i(s)$ are stable. It follows from either (2.3) or (2.4) and the Hermite-Biehler Theorem that $(h_+,g_+)$, $(h_+,g_-)$, $(h_-, g_+)$ and $(h_-,g_-)$ are positive pairs of polynomials; see Gantmacher [4] for details. Consequently, the zeroes of $h_+(u)$, $h_-(u)$, $g_+(u)$ and $g_-(u)$ are negative and distinct and the zeroes of $h_+(u)$ and $g_+(u)$ alternate. Similarly, the zeroes of each of the pairs $(h_+,g_-)$, $(h_-,g_+)$ and $(h_-,g_-)$ alternate. Finally, recall (see (2.2)) that $h_+$ and $g_+$ dominate $h_-$ and $g_-$ respectively. These facts constrain the zeroes of $h_+(u)$, $h_-(u)$, $g_+(u)$ and $g_-(u)$ to a particular arrangement on the negative real axis. A typical pattern is illustrated in Figure 1.

Now, using the zeroes of $h_+$, $h_-$, $g_+$ and $g_-$ we partition the negative u-axis. For n = even, let m = n/2 while for n = odd let m = (n-1)/2. In either case order the zeroes of $h_+$ and $h_-$ as

$$\begin{aligned} h_+&: \alpha_m^+ < \dots < \alpha_2^+ < \alpha_1^+ < 0 \\ h_-&: \alpha_m^- < \dots < \alpha_2^- < \alpha_1^- < 0 \end{aligned} \tag{2.6a}$$

while for n = odd

$$\begin{aligned} g_+&: \beta_m^+ < \dots < \beta_2^+ < \beta_1^+ < 0 \\ g_-&: \beta_m^- < \dots < \beta_2^- < \beta_1^- < 0 \end{aligned} \tag{2.6b}$$

and for n = even

$$\begin{aligned} g_+&: \beta_{m-1}^+ < \dots < \beta_2^+ < \beta_1^+ < 0 \\ g_-&: \beta_{m-1}^- < \dots < \beta_2^- < \beta_1^- < 0. \end{aligned} \tag{2.6c}$$

For $n \geq 2$ and for n = even let p = 4m-1; if n = odd, let p = 4m + 1. Partition the negative u-axis into the intervals $U_1, U_2, \dots, U_p$ described by

$$U_1 = [\alpha_1^-, 0];\ U_2 = [\alpha_1^+, \alpha_1^-];\ U_3 = [\beta_1^-, \alpha_1^+];\ U_4 = [\beta_1^+, \beta_1^-];\ U_5 = [\alpha_2^+, \beta_1^+]$$

$$U_6 = [\alpha_2^-, \alpha_2^+];\ U_7 = [\beta_2^+, \alpha_2^-];\ U_8 = [\beta_2^-, \beta_2^+];\ U_9 = [\alpha_3^-, \beta_2^-];\ U_{10} = [\alpha_3^+, \alpha_3^-]$$

$$\vdots \tag{2.7}$$

Observe that the pattern in intervals $U_1 - U_8$ is repeated in intervals $U_9 - U_{16}$, $U_{17} - U_{24}$ and so forth. The cycle is evident from Figure 1 where $\bar{n} = 5$, $\bar{m} = 2$ and where p = 9 such intervals are required to describe complete cycles of $h_+$, $h_-$, $g_+$ and $g_-$.

**Lemma 2.2: Consider n $\geq$ 2 and N an interval of polynomials as in (1.1). Suppose there exists a d(s) which satisfies the following conditions:**

**(i) For n = even:**

$$\begin{array}{llll}
\text{Real}[K_1(j\sqrt{-u})/d(j\sqrt{-u})] > 0 & \text{and} & \text{Real}[K_2(j\sqrt{-u})/d(j\sqrt{-u})] > 0; & u \in U_1; \\
\text{Real}[K_2(j\sqrt{-u})/d(j\sqrt{-u})] > 0 & \text{and} & \text{Real}[K_3(j\sqrt{-u})/d(j\sqrt{-u})] > 0; & u \in U_2; \\
\text{Real}[K_3(j\sqrt{-u})/d(j\sqrt{-u})] > 0 & \text{and} & \text{Real}[K_4(j\sqrt{-u})/d(j\sqrt{-u})] > 0; & u \in U_3; \\
\text{Real}[K_4(j\sqrt{-u})/d(j\sqrt{-u})] > 0 & \text{and} & \text{Real}[K_2(j\sqrt{-u})/d(j\sqrt{-u})] > 0; & u \in U_4; \\
\text{Real}[K_2(j\sqrt{-u})/d(j\sqrt{-u})] > 0 & \text{and} & \text{Real}[K_1(j\sqrt{-u})/d(j\sqrt{-u})] > 0; & u \in U_5; \\
\text{Real}[K_1(j\sqrt{-u})/d(j\sqrt{-u})] > 0 & \text{and} & \text{Real}[K_4(j\sqrt{-u})/d(j\sqrt{-u})] > 0; & u \in U_6; \\
\text{Real}[K_4(j\sqrt{-u})/d(j\sqrt{-u})] > 0 & \text{and} & \text{Real}[K_3(j\sqrt{-u})/d(j\sqrt{-u})] > 0; & u \in U_7; \\
\text{Real}[K_3(j\sqrt{-u})/d(j\sqrt{-u})] > 0 & \text{and} & \text{Real}[K_1(j\sqrt{-u})/d(j\sqrt{-u})] > 0; & u \in U_8; \\
\text{Real}[K_1(j\sqrt{-u})/d(j\sqrt{-u})] > 0 & \text{and} & \text{Real}[K_2(j\sqrt{-u})/d(j\sqrt{-u})] > 0; & u \in U_9; \\
\text{Real}[K_2(j\sqrt{-u})/d(j\sqrt{-u})] > 0 & \text{and} & \text{Real}[K_3(j\sqrt{-u})/d(j\sqrt{-u})] > 0; & u \in U_{10}; \\
 & \vdots & &
\end{array} \tag{2.8a}$$

**(ii) For n = odd:**

$$\begin{array}{llll}
\text{Real}[K_2(j\sqrt{-u})/d(j\sqrt{-u})] > 0 & \text{and} & \text{Real}[K_1(j\sqrt{-u})/d(j\sqrt{-u})] > 0; & u \in U_1; \\
\text{Real}[K_1(j\sqrt{-u})/d(j\sqrt{-u})] > 0 & \text{and} & \text{Real}[K_4(j\sqrt{-u})/d(j\sqrt{-u})] > 0; & u \in U_2; \\
\text{Real}[K_4(j\sqrt{-u})/d(j\sqrt{-u})] > 0 & \text{and} & \text{Real}[K_3(j\sqrt{-u})/d(j\sqrt{-u})] > 0; & u \in U_3; \\
\text{Real}[K_3(j\sqrt{-u})/d(j\sqrt{-u})] > 0 & \text{and} & \text{Real}[K_1(j\sqrt{-u})/d(j\sqrt{-u})] > 0; & u \in U_4; \\
\text{Real}[K_1(j\sqrt{-u})/d(j\sqrt{-u})] > 0 & \text{and} & \text{Real}[K_2(j\sqrt{-u})/d(j\sqrt{-u})] > 0; & u \in U_5; \\
\text{Real}[K_2(j\sqrt{-u})/d(j\sqrt{-u})] > 0 & \text{and} & \text{Real}[K_3(j\sqrt{-u})/d(j\sqrt{-u})] > 0; & u \in U_6; \\
\text{Real}[K_3(j\sqrt{-u})/d(j\sqrt{-u})] > 0 & \text{and} & \text{Real}[K_4(j\sqrt{-u})/d(j\sqrt{-u})] > 0; & u \in U_7; \\
\text{Real}[K_4(j\sqrt{-u})/d(j\sqrt{-u})] > 0 & \text{and} & \text{Real}[K_2(j\sqrt{-u})/d(j\sqrt{-u})] > 0; & u \in U_8;
\end{array}$$

**$\mathrm{Real}[K_2(j\sqrt{-u})/d(j\sqrt{-u})] > 0$ and $\mathrm{Real}[K_1(j\sqrt{-u})/d(j\sqrt{-u})] > 0$; $u \in U_9$;**

**$\mathrm{Real}[K_1(j\sqrt{-u})/d(j\sqrt{-u})] > 0$ and $\mathrm{Real}[K_4(j\sqrt{-u})/d(j\sqrt{-u})] > 0$; $u \in U_{10}$;**

$$\vdots \qquad \textbf{(2.8b)}$$

**Then, $\Gamma(N,d)$ is SPR-invariant.**

Observe the the conditions in (2.8) for intervals $U_1$ - $U_8$ are repeated in intervals $U_9$ - $U_{16}$, $U_{17}$ - $U_{24}$ and so on.

Proof of Lemma 2.2: The proof is geometric and leans of the fact that the so-called value set $V_N(jw)$ (for polynomial family N at frequency ω), defined by

$$V_N(j\omega) = \{n(j\omega): n(s) \in N\},$$

is a level rectangle as depicted in Figure 2. This fact was reported in Dasgupta [7] and Minnichelli, Anagnost and Desoer [6]. The vertices $V_N^i$ of this level rectangle are the Kharitonov polynomials identified for n = even by

$$V_N^1 = K_2;\ V_N^2 = K_4;\ V_N^3 = K_1;\ V_N^4 = K_3$$

and for n = even by

$$V_N^1 = K_1;\ V_N^2 = K_3;\ V_N^3 = K_2;\ V_N^4 = K_4.$$

For simplicity, assume n = even. From (2.5), (2.6a) and (2.6c) it follows from Figure 3 that

$u \in U_1 \Rightarrow V_N(j\sqrt{-u})$ is contained in first quadrant;

$u \in U_2 \Rightarrow V_N(j\sqrt{-u})$ is contained in first and second quadrants;

$u \in U_3 \Rightarrow V_N(j\sqrt{-u})$ is contained in second quadrant;

$u \in U_4 \Rightarrow V_N(j\sqrt{-u})$ is contained in second and third quadrants;

and so forth. By definition, $\Gamma(N,d)$ is SPR-invariant if and only if Condition (SPR) holds for some d(s) and for all $n(s) \in N$. This is equivalent to the existence of a d(s) such that

$$\mathrm{Real}\ [n(j\omega)d^*(j\omega)] > 0, \qquad \text{for all } \omega \geq 0 \tag{2.9}$$

and for all $n(s) \in N$ where $d^*(j\omega)$ is the complex conjugate of $d(j\omega)$. Suppose $u \in U_1$. From Figure 3a we see that

$$\mathrm{Real}\ [n(j\sqrt{-u})d^*(j\sqrt{-u})] > 0 \tag{2.10}$$

for all $n(s) \varepsilon N$ if and only if this condition holds for all $n(j\sqrt{-u}) \varepsilon V_N(j\sqrt{-u})$. This is true if and only if

$$\mathrm{Real}[K_1(j\sqrt{-u})/d^*(j\sqrt{-u})] > 0 \quad \text{and} \quad \mathrm{Real}[K_2(j\sqrt{-u})/d^*(j\sqrt{-u})] > 0. \qquad (2.11)$$

If $u \varepsilon U_2$, then from Figure 3b we see that (2.10) hold if and only if

$$\mathrm{Real}[K_2(j\sqrt{-u})/d^*(j\sqrt{-u})] > 0 \quad \text{and} \quad \mathrm{Real}[K_3(j\sqrt{-u})/d^*(j\sqrt{-u})] > 0.$$

Continuing in this fashion it's clear from (2.9) and the preceding conditions that (2.8) are sufficient for $\Gamma(N,d)$ to be SPR-invariant for some polynomial d(s). This proves Lemma 2.3. ∇∇∇

We are now in a position to state and prove the main result.

**Theorem 2.1: Consider N an interval polynomial as in (1.1) which satisfies**

**(i) $\underline{a}_i = \bar{a}_i$ for all i = even or all i = odd.**

**or**

**(ii) n = 1, 2, or 3.**

**Then, there exists a d(s) such that $\Gamma(N,d)$ is SPR-invariant if and only if the Kharitonov polynomials $K_1$, $K_2$, $K_3$ and $K_4$ are all stable. Furthermore, there exists a $k \geq 0$ such that this d(s) can be written as**

$$d(s) = h_-(s^2) + ksg_+(s^2). \qquad (2.12)$$

Proof of Theorem 2.1: (Necessity) For $\Gamma(N,d)$ to be SPR-invariant it is clear that all of N be stable. Since $K_i(s) \varepsilon N$, i=1,2,3,4, then these Kharitonov polynomials $K_i(s)$ must be stable.

(Sufficiency) Assume that the Kharitionov polynomials $K_i(s)$ are stable. If Condition (i) or (ii) of the theorem statement holds, then we must find a d(s) such that $\Gamma(N,d)$ is SPR-invariant; that is, Condition (SPR) holds for all $n(s) \varepsilon N$.

For $n(s) \varepsilon N$ and a candidate d(s), write

$$n(s) = h_n(s^2) + sg_n(s^2); \quad d(s) = h_d(s^2) + sg_d(s^2). \qquad (2.13)$$

For this n(s) and d(s) Condition (SPR) is equivalent to

$$\mathrm{Real}\ [n(j\omega)d^*(j\omega)] > 0, \quad \text{for all } \omega \geq 0 \qquad (2.14)$$

where $d^*(j\omega)$ is the complex conjugate of $d(j\omega)$. Substituting (2.13) into the left hand side of inequality (2.14) gives

$$\begin{aligned}\text{Real } [n(j\omega)d^*(j\omega)] &= \text{Real } [(h_n(-\omega^2)+j\omega g_n(-\omega^2))(h_d(-\omega^2)-j\omega g_d(-\omega^2)]\\ &= h_n(-\omega^2)h_d(-\omega^2) + \omega^2 g_n(-\omega^2)g_d(-\omega^2)\\ &= h_n(u)h_d(u) - ug_n(u)g_d(u) \qquad (2.15)\end{aligned}$$

where $u = -\omega^2$. From (2.14) and (2.15) we see that Condition (SPR) is equivalent to

$$h_n(u)h_d(u) - ug_n(u)g_d(u) > 0 , \quad \text{for all } u \in (-\infty, 0]. \qquad (2.16)$$

From Lemma 2.2 it suffices to show Condition (SPR) holds for only the Kharitonov polynomials $K_i(s)$. Furthermore, we'll restrict d(s) to the form in (2.5); i.e., in (2.13) we take $h_d \equiv h_-$ and $g_d \equiv kg_+$. Therefore, to complete the proof, we use Lemma 2.2, (2.3)-(2.5) and (2.14) and seek a $k \geq 0$ satisfying

$$h_-(u)h_-(u) - kug_+(u)g_+(u) > 0, \qquad (2.17a)$$
$$h_-(u)h_-(u) - kug_-(u)g_+(u) > 0, \qquad (2.17b)$$
$$h_-(u)h_+(u) - kug_-(u)g_+(u) > 0 \qquad (2.17c)$$

and

$$h_-(u)h_+(u) - kug_+(u)g_+(u) > 0 \qquad (2.17d)$$

for all $u \in (-\infty, 0]$. Later in the proof we'll make use of root locus techniques to find a suitable k satisfying (2.17). For the moment we reconsider Figure 1 and make two observations concerning the inequalities in (2.17). First, $h_-(u)$ and $g_+(u)$ have no common roots. Hence, (2.17a) holds for all $k \geq 0$ and for all $u \in (-\infty, 0]$. Secondly, $h_+(0)$ and $h_-(0)$ are positive. Thus, the requirements in (2.17a-c) are equivalent to finding a $k \geq 0$ such that

$$h_-(u)h_-(u) - kug_-(u)g_+(u) = 0, \qquad (2.18a)$$
$$h_-(u)h_+(u) - kug_-(u)g_+(u) = 0 \qquad (2.18b)$$

and

$$h_-(u)h_+(u) - kug_+(u)g_+(u) = 0 \qquad (2.18c)$$

possess no solutions u in $(-\infty, 0]$. We are now in a position to consider conditions (i) and (ii) in the theorem statement.

<u>Condition (i) holds</u>: (Case 1: $\underline{a}_i = \bar{a}_i$ for i = even) In this situation $h_+ \equiv h_-$, thus the requirements on (2.18a-c) become one of finding a $k \geq 0$ such that

$$h_-(u)h_-(u) - kug_-(u)g_+(u) = 0 \qquad (2.19)$$

has no solutions u in $(-\infty, 0]$. The problem of finding such a k can be resolved using root locus techniques. Indeed, (2.19) is equivalent to

$$1 - k\,\frac{ug_-(u)g_+(u)}{h_-(u)h_-(u)} = 0. \qquad (2.20)$$

It's evident from the root locus diagram in Figure 4 that (2.19) has no roots in $(-\infty, 0]$ for sufficiently small $k \geq 0$. Thus, there exists a (sufficiently small) $k \geq 0$ such that (2.19) possess no roots u in $(-\infty, 0]$. This completes the proof of Case 1.

(Case 2: $\underline{a}_i = \bar{a}_i$ for i = odd) The proof for this case is dual to the proof of Case 1. We have $g_+ \equiv g_-$ so that equations (2.18a-c) reduce to

$$h_-(u)h_+(u) - kug_+(u)g_+(u) = 0. \tag{2.21}$$

Again, using root locus arguments, one sees that (2.14) possesses no roots u in $(-\infty, 0]$ for sufficiently large k; see Figure 3. Therefore, there exists a (sufficiently large) $k \geq 0$ such that (2.12) has no roots in $(-\infty, 0]$. This completes the proof of Case 2.

<u>Condition (ii) holds</u>: We consider the three cases n = 1,2, and 3 separately.

(Case 1: n = 1) Here, $h_+ \equiv \bar{a}_1$, $h_- \equiv \underline{a}_1$, $g_+ \equiv \bar{a}_0$ and $g_- \equiv \underline{a}_0$ which are all positive numbers. If we take $k = 0$, then (2.18) never has non-positive real roots u.

(Case 2: n = 2) The polynomials $h_+(u)$ and $h_-(u)$ are first order, while $g_-(u) = \underline{a}_1$ and $g_+(u) = a_1$. Equations (2.18) become

$$h_-(u)h_-(u) - ku\underline{a}_1\underline{\bar{a}}_1 = 0 \tag{2.22a}$$
$$h_-(u)h_+(u) - ku\underline{a}_1 a_1 = 0 \tag{2.22b}$$

and

$$h_-(u)h_+(u) - ku\bar{a}_1^2 = 0. \tag{2.22c}$$

For all $k \geq 0$, (2.22a) possesses no real, non-positive roots u. Since $-ku\bar{a}_1^2 \geq -ku\underline{a}_1\bar{a}_1$ for all $u \leq 0$, then the requirements on (2.22b) and (2.22c) are met if just (2.22b) holds. However, (2.22b) is just a special case of (2.21). Thus, for sufficiently large $k > 0$, (2.22b) possesses no roots in $(-\infty, 0]$.

(Case 3: n=3) Polynomials $h_+, h_-, g_+$ and $g_-$ are first order with zeroes $\alpha_1^+ < 0$, $\alpha_1^- < 0$, $\beta_1^+ < 0$ and $\beta_1^- < 0$ respectively. To make use of Lemma 2.2 we identify the five intervals (p = 4m+1 = 5)

$$U_1 = [\alpha_1^-, 0];\ U_2 = [\alpha_1^+, \alpha_1^-];\ U_3 = [\beta_1^-, \alpha_1^+];\ U_4 = [\beta_1^+, \beta_1^-];\ U_5 = [-\infty, \beta_1^+]. \tag{2.24}$$

With the form of d(s) taken as in (2.12) we now find a $k > 0$ such that conditions (2.8b) of Lemma 2.2 are satisfied.

From the preceding development, (2.13)-(2.18), we see using (2.4) and (2.18a-c) that the requirements in (2.8b) are equivalent to the existence of a $k > 0$ such that

(2.18b) has no solutions $u \varepsilon U_1$; (2.25a)
(2.18a) and (2.18b) have no solutions $u \varepsilon U_2$; (2.25b)
(2.18a) and (2.18c) have no solutions $u \varepsilon U_3$; (2.25c)
(2.18b) and (2.18c) have no solutions $u \varepsilon U_4$; (2.25d)
and
(2.18b) has no solutions $u \varepsilon U_5$. (2.25e)

The solutions to (2.18a-c), as a function of k, can be determined from the three root loci in Figure 6. Now, choose $k > 0$ such that the solutions to (2.18b) are neither real nor negative. It's clear from Figure 6b that this is possible; e.g., take $k > 0$ so that the roots of (2.18b) coincide with the "*'s" in Figure 6b. For this value of $k = k_*$ all conditions in (2.25) are satisfied. This is clear from (2.24), (2.25) and Figure 6. For instance, conditions (2.25a-e) are satisfied since the root locus in Figure 6b does not touch $U_1$ or $U_5$. Finally, observe that even though the root loci associated with (2.18a) and (2.18c), see Figures 6a and 6b respectively, touch intervals $U_2$ and $U_4$, none of the conditions (2.25) are violated. Therefore, for $k = k_*$, all of the conditions (2.25) are satisfied. This implies that $d(s) = h(s^2) + sk_*g_+(s^2)$ satisfies the conditions of Lemma 2.2. Hence, $\Gamma(N,d)$ is SPR-invariant which proves Case 3 and the theorem. ∇∇∇

## 3. EXAMPLE

Consider an interval polynomial family N of degree n = 3 described by

$$a_0s^3 + a_1s^2 + a_2s + a_3; \quad a_i \varepsilon [\underline{a}_i, \bar{a}_i]$$

where $\underline{a}_0 = \bar{a}_0 = \underline{a}_1 = \bar{a}_1 = 1$; $\underline{a}_2 = 3$; $\bar{a}_2 = 4$; $\underline{a}_3 = 1$; $\bar{a}_3 = 2$. The four Kharitonov polynomials in (1.3) are

$$K_1(s) = s^3 + s^2 + 3s + 2; \quad K_2(s) = s^3 + s^2 + 4s + 1;$$
$$K_3(s) = s^3 + s^2 + 4s + 2; \quad K_4(s) = s^3 + s^2 + 3s + 1$$

which are all stable. We conclude from Theorem 2.1 that there exists a polynomial d(s) such that the family of transfer functions $\Gamma(N,d)$ is SPR-invariant. Moreover, a suitable d(s) is given by

$$d(s) = h_+(s^2) + ksg_+(s^2)$$

where $h_+$ and $g_+$ are defined in (2.2) and given by

$$h_+(s) = s + 2; \quad g_+(s) = s + 4$$

and where K is some non-negative number. In fact, a class of admissible d(s) is

$$d(s) = ks^3 + s^2 + 4ks + 2; \quad k \varepsilon [.05, 4.1].$$

The research of the first author was supported in part by NSF Grant ECSE-8858366. The second author's research was supported in part by the National Natural Science Foundation of China.

REFERENCES

[1] B.D.O Andersen, R.R. Bitmead, C.R. Johnson, P.V. Kokotovic, R.L. Kosut, I.M.Y. Mareels, L. Praly and B.D. Riedle, Stability of Adaptive Systems: Passivity and Averaging Analysis, MIT Press Cambridge, 1986.

[2] S. Dasgupta and A.S. Bhagwat, "Conditions for Designing Strictly Positve Real Transfer Functions for Adaptive Output Error Identification," IEEE Transactions on Circuits and Systems, Vol. CAS-34, no. 7, pp. 731 - 736, 1987.

[3] V.L. Kharitonov, "Asymptotic Stability of an Equilibrium Position of a Family of Linear Differential Equations," Differential'nye Uravneniya, Vol. 14, pp. 2086 2088, 1978.

[4] F.R. Gantmacher, Matrix Theory, Vol. 2, Chelsea, New York, 1959.

[5] S. Dasgupta, B.D.O Anderson and R.J. Kaye, "Output-Error Identification Methods for Partially Known Systems," International Journal of Control, Vol. 43, no. 1, pp. 177 - 191, 1986.

[6] R. Minnichelli, J.J. Anagnost, and C.A. Desoer, "An Elementary Proof of Kharitonov's Theorem with Extensions," to appear IEEE Transactions on Automatic Control, August, 1989.

[7] S. Dasgupta, "Perspective on Kharitonov's Theorem: A View from the Imaginary Axis," University of Iowa, Iowa City, IA, 52242, 1987.

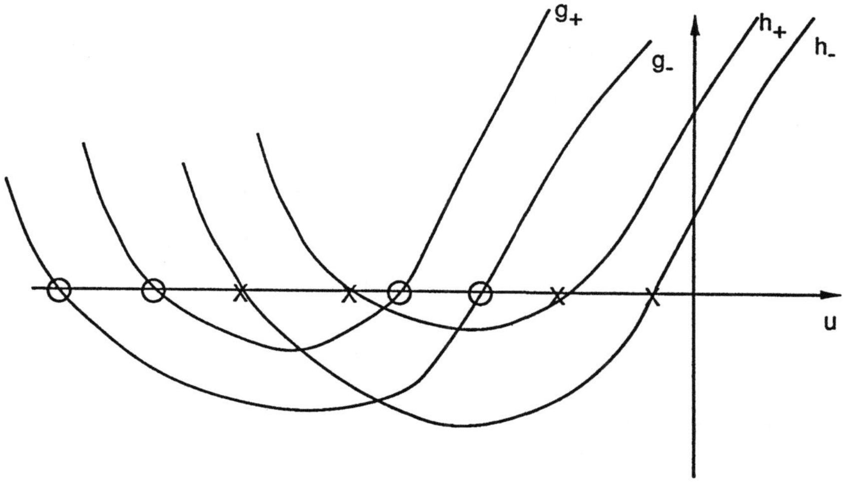


Figure 1: The pairs $(h_+, g_+)$, $(h_+,g_-)$, $(h_-,g_+)$ and $(h_-,g_-)$ are all positive pairs, see Gantmacher [4]. Furthermore, h+ and g+ dominate h- and g- respectively.

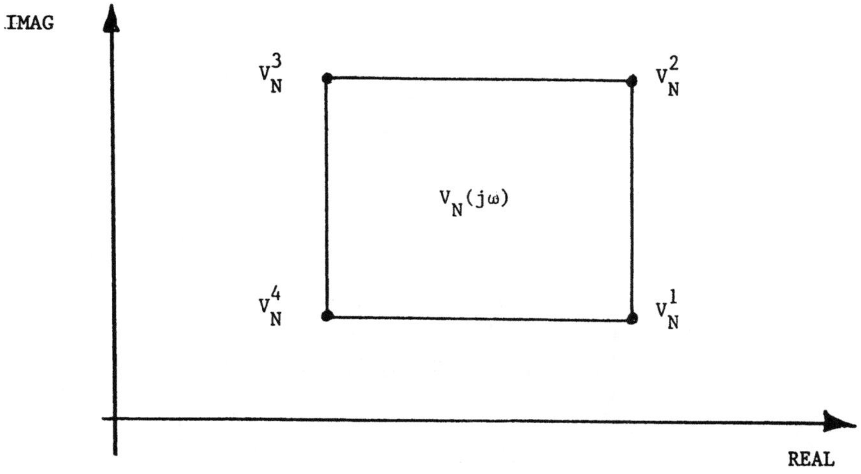


Figure 2: The value set of an interval of polynomials is a level rectangle.

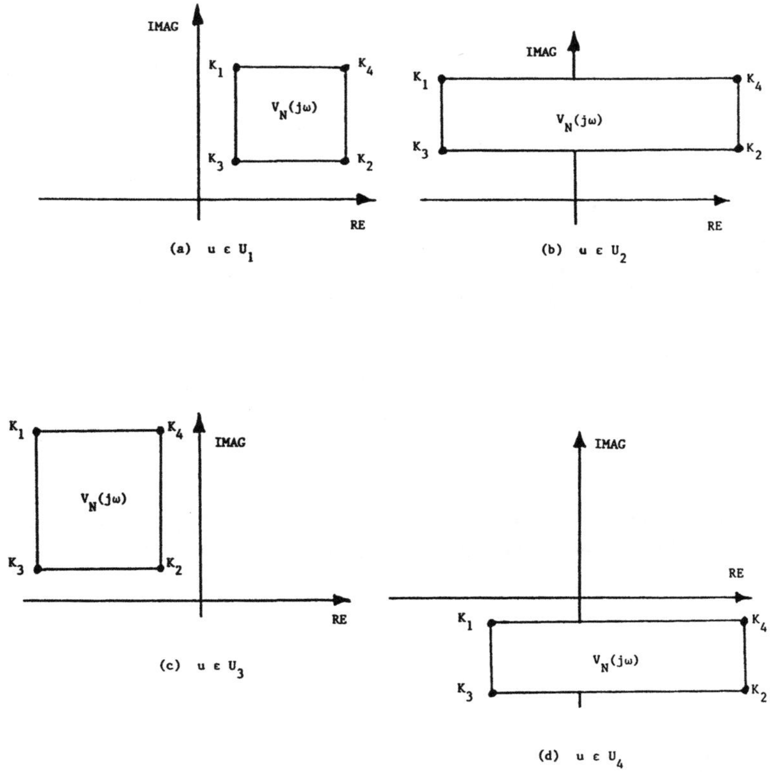


Figure 3: The value set $V_N(j\omega)$, where N is a stable, level rectangle of polynomials, traces out a distinctive path in the complex plane.

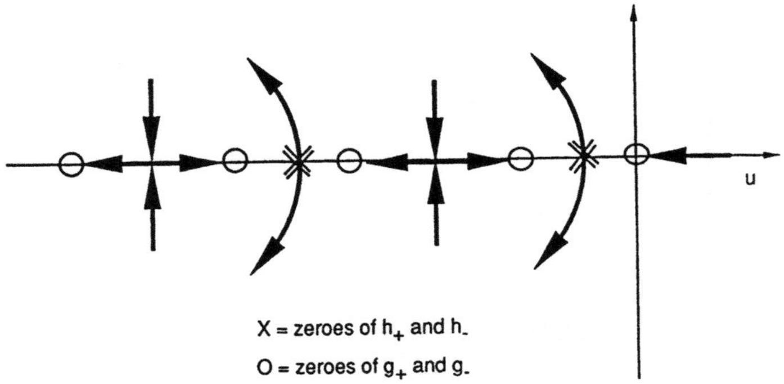


Figure 4: From the root locus we see that for sufficiently small values of the variable k that (2.19) has no real roots $\mu$ in $(-\infty,0]$.

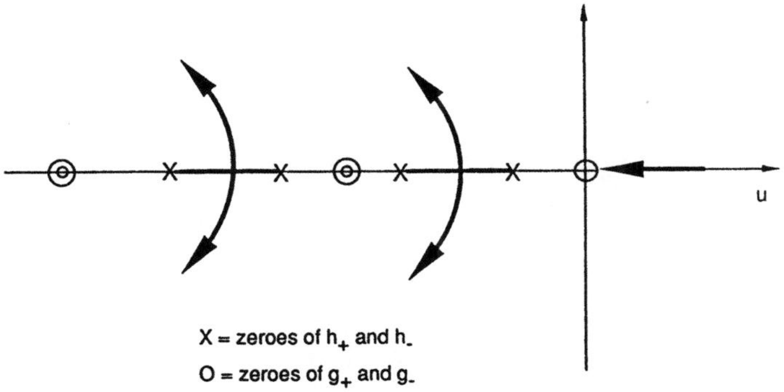


Figure 5: This root locus possesses no solutions on the negative real axis as long as the root locus variable k is sufficiently large. This locus describes the roots of (2.21).

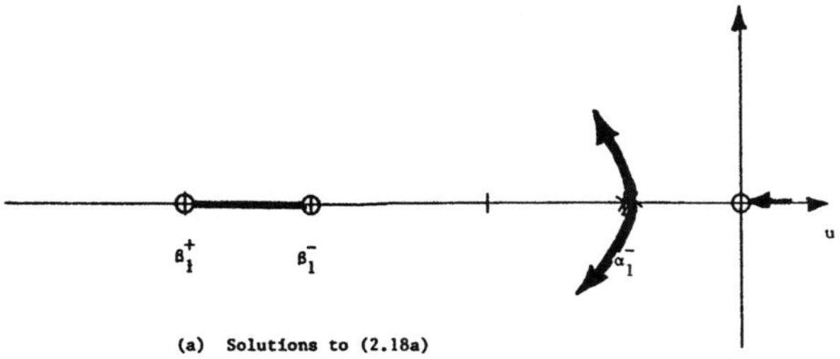


(a) Solutions to (2.18a)

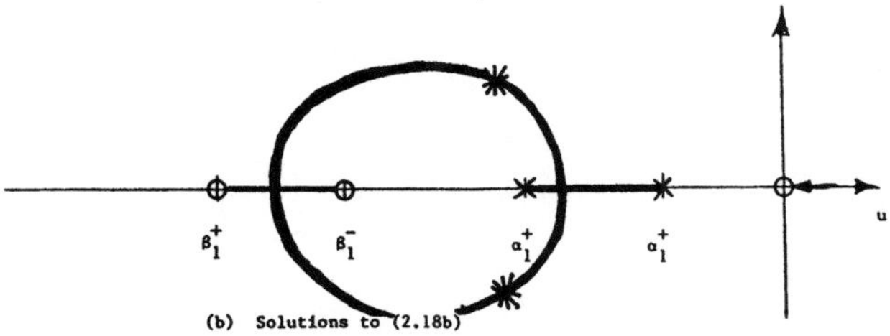


(b) Solutions to (2.18b)

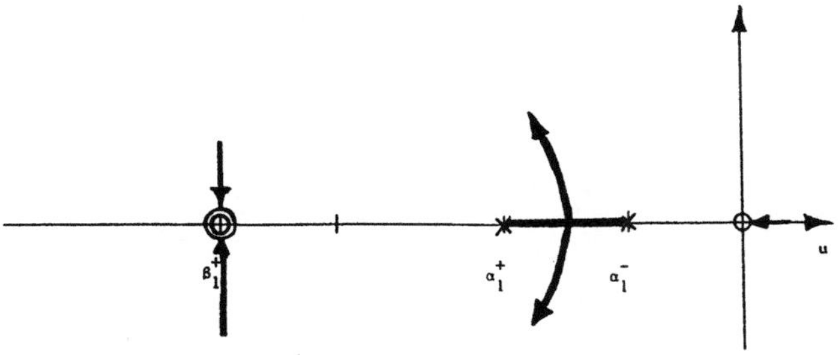


(c) Solutions to (2.18c)

Figure 6: For n = 3, equations (2.18a-c) have solutions as depicted by these root loci. These results are used in finding a k > 0 such that (2.25a-e) are satisfied.

# QUADRATIC STABILIZABILITY OF LINEAR SYSTEMS WITH STRUCTURAL INDEPENDENT TIME VARYING UNCERTAINTIES

Kehui Wei

Abstract

This paper investigates the problem of designing a linear state feedback control to stabilize a class of single-input uncertain linear dynamical systems. The systems under consideration contain time-varying uncertain parameters whose values are unknown but bounded in given compact sets. The method used to establish asymptotical stability of the closed loop system (obtained when the feedback control is applied) involves the use of a quadratic Lyapunov function. Under the assumption that each entry of system matrices independently varies in a sufficient large range we first show that to insure a system stabilizable some entries of the system matrices must be sign invariant, more precisely, the number of the least required sign invariant entries is equal to the system order. Then, for a class of systems containing both the least required sign invariant entries and sign varying structural uncertainties we provide the necessary and sufficient conditions under which the system can be quadratically stabilized by a linear control for all admissible variations of uncertainties. The conditions show that all uncertainties can only enter the system matrices in a way to form a particular geometrical pattern called "anti-symmetry stepwise configuration".

## 1. Introduction.

In recent years, the problem of designing a feedback control law to stabilize an uncertain dynamical system has received considerable attention; e.g., see [1]-[20] and their bibliographies. In this paper the uncertain dynamical systems under consideration are described by state equations containing time-varying uncertain parameters which are unknown but bounded in a prescribed arbitrary compact sets. In order to establish the stability of the closed loop time varying uncertain system, a quadratic Lyapunov function is used. The quadratic stabilization problem can be roughly stated as follows: Provide conditions under which it is possible to find a suitable quadratic Lyapunov function and design a continuous feedback controller which guarantees uniform asymptotic stability of the origion for all admissible variations of uncertainties.

It is convenient to classify the existing results on robust stabilization into two categories. First, there are a number of results which treate the uncertain system as a nominal system with uncertain perturbations. No special assumptions on the location of uncertain entries are required. By using the knowledge of the nominal system, one can construct a feedback control and a related Lyapunov function to prove uniform asymptotic stability for all admissible uncertainties; e.g., see [1], [9],[13] and [15]. With this method, usually only sufficiently small "size" of perturbation is allowable.

The second category of results are applicable to systems having some arbitrarily large varying terms. In this case, the locations of uncertainties in the system matrices play a crucial role. It has been found that if sufficiently large uncertainties enter some entries, the system may not be stabilizable. In other words, to guarantee robust stabilizability of a system, one has to propose some restrictions on which entries of system matrices are permitted to be uncertain. In [4], [7]-[8] and [11]-[12], the uncertainty in the systems is assumed to satisfy the so-called "matching conditions". In view of the fact that matching condition as sufficient conditions are often too conservative, generalized matching conditions are found in [16], provided the nominal system is of cannonical form. This result is further generalized to obtain so-called "admissible shuffle" structures for single-input systems in [2]. In [10], a sufficient condition on multi-input systems is proposed, also based on a canonical form assumption.

This paper is fall into the second category. The main aim of this paper is to examine the following question: When each entry of system matrices varies independently in a sufficient large range, under what conditions can the system be quadratically stabilized by a linear control. It is not hard to see that if every entry is allowed to vary independently through zero, the system can not be stabilizable. In other words, in order to guarantee the stabilizability for an uncertain system some entries of the system matrices must be sign invariant (including a constant number). We first show that to insure a system stabilizable via linear control the least number of sign invariant uncertain entries in the system matrices is equal to the system order, provided that those sign invariant uncertain entries locate in proper places. Then, under the assumption that a system containing the least required sign invariant uncertainties in proper entries and structural uncertainties (i.e., uncertainties which vary in arbitrary bounding sets) in some of the rest entries we derive necessary and sufficient conditions under which the uncertain system is quadratically stabilizable via linear control. Roughly speaking, those structural uncertainties can only locate in such places which form a certain geometrical pattern called anti-symmetry stepwise configuration. This paper has some siliant features: First, our stabilizability conditions can be easily checked by just examining the uncertainty locations in the system matrices. Second, once a system satisfies the stabilizability conditions, a suitable quadratic Lyapunov function and a desired linear stabilizer can be computed following a constructive procedure. The control gain will depend on the given varying ranges of all uncertain parameters. Thirdly, as a necessary and sufficient condition, our criterion captures all the stabilizable systems which satisfying the forementioned sufficient conditions in the same category.

Due to space constraints, all proofs of preliminary lemmas and main theorems has been omitted and can be found in [19].

## 2. Systems, assumptions and definitions.

We consider a <u>linear time varying uncertain system</u> $\Sigma(A(q(t)), b(q(t)))$ (or <u>uncertain system</u> $\Sigma(A(q), b(q))$ for short) described by the state equation

$$\dot{x}(t) = A(q(t))x(t) + b(q(t))u(t) \; ; \; t \geq 0$$

where $x(t) \in R^n$ is the state; $u(t) \in R$ is the control; $q(t) \in Q \subset R^p$ is the model uncertainty which is restricted to a prescribed bounding set Q. Within this framework, $A(\cdot)$ and $b(\cdot)$ must be n×n and n×1 dimensional matrix functions on the set Q, respectively. Hence for fixed $q \in Q$, A(q) and b(q) are the model matrices which result.

In this paper, unless otherwise stated, we assume that A(q) and b(q) each depends on different components of q; that is, we have $q=[r : s]'$, where $A(\cdot)$ depends solely on r and $b(\cdot)$ on s.

Throughout the remainder of this paper, the following assumptions are taken as standard:

Assumption 2.0.1 (continuity): The matrices $A(\cdot)$ and $b(\cdot)$ depend continuously on their arguments.

Assumption 2.0.2 (compactness): The bounding set Q is compact.

Assumption 2.0.3 (measurability): The uncertainty $q(\cdot)$: $[0,\infty) \rightarrow Q$ is required to be Lebesgue measurable.

Definition 2.1: An uncertain system $\Sigma(A(q(t)), b(q(t)))$ is said to be quadratically stabilizable (with respect to Q) if there exists a continuous (feedback control) mapping $p(\cdot)$: $R^n \rightarrow R$ with $p(0)=0$, an n×n positive-definite symmetric matrix P and a constant $\alpha > 0$ leading to the satisfaction of the following condition: Given any admissible uncertainties $q(\cdot)$, it follows that

$$L(x,t) \triangleq x'[A'(q(t))P + PA(q(t))]x + 2x'Pb(q(t))p(x) \leq -\alpha\|x\|^2 \qquad (2.1.1)$$

for all pairs $(x,t) \in R^n \times [0, +\infty)$. L(x,t) is the so-called Lyapunov derivative associated with the quadratic Lyapunov function $V(x) \triangleq x'Px$. Furthermore, $\Sigma(A(q(t), b(q(t)))$ is said to be quadratically stabilizable via linear control (with respect to Q) if $p(x)=Kx$ where K is a 1×n constant matrix.

Definition 2.2: An n×n uncertain matrix $M(q) \triangleq \{m_{i\,j}(q)\}$ is said to be in the standard form if for each $i = 1, 2, \ldots, n-1$, $m_{i\,i+1}(q)$ is an independent sign invariant function of q (including a constant function).

In the sequel, for notational simplicity, we always use $\theta$ to denote an entry which is a sign invariant uncertainty. Note that $\theta$ in different entries are not necessarily a same function of q.

Definition 2.3: An uncertain system $\Sigma(A(q), b(q))$ is said to be in the standard form with structural independent uncertainties if its corresponding matrix M(q) defined as

$$M(q) \triangleq \begin{bmatrix} A(q) & : & b(q) \\ \hline 0 & : & 0 \end{bmatrix} \triangleq \{m_{i\,j}(q)\} \tag{2.3.1}$$

is in the standard form and every non-supperdiagonal entry $m_{i\,j}(q)$ of M(q) is zero or varies independently in $[-r_{i\,j} \;\; r_{i\,j}]$ where $r_{i\,j}>0$ is allowed to be arbitrarily large.

Definition 2.4: An uncertain system $\Sigma(A(q), b(q))$ is said to have an anti-symmetric stepwise configuration if its corresponding matrix M(q) as in (2.3.1) satisfies the following conditions:

i) M(q) is in the standard form as in Definition 2.2;

ii) If $p \geq k+2$ and $m_{k\,p}(q) \not\equiv 0$, then $m_{u\,v}(q) \equiv 0$ for all $u \geq v$, $u \leq p-1$ and $v \leq k+1$.

Remark 2.5: A roughly geometric interpretation of an anti-symmetric stepwise configuration is shown in Figure 2.1 where the shaded regions denote permissible uncertain entries and the empty regions are composed entirely of zero entries. Note that in accordence with Definition 2.4 a precise geometric interpretation of an anti-symmetric stepwise configuration is easy to determine.

3. Preliminary lemmas for proving main theorems.

Lemma 3.1: (see [6] for proof): An uncertain system $\Sigma(A(q), b(q))$ is quadratically stabilizable if and only if there exists an $n \times n$ positive-definite matrix S such that

$$x'(A(q)S + SA'(q))x < 0$$

for all pairs $(x,q) \in N \times Q$ with $x \neq 0$ where $N \triangleq \{x \in R^n : b'x=0$ for some $b \in \text{conv}\,\{b(q) : q \in Q\}$.

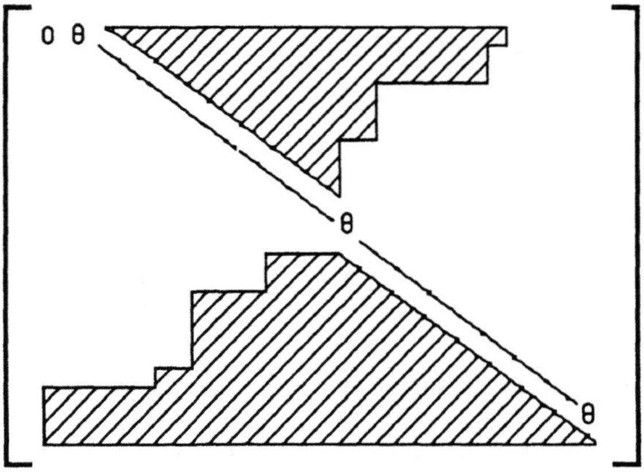


Figure 2.1: An anti-symmetric stepwise configuration.

Corollary 3.2: An uncertain system $\Sigma(A(q), b(q))$ with

$$b(q) \triangleq \begin{bmatrix} 0 \\ \hdashline \theta \end{bmatrix}$$

where $\theta$ is a sign invariant uncertain function of q is quadratically stabilizable if and only if there exists an n×n positive-definite matrix S and $\Theta =[\, I_{n-1} : 0\,]'$ such that

$$\pi(S,q) \triangleq \Theta'(A(q)S + SA'(q))\Theta \tag{3.2.1}$$

is negative-definite for all $q \in Q$.

The pair $(S, \pi)$ satisfying Corollary 3.2 is called an admissible pair for the system $\Sigma(A(q), b(q))$.

Lemma 3.3: (see [9] for proof): Consider an uncertain system $\Sigma(A(q), b)$ where b is a constant vector. Let $\Theta$ be any n×(n-1) orthonormal matrix whose columns span N(b'). Then the system $\Sigma(A(q), b)$ is quadratically stabilizable via linear control if and only if there exists an n×n positive-definite matrix S such that

$$\pi(s,q) \triangleq \Theta'(A(q)S + SA'(q))\Theta \tag{3.3.1}$$

is negative-definite for all $q \in Q$.

Lemma 3.4: (see [2] for proof): Consider an uncertain system $\Sigma(A(q(t)), b(q(t)))$. Define the (n+1)-dimentional system $\Sigma^+(A^+(q(t)), b^+(q(t)))$

$$\dot{x}^+(t) = A^+(q(t))x^+(t) + b^+(q(t))u^+(t) \; ; \; t \geq 0$$

where

$$A^+(q) \triangleq \begin{bmatrix} A(q) & : & b(q) \\ \hdashline 0 & : & 0 \end{bmatrix} \; ; \quad b^+(q) \triangleq \begin{bmatrix} 0 \\ \hdashline \theta \end{bmatrix}.$$

Then $\Sigma(A(q(t)), b(q(t)))$ is quadratically stabilizable via linear control if and only if $\Sigma^+(A^+(q(t)), b^+(q(t)))$ is quadratically stabilizable via linear control.

Note that in [2], only $b^+=[0\ 0\ \dots\ 0\ 1]'$ is considered. However, the same proof is also valid when $b^+$ is replaced by $b^+(q(t))=[0\ 0\ \dots\ 0\ \theta]'$.

Definition 3.5: Consider an uncertain system $\Sigma(A(q), b(q))$ where

$$b(q) \triangleq \begin{bmatrix} 0 \\ \hdashline \theta \end{bmatrix}$$

where $\theta$ is a sign invariant uncertainty. A down-augmentation system $\Sigma^+(A^+(q), b^+(q))$ of $\Sigma(A(q), b(q))$ is defined as follows:

$$A^+(q) \triangleq \begin{bmatrix} A(q) & : & b(q) \\ \hdashline * & : & * \end{bmatrix} \; ; \quad b^+(q) \triangleq \begin{bmatrix} 0 \\ \hdashline \theta \end{bmatrix}$$

where $\theta$ is a sign invariant uncertainty.

Lemma 3.6: Consider an uncertain system $\Sigma(A(q), b(q))$ and its down-augmentation system $\Sigma^+(A^+(q), b^+(q))$ as in Definition 3.5. Then, if $\Sigma(A(q), b(q))$ is quadratically stabilizable, $\Sigma^+(A^+(q), b^+(q))$ is also quadratically stabilizable.

Definition 3.7: Consider an uncertain system $\Sigma(A(q), b(q))$ where

$$A(q) \triangleq \left[\begin{array}{c:c} 0 & A^-(q) \\ \hdashline * & * \end{array}\right] \; ; \quad b(q) \triangleq \left[\begin{array}{c} 0 \\ \hdashline \theta \end{array}\right].$$

An up-augmentation system $\Sigma^+(A^+(q), b^+(q))$ of $\Sigma(A(q), b(q))$ is defined as follows:

$$A^+(q) \triangleq \left[\begin{array}{c:c:c} 0 & \theta & * * \cdots * \\ \hdashline 0 & 0 & A^-(q) \\ \hdashline * & * & * * \cdots * \end{array}\right] \; ; \quad b^+(q) \triangleq \left[\begin{array}{c} 0 \\ \hdashline \\ b(q) \end{array}\right].$$

Lemma 3.8: Consider an uncertain system $\Sigma(A(q), b(q))$ and its up-augmentation system $\Sigma^+(A^+(q), b^+(q))$ as in Definition 3.7. Then, if $\Sigma(A(q), b(q))$ is quadratically stabilizable, $\Sigma^+(A^+(q), b^+(q))$ is also quadratically stabilizable.

Observation 3.9: If an uncertain system $\Sigma(A(q), b(q))$ satisfies the following condition:

$$A^+(q) \triangleq \left[\begin{array}{c:c} A(q) & b(q) \\ \hdashline 0 & 0 \end{array}\right]$$

has an anti-symmetric stepwise configuration. Then the system $\Sigma^+(A^+(q), b^+)$

$$A^+(q) \triangleq \left[\begin{array}{c:c} A(q) & b(q) \\ \hdashline 0 & 0 \end{array}\right] \; ; \quad b^+ \triangleq \left[\begin{array}{c} 0 \\ \hdashline 1 \end{array}\right]$$

can always be generated from $A_0=[\,*\,]$, $b_0=[\,\theta\,]$ or $A_0=[\,0\,]$, $b_0=[\,\theta\,]$ via a sequence of augmentation (either down or up).

Lemma 3.10: Consider the free system $\Sigma(A_c, b_c)$ where

$$A_c \triangleq \begin{bmatrix} 0 & 1 & 0 & \cdots & 0 \\ 0 & 0 & 1 & \cdots & 0 \\ \vdots & & & & \vdots \\ \vdots & & & & 0 \\ \vdots & & & & 1 \\ 0 & 0 & & \cdots & 0 \end{bmatrix} \; ; \quad b_c \triangleq \begin{bmatrix} 0 \\ 0 \\ \vdots \\ \vdots \\ 0 \\ 1 \end{bmatrix}. \tag{3.10.1}$$

If there is an admissible pair $(S,\pi)$ for the system $\Sigma(A_c, b_c)$, then the entries $s_{i\,j}$ of S have the following properties:

1) $s_{i\,i} > 0$ for all $i = 1, 2, \ldots, n$ and $s_{i\,i+1} < 0$ for all $i = 1, 2, \ldots, n-1$.

2) If $s_{i+1\,i+1} \geq \alpha^2 s_{i\,i}$, then for $i+1 \leq k \leq n$

$$s_{k\,k} > \alpha^2 4^{-4(k-i-1)} s_{k-1\,k-1}.$$

When $\alpha^2$ is large enough, then

$$s_{i\,i} < s_{i+1\,i+1} < \ldots < s_{n\,n}$$

3) if $s_{i-1\,i-1} \geq \beta^2 s_{i\,i}$, then for $i-1 \geq k \geq 1$

$$s_{k\,k} > \beta^2 4^{-4(i-k-1)} s_{k+1\,k+1}.$$

When $\beta^2$ is large enough, then

$$s_{i\,i} < s_{i-1\,i-1} < \cdots < s_{1\,1}.$$

Lemma 3.11: Consider an uncertain system $\Sigma(A_c(q), b_c)$

$$A_c(q) \triangleq \begin{bmatrix} 0 & 1 & 0 & \ldots & 0 \\ 0 & 0 & 1 & \ldots & 0 \\ \vdots & & & & \vdots \\ \vdots & & & & \vdots \\ \vdots & & a_{u\,v} & & 0 \\ \vdots & & & & 1 \\ 0 & 0 & \ldots & & 0 \end{bmatrix}; \quad b_c \triangleq \begin{bmatrix} 0 \\ 0 \\ \vdots \\ \vdots \\ \vdots \\ 0 \\ 1 \end{bmatrix}$$

having one structural uncertainty $a_{u\,v}$ where $1 \leq v \leq u \leq n-1$, $|a_{u\,v}(q)| \leq r$ and r is sufficiently large. If there is an admissible pair $(S, \pi)$ for the system $\Sigma(A_c(q), b_c)$, then the entries $s_{i\,j}$ of S have following properties:

1) $|s_{u\,u+1}| > |s_{v-1\,v}|$ when $v \neq 1$;

2) $|s_{u\,u+1}|\,|s_{v\,v+1}| > s^2_{v\,v}$;

3) $s_{u\,u} < s_{u+1\,u+1} < \cdots < s_{n\,n}$.

Lemma 3.12: Consider an uncertain system $\Sigma(A_c, b_c(q))$

$$A_c \triangleq \begin{bmatrix} 0 & 1 & 0 & \ldots & 0 \\ 0 & 0 & 1 & \ldots & 0 \\ \vdots & & & & \vdots \\ \vdots & & & & \vdots \\ \vdots & & & & 0 \\ \vdots & & & & 1 \\ 0 & 0 & \ldots & & 0 \end{bmatrix}; \quad b_c(q) \triangleq \begin{bmatrix} 0 \\ \vdots \\ a_{k\,p} \\ \vdots \\ 0 \\ 1 \end{bmatrix}$$

where $a_{k\,p}$ is a structural uncertainty : $|a_{k\,p}(q)| \leq r$ and r is sufficiently large. If there exists an admissible pair $(S, \pi)$ for the system $\Sigma(A_c, b_c(q))$, then the entries $s_{i\,j}$ of S have the following properties:

1) $|s_{k\,k+1}|\,|s_{n\,n-1}| > s^2_{n\,n}$;

2) $s_{1\,1} > s_{2\,2} > \cdots > s_{k\,k} > s_{k+1\,k+1}$.

Lemma 3.13: Consider an uncertain system $\Sigma(A_c(q, b_c(q))$

$$A_c(q) \triangleq \begin{bmatrix} 0 & 1 & 0 & \dots & 0 \\ 0 & 0 & 1 & \dots & 0 \\ \vdots & & & & \vdots \\ \vdots & & & & \vdots \\ \vdots & & & & 0 \\ \vdots & & & & 1 \\ 0 & 0 & a_{u\,v} & \dots & 0 \end{bmatrix} ; \quad b_c(q) \triangleq \begin{bmatrix} 0 \\ \vdots \\ a_{k\,p} \\ \vdots \\ 0 \\ 1 \end{bmatrix}$$

where $a_{u\,v}$ and $a_{k\,p}$ are independent structural uncertainties. If $k \geq v-1$, then the system is not quadratically stabilizable.

## 4. Main results.

We now state our main results.

Theorem 4.1: Consider an uncertain system $\Sigma(A(q(t)), b(q(t)))$. If every entry of system matrices $A(q)$ and $b(q)$ is an independent uncertain function of $q$, then the system is quadratically stabilizable via linear control only if the following condition holds: There are sign invariant entries in every row of the first $n$ rows and every column of $n$ columns of the $(n+1)\times(n+1)$ matrix $M(q)$ where

$$M(q) \triangleq \begin{bmatrix} A(q) & \vdots & b(q) \\ \hline 0 & \vdots & 0 \end{bmatrix}.$$

Theorem 4.1 implies that if a system having independent varing entries is stabilizable, the number of the least required sign invariant entries is equal to the system order, provided they are in proper locations. In the follows, we only consider the systems having the lowest number of sign invariant uncertain entries. Obviously, the supper diagonal of $M(q)$ is a proper location for sign invariant entries in order to satisfy the requirements of Theorem 4.1.

Theorem 4.2: An uncertain system $\Sigma(A(q(t)), b(q(t)))$ in the standard form with structural independent uncertainties is quadratically stabilizable via linear control if and only if the system has an anti-symmetric stepwise configuration.

This research was supported by Deutsche Forschungsanstalt fuer Luft- und Raumfahrt.

References


[1] Barmish, B. R. and G. Leitmann (1982). On ultimate boundedness control of uncertain systems in the absence of matching conditions. IEEE Trans. Aut. Control, **AC-27**, 153.

[2] Barmish, B. R. (1982). On enlarging the class of stabilizable linear systems with uncertainty. Proc. IEEE Asilomar Conf. Comput., Cts Syst., Monterey.

[3] Barmish, B. R., I. R. Petersen and A. Feuer (1983). Linear ultimate boundedness control of uncertain dynamical systems. Automatica, **19**, 523.

[4] Barmish, B. R., M. Corless and G. Leitmann (1983). A new class of stabilizing controllers for uncertain dynamical systems. SIAM J. Control & Opt., **21**, 246.

[5] Barmish, B. R. (1983). Stabilization of uncertain systems via linear control. IEEE Trans. Aut.Control, **AC-28**, 848.

[6] Barmish, B. R. (1985). Necessary and sufficient conditions for quadratic stabilizability of an uncertain system. J. Opt. Theory Applic., **46**, 399.

[7] Corless, M. and G. Leitmann (1981). Continuous state feedback guaranteeing uniform ultimate boundedness of uncertain dynamic systems. IEEE Trans. Aut. Control, **AC-26**, 1139.

[8] Gutman, S. and Z. Palmor (1982). Properties of min-max controllers in uncertain dynamical systems. SIAM J. Control & Opt., **20**, 850.

[9] Hollot, C. V. and B. R. Barmish (1980). Optimal quadratic atabilizability of uncertain linear systems. Proc. 18th Allerton Conf. Communic., Control Comput., University of Illinois, Monticello.

[10] Hollot, C. V. (1985). Matrix uncertainty structures for robust stabilizability. Proc. of the American Control Conf.. Boston.

[11] Leitmann, G. (1979). Guaranteed asymptotic stability for some linear systems with bounded uncertainties. J. Dyn. Syst., Meas. Control, **101**, 212.

[12] Leitmann, G. (1981). On the efficacy of nonlinear control in uncertain linear systems. J. Dyn. Syst., Meas. Control, **103**, 95.

[13] Patel, R. V., M. Toda and B. Sridhar (1977). Robustness of linear quadratic state feedback designs in the presence of system uncertainty. IEEE Trans. Aut. Control, **AC-22**, 290.

[14] Petersen, I. R. (1985). Structural stabilization of uncertain systems: necessity of the matching condition. SIAM J. Control & Opt., **23**, 286.

[15] Petersen, I. R. and C. V. Hollot (1986). A riccati equation approach to the stabilization of uncertain linear systems. Automatica, **22**, 397.

[16] Petersen, I. R. (1988). Quadratic stabilizability of uncertain linear systems containing both constant and time-varying uncertain parameters. J. Opt. Theory Applic., **57**, 439.

[17] Thorp, J. S. and B. R. Barmish (1981). On guaranteed stability of uncertain linear systems via linear control. J. Opt. Theory Applic., **35**, 559.

[18] Wei, K. (1989). Robust stabilizability of linear time varying uncertain dynamical systems via linear feedback control. Proc. IEEE International Conference on Control and Applications. Jerusalem.

[19] Wei, K. The full version of this paper, to be published in IEEE Trans. Aut. Control.

[20] Zhou, K. and P. Khargonekar (1988). Robust stabilization of linear systems with norm-bounded time-varying uncertainty. Systems & Control Letters, **10**, 17.

# A Finite Zero Exclusion Principle

**Anders Rantzer**

*Abstract.* *The paper shows that the "frequency sweep" in the well known "zero exclusion principle" for checking robust stability of linear systems, can be avoided. In fact, we derive a simple sufficient condition for zero exclusion in an entire frequency interval. The main idea is that a polynomial $p$ of degree $n$ is Hurwitz if and only if $\sum_{k=2}^{N} \arg(p(i\omega_k)/p(i\omega_{k-1})) > (n-1)\pi$ for some $\omega_1 < \ldots < \omega_N$.*

*As an application, we consider polynomials of degree $n$ with coefficients depending linearly on $m$ parameters in the interval $[0,1]$. The number of calculations for checking Hurwitz stability of the complete family then grows only as $n^2 m \log m$, but depends also on the "stability margin" of the family.*

*The test is also applied to problems with multilinear parameter dependence, in particular checking positive realness of a rational function, whose numerator and denominator depend linearly on interval bounded parameters.*

## Introduction

Considering stability robustness of linear systems raises the question whether the characteristic polynomial is Hurwitz for every possible value of some uncertain parameters. (A Hurwitz polynomial is usually defined as a polynomial with real coefficients, having no zero $z$ with Re $z \geq 0$. For convenience, we shall use the same notation for polynomials with complex coefficients and for families of such objects.) A celebrated theorem by V.L. Kharitonov [6] [7], states that a family of polynomials, defined by letting the coefficients vary independently of each other in specified intervals, is Hurwitz if and only if four special "corner polynomials" of the family are Hurwitz.

Other methods have been proposed for treating other stability regions and letting several coefficients of the polynomial depend on one common set of uncertain variables. In particular, Bartlett, et.al. [3], proved that the convex hull of a set of polynomials $p_1(s), \ldots, p_m(s)$, is $\Omega$-stable (has all its zeros within the pathwise connected region $\Omega \subset \mathbf{C}$), if all the edges $\lambda p_i(s) + (1-\lambda)p_j(s)$, $1 \leq i, j \leq m$, $0 \leq \lambda \leq 1$, are $\Omega$-stable. Edge stability may be checked using the criterions in [4] and [5].

An older result, recently developed by Barmish [2] and Anagnost, et.al. [1], is the "zero exclusion principle", stating that a connected family of polynomials is $\Omega$-stable if it contains a stable polynomial and no polynomial of the family has a zero on the boundary of $\Omega$. For example, Anagnost, et.al. sweep the boundary of $\Omega$, plotting, for a sufficient number of points, the value closest to zero that is taken in that point by a polynomial of the family. The family is stable, unless the plot intersects zero. Considering Hurwitz stability, this paper shows that the boundary sweep can actually be replaced by calculations for a much smaller number of frequencies. The same method can be applied to any circular stability region $D \subset \mathbf{C}$, by using an appropriate transformation $p(s) \mapsto (as+b)^n p((cs+d)/(as+b))$ of the polynomials.

The argument principle implies that a polynomial $p$ of degree $n$ is Hurwitz if and only if the argument variation of $p$ along the imaginary axis equals $n\pi$. Our first result is a *finite argument principle* (f.a.p.), stating that $p$ is Hurwitz if $\sum_{k=2}^{N} \arg\big(p(i\omega_k)/p(i\omega_{k-1})\big) > (n-1)\pi$ for some $\omega_1 < \ldots < \omega_N$.

When the polynomial $p(s,\lambda)$ also depends on $\lambda \in \mathbb{R}^m$, we use the notation $p(s,\Lambda)$ for the set $\{p(s,\lambda) \mid \lambda \in \Lambda\}$. If the f.a.p. is satisfied by $p(\cdot,0)$, the additional conditions $0 \notin \operatorname{conv}\big[p(i\omega_{k-1},\Lambda) \cup p(i\omega_k,\Lambda)\big]$, $k = 2,\ldots,N$, are sufficient to prove that every $p(\cdot,\lambda)$, $\lambda \in \Lambda$, satisfies the f.a.p. This is called the *finite zero exclusion principle* (f.z.e.p.).

After developing this general stability criterion, we devote one section to the particular case

$$p(\cdot,\lambda) = p_0 + \textstyle\sum_{j=1}^{m} \lambda_j p_j, \quad \lambda = (\lambda_1,\ldots,\lambda_m) \in [0,1]^m \tag{1}$$

where $p_0,\ldots,p_m$ are polynomials and $\deg p_j \leq \deg p_0 = n$, $j = 1,\ldots m$. One drawback of the "edge criterion" mentioned above, is that stability has to be checked separately for each edge of the polytope and that the number of edges grows with $m$ as $2^m$. This makes the method inconvenient when the number of uncertain parameters is large. For the same problem, the complexity of the f.z.e.p. grows with $m$ and $n$ only as $n^2 m \log m$.

A more general problem of great importance is to check Hurwitz stability of a polynomial which depends *multilinearly* on a number of unknown independent parameters, each bounded in an interval. For example, the characteristic polynomial of a matrix depends multilinearly on the entries of the matrix. Our algorithm is generalized to treat this problem as well, however with exponential complexity. Let the set of unknown parameters be divided into two categories with $l$ and $m$ parameters each, such that the polynomial depends linearly on the parameters of the first category, as the second is kept fixed. Then, the computational complexity for problems with fixed stability margin grows as $n^2 2^m l \log l$.

As an application, we show that the problem of checking positive realness of a rational function, whose numerator and denominator depend linearly on $m$ independent unknown interval bounded paramters, can be reduced to the multilinear polynomial case with $m = 2$.

## A Finite Zero Exclusion Principle

The principal branch of arguments of complex numbers is used, i.e. $-\pi < \arg z \leq \pi$ for $z \neq 0$. The notation $\arg p(\pm i\infty)$ denotes $\lim_{\omega \to \pm\infty} \arg p(i\omega)$.

THEOREM 1 (FINITE ARGUMENT PRINCIPLE, F.A.P.) *Suppose* $-\infty = \omega_1 < \ldots < \omega_N = +\infty$. *If $p$ is a polynomial of degree $n$ with complex coefficients, and*

$$\begin{cases} p(i\omega_k) \neq 0, \quad k = 1,\ldots,N \\ \arg\big(p(i\omega_k)/p(i\omega_{k-1})\big) < \pi, \quad k = 2,\ldots,N \\ \displaystyle\sum_{k=2}^{N} \arg\frac{p(i\omega_k)}{p(i\omega_{k-1})} = \pi n, \end{cases} \tag{2}$$

*then $p$ is Hurwitz.*

*Conversely, if $p$ is a Hurwitz polynomial of degree $n$, there are frequencies $\{\omega_k\}_{k=1}^{2n}$ that satisfy* (2) .

*Proof:* Suppose $p(s) = p_0(s-\alpha_1)\cdots(s-\alpha_n)$. The first two conditions immediately imply that $p$ has no zeros on the imaginary axis, so the expression $\arg\big((i\omega_k-\alpha_j)/(i\omega_{k-1}-\alpha_j)\big)$ is $>0$ if $\operatorname{Re}\alpha<0$ and $<0$ if $\operatorname{Re}\alpha>0$. Since $\arg(u_1\cdots u_n)\le|\arg u_1|+\cdots+|\arg u_n|$ with strict inequality if $\arg u_k<0$ for some $k$, the third condition implies that

$$n\pi=\sum_{k=2}^{N}\arg\frac{p(i\omega_k)}{p(i\omega_{k-1})}=\sum_{k=2}^{N}\arg\prod_{j=1}^{n}\frac{(i\omega_k-\alpha_j)}{(i\omega_{k-1}-\alpha_j)}$$
$$\le\sum_{k=2}^{N}\sum_{j=1}^{n}\left|\arg\frac{(i\omega_k-\alpha_j)}{(i\omega_{k-1}-\alpha_j)}\right|=\sum_{j=1}^{n}\left|\sum_{k=2}^{N}\arg\frac{(i\omega_k-\alpha_j)}{(i\omega_{k-1}-\alpha_j)}\right|=n\pi$$

Strict inequality is impossible, so $p$ must be Hurwitz.

If $p$ is Hurwitz, then we can choose $\{\omega_k\}_{k=1}^{2n}$ with $\omega_1=-\infty$, $\omega_{2n}=\infty$ such that each of the intervals $]\omega_{k-1},\omega_k[$ contains exactly one zero of either $\operatorname{Re}p(i\omega,0)$ or $\operatorname{Im}p(i\omega,0)$. This proves the second part. ∎

The main result of this paper now follows as a natural generalization of the f.a.p. to families of polynomials:

THEOREM 2 (FINITE ZERO EXCLUSION PRINCIPLE, F.Z.E.P.) *Suppose $p(s,\lambda)$ is a polynomial in $s$ which depends continuosly on the $\lambda\in\mathbb{R}^m$. Let $\Lambda\subset\mathbb{R}^m$ be pathwise connected, and suppose $p(\cdot,\lambda_0)$, $\lambda_0\in\Lambda$, together with $\{\omega_k\}_{k=1}^{N}$ satisfies the f.a.p. If $\Lambda=\cup_{j=1}^{J}\Lambda_j$ and*

$$0\notin\operatorname{conv}\big[p(i\omega_{k-1},\Lambda_j)\cup p(i\omega_k,\Lambda_j)\big]\ \ \textit{for } j=1,\ldots,J,\quad k=2,\ldots,N,\tag{3}$$

*then every polynomial in $p(\cdot,\Lambda)$ is Hurwitz.*

*Conversely, if $p(\cdot,\Lambda)$ is Hurwitz, then there is a partition $\Lambda=\cup_{j=1}^{J}\Lambda_j$ and a sequence $(\omega_k)_{k=1}^{N}$ such that all conditions above are fulfilled.*

*Remark.* The partition $\Lambda=\cup_{j=1}^{J}\Lambda_j$ may be necessary if $0\in\operatorname{conv}p(i\omega,\Lambda)\setminus p(i\omega,\Lambda)$ for some $\omega$.

*Proof:* Conditions (3) imply that $\sum_{k=2}^{n}\arg\big(p(i\omega_k,\lambda)/p(i\omega_{k-1},\lambda)\big)/\pi$ is welldefined for $\lambda\in\Lambda$. The sum depends continuously on $\lambda$ but takes only integer values. Since $\Lambda$ is pathwise connected, every $\lambda\in\Lambda$ must give the same value as $\lambda_0$, so the f.a.p. completes the proof of the first part.

The second part is evident, since $\cup_{j=1}^{J}\operatorname{conv}\big[p(i\nu,\Lambda_j)\cup p(i\omega,\Lambda_j)\big]\to p(i\omega,\Lambda)$ as the refinement of the partition increases and $\nu\to\omega$. ∎

To apply this criterion one gradually refines the partition $\Lambda=\cup_{j=1}^{J}\Lambda_j$ and adds new frequencies until either the conditions of the theorem are fulfilled, or $0\in p(i\omega_k,\Lambda)$ for some $k$. Unfortunately, the algorithm may not stop if $p(\cdot,\Lambda)$ just touches the boundary of the set of Hurwitz polynomials. However, the successive values of new frequencies reveals what is going on.

Another complication may be the computation of $p(i\omega,\Lambda_j)$. In the following sections, this problem is analysed in two important cases.

## Polynomials with linear parameter dependence

For the case (1), no partition $\Lambda = \cup_{j=1}^{J} \Lambda_j$ is necessary, since $\operatorname{conv} p(i\omega, \Lambda) = p(i\omega, \Lambda)$. The following theorem shows how to calculate $p(i\omega, I)$ for a given $\omega$ when $I = [0,1]^m$.

THEOREM 3 *Let* $p(\cdot, \lambda) = p_0 + \sum_{j=1}^{m} \lambda_j p_j$ *where* $\lambda = (\lambda_1, \ldots, \lambda_m) \in I = [0,1]^m$ *and suppose that* $\arg p_1(i\omega) \leq \arg p_2(i\omega) \leq \ldots \leq \arg p_m(i\omega) \leq \arg p_1(i\omega) + \pi$ *for some* $\omega \in [-\infty, \infty]$. *Let* $p_{m+j} = -p_j$ *for* $j = 1, \ldots, m$ *and* $q_l = p_0 + \sum_{j=1}^{l} p_j$, $l = 1, \ldots, 2m$. *Then*

$$p(i\omega, I) = \operatorname{conv}\{q_1(i\omega), \ldots, q_{2m}(i\omega)\}. \tag{4}$$

*Remark.* It should be noted that for any family of the form (1) and any given $\omega \in \mathbb{R}$, the argument condition can be fulfilled by replacing $p_0$ with $(p_0 + \sum_{j \in B} p_j)$ and $p_j$ with $-p_j$ for $j \in B = \{j \,|\, \arg p_j(i\omega) < 0\}$, then renumbering the polynomials.

*Proof:* Let $\Phi_j = [\arg p_j(i\omega), \arg p_{j+1}(i\omega)[$ for $j = 1, \ldots, m$. The set $p(i\omega, I)$ is obviously a convex polygon with finitely many vertices of the form $\xi = (p_0 + \sum_{l \in A} p_l)(i\omega)$. Each of these has a supporting line with an argument $\phi \in \Phi_j$ for some $j$. It follows that either $\xi = (p_0 + \sum_{l=1}^{j} p_l)(i\omega) = q_j(i\omega)$ or $\xi = (p_0 + \sum_{l=j+1}^{m} p_l)(i\omega) = q_{m+j}(i\omega)$. ∎

Combining this result with the finite zero exclusion principle gives the following Hurwitz test for polynomial families of the form (1).

ALGORITHM 1.

1. Check Hurwitz stability of the nominal polynomial $p(\cdot, 0)$, using for example Routh-Hurwitz criterion.
2. Choose $2n$ frequencies $-\infty = \omega_1 < \ldots < \omega_{2n} = +\infty$ such that each of the intervals $[\omega_{k-1}, \omega_k]$ contains exactly one zero of either Re $p(i\omega, 0)$ or Im $p(i\omega, 0)$. Then the f.a.p. is satisfied by $p(\cdot, 0)$ together with $\{\omega_k\}_{k=1}^{2n}$. (For real polynomials, it is sufficient to consider positive frequencies.)
3. Calculate $p(i\omega_k, I)$ for $k = 1, \ldots, 2n$, using Theorem 3, and add new frequencies until either $0 \notin \operatorname{conv}[p(i\omega_{k-1}, I) \cup p(i\omega_k, I)]$ for $k = 2, \ldots, 2n$ or $0 \in p(i\omega_k, I)$ for some $k \in \{1, \ldots, 2n\}$. In the first case, the entire family $p(\cdot, I)$ is Hurwitz by the f.z.e.p. In the second case, it obviously contains a non-Hurwitz polynomial.

*Complexity.* The complexity of this algorithm grows with $m$ and $n$ essentially as $n^2 m \log m$. For step 3, the quadratic dependence on $n$ follows since the number of frequencies is proportional to $n$ and for each frequency, the evaluation of the polynomials also has complexity $n$. The factor $m \log m$ is due to the renumbering of polynomials, necessary before applying Theorem 3. It should be noted that computation of the zeros of Re $p(i\omega, 0)$ and Im $p(i\omega, 0)$ in step 2 would be of complexity $n^3$, so for large $n$ the initial frequencies had better be determined from a Nyquist plot of the nominal polynomial $p$. This "frequency sweep" is exceedingly simple compared to the direct application of the classical zero exclusion principle and it grows as $n^2$.

We shall now illustrate our stability test with an example.

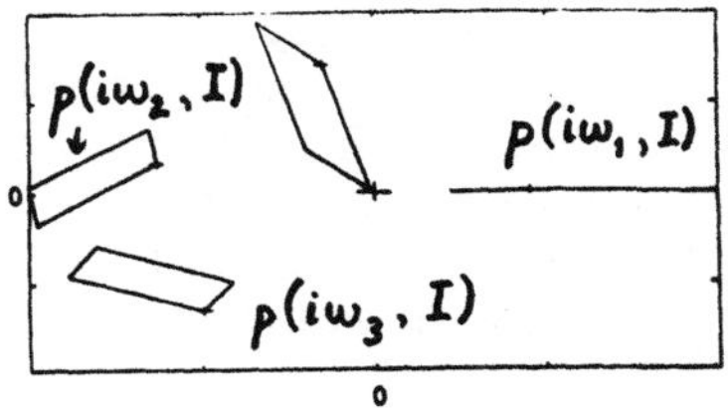


Figure 1.

**EXAMPLE.** Suppose a discrete time linear timeinvariant system has the transfer function $B(z)/A(z) = ((2+\lambda_1)z + 4 + \lambda_2)/(z^3 - 1)$, where $0 \le \lambda_1 \le 1$ and $0 \le \lambda_2 \le 2.4$ are unknown parameters. We apply feedback with transfer function $D(z)/C(z) = (z+2)/(8z+8)$ and get the closed loop characteristic polynomial

$$A(z)C(z) + B(z)D(z) = 8z^4 + 8z^3 + 2z^2 + \lambda_1(z^2 + 2z) + \lambda_2(z+2).$$

We would like to decide whether the closed loop system is stable, i.e. if this polynomial has all its zeros in the unit disk.

A Moebius transformation shows that this happens if and only if the polynomial $p(s,\lambda) = p_0(s) + \lambda_1 p_1(s) + \lambda_2 p_2(s)$ is Hurwitz for $\lambda_1, \lambda_2 \in [0,1]$ when

$$\begin{aligned} p_0(s) &= 8(s+1)^4 + 8(s+1)^3(s-1) + 2(s+1)^2(s-1)^2 \\ p_1(s) &= (s+1)^2(s-1)^2 + 2(s+1)(s-1)^3 \\ p_2(s) &= 2.4((s+1)(s-1)^3 + 2(s-1)^4). \end{aligned}$$

Algorithm 1 gives the following calculations.

1. The nominal polynomial $p(\cdot,0) = p_0$ is Hurwitz.
2. $\operatorname{Re} p_0(i\omega) = 18\omega^4 - 44\omega^2 + 2 = (\omega^2 - 0.046)(\omega^2 - 2.4)$ and $\operatorname{Im} p_0(i\omega) = -48(\omega^2 - 0.33)$, so $\sum_{k=2}^{4} \arg(p(i\omega_k, 0)/p(i\omega_{k-1}, 0)) = 2\pi$ when $\omega_1 = 0$, $\omega_2 = 0.5$, $\omega_3 = 1$ and $\omega_4 = +\infty$.
3. Theorem 3 is used for calculation of $p(i\omega_k, I)$, $k = 1, \ldots, 4$. In fact, we have $\arg p_1(i\omega_k) \le \arg p_2(i\omega_k) \le \arg p_1(i\omega_k) + \pi$ for all $k$ without redefinition of $p_1$ and $p_2$, so

$$p(i\omega_k, I) = p_0(i\omega_k) + \operatorname{conv}\{0, p_1(i\omega_k), (p_1+p_2)(i\omega_k), p_2(i\omega_k)\}, \quad k = 1, \ldots, 4.$$

It turns out that $0 \notin \operatorname{conv}[p(i\omega_{k-1}, I) \cup p(i\omega_k, I)]$ for all $k$ except for $k = 2$. We therefore add another frequency midway between $\omega_1$ and $\omega_2$ and check the f.z.e.p. again. This time all conditions are satisfied, so the closed loop system is stable for all parameter values (Figure 1).

□

## Polynomials with multilinear parameter dependence.

In this section we let $p(s,\lambda)$ depend multilinearly on $\lambda = (\lambda_1, \ldots, \lambda_m) \in I = [0,1]^m$. This means linear dependence on each $\lambda_j$ as other coefficients are kept fixed. Then the partition $\Lambda = \cup_{j=1}^{J} \Lambda_j$ in the f.z.e.p. becomes necessary. For computation of $\operatorname{conv} p(i\omega, I)$ we use the following theorem by Zadeh and Desoer [8].

THEOREM 4 *Let $p(s,\lambda)$ depend multilinearly on $\lambda \in I = [0,1]^m$. Define the finite set $\Delta I = \{0,1\}^m$. Then $\operatorname{conv} p(s,I) = \operatorname{conv} p(s,\Delta I)$ for all $s \in \mathbf{C}$.*

Often the computation of $\operatorname{conv} p(i\omega, I)$ can be further simplified using Theorem 3. Suppose, possibly with renumbered coefficients of $\lambda$, that $I = M \times L$ and $p(\cdot,\lambda)$ depends linearly on $(\lambda_{m+1},\dots,\lambda_{m+l}) \in L$. Then $\operatorname{conv} p(i\omega, I) = p(i\omega, \Delta M \times L)$, and $p(i\omega, \{(\lambda_1,\dots,\lambda_m)\} \times L)$ can be computed from Theorem 3 for each $(\lambda_1,\dots,\lambda_m)$ in the finite set $\Delta M$.

The following stability test now falls out for polynomials with multilinear parameter dependence.

ALGORITHM 2.

1. Check Hurwitz stability of the nominal polynomial $p(\cdot,0)$, using for example Routh-Hurwitz criterion.
2. Choose $2n$ frequencies $-\infty = \omega_1 < \ldots < \omega_{2n} = +\infty$ such that each of the intervals $[\omega_{k-1},\omega_k]$ contains exactly one zero of either Re $p(i\omega,0)$ or Im $p(i\omega,0)$. Then the f.a.p. is satisfied by $p(\cdot,0)$ together with $\{\omega_k\}_{k=1}^{2n}$. Let $J = 1$, $M = M_1$.
3. Calculate $\operatorname{conv} p(i\omega_k, M_j \times L) = p(i\omega_k, \Delta M_j \times L)$ for all $(j,k)$ using the method above. Refine the partition $M = \cup_{j=1}^{J} M_j$ until either $0 \notin \operatorname{conv} p(i\omega_k, \Delta M_j \times L)$ for all $(j,k)$ or $0 \in p(i\omega_k, \Delta M_j \times L)$ for some $(j,k)$. In the second case, stop the algorithm. The family $p(\cdot, I)$ obviously contains a polynomial with a zero on the imaginary axis.
4. If $0 \in \operatorname{conv}\big[p(i\omega_{k-1}, \Delta M_j \times L) \cup p(i\omega_k, \Delta M_j \times L)\big]$ for some $(j,k)$, either split $M_j$ as in step 3, or add a new frequency between $\omega_{k-1}$ and $\omega_k$, then return to step 3. If not, the conditions of Theorem 2 are fulfilled, so we may conclude that the family $p(\cdot, I)$ is Hurwitz.

The choice in 4, whether to add a new frequency or split $M_j$ in one of its directions, can be made in different ways. One way, that has been implemented by the author, is to consider the value $p(i\omega, (\lambda_1,\dots,\lambda_m,0,\dots,0))$ for some $(\lambda_1,\dots,\lambda_m) \in \Delta M_j$, $\omega \in \{\omega_{k-1},\omega_k\}$ then check which coefficient $\lambda_1,\dots,\lambda_m$ or frequency $\omega$ should be replaced by its opposite limit, to cause the biggest possible change of complex argument. If it turns out to be one of the coefficients, then split $M_j$ in the corresponding direction, otherwise add a new frequency between $\omega_{k-1}$ and $\omega_k$.

*Complexity.* The complexity $n^2 2^m l \log l$ is obtained as in the linear case, with the only difference that asymptotically each "frequency split" is accompanied by $2^m$ other "splits".

## Strictly Positive Real (SPR) Rational Functions

The real function $p/q$ is called SPR if for some $\varepsilon > 0$, we have Re $(p/q)(s) > 0$ for all $s$ with Re $s > -\varepsilon$. Such functions are important, e.g. in stochastic realization theory and adaptive control theory. The next lemma shows that the multilinear case with $m = 2$, can be used to test the SPR property of families of rational functions of the form

$$\frac{p_0 + \sum_{j=1}^{m} \lambda_j p_j}{q_0 + \sum_{j=1}^{m} \lambda_j q_j} \quad \text{where} \quad 0 \le \lambda_j \le 1.$$

LEMMA 5 *The real rational function $p/q$ is SPR if and only if the polynomial $\mu_1\mu_2 p + \big(\mu_2(1-i) + i - \mu_1\mu_2\big)q$ is Hurwitz for all $\mu_1,\mu_2 \in [0,1]$. ($i$ is the imaginary unit.)*

This lemma also shows why it is sometimes interesting to consider polynomials with complex coefficients.

*Proof:* First note that

$$\frac{\mu_1(1-i)+i-\mu_2\mu_1}{\mu_2\mu_1} = \frac{i(\frac{1}{\mu_1}-1)+1}{\mu_2} - 1.$$

Hence, the condition of the lemma is equivalent to stating that $p+\alpha q$ is Hurwitz when $0 \leq \arg\alpha \leq \pi/2$. Since $p$ and $q$ are supposed to be real, all solutions of the equation $(p/q)(s) = -\alpha$ must belong to the open left half plane when $\operatorname{Re}\alpha \geq 0$. The solutions depend continuously on $\alpha$, so it follows that their real parts remain smaller than some $-\varepsilon < 0$. This completes the proof. ∎

## References


1. ANAGNOST, J.J., DESOER, C.A, MINNICHELLI, R.J. (1988): "Graphical Stability Robustness Tests for Linear Time-Invariant Systems: Generalizations of Kharitonov's Stability Theorem", *Proc. of IEEE Conf. on Decision and Control.*

2. BARMISH, B.R. (1989): "A Generalization of Kharitonov's Four-Polynomial Concept for Robust Stability Problems with Linearly Dependent Coefficient Pertubations", *IEEE Trans. Autom. Control* **34**, 157-165.

3. BARTLETT, A.C., HOLLOT, C.V. AND LIN, H. (1988): "Root Locations of an Entire Polytope of Polynomials: It Suffices to Check the Edges", *Mathematics of Control, Signals and Systems* **1**, 61-71.

4. BIALAS, S. (1985): "A Necessary and Sufficient Condition for the Stability of Convex Combinations of Stable Polynomials and Matrices", *Bull. Polish Acad. Sci. Tech. Sci.* **33**, 473-480.

5. FU, M. AND BARMISH, B.R., "Stability of Convex and Linear Combinations of Polynomials and Matrices Arising in Robustness Problems", *Proc. 1987 Conf. Information Sciences & Systems.*

6. KHARITONOV, V.L. (1979): "Asymptotic Stability of an Equilibrium Position of a Family of Systems of Linear Differential Equations.", *Differential Equations* **14**, 1483-1485.

7. MINNICHELLI, R.J., ANAGNOST, J.J., DESOER, C.A. (Aug 1989, to appear): "An Elementary Proof of Kharitonov's Theorem with Extensions", *IEEE Transactions on Automatic Control.*

8. ZADEH, L.A. AND DESOER, C.A. (1963): *Linear System Theory*, McGraw-Hill Book Company, New York, San Francisco, Toronto, London, p. 476.

# DESIGN OF CONTROLLER WITH ASYMPTOTIC DISTURBANCE ATTENUATION

**Kenko Uchida, Masayuki Fujita, and Etsujiro Shimemura**

**Abstract**: This paper formulates a generalized disturbance attenuation problem, in which two separate disturbance attenuations are required, for linear time-invariant systems having a direct transmission of controls in controlled outputs or considering disturbances in observations. We propose a design method of controllers which attain asymptotically the generalized disturbance attenuation based on the perfect regulation and the perfect observation under certain minimum phase conditions.

**Key words**: Disturbance attenuation; Riccati equation; stability margin; perfect regulation; perfect observation; $H^\infty$ control.

## 1. Introduction

To reduce the effect of external disturbances is a main objective of designing control systems. Petersen [9] recently considered a problem, which he called the disturbance attenuation problem, of finding feedback controls which reduce the effect of the disturbances to a prespecified level, and he presented a state feedback solution in terms of an algebraic Riccati equation arising in linear quadratic differential games. His result also suggested new state space solutions to $H^\infty$ control problem [1][5]. Petersen and Hollot [10] attempted to solve the disturbance attenuation problem in the case when the state information is not available; their idea is to use the accurate optimal observer [2][3], which is the dual of the cheap optimal regulator [8], to recover asymptotically the norm of a certain transfer function. Another approach to output feedback case is, of course, to treat the problem within the framework of the standard $H^\infty$ control problem [1].

In this paper, we formulate a generalized disturbance attenuation problem, in which two separate disturbance attenuations are required and a direct transmission of the control in the controlled output or an observation disturbance is taken into account, and we propose a design method of controllers which attain asymptotically the generalized disturbance attenuation under certain minimum phase conditions, based on the perfect observation or the perfect regulation posed by Kimura and Sugiyama [6][7]. The idea of asymptotic attenuation is the same to that of Petersen and Hollot [10]; the use of the perfect observation and the perfect regulation in this paper, however, makes it possible to consider the generalized disturbance attenuation problem and to delete a matching condition which is required in their argument [10] (concerning the matching condition, see Remark 2).

## 2. Problem formulation

We consider two types of linear time-invariant systems. One is described by

$$( \Sigma ) \qquad \dot{x} = Ax + Bu + Dv, \quad y = Cx, \quad z = Fx, \quad g = \begin{bmatrix} z \\ u \end{bmatrix}$$

and the other is described by

$$( \Sigma^* ) \qquad \dot{x} = Ax + Bu + Dv, \quad y = Cx + w, \quad z = Fx$$

where $x \in R^n$ is the state; $u \in R^r$ is the control input; $y \in R^m$ is the observed output; in the system ( $\Sigma$ ), $v \in R^p$ is the disturbance and $g = [z' \ u']' \in R^{q+r}$ is the controlled output; in the system ( $\Sigma^*$ ), $h := [v' \ w']' \in R^{p+m}$ is the disturbance and $z \in R^q$ is the controlled output. $B$, $C$, $D$ and $F$ are assumed to be full rank.

Let the controller for the system ( $\Sigma$ ) or ( $\Sigma^*$ ) be restricted to the form consisting of a constant feedback gain $K$ and a state observer with a constant observer gain $L$:

$$( \Gamma ) \qquad u = K\mathbf{x} , \qquad \dot{\mathbf{x}} = A\mathbf{x} + Bu + L(C\mathbf{x} - y).$$

In the closed-loop system formed by the system ( $\Sigma$ ) with the controller ( $\Gamma$ ), denote by $S(s) = [\ S_z(s)' \ S_u(s)' \ ]'$ the transfer function from the disturbance $v$ to the controlled output $g = [z' \ u']'$. In the closed-loop system formed by the system ( $\Sigma^*$ ) with the controller ( $\Gamma$ ), denote by $T(s) = [\ T_v(s) \ T_w(s) \ ]$ the transfer function from the disturbance $h = [v' \ w']'$ to the controlled output $z$.

Our concern is to find a controller with the parameter ( $K$, $L$ ) which makes the system ( $\Sigma$ ) or ( $\Sigma^*$ ) internally stable (i.e., $A + BK$ and $A + LC$ are stable) and realizes the disturbance attenuation: for the system ( $\Sigma$ )

$$|| \ S_z(j\omega) \ || \leq \kappa, \quad || \ S_u(j\omega) \ || \leq \mu \tag{2.1}$$

( then $|| \ S(j\omega) \ || \leq \sqrt{2} \max(\kappa, \mu)$ ) for given constants $\kappa > 0$, $\mu > 0$ and all $\omega \in R$, or for the system ( $\Sigma^*$ )

$$|| \ T_v(j\omega) \ || \leq \nu, \quad || \ T_w(j\omega) \ || \leq \rho \tag{2.2}$$

( then $|| \ T(j\omega) \ || \leq \sqrt{2} \max(\nu, \rho)$ ) for given constants $\nu > 0$, $\rho > 0$ and all $\omega \in R$. In this paper we will present an asymptotic design method for such controllers.

## 3. State feedback and state estimation with disturbance attenuation

As preliminaries, we discuss here two special cases of the design problem. For the system ( $\Sigma$ ) we consider the state feedback problem; if we adopt the feedback gain $K$, the closed-loop system has the form

$$\dot{x} = (A + BK)x + Dv, \quad g = \begin{bmatrix} z \\ u \end{bmatrix} = \begin{bmatrix} F \\ K \end{bmatrix} x; \tag{3.1}$$

then the transfer function from the disturbance $v$ to the controlled output $g = [z' \ u']'$, denoted by $G(s) = [\ G_z(s)' \ \ G_u(s)'\ ]'$, is given by

$$G(s) = \begin{bmatrix} G_z(s) \\ G_u(s) \end{bmatrix} = \begin{bmatrix} F \\ K \end{bmatrix} (sI_n - A^{\#})^{-1} D \tag{3.2}$$

where $A^{\#} = A + BK$. Now, as state feedback gain $K$, we choose

$$K_o = -\mu^2 B' M$$

where $M$ is a positive definite solution to the Riccati equation

$$A' M + MA + \kappa^{-2} F' F - M(\mu^2 BB' - DD')M = 0 \tag{3.3}$$

with parameters $\kappa > 0$ and $\mu > 0$.

**Theorem 1.** Assume that the following conditions are satisfied:

($A_1$) $(\ F,\ A\ )$ is detectable.

($A_2$) For given constants $\kappa > 0$ and $\mu > 0$, the Riccati equation (3.3) has a positive definite solution $M$.

Then, for the state feedback gain $K_o$, the closed-loop system (3.1) is internally stable, i.e., $A + BK_o$ is stable, and has the following disturbance attenuation property:

$$G(j\omega) G(-j\omega)' \le \begin{bmatrix} \kappa^2 I_q & 0 \\ 0 & \mu^2 I_r \end{bmatrix} \tag{3.4}$$

for all $\omega \in R$; in particular,

$$G_z(j\omega) G_z(-j\omega)' \le \kappa^2 I_q, \qquad G_u(j\omega) G_u(-j\omega)' \le \mu^2 I_r \tag{3.5}$$

for all $\omega \in R$.

For the system $(\ \Sigma^*\ )$ we consider the state estimation problem; if we adopt the state observer with the observer gain $L$, the estimation error $e := x - \mathbf{x}$ obeys the equation

$$\dot{e} = (A + LC)e + [D \ \ L] \begin{bmatrix} v \\ w \end{bmatrix}; \tag{3.6}$$

then the transfer function from the disturbance $h = [v' \ \ w']'$ to the estimation error $Fe$, denoted by $H(s) = [\ H_v(s)\ H_w(s)\ ]$, is given by

$$H(s) = [\ H_v(s)\ H_w(s)\ ] = F(sI_n - A_{\#})^{-1} [D \ \ L] \tag{3.7}$$

where $A_{\#} = A + LC$. Now, as state observer gain $L$, we choose

$$L_o = -\rho^2 P C'$$

where $P$ is a positive definite solution to the Riccati equation

$$AP + PA' + \nu^{-2}DD' - P(\rho^2 C' C - F' F)P = 0 \tag{3.8}$$

with parameters $\nu > 0$ and $\rho > 0$.

**Theorem 2.** [13] Assume that the following conditions are satisfied:

($A_3$) ( $A$, $D$ ) is stabilizable.

($A_4$) For given constants $\nu > 0$ and $\rho > 0$, the Riccati equation (3.8) has a positive definite solution $P$.

Then, for the state observer gain $L_o$, the error system (3.6) is internally stable, i.e., $A + L_o C$ is stable, and has the following disturbance attenuation property:

$$H(-j\omega)' H(j\omega) \leq \begin{bmatrix} \nu^2 I_p & 0 \\ 0 & \rho^2 I_m \end{bmatrix} \tag{3.9}$$

for all $\omega \in R$; in particular,

$$H_v(-j\omega)' H_v(j\omega) \leq \nu^2 I_p, \qquad H_w(-j\omega)' H_w(j\omega) \leq \rho^2 I_m \tag{3.10}$$

for all $\omega \in R$.

**Remark 1.** It is shown in [4] that the state feedback gain $K_o$ has a large stability margin at the input side, which covers the well-known gain and phase margins of the linear quadratic optimal regulator [12]. It can be also shown that the observer gain $L_o$ guarantees a similar stability margin at the input side in the dual system.

## 4. Asymptotic disturbance attenuation

In this section, using the perfect observation or the perfect regulation posed by Kimura and Sugiyama [6][7], we design a controller guaranteeing asymptotically the internal stability and the disturbance attenuation for the system ( $\Sigma$ ) or ( $\Sigma^*$ ). We summarize here materials about the perfect observation and the perfect regulation [6][7], which are necessary to our discussion. Let the observer gain $L_f$ ( feedback gain $K_f$ ) be rational in scalar $f > 0$ and assume that the all the eigenvalues of $A_{\#f} := A + L_f C$ ($A^{\#f} := A + BK_f$), denoted by $\lambda_i(A_{\#f})$ ( $\lambda_i(A^{\#f})$ ), $i = 1,\ldots,n$, satisfy either

$$\lambda_i(A_{\#f}) \rightarrow \alpha_i \qquad ( \lambda_i(A^{\#f}) \rightarrow \beta_i )$$

or

$$f^{-1}\lambda_i(A_{\#f}) \rightarrow \alpha_i \qquad ( f^{-1}\lambda_i(A^{\#f}) \rightarrow \beta_i )$$

as $f \rightarrow \infty$, where $\alpha_i$ ( $\beta_i$ ) is a complex number with negative real part. Then, we call that the observer gain $L_f$ ( feedback gain $K_f$ ) attains the perfect observation of the system ( $C$, $A$, $D$ ) ( perfect regulation of the system

( $F$, $A$, $B$ ) ) if

$$\int_0^\infty \| (\exp A_{\#f}t)D \|^2 dt \to 0 \qquad ( \int_0^\infty \| F(\exp A^{\#f}t) \|^2 dt \to 0 )$$

as $f \to \infty$.

**Lemma.** [6][7] (i) There exits an observer gain $L_f$ ( feedback gain $K_f$ ) which attains the perfect observation of the system ( $C$, $A$, $D$ ) ( perfect regulation of the system ( $F$, $A$, $B$ ) ) if and only if the conditions ($A_5$) and ($A_6$) ( ($A_7$) and ($A_8$) ) holds.

($A_5$) ( $C$, $A$ ) is observable.

($A_6$) ( $C$, $A$, $D$ ) is left-invertible and minimum phase.

($A_7$) ( $A$, $B$ ) is controllable.

($A_8$) ( $F$, $A$, $B$ ) is right-invertible and minimum phase.

(ii) If the observer gain $L_f$ ( feedback gain $K_f$ ) attains the perfect observation of the system ( $C$, $A$, $D$ ) (perfect regulation of the system ( $F$, $A$, $B$ )), then, for each $s$,

$$(sI_n - A_{\#f})^{-1}D \to 0 \qquad ( F(sI_n - A^{\#f})^{-1} \to 0 ) \qquad (4.1)((4.2))$$

as $f \to \infty$.

For the computational procedures of the perfect observation gain $L_f$ and the perfect regulation gain $K_f$, see [6][7].

Now we turn to the original design problem formulated in Section 2. First consider the system ( $\Sigma$ ); in the closed-loop system given by the controller ( $\Gamma$ ), the transfer function $S(s)$ from the disturbance to the controlled output is written as

$$S(s) = \begin{bmatrix} F \\ K \end{bmatrix} (sI_n - A^{\#})^{-1} LC(sI_n - A_{\#})^{-1}D + \begin{bmatrix} F \\ 0 \end{bmatrix} (sI_n - A_{\#})^{-1}D. \quad (4.3)$$

Here, as observer gain $L$, we choose the perfect observation gain $L_f$ of the system ( $C$, $A$, $D$ ). From (4.1), $(sI_n - A_{\#f})^{-1}D \to 0$, which implies

$$L_f C(sI_n - A_{\#f})^{-1}D = (sI_n - A)(sI_n - A_{\#f})^{-1}D - D \to -D$$

as $f \to \infty$. Applying these relations to the formula (4.3), we have the following result:

**Theorem 3.** Assume that ($A_5$) and ($A_6$) are satisfied. Then, for the system ( $\Sigma$ ), the controller ( $\Gamma$ ) with an arbitrary feedback gain $K$ and the observer gain $L_f$ attaining the perfect observation of the system ( $C$, $A$, $D$ ) ensures that, for each $s$,

$$S(s) = \begin{bmatrix} S_z(s) \\ S_u(s) \end{bmatrix} \to -G(s) = -\begin{bmatrix} G_z(s) \\ G_u(s) \end{bmatrix}$$

as $f \to \infty$, where $G(s)$ is defined as (3.2).

Thus, from Theorem 1 and Theorem 3, we see that the controller ( $\Gamma$ ) with the parameter ( $K_o$, $L_f$ ) guarantees asymptotically the internal stability and the disturbance attenuation for the system ( $\Sigma$ ).

Next we consider the system ( $\Sigma^*$ ); in the closed-loop system given by the controller ( $\Gamma$ ), the transfer function $T(s)$ from the disturbance to the controlled output is written as

$$T(s) = F(sI_n - A^{\#})^{-1}[D \;\; 0] + F(sI_n - A^{\#})^{-1}BK(sI_n - A_{\#})^{-1}[D \;\; L].$$

Then, by the dual argument together with the property (4.2) for the perfect regulation gain $K_f$ of the system ( $F$, $A$, $B$ ), we have the following result:

**Theorem 4.** Assume that ($A_7$) and ($A_8$) are satisfied. Then, for the system ( $\Sigma^*$ ), the controller ( $\Gamma$ ) with the feedback gain $K_f$ attaining the perfect regulation of the system ( $F$, $A$, $B$ ) and an arbitrary observer gain $L$ ensures that, for each $s$,

$$T(s) = [\; T_v(s) \; T_w(s) \;] \to -\, H(s) = -\,[\; H_v(s) \; H_w(s) \;]$$

as $f \to \infty$, where $H(s)$ is defined as (3.7).

Thus, from Theorem 2 and Theorem 4, we see that the controller ( $\Gamma$ ) with the parameter ( $K_f$, $L_o$ ) guarantees asymptotically the internal stability and the disturbance attenuation for the system ( $\Sigma^*$ ).

**Remark 2.** By using the accurate optimal observer [2][3], which is the dual of the cheap optimal regulator [8], Petersen and Hollot [10] shows that $||S_z(j\omega)|| \to ||G_z(j\omega)||$ for each $\omega$; compared with Theorem 3, their result requires an additional assumption that $\mathrm{Im}\,C'$ includes $\mathrm{Im}\,F'$, called the "matching" condition. After submitting this paper, the authors learned that Petersen and Hollot in their recent paper [11] succeeded in deleting the matching condition by generalizing the algebraic Riccati equation, on which the accurate optimal observer is based, to an indefinite type of linear quadratic differential games; however, their method and result seem still less simple and less general than ours. Note also that the computation [6][7] of the perfect observation gain is in general easier than that of the accurate optimal observer gain.

**Remark 3.** In [7] it is shown that the perfect observation ( the perfect regulation ) of the system ( $C$, $A$, $B$ ) recovers asymptotically the return difference at the input ( the output ) side of the system. Therefore, if $C = F$ ( $B = D$ ), the perfect observation in Theorem 3 ( the perfect regulation in Theorem 4 ) recovers also such a large stability margin as stated in Remark 1 at the input ( the output ) side of the system ( $\Sigma$ ) ( ( $\Sigma^*$ ) ).

## References


[1] J.C. Doyle, K. Glover, P.P. Khargonekar, B.A. Francis, "State-space solutions to standard $H_2$ and $H_\infty$ control problems", in *Proc. 1988 American Contr. Conf.*, Atlanta, 1988.

[2] J.C. Doyle, G. Stein, "Robustness with observers", *IEEE Trans. Automat. Contr.* **AC-24** (1979), 607-611.

[3] B.A. Francis, "The optimal linear-quadratic time-invariant regulator with cheap control", *IEEE Trans. Automat. Contr.* **AC-24** (1979), 616-621.

[4] M. Fujita, K. Uchida, F. Matsumura, "Gain perturbation tolerance in $H^\infty$ state feedback control", submitted for publication, 1989.

[5] P.P. Khargonekar, I.R. Petersen, M.A. Rotea, "$H_\infty$-optimal control with state-feedback", *IEEE Trans. Automat. Contr.* **AC-33** (1988), 786-788.

[6] H. Kimura, "A new approach to the perfect regulation and the bounded peaking in linear multivariable control systems", *IEEE Trans. Automat. Contr.* **AC-26** (1981), 253-270.

[7] H. Kimura, O. Sugiyama, "Design of robust controller using perfect regulation and perfect observation (in Japanese)", *Trans. SICE* **18** (1982), 955-960.

[8] H. Kwakernaak, R. Sivan, "The maximally achievable accuracy of linear optimal regulators and linear optimal filters", *IEEE Trans. Automat. Contr.* **AC-17** (1972), 79-86.

[9] I.R. Petersen, "Disturbance attenuation and $H^\infty$ optimization: A design method based on the algebraic Riccati equation", *IEEE Trans. Automat. Contr.* **AC-32** (1987), 427-429.

[10] I.R. Petersen, C.V. Hollot, "Using observers in the stabilization of uncertain linear systems and in disturbance rejection problems", in *Proc. 25th IEEE Conf. Decision and Contr.*, Athens, 1986.

[11] I.R. Petersen, C.V. Hollot, "High gain observers applied to problems in the stabilization of uncertain linear systems, disturbance attenuation and $H^\infty$ optimization", *Int. J. Adaptive Contr. and Signal Processing* **2** (1988), 347-369.

[12] M.G. Safonov, M. Athans, "Gain and phase margin for multiloop LQG regulators", *IEEE Trans. Automat. Contr.* **AC-22** (1977), 173-179.

[13] K. Uchida, "Properties of a Kalman-Bucy type robust filter (in Japanese)", in *Preprints 31st Japan Joint Automat. Contr. Conf.*, Osaka, 1988.


## Appendices

**Proof of Theorem 1.** We first rewrite the Riccati equation (3.3) as

$$A^{o\prime} M + MA^o + \kappa^{-2}F'F + \mu^{-2}K_o{}'K_o + MDD'M = 0 \tag{A.1}$$

where $A^o = A + BK_o$ and $K_o = -\mu^2 B'M$. The stability of $A^o$ follows from the identity (A.1) and a standard argument under the assumptions ($A_1$) and ($A_2$). If we substitute $M = -N^{-1}$ into (A.1) and add $-sI_nN^{-1} + N^{-1}sI_n$ to the left hand side, we have

$$N(-sI_n - A^{o\prime}) + (sI_n - A^o)N + N(\kappa^{-2}F'F + \mu^{-2}K_o{}'K_o)N + DD' = 0. \tag{A.2}$$

Furthermore, multiplying the both sides of (A.2) by $(F' K_o')'(sI_n - A^o)^{-1}$ from the left and by $(-sI_n - A^{o\prime})^{-1}(F' K_o')$ from the right and rearranging terms, we have

$$G(s)G(-s)' - \begin{bmatrix} \kappa^2 I_q & 0 \\ 0 & \mu^2 I_r \end{bmatrix} = -V(s)V(-s)' \tag{A.3}$$

where

$$G(s) = \begin{bmatrix} F \\ K_o \end{bmatrix} (sI_n - A^o)^{-1} D$$

$$V(s) = I_{q+r} + \begin{bmatrix} F \\ K_o \end{bmatrix} (sI_n - A^o)^{-1} N (F' K_o') \begin{bmatrix} \kappa^2 I_q & 0 \\ 0 & \mu^2 I_r \end{bmatrix}.$$

The inequality (3.4), which implies the inequalities (3.5), follows from the identity (A.3) with $s = j\omega$. □

**Proof of Theorem 2.** The proof of the theorem is completely dual to that of Theorem 1. □

# Gas Turbine Control Using Mixed Sensitivity H∞-Optimisation

D.Biss and K.G.Woodgate

Abstract: The purpose of this paper is to present an industrial application of the polynomial approach to multivariable mixed sensitivity H∞ optimisation for feedback systems introduced by Kwakernaak [8].The software implemented was supplied by the Dept. of Applied Mathematics,Twente University [7] based on the design algorithm of [2].The multivariable process model was derived from non-linear simulation data for a 1.5 MW gas turbine supplied by Hawker Siddeley Dynamics [6] as part of a collaborative research project at the Industrial Control Unit,University of Strathclyde.A previous scalar design study relating to the gas turbine problem has been presented by Biss [1] .The design results for the MISO gas turbine problem with respect to the controller will demonstrate the robustness properties which can be produced by judicious choice of weighting function matrices $V, W_1$ and $W_2$ within a criterion to be minimized,with respect to the stabilising compensator transfer matrix G, of the form $||Z(s)||_\infty$ where

$$Z := V^*(S^*W_1^*W_1S + T^*W_2^*W_2T)V$$

and $S := (I+HG)^{-1}$ and $T := GS$

## 1.0 Industrial Application -Gas Turbine Control

### 1.1 Introduction

The design of a modern gas turbine control system has a usual sequence of events from specifying a set of performance specifiçations for the engine to completion of adequate simulation and testing of the control system .The objectives of this section are to present the linear modelling of the gas turbine system obtained from non-linear simulation data for a 1.5 MW gas turbine provided by H.S.D.E [6] and outline the usual disturbances which affect the performance of the system .The latter sections will discuss the actual control design and the simulation study .

## 1.2 Modelling of the Gas Turbine

Non-linear simulation data for a gas turbine has been used to develop a mathematical model of the system which can be used for design and simulation purposes The resultant system requires the use of MIMO compensator design therefore the use of a modern/advanced control technique is preferable to a classical design approach .

The gas turbine is a prime mover,its purpose is to deliver power and the primary control requirement over this developed power is the fuel input .For a simple single shaft turbojet problem (SISO design),the thrust developed by the engine can be measured by use of the gas generator speed as an indirect method of power measurement .This is the starting point of the modelling and control design process .

### 1.2.1 Gas Generator

From Fig.2 ,it can be seen that a SISO loop for simple power control can be established where the gas generator speed,$N_g$,which is a function of Power,is the control input and the fuel flow into the engine ,$F_{t-Td}$,which is a function of the actuator valve angle,$\theta_v$,is the control output [1] .

From the non-linear simulation data,a block diagram representing the gas generator characteristics was reduced to a general continuous-time plant model of the form :

$$H_1(s) = \frac{K_1 e^{-Tds}}{(\tau_1 s+1)(\tau_2 s+1)} = \frac{N_g \text{ (r.p.m)}}{\theta_v \text{ (deg)}} \tag{1}$$

where $K_1, \tau_1$ and $\tau_2$ are determined from the linearised gains of the system and the combustion dead time, $T_d$, is 0.04 secs .The gains of the gas generator model vary over the fuel range and four models have been determined for the range of fuel input .

### 1.2.2 Free Turbine

The inclusion of the free turbine dynamics into the description of the system requires further linearisation of the simulation data provided by H.S.D.E [6] .The extended problem for the gas turbine can be represented by the multi-loop configuration given in Fig.2 .

The derivation of a linear model for the free turbine involves three stages - (i) derivation of the relationship for the total driving output torque of the free turbine shaft,(ii) the discussion of the complex load and the derivation of the mathematical representation and (iii) formulation of the general linear model .

$$(2) \qquad H_2 = \frac{(\tau_1 s + 1)(K_3 + K_4 \delta N_{f1})}{(J_f s/\beta + K_5 + K_6)} = \frac{N_f \ (\text{r.p.m})}{N_g \ (\text{r.p.m})}$$

where $K_3, K_4$ linearised gains, function of total steady state driving torque

$J_f$ total free turbine inertia (3.7 kgm^2/rad)

$\tau_1$ time constant derived for gas generator model

$\delta N_{f1}$ $= N_f - N_{fn}$

$N_f$ free turbine speed

$N_{fn}$ lowest free turbine speed

$\beta$ $= 60/2*\pi$

$K_5, K_6$ gradients of complex load relationships

### 1.2.3 Gas Turbine Model for Control Design

For the MISO compensator design problem, the plant

models of the gas generator, $H_1$, and the free turbine, $H_2$, need to be formulated in a left coprime polynomial matrix representation. This can be achieved by consideration of Fig.2 :

(3) $$\begin{bmatrix} N_f \\ N_g \end{bmatrix} = \begin{bmatrix} H_1 & 0 \\ H_1 H_2 & 0 \end{bmatrix} \begin{bmatrix} F_g \\ 0 \end{bmatrix} \qquad H_1 = \Phi_1/\theta_1,\ H_2 = \Phi_2/\theta_2$$

(4) $$\text{therefore} \quad H = \begin{bmatrix} H_1 & 0 \\ H_1 H_2 & 0 \end{bmatrix} = \begin{bmatrix} \Phi_1/\theta_1 & 0 \\ \Phi_1\Phi_2/\theta_1\theta_2 & 0 \end{bmatrix}$$

Later the left coprime matrix form will be utilised :

(5) $$H = D^{-1}N \quad \text{where } D = \begin{bmatrix} -\Phi_2 & \theta_2 \\ \theta_1 & 0 \end{bmatrix} \text{ and } N = \begin{bmatrix} 0 & 0 \\ \Phi_1 & 0 \end{bmatrix}$$

The corresponding right coprime matrix form is :

(6) $$H = N_1 D_1^{-1} \quad \text{where } D_1 = \begin{bmatrix} \theta_1\theta_2 & 0 \\ 0 & 1 \end{bmatrix} \text{ and } N_1 = \begin{bmatrix} \Phi_1\theta_2 & 0 \\ \Phi_1\Phi_2 & 0 \end{bmatrix}$$

The polynomial theory of Kwakernaak [8] is for square systems therefore the use of a dummy (zero) input is required in the plant matrix H. The use of a dummy input does not affect the squareness or invertibility of H in the polynomial compensator design method [9].

1.3 Disturbances

A typical disturbance associated with the scalar control design for the gas generator loop is a high frequency disturbance due to noise which can be represented by high pass functions for V. For the free turbine loop, load change in the demand represents the largest disturbance therefore the system requires a large degree of stability robustness .

## 2.0 Control Design Algorithm - H∞ Optimisation

### 2.1 Problem Formulation

The multivariable linear feedback control scheme considered here is shown in Fig.1, where the multivariable plant with transfer matrix H(s) corresponds to the gas turbine model derived in §1.2.3 and the compensator to be designed is denoted by the transfer function matrix G(s) .

The optimisation problem can be defined as the minimisation with respect to stabilising compensator G :

(7) $$\|Z(s)\|_\infty = \sup_{\Omega \in \mathbb{R}} \|Z(i\Omega)\|_2$$

and the system equations are :

(8) $$H = D^{-1}N = N_1 D_1^{-1} \quad \text{where } D = \begin{bmatrix} -\Phi_2 & \theta_2 \\ \theta_1 & 0 \end{bmatrix} \quad N = \begin{bmatrix} 0 & 0 \\ \Phi_1 & 0 \end{bmatrix}$$

(9) $$V = D^{-1}M \quad \text{where } M = \begin{bmatrix} \mu_1(s) & 0 \\ 0 & \mu_2(s) \end{bmatrix} \quad \begin{matrix} \deg(\mu_1)=\deg(\theta_2) \\ \deg(\mu_2)=\deg(\theta_1) \end{matrix}$$

where M(s) a rational matrix function, $M^*(s)=M^T(-s)$ and T denotes the transpose

(10) $$W_1 = A_1 B_1^{-1} \quad \text{where } A_1 = B_1 = I$$

(11) $$W_2 = A_2 B_2^{-1} \quad \text{where } A_2 = \begin{bmatrix} \Gamma s^k & 0 \\ 0 & \alpha(s) \end{bmatrix} \quad B_2 = I$$

where $\alpha(s)$ a asymptotically stable scalar polynomial

$\Gamma$ a scalar constant

k some non-negative integer

The restrictions on the degrees of $\mu_1$ and $\mu_2$ are necessary because V must be bi-proper .

(12) $G = YX^{-1} = B_2Q(B_1P)^{-1}$ compensator

In order to simplify the solution procedure a so-called equalising solution is sought i.e stabilising compensators G which satisfy the following :

(13) $Z(i\Omega) = \lambda^2 I \ , \lambda \in \mathbb{R}, \Omega \in \mathbb{R}$ equalising solution

To achieve this it is necessary to choose $X=B_1P$ and $Y=B_2Q$,where P and Q are square polynomial equations to be determined .Minimisation of $\|Z\|_\infty$ then becomes that of minimising $\lambda^2$. The generalised closed loop characteristic polynomial is given by :

(14) $F = DX + NY = DB_1P + NB_2Q$

where F is a polynomial matrix .The resulting polynomial equations which need to be solved to obtain an equalising solution for the mixed sensitivity $H^\infty$ problem are given below,for further details of their derivation see [1] .

(15) $1/\lambda M\_R = DB_1P + NB_2Q$

where M_ is defined by $M\_M\_^* = MM^*$ with M_ an asymptotically stable polynomial matrix and R is a polynomial matrix such that

(16) $$R^{*}R = P^{*}A_1^{*}A_1P + Q^{*}A_2^{*}A_2Q$$

By inspection of (15) and (16) stability of the closed loop system is equivalent to that of $M\_R$ which is in turn equivalent to the stability of R .

2.2 Design Algorithm

The software which will be discussed in this section was developed at Twente University [2] .

Step 1. Specify the plant model,disturbances and weighting function matrices ,equations (8)-(11) .

Step 2.Calculate a particular equalising solution at $\lambda=\infty$ by determining $R_\infty$,$P_\infty$ and $Q_\infty$ from the equations (15) and (16) .For the special case $B_1=B_2=I$,$P_\infty$ and $Q_\infty$ correspond to the right-coprime representation of the plant transfer function H ,i.e $P_\infty = -N_1$ and $Q_\infty = D_1$, which can be computed using a standard algorithm [3].The matrix $R_\infty$ can be calculated using $P_\infty$ and $Q_\infty$ in equation (16) by spectral factorisation,though two assumptions must be satisfied :

(i) $R_\infty$ has no roots on the imaginary axis i.e $|R^{*}_\infty(i\Omega)R_\infty(i\Omega)| > 0$ for all $\Omega \in \mathbb{R}$

(ii)the column degrees of $R_\infty$ must be equal to those of $A_2Q_\infty$

In general these assumptions are non-restrictive .

Step 3. Calculate the degrees of P,Q and R for general $\lambda$ from those of $P_\infty$,$Q_\infty$ and $R_\infty$ [8] .For the special case:

(17) $$A_2 = \begin{bmatrix} \Gamma s^k & 0 \\ 0 & \alpha(s) \end{bmatrix} \quad \text{if } k=0 \text{ ,deg}(\alpha(s))=2$$

This choice of deg($\alpha(s)$) will also minimise the total number of coefficients of the unknown polynomials in R,P and Q [9] . Step 4.Having determined the degrees of all unknown polynomials, the equations (15) and (16) may be solved by equating coefficients in like powers of the Laplace variable 's' and using the REDUCE symbolic language package [4] a solution can be obtained. Further details of this procedure may be found in [2] where the standard $H_\infty$ problem is solved using the same approach.

Step 5. An optimal compensator is obtained from the solution using $P_{opt}$ and $Q_{opt}$ in equation (12).

## 2.4 Choice of Weighting Functions

The choice of the weighting functions $V, W_1$ and $W_2$ can be categorised according to the desired performance for S and T and is discussed in detail in [8].

## 3.0 Design Results

Consider the plant transfer matrix of equation (18), this represents the midway power range model for the gas turbine system :

$$(18)\quad H(s) = \begin{bmatrix} 208.23/0.03439s^2+1.829s+1 & 0 \\ 227.2s+436.9/0.034s^3+1.8s^2+1.05s+0.027 & 0 \end{bmatrix}$$

The design specifications are :

(a) to achieve performance robustness with respect to variations in the plant dynamics by keeping the magnitude of the Sensitivity matrix elements small and the closed loop transfer function matrix T' ($T'=HG(I+HG)^{-1}$) small at low frequency and

(b) to ensure disturbance rejection at high frequency of the high frequency noise represented by V.

The noise model V is defined to be :

$$(19)\qquad V=D^{-1}M \quad \text{where } D = \begin{bmatrix} -0.073s-0.04 & s+0.0301 \\ 0.034s^2+1.8s+1 & 0 \end{bmatrix}$$

$$\text{and} \quad M = \begin{bmatrix} s+1 & 0 \\ 0 & s^2+2s+1 \end{bmatrix}$$

The weighting functions $W_1$ and $W_2$ are chosen to be :

$$(20) \qquad A_1 = B_1 = I \quad \text{and} \quad A_2 = \begin{bmatrix} 1 & 0 \\ 0 & s^2+2s+1 \end{bmatrix} \; B_2 = I$$

An equalising solution was obtained for the system using the steps outlined in $2.2 .

$$(21) \qquad G_{opt} = Q_{opt} \, P_{opt}^{-1} \quad \text{and} \quad |\lambda_{opt}| = 27.12$$

$$(22) \; P_{opt} = \begin{bmatrix} -0.09s^3-0.75s^2-208.7s-6.8 & 0.65s^2-3.39s-0.11 \\ 0.008s^3+0.21s^2-7.4s-0.81 & 0.006s^2-0.12s-0.01 \end{bmatrix}$$

$$(23) \; Q_{opt} = \begin{bmatrix} 0.04s^3+1.83s^2+1.06s+0.03 & 0.029s^2+0.017s+0.052 \\ -0.23 & -0.0017 \end{bmatrix}$$

The corresponding Sensitivity ,$S_{opt}$, and Control Sensitivity ,$T_{opt}$, plots are given in Figs 3 and 4 determined using the MATLAB multivariable frequency domain toolbox [4] .The results have satisfied the design specifications since both S and T' are small at low frequency providing robust performance and the elements of S ,Fig.4, are less than unity over the entire frequency range therefore providing good disturbance rejection .

## 4.0 Conclusions

The design results presented show that robustness can be achieved for the gas turbine engine by use of the polynomial $H_\infty$ control design method due to Kwakernaak [8].

Acknowledgements :

This research was supported by the Science and Engineering Research Council,U.K and H.S.D.E,U.K .

References

[1] Biss,D ,M.J.Grimble,M.A.Johnson and M.L.G,Hill '$H_\infty$ Robust Control Algorithm and Application ',MTNS Proceedings 1987

[2] Boekhoudt,P and H,Kwakernaak , 'A polynomial approach to the $H_\infty$ control design method' ,A.C.C Proceedings,June 1988

[3] Chang,B.C and J.B,Pearson, "Algorithms for the solution of polynomial equations arising in multivariable control theory " Technical Report # 8208,Dept. of Electrical Engineering,Rice University,May 1982 .

[4] Ford,M.P ,J.M Maciejowski and J.M Boyle , 'Multivariable Frequency Domain Toolbox ',GEC Engineering Research Centre and Cambridge Control Limited .

[5] Hearn,A.C 'REDUCE User's Manual ' Rand Corporation , Santa Monica,CA 90406,1983

[6] Hill,M.L.G ., 'H.S.D.E Non-linear simulation data for a 1.5 MW gas turbine engine and load', $H_\infty$ Controller Research Project at Strathclyde University,TPL 2179 Issue No. 2 , March 1986

[7] Kaspers,M.A ., 'Het oplossen van polynoommatrix-ver gelijkingen bij $H_\infty$-optimalisatieproblemen',(in Dutch) Student project report ,Dept. of Applied Math., University of Twente,Enschede,August 1987.

[8] Kwakernaak,H ., 'A polynomial approach to minimax frequency domain optimisation of multivariable feedback systems',Int.J.Control,1986,Vol.44 No.1,117-156

[9] Woodgate,K.G , "Robust stabilisation of an inverted double pendulum via a polynomial approach to $H_\infty$ optimal control" Internal Memorandum #742,Dept. of Maths,University of Twente,October 1988 .

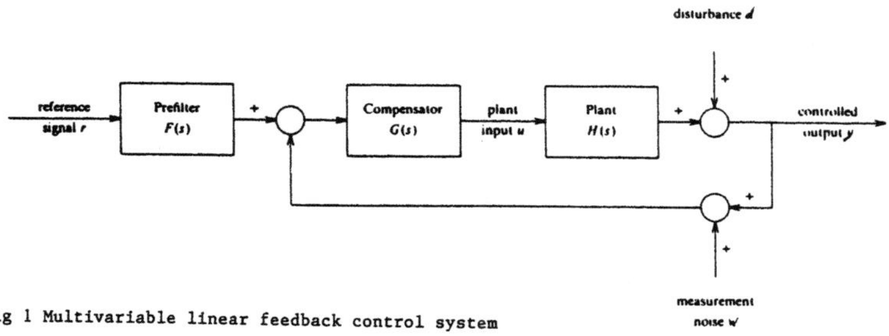


Fig 1 Multivariable linear feedback control system

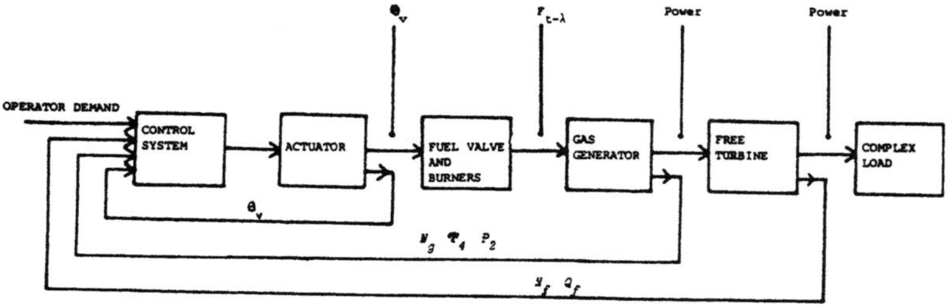


Fig 2 Gas Turbine Control Configuration for MIMO System

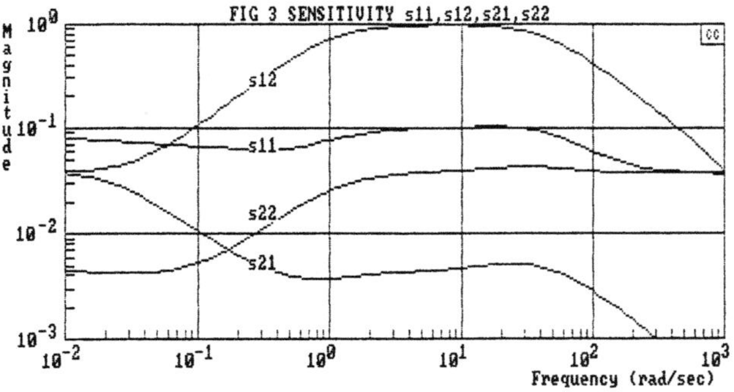


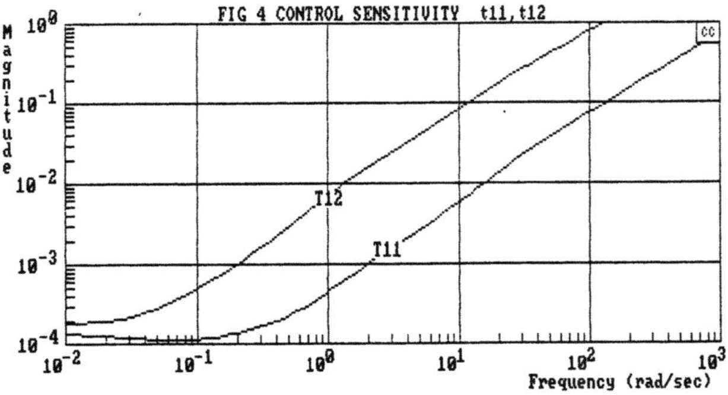

# NONLINEAR $H^\infty$ THEORY

Ciprian Foias Allen Tannenbaum

**Abstract**

In this paper we discuss a natural nonlinear extension of $H^\infty$ synthesis theory. We base our results on our previous papers [9] and [11].

## 1 Introduction

This note will be concerned with nonlinear extensions of the $H^\infty$ design theory. In the papers [2], [3] an extension of the commutant lifting theorem to a local nonlinear setting was given, together with a discussion of how this result could be used to develop a design procedure for nonlinear systems. In the present paper, we continue this line of research with a constructive extension of the linear $H^\infty$ theory to nonlinear systems. We should note that our colleagues Joe Ball and Bill Helton [5] have developed a completely different novel approach to this problem based on a nonlinear version of Ball-Helton theory.

In the theory presented below, we will consider majorizable input/output operators (see Section 2 for the precise definition). In particular, these operators are analytic in a ball around the origin in a complex Hilbert space, and it turns out that it is possible to express each $n$-linear term of the Taylor expansion of such an operator as a linear operator on a certain tensor space. (Our class of operators also include Volterra series of fading memory [7].) This allows us to iteratively apply the classical commutant lifting theorem in designing a compensator. (The general technique we call the "iterative commutant lifting procedure." See Section 6.) For single input/single output (SISO) systems, this leads to the construction of a compensator which is optimal relative to a certain sensitivity function which will be defined below. Moreover in complete generality (i.e. for multiple input/multiple ouput (MIMO) systems), our procedure will ameliorate (in the sense of our nonlinear weighted sensitivity criterion), any given design. We note that for linear systems, our method reduces to the standard $H^\infty$ design technique as discussed for example in [13] and [16].

In developing the present theory, we have had to extend some of the skew Toeplitz techniques of [6], [10], and [12] to linear operators defined on certain tensor spaces. This has lead to several novel results in computational operator theory, and for example provides a way of iteratively constructing the nonlinear intertwining dilation of the nonlinear commutant lifting theorem considered in [2] and [3].

## 2 Analytic Mappings on Hilbert Space

In order to carry out our extension of $H^\infty$ synthesis theory to nonlinear systems, we will need to first discuss a few standard results about analytic mappings on Hilbert spaces. We are essentially following the treatments of [3], [4] to which the reader may refer for all of the details. In particular, input/output operators which admit Volterra expansions are special cases of the operators which we study here. See [7], [14].

Let $G$ and $H$ denote complex Hilbert spaces. Set

$$B_{r_o}(G) := \{g \in G : \|g\| < r_o\}$$

(the open ball of radius $r_o$ in $G$ about the origin). Then we say that a mapping $\phi : B_{r_o}(G) \to H$ is *analytic* if the complex function $(z_1, \ldots, z_n) \mapsto \langle (z_1 g_1 + \ldots + z_n g_n), h \rangle$ is analytic in a neighborhood of $(1, 1, \ldots, 1) \in \mathbf{C^n}$ as a function of the complex variables $z_1, \ldots, z_n$ for all $g_1, \ldots, g_n \in G$ such that $\|g_1 + \ldots + g_n\| < r_o$, for all $h \in H$, and for all $n > 0$. (Note that we denote the Hilbert space norms in $G$ and $H$ by $\| \quad \|$ and the inner products by $\langle , \rangle$.)

We will now assume that $\phi(0) = 0$. It is easy to see that if $\phi : B_{r_o}(G) \to H$ is analytic, then $\phi$ admits a convergent Taylor series expansion, i.e.

$$\phi(g) = \phi_1(g) + \phi_2(g, g) + \cdots + \phi_n(g, \cdots, g) + \cdots$$

where $\phi_n : G \times \cdots \times G \to H$ is an $n$-linear map. Clearly, without loss of generality we may assume that the $n$-linear map $(g_1, \cdots, g_n) \to \phi(g_1, \cdots, g_n)$ is symmetric in the arguments $g_1, \cdots, g_n$. This assumption will be made throughout this paper for the various analytic maps which we consider. For $\phi$ a Volterra series, $\phi_n$ is basically the $n^{th}$-Volterra kernel.

Now set

$$\hat{\phi}_n(g_1 \otimes \cdots \otimes g_n) := \phi_n(g_1, \cdots, g_n).$$

Then $\hat{\phi}_n$ extends in a unique manner to a dense set of $G^{\otimes n} := G \otimes \ldots \otimes G$ (tensor product taken $n$ times). Notice by $G^{\otimes n}$ we mean the Hilbert space completion of the algebraic tensor product of the $G$'s. Clearly if $\hat{\phi}_n$ has finite norm on this dense set, then $\hat{\phi}_n$ extends by continuity to a bounded linear operator $\hat{\phi}_n : G^{\otimes n} \to H$. By abuse of notation, we will set $\phi_n := \hat{\phi}_n$.

We now conclude this section with two key definitions.

**Definitions 1.**

(i) Notation as above. By a *majorizing sequence* for the holomorphic map $\phi$, we mean a positive sequence of numbers $\alpha_n$ $n = 1, 2, \ldots$ such that $\|\phi_n\| < \alpha_n$ for $n \geq 1$. Suppose that $\rho := \limsup {\alpha_n}^{1/n} < \infty$. Then it is completely standard ([8]) that the Taylor series expansion of $\phi$ converges at least on the ball $B_r(G)$ of radius $r = 1/\rho$.

(ii) If $\phi$ admits a majorizing sequence as in (i), then we will say that $\phi$ is *majorizable*.

We will see in the next section that a very important class of input/output operators from systems and control theory are in point of fact majorizable.

## 3 Operators with Fading Memory

In this section, we will show that perhaps the most natural class of input/output operators from the systems standpoint are majorizable. Moreover for this class of operators we will even derive *a priori* majorizing sequence. We begin with the following key definition:

**Definition 2.**

An analytic map $\phi : B_{r_o}(G) \to H$, $\phi(0) = 0$ has *fading memory* if its nonlinear part $\phi - \phi'(0)$ admits a factorization

$$\phi - \phi'(0) = \hat{\phi}_o W$$

where $\hat{\phi}$ is an analytic defined in some neighborhood of $0 \in G$, and $W$ is a linear Hilbert-Schmidt operator. (In this case, one can assume that there exists an orthonormal basis of eigenvectors for $W$ in $G$, $\{e_k\}, k = 1, 2, \ldots$ such that $We_k = \lambda_k e_k$ with

$$\|W\|_2^2 := \sum_{k=1}^{\infty} |\lambda_k|^2 < \infty.$$

$\|W\|_2$ is called the *Hilbert-Schmidt norm* of $W$. )

**Remark 1.** System-theoretically fading memory input/output operators have the property that any two input signals which are close in the recent past but not necessarily close in the remote past will yield present outputs which are close. For more details about this important class of operators see [7].

For fading memory operators, we can construct an explicit majorizing sequence:

**Lemma 1** *Let $\phi : B_{r_o}(G) \to H, \phi(0) = 0$, have fading memory. Suppose moreover that if we write*

$$\phi - \phi'(0) = \hat{\phi}_o W$$

*as in (3.1), then $\hat{\phi} : B_{r_1}(G) \to B_{r_2}(H)$. Then the sequence*

$$\alpha_1 := \|\phi'(0)\|$$

$$\alpha_n := \frac{r_2 e^n \|W\|_2^n}{r_1^n}$$

*for $n \geq 2$, is a majorizing sequence for $\phi$.*

**Proof.** See [2], Lemma (3.5). □

In what follows, we will assume that all of the input/output operators we consider are causal and have fading memory. An interesting and useful property of fading memory operators is the following:

**Proposition 1** *Notation and hypotheses as in (1). Then each $\phi_n$ (regarded as a linear operator on $G^{\otimes n}$) is compact for $n \geq 2$.*

**Proof.** See (3.5) of [11]. □

# 4 Control Theoretic Preliminaries

We start here with the control problem definition. First, we will need to consider the precise kind of input/output operator we will be considering. See [9], [11] for closely related discussions. We will assume that all of the operators we consider are causal and majorizable. Throughout this paper $H^2(\mathbf{C^k})$ will denote the standard Hardy space of $\mathbf{C^k}$-valued functions on the unit circle ($k$ may be infinite, i.e., in this case $\mathbf{C^k}$ is replaced by $h^2$, the space of one-sided square summable sequences). We now make the following definition:

**Definition 2.**

Let $S : H^2(\mathbf{C^k}) \to H^2(\mathbf{C^k})$ denote the canonical unilateral right shift. Then we say an input/output operator $\phi$ is *locally stable* if it is causal and majorizable, $\phi(0) = 0$, and if there exists an $r > 0$ such that $\phi : B_r(H^2(\mathbf{C^k})) \to H^2(\mathbf{C^k})$ with $S\phi = \phi \circ S$ on $B_r(H^2(\mathbf{C^k}))$. We set

$$C_l := \{\text{space of locally stable operators}\}.$$

Since the theory we are considering is local, the notion of local stability is sufficient for all of the applications we have in mind. The interested reader can compare this notion, with the more global notions of stability as for example discussed in [5].

The theory we are about to give holds for all plants which admit coprime locally stable factorizations. However, for simplicity we will assume that our plant is also locally stable. Accordingly, let $P, W$ denote locally stable operators, with $W$ invertible. In a typical feedback system [16], $P$ represents the plant, and $W$ the weight or filter on the set of disturbances whose energy is bounded by 1. Now we say that the feedback compensator $C$ *locally stabilizes* the closed loop if the operators $(I + P \circ C)^{-1}$ and $C \circ (I + P \circ C)^{-1}$ are well-defined and locally stable. By a result of [1], $C$ locally stabilizes the closed loop if and only if

$$C = \hat{q} \circ (I - P \circ \hat{q})^{-1} \tag{1}$$

for some $\hat{q} \in C_l$. Notice then that the weighted sensitivity (see [13] and [16] for all the relevant engineering definitions and motivation), $(I + P \circ C)^{-1} \circ W$ can be written as $W - P \circ q$, where $q := \hat{q} \circ W$. (Since $W$ is invertible, the data $q$ and $\hat{q}$ are equivalent.) In this context, we will call such a $q$, a *compensating parameter*. Note that from the compensating parameter $q$, we get a locally stabilizing compensator $C$ via the formula (1).

The problem we would like to solve here, is a version of the classical disturbance attenuation problem of [13], [16]. This of course corresponds to the "minimization" of the "sensitivity" $W - P \circ q$ taken over all locally stable $q$. In order to formulate a precise mathematical problem, we need to say in what sense we want to minimize $W - P \circ q$. This we will do in the next section where we will propose a notion of "sensitivity minimization" which seems quite natural to analytic input/output operators.

# 5 Sensitivity Function

In this section we define a fundamental object, namely a nonlinear version of *sensitivity*. We will see that while the optimal $H^\infty$ sensitivity is a real number in the linear case, the measure of performance which seems to be more natural in this nonlinear setting is a certain function defined in a real interval.

In order to define our notion of sensitivity, we will first have to partially order germs of analytic mappings. All of the input/output operators here will be locally stable. We also follow here our convention that for given $\phi \in C_l$, $\phi_n$ will denote the bounded linear map on the tensor space $(H^2(\mathbf{C^k}))^{\otimes n}$ associated to the $n$-linear part of $\phi$ which we also denote by $\phi_n$ (and which we always assume without loss of generality is symmetric in its arguments). The context will always make the meaning of $\phi_n$ clear.

We can now state the following key definitions:

**Definitions 3.**

(i) For $W, P, q \in C_l$ ($W$ is the weight, $P$ the plant, and $q$ the compensating parameter), we define the *sensitivity function* $S(q)$,

$$S(q)(\rho) := \sum_{n=1} \rho^n \|(W - P \circ q)_n\|$$

for all $\rho > 0$ such that the sum converges. Notice that for fixed $P$ and $W$, for each $q \in C_l$, we get an associated sensitivity function.

(ii) We write $S(q) \preceq S(\tilde{q})$, if there exists a $\rho_o > 0$ such that $S(q)(\rho) \leq S(\tilde{q})(\rho)$ for all $\rho \in [0, \rho_o]$. If $S(q) \preceq S(\tilde{q})$ and $S(\tilde{q}) \preceq S(q)$, we write $S(q) \cong S(\tilde{q})$. This means that $S(q)(\rho) = S(\tilde{q})(\rho)$ for all $\rho > 0$ sufficiently small, i.e. $S(q)$ and $S(\tilde{q})$ are equal as germs of functions.

(iii) If $S(q) \preceq S(\tilde{q})$, but $S(\tilde{q}) \not\preceq S(q)$, we will say that $q$ *ameliorates* $\tilde{q}$. Note that this means $S(q)(\rho) < S(\tilde{q})(\rho)$ for all $\rho > 0$ sufficiently small.

Now with Definitions 3, we can define a notion of "optimality" relative to the sensitivity function:

**Definitions 4.**

(i) $q_o \in C_l$ is called *optimal* if $S(q_o) \preceq S(q)$ for all $q \in C_l$.

(ii) We say $q \in C_l$ is *optimal with respect to its n-th term* $q_n$, if for every $n$-linear $\hat{q}_n \in C_l$, we have

$$S(q_1 + \ldots + q_{n-1} + q_n + q_{n+1} \ldots) \preceq S(q_1 + \ldots + q_{n-1} + \hat{q}_n + q_{n+1} + \ldots).$$

If $q \in C_l$ is optimal with respect to all of its terms, then we say that it is *partially optimal.*

Clearly, if $q$ is optimal, then it is partially optimal, but the converse may not hold. Notice moreover that if $\phi$ is a Volterra series, then our definition of sensitivity measures in a precise sense the amplification of energy of each Volterra kernel on signals whose energy is bounded by a given $\rho$. For this reason, it appears that in this context the Definitions 3 of the sensitivity function $S(q)$ seems physically natural. In the next section, we will discuss a procedure for constructing partially optimal compensating parameters, and then in Section 7 we will show how this procedure leads to the construction of optimal compensating parameters for SISO systems. Of course, from formula (1) above, one can derive the corresponding partially optimal (resp., optimal) compensator from the partially optimal (resp., optimal) compensating parameter.

# 6 Iterative Commutant Lifting Method

In this section, we discuss the main construction of this paper from which we will derive both partially optimal and optimal compensators relative to the sensitivity function given in Definitions 3 above. As before, $P$ will denote the plant, and $W$ the weighting operator, both of which we assume are locally stable. As in the linear case, we always suppose that $P_1$ is an isometry, i.e. $P_1$ is **inner.** In order to state our results, we will need a few preliminary remarks and to set-up some notation. We refer the interested reader to [11] for the precise proofs of the various results in this section.

We begin by noting the following key relationship:

$$(W - P \circ q)_k = W_k - \sum_{1 \leq j \leq k} \sum_{i_1 + .... + i_j = k} P_j(q_{i_1} \otimes \ldots \otimes q_{i_j})$$

Note that once again for $\phi$ of fading memory, $\phi_n$ denotes the $n$-linear part of $\phi$, as well as the associated linear operator on the appropriate tensor space.

We are now ready to formulate the *iterative commutant lifting procedure*. Let $\Pi$ : $H^2(\mathbf{C}^\mathbf{k}) \to H^2(\mathbf{C}^\mathbf{k}) \ominus P_1 H^2(\mathbf{C}^\mathbf{k})$ denote orthogonal projection. Using the linear commutant lifting theorem (CLT) (see [15] for the details), we may choose $q_1$ such that

$$\|W_1 - P_1 q_1\| = \|\Pi W_1\|.$$

Now given this $q_1$, we choose (using CLT) $q_2$ such that

$$\|W_2 - P_2(q_1 \otimes q_1) - P_1 q_2\| = \|\Pi(W_2 - P_2(q_1 \otimes q_1))\|.$$

Inductively, given $q_1, \ldots, q_{n-1}$, set

$$A_n := (W_n - \sum_{2 \leq j \leq n} \sum_{i_1 + .... + i_j = n} P_j(q_{i_1} \otimes \ldots \otimes q_{i_j}))$$

for $n \geq 2$. Then from the CLT, we may choose $q_n$ such that

$$\|A_n - P_1 q_n\| = \|\Pi A_n\|. \tag{2}$$

We now come to the key point on the convergence of the iterative commutant lifting method.

**Proposition 2** *With the above notation, let* $q^{(1)} := q_1 + q_2 + \ldots$. *Then* $q^{(1)} \in C_l$.

Note that given any $q \in C_l$, we can apply the iterative commutant lifting procedure to $W - P \circ q$. Now set

$$S_\Pi(q)(\rho) := \sum_{n=1} \rho^n \|\Pi(W - P \circ q)_n\|.$$

Clearly, $S_\Pi(q) \leq S(q)$ (as functions). We can now state the following result whose proof is immediate from the above discussion:

**Proposition 3** *Given* $q \in C_l$, *there exists* $\tilde{q} \in C_l$, *such that* $S(\tilde{q}) \equiv S_\Pi(q)$. *Moreover* $\tilde{q}$ *may be constructed from the iterated commutant lifting procedure.*

Moreover, we easily have the following result:

**Proposition 4** $q$ *is partially optimal if and only if* $S(q) \cong S_\Pi(q)$.

We can now summarize the above discussion with the following:

**Theorem 1** *For given* $P$ *and* $W$ *as above, any* $q \in C_l$ *is either partially optimal or can be ameliorated by a partially optimal compensating parameter.*

**Proof.** Immediate from Propositions 2, 3, and 4. □

It is important to emphasize that a partially optimal compensating parameter need not be optimal in the sense of Definition 4(i). Basically, what we have shown here is that using the iterated commutant lifting procedure, we can ameliorate any given design. The question of optimality will be considered in the next section.

# 7 Optimal Compensators

In this section we will derive our main results about optimal compensators. Basically, we will show that in the single input / single output setting, the iterated commutant lifting procedure leads to an optimal design. We begin with the following:

**Theorem 2** *There exist optimal compensators.*

**Proof.** See (7.1) of [11]. □

For the construction of the optimal compensator in Theorem 3 below, we will need one more technical result. Accordingly, we will need to set-up a bit more notation. First set $H^2 := H^2(\mathbf{C})$, and $H^\infty := H^\infty(\mathbf{C})$ (the space of bounded analytic complex-valued functions on the unit disc). Let $m \in H^\infty$ be a nonconstant inner function, let $\Pi_1 : H^2 \to H^2 \ominus mH^2 =: H(m)$ denote orthogonal projection, and set $T := \Pi_1 S|H(m)$, where $S$ is the canonical unilateral shift on $H^2$. ($T$ is the compressed shift.) For $H$ a complex separable Hilbert space, let $S_\infty : H \to H$ denote a unilateral shift, i.e. an isometric operator with no unitary part. This means that $S_\infty^{*n} \to 0$ for all $h \in H$ as $n \to \infty$. (See [15].) We can now state the following generalization of a nice result due to Sarason:

**Lemma 2** *Notation as above. Let $A : H \to H^2 \ominus mH^2$ be a bounded linear operator which attains its norm, i.e. such that there exists $h_o \in H$ with $\|Ah_o\| = \|A\|\|h_o\| \neq 0$. Suppose moreover that*

$$AS_\infty = TA.$$

*Then there exists a unique minimal intertwining dilation $B$ of $A$, i.e. an operator $B : H \to H^2$ such that $BS_\infty = SB$, $\|A\| = \|B\|$, and $\Pi_1 B = A$.*

**Proof.** See (7.2) of [11]. □

We now come to the main result of this section:

**Theorem 3** *Let $W$ and $P$ be single SISO locally stable operators, with $W$ the weight and $P$ the plant. Suppose that $\Pi W_j$ is compact for $j \geq 1$ and $\Pi P_k$ is compact for $k \geq 2$. ($\Pi : H^2 \to H^2 \ominus P_1 H^2$ denotes orthogonal projection.) Let $q_{opt}$ be a partially optimal compensating parameter as constructed by the iterated commutant lifting procedure. Then $q_{opt}$ is optimal.*

**Proof.** First of all, since $\Pi W_1$ attains its norm, from Lemma 2 we have that the optimal $q_1$ constructed relative to $W_1$ and $P_1$ is unique. Now from our above hypotheses, each $\Pi A_k$ is compact for $k \geq 2$, and hence each $\Pi A_k$ attains its norm. Therefore by Lemma 2 each optimal $q_k$ constructed by the iterated commutant lifting procedure is unique. Theorem 3 now follows immediately from Theorem 1. □

**Corollary 1** *Let $P$ and $W$ be locally stable and SISO, with linear part $P_1$ rational. Then the partially optimal compensating parameter $q_{opt}$ constructed by the iterated commutant lifting procedure is optimal.*

**Proof.** Indeed, since $P_1$ is SISO rational (recall that we also always assume that $P_1$ is inner), $H^2 \ominus P_1 H^2$ is finite dimensional, and so we are done by Theorem 3. □

**Remark 2.** Corollary 1 gives a constructive procedure for finding the optimal compensator under the given hypotheses. Indeed, when $P_1$ is SISO rational, the iterative commutant lifting procedure can be reduced to *finite dimensional matrix calculations.* In our paper [9], we have shown that when the hypotheses of Theorem 3 are satisfied, the skew Toeplitz theory of [6] provides an algorithmic design procedure for distributed nonlinear systems as well.

This research was supported in part by grants from the Research Fund of Indiana University, Department of Energy DE-FG02-86ER25020, NSF (ECS-8704047), NSF (DMS-8811084), and the Air Force Office of Scientific Research AFOSR-88-0020, AFOSR-90-0024.

## References

[1] V. Anantharam and C. Desoer, "On the stabilization of nonlinear systems," *IEEE Trans. Automatic Control* **AC-29** (1984), pp. 569–573.

[2] J. Ball, C. Foias, J. W. Helton, and A. Tannenbaum, "On a local nonlinear commutant lifting theorem," *Indiana J. Mathematics* **36** (1987), pp. 693–709.

[3] J. Ball, C. Foias, J. W. Helton, and A. Tannenbaum, "Nonlinear interpolation theory in $H^\infty$," in *Modelling, Robustness, and Sensitivity in Control Systems* (edited by Ruth Curtain), NATO-ASI Series, Springer-Verlag, New York, 1987.

[4] J. Ball, C. Foias, J. W. Helton, and A. Tannenbaum, "A Poincare-Dulac approach to a nonlinear Beurling-Lax-Halmos theorem," to appear *Journal of Math. Anal. and Applications.*

[5] J. Ball and J. W. Helton, "Sensitivity bandwidth optimization for nonlinear feedback systems," Technical Report, Department of Mathematics, University of California at San Diego, 1988.

[6] H. Bercovici, C. Foias, and A. Tannenbaum, "On skew Toeplitz operators. I," *Operator Theory: Advances and Applications* **29** (1988), pp. 21–44.

[7] S. Boyd and L. Chua, "Fading memory and the problem of approximating nonlinear operators with Volterra series," *IEEE Trans. Circuits and Systems* **CAS-32** (1985), pp. 1150–1161.

[8] A. Feintuch and B. Francis, "Uniformly optimal control of linear systems," *Automatica* **21** (1986), pp. 563-574.

[9] C. Foias and A. Tannenbaum, "Iterated cummutant lifting for systems with rational symbol," to appear *Operator Theory: Advances and Applications.*

[10] C. Foias and A. Tannenbaum, "On the four block problem, II: the singular system," *Integral Equations and Operator Theory* **11** (1988), pp. 726–767.

[11] C. Foias and A. Tannenbaum, "Weighted optimization theory for nonlinear systems," to appear in *SIAM J. on Control and Optimization.*

[12] C. Foias, A. Tannenbaum, and G. Zames, "Some explicit formula for the singular values of certain Hankel operators with factorizable symbol," *SIAM J. on Math. Analysis* **19** (1988), pp. 1081–1089.

[13] B. Francis, *A Course in $H^\infty$ Control Theory*, McGraw-Hill, New York, 1981.

[14] E. Hille and R. S. Phillips, *Functional Analysis and Semigroups*, AMS Colloquium Publications XXIII, Providence, 1957.

[15] B. Sz.-Nagy and C. Foias, *Harmonic Analysis of Operators on Hilbert Space*, North-Holland Publishing Company, Amsterdam, 1970.

[16] G. Zames, "Feedback and optimal sensitivity: model reference transformations, multiplicative seminorms, and approximate inverses," *IEEE Trans. Auto. Control* **AC-26** (1981), pp. 301–320.

# A J-SPECTRAL FACTORIZATION APPROACH TO $H_\infty$ CONTROL

Michael Green[1], Keith Glover[2], David Limebeer[1] and John Doyle[3]

Abstract

It is shown that necessary and sufficient conditions for the existence of sub-optimal solutions to the model matching problem associated with $\mathcal{H}_\infty$ control are that two coupled J-spectral factorizations exist. The second J-spectral factor provides a parameterization of all solutions to the model matching problem. The existence of the J-spectral factors is then shown to be equivalent to the existence of stabilizing, non-negative definite solutions to two algebraic Riccati equations, allowing a state-space formula for a linear fractional representation of all controllers to be given.

## 1. Introduction

In a recent paper [1] a general class of $\mathcal{H}_\infty$ control problems was solved via several spectral and J-spectral factorizations. The resulting algorithm is not computationally optimal, since the solution to the $\mathcal{H}_\infty$ problem presented in [9] requires just two algebraic Riccati equations to be solved. It was also observed that these Riccati equations could be associated with two J-factorizations.

Here, we re-analyse the work in [1], showing that all the spectral and J-spectral factorizations can be subsumed into just two J-spectral factorizations. The BGK factorization theory [4] can then be used show that J-spectral factorization is equivalent to the soluability of indefinite algebraic Riccati equations, enabling a generator of all solutions to the model matching and $\mathcal{H}_\infty$ control problems to be given.

Concurrent with this work, several of the other approaches to $\mathcal{H}_\infty$ control have been generalized and entirely new connections uncovered. The following remarks, which are in no way a complete survey, are intended to connect this paper with these other developments. The four block distance problem has been solved [9,10,14] using all-pass embedding. Connections between maximum entropy $\mathcal{H}_\infty$ control and risk sensitive optimal control have been established [9], a connection observed also in [5]. Moreover, in [7], a state space approach which is reminiscent of classical LQG theory is developed. [12] also considers a state feedback approach, observing a connection with LQ game theory. The connection between game theory and J-spectral factorization is long standing [3]. A conjugation approach developed in [13] is also closely related to the J-spectral factorization method pursued here.

[1] Dept. Elec. Eng., Imperial College, London SW7 2BT, U.K.

[2] Engineering Dept., University of Cambridge, Trumpington St., Cambridge CB2 1PZ U.K.

[3] Dept. Elec. Eng., California Institute of Technology, Pasedena CA91125 U.S.A.

Notation throughout is standard. The prefix $\mathcal{R}$ (eg., $\mathcal{RH}_\infty^{pxq}$) denotes rational, whilst the prefix $\mathcal{G}$ denotes units ($\mathcal{GH}_\infty^{p} = \{M \in \mathcal{RH}_\infty^{pxp}: M^{-1} \in \mathcal{RH}_\infty^{pxp}\}$). $M^{\sim}(s) = [M(-\bar{s})]^*$. The indefinite matrix $\begin{bmatrix} I_p & 0 \\ 0 & -\gamma^2 I_q \end{bmatrix}$ we denote by $J_{pq}(\gamma)$, but we freqently abreviate this to J. $\Gamma_R$ is the Hankel operator with symbol R.

Recall [6,8,15] that by the use of doubly coprime factorizations and the Youla paramentrization, the $\mathcal{H}_\infty$ control problem can be posed as a model matching problem: Given the $T_{ij}$'s, find necessary and sufficient conditons for the existence of $Q \in \mathcal{RH}_\infty$ such that $\|T_{11} + T_{12}QT_{21}\|_\infty < \gamma$, and when such conditions hold, parameterize all solutions.

## 2. The Nehari problem

J-spectral factorization has been associated with the Nehari problem for some time [2,8]. Here, a new condition on the J-spectral factor—namely that it have outer 1,1 block—allows solutions for more general model matching problems to be developed by boot-strapping from the Nehari problem.

Theorem 2.1: Let $R \in \mathcal{RL}_\infty^{pxq}$. Then $\|\Gamma_R\| < \gamma \Leftrightarrow$ there exists $Q \in \mathcal{RH}_\infty^{pxq}$ such that $\|R + Q\|_\infty < \gamma \Leftrightarrow$ there exists $W \in \mathcal{GH}_\infty^{p+q}$ with $W_{11} \in \mathcal{GH}_\infty^{p}$ satisfying

$$G^{\sim}J_{pq}(\gamma)G = W^{\sim}J_{pq}(\gamma)W, \qquad G = \begin{bmatrix} I_p & R \\ 0 & I_q \end{bmatrix}. \tag{2.1}$$

Proof: 1 $\Leftrightarrow$ 2 is Nehari's Theorem. We shall prove that 1 $\Rightarrow$ 3 and that 3 $\Rightarrow$ 2.

3 $\Rightarrow$ 2: Let $V = W^{-1}$ and note that $W_{11} \in \mathcal{GH}_\infty^{p} \Leftrightarrow V_{22} \in \mathcal{GH}_\infty^{q}$. Set $Q = V_{12}(V_{22})^{-1} \in \mathcal{RH}_\infty$. It follows from $(GV)^{\sim}J(GV) = J$ that $(R+Q)^{\sim}(R+Q) - \gamma^2 I = -\gamma^2(V_{22}V_{22}^{\sim})^{-1} < 0$.

1 $\Rightarrow$ 3: Decompose R as $R = R_+ + R_-$, with $R_-(-s) \in \mathcal{RH}_\infty$ and strictly proper, $R_+ \in \mathcal{RH}_\infty$. Using the state-space construction of [8, Chapter 7], construct $X \in \mathcal{GH}_\infty$ such that $G_-^{\sim}JG_- = XJX$, $G_- = \begin{bmatrix} I_p & R_- \\ 0 & I_q \end{bmatrix}$. A Lyapunov equation argument shows that $X_{11} \in \mathcal{GH}_\infty$. Define $W = X\begin{bmatrix} I & R_+ \\ 0 & I \end{bmatrix}$. □

The next result provides a characterization of all solutions to sub-optimal Nehari extension problems.

Theorem 2.2: Let $R \in \mathcal{RL}_\infty^{pxq}$, $\|\Gamma_R\| < \gamma$. With W as in Theorem 2.1, the set of all $Q \in \mathcal{RH}_\infty^{pxq}$ such that $\|R + Q\|_\infty \le \gamma$ is given by

$$Q = Q_1Q_2^{-1}, \qquad \begin{bmatrix} Q_1 \\ Q_2 \end{bmatrix} = W^{-1}\begin{bmatrix} U \\ I_q \end{bmatrix} \quad U \in \mathcal{RH}_\infty^{pxq} \text{ with } \|U\|_\infty \le \gamma. \tag{2.2}$$

Proof: Let $V = W^{-1}$ and recall $V_{22} \in \mathcal{GH}_\infty$. By using the fact that $VJ^{-1}V^\sim = G^{-1}J^{-1}(G^{-1})^\sim$, we see that $\|V_{22}^{-1}V_{21}\|_\infty < \gamma^{-1}$. It follows that $(V_{22}^{-1}V_{21}U + I) \in \mathcal{GH}_\infty$, so $Q \in \mathcal{RH}_\infty$. From $(GV)^\sim J(GV) = J$ it follows that $(R+Q)^\sim(R+Q) - \gamma^2 I = (Q_2^{-1})^\sim[U^\sim U - \gamma^2 I]Q_2^{-1} \leq 0$.

Conversely, suppose $Q \in \mathcal{RH}_\infty$ is such that $\|R + Q\|_\infty \leq \gamma$. Define

$$\begin{bmatrix} U_1 \\ U_2 \end{bmatrix} = W\begin{bmatrix} Q \\ I \end{bmatrix} = WG^{-1}\begin{bmatrix} R+Q \\ I \end{bmatrix} \in \mathcal{RH}_\infty.$$

Thus $U_1^\sim U_1 - \gamma^2 U_2^\sim U_2 \leq 0$. Since $U_1$, $U_2$ are coprime we have that $U_2$ is invertible in $\mathcal{RL}_\infty$, and that $U = U_1U_2^{-1} \in \mathcal{RL}_\infty$ with $\|U\|_\infty \leq \gamma$. Hence (2.2) holds, with $Q_2 = U_2^{-1}$, and $Q_1 = QQ_2$. To show that $U \in \mathcal{RH}_\infty$, we show that $U_2 \in \mathcal{GH}_\infty$. To see this, observe that $V_{22}^{-1} = (V_{22}^{-1}V_{21}U + I)U_2 \in \mathcal{GH}_\infty$. Also, since $\|V_{22}^{-1}V_{21}U\|_\infty \leq \|V_{22}^{-1}V_{21}\|_\infty\|U\|_\infty < 1$, we have that the winding number (around the origin) of $\det\{(V_{22}^{-1}V_{21}U + I)(j\omega)\}$ is zero. Thus the winding number of $\det(U_2(j\omega))$ is zero, giving $U_2 \in \mathcal{GH}_\infty$, since $U_2 \in \mathcal{RH}_\infty$. □

## 3. The unilateral model matching problem

We now seek $Q \in \mathcal{RH}_\infty$ such that $\|A + BQ\|_\infty < \gamma$, where B is "tall" and the relevant "G" is now also "tall". A related theoren is given in [11, page 58].

Theorem 3.1: Suppose $G = \begin{bmatrix} B & A \\ 0 & I_q \end{bmatrix} \in \mathcal{RL}_\infty^{(\ell+q)\times(p+q)}$ has a left inverse in $\mathcal{RL}_\infty$. Then the following are equivalent:

1. There exists a $Q \in \mathcal{RH}_\infty^{p\times q}$ such that $\|A + BQ\|_\infty < \gamma$.
2. There exists a $W \in \mathcal{GH}_\infty^{p+q}$ with $W_{11} \in \mathcal{GH}_\infty^{p}$ satisfying $G^\sim J_{\ell q}(\gamma)G = W^\sim J_{pq}(\gamma)W$.

The set of all $Q \in \mathcal{RH}_\infty$ satisfying $\|A + BQ\|_\infty \leq \gamma$ is given by

$$Q = Q_1Q_2^{-1}, \quad \begin{bmatrix} Q_1 \\ Q_2 \end{bmatrix} = W^{-1}\begin{bmatrix} U \\ I_q \end{bmatrix} \quad U \in \mathcal{RH}_\infty^{p\times q} \text{ with } \|U\|_\infty \leq \gamma.$$

Proof: Reduce to the Nehari problem as follows:

Let $B_o \in \mathcal{GH}_\infty$ satisfy $B_o^\sim B_o = B^\sim B$, define $B_i = BB_o^{-1}$ and note $B_i^\sim B_i = I$. Let $B_\perp$ be such that $\begin{bmatrix} B_i & B_\perp \end{bmatrix}$ is all-pass. Then $\|A + BQ\|_\infty < \gamma$

$$\Leftrightarrow \left\|A + \begin{bmatrix} B_i & B_\perp \end{bmatrix}\begin{bmatrix} B_oQ \\ 0 \end{bmatrix}\right\|_\infty < \gamma$$

$$\Leftrightarrow \left\|\begin{bmatrix} R_1 \\ R_2 \end{bmatrix} + \begin{bmatrix} B_oQ \\ 0 \end{bmatrix}\right\|_\infty < \gamma, \quad R = \begin{bmatrix} B_i & B_\perp \end{bmatrix}^\sim A$$

$$\Leftrightarrow \|R_2\|_\infty < \gamma \text{ and } (R_1+B_oQ)^\sim(R_1+B_oQ) + R_2^\sim R_2 < \gamma^2 I.$$

Thus $\exists\, Q \in \mathcal{RH}_\infty$ such that $\|A + BQ\|_\infty < \gamma$ if and only if:

a) $\exists\, N \in \mathcal{GH}_\infty$ with $\gamma^2 N^\sim N = \Phi = \gamma^2 I - R_2^\sim R_2 = \gamma^2 I_q - A^\sim[I - B(B^\sim B)^{-1}B^\sim]A$.

b) $\exists\, \hat{Q}\ (= B_oQN^{-1}) \in \mathcal{RH}_\infty$ such that $\|R_1N^{-1} + \hat{Q}\|_\infty < \gamma$.

By Theorem 2.1, (b) holds $\Leftrightarrow \exists\, X \in \mathcal{GH}_\infty$ with $X_{11} \in \mathcal{GH}_\infty$ such that

$$\begin{bmatrix} I & 0 \\ (N^{-1})^\sim R_1^\sim & I \end{bmatrix} J \begin{bmatrix} I & R_1N^{-1} \\ 0 & I \end{bmatrix} = X^\sim JX.$$

Note also that $R_1 = (B_o^{\sim})^{-1}B^{\sim}A$. Now observe that

$$G^{\sim}JG = \begin{bmatrix} B_o^{\sim} & 0 \\ A^{\sim}BB_o^{-1} & I \end{bmatrix}\begin{bmatrix} I & 0 \\ 0 & -\Phi \end{bmatrix}\begin{bmatrix} B_o & (B_o^{\sim})^{-1}B^{\sim}A \\ 0 & I \end{bmatrix}.$$

It follows that W exists $\Leftrightarrow$ X and N exist ($X = W\begin{bmatrix} B_o & 0 \\ 0 & N \end{bmatrix}^{-1}$). □

The condition that G have a left inverse in $\mathcal{RL}_\infty$ is not necessary for there to exist a solution to the model matching problem. It is, however, a necessary condition for the existence of a $W \in \mathcal{GH}_\infty$ such that $G^{\sim}JG = W^{\sim}JW$.

## 4. The bilateral model matching problem

We seek $Q \in \mathcal{RH}_\infty$ such that $\|A + BQC\|_\infty < \gamma$, with B "tall" and C "wide". The technique is based on reduction to the unilateral case, and the result involves two J-spectral factorizations.

Theorem 4.1: Suppose $A \in \mathcal{RL}_\infty^{\ell x p}$, $B \in \mathcal{RL}_\infty^{\ell x q}$ and $C \in \mathcal{RL}_\infty^{m x p}$. Suppose also that B has a left inverse and C has a right inverse in the appropriate $\mathcal{RL}_\infty$ spaces. Let $B = B_a B_s$ in which $B_a \in \mathcal{RL}_\infty^{\ell x \ell}$ is all-pass and $B_s \in \mathcal{RH}_\infty^{\ell x q}$.

There exists a $Q \in \mathcal{RH}_\infty^{q x m}$ such that $\|A + BQC\|_\infty < \gamma$ if and only if:

1. There exists a $V \in \mathcal{GH}_\infty^{m+\ell}$ with $V_{11} \in \mathcal{GH}_\infty^{m}$ satisfying

$$HJ_{p\ell}(\gamma)H^{\sim} = VJ_{m\ell}(\gamma)V^{\sim}, \quad H = \begin{bmatrix} C & 0 \\ B_a^{\sim}A & I_\ell \end{bmatrix}. \tag{4.1}$$

2. There exists a $W \in \mathcal{GH}_\infty^{q+m}$ with $W_{11} \in \mathcal{GH}_\infty^{q}$ satisfying

$$G^{\sim}J_{\ell m}(\gamma)G = W^{\sim}J_{qm}(\gamma)W, \qquad G = \hat{J}V^{-1}\hat{J}^{*}\begin{bmatrix} B_s & 0 \\ 0 & I_m \end{bmatrix}, \quad \hat{J} = \begin{bmatrix} 0 & -I_\ell \\ I_m & 0 \end{bmatrix}. \tag{4.2}$$

The set of all $Q \in \mathcal{RH}_\infty^{q x m}$ such that $\|A + BQC\|_\infty \le \gamma$ is given by

$$Q = Q_1Q_2^{-1}, \quad \begin{bmatrix} Q_1 \\ Q_2 \end{bmatrix} = W^{-1}\begin{bmatrix} U \\ I_m \end{bmatrix} \quad U \in \mathcal{RH}_\infty^{q} \text{ with } \|U\|_\infty \le \gamma. \tag{4.3}$$

Proof: We may assume, without loss of generality, that $B \in \mathcal{RH}_\infty$, since $\|A + BQC\|_\infty \le \gamma$ $\Leftrightarrow \|B_a^{\sim}A + B_sQC\|_\infty \le \gamma$.

With $B \in \mathcal{RH}_\infty$ we see that 1 is necessary by applying Theorem 3.1 to the problem $A^*$ $+ C^*\hat{Q}$, where $\hat{Q} = (BQ)^*$.

Let $C_o \in \mathcal{GH}_\infty$ be such that $CC^{\sim} = C_oC_o^{\sim}$ and define $C_i = C_o^{-1}C$. Let $C_\perp$ be such that $\begin{bmatrix} C_i^{\sim} & C_\perp^{\sim} \end{bmatrix}$ is all-pass. Define R by $R = \begin{bmatrix} R_1 & R_2 \end{bmatrix} = A\begin{bmatrix} C_i^{\sim} & C_\perp^{\sim} \end{bmatrix}$.

As in the proof of Theorem 3.1, the existence of V satisfying (4.1) implies $\exists$ $M \in \mathcal{GH}_\infty$ such that $\gamma^2MM^{\sim} = \gamma^2I - R_2R_2^{\sim}$. So $Q \in \mathcal{RH}_\infty$ satisfies $\|A + BQC\|_\infty < \gamma \Leftrightarrow$ V exists and $\|M^{-1}R_1 + M^{-1}BQC_O\|_\infty < \gamma$. But, since $C_O \in \mathcal{GH}_\infty$, this is just a unilateral model matching problem. Therefore, by Theorem 3.1, we know that Q exists iff $\exists$ $Y \in \mathcal{GH}_\infty$ with $Y_{11} \in \mathcal{GH}_\infty$ such that

$$Y^{\sim}JY = P_1^{\sim}JP_1, \quad P_1 = \begin{bmatrix} M^{-1}B & M^{-1}R_1 \\ 0 & I \end{bmatrix}$$

and that $Y^{-1}$ "generates" all $QC_O$'s. But such a Y exists iff $\exists$ $W \in \mathcal{GH}_\infty$ with $W_{11} \in \mathcal{GH}_\infty$ satisfying

$$W^{\sim}JW = P^{\sim}JP, \quad P = P_1\begin{bmatrix} I & 0 \\ 0 & C_O^{-1} \end{bmatrix}$$

and that $W^{-1}$ "generates" all Q's. It remains to show that $P^{\sim}JP = G^{\sim}JG$, with G as in (4.2):

$$P = \begin{bmatrix} M^{-1} & M^{-1}R_1C_O^{-1} \\ 0 & C_O^{-1} \end{bmatrix}\begin{bmatrix} B & 0 \\ 0 & I \end{bmatrix} = \hat{J}\begin{bmatrix} C_O & 0 \\ R_1 & M \end{bmatrix}^{-1}\hat{J}^*\begin{bmatrix} B & 0 \\ 0 & I \end{bmatrix}.$$

Observe that $\hat{J}^*J\hat{J} = -\gamma^2J^{-1}$, that $\hat{J}\hat{J}^* = I$ and that $\begin{bmatrix} C_O & 0 \\ R_1 & M \end{bmatrix}J\begin{bmatrix} C_O & 0 \\ R_1 & M \end{bmatrix}^{\sim} = HJH^{\sim} = VJV^{\sim}$, from which it is easy to check that $G^{\sim}JG = P^{\sim}JP$. □

## 5. J-spectral factorization theory

In the last section we solved the model matching problem in terms of J-spectral factorization. We now show how such factorizations can be calculated by solving an algebraic Riccati equation. The main tool for this work is the state space factorization theory of Bart, Gohberg and Kaashoek [4], but some care needs to be taken to avoid minimality assumptions.

A matrix $H \in \mathbb{C}^{2n\times 2n}$ is a Hamiltonian matrix if $\hat{J}H = H^*\hat{J}^*$, $\hat{J} = \begin{bmatrix} 0 & -I_n \\ I_n & 0 \end{bmatrix}$. If $H \in \mathbb{C}^{2n\times 2n}$ is a Hamiltonian matrix, we say $H \in \mathrm{dom(Ric)}$ if there exists a matrix $Q \in \mathbb{C}^{n\times n}$ such that $H\begin{bmatrix} I_n \\ Q \end{bmatrix} = \begin{bmatrix} I_n \\ Q \end{bmatrix}\Lambda$, with $\Lambda$ asymptotically stable. If $H \in \mathrm{dom(Ric)}$, then $Q = \mathrm{Ric}(H)$ is Hermitian and $QH_{11} + H_{11}^*Q + QH_{12}Q - H_{21} = 0$.

<u>Theorem 5.1</u>: Suppose $G \in \mathcal{RH}_\infty^{(p+q)\times(m+\ell)}$ is given by the realization $G(s) = D + C(sI - A)^{-1}B$, with A asymptotically stable. Then there exists a $W \in \mathcal{GH}_\infty^{m+\ell}$ such that $G^{\sim}J_{pq}(\gamma)G = W^{\sim}J_{m\ell}(\gamma)W$ if and only if:

1. There exists a non-singular matrix $W_\infty \in \mathbb{C}^{(m+\ell)\times(m+\ell)}$ such that $D^*J_{pq}(\gamma)D = W_\infty^*J_{m\ell}(\gamma)W_\infty$.

2. $H \in \mathrm{dom(Ric)}$, where, with $J = J_{pq}(\gamma)$,

$$H = \begin{bmatrix} A & 0 \\ -C^*JC & -A^* \end{bmatrix} - \begin{bmatrix} B \\ -C^*JD \end{bmatrix}(D^*JD)^{-1}\begin{bmatrix} D^*JC & B^* \end{bmatrix}. \tag{5.1}$$

In this case W is given by $W(s) = W_\infty + L(sI - A)^{-1}B$, where $L = J_{m\ell}^{-1}(\gamma)W_\infty^{-*}[D^*J_{pq}(\gamma)C +$

$B^*Q]$ and $Q = \mathrm{Ric}(H)$.

Proof: Suppose 1 and 2 hold. Then $Q = \mathrm{Ric}(H)$ implies that $A - B(D^*JD)^{-1}[D^*JC + B^*Q] = A - BW_\infty^{-1}L$ is asymptotically stable. It follows that $W \in \mathcal{G}\mathcal{H}_\infty$. Now note that the Riccati equation for Q can be written as $C^*JC - Q(sI-A) - (-sI-A^*)Q = L^*JL$. A standard calculation, substituting for $L^*JL$, shows that $W^\sim JW = G^\sim JG$.

Conversely, suppose W exists. Then 1 follows by setting $W_\infty = W(\infty)$. Observe that $G^\sim JG = (W^\sim J)W$, $W \in \mathcal{G}\mathcal{H}_\infty$ is a canonical Wiener-Hopf factorization (see [4]; also [8]).

Temporary assumtion: (A,B) is controllable. Let $\exists\ P = P^*$ satisfy $PA + A^*P + C^*JC = 0$. It follows that

$$G^\sim JG = \begin{bmatrix} A & 0 & B \\ 0 & -A^* & -K^* \\ K & B^* & D^*JD \end{bmatrix}, \quad K = D^*JC + B^*P.$$

The unobservable/uncontrollable modes are the unobservable modes of of (K,A)/uncontrollable mdles of $(-A^*,-K^*)$. Therefore, w.l.o.g, suppose A,B,C are such that

$$A = \begin{bmatrix} A_{11} & 0 \\ A_{21} & A_{22} \end{bmatrix}, \quad B = \begin{bmatrix} B_1 \\ B_2 \end{bmatrix}, \quad K = \begin{bmatrix} K_1 & 0 \end{bmatrix} \quad (K_1,A_{11}) \text{ observable.}$$

A minimal realization of $G^\sim JG$ is given by

$$G^\sim JG = \begin{bmatrix} A_{11} & 0 & B_1 \\ 0 & -A_{11}^* & -K_1^* \\ K_1 & B_1^* & D^*JD \end{bmatrix} = \begin{bmatrix} \hat{A} & \hat{B} \\ \hat{C} & \hat{D} \end{bmatrix}.$$

Since $G^\sim JG$ has a canonical factorization, $\hat{A}^X = \hat{A} - \hat{B}\hat{D}^{-1}\hat{C}$ has no imaginary axis eigenvalues. Hence, since $\hat{A}^X$ is Hamiltonian, $\exists$ a non-singular matrix $\hat{X}$ such that $\hat{A}^X\hat{X} = \hat{X}T$, $T = \begin{bmatrix} T_1 & T_2 \\ 0 & -T_1^* \end{bmatrix}$ with $T_1$ asymptotically stable. Hence $X_+(\hat{A}) = \mathrm{Im}\begin{bmatrix} 0 \\ I \end{bmatrix}$ and $X_-(\hat{A}^X) = \mathrm{Im}\begin{bmatrix} \hat{X}_{11} \\ \hat{X}_{21} \end{bmatrix}$. By the BGK theorem, $X_+(\hat{A})$ and $X_-(\hat{A}^X)$ are complementary, i.e., $\hat{X}_{11}$ is non-singular, so define $\hat{Q} = \hat{X}_{21}\hat{X}_{11}^{-1}$. It is now straightforward to check that $\mathrm{Ric}(H) = P + \begin{bmatrix} \hat{Q} & 0 \\ 0 & 0 \end{bmatrix}$.

Removal of the controllability assumtion: Suppose (A,B,C) is in controllable canonical form and let $H_{\mathrm{cont}}$ denote the Hamiltonian analogous to (5.1) constructed from the controllable part. Apply the above result (i.e., with the controllability assumption), and observe that Ric(H) can be constructed from $\mathrm{Ric}(H_{\mathrm{cont}})$ and the solutions of 3 linear equations, which the stability conditions ensure have solutions. □

## 6. A state-space formula for all solutions to the $\mathcal{H}_\infty$ control problem

Theorem 6.1: Suppose $P(s) = D + C(sI - A)^{-1}B$, $(\ell+m)x(p+q)$ satisfies:

a) $(A, B_2)$ stabilizable and $(C_2, A)$ detectable.

b) $\begin{bmatrix} A-sI & B_2 \\ C_1 & D_{12} \end{bmatrix}, \begin{bmatrix} A-sI & B_1 \\ C_2 & D_{21} \end{bmatrix}$ full column and row rank respectively for all $s+\bar{s} = 0$.

c) $D_{12}^*D_{12} = I$, $D_{21}D_{21}^* = I$, $D_{11} = 0$, $D_{22} = 0$ (w.l.o.g. by [16]).

Then there exists an internally stabilizing controller K such that $\|P_{11} + P_{12}K(I-P_{22}K)^{-1}P_{21}\|_\infty < \gamma$ if and only if:

1. $H_Y \in \text{dom(Ric)}$ and $\text{Ric}(H_Y) \geq 0$, where

$$H_Y = \begin{bmatrix} A^* & 0 \\ -B_1B_1^* & -A \end{bmatrix} - \begin{bmatrix} C_2^* & C_1^* \\ -B_1D_{21}^* & 0 \end{bmatrix} (J_{m\ell}(\gamma))^{-1} \begin{bmatrix} D_{21}B_1^* & C_2 \\ 0 & C_1 \end{bmatrix}.$$

2. $H_Z \in \text{dom(Ric)}$ and $\text{Ric}(H_Z) \geq 0$, where, with $J = J_{qm}(\gamma)$, $M = \begin{bmatrix} M_1 & M_2 \end{bmatrix} = \begin{bmatrix} Y_\infty C_2^* + B_1D_{21}^* & -\gamma^{-2}Y_\infty C_1^* \end{bmatrix}$ and $Y_\infty = \text{Ric}(H_Y)$, $H_Z$ is defined by

$$H_Z = \begin{bmatrix} A-M_2C_1 & 0 \\ -C_1^*C_1 & -(A-M_2C_1)^* \end{bmatrix}$$
$$- \begin{bmatrix} B_2-M_2D_{12} & M_1 \\ -C_1^*D_{12} & 0 \end{bmatrix} J^{-1} \begin{bmatrix} D_{12}^*C_1 & (B_2-M_2D_{12})^* \\ 0 & M_1^* \end{bmatrix}$$

All controllers are given by $K = -K_1K_2^{-1}$, $\begin{bmatrix} K_1 \\ K_2 \end{bmatrix} = W_1^{-1} \begin{bmatrix} U \\ I \end{bmatrix}$, $U \in \mathcal{RH}_\infty$ with $\|U\|_\infty \leq \gamma$, where

$$W_1 \stackrel{s}{=} \begin{bmatrix} A-M_1C_2-M_2C_1 & B_2-M_2D_{12} & M_1 \\ L_1 & I & 0 \\ L_2 & 0 & I \end{bmatrix}$$

in which

$$L = \begin{bmatrix} L_1 \\ L_2 \end{bmatrix} = \begin{bmatrix} D_{12}^*C_1 + (B_2 - M_2D_{12})^*Z_\infty \\ -(C_2 + \gamma^{-2}M_1^*Z_\infty) \end{bmatrix}, \quad Z_\infty = \text{Ric}(H_Z).$$

<u>Proof</u>: Use the Youla parameterization to reduce the $\mathcal{H}_\infty$ control problem to the model matching problem $\|T_{11} + T_{12}QT_{21}\|_\infty < \gamma$ (see [6,8,15]). Assumption (a) is required here. Now apply Theorem 4.1—assumtion (b) is required for the left/right invertibility constraints on $T_{12}$ and $T_{21}$. Using Theorem 5.1, conditions 1 and 2 of the theorem are equivalent to conditions 1 and 2 of Theorem 4.1, with a Lyapunov equation argument required to show that outer constriants on the 1,1 blocks hold if and only if $\text{Ric}(H_Y)$ and $\text{Ric}(H_Z) \geq 0$.

Obtain a generator W for all Q from Theorems 4.1 and 5.1 and observe that

$$W = W_1 \begin{bmatrix} D_r & -U_\ell \\ N_r & V_\ell \end{bmatrix}$$

It follows that all controllers are generated by $W_1^{-1}$.

A few tricks can be used to avoid 2n dimensional state-space calculations in the procedure above. Also, the formula for all controllers can easily be turned in to an equivalent feedback form. □

## References

1] J.A. Ball and N. Cohen, "Sensitivity minimization in an $H^\infty$ norm: parameterization of all sub-optimal solutions", Int. J. Control, vol 46, pp 785-816, 1987.

2] J.A. Ball and A.C.M. Ran, "Optimal Hankel norm model reductions and Weiner-Hopf factorization, I: The canonical case", SIAM J Control and Optimization, 25, pp 362-383, 1987.

3] M. Banker, "Linear stationary quadratic games", Proc. IEEE C.D.C, pp 193-197, 1973.

4] H. Bart, I. Gohberg and M.A. Kaashoek, Minimal factorization of matrix and operator functions, Birkhauser Verlag, Basel, 1979.

5] D.S. Bernstein and W.M. Haddad, "LQG control with an $H_\infty$ performance bound: A Riccati equation approach", submitted to IEEE Trans. Auto Control.

6] J.C. Doyle et al, "Lecture notes in advances in multivariable control" ONR/Honeywell Workshop, Minneapolis, 1984.

7] J.C. Doyle, K. Glover, P. Khargonekar and B. Francis, "State-space solutions to standard $H_2$ and $H_\infty$ control problems", Proc. IEEE A.C.C, 1988; an extended version has been submitted to IEEE Trans Auto Control.

8] B.A. Francis, A course in $H_\infty$ control theory, Lecture notes in Control and Information Sciences number 88, Springer Verlag, 1987, 2nd ed.

9] K. Glover and J.C. Doyle, "State-space formulae for all stabilizing controllers that satisfy a $H^\infty$ norm bound and relations to risk sensitivity", Systems & Control Letters, 11, pp 167-172, 1988.

10] K. Glover, D.J.N. Limebeer, J. Dolye, E.M. Kasenally and M.G. Safonov, "A characterization of all the solutions to the four block general distance problem", under revision.

11] J.W. Helton et al, Operator theory, analytic functions, matrices and electrical engineering, Conference Board of the Mathematical Sciences, Regional Conference Series in Mathematics number 68, American Mathematical Society, 1987.

12] P.P. Khargonekar, I.R. Petersen and M.A. Rotea, "$H_\infty$ optimal control with state feedback", IEEE Trans. Auto. Control, 33, 1988, pp 783-786.

13] H. Kimura and R. Kawatani, "Synthesis of $H^\infty$ controllers based on conjugation", Proc. IEEE C.D.C, 1988.

14] D.J.N. Limebeer, E.M. Kasenally, M.G. Safonov and I. Jaimoukha, "A characterization of all the solutions to the four block general distance problem", Proc. IEEE C.D.C., 1988.

15] M.G. Safonov, E.A. Jonckheere, M. Verma and D.J.N. Limebeer, "Synthesis of positive real multivariable feedback systems", Int. J. Control., Vol 45, pp 817–842, 1987.

16] M.G. Safonov and D.J.N. Limebeer, "Simplifying the $H^\infty$ theory via loop shifting", Proc. IEEE C.D.C, 1988.

# VECTOR INTERPOLATION, $\mathcal{H}_\infty$ CONTROL and MODEL REDUCTION

Michael Green and David J.N. Limebeer

## 1. Introduction

The vector interpolation problem, posed by Ball and Helton [1,2], is the most general version of such Nevanlinna-Pick interpolation problems, containing the matrix interpolation problem as a special case. The problem has been solved by several methods, all of which rely on deep and abstract mathematics. Yet it is possible to solve the vector interpolation problem in a very straightforward way, as was done in [3,8] for the scalar case and in [11] for the vector case.

It is our purpose in this paper to detail the basic connections between $\mathcal{H}_\infty$-control, model reduction and vector interpolation. Given these connections it is clearly necessary to develop a solution for one of these problems for the solutions to the others to follow. We focus attention on the interpolation problem, showing that a solution of this problem for the rational case can be developed using no more than the classical Schur construction based on elementary linear fractional transformations and results on the inertia of matrices. The connections also allow a state-space solution to the interpolation problem to be given by exploiting Glover's state-space solution to the model matching problem.

As applications, we look in detail at the one-block model matching problem from $\mathcal{H}_\infty$ control and the model reduction problem, explicity constucting the associated Pick matrices.

Notation is standard. $\mathcal{RH}_\infty(k)$ denotes rational matrix functions with at most k poles in the closed right half plane. $\mathcal{RH}_\infty = \mathcal{RH}_\infty(0)$ and $M \in \mathcal{RH}_\infty^-$ if $M(-s) \in \mathcal{RH}_\infty$. $\mathcal{BH}_\infty$ and $\mathcal{BL}_\infty$ are the closed unit balls of $\mathcal{RH}_\infty$ and $\mathcal{RL}_\infty$. $\mathcal{F}(H,U)$ denotes the linear fractional map $\mathcal{F}(H,U) = H_{11} + H_{12}U(I-H_{22}U)^{-1}H_{21}$.

## 2. Vector interpolation and $\mathcal{H}_\infty$ control

To see the connection between $\mathcal{H}_\infty$-control and interpolation, recall [5,7,12] that a class of $\mathcal{H}_\infty$-control problems may be posed as the model matching problem: Find $Q \in \mathcal{RH}_\infty$ such that $\|T_{11} - T_{12}QT_{21}\|_\infty \leq \gamma$. Here, the $T_{ij}$'s are stable, $T_{12}$ and $T_{21}$ are square and, without loss of generality, may be chosen inner (stable and all-pass). Since Q is required to be stable, every right half plane zero of either $T_{12}$ or $T_{21}$ is a zero of $T_{12}QT_{21}$. Therefore, let $\{s_i : i=1,\cdots,n_r\}$ and $\{s_i : i=n_r+1,\cdots,n\}$ be the right half plane zeros of $T_{21}$ and $T_{12}$ respectively. Also, let $a_i \neq 0$ be such that $T_{21}(s_i)a_i = 0$, $i = 1,\cdots,n_r$ and $a_i^*T_{12}(s_i)$

$= 0$, $i = n_r+1,\cdots,n$. Denote the closed-loop by $\gamma R$, ie, $\gamma R := T_{11} - T_{12}QT_{21}$ where $\|R\|_\infty \leq 1$ and $\gamma$ is a gain parameter. Then $R \in \mathcal{BH}_\infty$ and must satisfy the interpolation constraints

$$R(s_i)a_i = \frac{T_{11}(s_i)a_i}{\gamma},\ i = 1,\cdots,n_r \text{ and } a_i^*R(s_i) = \frac{a_i^*T_{11}(s)}{\gamma},\ i = n_r+1,\cdots,n. \tag{2.1}$$

Conversely, if $R \in \mathcal{BH}_\infty$ satisfies (2.1), then Q defined by $Q = [T_{12}^*(T_{11} - \gamma R)T_{21}^*]$ is stable and $\|T_{11} - T_{12}QT_{21}\|_\infty = \gamma\|R\|_\infty \leq \gamma$. Essentially, the interpolation constraints (2.1) ensure that Q and R share the same stability properties.

## 3. Vector interpolation and model reduction

For consistency with the conventional set-up, where the interpolation points are in the right half plane, we consider the approximation of completely unstable systems.

Suppose $G \in \mathcal{RH}_\infty(m) \cap \mathcal{RH}_\infty^-$ (i.e., is anti-stable of degree $\leq m$). We seek $\hat{G} \in \mathcal{RH}_\infty(k)$, with $k \leq m$, such that $\|G - \hat{G}\|_\infty \leq \gamma$. We call $\hat{G}_-$, the unstable part of $\hat{G}$, a reduced order model of G.

Factorize G as $G = G_xG_i$, with $G_x \in \mathcal{RH}_\infty$, $G_i \in \mathcal{RH}_\infty^-$ all-pass and let $\gamma R := G_x - \hat{G}G_i^*$. Then, provided no pole of $\hat{G}$ cancels a zero of $G_i^*$ in the product $\hat{G}G_i^*$, we see that $G_i^*(s_j)a_j = 0$ $j = 1,\cdots,n$ implies

$$\gamma R(s_j)a_j = b_j := G_x(s_j)a_j \tag{3.1}$$

Conversely, $\hat{G}$ is recovered from R by $\hat{G} = (G_x - \gamma R)G_i = G - \gamma R$. Again, the interpolation constraints ensure that $\hat{G}$ and R share the same right half plane poles.

The cancellation condition above does not permit approximate models to share the same poles as G. This is perhaps a perverse condition, but one which an analysis of the Hankel norm model reduction formulae of [9] shows is met except in the trivial case k=m and for isolated values of $\gamma$ in problems having a particular structure.

## 4. Single point interpolation

Lemma 4.1 (elementary interpolants): Suppose $s_1$ is a point in the open right half plane, that $\gamma$ is a real parameter and that a and b are complex vectors. Define

$$H_\gamma(s) := \begin{bmatrix} 0 & I \\ I & 0 \end{bmatrix} + \begin{bmatrix} \phi_\gamma b \\ -\gamma\phi_\gamma a \end{bmatrix}(s - (-\bar{s}_1+\phi_\gamma b^*b))^{-1}\begin{bmatrix} -\gamma a^* & b^* \end{bmatrix} \tag{4.2}$$

in which $\phi_\gamma = \dfrac{-(s_1+\bar{s}_1)}{(\gamma^2 a^*a - b^*b)}$.

1. $H_\gamma$ is all-pass.
2. $\phi_\gamma < 0$ implies $H_\gamma \in \mathcal{RH}_\infty$ and $\phi_\gamma > 0$ implies $H_\gamma \in \mathcal{RH}_\infty^-$.

3. If $\phi_\gamma=0$ or $\infty$ (i.e., $|\gamma^2 a^*a-b^*b| \to \infty$ or $|\gamma| \to \frac{|b|}{|a|}$) then $H_\gamma(s)$ is non dynamic, i.e., independent of s.

Futhermore, provided $\phi_\gamma \neq 0$,

4. $R_\gamma(s)$ satisfies $R_\gamma(s_1)a = \frac{b}{\gamma}$ if and only if $R_\gamma = \mathcal{F}(H_\gamma,U)$, $U \in \mathcal{RL}_\infty$. Futhermore, $\|R_\gamma\|_\infty \leq 1$ if and only if $U \in \mathcal{BL}_\infty$.

Proof: Direct observation and calculation. Note that the only if part of 4 comes from defining

$$U = \mathcal{F}(\begin{bmatrix} H_{22}^* & -H_{12}^* \\ -H_{21}^* & H_{11}^* \end{bmatrix}, R). \qquad \square$$

## 5. Parametric Interpolation

Suppose we are given two data sets $\{s_i, a_i \in C^q, b_i \in C^p : i=1,\cdots,n_r\}$ and $\{s_i, a_i \in C^p, b_i \in C^q : i=n_r+1,\cdots,n\}$. The *standard assumptions* will be: $s_i+\bar{s}_i>0$ for $i=1,\cdots,n$; $s_i \neq s_j$ for $i \leq n_r$, $j>n_r$; and if $\mathcal{I}$ is an index set such that $s_i=s_j$, $i,j \in \mathcal{I}$ then $\{a_i : i \in \mathcal{I}\}$ are linearly independent. We seek a characterization of all $R_\gamma \in \mathcal{BL}_\infty$ such that

$$R_\gamma(s_i)a_i = \frac{b_i}{\gamma}, \quad i = 1,\cdots,n_r \quad \text{and} \quad a_i^* R_\gamma(s_i) = \frac{b_i^*}{\gamma}, \quad i = n_r+1,\cdots,n. \tag{5.1}$$

We call this the *parametric vector interpolation problem* (*PVIP*). Unlike the standard Nevanlinna-Pick problem, we have said nothing about the stability properties of the interpolating function at this stage.

## 6. The Schur construction

In the Schur construction, for iteration i, step 1 builds an elementary interpolant $H_\gamma^i(s)$ for the data $s_i$, $a_i^i$, $b_i^i$. Thus, provided $\phi_\gamma^i \neq 0$, $\gamma[\mathcal{F}(H_\gamma^i,U^i)(s_i)]a_i^i = b_i^i \ \forall U^i \in \mathcal{BL}_\infty$. We then need to choose $U^i(s)$ so that the other constraints are satisfied. To do this, we feed the remaining constraints "down through $H_\gamma^i$" to give an interpolation problem for $U^i$. The algorithm returns to step one, thus satisfying another constraint. After all the all the constraints are met, the final contraction U(s) is free:

Initial data: $\{s_i, a_i^1 \in C^q, b_i^1 \in C^p : i=1,\cdots,n_r\}$, $\{s_i, a_i^1 \in C^p, b_i^1 \in C^q : i=n_r+1,\cdots,n\}$, $\gamma$

Initialize count $i=1$.

1. Let $H_\gamma^i(s)$ be the elementary interpolant, where $a=a_i^i$ and $b=b_i^i$.

2. If $i \leq n_r-1$, update the right constraints by $\begin{bmatrix} b_j^{i+1} \\ a_j^{i+1} \end{bmatrix} = S_R^i(s_j) \begin{bmatrix} a_j^i \\ b_j^i \end{bmatrix}$, $j = i+1,\cdots,n_r$.

3. Update the left constraints by $\begin{bmatrix} a_j^{i+1} \\ b_j^{i+1} \end{bmatrix} = S_L^i(\bar{s}_j) \begin{bmatrix} b_j^i \\ a_j^i \end{bmatrix}$, with $j = n_r+1,\cdots,n$ if $i \leq n_r$, otherwise $j=i+1,\cdots,n$.

$$S_R(s) := \begin{bmatrix} I & 0 \\ 0 & I \end{bmatrix} + \begin{bmatrix} \gamma\phi_\gamma a \\ \phi_\gamma b \end{bmatrix} (s-(-\bar{s}_1)^{-1} \begin{bmatrix} \gamma a^* & -b^* \end{bmatrix} = S_L(-s) \tag{6.1}$$

4. If $i \leq n-1$, set $i=i+1$ and return to 1.

Theorem 6.1: Suppose that $\phi_\gamma^i \neq 0$, i=1,···,n in the Schur construction. Then $\mathcal{R}_\gamma$ given by $\mathcal{R}_\gamma = \mathcal{F}(H_\gamma^1, \mathcal{F}(H_\gamma^2, \cdots, \mathcal{F}(H_\gamma^n)\cdots)$ is a generator of all solutions to the PVIP. That is, $R_\gamma$ is a solution to the PVIP if and only if $R_\gamma = \mathcal{F}(\mathcal{R}_\gamma, U)$ for some $U \in \mathcal{BL}_\infty$. □

Remark 6.2: The condition $\phi_\gamma^i \neq 0$ ∀i is affected by $\gamma$ and the ordering of the data. For given $\gamma$, a generator of all solutions to the PVIP can be found by the Schur construction if *for some ordering of the data* $\phi_\gamma^i \neq 0$ ∀i. We call these values of $\gamma$ *regular values.* It can be shown that $\gamma$ is non-regular if and only if every ixi principal minor of the Pick matrix is zero. (In the 2x2 case, this means both diagonal entries are zero).

Genin & Kung [8] claim that the Schur construction always solves the (scalar) interpolation problem. This claim is based on the fact that since $R(\gamma,s) := R_\gamma(s)$ is a rational function of $\gamma$ and s there can only be finitely many $\gamma$ such that, for some i, $\phi_\gamma^i = 0$. It is therefore true that for any $\gamma_0$ and k, $\lim_{\gamma\to\gamma_0}\{\lim_{s\to s_k} R_\gamma(s)\}a_k = \frac{b_k}{\gamma_0}$. We know by examining the elementary interpolants, however, that if $\phi_\gamma^i \to 0$ as $\gamma \to \gamma_0$, then

$$\lim_{s\to s_{i+1}}\{\lim_{\gamma\to\gamma_0} R_\gamma(s)\}a_i \neq \frac{b_i}{\gamma_0}.$$

That is, the two variable rational function $R(\gamma,s)$ has a discontinuity at $(\gamma_0, s_{i+1})$.

## 7. The Pick matrix and stability properties of interpolating functions

The Pick matrix associated with our PVIP is:

$$\Pi(\gamma) = \begin{bmatrix} \Pi_{11} & \Pi_{21}^* \\ \Pi_{21} & \Pi_{22} \end{bmatrix}, \quad \text{with } \Pi_{11} = \left\{\frac{\gamma^2 a_i^* a_k - b_i^* b_k}{\bar{s}_i + s_k}\right\}_{k=1,n_r}^{i=1,n_r}$$

$$\Pi_{21} = \left\{\frac{\gamma b_i^* a_k - \gamma a_i^* b_k}{s_k - s_i}\right\}_{k=1,n_r}^{i=n_r+1,n} \qquad \Pi_{22} = \left\{\frac{\gamma^2 a_i^* a_k - b_i^* b_k}{\bar{s}_k + s_i}\right\}_{k=n_r+1,n}^{i=n_r+1,n}$$

Notice from Lemma 4.1 that the stability properties of the elementary interpolant which solves a 1 point problem are determined by the sign of $\phi_\gamma$, and that $-\phi_\gamma^{-1} = \Pi(1,1)$. For the general case, the inertia of the Pick matrix determines the stability properties of the generator of all interpolating functions.

Lemma 7.1: Let $\Pi^{i+1}(\gamma)$ be the Pick matrix for the n−i point interpolation problem obtained after i iterations of the Schur construction. Partition $\Pi^i(\gamma)$ as

$$\Pi^i(\gamma) = \begin{bmatrix} \pi_{11}^i & \pi_{21}^{i*} \\ \pi_{21}^i & \Pi_{22}^i \end{bmatrix}, \text{ with } \pi_{11}^i \text{ scalar. Then } \Pi^{i+1} = \Pi_{22}^i - \pi_{21}^i \pi_{21}^{i*}/\pi_{11}^i.$$

Proof: Calculation. □

Theorem 7.2: Suppose that $\gamma_0$ is a regular value and that the Pick matrix $\Pi(\gamma_0)$ has k strictly negative eigenvalues. Then the generator $\mathcal{R}_{\gamma_0}$ obtained from the Schur construction

$\in \mathcal{RH}_\infty(k)$, i.e., has no more than k right half plane poles.

Proof: By Lemma 4.1, the numer of unstable $H^i_\gamma$ in the Schur construction is the number of positive $\phi^i_\gamma$. The result follows from Lemma 7.1, since $-(\phi^i)^{-1} = \Pi^i(1,1)$. □

Theorem 7.3: If there exists a solution $R\gamma_0$ to the PVIP such that $R\gamma_0 \in \mathcal{RH}_\infty(k)$, then $\Pi(\gamma_0)$ has no more than k negative eigenvalues.

Proof: Write the Pick matrix as $\Pi(\gamma_0) = Z^*BZ$, $Z = \mathrm{diag}\{\gamma_0 a_i\}$, $i = 1, \cdots, n$, in which

$$B_{11} = \left\{\frac{I - R^*(s_i)R(s_k)}{\bar{s}_i + s_k}\right\}_{k=1,\cdots,n_r}^{i=1,\cdots,n_r} \qquad B_{12} = \left\{\frac{R^*(s_k) - R^*(s_i)}{\bar{s}_i - \bar{s}_k}\right\}_{k=n_r+1,\cdots,n}^{i=1,\cdots,n_r}$$

$$B_{22} = \left\{\frac{I - R(s_i)R^*(s_k)}{s_i + \bar{s}_k}\right\}_{k=n_r+1,\cdots,n}^{i=n_r+1,\cdots,n}$$

By the standard assumptions, Z has full column rank, from which it follws that $\nu(\Pi) \leq \nu(B)$ (i.e. B has at least as many negative eigenvalues as $\Pi$).

Let [A,B,C,D] be a minimal realization of R. Since $R \in \mathcal{BL}_\infty$, we can construct

$$\mathcal{R}_a(s) = \begin{bmatrix} D & D_{12} \\ D_{21} & D_{22} \end{bmatrix} + \begin{bmatrix} C \\ C_a \end{bmatrix}(sI - A)^{-1}\begin{bmatrix} B & B_a \end{bmatrix}$$

such that $R_a(s)$ is all-pass [9, Theorem 5.2]. Now let P and Q be the controllability and observability gramians of $R_a$. Calculation, using the all-pass equations [9], gives:

$$B_{11}(i,k) = \frac{\Delta_d^*(s_i)\Delta_d(s_k)}{\bar{s}_i + s_k} + B^*(\bar{s}_iI - A^*)^{-1}Q(s_kI - A)^{-1}B, \qquad \Delta_d(s) \overset{s}{=} [A,B,C_a,D_{21}]$$

$$B_{22}(i,k) = \frac{\Delta(s_i)\Delta^*(s_k)}{s_i + \bar{s}_k} + C(s_iI - A)^{-1}P(\bar{s}_kI - A^*)^{-1}C^*, \qquad \Delta(s) \overset{s}{=} [A,B_a,C,D_{12}]$$

$$B_{12}(i,k) = B^*(\bar{s}_kI - A^*)^{-1}(\bar{s}_iI - A^*)^{-1}C^*.$$

Hence

$$B = T_1NT_1^* + \begin{bmatrix} T_2 & 0 \\ 0 & T_3 \end{bmatrix}\begin{bmatrix} Q & I \\ I & P \end{bmatrix}\begin{bmatrix} T_2 & 0 \\ 0 & T_3 \end{bmatrix}^*$$

$$= T_1NT_1^* + \begin{bmatrix} T_2 & 0 \\ T_3Q^{-1} & T_3 \end{bmatrix}\begin{bmatrix} Q & 0 \\ 0 & 0 \end{bmatrix}\begin{bmatrix} T_2 & 0 \\ T_3Q^{-1} & T_3 \end{bmatrix}^* \qquad \text{since } P = Q^{-1}.$$

The various matrices above are identified as $N = \begin{bmatrix} N_1 & 0 \\ 0 & N_2 \end{bmatrix}$ with $N_1 = \{\frac{1}{\bar{s}_i + s_k}\}_{k=1,\cdots,n_r}^{i=1,\cdots,n_r}$, $N_2 = \{\frac{1}{\bar{s}_i + s_k}\}_{k=n_r+1,\cdots,n}^{i=n_r+1,\cdots,n}$, $T_1 = \mathrm{blockdiag}\{\Delta_d^*(s_i)_{i=i,n_r}, \Delta(s_i)_{i=n_r+1,n}\}$, $T_2 = [(s_1 - A)^{-1}B, (s_2 - A)^{-1}B, \cdots, (s_{n_r} - A)^{-1}B]^*$ and $T_3 =$

$[(\bar{s}_{n_r+1}-A^*)^{-1}C^*, (\bar{s}_{n_r+2}-A^*)^{-1}C^*, \cdots, (\bar{s}_n-A^*)^{-1}C^*]^*$. A Lyapunov equation argument establishes that $N \geq 0$, and consequently $\nu(\Pi) \leq \nu(B) \leq \nu(Q) \leq \pi(A) \leq k$. □

## 8. Pick matrices for the model matching and model reduction problems

From [11,12], we obtain the following state-space realization for the $T_{ij}$'s:

$$\begin{bmatrix} T_{11} & T_{12} \\ T_{21} & 0 \end{bmatrix} \stackrel{s}{=} \left[\begin{array}{cc|cc} A-B_2F & B_2F & B_1 & B_2 \\ 0 & A-HC_2 & -YC_2^* & 0 \\ \hline -B_2^*X & F & D_{11} & I \\ 0 & C_2 & I & 0 \end{array}\right] \tag{8.1}$$

$(A,B_2)$ and $(A,C_2)$ are assumed stabilizable and detectable respectively. The matices H and F are defined by $F = C_1 + B_2^*X$, $H = B_1 + YC_2^*$ in which X and Y are the unique positive semi-definite solutions to the Riccati equations

$$X(A-B_2C_1) + (A-B_2C_1)^*X - XB_2B_2^*X = 0 \tag{8.2}$$

$$(A-B_1C_2)Y + Y(A-B_1C_2)^* - YC_2^*C_2Y = 0 \tag{8.3}$$

Note that $A-B_2F$ and $A-HC_2$ are asymptotically stable, and that the matrices $D_{21}$ and $D_{12}$ which appear in [11,12] have been assumed to be scaled to the identity.

Lemma 8.1: If $\lambda+\bar{\lambda}>0$, then there exists $a\neq 0$ such that $T_{21}(\lambda)a = 0$ if and only if there exists $x\neq 0$ such that

$$[\lambda I-(A-B_1C_2)]x = 0. \tag{8.4}$$

In this case, $a = -C_2x$ and $T_{11}(\lambda)a = b = (C_1-D_{11}C_2)x$.

Proof: Since $\lambda+\bar{\lambda}>0$, the realization of $\begin{bmatrix} T_{11} \\ T_{21} \end{bmatrix}$ obtained from (8.1) is controllable and observable at $s=\lambda$. So $\begin{bmatrix} T_{11} \\ T_{21} \end{bmatrix}(\lambda)a = \begin{bmatrix} b \\ 0 \end{bmatrix}$ if and only if

$$\begin{bmatrix} \lambda I-(A-B_2F) & -B_2F & -B_1 \\ 0 & \lambda-(A-HC_2) & YC_2^* \\ -B_2^*X & C_1+B_2^*X & D_{11} \\ 0 & C_2 & I \end{bmatrix} \begin{bmatrix} y \\ x \\ a \end{bmatrix} = \begin{bmatrix} 0 \\ 0 \\ b \\ 0 \end{bmatrix}.$$

which is equivalent to $a=-C_2x$ where x satisfies (8.4), $y = x$ and $b=(C_1-D_{11}C_2)x$. □

Lemma 8.2: There exists $x\neq 0$ satisfying (8.4) with $\lambda+\bar{\lambda}>0$ if and only if there exists z, with $Yz\neq 0$, such that $[\lambda I+(A-HC_2)^*]z = 0$. In this case $x = Yz$.

Proof: Follows from the equation $Y(A-HC_2)^* + (A-B_1C_2)Y = 0$. □

Theorem 8.3: Let $s_i$, $z_i$, $Yz_i\neq 0$ for $i=1,\cdots,n_r$ with $n_r=\text{rank}(Y)$ satisfy $[s_iI+(A-HC_2)^*]z_i = 0$. Let $s_{n_r+i}$, $w_i$, $Xw_i\neq 0$ for $i=1,\cdots,n_\ell$, with $n_\ell=\text{rank}(X)$ satisfy $w_i^*[s_{i+n_r}I+(A-B_2F)^*] = 0$. Define $a_i = -C_2Yz_i$, $b_i = (C_1-D_{11}C_2)Yz_i$, $i=1,\cdots,n_r$. With $n=n_r+n_\ell$ define $a_i = -B_2^*Xw_{i-n_r}$, $b_i = (B_1-B_2D_{11})^*Xw_{i-n_r}$, $i=n_r+1,\cdots,n$.

The interpolation data associated with the model matching problem $T_{11} + T_{12}QT_{21}$, is $\{s_i,a_i,b_i\ i=1,\cdots n_r\}$ and $\{s_i,a_i,b_i\ i=n_r+1,\cdots n\}$. The interpolation data satisfy the standard assumptions if and only if $s_i \neq s_j$ for $i \leq n_r$, $j > n_r$ and $Z = \begin{bmatrix} z_1 & z_2 & \cdots & z_{n_r} \end{bmatrix}$, $W = \begin{bmatrix} w_1 & w_2 & \cdots & w_{n_\ell} \end{bmatrix}$ have full column rank. In this case, the Pick matrix is given by

$$\Pi(\gamma) = \begin{bmatrix} Z^* & 0 \\ 0 & W^* \end{bmatrix} \begin{bmatrix} \gamma^2 Y - \tilde{Y} & \gamma YX \\ \gamma XY & \gamma^2 X - \tilde{X} \end{bmatrix} \begin{bmatrix} Z & 0 \\ 0 & W \end{bmatrix}$$

where $\tilde{Y}$, $\tilde{X}$ are the solutions of the Lyapunov equations

$$(A-HC_2)\tilde{Y} + \tilde{Y}(A-HC_2)^* + Y(C_1-D_{11}C_2)^*(C_1-D_{11}C_2)Y = 0$$

$$(A-B_2F)^*\tilde{X} + (A-B_2F)\tilde{X} + X(B_1-B_2D_{11})(B_1-B_2D_{11})^*X = 0. \qquad \square$$

Remark 8.4: It is possible to find the least $\gamma$ such that $\Pi(\gamma) \geq 0$ by an eigenvalue calculation: Assume for simplicity that X and Y are non-singular and that $Z=I$, $W=I$. Then

$$\Pi(\gamma) = \begin{bmatrix} Y & Y \\ 0 & \gamma \end{bmatrix} \left\{ \gamma^2 \begin{bmatrix} I & 0 \\ 0 & I \end{bmatrix} - \begin{bmatrix} Y^{-1}\tilde{Y}+XY & -\tilde{X}X^{-1} \\ -XY & \tilde{X}X^{-1} \end{bmatrix} \right\} \begin{bmatrix} I & 0 \\ 0 & \gamma^{-1}X \end{bmatrix}. \qquad \square$$

For the model reduction problem, suppose we are given a controllable realization [A,B,C,D] of the system G in which A has all it's eigenvalues in the open right-half plane. Let $P=P^*<0$ and $Q=Q^*\leq 0$ be the controllability and observability gramians: $AP + PA^* + BB^* = 0$, $A^*Q + QA + C^*C = 0$. A straightforward calculation shows that $G_i(s) = I - B^*P^{-1}(sI - A)^{-1}B$ and $G_X(s) = D + [DB^* + CP](sI + A^*)^{-1}P^{-1}B$ satisfy $G=G_XG_i$, $G_i \in \mathcal{RH}_\infty^-$ all-pass and $G_X \in \mathcal{RH}_\infty$. The interpolation data $G_i^*(s_j)a_j = 0$, $G_X(s_j)a_j = b_j$ is given by $a_j = -B^*P^{-1}w_j$, $b_j = Cw_j$, where $(s_jI - A)w_j = 0$,. The standard assumptions are equivalent to the non-singularity of $W = \begin{bmatrix} w_1 & w_2 & \cdots & w_n \end{bmatrix}$ and

$$\Pi(\gamma) = \left\{ \frac{\gamma^2 a_i^* a_k - b_i^* b_k}{\bar{s}_i + s_k} \right\}_{k=1,\cdots,n}^{i=1,\cdots,n} = W^*(Q - \gamma^2 P^{-1})W \tag{8.5}$$

Since $\lambda^{\frac{1}{2}}(PQ)$ are the Hankel singular values of G(s) [9], it follows that the (positive) zeros of the Pick matrix are precisely the Hankel singular values of G. Furthermore, the Nehari/AAK theorem of optimal Hankel norm approximation follows directly from the interpolation results: Every positive zero $\gamma_k$ of $\det(\Pi(\gamma))$ is a singular value of $\Gamma_G$. Consequently $\inf_{\hat{G} \in \mathcal{RH}_\infty(k)} \|G - \hat{G}\|_\infty = \gamma_{k+1}$. $\square$

## 9. A generator of solutions to PVIP

Suppose we are given $\{s_j,\ a_j \in \mathbb{C}^q,\ b_j \in \mathbb{C}^p,\ j=1,\cdots,n\}$ for a right-sided PVIP. We use the state space connection of §8 'backwards' to construct the G of an associated model reduction problem: Define

$$A = \mathrm{diag}\{s_i\ i=1,\cdots,n\},\ C = [b_1\ b_2 \cdots b_n],\ X = [a_1\ a_2 \cdots a_n] \tag{9.6}$$

Let $F=F^*$ be a solution of $FA + A^*F + X^*X = 0$. The standard assumptions imply (X,A) is observable, so $F < 0$. Define

$$P = F^{-1},\ B = -PX^* \tag{9.7}$$

An appropriate G is given by the realization [A,B,C,D], for any $D \in C^{pxq}$. That is, with the realization $G(s) \overset{s}{=} [A,B,C,D]$, the PVIP obtained via the procedure in §8 is our original PVIP defined by the data $\{s_i, a_i, b_i\ i=1,\cdots,n\}$.

Recall [9] that, provided $R = (PQ-\gamma^2 I)$ is non-singular,

$$\mathcal{H}_\gamma(s) = \begin{bmatrix} D & \gamma I \\ \gamma I & 0 \end{bmatrix} + \begin{bmatrix} CP \\ -\gamma B^* \end{bmatrix}(sI - R^{-*}(\gamma^2 A^* + QAP))^{-1} R^{-*} \begin{bmatrix} QB & -\gamma C^* \end{bmatrix}$$

is a generator of all solutions to the model reduction problem, i.e., all reduced order models $\hat{G}_\gamma \in \mathcal{RH}_\infty(k)$ such that $\|G-\hat{G}_\gamma\|_\infty \le \gamma$ are given by $\hat{G}_\gamma = F(\mathcal{H}_\gamma, U)$, $U \in \mathcal{BH}_\infty$. Hence a generator of all solutions to PVIP is given by $\mathcal{R}_\gamma(s) = \frac{1}{\gamma}\left\{\begin{bmatrix} G_X & 0 \\ 0 & 0 \end{bmatrix} - \mathcal{H}_\gamma \begin{bmatrix} G_i^* & 0 \\ 0 & I \end{bmatrix}\right\}(s)$

$$= \begin{bmatrix} 0 & -I \\ -I & 0 \end{bmatrix} + \begin{bmatrix} -CP \\ \gamma B^* \end{bmatrix}(sI - R^{-*}(\gamma^2 A^* + QAP))^{-1} R^{-*} \begin{bmatrix} P^{-1}B & -\gamma C^* \end{bmatrix}.$$

## References


1] Ball, J.A., "Interpolation problems of the Pick-Nevanlinna and Loewner types for meromorphic matrix functions", Integral equations and operator theory, 6, 1983, 804-840 and 9, 1986, 155-203.

2] Ball, J.A. and Helton, J.W., "Beurling-Lax representations using classical Lie groups with many applications III: Groups preserving two bilinear forms", Amer. J. Maths, 1983, 95-174.

3] Delsarte, Ph., Genin Y.V. and Kamp, Y., "On the role of the Nevanlinna-Pick problem in circuit and system theory", Circuit theory and applications, 9, 1981, 177-187.

4] Dewilde, P. and Dym, H., "Lossless chain scattering matrices and optimum linear prediction: the vector case", Circuit theory and applications, 9, 1981, 135-175.

5] Doyle, J.C. et al, "Advances in multivariable control", ONR/Honeywell workshop, 1984.

6] Dym, H., "J contractive matrix functions, reproducing Hilbert spaces and interpolation", CBMS Regional Conference Series, vol 71, Amer. Math. Soc., Providence RI, 1989.

7] Francis, B.A., "A course in $H^\infty$ control theory, Lecture notes in Control and Information Sciences number 88, Springer Verlag, 1987, 2nd Ed.

8] Genin, Y.V. and Kung, S.Y., "A two-variable approach to the model reduction problem with Hankel norm criterion", IEEE Trans. Circuits and Systems, CS-28, 1981, 912-924.

9] Glover, K., "All optimal Hankel-norm approximations of linear multivariable systems and their $L^\infty$ error bounds", Int. J. Control, 39, 1984, 1115-1193.

10] Kimura, H., "Directional interpolation approach to $H^\infty$ optimization and robust stabilization", IEEE Trans. Auto Control, AC-32, 1987, 1085-1093.

11] Limebeer, D.J.N. and Anderson, B.D.O., "An interpolation theory approach to $H^\infty$ controller degree bounds", Linear Algebra and its applications, 98, 1988, 347-386.

12] Safonov, M.G., Jonckheere, E.A., Verma, M., and Limebeer, D.J.N., "Synthesis of positive real multivariable feedback systems", Int. J. Control, 45, 1987, 817-842.

# Sensitivity Minimization and Robust Stabilization by Stable Controller

S. Hara and M. Vidyasagar

This paper is concerned with the problems of stabilizing an SISO system using a stable controller. Two interpolation-minimization problems for unit functions are investigated, and necessary and/or sufficient conditions for solvability are obtained using the logarithm function. A necessary and sufficient conditions is also derived for the sensitivity reduction by a stable controller. We obtain a lower bound and two upper bounds on the uncertainty in the plant which can be stabilized by a stable controller. A procedure for designing a stable robust controller is presented, based on solving a nonlinear min-max optimization problem.

# 1 Introduction

This paper is concerned with the problems of stabilizing an SISO system using a stable controller. Stable controllers are desirable from the standpoint of the integrity of the closed-loop system and they are commonly used in classical control system design. However, it is well known that in a recent approach called $H_\infty$ norm optimization (e.g. sensitivity minimization [13] and robust stabilization [5]) the optimal controller or robust compensator often unstable if the plant has unstable zeros or unstable poles.

A necessary and sufficient condition for stabilizability by a stable controller has been derived by Youla et al. [12]. The condition is called the parity interlacing property (*p.i.p.*). A parametrization of all stable stabilizing compensators has been developed by Vidyasagar [9]. Several design problems with stable controllers such as the gain margin problem and the sensitivity minimization problem have been discussed in the literature [7], [8] , [2], [3], [6]. However, there is no research on robust stabilization by a stable controller. It is also noted that the results in [8] and [2] are not complete from the practical design point of view. The results in [8] and [2] were obtained by means of a modified Nevanlinna-Pick interpolation problem first discussed by Ball and Helton [1]. The modified interpolation problem is to find a unit (rather than stable) function such that it satisfies given interpolation conditions and its infinity-norm is less than 1. However the class of functions treated in [1] is not restricted to symmetric. Their results are therefore inadequate for problems of control system design, since the resulting controller may be a rational function with complex coefficients.

In this paper, we consider problems of sensitivity minimization and robust stabilization by a real rational stable controller. The paper is organized as follows: In Section 2, we propose two interpolation-minimization problems for unit functions. The first and second problems are related to the sensitivity minimization problem and robust stabilization problem, respectively. A necessary and sufficient condition for the solvability of the first problem is obtained using the logarithm function. The problem and the derived condition are slightly different from those in [1]. Two necessary conditions and a sufficient condition are developed for the second problem. We investigate the sensitivity minimization problem in Section 3. A necessary and sufficient condition is derived for the sensitivity reduction by a stable controller. Section 4 is devoted to the robust stabilization problem. It is shown that the necessary condition (resp. sufficient condition) for the second interpolation problem leads to an upper bound (resp. lower bound) on the plant uncertainties such that there exists a stable controller which internally stabilizes all plants in the prescribed bound. We also show that the problem can be approximated by a nonlinear min-max optimization problem.

Since the problems for continuous-time systems can be converted to those for discrete-time systems by an appropriate bilinear transformation, we only consider the discrete-time case in this paper and all the proofs are omitted.

We use the following notation. $\mathbf{D}$ (resp. $\bar{\mathbf{D}}$) denotes the open (resp. closed) unit disk. The set $\mathbf{A}$ consists of all functions mapping $\bar{\mathbf{D}}$ into the complex numbers such that they are analytic on $\mathbf{D}$ and continuous on $\bar{\mathbf{D}}$. $\mathbf{A}_s$ denotes the subset of $\mathbf{A}$ consisting of all symmetric functions, i.e.

$$\mathbf{A}_s = \{f \in \mathbf{A} : \bar{f}(\bar{z}) = f(z), \forall z \in \mathbf{D}\}$$

where the bar denotes complex conjugation. The set of all rational functions is denoted by $\mathbf{R(z)}$ and we define two subsets of $\mathbf{R(z)}$:

$$\mathbf{RH}_\infty = \{f \in \mathbf{A}_s \cup \mathbf{R(z)}\}, \quad \mathbf{B}_\infty = \{f \in \mathbf{RH}_\infty : \|f\|_\infty < 1\}$$

where $\|f\|_\infty$ denotes the $H_\infty$ norm of $f$ defined by

$$\|f\|_\infty := \sup_{r \to 1^-, r \in [0,1)} \max_{\theta \in [0,2\pi]} |f(re^{j\theta})|$$

## 2 Interpolation by unit function

In this section, we consider the following two interpolation problems for unit functions:

**Problem 1:** Given complex numbers $\alpha_i, \beta_i$ $(i = 1 \sim n)$ with $|\alpha_i| < 1$ and $|\beta_i| < 1$ for all $i$, find a function $f$ in $\mathbf{A}_s$ satisfying

a) $f$ is a unit of $\mathbf{A}_s$, i.e., $f(z) \neq 0, \forall z \in \bar{\mathbf{D}}$

b) $f(\alpha_i) = \beta_i \quad (i = 1 \sim n := 2s + r)$

c) $\|f\|_\infty < 1$

In order for b) to be satisfied by a symmetric function, the $\alpha_i$'s and $\beta_i$'s occur in complex conjugate pairs. To be specified, suppose

i) $\alpha_i (i = 1 \sim s)$ are nonreal with $\alpha_{i+s} = \bar{\alpha}_i$ and $\beta_{i+s} = \bar{\beta}_i$ $(i = 1 \sim s)$

ii) $\alpha_i (i = 2s + 1 \sim n)$ are real and $\beta_i (i = 2s + 1 \sim n)$ are real and positive

**Problem 2:** Given complex numbers $\alpha_i, \beta_i (i = 1 \sim n)$ with $|\alpha_i| < 1$ and $|\beta_i| < 1$ for all $i$ and a function $h \in \mathbf{A_s}$, find a function $f$ in $\mathbf{A_s}$ satisfying

a) $f$ is a unit of $\mathbf{A_s}$, i.e., $f(z) \neq 0, \forall z \in \bar{\mathbf{D}}$

b) $f(\alpha_i) = \beta_i \quad (i = 1 \sim n := 2s + r)$

d) $\|f - h\|_\infty < 1$

Problems 1 and 2 are closely related to the problems of sensitivity minimization and robust stabilization by a stable controller, respectively. This is shown in Sections 3 and 4.

A necessary and sufficient condition for Problem 1 to have a solution is given next:

**Theorem 2.1:** There exists a function $f$ in $\mathbf{A_s}$ satisfying a), b) and c), if and only if there exists a set of integers $\{m_k\}_{k=1\sim s}$ satisfying

$$m_{k+s} = -m_k \ (k = 1 \sim s) \ \text{ and } \ m_{2s+\ell} = 0 \ (\ell = 1 \sim r) \tag{2.1}$$

such that the Pick matrix

$$Q(\{m_k\}) := \left[\frac{-\ln \beta_i - \ln \bar{\beta}_k - j2\pi(m_i - m_k)}{1 - \alpha_i \bar{\alpha}_k}\right]_{i,k=1}^{n} \tag{2.2}$$

is positive definite.

**Corollary 2.1:** Suppose $s = 0$ in Problem 1, i.e., there exist no nonreal interpolation points. There exists a function $f$ in $\mathbf{A_s}$ satisfying a), b) and c), if and only if

$$Q := \left[\frac{-\ln \beta_i - \ln \beta_k}{1 - \alpha_i \bar{\alpha}_k}\right]_{i,k=1}^{r} > 0 \tag{2.3}$$

*Remark 2.1:* The above results can be derived by using a logarithm function $-\ln f$. A result similar to Theorem 2.1 was developed by Ball and Helton [1] and their result was used to solve the sensitivity minimization problem and gain margin problem by a stable controller by Genesh and Pearson [2] and Tannenbaum [8], respectively. However, the class of functions considered in [1] is $\mathbf{A}$ rather than $\mathbf{A_s}$. Therefore, we may not obtain a rational function with real coefficients even if the condition in [1] holds. This means that the result in [1] is not adequate for control problems. The only difference between Theorems 2.1 and the result in [1] is in the choice of the set of integers $\{m_k\}$. We must choose $\{m_k\}$ in Theorem 2.1 such that the consistency condition (2.1) holds, while $m_k$ $(k = 1 \sim n)$ are arbitrary integers in [1].

*Remark 2.2:* Since the numbers of possible choice of $\{m_k\}$ is finite as shown in [2], we can check the solvability condition in Theorem 2.1 in finitely many steps.

Next, we investigate Problem 2. Unfortunately, the problem is quite difficult, so we only give two necessary conditions and a sufficient condition for the solvability (all the proofs are omitted). The conditions lead to upper and lower bounds on the maximum plant uncertainty for which there exists a stable controller that achieves robust stabilization.

Let $\hat{h}$ is a unit in $\mathbf{A_s}$ such that

$$|\hat{h}(e^{j\theta})| = 1 + |h(e^{j\theta})|, \quad \forall\theta \tag{2.4}$$

then we have the necessary condition.

**Theorem 2.2:** Given $h \in \mathbf{A_s}$, choose a unit $\hat{h}$ of $\mathbf{A_s}$ such that (2.4) holds. Under these conditions, if there exists a function $f$ in $\mathbf{A_s}$ satisfying a), b) and d) in Problem 2, then there exists a set of integers $\{m_k\}_{k=1\sim s}$ satisfying the consistency condition (2.1) such that the Pick matrix

$$\hat{Q}_h(\{m_k\}) := \left[\frac{-\ln\beta_i - \ln\bar{\beta}_k + \ln\nu_i + \ln\bar{\nu}_k - j2\pi(m_i - m_k)}{1 - \alpha_i\bar{\alpha}_k}\right]_{i,k=1}^{n} \tag{2.5}$$

is positive definite.

We can obtain another necessary condition by replacing $\beta_i$ and $\nu_i$ $(i = 1 \sim n)$ by $1/\beta_i$ and

$$\mu_i := 1/\bar{h}(\alpha_i) \quad (i = 1 \sim n) \tag{2.6}$$

respectively, where $\bar{h}$ is a unit in $\mathbf{A_s}$ satisfying

$$|\bar{h}(e^{j\theta})| \geq \max\{|h(e^{j\theta})| - 1, 0\}, \quad \forall\theta \tag{2.7}$$

Note that $\bar{h}$ is not unique but it can be chosen such that $|\bar{h}(e^{j\theta})|$ is as close as possible to $max$ $\{|h(e^{j\theta})| - 1, 0\}$.

**Theorem 2.3:** Suppose $\mu_i$ $(i = 1 \sim n)$ are defined by (2.6). If there exists a function $f$ in $\mathbf{A_s}$ satisfying a), b) and d) in Problem 2, then there exists a set of integers $\{m_k\}_{k=1\sim s}$ satisfying the consistency condition (2.1) such that the Pick matrix

$$\bar{Q}_h(\{m_k\}) := \left[\frac{\ln\beta_i + \ln\bar{\beta}_k + \ln\mu_i + \ln\bar{\mu}_k - j2\pi(m_i - m_k)}{1 - \alpha_i\bar{\alpha}_k}\right]_{i,k=1}^{n} \tag{2.8}$$

is positive definite.

Finally, we will derive a sufficient condition for the solvability of Problem 2:

**Theorem 2.4:** Suppose that $\|h\|_\infty < 1$ and select a unit $h_s$ in $\mathbf{A_s}$ such that

$$|h_s(e^{j\theta})| = 1 - |h(e^{j\theta})|, \quad \forall\theta \tag{2.9}$$

Define

$$\delta_i := h_s(\alpha_i) \quad (i = 1 \sim n) \tag{2.10}$$

Then there exists a function $f$ in $\mathbf{A_s}$ satisfying a), b) and d) in Problem 2, if there exists a set of integers $\{m_k\}_{k=1\sim s}$ satisfying the consistency condition (2.1) such that the Pick matrix

$$Q_{h_s}(\{m_k\}) := \left[\frac{-\ln\beta_i - \ln\bar{\beta}_k + \ln\delta_i + \ln\bar{\delta}_k - j2\pi(m_i - m_k)}{1 - \alpha_i\bar{\alpha}_k}\right]_{i,k=1}^{n} \tag{2.11}$$

is positive definite.

# 3 Sensitivity minimization by stable controller

Consider a feedback control system shown in *Fig*.1, where $P(z) \in \mathbf{R}(z)$ is the plant to be controlled and $C(z) \in \mathbf{RH}_\infty$ is the controller to be designed, i.e., the controller is itself stable. The closed-loop system is said to be *stable* if

$$H(P,C) := \begin{bmatrix} 1/(1+PC) & -P/(1+PC) \\ C/(1+PC) & 1/(1+PC) \end{bmatrix} \in \mathbf{RH}_\infty \tag{3.1}$$

We note that a constraint so called *p.i.p* (parity interlacing property) is required for the plant to be stabilized by a stable controller [12], [9].

Let $p_k(k = 1 \sim m)$ be the distinct poles of $P(z)$ in $\mathbf{D}$, and let $p_k(k = m+1 \sim m+\mu)$ be the distinct poles of $P(z)$ on the unit circle. Let $z_k(k = 1 \sim n)$ be the distinct zeros of $P(z)$ in $\mathbf{D}$, and let $z_k(k = n+1 \sim m+\nu)$ be the distinct zeros of $P(z)$ on the unit circle. Then, the stability condition requires that the sensitivity function defined by

$$S(z) := 1/(1 + P(z)C(z)) \in \mathbf{RH}_\infty \tag{3.2}$$

should have the following properties:

S1) $S(z)$ is real rational and analytic, i.e., $S(z) \in \mathbf{RH}_\infty$

S2) For each $k = 1 \sim \mu$, $p_k$ is a zero of $S(z)$; moreover, its multiplicities as a zero of $S(z)$ is at least equal to its degree as a pole of $P(z)$.

S3) For each $k = 1 \sim \nu$, $z_k$ is a zero of $S(z) - 1$; moreover, its multiplicities as a zero of $S(z) - 1$ is at least equal to its degree as a zero of $P(z)$.

We now consider the problem of weighted sensitivity minimization by a stable controller, i.e., selecting a stable controller $C(z) \in \mathbf{RH}_\infty$ such that (3.1) and

$$\|WS\|_\infty < \gamma \tag{3.3}$$

hold, where $W(z) \in \mathbf{R}(z)$ is a given unit.

For simplicity we first assume that the plant $P(z)$ has no unit circle poles or zeros, i.e., $\mu = 0$ and $\nu = 0$, and that all poles and zeros in $\mathbf{D}$ are simple. Let

$$f(z) := (W(z)S(z)/d_i(z))/\gamma = W(z)u(z)/\gamma \tag{3.4}$$

where $d_i(z) := \prod_{k=1}^{m}(z - p_k)/(1 - \bar{p}_k z)$.

Then $f$ must satisfy

a) $f$ is a unit of $\mathbf{RH}_\infty$

b) $f(z_k) = W(z_k)/(\gamma d_i(z_k)), \quad k = 1 \sim n$

c) $\|f\|_\infty < 1$

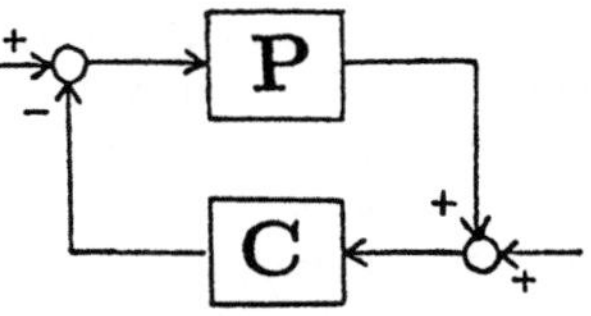


*Fig*.1 FEEDBACK SYSTEM

We can readily show that the existence of such an $f(z)$ is the sufficient condition. Consequently, sensitivity minimization by a stable controller is reduced to a problem of interpolation by a unit function (Problem 1 proposed in Section 2). Hence, we can apply Theorem 2.1 to solve the problem.

**Theorem 3.1:** Suppose all poles and zeros of $P(z)$ in $\mathbf{D}$ are simple (unit circle poles or zeros of $P(z)$ may not be simple). Let $\alpha_k(k = 1 \sim 2s)$ be the nonreal zeros of $P(z)$ in $\mathbf{D}$ with $\alpha_{k+s} = \bar{\alpha}_k$ and $\alpha_k(k = 2s+1 \sim n = 2s+r)$ be the real zeros of $P(z)$ in $\mathbf{D}$, and let $\alpha_k(k = n+1 \sim n+\nu)$ be the distinct unit circle zeros of $P(z)$.

Under these conditions, there exists a stable controller $C(z) \in \mathbf{RH}_\infty$ satisfying (3.1) and (3.3) if and only if

1) $P(z)$ has the parity interlacing property

2) $|W(\alpha_k)| < \gamma \quad (k = n+1 \sim n+\nu)$

3) there exists a set of integers $\{m_k\}_{k=1\sim s}$ satisfying the consistency condition (2.1) such that $Q(\{m_k\}) > 0$ holds, where $Q(\{m_k\})$ is defined by (2.2) with

$$\beta_k := \hat{\beta}_k/\gamma, \quad \hat{\beta}_k := W(\alpha_k)/d_i(\alpha_k) \quad (k = 1 \sim n) \tag{3.5}$$

*Remark 3.1:* The condition 2) in Theorem 3.1 is related to the unit circle zeros. The case where $P(z)$ has zeros with multiplicities in **D** can be also treated by using the corresponding Pick matrix made up of interpolation data on derivatives.

*Remark 3.2:* The same technique is valid for the gain margin problem [7], [8], [4] via an appropriate conformal mapping. In the MIMO case, if all zeros of $P(z)$ in **D** are real and blocking zeros, the problem of sensitivity minimization by a stable controller can be solved by using the corresponding Pick matrix proposed by Sideris and Safonov [6] without searching.

# 4 Robust stabilization by stable controller

The robust stabilization problem [5] for multiplicative (or additive) perturbations can be reduced to an $H_\infty$-norm optimization problem of the form

$$\inf \|WT\|_\infty < 1 \tag{4.1}$$

where

$$T := C/(1+PC) \tag{4.2}$$

and $W$ is an appropriate $\mathbf{RH}_\infty$ function which is determined by the frequency shaped uncertainty bound [10]. Let $P = n_p/d_p$ and $C = n_c/d_c$ be coprime factorizations over $\mathbf{RH}_\infty$; then $T$ can be rewritten as

$$T = d_p n_c/(d_p d_c + n_p n_c) \tag{4.3}$$

We can see from (4.3) that the problem of robust stabilization by a minimum-phase controller is dual to sensitivity minimization by a stable controller. Hence, problem of robust stabilization by a minimum-phase controller can be solved as in Section 3 by interchanging the roles of the unstable poles and zeros of $P(z)$.

Unfortunately, the technique used in Section 3 cannot be adopted to solve the problem of robust stabilization by a stable controller, since $n_c(z)$ may have unstable zeros even if $C(z)$ is stable. Therefore, we must modify the problem.

It is well known that the class of all robustly stabilizing controllers can be expressed as

$$C = (\hat{a} + q\hat{b})/(a + qb), \quad q \in \mathbf{B}_\infty \tag{4.4}$$

where $\hat{a}$, $\hat{b}$, $a$ and $b$ are appropriate $\mathbf{RH}_\infty$ functions [5]. Hence, the problem of robust stabilization by a stable controller can be reduced to finding a $q \in \mathbf{B}_\infty$ such that

$$a + qb \text{ is a unit} \tag{4.5}$$

Note that there exists a $q(z) \in \mathbf{RH}_\infty$ satisfying (4.5) if and only if $b/a$ has the *p.i.p*, or equivalently $P(z)$ has the *p.i.p*.

Let $v := a + qb$ and $b = b_i b_o$ is the inner-outer factorization; then $q = (v-a)/b = (v-a)/(b_i b_o)$ and hence we have

$$\|q\|_\infty = \|(v-a)/b_o\|_\infty < 1 \tag{4.6}$$

If we set

$$f := v/b_o, \quad h := a/b_o \tag{4.7}$$

then the problem is reduced to Problem 2 stated in Section 2 and hence Theorems 2.2 (or 2.3) and 2.4 can be applied to obtain a necessary condition and a sufficient condition for the solvability (note that the technique in the proof of Theorem 3.1 is still valid for the case where $b_o$ is not a unit, i.e., $b_o$ has unit circle zeros). In other words, Theorems 2.2 (or 2.3) and 2.4 give an upper bound and a lower bound on the plant uncertainty for which there exists a stable robustly stabilizing controller. Further discussion is omitted in the interests of brevity.

Since one cannot expect to find a closed form solution to this problem, we now present an outline of a procedure to solve the problem numerically.

Consider the following problem:

Find (if exists) a $q(z) \in \mathbf{A_s}$ such that $a(z) + q(z)b(z)$ is a unit and $\|q\|_\infty < 1$, where $a(z) \in \mathbf{A_s}$ and $b(z) \in \mathbf{A_s}$.

Let $q_o(z)$ be a function in $\mathbf{A_s}$ such that $u := a + q_o b$ is a unit (note that such a $q_o$ exists if $b/a$ has the *p.i.p.*). Then the set of all units of the form $a + qb$ can be expressed as

$$a + qb = u\exp\{hb + \sum_{k=1}^{s} 2\pi m_k \phi_k\} = u\exp(v) \tag{4.8}$$

where $h(z) \in \mathbf{A_s}$ is arbitrary, $m_k (k = 1 \sim s)$ are arbitrary integers and $\phi_k(z)(k = 1 \sim s)$ defined as in (3.3.6) in [9] correspond to the nonreal unstable zeros of $b(z)$. This implies that

$$q = \frac{ue^v - a}{b} = \frac{u-a}{b} + \frac{u(e^v - 1)}{b} = q_o + \frac{u(e^v-1)}{b} \tag{4.9}$$

and hence the problem is reduced to an infinite-dimensional nonlinear optimization problem:

$$\min_{h(z),\{m_k\}} \|q_o + \tfrac{u}{b}(\exp(hb + \Sigma 2\pi m_k \phi_k) - 1)\|_\infty < 1 \tag{4.10}$$

We now assume $b(z)$ has no nonreal unstable zeros for simplicity and propose an algorithm for solving the optimization problem by a discretizing method.

Let

$$h(z) := \sum_{i=0}^{\ell} h_i z^i \tag{4.11}$$

and evaluate the norm at some points $e^{\theta_i}(i = 1 \sim N)$ on the unit circle. Then, the approximate value of the minimum norm of (4.10) can be calculated by solving a nonlinear min-max optimization problem:

$$\min_{h_1 \sim h_\ell} \max_i |[q_o + \frac{u}{b}(e^{bh} - 1)](e^{j\theta_i})| \tag{4.12}$$

# 5 Conclusion

We proposed two interpolation- minimization problems for unit functions which are related to the problems of sensitivity minimization and robust stabilization.

A necessary and sufficient condition has been obtained for the problem of sensitivity reduction by a stable controller. On the other hand, it is very hard to derive a closed form solution for the problem of robust stabilization by a stable controller. We showed that the problem can be approximately reduced to a nonlinear min-max optimization problem.

This research was carried out while the first author was visiting the Dept. of Electrical Engineering, University of Waterloo, and was supported by the Manufacturing Research Corporation of Ontario.

# References

[1] J.A. Ball and J.W. Helton, "Interpolation with outer functions and gain equalization in amplifiers", Proc. Int. Symp. on Mathematical Theory of Networks Systems, p.41-49, Delft (1979).

[2] C. Ganesh, J.B. Pearson, "Design of optimal control systems with stable feedback", Technical report No.8514, Rice University (1985).

[3] C. Ganesh, J.B. Pearson, "On $H^2$-optimization with stable controllers," to appear in Automatica (1989).

[4] P.P. Khargonekar, and A. Tannenbaum, "Non-euclidian metrics and the robust stabilization of systems with parameter uncertainty", IEEE Trans. Automat. Control, AC-30, pp.1005-1013 (1985).

[5] H. Kimura, "Robust stabilization for a class of transfer functions", IEEE Trans. Automat. Control, AC-29, pp.788-793 (1984).

[6] A. Sideris, and M.G. Safonov, "Infinity-norm optimization with a stable controller," Proc. American Control Conf., pp.804-805 (1985).

[7] A. Tannenbaum, "Feedback stabilization of linear dynamical plants with uncertainty in the gain factor", Int. J. Control, 32, pp.1-16 (1980).

[8] A. Tannenbaum, "Modified Nevanlinna-Pick interpolation and feedback stabilization of linear plants with uncertainty in the gain factor", Int. J. Control, 36, pp.331-336 (1982).

[9] M. Vidyasagar, "Control System Synthesis: A Factorization Approach", Cambridge, MA: MIT Press (1985).

[10] M. Vidyasagar and H. Kimura, "Robust controllers for uncertain linear multivariable systems", Automatica, 22, pp.85-94 (1986).

[11] J. Walsh, "Interpolation and Approximation by Rational Functions in the Complex Domain", AMS Colloquium Publications, 20, Fourth Edition (1965).

[12] D.C. Youla, J.J. Bongiorno, Jr. and C.N. Lu, "Single-loop feedbavk stabilization of linear multivariable dynamical plants", Automatica, 10, pp.159-173 (1974).

[13] G. Zames, "Feedback and optimal sensitivity: Model reference transformations, multiplicative seminorms, and approximate inverses", IEEE Trans. Automat. Control, AC-26, pp.301-320 (1981).

# CONJUGATION AND $H^\infty$ CONTROL

HIDENORI KIMURA

**ABSTRACT**

This paper proposes a new approach to the $H^\infty$ control problem based on the notion of conjugation. A method of controller augmentation is introduced which leads to the formulation of $H^\infty$ control problem as a J-lossless conjugation. The complete characterization of the class of desired controllers is given in the state space for the so-called one-block problem. An extension of this result to the general four-block problem is briefly discussed.

## 1. INTRODUCTION

Let

$$\begin{bmatrix} z(s) \\ y(s) \end{bmatrix} = G(s) \begin{bmatrix} w(s) \\ u(s) \end{bmatrix} = \begin{bmatrix} G_{11}(s) & G_{12}(s) \\ G_{21}(s) & G_{22}(s) \end{bmatrix} \begin{bmatrix} w(s) \\ u(s) \end{bmatrix} \tag{1}$$

be the input-output description of the plant to be controlled, where z is the error vector of dimension m, y the observation vector of dimension q, w the disturbance vector of dimension r and u the control input vector of dimension p. The purpose of $H^\infty$ control is to find a control law

$$u(s) = K(s)\, y(s) \tag{2}$$

such that the closed-loop system is internally stable and satisfies the norm bound

$$\|\Phi\|_\infty < \gamma \tag{3}$$

for some $\gamma$, where $\Phi$ is the closed-loop transfer function from w to z given explicitly by

$$\Phi := G_{11} + G_{12}K(I - G_{22}K)^{-1}G_{21} \tag{4}$$

Remarkable progress is now being made in the field of $H^\infty$ control concerning the existence condition of the desired $H^\infty$ control law and the method of synthesizing such control law ([1]-[4]). Particularly, the method of J-spectral factorization [1][2] was shown to be effective for solving the case of output feedback and the game-theoretic approach is useful for attacking the case of state feedback [3][4]. Other approaches have also been proposed and successfully used ([5][6]). However, it is the author's impression that the unified framework of $H^\infty$ control is yet to be established which covers various problems ranging from "one-block" cases to "four-block" cases and from the case of output feedback to that of state feedback in their full generality. For instance, game-theoretic approach is not very convenient for parameterizing all $H^\infty$ state feedback control laws, while the method of J-spectral factorization is not appropriate to treat the case of state feedback.

In this paper, we propose a new method of $H^\infty$ control based on the controller augmentation and the J-lossless conjugation. This method can solve various $H^\infty$ control problems in their full generality in a unified and systematic way. The formulation of the $H^\infty$ control problem rendered in this framework is conceptually simple and gives a deep insight on the fundamental structure of the $H^\infty$ control system.

*Notations:*

$\mathbf{R}_{m\times r}$; the set of constant real matrices of size m × r.

$\mathbf{RL}^\infty_{m\times r}$; the set of rational proper matrices with size m × r.

$\mathbf{RH}^\infty_{m\times r}$; the subset of stable matrices in $\mathbf{RL}^\infty_{m\times r}$.

$\mathbf{BH}^\infty_{m\times r}$; the set of all $S\in\mathbf{RH}^\infty_{m\times r}$ satisfying $\|S\|_\infty < 1$.

$\tilde{A}(s) := A^T(-s)$, $A^*(s) := \overline{A}^T(s)$.

$G(s) = \{A, B, C, D\} := D + C(sI - A)^{-1}B$.

## 2. LINEAR FRACTIONAL TRANSFORMATIONS AND J-LOSSLESS SYSTEMS

Let U and W be two square matrices in $\mathbf{RL}^\infty_{(m+r)\times(m+r)}$ which are represented in the partitioned forms

$$U = \begin{bmatrix} U_{11} & U_{12} \\ U_{21} & U_{22} \end{bmatrix}, \quad W = \begin{bmatrix} W_{11} & W_{12} \\ W_{21} & W_{22} \end{bmatrix}, \tag{5}$$

where the (1,1) elements denote m × m blocks. Associated with (5), we define two types of linear fractional transformations

$$F(U, S) := (U_{11}S + U_{12})(U_{21}S + U_{22})^{-1}, \tag{6}$$

$$F'(W, S) := (W_{11} + SW_{21})^{-1}(W_{12} + SW_{22}), \tag{7}$$

where $S \in \mathbf{RL}^\infty_{m\times r}$. The following lemma demonstrates a useful property of these transformations.

**LEMMA 1.** For each $U_i, W_i$ (i = 1,2) and S of compatible sizes, the following identities hold:

$$F(U_1, F(U_2, S)) = F(U_1U_2, S)$$

$$F'(W_1, F'(W_2, S)) = F'(W_2W_1, S)$$

If (6) respresents a right coprime fractional representation of F(U,S), the corresponding left coprime fractional representation can be represented in the form of (7) for some W. The following lemma establishes a condition under which the two transformations (6) and (7) are identical.

**LEMMA 2.** F(U, S) = F' (W, S) for each S, if and only if

$$WJU = \alpha J \tag{8}$$

for some scalar α, where J is given by

$$J := \begin{bmatrix} I_m & 0 \\ 0 & -I_r \end{bmatrix}. \tag{9}$$

The proof is elementary and omitted here.

A matrix $\Theta(s) \in \mathbf{RL}^{\infty}_{(m+r)\times(m+r)}$ is said to be J-unitary, if

$$\Theta^{\sim}(s)\, J\Theta(s) = J, \tag{10}$$

holds for each s, where J is given by (9). A J-unitary matrix $\Theta(s)$ is said to be J-lossless, if

$$\Theta^{*}(s)\, J\Theta(s) < J \tag{11}$$

for each $\mathrm{Re}[s] > 0$.

Let S be an arbitrary matrix in $\mathbf{BH}^{\infty}_{m\times r}$. If $\Theta(s) \in \mathbf{RL}^{\infty}_{(m+r)\times(m+r)}$ is J-lossless, we have $\begin{bmatrix} S^{*} & I \end{bmatrix} \Theta^{*}(s)\, J\, \Theta(s) \begin{bmatrix} S \\ I \end{bmatrix} < \begin{bmatrix} S^{*} & I \end{bmatrix} J \begin{bmatrix} S \\ I \end{bmatrix} < 0$ for each $\mathrm{Re}[s] > 0$. This obviously implies that $F(\Theta, S) \in \mathbf{BH}^{\infty}_{m\times r}$. In exactly the same way, we can show that $F(\Theta^{-1}, S) \in \mathbf{BH}^{\infty}_{m\times r}$. Thus, the following lemma holds:

**LEMMA 3.** If $\Theta(s) \in \mathbf{RL}^{\infty}_{(m+r)\times(m+r)}$ is J-lossless, then, both $F(\Theta, S)$ and $F(\Theta^{-1}, S)$ are in $\mathbf{BH}^{\infty}_{m\times r}$ for each $S \in \mathbf{BH}^{\infty}_{m\times r}$.

## 3. CONJUGATION

Let $G(s) = \{A_0, B_0, C_0, D_0\}$. A system

$$V(s) = \{-A_0^{T}, B_c, C_c, D_c\} \tag{12}$$

is said to be a *conjugator* of G(s), if the equations

$$A_0X + XA_0^{T} + B_0C_c = 0 \tag{13}$$

$$B_0D_c - XB_c = 0 \tag{14}$$

holds for some X. Based on (13) (14), we can easily show that

$$G(s)\, V(s) = \{-A_0^{T}, B_c, C_0X + D_0C_c, D_0D_c\} \tag{15}$$

Notice that the A-matrix $A_0$ of G(s) is replaced by its conjugate $-A_0^{T}$ by the postmultiplication of V(s). The operation of multiplying a conjugator is said to be the *conjugation*.

Many interesting properties of conjugations were extensively discussed in [7] including the structure of pole-zero cancellation and the relation to the inner-outer factorization. It should be noted that the defining equations (13) (14) of conjugators depend only on the pair $(A_0, B_0)$. Therefore, we sometimes call it a conjugator of $(A_0, B_0)$.

An important class of conjugations is the J-lossless conjugation, the cojugation by a J-lossless conjugator. In [7], the existence condition for the J-lossless conjugator was derived. Also the J-lossless conjugation was shown to be equivalent to the Nevanlinna-Pick interpolation represented in the state space.

**LEMMA 4.** A J-lossless conjugator $\Theta(s)$ of a pair $(A_0, B_0)$, $A_0 \in \mathbf{R}_{n\times n}$, $B_0 \in \mathbf{R}_{n\times(m+r)}$, exists, if and only if the equation

$$A_0X + XA_0^{T} = B_0JB_0^{T} \tag{16}$$

has a positive definite solution X. In that case, a J-lossless conjugator of $(A_0, B_0)$ is given by

$$V(s) = \{-A_0^{T}, X^{-1}B_0D_c, -JB_0^{T}, D_c\} \tag{17}$$

where $D_c$ is any matrix satisfying $D_c^T J D_c = J$.

## 4. STRUCTURE OF $H^\infty$ CONTROLLERS

The input-output description (1) of the plant is represented in the state-space form

$$\dot{x} = Ax + B_1 w + B_2 u \tag{18a}$$

$$z = C_1 x + D_{11} w + D_{12} u \tag{18b}$$

$$y = C_2 x + D_{21} w \tag{18c}$$

where x is the state vector of dimension n. We assume that $(A, B_2)$ is stabilizable and $(C_2, A)$ is detectable. Hence, there exists a matrix $F \in \mathbf{R}_{n\times p}$ and $H \in \mathbf{R}_{q\times n}$ such that the two matrices

$$A_F := A + B_2 F, \quad A_H := A + HC_2 \tag{19}$$

are both stable. It is well-known that each stabilizing control law (2) is given by

$$K(s) = F(Z, Q) \tag{20}$$

for some $Q \in \mathbf{RH}^\infty_{p\times q}$, where

$$Z = \begin{bmatrix} Z_{11} & Z_{12} \\ Z_{21} & Z_{22} \end{bmatrix} = \left\{ A_F, \ -\begin{bmatrix} B_2 & H \end{bmatrix}, \begin{bmatrix} F \\ C_2 \end{bmatrix}, \begin{bmatrix} -I_p & 0 \\ 0 & I_q \end{bmatrix} \right\} \tag{21}$$

Substituting (20) in (4) yields

$$\Phi = T_1 - T_2 Q T_3, \tag{22}$$

where

$$\begin{bmatrix} T_1 & T_2 \\ T_2 & 0 \end{bmatrix} = \left\{ \begin{bmatrix} A_F & -B_2 F \\ 0 & A_H \end{bmatrix}, \begin{bmatrix} B_1 & B_2 \\ B_1 + HD_{21} & 0 \end{bmatrix}, \begin{bmatrix} C_1 + D_{12}F & -D_{12}F \\ 0 & C_2 \end{bmatrix}, \begin{bmatrix} D_{11} & D_{12} \\ D_{21} & 0 \end{bmatrix} \right\} \tag{23}$$

For details, see [9]. Without loss of generality, we can normalize $\gamma$ in (3) to 1. Hence our problem is to find a $Q \in \mathbf{RH}^\infty_{p\times q}$ such that

$$\| \Phi \|_\infty = \| T_1 - T_2 Q T_3 \|_\infty < 1 \tag{24}$$

Now, we make the following assumption :

($A_1$) Both $T_2^{-1}$ and $T_3^{-1}$ exist.

This implies that both $D_{12}^{-1}$ and $D_{21}^{-1}$ exist, which, of course, requires m = p and r = q. Under this assumption, (22) can be rewritten as $\Phi = (T_1 T_3^{-1} - T_2 Q)T_3 = T_2(T_2^{-1} T_1 - QT_3)$. Therefore, we have

$$\Phi = F(U_a, Q) = F'(U_b, Q) \tag{25}$$

$$U_a := \begin{bmatrix} -T_2 & T_1 T_3^{-1} \\ 0 & T_3^{-1} \end{bmatrix}, \quad U_b := \begin{bmatrix} T_2^{-1} & T_2^{-1} T_1 \\ 0 & -T_3 \end{bmatrix} \tag{26}$$

Obviously, $U_b J U_a = -J$.

The design specification (24) can be satisfied if there exists a $\Pi_a \in \mathbf{RH}^\infty_{(m+r)\times(m+r)}$ such that $\Pi_a^{-1}$ is stable and

$$U_a \Pi_a = \Theta \tag{27}$$

is J-lossless. In this case,

$$Q = F(\Pi_a, S) \tag{28}$$

satisfies (24) for each $S \in RH^{\infty}_{m \times r}$. Indeed, substituting (28) in (25) yields $\Phi$ = $F(U_a, F(\Pi_a, S)) = F(U_a \Pi_a, S) = F(\Theta, S)$. Due to Lemma 3, $\Phi \in RH^{\infty}_{m \times r}$. Let

$$\Pi_a = \begin{bmatrix} \Pi_{11} & \Pi_{12} \\ \Pi_{21} & \Pi_{22} \end{bmatrix}, \quad \Theta = \begin{bmatrix} \Theta_{11} & \Theta_{12} \\ \Theta_{21} & \Theta_{22} \end{bmatrix} \tag{29}$$

be the partitioned representations of $\Pi_a$ and $\Theta$. Since both $\Pi_a = U_a^{-1}\Theta$ and $\Pi_a^{-1} = \Theta^{-1}U_a$ are stable, all the unstable zeros of $T_2$ and $T_3$ are cancelled out by the zeros and the poles of $\Theta$. Therefore, $(\Pi_{21}S + \Pi_{22})^{-1} = (\Theta_{21}S + \Theta_{22})^{-1}T_3^{-1}$ is stable. Hence, Q is stable. The structure of the closed-loop system is illustrated in Fig. 1.

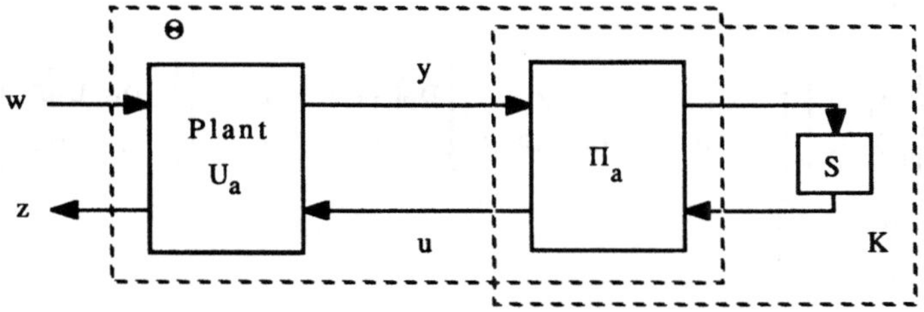


Fig. 1 Structure of Closed-loop System

The dual argument shows that if there exists a $\Pi_b \in RH^{\infty}_{(m+r)\times(m+r)}$ such that $\Pi_b^{-1}$ is stable and $\Pi_b U_b = \Theta_1^{-1}$ for some J-lossless matrix $\Theta_1$, then,

$$Q = F'(\Pi_b, S) \tag{30}$$

satisfies the specification (24). According to Lemma 2, we choose $\Pi_b$ satisfying $\Pi_b J \Pi_a = -J$ so that (28) and (30) are identical. In this case, $\Pi_b U_b J\Theta = \Pi_b U_b J U_a \Pi_a = -\Pi_b J \Pi_a = J$. Hence, we have

$$\Pi_b U_b = \Theta^{\sim}. \tag{31}$$

Due to (20) (28), a desired controller is parameterized as

$$K(s) = F(Z\Pi_a, S) \tag{32}$$

for each $S \in BH^{\infty}_{m \times r}$. We now summarize the above reasoning as follows:

**LEMMA 5.** There exists a stabilizing controller satisfying $\|\Phi\|_{\infty} < 1$, if there exist stable matrices $\Pi_a$ and $\Pi_b$ such that (27) and (31) hold for some J-lossless $\Theta$. In this case, desired controllers are given by (32) for each $S \in BH^{\infty}_{m \times r}$.

## 5. CALCULATION OF H$^{\infty}$ CONTROLLER

The relations (27) and (31) are the basis of calculating a desired H$^{\infty}$ controller. From these relations, we have

$$\Pi_a = U_a^{-1}\Theta, \quad \Pi_b^{\sim} = (U_b^{\sim})^{-1}\Theta. \tag{33}$$

Since $\Pi_a$ is stable, $\Theta$ must conjugate the anti-stable part of $U_a^{-1}$. Also, since $\tilde{\Pi}_b$ is anti-stable, $\Theta$ must conjugate the stable part of $(\tilde{U}_b)^{-1}$. Therefore, $\Theta$ must be a J-lossless conjugator of the unstable part of $U_a^{-1}$ and the stable part of $(\tilde{U}_b)^{-1}$.

In order to make the subsequent argument simple, we make the following assumption:

($A_2$) Both $T_2^{-1}$ and $T_3^{-1}$ are anti-stable.

From (23), it follows that $T_2^{-1} = \{A - B_2D_{12}^{-1}C_1, -B_2D_{12}^{-1}, D_{12}^{-1}C_1 + F, D_{12}^{-1}\}$ and $T_3^{-1} = \{A - B_1D_{21}^{-1}C_2, -B_1D_{21}^{-1} - H, D_{21}^{-1}C_2 + F, D_{21}^{-1}\}$. The assumption ($A_2$) implies that both $A_a := A - B_2D_{12}^{-1}C_1$ and $A_b := A - B_1D_{21}^{-1}C_2$ are anti-stable. Simple manipulations yield

$$U_a^{-1} = \begin{bmatrix} -T_2^{-1} & T_2^{-1}T_1 \\ 0 & T_3 \end{bmatrix}$$

$$= \left\{ \begin{bmatrix} A_a & 0 \\ 0 & A_H \end{bmatrix}, \begin{bmatrix} L_a & -M_a \\ 0 & -(B_1 + HD_{21}) \end{bmatrix}, \begin{bmatrix} D_{12}^{-1}C_1 + F & F \\ 0 & -C_2 \end{bmatrix}, \begin{bmatrix} -D_{12}^{-1} & D_{12}^{-1}D_{11} \\ 0 & D_{21} \end{bmatrix} \right\}$$

$$U_b^{-1} = \begin{bmatrix} T_2 & T_1T_3^{-1} \\ 0 & -T_3^{-1} \end{bmatrix}$$

$$= \left\{ \begin{bmatrix} A_b & 0 \\ 0 & A_F \end{bmatrix}, \begin{bmatrix} 0 & B_1D_{21}^{-1} + H \\ B_2 & -H \end{bmatrix}, \begin{bmatrix} M_b^T & C_1 + D_{12}F \\ -L_b^T & 0 \end{bmatrix}, \begin{bmatrix} D_{12} & D_{11}D_{21}^{-1} \\ 0 & -D_{21}^{-1} \end{bmatrix} \right\}$$

where $L_a = B_2D_{12}^{-1}$, $L_b^T = D_{21}^{-1}C_2$, $M_a = L_aD_{11} - B_1$ and $M_b^T = D_{11}L_b^T - C_1$ . From the assumption, the anti-stable portion of $U_a^{-1}$ is $(A_a, [L_a \ \ -M_a])$ and the stable portion of $(\tilde{U}_b)^{-1}$ is $(-A_b^T, [M_b \ \ -L_b])$. Therefore, $\Theta$ must be a J-lossless conjugator of the pair

$$\left( \begin{bmatrix} A_a & 0 \\ 0 & -A_b^T \end{bmatrix}, \begin{bmatrix} L_a & -M_a \\ M_b & -L_b \end{bmatrix} \right) \tag{34}$$

Due to Lemma 4, a J-lossless conjugator of the pair (34) exists, if and only if the equation

$$\begin{bmatrix} A_a & 0 \\ 0 & -A_b^T \end{bmatrix} P + P \begin{bmatrix} A_a^T & 0 \\ 0 & -A_b \end{bmatrix} = \begin{bmatrix} L_a & M_a \\ M_b & L_b \end{bmatrix} J \begin{bmatrix} L_a & M_a \\ M_b & L_b \end{bmatrix}^T \tag{35}$$

has the positive definite solution P. It is not difficult to show that solution P of (35) is written as

$$P = \begin{bmatrix} P_a & I \\ I & P_b \end{bmatrix}, \tag{36}$$

where $P_a$ and $P_b$ are the solutions of

$$A_aP_a + P_aA_a^T = L_aL_a^T - M_aM_a^T, \quad A_b^TP_b + P_bA_b = L_bL_b^T - M_bM_b^T . \tag{37}$$

If $P > 0$, a J-lossless conjugator of the pair (34) is calculated to be

$$\Theta(s) = \left\{ \begin{bmatrix} -A_a^T & 0 \\ 0 & A_b \end{bmatrix}, P^{-1}\begin{bmatrix} L_a & -M_a \\ M_b & -L_b \end{bmatrix}, -\begin{bmatrix} L_a & M_a \\ M_b & L_b \end{bmatrix}^T, \begin{bmatrix} I_r & 0 \\ 0 & I_m \end{bmatrix} \right\} \tag{38}$$

due to Lemma 4. Thus, the closed-loop transfer function is given by $\Phi = F(\Theta, S)$ with $\Theta$ being given by (38).

According to (27) (32), the controller is given by

$$K = F(ZU_a^{-1}\Theta, S) \tag{39}$$

Lengthy but simple calculation yields

$$ZU_a^{-1}\Theta = \left\{ -A_a^T, [I\ 0]P^{-1}\begin{bmatrix} L_a M_a \\ M_b L_b \end{bmatrix}, \begin{bmatrix} D_{12}^{-1}(D_{11}M_a^T - L_a^T + C_1P_a) \\ D_{21}(M_a^T - L_a^T P_a) \end{bmatrix}, \begin{bmatrix} -D_{12}^{-1} & D_{12}^{-1}D_{11} \\ 0 & D_{21} \end{bmatrix} \right\} \tag{40}$$

These results are summed up as follows :

**THOREM 1.** There exists a $H^\infty$ controller satisfying (24), if and only if P given in (36) is positive definite. In that case, a desired controller is parameterized as (39) (40).

An extension of the above result to the general case where the assumption $(A_2)$ does not hold is found in [8] .

We briefly discuss how to extend the above results to the general "four-block" case, where the assumption $(A_1)$ no longer holds. The key idea is to find a generalization of (33) to the case where $T_2$ and $T_3$ are no longer invertible. For that purpose, we must introduce the notion of (J, J')-lossless system, which is a generalization of J-lossless system to non-square matrices and is defined as the matrix $\Theta_1 \in RL^\infty_{(m+r)\times(p+q)}$ satisfying

$$\tilde{\Theta}_1 J \Theta_1 = J', \ \forall s, \quad \Theta_1^* J \Theta_1 \le J', \ Re[s] \ge 0$$

$$J' = \begin{bmatrix} I_p & 0 \\ 0 & -I_q \end{bmatrix}.$$

We seek a (J, J')-lossless matrix such that

$$\begin{bmatrix} -T_2 & 0 \\ 0 & I_q \end{bmatrix}\Pi_a = \begin{bmatrix} I_m & -T_1 \\ 0 & T_3 \end{bmatrix}\Theta_1, \quad \Pi_b\begin{bmatrix} I_p & 0 \\ 0 & -T_3 \end{bmatrix}\tilde{\Theta}_1\begin{bmatrix} T_2 & -T_1 \\ 0 & I_p \end{bmatrix}, \tag{41}$$

for stable $\Pi_a$ and $\Pi_b$. It is easy to see that the above relations are identical to (38), if $T_2^{-1}$ and $T_3^{-1}$ exist. This is again a variation of the J-lossless conjugation, and the method used for the "one-block" case can be applied to this case. The existence condition is represented in terms the two algebraic Riccati equations, instead of the two Lyapunov equations (37). Due to the space limitation we must omit the details here [8].

## 6. CONCLUSION

A controller augmentation is introduced through a linear fractional transformation, which is similar to the Youla parameterization of stabilizing controllers. This augmentation leads naturally to the formulation of $H^\infty$ control problem as a J-lossless conjugation. A parameterization of all desired controllers is obtained in the state space, as well as the existence condition represented in term of two Lyapunov-type equations for the "one-block" problem. Extension to the general "four-block" problem is briefly discussed.

## REFERENCES

[1] J. A. Ball and N. Cohen, " Sensitivity minimization in an $H^\infty$ norm parameterization of all suboptimal solutions," Int. J. Control, vol. 46, pp. 785-816, 1987.

[2] M. Green, K. Glover, D. Limebeer and J. Doyle, "A J-spectral factorization approach to $H^\infty$ control," Submitted for publication.

[3] J. Doyle, K. Glover, P. Khargonekar and B. Francis, "State-space solutions to standard $H_2$ and $H^\infty$ control problems," to appear in IEEE Trans. AC.

[4] G. Tadmor," $H^\infty$ in the time domain : The standard four blocks problem," Reprint, 1988.

[5] D. Limebeer, E. Kasenally and M. Safonov, "A characterization of all solutions to the four block general distance problem," submitted for publication.

[6] A. A. Stoorvogel, " $H^\infty$ control with state feedback," Technical Univ. Eindhoven, 1989.

[7] H. Kimura, "Conjugation, interpolation and model-matching in $H^\infty$," Int. J. Control, vol.49, pp.269-307, 1989.

[8] H. Kumira and R. Kawatani, " Synthesis of $H^\infty$ controllers based on conjugation," Proc. of the 27th IEEE Conf. on Decision and Control, pp.7-13, 1988.

[9] B. A. Francis, *A Course in $H^\infty$ Control*, Springer-Verlag, New York, 1987.

Necessary and sufficient conditions for the existence of $H^\infty$-optimal controllers: An interpolation approach.

D.J.N. Limebeer and E.M. Kasenally

Abstract

A recent advance in $H^\infty$-optimal control has shown that it is possible to parameterize all controllers in terms of two n-dimensional Riccati equations—n is the dimension of the problem [4,6,10]. If $X_\infty$ and $Y_\infty$ are the solutions to these equations, then it has been shown that (1) $X_\infty \geq 0$, (2) $Y_\infty \geq 0$ and (3) $\lambda_{max}(X_\infty Y_\infty) \leq \gamma^2$ are necessary and sufficient conditions for the existence of a solution such that $\|F_u(P,K)\|_\infty \leq \gamma$; $\gamma$ is the target $H^\infty$-norm. Doyle et al [4] have proved this result in the sub-optimal case using classical time domain arguments. The aim of this paper is to prove this result using interpolation theory; this proof addresses the optimal cases in which $X_\infty$ and $Y_\infty$ are bounded. As was mentioned in [9], we will prove that the four block general distance problem has an associated Pick matrix which is congruent to

$$\Pi(\gamma) = \begin{bmatrix} X_\infty^{-1} & \gamma^{-1}I \\ \gamma^{-1}I & Y_\infty^{-1} \end{bmatrix} \tag{0.1}$$

when the indicated inverses exist. It will be shown that the necessary and sufficient condition for a solution to exist is $\Pi(\gamma) \geq 0$, which by taking Schur complements, is equivalent to $X_\infty > 0$, $Y_\infty > 0$ and $\lambda_{max}(X_\infty Y_\infty) \leq \gamma^2$. If either $X_\infty^{-1}$ or $Y_\infty^{-1}$ do not exist, a routine balancing argument will reduce the dimension of the interpolation problem; $X_\infty^{-1}$ and $Y_\infty^{-1}$ will always exist for the smaller problem[6].

## 1. Introduction

A recent advance has shown that it is possible to express all the solutions to a general class of $H^\infty$-optimal control problems in terms of the solutions of two n-dimensional Riccati equations. The sub-optimal case is dealt with in [4], while the optimal cases may be addressed via generalized state-space theory [6,10]. In these papers it is stated that necessary and sufficient conditions for a solution to exist may be given in terms of the Riccati equation solutions: $X_\infty \geq 0$, $Y_\infty \geq 0$, and $\lambda_{max}(X_\infty Y_\infty) \leq \gamma^2$ are all required. Once the pair $(X_\infty, Y_\infty)$ with the desired properties has been found, the solutions may be substituted into a representation formula which parameterizes all the solutions [4,6,10].

The purpose of this paper is to establish the three necessary and sufficient conditions by vector interpolation theory. When treating the optimalities we assume that $\|X_\infty\| < \infty$ and $\|Y_\infty\| < \infty$. There are certain problems in which $X_\infty$ and/or $Y_\infty$ do not exist at optimality[8].

It is well known that $H^\infty$ control problems are equivalent to finding all Q's$\in \mathcal{H}_+^\infty$ (if any exist) such that[3]:

$$\|(T_{11}+T_{12}QT_{21}=T_{11}+[T_{12}\ T_\perp]\begin{bmatrix} Q & 0 \\ 0 & 0 \end{bmatrix}\begin{bmatrix} T_{21} \\ \tilde{T}_\perp \end{bmatrix})(s)\|_\infty \leq \gamma \tag{1.1}$$

in which $[T_{12}\ T_\perp](s)$ and $[T_{21}^\sim\ \tilde{T}_\perp^\sim]^\sim(s)$ are inner. The problem in (1.1) is cleary equivalent to finding all Q's$\in\mathcal{H}_+^\infty$ such that

$$\left\| \begin{matrix} R_{11}+Q & R_{12} \\ R_{21} & R_{22} \end{matrix} \right\|_\infty \leq \gamma \tag{1.2}$$

in which

$$\begin{bmatrix} R_{11} & R_{12} \\ R_{21} & R_{22} \end{bmatrix}(s) = \begin{bmatrix} T_{12}^\sim \\ T_\perp^\sim \end{bmatrix} T_{11}\, [T_{21}^\sim\ \tilde{T}_\perp^\sim](s) \in \mathcal{H}_-^\infty \tag{1.3}$$

Four block problems of the type shown in (1.2) may be solved by interpolation [5], and we have shown that they may also may be solved by inner embedding [6,10]. Lets suppose that (1.2) is embedded in the all-pass matrix

$$\mathcal{A}(s) = \begin{bmatrix} R_{00}+Q_{00} & R_{01}+Q_{01} & R_{02} \\ R_{10}+Q_{10} & R_{11}+Q_{11} & R_{12} \\ R_{20} & R_{21} & R_{22} \end{bmatrix}(s) = R_a(s)+Q_a(s) \tag{1.4}$$

such that $\mathcal{A}(s)\mathcal{A}^\sim(s)=\gamma^2 I$. This is always possible[6]. Next, we unwrap (1.4) by writing

$$\mathcal{R}(s) = \begin{bmatrix} I & 0 & 0 \\ 0 & T_{12} & T_\perp \end{bmatrix} \mathcal{A}(s) \begin{bmatrix} I & 0 \\ 0 & T_{21} \\ 0 & \tilde{T}_\perp \end{bmatrix}(s) \qquad Q\in\mathcal{H}_+^\infty$$

$$= \left( \begin{bmatrix} T_{00} & T_{01} \\ T_{10} & T_{11} \end{bmatrix} + \begin{bmatrix} I & 0 & 0 \\ 0 & T_{12} & T_\perp \end{bmatrix} \begin{bmatrix} Q_{00} & Q_{01} & 0 \\ Q_{10} & Q_{11} & 0 \\ 0 & 0 & 0 \end{bmatrix} \begin{bmatrix} I & 0 \\ 0 & T_{21} \\ 0 & \tilde{T}_\perp \end{bmatrix} \right)(s) \tag{1.5}$$

which is a dilated version of the original problem posed in (1.1). The necessary and sufficient conditions are established by studying the vector interpolation problem associated with (1.5). Use will be made of the interpolation theory described in [1,2,7]. In particular, we show that the Pick matrix associated the problem in (1.5) is congruent to (0.1) if X and Y are nonsingular (X and Y are the standard $H_2$ Riccati equation solutions). Otherwise, a smaller Pick matrix with dimension Rank(X)+Rank(Y) must be constructed[6]. We will use the standard notation as detailed in [3,4,6].

## 2. Preliminaries

We will use this section as a repository for a number of state-space models. To begin, we assume that the standard problem matrix is given by [3,6]

$$P(s) \stackrel{s}{=} \left[\begin{array}{c|cc} A & B_1 & B_2 \\ \hline C_1 & 0 & D_{12} \\ C_2 & D_{21} & 0 \end{array}\right] \tag{2.1}$$

in which $D_{11}=0$ and $D_{22}=0$ (this may be assumed without loss of generality[6]), and that $[D_{12}\ D_\perp]$ and $[D_{21}'\ \tilde{D}_\perp']'$ are orthogonal. Next, we need the $H_2$ Riccati equation solutions which are given by

$$X = \mathrm{Ric}\begin{bmatrix} A-B_2D_{12}'C_1 & -B_2B_2' \\ -C_1'D_\perp D_\perp'C_1 & -(A-B_2D_{12}'C_1)' \end{bmatrix} \tag{2.2a}$$

$$Y = \mathrm{Ric}\begin{bmatrix} (A-B_1D_{21}'C_2)' & -C_2'C_2 \\ -B_1\tilde{D}_\perp'\tilde{D}_\perp B_1' & -(A-B_1D_{21}'C_2) \end{bmatrix} \tag{2.2b}$$

See [4] for an explaination of the Ric[·] notation. The $T_{ij}(s)$'s in (1.1) are given by

$$\begin{bmatrix} T_{11} & T_{12} & T_\perp \\ T_{21} & 0 & 0 \\ \tilde{T}_\perp & 0 & 0 \end{bmatrix}(s) \stackrel{s}{=} \left[\begin{array}{cc|ccc} A-B_2F & B_2F & B_1 & B_2 & -X^{-1}C_1'D_\perp \\ 0 & A-HC_2 & B_1-HD_{21} & 0 & 0 \\ \hline C_1-D_{12}F & D_{12}F & 0 & D_{12} & D_\perp \\ 0 & C_2 & D_{21} & 0 & 0 \\ 0 & -\tilde{D}_\perp B_1'Y^{-1} & \tilde{D}_\perp & 0 & 0 \end{array}\right] \tag{2.3}$$

if $X^{-1}$ and $Y^{-1}$ exist. If this is not the case, a lower dimension problem may be found for which these inverses exist[6]. The $H_2$ state feedback and output injection matrices are given by

$$F = D_{12}'C_1 + B_2'X \qquad\qquad H = B_1D_{21}' + YC_2' \tag{2.4}$$

and with this particular choice of F and H, $[T_{12} \mid T_\perp](s)$ and $[T_{21}^{\sim} \mid \tilde{T}_\perp^{\sim}]^{\sim}(s)$ are inner [3]. In addition, it is easy to see that $X_\infty^{-1}$ and $Y_\infty^{-1}$ exist since they are given by $X_\infty=(I-\gamma^{-2}\tilde{X}X^{-1})^{-1}X$ and $Y_\infty=Y(I-\gamma^{-2}Y^{-1}\tilde{Y})^{-1}$ in which $\tilde{X}$ and $\tilde{Y}$ are the solutions to spectral factorization Riccati equations associated with the embedding process[6]. Finally

$$\begin{bmatrix} T_{00} & T_{01} \\ T_{10} & T_{11} \end{bmatrix} \stackrel{s}{=} \left[\begin{array}{cc|cc} A-B_2F & B_2F & -\gamma X_\infty^{-1}C_1'D_\perp & B_1 \\ 0 & A-HC_2 & 0 & B_1-HD_{21} \\ \hline 0 & -\gamma\tilde{D}B_1'Y_\infty^{-1} & 0 & \gamma\tilde{D}_\perp \\ C_1-D_{12}F & D_{12}F & \gamma D_\perp & 0 \end{array}\right] \tag{2.5}$$

$X_\infty$ and $Y_\infty$ may be found directly from

$$X_\infty = \text{Ric}\begin{bmatrix} A-B_2D_{12}'C_1 & -(B_2B_2'-\gamma^{-2}B_1B_1') \\ -C_1'D_\perp D_\perp' C_1 & -(A-B_2D_{12}'C_1)' \end{bmatrix} \tag{2.6a}$$

and

$$Y_\infty = \text{Ric}\begin{bmatrix} (A-B_1D_{21}'C_2)' & -(C_2'C_2-\gamma^{-2}C_1'C_1) \\ -B_1\tilde{D}_\perp'\tilde{D}_\perp B_1' & -(A-B_1D_{21}'C_2) \end{bmatrix} \tag{2.6b}$$

This completes the state-space specification of the interpolation problem in (1.5)

## 3 Interpolation theory

We will prove the necessary and sufficient conditions in two steps. Firstly, we will establish that a solution to the $H^\infty$ control problem exists if and only if the Pick matrix associated with the interpolation problem in (1.5) is positive semi-definite. Secondly, we will show by state-space calculation that the Pick matrix is congruent to $\Pi(\gamma)$ in (0.1).

Theorem 3.1 The $H^\infty$ control problem specified by P(s) in (2.1) has a solution if and only if the Pick matrix associated with (1.5) is non-negative.

Necessity. Suppose there exists an internally stabilizing controller K(s) such that $\|F_l(P(s),K(s))\|_\infty \leq \gamma$; P(s) is the design problem matrix given in (2.1). This assumption implies that there exists a $Q(s) \in \mathcal{H}_+^\infty$ such that

$$\left\| \begin{bmatrix} R_{11}+Q & R_{12} \\ R_{21} & R_{22} \end{bmatrix}(s) \right\|_\infty \leq \gamma \tag{3.1}$$

By dilation[6,10], one may construct the all pass matrix $\mathcal{A}(s)$ given in (1.4) in such a way that the last block row and column are parts of inner matrices. Finally,

$\begin{bmatrix} I & 0 \\ 0 & T_{12}^{\sim} \\ 0 & T_\perp^{\sim} \end{bmatrix} \mathcal{A}(s) \begin{bmatrix} I & 0 & 0 \\ 0 & T_{21}^{\sim} & \tilde{T}_\perp^{\sim} \end{bmatrix}(s)$ gives the dilated closed loop

$$\mathcal{R}(s) = \left( \begin{bmatrix} T_{00} & T_{01} \\ T_{10} & T_{11} \end{bmatrix} + \begin{bmatrix} I & 0 & 0 \\ 0 & T_{12} & T_\perp \end{bmatrix} \begin{bmatrix} Q_{00} & Q_{01} & 0 \\ Q_{10} & Q_{11} & 0 \\ 0 & 0 & 0 \end{bmatrix} \begin{bmatrix} I & 0 \\ 0 & T_{21} \\ 0 & \tilde{T}_\perp \end{bmatrix} \right)(s) \quad \mathcal{R}(s) \in H_+^\infty \tag{3.2}$$

Since $\mathcal{R}(s)$ is an interpolating matrix function, it is necessarily the case that $\Pi(\gamma) \geq 0$ [1,2,7].

Sufficiency Suppose $\Pi(\gamma) \geq 0$. Then there exists an interpolating matrix function such that[1,2,7]

$$\| \mathcal{R}(s) = \left( \begin{bmatrix} T_{00} & T_{01} \\ T_{10} & T_{11} \end{bmatrix} + \begin{bmatrix} I & 0 & 0 \\ 0 & T_{12} & T_\perp \end{bmatrix} \begin{bmatrix} Q(s) \in \mathcal{H}_+^\infty \end{bmatrix} \begin{bmatrix} I & 0 \\ 0 & T_{21} \\ 0 & \tilde{T}_\perp \end{bmatrix} \right)(s) \|_\infty \leq \gamma \quad \mathcal{R}(s) \in H_+^\infty \tag{3.3}$$

What remains to be shown is that Q(s) may always be chosen with the form

$$\begin{bmatrix} Q_{00} & Q_{01} & 0 \\ Q_{10} & Q_{11} & 0 \\ 0 & 0 & 0 \end{bmatrix}(s) \tag{3.4}$$

To demonstrate that this is the case, we note that (1)$\Pi(\gamma)\geq 0 \Leftrightarrow \|R_a\|_H \leq \gamma$[1,2,7], (2) we invoke the state-space construction in [6] to show that (3.3) always admits a solution of the form indicated in (3.4). □

## 3.2 The state-space theory

In this section we derive state-space expressions for the left and right interpolation constraints which are associated with (1.5). Following that, we go on to verify that the Pick matrix associated with (1.5), is (0.1). $\begin{bmatrix} T_{21} \\ \tilde{T}_{\perp} \end{bmatrix}(s)$ will have n right half plane zeros $\{s_i: i= 1,2,...,n\}$ (which are assumed distinct) together with a sequence of vectors $a_i$ which satisfy $\{\begin{bmatrix} T_{21} \\ \tilde{T}_{\perp} \end{bmatrix}(s_i)a_{i2}=0 : i=1,2,...,n\}$[1]. Further, we define $\{b_i=\begin{bmatrix} T_{01} \\ T_{11} \end{bmatrix}(s_i)a_{i2}: i=1,2,...,n\}$ and any interpolating matrix function associated with (1.5) must satisfy $\{\mathcal{R}(s_i)a_i=b_i\ i=1,2,...,n\}$ [1,2,7]. By duality, there exist a set of zeros and zero vectors such that $a'_{i2}[\ T_{12}\ T_{\perp}](s_i)=0: i=n+1,...,2n\}$. If we define $\{b'_i=a'_{i2}[\ T_{10}\ T_{11}](s_i): i=n+1,...,2n\}$, then it is necessarily the case that $\{a'_i\mathcal{R}(s_i)=b'_i: i=n+1,...,2n\}$ if $\mathcal{R}(s)$ is interpolating.

The Pick matrix associated with the above interpolation data is given by [7]

$$\Pi(\gamma) = \begin{bmatrix} \Pi_{11} & \Pi_{12} \\ \Pi'_{12} & \Pi_{22} \end{bmatrix}(\gamma) \tag{3.5}$$

in which

$$\Pi_{11} = \left\{\frac{\gamma^2 a'_i a_k - b'_i b_k}{\bar{s}_i + s_k}\right\}_{k=1,...,n}^{i=1,...,n} \qquad \Pi_{12} = \left\{\frac{\gamma a'_i b_k - \gamma b'_i a_k}{\bar{s}_i - \bar{s}_k}\right\}_{k=n+1,...,2n}^{i=1,...,n}$$

$$\Pi'_{12} = \left\{\frac{\gamma b'_i a_k - \gamma a'_i b_k}{s_k - s_i}\right\}_{k=1,...,2n}^{i=n+1,...,2n} \qquad \Pi_{22} = \left\{\frac{\gamma^2 a'_i a_k - b'_i b_k}{\bar{s}_k + s_i}\right\}_{k=n+1,...,2n}^{i=n+1,..,2n}$$

We will now find explicit expressions for the $s_i$'s, $a_{i2}$'s and $b_i$'s described above. To begin with, we note that the $s_i$'s associated with the right constraints are given by $-\bar{s}_i=\lambda_i(A-HC_2)$ since

---

1 In order that all the dimensions match, we should write $\begin{bmatrix} I & 0 \\ 0 & T_{21} \\ 0 & \tilde{T}_{\perp} \end{bmatrix}(s_i)\begin{bmatrix} a_{i1} \\ a_{i2} \end{bmatrix}=0$. This clearly requires that $a_{i1}=0\ i=1,...n$.

A similar remark applies in the case of the left vector constraints.

$\begin{bmatrix} T_{21} \\ \tilde{T}_{\perp} \end{bmatrix}(s)$ is all-pass. To obtain state-space models for the $a_i$'s, we study the solutions to the equation

$$\left[\begin{array}{c|c} s_iI-A+(B_1D_{21}'+YC_2')C_2 & -B_1+(B_1D_{21}'+YC_2')D_{21} \\ \hline C_2 & D_{21} \\ -\tilde{D}_{\perp}B_1'Y^{-1} & \tilde{D}_{\perp} \end{array}\right]\begin{bmatrix} Yx_i \\ a_{i2} \end{bmatrix} = \begin{bmatrix} 0 \\ \hline 0 \\ 0 \end{bmatrix} \qquad (3.6)$$

The (2,1) and (3,1) blocks of (3.6) $\Rightarrow$

$$a_{i2}=[D_{21}' \ \tilde{D}'_{\perp}]\begin{bmatrix} -C_2Y \\ \tilde{D}_{\perp}B_1' \end{bmatrix}x_i \qquad i=1,2,\dots,n \qquad (3.7)$$

Substituting this into the (1,1) partition of (3.6) yields

$$(s_iI+A'-C_2'H')x_i=0 \qquad i=1,2,\dots,n \qquad (3.8)$$

after substituting from (2.2b).

A dual sequence of arguments shows that that

$$a_{i2}'=x_i'[-XB_2 \ \ C_1'D_{\perp}]\begin{bmatrix} D_{12}' \\ D'_{\perp} \end{bmatrix} \qquad i=n+1,\dots,2n \qquad (3.9)$$

where

$$x_i'(s_iI+A'-F'B_2')=0 \qquad i=n+1,\dots,2n \qquad (3.10)$$

A direct calculation using (2.5), (3.7) and (3.8) shows that

$$\left[\begin{array}{cc|c} s_iI-A+B_2F & -B_2F & -B_1 \\ 0 & s_iI-A+HC_2 & HD_{21}-B_1 \\ \hline 0 & -\gamma\tilde{D}_{\perp}B_1'Y_{\infty}^{-1} & \gamma\tilde{D}_{\perp} \\ C_1-D_{12}F & D_{12}F & 0 \end{array}\right]\begin{bmatrix} Yx_i \\ Yx_i \\ a_{i2} \end{bmatrix} = \begin{bmatrix} 0 \\ 0 \\ b_{i1} \\ b_{i2} \end{bmatrix} \qquad i=1,2,\dots,n \qquad (3.11)$$

which gives

$$b_i=\begin{bmatrix} b_{i1} \\ b_{i2} \end{bmatrix}=\begin{bmatrix} \gamma\tilde{D}_{\perp}B_1'(Y^{-1}-Y_{\infty}^{-1})Y \\ C_1Y \end{bmatrix}x_i \qquad i=1,2,\dots,n \qquad (3.12)$$

In the same way it is not difficult to verify that

$$[x_i'X \quad 0 \quad a_{i2}'] \left[\begin{array}{cc|cc} s_iI-A+B_2F & -B_2F & -X_\infty^{-1}C_1'D_\perp\gamma & B_1 \\ 0 & s_iI-A+HC_2 & 0 & B_1-HD_{21} \\ \hline -(C_1-D_{22}F) & -D_{12}F & \gamma D_\perp & 0 \end{array}\right] = [0 \quad 0 \quad b_{i1}' \quad b_{i2}'] \qquad i=n+1,\ldots,2n \tag{3.13}$$

in which

$$b_i'=[b_{i1}' \quad b_{i2}'] = x_i'[\ \gamma X(X^{-1}-X_\infty^{-1})C_1'D_\perp \qquad XB_1] \qquad i=n+1,\ldots,2n \tag{3.14}$$

In the last phase of our calculations we substitute the state-space expressions for the interpolation constraints into (3.5). We begin with the (1,1) block

$$\Pi_{11}(\gamma) = \left\{\frac{\gamma^2 a_i'a_k - b_i'b_k}{\bar{s}_i + s_k}\right\}_{k=1,\ldots,n}^{i=1,\ldots,n} \tag{3.15}$$

Substituting from (3.7) and (3.12) gives

$$\Pi_{11}(\gamma)=\gamma^2x_i'Y\Big\{[Y^{-1}B_1\tilde{D}_\perp'\tilde{D}_\perp-C_2'D_{21}][\tilde{D}_\perp'\tilde{D}_\perp B_1'Y^{-1}-D_{21}'C_2]-C_1'C_1\gamma^{-2}$$
$$-[Y^{-1}-Y_\infty^{-1}]B_1\tilde{D}_\perp'\tilde{D}_\perp B_1'[Y^{-1}-Y_\infty^{-1}]\Big\}Yx_k\Big/(\bar{s}_i-s_k) \tag{3.16}$$

Eliminating terms and substituting

$$Y_\infty^{-1}(A-B_1D_{21}'C_2)+(A-B_1D_{21}'C_2)'Y_\infty^{-1}-C_2'C_2+\gamma^{-2}C_1'C_1+Y_\infty^{-1}B_1\tilde{D}_\perp'\tilde{D}_\perp B_1'Y_\infty^{-1}=0 \tag{3.17}$$

yields:

$$\Pi_{11}(\gamma)=\gamma^2x_i'Y\Big\{[Y_\infty^{-1}[A-B_1D_{21}'C_2+B_1\tilde{D}_\perp'\tilde{D}_\perp B_1'Y^{-1}]+$$
$$[A-B_1D_{21}'C_2+B_1\tilde{D}_\perp'\tilde{D}_\perp B_1'Y^{-1}]'Y_\infty^{-1}\Big\}Yx_k\Big/(\bar{s}_i-s_k) \tag{3.18}$$

Substituting from (2.2b) and (2.4b) allows us to write

$$\Pi_{11}(\gamma)=-\gamma^2x_i'Y\Big\{[Y_\infty^{-1}Y[A-HC_2]'Y^{-1}+Y^{-1}[A-HC_2]YY_\infty^{-1}\Big\}Yx_k\Big/(\bar{s}_i-s_k) \tag{3.19}$$

Finally we invoke (3.8) to obtain

$$\Pi_{11}(\gamma)=\left\{\gamma^2 x_i' Y Y_\infty^{-1} Y x_k\right\}_{k=1,\dots,n}^{i=1,\dots,n} \tag{3.20}$$

In the same way we get

$$\Pi_{12}(\gamma)=\left\{\gamma x_i' Y X x_k\right\}_{k=n+1,\dots,2n}^{i=1,\dots,n} \tag{3.21}$$

By a symmetrical set of calculations one may establish that

$$\Pi_{21}(\gamma)=\left\{\gamma x_i' X Y x_k\right\}_{k=1,\dots,n}^{i=n+1,\dots,2n} \tag{3.22}$$

and

$$\Pi_{22}(\gamma)=\left\{\gamma^2 x_i' X X_\infty^{-1} X x_k\right\}_{k=n+1,\dots,2n}^{i=n+1,\dots,2n} \tag{3.23}$$

Since the $x_i$'s are linearly independent, it follows from Sylvester's inertia theorem that the inertia of the Pick matrix is the same as that given by $\Pi(\gamma)$ in (0.1). □

## 4. References

1] J.A. Ball, "Interpolation problems of the Pick–Nevanlinna and Loewner types for meromorphic matrix functions" Integral equations and operator theory, vol 6, pp 804–840, 1983.

2] J.A. Ball and J.W. Helton, "Interpolation problems of the Pick–Nevanlinna and Loewner types for meromorphic matrix functions: parameterization of all the solutions" Integral equations and operator theory, vol 9, pp 155–203, 1986.

3] J.C. Doyle et al, "Lecture notes in advances in multivariable control" ONR/Honeywell Workshop, Minneapolis 1984.

4] J.C. Doyle, K. Glover, P. Khargonekar and B. Francis, "State–space solutions to standard $H_2$ and $H_\infty$ control problems" IEEE American control conference, Atlanta, June 1988; an extended version is to appear in IEEE Trans. Auto. Control.

5] C. Foias and A. Tannenbaum, "On the four block problem I", Operator theory: Adv and Appl., Vol. 32, pp 93-122, 1988 and "On the four block problem II: The singular system", Integral equations and operator theory, Vol. 11, pp 726-767, 1988.

6] K. Glover, D.J.N. Limebeer, J.C. Doyle, E.M. Kasenally and M.G. Safonov,"A characterization of all solutions to the four block general distance problem", Submitted for publication, 1988.

7] D.J.N Limebeer and B.D.O. Anderson, "An interpolation theory approach to $H^\infty$ controller degree bounds" Linear algebra and its application, Vol. 98, pp347–386, 1988.

8] D.J.N. Limebeer and E.M. Kasenally,"Closed formulae for a parametric mixed sensitivity problem", Systems and Control Letters, 12, pp 1-7, 1989.

9] D.J.N. Limebeer and E.M. Kasenally,"Vector interpolation: Applications to $H^\infty$ control", IMA conference on matrix theory, Bradford, 1988.

10] D.J.N. Limebeer, E.M. Kasenally, E. Jaimouka and M.G. Safonov,"A characterization of all solutions to the four block general distance problem" Proc, IEEE CDC, Austin, 1988.

# On $\mathcal{H}_\infty$ Control, LQG Control and Minimum Entropy

Denis Mustafa

It is shown that the usual $\mathcal{H}_\infty$-optimal and LQG control problems are limiting cases of the minimum entropy/$\mathcal{H}_\infty$ control problem. It is explained how, in general, the minimum entropy/$\mathcal{H}_\infty$ problem may be seen as a link between the $\mathcal{H}_\infty$ and LQG problems. The results are illustrated using a particular normalized problem, for which a numerical example is given.

## 1 Introduction

In this paper we discuss an $\mathcal{H}_\infty$ control problem with a *minimum entropy* criterion, which contains both the Linear Quadratic Gaussian and $\mathcal{H}_\infty$-optimal control problems as limiting cases. This is important and interesting, because both the LQG and $\mathcal{H}_\infty$-optimal problems are well-established and have attracted considerable attention in their own right; details may be found in [10] for the LQG approach and [5] for the $\mathcal{H}_\infty$-optimal approach. Briefly, what LQG guarantees is system stability and good performance in the face of stochastic disturbance signals and noise. This is achieved by minimizing a quadratic cost criterion subject to a closed-loop stability constraint. Stability and good performance are also provided by the $\mathcal{H}_\infty$ approach, where the $\mathcal{H}_\infty$-norm of a closed-loop transfer function is minimized subject to a closed-loop stability constraint. However, unlike the LQG approach, the $\mathcal{H}_\infty$ approach guarantees certain robust stability properties.

We will show that the minimum entropy/$\mathcal{H}_\infty$ problem provides a link between the $\mathcal{H}_\infty$ and LQG problems. We will discuss this link and explain the tradeoff between the $\mathcal{H}_\infty$ and LQG criteria. Essentially, less emphasis on $\mathcal{H}_\infty$ performance and/or robust stability implies improved LQG performance.

A more detailed treatment of the minimum entropy approach may be found in [13,8,11]. There are a number of related problems. For example, working from a different viewpoint, in [1] a particular combined $\mathcal{H}_\infty$/LQG problem is considered, which turns out to be one of entropy minimization. This link is described in [12]. Also, [7] establishes equivalence with the risk-sensitive LQG problem [15], in which an exponential-of-quadratic cost criterion is minimized.

The layout of the paper is as follows. In the next section the $\mathcal{H}_\infty$, LQG and minimum entropy/$\mathcal{H}_\infty$ problems are formulated in a consistent framework. Then, by summarizing some properties of entropy, it is made clear in what sense we can interpret the minimum entropy/$\mathcal{H}_\infty$ problem as a combination of $\mathcal{H}_\infty$ and LQG objectives. In doing so, we explain the inherent tradeoff between $\mathcal{H}_\infty$ and LQG which is implied by the minimum entropy criterion. To illustrate the results, and to highlight some general properties of minimum entropy controllers, in the final section we give the solutions to the $\mathcal{H}_\infty$, LQG and minimum entropy/$\mathcal{H}_\infty$ problems for a particular configuration. The configuration chosen corresponds to the 'normalized' LQG problem as in [9]. The results are particularly transparent in this case and a numerical example is given.

**Notation.** In this paper all systems are linear and all transfer functions are real-rational. The right-half complex plane is written $C_+$, the real numbers are denoted by $\mathbb{R}$. The usual Hardy spaces $\mathcal{H}_\infty$ and $\mathcal{H}_2$, defined on $C_+$, are used, with norms

$$\|H\|_\infty := \sup_{\omega \in \mathbb{R}} \{\sigma_{max}(H(j\omega))\}$$

and

$$\|H\|_2 := \left\{ \frac{1}{2\pi} \int_{-\infty}^{\infty} \text{trace}[H(j\omega)H^*(j\omega)]dw \right\}^{1/2},$$

respectively. Here $\sigma_{max}(\,\cdot\,)$ denotes maximum singular value and $H^*(s) := H^T(-s)$. The symbol $\mathbf{E}$ is used to denote expectation. For a matrix $Q = Q^T$, the notation $Q > 0$ (*respectively* $Q \geq 0$) means that $Q$ is positive definite (*resp.* positive semi-definite). Spectral radius is denoted by $\rho(\,\cdot\,)$. The Laplace transform variable $s$ and the time variable $t$ will usually be omitted if no confusion can occur. The subscripts 'ME' and 'LQG' denote 'minimum entropy' and 'LQG-optimal', respectively. A (lower) linear fractional map of appropriately partitioned transfer function matrices $P$ and $K$ is defined by $\mathcal{F}(P,K) := P_{11} + P_{12}K(I - P_{22}K)^{-1}P_{21}$. Finally, a state space realization $G = D + C(sI - A)^{-1}B$ of a transfer function matrix will be written

$$G = (A, B, C, D) \qquad \text{or} \qquad G = \left[\begin{array}{c|c} A & B \\ \hline C & D \end{array}\right].$$

# 2 Problem Formulation and Connections

It is the purpose of this section to state the problems of interest in a consistent way, to describe their connections, and to examine the implied tradeoff. The framework used is standard, and is convenient because a large number of sensible problems may be rearranged to fit it. Consider an $n$-state plant $P$ with state-space description

$$\begin{aligned} \dot{x} &= Ax + B_1 w + B_2 u \\ z &= C_1 x + D_{12} u \\ y &= C_2 x + D_{21} w \end{aligned} \tag{1}$$

where $w \in \mathbb{R}^{m_1}$ is the disturbance vector; $u \in \mathbb{R}^{m_2}$ is the control input vector; $z \in \mathbb{R}^{p_1}$ is the error vector; $y \in \mathbb{R}^{p_2}$ is the measurement vector and $x \in \mathbb{R}^n$ is the state-vector. Here, as with other $\mathcal{H}_\infty$ problems, we assume that $m_1 \geq p_2$, $p_1 \geq m_2$. The transfer function $P$ from $[w^T\ u^T]^T$ to $[z^T\ y^T]^T$ is given by

$$P := \begin{bmatrix} P_{11} & P_{12} \\ P_{21} & P_{22} \end{bmatrix} = \left[\begin{array}{c|cc} A & B_1 & B_2 \\ \hline C_1 & 0 & D_{12} \\ C_2 & D_{21} & 0 \end{array}\right] \tag{2}$$

Note that $P$ is often called the 'standard plant' to distinguish it from the actual plant, which is embedded in $P$ together with the weighting functions and interconnections appropriate for the problem in hand.

Let a feedback controller $K = (\hat{A}, \hat{B}, \hat{C}, 0)$ be connected from $y$ to $u$. Then the overall closed-loop transfer function $H$ is just $H = \mathcal{F}(P, K)$; see Figure 1 for a block-diagram.

A basic aim of any control problem is to stabilize the system. We assume that $(A, B_2, C_2)$ is both stabilizable and detectable; this guarantees [5] that the set of controllers which (internally) stabilize $P$ is non-empty. Now suppose we specify that $\|H\|_\infty < \gamma$, where $\gamma \in \mathbb{R}$ is given. It is standard that this implies that the $\mathcal{H}_2$-gain from $w$ to $z$ is strictly less than $\gamma$. Here the disturbance $w$ is taken to be in $\mathcal{H}_2$ i.e., of bounded energy. Furthermore, consider a stable perturbation matrix $\Delta$ connected from $z$ to $w$ as shown in Figure 2. Then by the Small Gain Theorem [18], this closed-loop is stable for all such perturbations satisfying $\|\Delta\|_\infty \leq \gamma^{-1}$. Thus, specifying $\|H\|_\infty < \gamma$ leads to both performance and robust stability guarantees.

Although this is a reasonable problem, the solution is non-unique. To deal with this in a sensible way, a minimum entropy criterion has been developed in [13,8,11]. The *entropy* of $H$, where $\|H\|_\infty < \gamma$, is defined by

$$I(H,\gamma) := \lim_{s_0 \to \infty} \left\{ -\frac{\gamma^2}{2\pi} \int_{-\infty}^{\infty} \ln |\det(I - \gamma^{-2} H(j\omega) H^*(j\omega))| \left[ \frac{s_0}{|s_0 - j\omega|} \right]^2 d\omega \right\}.$$

It is easy to see that the entropy is well-defined and non-negative; the standing assumption that $P_{11}$ is strictly proper (see (1) and (2)) is enough to guarantee that the minimum value of the entropy is finite. We can now state the minimum entropy/$\mathcal{H}_\infty$ ('ME/$\mathcal{H}_\infty$') problem:

**Problem 2.1 (The ME/$\mathcal{H}_\infty$ Problem)** *Minimize $I(H,\gamma)$ over all stabilized closed-loops $H$ which satisfy $\|H\|_\infty < \gamma$.*

This problem was solved in [13] and [8], by exploiting recent state-space formulae of [7]. In particular it was shown in [8] that the minimum entropy controller

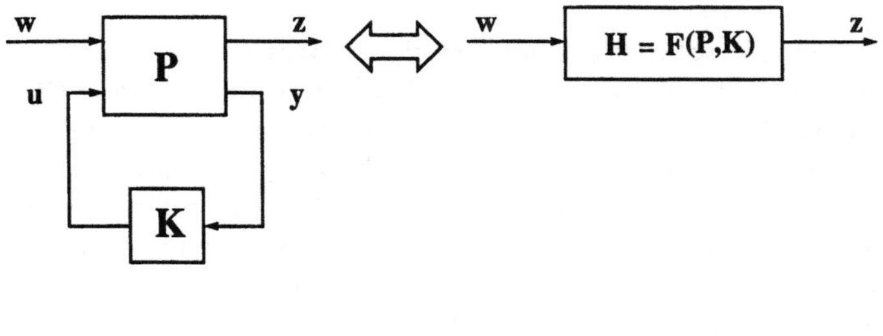


Figure 1: The closed-loop system

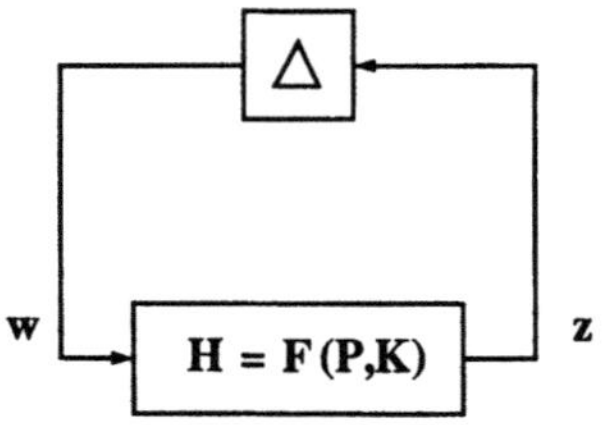


Figure 2: Block diagram for robust stability analysis

is just the central controller from the parametrization in [7] of all stabilizing controllers which satisfy the closed-loop $\mathcal{H}_\infty$-norm bound.

Denote by $\gamma_o$ the infimum of those $\gamma$ for which $\|H\|_\infty < \gamma$ where $H$ is a stabilized closed-loop. Finding this $\gamma_o$ and the controller which achieves it is just the $\mathcal{H}_\infty$-optimal control problem:

**Problem 2.2 (The $\mathcal{H}_\infty$-optimal Problem)** *Minimize $\|H\|_\infty$ over all stabilized closed-loops $H$.*

Clearly, the $\mathcal{H}_\infty$-optimal problem is just the limit of the ME/$\mathcal{H}_\infty$ problem as $\gamma \to \gamma_o$. Finding $\gamma_o$ usually requires an iterative search, ('$\gamma$-iteration,' see, e.g., [5]) using e.g., the state-space formulae of [7]. Note that the $\mathcal{H}_\infty$-optimal solution allows the *largest* permissible perturbation $\Delta$ to be tolerated in the system of Figure 2.

The final problem of interest here is the LQG problem associated with the system in (1). Define the LQG *quadratic cost* by

$$C(H) := \lim_{t_f \to \infty} \mathbf{E}\left\{\frac{1}{t_f}\int_0^{t_f} z^T(t)z(t)\,dt\right\}. \tag{3}$$

Of course, the disturbance signal $w$ is now interpreted as normalized Gaussian white noise. The LQG problem is then to stabilize the system from $u$ to $y$ and minimize $C(H)$:

**Problem 2.3 (The LQG Problem)** *Minimize $C(H)$ over all stabilized closed-loops $H$.*

It is a fact that the LQG problem may be obtained from the ME/$\mathcal{H}_\infty$ problem by relaxing the $\mathcal{H}_\infty$ constraint completely i.e., by letting $\gamma \to \infty$. A proof is based the following result [13]; that

$$I(H,\infty) := \lim_{\gamma \to \infty}\{I(H,\gamma)\} = C(H).$$

So minimizing $I(H,\infty)$ over all stabilized closed-loops $H$ is equivalent to minimizing $C(H)$ over all stabilized closed-loops $H$, which is precisely the LQG problem.

Having seen that the ME/$\mathcal{H}_\infty$ problem becomes the $\mathcal{H}_\infty$-optimal problem when $\gamma \to \gamma_o$, and becomes the LQG problem when $\gamma \to \infty$, let us now turn to the intermediate case when $\gamma_o < \gamma < \infty$. Then, of course, it is guaranteed that $\|H\|_\infty < \gamma$; but not only that, the entropy gives a guaranteed upper bound on $C(H)$: from [13], the entropy satisfies $I(H,\gamma) \geq C(H)$. The following bounds on the $\mathcal{H}_\infty$-norm and the LQG cost of the solution to the ME/$\mathcal{H}_\infty$ problem are then evident:

**Proposition 2.4 (Guaranteed Bounds from the ME/$\mathcal{H}_\infty$ Problem)** *The minimum entropy closed-loop, $H_{ME}$, satisfies*

$$\|H_{ME}\|_\infty < \gamma$$
$$C(H_{ME}) \leq I(H_{ME}, \gamma).$$

It is these bounds which enable us to interpret the ME/$\mathcal{H}_\infty$ problem as a combination of $\mathcal{H}_\infty$ and LQG criteria. Now, if $\gamma_u > \gamma_l$ then (with obvious notation) it can be shown [11] that $I(H_{ME_l}, \gamma_l) \geq I(H_{ME_l}, \gamma_u)$. But $I(H_{ME_l}, \gamma_u) \geq I(H_{ME_u}, \gamma_u)$, because the set of stabilized closed-loops satisfying $\|H\|_\infty < \gamma_l$ is a subset of those which satisfy $\|H\|_\infty < \gamma_u$. Therefore $I(H_{ME_l}, \gamma_l) \geq I(H_{ME_u}, \gamma_u)$ and the conclusion is:

**Proposition 2.5 (The $\mathcal{H}_\infty$/LQG Tradeoff.)** *The minimum value of the closed-loop entropy, $I(H_{ME}, \gamma)$, is a monotonically decreasing function of $\gamma$.*

This is a simple and direct tradeoff between the (upper bounds on the) $\mathcal{H}_\infty$ and LQG objectives. The $\mathcal{H}_\infty$ objectives reflect both robust stability and performance requirements, where noise is taken to be of bounded energy. The tradeoff is against the LQG measure of performance where noise is taken to be Gaussian and white. This tradeoff may also be found in the independent work of [1]. Although the tradeoff is in terms of the *upper bounds* on $\|H_{ME}\|_\infty$ and $C(H_{ME})$, we will see in the example in the next section that the same tradeoff may be exhibited by the quantities $\|H_{ME}\|_\infty$ and $C(H_{ME})$ themselves.

# 3 A Normalized $\mathcal{H}_\infty$ Problem

In this section we will consider a particular problem. We choose the 'normalized' LQG problem [9] and look at the ME/$\mathcal{H}_\infty$ and $\mathcal{H}_\infty$-optimal problems implied by it. We find that the implied $\mathcal{H}_\infty$ problem is sensible and the associated ME/$\mathcal{H}_\infty$ problem is useful for our purposes because the dependence of the solution on $\gamma$ is fairly clear, which aids the illustration of the results of the previous section.

The normalized LQG problem is based on a system

$$\dot{x} = Ax + Bw_1 + Bu$$
$$z_1 = Cx, \qquad z_2 = u$$
$$y = Cx + w_2,$$

where $w_1$ and $w_2$ are each normalized Gaussian white noise signals, and the given plant $G = (A, B, C, 0)$ is stabilizable and detectable. Putting $z := [z_1^T \; z_2^T]^T$ and $w := [w_1^T \; w_2^T]^T$, the LQG cost (3) is just

$$C(H) = \lim_{t_f \to \infty} \mathbf{E}\left\{\frac{1}{t_f}\int_0^{t_f} x^T(t)C^TCx(t) + u^T(t)u(t)\,dt\right\}.$$

We can see that this LQG problem has a standard plant (in the sense of Figure 1) given by

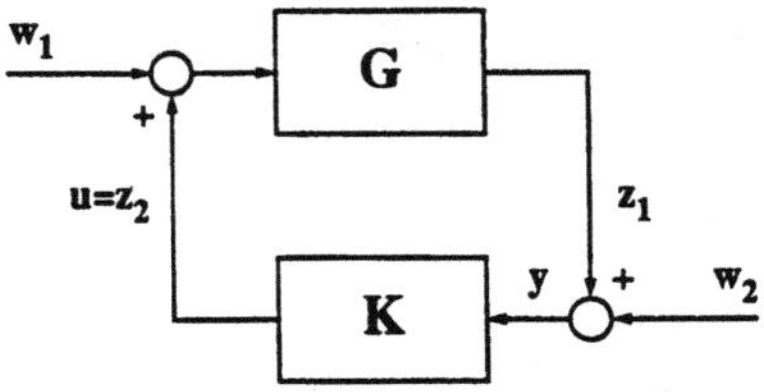


Figure 3: Block diagram for the normalized problem

$$P = \left[\begin{array}{c|cc} A & \begin{bmatrix} B & 0 \end{bmatrix} & \begin{bmatrix} B \end{bmatrix} \\ \hline \begin{bmatrix} C \\ 0 \\ C \end{bmatrix} & \begin{bmatrix} 0 & 0 \\ 0 & 0 \\ 0 & I \end{bmatrix} & \begin{bmatrix} 0 \\ I \\ 0 \end{bmatrix} \end{array}\right] \tag{4}$$

The solution to this LQG problem follows from standard results (see, e.g., [10]) and is stated immediately. For a more recent treatment, which develops the solution as an $\mathcal{H}_2$ problem in parallel with an $\mathcal{H}_\infty$ problem, see [3].

**Proposition 3.1 (Solution to the Normalized LQG Problem)** *The controller is*

$$K_{LQG} = \left[\begin{array}{c|c} A - Y_2C^TC - BB^TX_2 & Y_2C^T \\ \hline -B^TX_2 & 0 \end{array}\right] \tag{5}$$

*where $X_2 \geq 0$ and $Y_2 \geq 0$ are the stabilizing solutions to the algebraic Riccati equations*

$$0 = X_2A + A^TX_2 - X_2BB^TX_2 + C^TC \tag{6}$$

$$0 = Y_2A^T + AY_2 - Y_2C^TCY_2 + BB^T. \tag{7}$$

*The minimum value of $C(H)$ is*

$$C(H_{LQG}) = \text{trace}[X_2BB^T + X_2Y_2X_2BB^T]. \tag{8}$$

Note that the first and second terms in (8) are associated with the control and filtering, respectively, that is implicit in the LQG solution. Now let us examine the $\mathcal{H}_\infty$ problem implied by the standard plant $P$ given in (4) above. Some simple manipulations show that the closed-loop transfer function from $w$ to $z$ is

$$H = \begin{bmatrix} SG & SGK \\ KSG & KS \end{bmatrix}$$

where $S := (I - GK)^{-1}$ and $G = (A, B, C, 0)$. The block diagram of this system is given in Figure 3; we recognise this system to be the one used to analyse internal stability [14, p101]. Since, in addition, all four elements of $H$ are fairly familiar objective functions in $\mathcal{H}_\infty$ problems, this configuration is not unreasonable (although in general we would consider a frequency weighted version of this problem). Indeed, $SG$ corresponds to additive uncertainty on the controller, $SGK$ corresponds to output multiplicative uncertainty on the plant, $KSG$ corresponds to input multiplicative uncertainty on the plant and $KS$ corresponds to additive uncertainty on the plant. (See [4] for some details of various types of uncertainty.)

The solutions to both the ME/$\mathcal{H}_\infty$ and $\mathcal{H}_\infty$-optimal problems (using the state-space results of [7]), rely on the stabilizing solutions $X_\infty \geq 0$ and $Y_\infty \geq 0$, to two algebraic Riccati equations which in this case are

$$0 = X_\infty A + A^T X_\infty - (1-\gamma^{-2})X_\infty BB^T X_\infty + C^T C \tag{9}$$

$$0 = Y_\infty A^T + AY_\infty - (1-\gamma^{-2})Y_\infty C^T CY_\infty + BB^T. \tag{10}$$

We can now use the results of [8] to state the solution to the ME/$\mathcal{H}_\infty$ problem.

**Proposition 3.2 (Solution to the associated ME/$\mathcal{H}_\infty$ problem)** *The controller is*

$$K_{ME} = \left[\begin{array}{c|c} A-(1-\gamma^{-2})Y_\infty C^T C - BB^T X_\infty Z & Y_\infty C^T \\ \hline -B^T X_\infty Z & 0 \end{array}\right] \tag{11}$$

*where $X_\infty \geq 0$, $Y_\infty \geq 0$ are the stabilizing solutions to (9) and (10) and*

$$Z := (I - \gamma^{-2} Y_\infty X_\infty)^{-1}.$$

*The minimum value of the entropy is given by*

$$I(H_{ME},\gamma) = \text{trace}[X_\infty BB^T + X_\infty Z Y_\infty X_\infty BB^T]. \tag{12}$$

To solve the $\mathcal{H}_\infty$-*optimal* problem for this plant, we apply Theorem 1 of [7], which states that there exists a stabilizing controller such that $\|H\|_\infty < \gamma$, if and only if there exists $X_\infty \geq 0$ and $Y_\infty \geq 0$ as above, and $\rho(X_\infty Y_\infty) < \gamma^2$. Take the infimum over $\gamma$. In general a closed-form solution for $\gamma_o$ is not available, but $\gamma$-iteration can be used to isolate $\gamma_o$ to an arbitrary accuracy.

It is interesting to note that $\gamma_o < 1$ *only if the given plant $G = (A,B,C,0)$ is asymptotically stable.* Conversely, *if the given plant $G = (A,B,C,0)$ is not asymptotically stable, then $\gamma_o \geq 1$.* To prove this suppose $\gamma_o < \gamma \leq 1$. Then (10) can be written as a Lyapunov equation:

$$0 = Y_\infty A^T + AY_\infty + [\alpha Y_\infty C^T \ \ B][\alpha Y_\infty C^T \ \ B]^T,$$

where $\alpha^2 := \gamma^{-2} - 1 \geq 0$. The pair $(A, [\alpha Y_\infty C^T \ \ B])$ is stabilizable because, by assumption, $(A,B)$ is. This together with $Y_\infty \geq 0$ implies that $A$ is asymptotically stable, by a standard result on Lyapunov equations [17, Lemma 12.2].

Note that as $\gamma \to \gamma_o$, $Z$ tends to a singular matrix. A singular perturbation analysis of the controller in (11) yields the $\mathcal{H}_\infty$-optimal controller.

The guaranteed bounds provided by the ME/$\mathcal{H}_\infty$ problem are as follows:

**Proposition 3.3 (Guaranteed bounds from the ME/$\mathcal{H}_\infty$ problem)**
*The minimum entropy closed-loop, $H_{ME}$, satisfies*

$$\|H_{ME}\|_\infty < \gamma \tag{13}$$

$$C(H_{ME}) \leq \text{trace}[X_\infty BB^T + X_\infty Z Y_\infty X_\infty BB^T]. \tag{14}$$

To see the tradeoff between (13) and (14), we need to examine how $X_\infty$, $Y_\infty$ and $Z$ behave as $\gamma$ varies. Using [16] it can be shown that [6], as $\gamma$ increases monotonically to infinity, so $X_\infty$ decreases monotonically to $X_2$, $Y_\infty$ decreases monotonically to $Y_2$ and $Z$ decreases monotonically to $I$. The tradeoff of Proposition 2.5 between the $\mathcal{H}_\infty$-norm bound and LQG cost bound is then apparent: as the RHS of (13) increases (*resp.* decreases) so the RHS of (14) decreases (*resp.* increases). It is also clear that the LQG solution is obtained exactly when $\gamma \to \infty$: review (5)-(8) and (9)-(12) in that case.

## 3.1 A Numerical Example

Here we illustrate the main points with a simple numerical example. Take

$$G = \left[\begin{array}{c|c} A & B \\ \hline C & 0 \end{array}\right] = \left[\begin{array}{cc|c} 20 & -100 & 1 \\ 1 & 0 & 0 \\ \hline 1 & -0.1 & 0 \end{array}\right]$$

This has a $\mathcal{C}_+$-zero at 0.1 and two $\mathcal{C}_+$-poles at 10. Using $\gamma$-iteration gives $\gamma_o = 40.87$ (the minimum value of $\|H\|_\infty$ i.e., this corresponds to the $\mathcal{H}_\infty$-optimal solution). Then, solving the $X_2$ and $Y_2$ equations, we find from (8) that $C(H_{LQG}) = 64 \times 10^3$ (the minimum value of $C(H)$, i.e., this corresponds to the LQG solution).

For $\gamma_o < \gamma < \infty$, $X_\infty$ and $Y_\infty$ are found from (9) and (10), and the minimum entropy controller, $K_{ME}$, is found using (11). Then the minimum entropy closed-loop is $H_{ME} = \mathcal{F}(P, K_{ME}) := (\tilde{A}, \tilde{B}, \tilde{C}, 0)$, say. The minimum value of the entropy is evaluated using (12). To calculate $C(H_{ME})$, solve the Lyapunov equation for the closed-loop controllability Gramian $\mathcal{P}$; then $C(H_{ME}) = \mathrm{trace}[\mathcal{P}\tilde{C}^T\tilde{C}]$. Calculation of $\|H_{ME}\|_\infty$ can be performed to prespecified accuracy using the algorithm in [2]. These calculations were done for a number of values of $\gamma$, ranging over several orders of magnitude from close to $\gamma_o$. The results are illustrated in Figures 4-6.

Figure 4 is a plot of (13) as $\gamma$ varies i.e., of the upper bound $\gamma$ on $\|H_{ME}\|_\infty$, and the actual value of $\|H_{ME}\|_\infty$, against $\gamma$. Figure 5 is a plot of (14) as $\gamma$ varies i.e., of the upper bound $I(H_{ME}, \gamma)$ on $C(H_{ME})$, and the actual value of $C(H_{ME})$, against $\gamma$.

The graphs illustrate clearly the $\mathcal{H}_\infty$/LQG tradeoff: the upper curve in Figure 4 *increases* with increasing $\gamma$, whilst the upper curve in Figure 5 *decreases* with increasing $\gamma$. In fact, we notice that the tradeoff is even stronger than this—the *achieved* values (i.e., lower curves in Figures 4 and 5) exhibit the same behaviour as their upper bounds.

Notice that in Figure 4, as $\gamma$ becomes large, $\|H_{ME}\|_\infty$ tends to $\|H_{LQG}\|_\infty \approx 80$, a number slightly less than twice $\gamma_o$. The variation with $\gamma$ is more rapid in Figure 5. Both $C(H_{ME})$ and its upper bound $I(H_{ME}, \gamma)$ decrease quickly with $\gamma$ when $\gamma$ is close to $\gamma_o$. So large improvements in the LQG properties can be obtained with only modest degradation of $\mathcal{H}_\infty$ properties. Although theory predicts that $\gamma$ must be arbitrarily large before the ME/$\mathcal{H}_\infty$ problem reduces to the LQG problem, we see in Figure 5 that for $\gamma$ as small as twice $\gamma_o$, the LQG cost and its bound are very close to their minimum (LQG-optimal) values, and the upper bound provided by the entropy is quite tight.

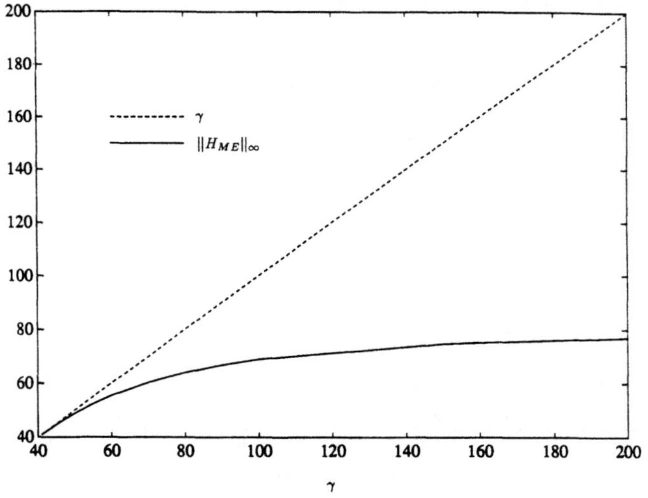


Figure 4: $\|H_{ME}\|_\infty$ and its upper bound $\gamma$, against $\gamma$

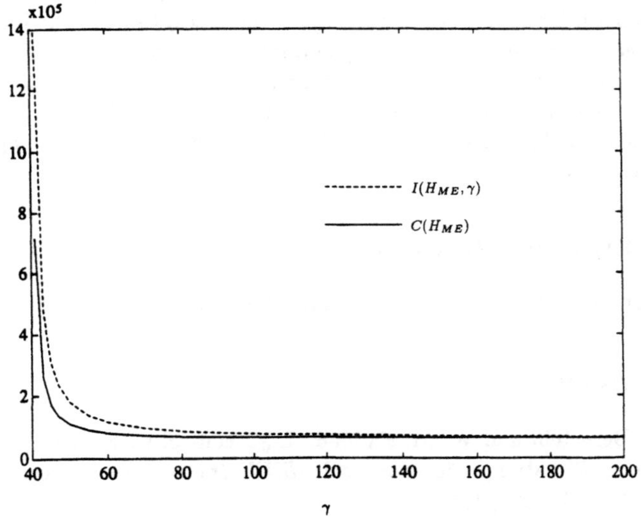


Figure 5: $C(H_{ME})$ and its upper bound $I(H_{ME},\gamma)$, against $\gamma$

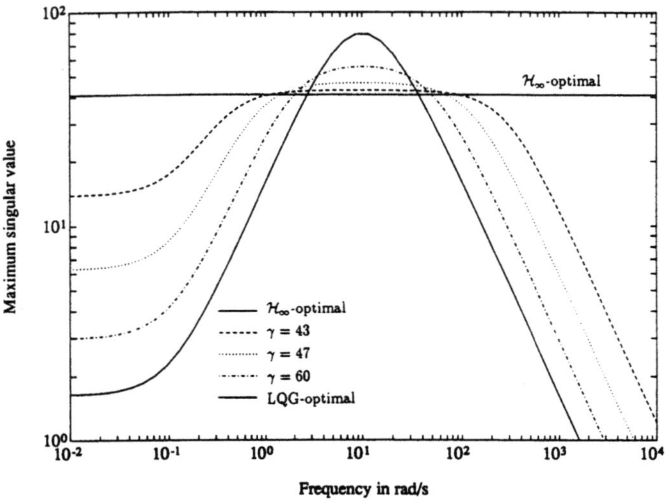


Figure 6: Maximum singular value of $H_{ME}$ against frequency, for various $\gamma$

The last graph, Figure 6, shows the maximum singular value of the minimum entropy closed-loop as a function of frequency. Five curves are plotted, corresponding to five representative values of $\gamma$: $\gamma = 40.87$ ($\mathcal{H}_\infty$-optimal), 43, 45, 60 and $\infty$ (LQG-optimal). The rapid convergence (as $\gamma$ increases) towards the LQG-optimal curve is clear, as is the fairly slow degradation of the peak around 10 $rad/s$.

We can conclude that allowing $\gamma$ to be approximately 60 ($\approx 1.5\gamma_o$) gives us nearly LQG-optimal performance without degrading the achieved $\mathcal{H}_\infty$-norm excessively. Examine Figures 4-6 for $\gamma = 60$ and compare with the $\mathcal{H}_\infty$-optimal and LQG-optimal cases.

# 4 Acknowledgement

The author would like to thank Dr K. Glover for useful discussions.

# References

[1] D. S. Bernstein and W. M. Haddad. LQG control with an $\mathcal{H}_\infty$ performance bound: a Riccati equation approach. *IEEE Transactions on Automatic Control*, 34(3):293–305, 1989.

[2] S. Boyd, V. Balakrishnan, and P. Kabamba. On computing the $\mathcal{H}_\infty$-norm of a transfer matrix. In *Proceedings of the American Control Conference*, Atlanta, GA, 1988.

[3] J. C. Doyle, K. Glover, P. P. Khargonekar, and B. A. Francis. State-space solutions to standard $\mathcal{H}_2$ and $\mathcal{H}_\infty$ control problems. In *Proceedings of the American Control Conference*, Atlanta, GA, 1988.

[4] J. C. Doyle, J. E. Wall, and G. Stein. Performance and robustness analysis for structured uncertainty. In *Proceedings of the Conference on Decision and Control*, Orlando, FL, 1982.

[5] B. A. Francis. *A Course in $\mathcal{H}_\infty$ Control Theory*. Volume 88 of *Lecture Notes in Control and Information Sciences*, Springer Verlag, 1987.

[6] K. Glover and J. C. Doyle. Report under preparation, 1988.

[7] K. Glover and J. C. Doyle. State-space formulae for all stabilizing controllers that satisfy an $\mathcal{H}_\infty$-norm bound and relations to risk sensitivity. *Systems and Control Letters*, 11:167–172, 1988.

[8] K. Glover and D. Mustafa. Derivation of the maximum entropy $\mathcal{H}_\infty$-controller and a state-space formula for its entropy. *International Journal of Control*, 50(3):899–916, 1989.

[9] E. A. Jonckheere and L. M. Silverman. A new set of invariants for linear systems—application to reduced order compensator design. *IEEE Transactions on Automatic Control*, 28(10):953–964, 1983.

[10] H. Kwakernaak and R. Sivan. *Linear Optimal Control Systems*. Wiley, 1972.

[11] D. Mustafa. *Minimum Entropy $\mathcal{H}_\infty$ Control*. PhD thesis, University of Cambridge, 1989.

[12] D. Mustafa. Relations between maximum entropy/$\mathcal{H}_\infty$ control and combined $\mathcal{H}_\infty$/LQG control. *Systems and Control Letters*, 12:193–203, 1989.

[13] D. Mustafa and K. Glover. Controllers which satisfy a closed-loop $\mathcal{H}_\infty$-norm bound and maximize an entropy integral. In *Proceedings of the Conference on Decision and Control*, Austin, TX, 1988.

[14] M. Vidyasagar. *Control Systems Synthesis: A Factorization Approach*. MIT Press, 1985.

[15] P. Whittle. Risk-sensitive Linear/Quadratic/Gaussian control. *Advances in Applied Probability*, 13:764–777, 1981.

[16] H. K. Wimmer. Monotonicity of maximal solutions of algebraic Riccati equations. *Systems and Control Letters*, 5:317–319, 1985.

[17] W. M. Wonham. *Linear Multivariable Control: A Geometric Approach*. Springer Verlag, 1979.

[18] G. Zames. On the input-output stability of time-varying nonlinear feedback systems, Part I. *IEEE Transactions on Automatic Control*, 11(2):228–238, 1966.

# Optimal $H_\infty$ - SISO-controllers with structural constraints

Seppo Pohjolainen

## Abstract

A modified Q-parameter method is used to improve the quality of SISO control under $H_\infty$- criterion. A necessary condition for the existence of optimal controllers under structural constraints will be given. An ascent algorithm to compute optimal controllers is presented.

## Introduction

A new field of control theory - called $H_\infty$-theory -has been developed during the 1980's. Modern $H_\infty$-theory provides rather sharp and useful results, such as parametrization of all stabilizing controllers, optimal $H_\infty$-controllers, and optimal robustness results [9] , [4] , [5] to name a few references. From the designer's point of view, one of the main advantages in the theory is that both the parameters and the structure of a controller may be varied. This provides an improvement to state space methods, where the dimension of a controller must first be fixed, after which the tuning may start. The new design freedom is useful, but one of its weak points is that the controller may turn out to be so complicated that it will be almost useless.

In this paper a different design method is proposed. We shall start with a simple stabilizing controller and demonstrate how to improve the quality of the closed loop system using $H_\infty$-theory, without losing sight on the controller's structure.The starting point of the method is Q-parametrization, and an obvious duality result which makes the process and the controller interchangeable in the theory. Once a stabilizing controller has been found, the existing robustness results can be used to estimate how much the controller may be varied without losing stability. Next an $H_\infty$ -criterion is set up to measure the quality of the control. A linearized version of the criterion is then derived, from which one easily sees how to change the controller within a specified class to improve the quality of the control. The problem of improving the quality of the control turns out to be a real minimax approximation problem, for which there exists a plenty of numerical methods. Finally a tuning method, based on the ascent algorithm, will be derived. The method may be used if the cost function attains its maximum value in a finite point set. Lack of passivity of a transfer function on the frequencies, where the cost function attains its maximum points, is shown to be a necessary condition for optimality.

A preliminary version of the paper - without proofs- has been published in [8] .The proposed method has been succesfully used to control heat exchangers [6] and flexible beams [7] .

## Notations and preliminaries

We shall assume that the **system** may be described by a real rational strictly proper transfer function P(s), and the **controller** C(s) as a proper real rational function function . A transfer function is called **stable** if it has all of its poles in the open left half complex plane. For stable transfer functions the $H_\infty$ norm is defined as

$$\| P \| = \sup_{\omega \in R} | P(i\omega) |$$

Denote the transfer function from the perturbation signal w to the output y (Fig.1) by Q(s).

$$Q(s)=\frac{P(s)}{1+C(s)P(s)}$$

The strict properness of P implies Q to be strictly proper. The definition of Q-parameter differs from that of [10] , where Q parameter was chosen to be transfer function from $y_{ref}$ to u. Our definition of Q-parameter can be recovered from this by interchanging the roles of the controller and process. This fact has useful consequences because it enables us to take over the necessary results from [10].

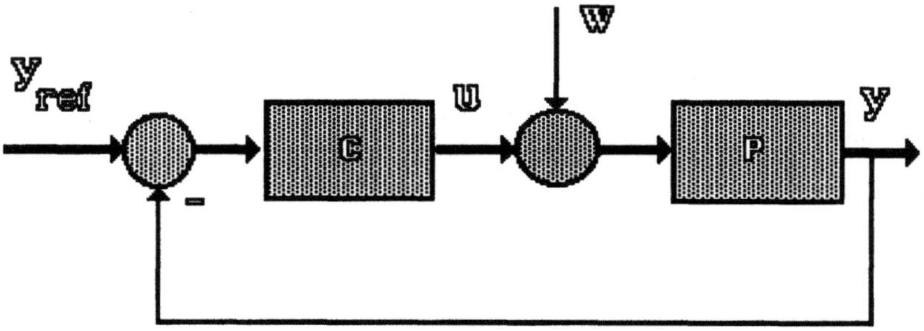


**Figure 1. The feedback system**

The overall transfer function is

$$\begin{pmatrix} y \\ u \end{pmatrix} = \begin{pmatrix} QC & Q \\ C(1-QC) & -CQ \end{pmatrix} \begin{pmatrix} y_{ref} \\ w \end{pmatrix}.$$

The above closed loop system is **stable** if all its component transfer functions are stable. This situation will be referred to by saying that C **stabilizes** P. It may be proved, as in [ 10 ], that the closed loop system is stable if and only if the Q-parameter is proper and stable and satisfies certain interpolation conditions on the unstable poles of C. We shall assume that a stabilizing controller C has been found and the problem is how to improve the quality of the control. The following theorem shows how much a given stabilizing controller may be varied without losing stability.

## Admissible controller variations

Assume that a stabilizing controller C has been found. If the controller is changed from C to C + ΔC , then the feedback system remains stable, if

(A1) C and C + ΔC have the same number of unstable poles

(A2) $\|Q\Delta C\| < 1$

The proof follows e.g. from [9 p.273-279] from the sufficiency part of the proof. If C has poles on the imaginary axis, a standard indentation technique may be used. However ,in this case we shall assume that (A2) holds on the imaginary axis. So if C has poles on the imaginary axis then C+ΔC must have the same imaginary axis poles.

## Cost function

In order to evaluate the quality of the control, a cost function will be defined as

$$J(\omega,C) = f(\omega)\,|\,1- Q(i\omega)C(i\omega)\,| + g(\omega)\,|\,Q(i\omega)\,|.$$

The component 1-QC measures the deviation between the desired output $y_{ref}$ and the measured output y. The second component measures the effect of the perturbation signal w on the output. The functions f and g are assumed to be nonnegative bounded continuous even functions in $\omega$ , and $\lim f(\omega) = 0$ as $\omega \to \infty$.

Let us denote the maximum value of the criterion as

$$J(C) = \max_{\omega} J(C,\omega) = \|J(\,.\,,C)\|.$$

The maximum exists, because all the functions are continuous , Q is strictly proper and C proper, so that $\lim J(C,\omega) = 0$ as $\omega \to \infty$. Further, let

$$Max=\{\,\omega\,|\,J(\omega,C)=J(C)\}$$

be the set in which the global maximums for $J(\omega,C)$ are obtained.

## Linearization of the cost function

**Theorem 1** . Let the controller C to be changed to C+ ΔC and assume $\|Q\Delta C\| < 1$ . Then the effect of the perturbation on the cost is bounded by

$$J(\omega,C+\Delta C) \le J(\omega,C)\,|\,1-Q(i\omega)\Delta C(i\omega)\,| + O(\|Q\Delta C\|^2),$$

where O(h) means M h.

**Proof**. If $\|Q\Delta C\| < 1$, then the perturbed Q-parameter may be written as

$$Q+\Delta Q = Q[1+Q\Delta C]^{-1} = Q- Q(Q\Delta C)[1+Q\Delta C]^{-1}$$

$$= Q-Q(Q\Delta C)+Q(Q\Delta C)^2[1+Q\Delta C]^{-1}.$$

$H_\infty$-bound for the second order term is

$$\|\,Q(Q\Delta C)^2[1+Q\Delta C]^{-1}\| \le \|Q\|\,\|Q\Delta C\|^2/(1-\|Q\Delta C\|\,) \le M\,\|Q\Delta C\|^2.$$

Because $g(\omega)$ is bounded, we have

$$g(\omega)\,|\,Q(i\omega)+\Delta Q(i\omega)\,| \le g(\omega)\,|\,Q(i\omega)\,|\;\;|\,1-Q(i\omega)\Delta C(i\omega)\,| + M\,\|Q\Delta C\|^2,$$

where M is a generic constant. An analogous reasoning with the first component of the criterion results in:

$$f(\omega)\,|\,1-(Q(i\omega)+\Delta Q(i\omega))(C(i\omega)+\Delta C(i\omega))\,|$$

$\leq f(\omega)\,|\,1\text{-}Q(i\omega)C(i\omega)\,|\ \ |\,1\text{-}Q(i\omega)\Delta C(i\omega)\,|\ + M\,\|Q\Delta C\|^2$

Adding up the components of the cost function finally gives the desired formula:

$$J(\omega,C+\Delta C) \leq J(\omega,C)\,|\,1-Q(i\omega)\Delta C(i\omega)\,| + O(\|Q\Delta C\|^2),$$

and completes the proof.

In what follows we shall assume that

(A3) $\quad J(C) > 0.$

for the obvious reason that if $J(C)=0$ then there is nothing to be done.

**Lemma 2.** If $J(C) > 0$, then Max is a compact set.

**Proof:** The closedness of Max follows because $J(\omega,C)$ is continuous in $\omega$. Max is bounded since $\lim J(\omega,C) = 0$, as $\omega \to \infty$.

The crucial step in improving the quality of the control is in finding a $\Delta C$ which satisfies the stability conditions (A1) and (A2) and, in addition

$$|\,1-Q(i\omega)\Delta C(i\omega)\,| < 1 \quad \text{for all } \omega \in \text{Max}, \tag{1}$$

for sufficiently small values of $\|Q\Delta C\|$. Next we shall find conditions, which guarantee (1) to hold

**Lemma 3.** If $|\,1-Q(i\omega)\Delta C(i\omega)\,| < 1$ for all $\omega \in$ Max, then $\text{Re}[Q(i\omega)\Delta C(i\omega)] > 0$ on the set Max.

**Proof:** $|\,1-Q(i\omega)\Delta C(i\omega)\,|^2 = 1-2\text{Re}[Q(i\omega)\Delta C(i\omega)] + |\,Q(i\omega)\Delta C(i\omega)\,|^2 < 1$

$$\Leftrightarrow |\,Q(i\omega)\Delta C(i\omega)\,|^2 < 2\,\text{Re}[Q(i\omega)\Delta C(i\omega)\,|.$$

Because $|\,Q(i\omega)\Delta C(i\omega)\,| > 0$, by assumption, the result follows.

A partial converse to the above lemma is given next.

**Lemma 4.** Let $\text{Re}[Q(i\omega)\Delta C(i\omega)] > 0$ on Max. Then for sufficiently small positive values of $\varepsilon$ $\ |\,1-\varepsilon\, Q(i\omega)\Delta C(i\omega)\,| < 1$ for all $\omega \in$ Max.

**Proof.** The transfer function $Q(i\omega)\Delta C(i\omega)$ is continuous on the imaginary axis. and so the functions $\text{Re}[Q(i\omega)\Delta C(i\omega)]$ and $|\,Q(i\omega)\Delta C(i\omega)\,|$ are continuous on the compact set Max. Thus there are positive constants k and K such that $\text{Re}[Q(i\omega)\Delta C(i\omega)] \geq k > 0$, and $|\,Q(i\omega)\Delta C(i\omega)\,| \leq K$ on Max. If $\varepsilon < k/(K^2)$, then

$$|\,1-\varepsilon\, Q(i\omega)\Delta C(i\omega)\,|^2 = 1-2\varepsilon\text{Re}[Q(i\omega)\Delta C(i\omega)] + \varepsilon^2\,|\,Q(i\omega)\Delta C(i\omega)\,|^2$$

$$< [1-(k/2)\,\varepsilon\,]^2, \text{ for all } \omega \text{ in Max.}$$

The next theorem gives the main result of the chapter by combining the previous results to prove that the improvement takes place on the whole imaginary axis instead of the subset Max.

**Theorem 5.** Select $\Delta C$ so that $\text{Re}[Q(i\omega)\Delta C(i\omega)] > 0$ for all $\omega \in$ Max. Then for sufficiently small positive values of $\varepsilon$

$$J(C+\varepsilon\,\Delta C) < J(C).$$

**Proof.** There are positive constants k, K such that $|\,Q(i\omega)\Delta C(i\omega)\,| \leq K$ and $\text{Re}[Q(i\omega)\Delta C(i\omega)] \geq k>0$ on Max. Because $\text{Re}[Q(i\omega)\Delta C(i\omega)]$ and $|\,Q(i\omega)\Delta C(i\omega)\,|$ are continuous, for every $\omega \in$ Max there is a open neighbourhood $U_r(\omega)$, such that

$Re[Q(i\omega)\Delta C(i\omega)] > k/2 > 0$ and $|Q(i\omega)\Delta C(i\omega)| < 2K$ on $U_r(\omega)$. Because Max is compact we may form a finite open subcover so that $Max \subset U = \cup U_{r_i}(\omega_i)$ . An analogous reasoning to Lemma 4 proves that

$$|1 - \varepsilon Q(i\omega)\Delta C(i\omega)| < (1 - \varepsilon k/4) ,$$

for all $\varepsilon$, $0 < \varepsilon < k/(8K^2)$ and for all $\omega \in U$. Hence

$$J(\omega,C)\,|1 - Q(i\omega)\Delta C(i\omega)| < (1-\varepsilon k/4)J(\omega,C) < (1-k^2/(32K^2))J(\omega,C) < J(C)$$

for all $\omega \in U$. Since $J(\omega,C) \to 0$, as $\omega \to \infty$ , the cost function $J(\omega,C)$ attains a maximum value on R-U.Let

$$\max_{\omega \in R-U.} |J(\omega,C)| = J_1 < J(C).$$

If $\varepsilon$ is small enough, then

$$J(\omega,C)\,|1 - \varepsilon\, Q(i\omega)\Delta C(i\omega)| < J_1\,(1 + \varepsilon\|Q\Delta C\|) < J(C),$$

for all $\omega \in$ R-U. The proof is complete, if we note that the second order term, $O(\|\varepsilon Q\Delta C\|^2)$ is uniform in $\omega$ and does not affect the reasoning provided that $\varepsilon$ is sufficiently small.

Because the conditions $|1 - \varepsilon Re[Q(i\omega)\Delta C(i\omega)]| < 1$ and $Re[Q(i\omega)\Delta C(i\omega)] > 0$ are equivalent for small values of $\varepsilon$ , we may replace the complex approximation problem by a real one: Find $\Delta C$ so that the stability condition (A1) is satisfied, and

$$|1 - \varepsilon\, Re[Q(i\omega)\Delta C(i\omega)| < 1 \text{ for all } \omega \in Max, \qquad (2)$$

or equivalently

$$Re[Q(i\omega)\Delta C(i\omega)] > 0 \qquad \text{for all } \omega \in Max. \qquad (3)$$

If $\varepsilon$ is small enough, then the selection $\varepsilon\Delta C$ is good for the original problem, too. The latter condition (3) means that the transfer function $Q(i\omega)\Delta C(i\omega)$ should be **passive** [2] on Max. The simple result (2), (3) above has useful consequences, because it enables us to apply real $L_\infty$-approximation methods in solving the original problem. Real linear minimax-problems may be solved in many ways, including the discrete ascent, descent and continuous Remez methods, [3] from which we start to elaborate with the discrete ascent method.

## Existence results

To simplify the problem a little bit further, we pose two simplifying assumptions. First that the controller parameters $\underline{k} \in R^n$ to be varied, will appear linearly :

(A4) $\qquad \Delta C(i\omega) = \underline{l}(i\omega)\underline{k},$

where the elements of $\underline{l}(i\omega) \in C^{1xn}$ are real proper rational functions in $\omega$, and second assumption states that the cost function attains its global maximum points in a finite set

(A5) $\qquad Max = \{\omega_1, \omega_2, \omega_3, \ldots\ldots, \omega_m\},$

where the number of global maximums may be different for different controllers.

In order to find better controller parameters $\underline{k}$ the condition (2) should be satisfied on the set Max. If we drop the $\varepsilon$ for a while, the problem can be written compactly as follows: Find the parameters $\underline{k}$ so that

$$\| A\underline{k} - \underline{1} \|_\infty < 1 \quad (4), \text{ or } \quad A\underline{k} > \underline{0}, \qquad (5)$$

where $A = (a_{hj})_{mxn} = ( \mathrm{Re}[Q(i\omega_h)I_j(i\omega_h)])_{mxn}$, and $\underline{1} = (1,1,\ldots 1)^T$, where $^T$ denotes the transpose of a vector, and

$$\|\underline{x}\|_\infty = \max |x_h|$$

is the maximum norm in $R^m$. The vector norm

$$\|\underline{x}\|_1 = \sum_h |x_h|$$

will also be used.

**Gordan's theorem** [1 p.50] states that either (5) or

$$\underline{p}^T A = \underline{0}^T, \qquad \underline{p} \geq 0, \underline{p} \neq \underline{0} \qquad (6)$$

is solvable.The condition (6) gives thus a necessary condition for the existence of optimal control parameters as the following theorem:

**Theorem 6.** Assume A1,...,A5 to be true. Then a necessary condition for a controller $C_o$ to be optimal is that there is a nonzero nonnegative vector $\underline{p}$ such that

$$\sum_{h=1}^{m} p_h \mathrm{Re}[Q_o(i\omega_h)I(i\omega_h)] = \underline{0}^T \qquad (7)$$

The criterion for optimality looks appealing, since it depends only on the system parameters. The weight functions in the cost function affect only the selection of the set Max of global maximum points.

## The Ascent algorithm

There are several methods available to solve the linear problem,

$$\min_{\underline{k}} \| A\underline{k} - \underline{b} \|_\infty . \qquad (8)$$

where $A \in R^{mxn}$, $\underline{b} \in R^m$ and $m > n$ .In this paper the ascent method [3] will be discussed. The presentation of [3] will be rewritten using matrix-vector notation to simplify some aspects in the algorithm The role of the important Haar condition can be clearly seen in this set-up.

Let us write the matrix A in the row vector form

$$A = [\underline{a}_1, \underline{a}_2, \ldots\ldots, \underline{a}_m]^T.$$

The components of the error vector $\underline{r}(\underline{k}) = A\underline{k} - \underline{b}$ are

$$r_h(\underline{k}) = \underline{a}_h^T \underline{k} - b_h \quad , h=1,2,..m$$

A well known result on minimax solutions [3 pp.35-36] states that every solution of the system (8) is the solution of an appropriate subsystem comprising of n+1 equations. This subsystem can be solved e.g. by the method of de La Vallée Poussin [3]. An important condition, which is needed to solve the minimax problem numerically effectively is the Haar condition:

**Definition:** The set of vectors $\{\underline{x}_1, \underline{x}_2, \ldots\ldots, \underline{x}_m\}$, m>n in $R^n$ is said to satisfy Haar condition if every selection of n vectors forms a linearly independent set.

(A6) The row vectors of matrix A are assumed to satisfy the Haar condition.

## The case m=n+1

**Theorem 7.** Let m=n+1. Necessary and sufficient conditions for the minimum of $\|A\underline{k} - \underline{b}\|$ are:

$$r_h(\underline{k}) = \gamma\sigma_h , \qquad h = 1,\dots,n+1 \tag{9}$$

$$\underline{0} \in H\{ \sigma_1 a_1^T, \sigma_2 a_2^T, \dots\dots\dots, \sigma_{n+1} a_{n+1}^T \}, \tag{10}$$

where $\sigma_h = \text{sgn}(r_h(\underline{k}))$ ( $\sigma_h > 0$ ,if $r_h(\underline{k})>0$, $\sigma_h< 0$ ,if $r_h(\underline{k})<0$ ,and $\sigma_h = 0$ ,if $r_h(\underline{k})=0$ ) and H{W} is the convex hull of the set W.

Proof. [3]

Let us find a $\underline{k}$ to satisfy the above conditions. Decompose the problem as

$$\begin{pmatrix} A_n \\ a_{n+1}^T \end{pmatrix} \underline{k} = \begin{pmatrix} \underline{b}_n \\ b_{n+1} \end{pmatrix} + \gamma \begin{pmatrix} \underline{\sigma}_n \\ \sigma_{n+1} \end{pmatrix} \tag{11}$$

Because the system $\{a_1^T, a_2^T, \dots\dots, a_{n+1}^T\}$ is linearly dependent, there is a nonzero vector $\underline{\lambda}^T = (\underline{\lambda}_n^T, \lambda_{n+1})^T$ such that $\underline{\lambda}^T A = \underline{0}^T$.The Haar condition implies that all the elements of $\underline{\lambda}$ are nonzero. If we select the vector $\underline{\sigma}$ so that $\sigma_i = \text{sgn}(\lambda_i)$ , then the condition (10) in Th.7 is fulfilled. Because of Haar condition $A_n$ is invertible and we may solve for the first n equations so that

$$\underline{k} = A_n^{-1}( \underline{b}_n + \gamma\underline{\sigma}_n ) .$$

Because $\lambda_{n+1}\, a_{n+1}^T = - \underline{\lambda}_n^T A_n$ the last equation in (11) may be written as

$$a_{n+1}^T \underline{k} = - \underline{\lambda}_n^T( \underline{b}_n + \gamma\underline{\sigma}_n )/ \lambda_{n+1} = b_{n+1} + \gamma\sigma_{n+1} .$$

This equation is fulfilled if we select $\gamma$ as

$$\gamma = -\underline{\lambda}^T\underline{b} /(\underline{\lambda}^T\underline{\sigma}) = - \underline{\lambda}^T\underline{b} /\|\underline{\lambda}\|_1 ,$$

and so the optimal solution is $(\underline{k}, \gamma)$.

## The case m> n+1

In this case the idea is to find the subsystem of n+1 equations, which gives the optimal solution. The subsystems will be selected by the ascent method so that the error $|\gamma|$ is strictly increasing until the algorithm terminates and the optimal solution has been found.

Let us assume the mxn matrix A to be so organized that we may start from the first n+1 equations. Let

$$\underline{\lambda}^{(1)T} = (\underline{\lambda}^{(1)}{}_{n+1}{}^T, \underline{0}^T) \in R^{n+1} \times R^{m-n-1}$$

be selected so that $\underline{\lambda}^{(1)\,T}A = 0$, and $\underline{\sigma}^{(1)}$ so that the elements $\sigma_h$ satisfy $\sigma_h = \mathrm{sgn}(\lambda_h)$. In the system

$$A\underline{k} = \underline{b} + \gamma\underline{\sigma}$$

the first n+1 equations may be solved as above, with

$$\gamma_1 = -\underline{\lambda}^{(1)\,T}\underline{b}/\|\underline{\lambda}^{(1)}\|_1 .$$

Let $(\underline{k}_1, \gamma_1)$ be the solution.The sign of $\gamma_1$ may be selected to be nonnegative by changing the sign of the vector $\underline{\lambda}^{(1)T}$, if necessary. The error in the remaining m-n-1 equations is

$$r_h(\underline{k}_1) = \underline{a}_h^T\underline{k}_1 - b_h \quad h=n+2,\ldots,m$$

Let us select the equation with the largest error in absolute value.For ease of notation assume that it is the equation n+2. Let $\sigma_{n+2}$ be the sign of $r_{n+2}(\underline{k}_1)$ so that $\sigma_{n+2}r_{n+2}(\underline{k}_1) = \gamma'$ If $\gamma' \le \gamma_1$ then we may stop, because the optimal solution has been found. Let $\gamma' > \gamma_1$. Select $\underline{\beta}^T = (\underline{\beta}_{n+1}{}^T, \sigma_{n+2}, \underline{0}^T)$ so that $\underline{\beta}^T A = \underline{0}^T$, and consider the vector $\underline{\lambda}^{(1)\,T} + \alpha\underline{\beta}^T$ .Then

$$(\underline{\lambda}^{(1)\,T} + \alpha\underline{\beta}^T)A = \sigma_{n+2}\underline{a}_{n+2}{}^T + \sum_{i=1}^{n+1}(\beta_i\sigma_i + \alpha\lambda_i\sigma_i)\sigma_i\,\underline{a}_i{}^T = 0$$

If $\alpha = \max\{-\beta_i\sigma_i/\lambda_i\sigma_i \mid i=1,\ldots,n+1\}$, then the Haar condition implies that exactly one of the coefficients will become zero, and rest of them will be positive. The condition (10) is fulfilled for a new set of basis vectors. Assume, for simplicity that the component n+1 will be first to become zero. Then the signs of the first n components of the vector

$$\underline{\lambda}^{(2)T} = \underline{\lambda}^{(1)T} + \alpha\underline{\beta}^T, \quad \alpha = -\beta_{n+1}/\lambda_{n+1}$$

$$= (\underline{\lambda}_n{}^{(2)T}, 0, \sigma_{n+2}, \underline{0}^T)$$

are retained, and thus the first n components in the vector $\underline{\sigma}^{(2)} = (\underline{\sigma}_n{}^T, 0, \sigma_{n+2}, \underline{0}^T)^T$ are the same as in $\underline{\sigma}^{(1)}$. Let $(\underline{k}_2, \gamma_2)$ be the solution in the new basis . Since

$$\underline{k}_1 = A_n{}^{-1}[\underline{b}_n + \gamma_1\underline{\sigma}_n], \text{ and } \underline{k}_2 = A_n{}^{-1}[\underline{b}_n + \gamma_2\underline{\sigma}_n] ,$$

we have

$$\underline{k}_1 - \underline{k}_2 = (\gamma_1 - \gamma_2)A_n{}^{-1}\underline{\sigma}_n ,$$

and

$$r_{n+2}(\underline{k}_1) = r_{n+2}(\underline{k}_2) + (\gamma_1 - \gamma_2)\,\underline{a}_{n+2}{}^T A_n{}^{-1}\underline{\sigma}_n.$$

Because $\underline{a}_{n+2}{}^T = -\underline{\lambda}_n{}^{(2)T}\underline{\sigma}_n/(\sigma_{n+2}) = -\|\underline{\lambda}_n{}^{(2)T}\|_1/(\sigma_{n+2})$ , then

$$\sigma_{n+2}r_{n+2}(\underline{k}_1) = \sigma_{n+2}r_{n+2}(\underline{k}_2) - (\gamma_1 - \gamma_2)\|\underline{\lambda}_n{}^{(2)T}\|_1 \text{ , i.e.}$$

$$\gamma' - \gamma_2 = -(\gamma_1 - \gamma_2)\|\underline{\lambda}_n{}^{(2)T}\|_1.$$

Because $\gamma > \gamma_1$ this results in the inequality

$$\gamma_1(1+\|\lambda_n^{(2)T}\|_1) < \gamma_2\,(1+\|\lambda_n^{(2)T}\|_1)$$

i.e. $\gamma_1 < \gamma_2$ .Note that this result guarantees $\gamma_2$ to be positive so that the algorithm may be continued further.

## The nonlinear ascent algorithm

The original nonlinear problem may now be numerically solved with the following algorithm:

1. Define process P(s), a stabilizing controller $C_0(i\omega)$, $\underline{I}(i\omega)$ and weight functions $f(\omega)$ and $g(\omega)$.

2. Set ind=0

3. Compute $Q_{ind}(i\omega)$, and $J(\omega,C_{ind})$

4. Max = $\{\omega \mid \omega$ is maximum point of $J(\omega,C_{ind})\}$

   It is useful to put some of the local maximum points in the set Max in addition to the global maximum points. Then the method can take into account those maximum points in which the value of the cost function is close to global maximum. Intuitively ,the numerical behaviour should become smoother.

5. Solve the linear problem

$$\min_{\underline{k}} \max_{h} \left| J(\omega_h,C_{ind})[1-\mathrm{Re}[Q(i\omega_h)\underline{I}(i\omega_h)]\,\underline{k} \right|$$

   with the ascent method.

6. Select $C_{ind+1}(i\omega) = C_{ind}(i\omega) + \varepsilon\,\underline{I}\,(i\omega)\underline{k}$, where $\varepsilon$ is the step length.

7. Set ind=ind+1 and go to 3 until termination criteria are satisfied.

## Conclusions

A modified Q-parameter method was used to improve the quality of the control under an $H_\infty$- criterion. Well-known robustness results were used to estimate how much a controller could be varied without losing closed loop stability. Then a linearized version of the criterion was derived and it was proved that the complex $H_\infty$ problem could be replaced by a real minimax problem, for which there exists plenty of numerical methods. As a by-product a necessary condition for the existence of an locally optimal $H_\infty$ -controller with structural constraints was given. A discrete ascent method was reviewed and used to solve the linearized problem. Finally a nonlinear ascent algorithm was presented to solve the original nonlinear $H_\infty$ -optimization problem. As the algorithm may stop to a local minimum, finding global minimums cannot be guaranteed at this stage.

## References


[1] Bazaraa M.S., Sfetty C.M.:"Nonlinear Programming: Theory and Algorithms", John Wiley & Sons, New York, 1979.

[2] Bode H.W. :"Network Analysis and Feedback Amplifier Design", D. Van Nostrand, Princeton New York, 1945.

[3] Cheney E. W. :"Introduction to Approximation Theory", Chelsea Publishing Company,New York, 1982.

[4] Francis B.: "A Course in $H_\infty$ Control Theory",Springer-Verlag, Berlin, 1986.

[5] Kimura H.:" Robust stabilizability for a class of transfer functions", IEEE Trans. Aut. Control, AC-29, No. 9, 1984, pp. 788-793.

[6] Lätti I., Pohjolainen S." How to tune a multivariable PI-controller for heat exchangers using $H_\infty$-methods", 5th IFAC Symposium on Control of Distributed Parameter Systems, Perpignan France, 26-29.6.1989.

[7] Piché R., Pohjolainen S. "The Q-parameter Method Applied to the Control of Flexible Beam" 5th IFAC Symposium on Control of Distributed Parameter Systems, Perpignan France, 26-29.6.1989.

[8] Pohjolainen S., Mäkelä T. "How to improve a given linear controller using $H_\infty$-methods", Proc. 12 IMACS World Congress on Scientific Computation, 1988, Paris, France,July 18-22.

[9] Vidyasagar M :"Control System Synthesis: A Factorization Approach", MIT Press, Cambridge, Massachusetts, 1985.

[10] Zames G., Francis B.A., "Feedback,minimax sensitivity, and optimal robustness," IEEE Trans.Aut.Control, vol. AC-28, No.5, May 1983, pp. 585-601.

# SUPER-OPTIMAL $H^\infty$ DESIGN

Ian Postlethwaite, Mi-Ching Tsai and Da-Wei Gu

## Abstract

In this paper, we examine the usefulness of super-optimal $H^\infty$ design in robust control. It is argued that the approach can lead to a more robust design than standard $H^\infty$ control, but that there is usually a cost to pay in complexity of controller.

We outline a recently developed state-space approach for solving 2-block problems, which enables realistic designs such as mixed sensitivity to be solved. This is then used to design the super-optimal controller for a simple example and the results are compared with those of a standard $H^\infty$ controller.

## 1 Introduction

In standard $H^\infty$ control only the maximum singular value ($H^\infty$ norm) of a cost function is minimized, whereas in super-optimal $H^\infty$ control all singular values are minimized. The purpose of this short paper is to examine the usefulness of this strengthened optimization in robust control.

State-space algorithms for super-optimal control are given in [7] [8] and [10] for "1-block" design problems, and a polynomial approach to the mixed sensitivity problem (a "2-block" problem) can be found in [6]. Recently, a state-space approach for the general 2-block problem was given in [4]. In this paper, we will review the latter and then use it to design a super-optimal controller for a simple example. The results of this are then compared with those for a standard $H^\infty$ optimal controller. It is argued that super-optimal $H^\infty$ control can lead to a more robust design than standard $H^\infty$ control albeit at the expense of a more complex (higher McMillan degree) controller.

## 2 Preliminaries

The standard $H^\infty$ design problem is equivalent to solving a model-matching problem (MMP) [1] [2]:

$$\inf_{\hat{Q}\in\Re\mathbf{H}^\infty} \|F-\hat{Q}\|_\infty \tag{1}$$

where $\hat{Q}$ may have zero rows and/or columns, and $F \in \Re\mathbf{H}_-^\infty$, *i.e.* it is anti-stable. When $F$ is a matrix there are in general many $\hat{Q}$'s satisfying (1). Thus for uniqueness Young [13] [14] proposed a stronger criterion, namely to minimize, in lexicographic order, the sequence

$$(s_1^\infty(F-\hat{Q}),\ s_2^\infty(F-\hat{Q}),\ \cdots) \text{ where } s_i^\infty\ (F-\hat{Q}) \triangleq \sup_\omega\ s_i\ [(F-\hat{Q})(j\omega)] \qquad (2)$$

and $s_i[A]$ denotes the *ith* largest singular value of $A$. The model-matching problem with this stronger minimization criterion is called the strengthened model-matching problem (SMMP), and its solution $Q_{sup}$ is called the super-optimal solution. We will call the optimal squence $(s_1^\infty(F-Q_{sup}),\ s_2^\infty(F-Q_{sup}),\ \cdots)$ the *s-numbers* of $F$. Note that for each $i$, $s_i[(F-Q_{sup})(j\omega)]$ is constant over all $\omega$.

To motivate super-optimal control we next give an interpretation of the s-numbers in terms of energy gains. Let the cost function corresponding to $F-Q$ have an input $d(t)$ and output $y(t)$, and let $d(t)$ be bounded in energy. Then

$$\sup_{d(t)\neq 0,\ d(t)\in\mathcal{B}} \|y(t)\|_2 = s_1^\infty(F-Q_{sup}) \qquad (3)$$

where

$$\mathcal{B} \triangleq \{d(t) : \|d(t)\|_2 \leq 1\}\ .$$

Now let $\hat{d}_1(s)$ be the Laplace transform of an input $d_1(t)$ which produces maximum energy in the output, then the direction (inner part) of $\hat{d}_1(s)$ is the direction (inner part) of a maximizing vector of the Hankel operator generated by $F$ [10] [5], and we have

$$\sup_{d(t)\neq 0,\ d(t)\in\mathcal{B}_1} \|y(t)\|_2 = s_2^\infty(F-Q_{sup}) \qquad (4)$$

where $\mathcal{B}_1$ is the subset of $\mathcal{B}$ whose elements have Laplace transforms which are pointwise orthogonal to $\hat{d}_1(s)$ for each $s$ on the $j\omega$-axis.

Similarly, let $\hat{d}_2(s)$ be the Laplace transform of an input $d_2(t)$ which produces the energy gain $s_2^\infty(F-Q_{sup})$. The inner part of $\hat{d}_2(s)$ is characterized by the inner part of a maximizing vector of a Hankel operator generated by a matrix which now is a function of $F$ but with dimensions each reduced by 1; details can be found in [10] [5]. We then have

$$\sup_{d(t)\neq 0,\ d(t)\in\mathcal{B}_2} \|y(t)\|_2 = s_3^\infty(F-Q_{sup}) \qquad (5)$$

where $\mathcal{B}_2$ is the subset of $\mathcal{B}_1$ whose elements have Laplace transforms which are pointwise orthogonal to $\hat{d}_2(s)$ for each $s$ on the $j\omega$-axis. Etc.

The s-numbers can therefore be interpreted as the largest energy gains from appropriately defined input spaces to the output. Intuitively then, if disturbance rejection is an objective reflected in the cost function it is better (more robust) to minimize all the singular values and not just the maximum.

Without loss of generality we only consider the variable $\hat{Q}$ in the 2-block form $\hat{Q} = \begin{bmatrix} Q \\ 0 \end{bmatrix}$ so that (1) can be rewritten as

$$\inf_{Q \in \Re \mathbf{H}^\infty} \left\| \begin{matrix} F_1 & - & Q \\ & F_2 & \end{matrix} \right\|_\infty . \tag{6}$$

We also assume the dimensions of $F_1$ and $F_2$ are $q_1 \times m$ and $q_2 \times m$ with $q_1 \geq m$ for compatibility with [10].

# 3 Solution to the 2-block standard MMP [1]

In general (6) has no closed-form solutions but the problem can be reduced to a succession of 1-block problems whose solutions approach as close as we like a solution. The method is called $\gamma$-iteration [1], and is based on the following theorem. (Note that $G^*(s)$ denotes $G(-s)^T$).

**Theorem 1** *[1] [12] For any positive number $\gamma > \|F_2\|_\infty$ and $Q \in R\mathbf{H}^\infty$, we have*

$$\left\| \begin{matrix} F_1 & - & Q \\ & F_2 & \end{matrix} \right\|_\infty \leq \gamma \textit{ if and only if } \| (F_1 - Q)\Phi_\gamma^{-1} \|_\infty \leq 1$$

*where $\Phi_\gamma(s)$ is a spectral factor of $(\gamma^2 I - F_2^* F_2)$.*

For a given $\gamma > \|F_2\|_\infty$, the value of $\min_Q \|(F_1 - Q)\Phi_\gamma^{-1}\|_\infty$ can be calculated and also the $\arg\min_Q \|(F_1 - Q)\Phi_\gamma^{-1}\|_\infty$ [9] [3]. Therefore we can try iteratively different values of $\gamma$ until the accuracy of $\gamma$ to the optimal value $\gamma_o$ is satisfactory, and hence obtain an approximation to the optimal $Q$ [1]. This is the so-called *$\gamma$-iteration* scheme.

# 4 Solution to the 2-block strengthened MMP [4]

Let

$$\gamma_{o,i} \triangleq \inf_Q s_i^\infty \left( \begin{bmatrix} F_1 - Q \\ F_2 \end{bmatrix} \right), \quad i = 1, ..., m$$

and make the following assumptions

- $\gamma_{o,i} > s_i^\infty(F_2), \; i = 1, ..., m.$
- $\gamma_{o,1} > \gamma_{o,2} > \cdots > \gamma_{o,m}.$

For the SMMP we will use a similar philosophy to that described in Section 3 for the standard problem. That is, we will work out successively the optimal singular values $\gamma_{o,i}$ by an iterative method and then synthesize the super-optimal solution $Q$. To do this, we need a criterion (analogous to Theorem 1) to judge whether or not we have reached (or are close enough to) the optimal singular values. Such a criterion is given by the following theorem.

**Theorem 2** *[4] For any $Q \in \Re\mathbf{H}^\infty$, let*

$$\gamma_{Q,i} \triangleq s_i^\infty\left(\begin{bmatrix} F_1 - Q \\ F_2 \end{bmatrix}\right), \quad i = 1, 2, \cdots, m.$$

*Then for any* m*-tuple* $\underline{\gamma} = (\gamma_1, \gamma_2, \cdots, \gamma_m)$ *where*

$$\gamma_1 \geq \gamma_2 \geq \cdots \geq \gamma_m \tag{7}$$

$$\gamma_i > s_i^\infty(F_2), \quad i = 1, \cdots, m \tag{8}$$

*and $Q \in \Re\mathbf{H}^\infty$, we have*

$$\gamma_{Q,i} \leq \gamma_i, \quad i = 1, \cdots, m \tag{9}$$

*if and only if there exists a square matrix $U(s)$ which is unitary for $s = j\omega$, $\omega \in [0, \infty]$, such that*

$$s_i^\infty\left((F_1 - Q)U\Phi_{\underline{\gamma}}^{-1}\right) \leq 1, \quad i = 1, \cdots, m \tag{10}$$

*where*

$$\Phi_{\underline{\gamma}}^*\Phi_{\underline{\gamma}} = \begin{bmatrix} \gamma_1^2 & & \\ & \ddots & \\ & & \gamma_m^2 \end{bmatrix} - U^*F_2^*F_2U \ . \tag{11}$$

For a given $\underline{\gamma}$, satisfying the assumptions of the theorem, we can solve $s_i^\infty\left[(F_1 - Q)U\Phi_{\underline{\gamma}}^{-1}\right]$ using a 1-block SMMP algorithm and hence condition (10) can be tested.

Comparing Theorem 2 with Theorem 1 one might initially be surprised to find the matrix $U(s)$ in the former. Its importance, however, is crucial to solving the SMMP. To see this suppose that in Theorem 2 we let $U(s)$ be the identity matrix, then even if assumption (8) holds the spectral factorization in (11) may still be unsolvable. $U(s)$ is required to rearrange the singular value structure of $F_2$ forcing the larger singular values up the diagonal and the smaller ones down. Since $U(j\omega)$ is square and unitary it represents an equivalence transformation of the SMMP. In our algorithm, we will need a stable $\Phi_{\underline{\gamma}}$ with stable inverse and a real-rational stable $U(s)$. For further details of the algorithm see [4] and [11].

# 5 Example

In this section, we assess the usefulness of the super-optimal approach by applying it to a simple example and making comparisons with a standard $H^\infty$ controller.

We will consider the design example given in [6] where the plant transfer function is

$$G(s) = \begin{bmatrix} \frac{1}{s^2} & \frac{1}{s+2} \\ 0 & \frac{1}{s+2} \end{bmatrix}$$

and the design objectives are : (a) disturbance attenuation on both channels up to 1 $rad/sec$; (b) rejection of constant disturbances; (c) compensator roll-off of 20 dB/decade starting at the lowest frequency without affecting (a).

To achieve specifications (b) and (c), we first form an augmented plant

$$G_a = GW_c = \begin{bmatrix} \frac{1}{s^2(s+1)} & \frac{1}{s(s+2)} \\ 0 & \frac{1}{s(s+2)} \end{bmatrix} \quad \text{where } W_c = \begin{bmatrix} \frac{1}{s+1} & 0 \\ 0 & \frac{1}{s} \end{bmatrix} .$$

We then consider the feedback configuration and design problem illustrated in Figure 1. Since the augmented plant is used the final optimal controller for $G$, denoted by $\hat{K}$, is implemented as $W_c K$ .

We will design a standard $H^\infty$ optimal controller $\hat{K}_\infty$ $(= W_c K_\infty)$, resulting from an equalizing solution (all singular values equal), and a super-optimal solution $\hat{K}_{sup}$ $(= W_c K_{sup})$ using the approach of Section 4. The same weights are used in each case. We choose $W_1 = I$ and $W_2 = 0.1\ I$ to trade-off the control signals and output errors. To satisfy the design goal (a), we choose

$$W_d = \begin{bmatrix} \frac{s^2+\sqrt{2}s+1}{s^2} & 0 \\ 0 & \frac{s+1}{s} \end{bmatrix}$$

which has the effect of moving the three undesired poles of $G_a$ at the origin to $\frac{1}{\sqrt{2}}(-1 \pm i)$ and $-1$. We choose $W_d$ so that its poles are included in the set of poles for $G_a$; for more details see [11].

The plant output responses for each of these controllers to unit step disturbances on each of the outputs (in turn) are shown in Figures 2 and 3. The plant input responses to the same inputs are shown in Figures 4 and 5. The performance of the super-optimal controller is better particularly in the cross channel coupling as might be expected from the discussion in section 2 and the assumption that $W_1 S W_d$ dominates the cost function at low frequencies while $W_2 K S W_d$ dominates at high frequencies.

As a further comparison the singular values of the loop gain are compared in Figure 6 for both $\hat{K}_\infty$ and $\hat{K}_{sup}$. $\hat{K}_{sup}$ is better in that at low frequencies it has higher gains for performance and at higher frequencies it has lower gains for robust stability.

Finally note that i) the McMillan degree of $K_{sup}$ $(\hat{K}_{sup})$ is 7 (8) and the McMillan degree of $K_\infty$ $(\hat{K}_\infty)$ is 4 (5), ii) $K_\infty$ has an unstable pole and $K_{sup}$ is stable. $K_{sup}$ will almost always be more complex than a standard equalizing $K_\infty$ and we conjecture that in general the McMillan degree bound of $K_{sup}$ (for the above problem with the special types of weights selected) is given by

$$n - m \leq deg(K_{sup}) \leq \sum_{i=1}^{m} (n - i)$$

where $n$ is the state dimension of $G_a$ and $G_a$ is an $m \times p$ strictly proper real-rational matrix with $p \geq m$.

## 6 Conclusions

The algorithm briefly outlined in Section 4 enables super-optimal controllers to be designed for realistic problems. It is argued that the super-optimal $H^{\infty}$ controller is better than a standard $H^{\infty}$ controller for some design objectives and this is illustrated in the example of Section 5. However, the super-optimal controller is generally significantly more complex than a standard $H^{\infty}$ controller suggesting that further studies are required to fully justify its use.

This work was supported in part by the UK Science and Engineering Research Council.

## References

[1] J. C. Doyle, Lecture Notes for ONR/Honeywell Workshop on advances in multivariable control, Minneapolis, MN (1984).

[2] B. A. Francis, *A Course in $H_{\infty}$ Control Theory* Springer-Verlag, (1987).

[3] K. Glover, "All optimal Hankel-norm approximations of linear multivariable system and their $L^{\infty}$-error bounds," *Int. J. Contr.* **39** (1984) 115-1193.

[4] D.-W. Gu, M. C. Tsai and I. Postlethwaite, "An algorithm for super-optimal $H^{\infty}$ design : the 2-block case," *Automatica* **25** (1989) 215-221.

[5] D.-W. Gu, M. C. Tsai and I. Postlethwaite, "A frame approach to the $H^{\infty}$ super-optimal solution," *Oxford University Report No. OUEL 1761/89.*

[6] H. Kwakernaak, "A polynomial approach to minimax frequency domain optimization of multivariable feedback systems,," *Int. J. Control* **44** (1986) 117-156.

[7] D. J. N. Limebeer, G. D. Halikias and K. Glover, "A state-space algorithm for the computation of superoptimal matrix interpolating functions," Proc. MTNS conference on linear circuits, systems and signal processing: Theory and applications. North-Holland, Amsterdam, (1988).

[8] I. Postlethwaite, M. C. Tsai and D.-W. Gu, "A state-space approach to discrete-time super-optimal $H^{\infty}$ control problems,," *Int. J. Control* **49** (1989) 247-268.

[9] M. G. Safonov and M. S. Verma, "Multivariable $L^{\infty}$ sensitivity optimization and Hankel approximation," *Proc. Amer. Contr. Conf.*, San Francisco, CA (1983).

[10] M. C. Tsai, D.-W. Gu and I. Postlethwaite, "A state-space approach to super-optimal $H^{\infty}$ control problems," *IEEE Trans. Automat. Contr.* **33** (1988) 833-843.

[11] M. C. Tsai, "Super-optimal control system design for multivariable plants," *D.Phil Thesis*, University of Oxford (1989).

[12] M. S. Verma and E.A. Jonckheere, "$L_{\infty}$ compensator with mixed-sensitivity as a broadband matching problem," *Syst. Contr. Lett.*, **4** (1984) 125.

[13] N. J. Young, "The Nevanlinna-pick problem for matrix-valued functions," *J. Operator Theory*, **15** (1986) 239-265.

[14] N. J. Young, "An algorithm for the super-optimal sensitivity minimising controller," *Proc. Workshop on New Perspectives in Industrial Control System Design using $H^{\infty}$ Methods*, Oxford, (1986).

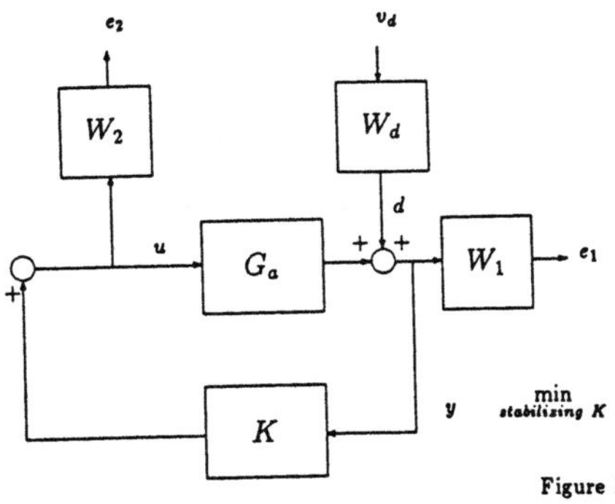


$$\min_{\text{stabilizing } K} \left\| \begin{matrix} W_1 S W_d \\ W_2 K S W_d \end{matrix} \right\|_\infty \qquad (S = (I - G_a K)^{-1})$$

**Figure 1: Disturbance Rejection Problem**

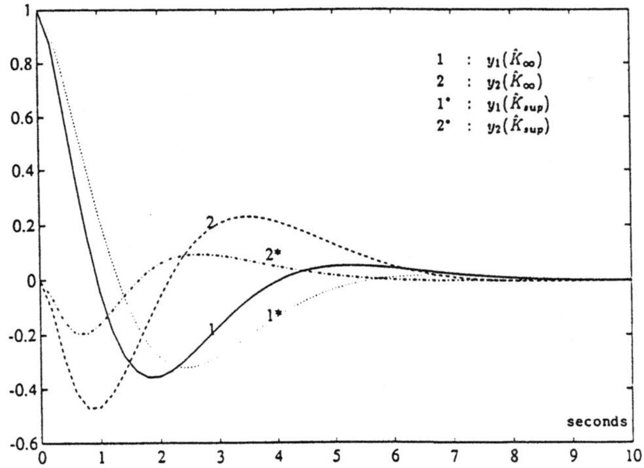


Figure 2: Plant output response to unit step disturbance in output channel 1

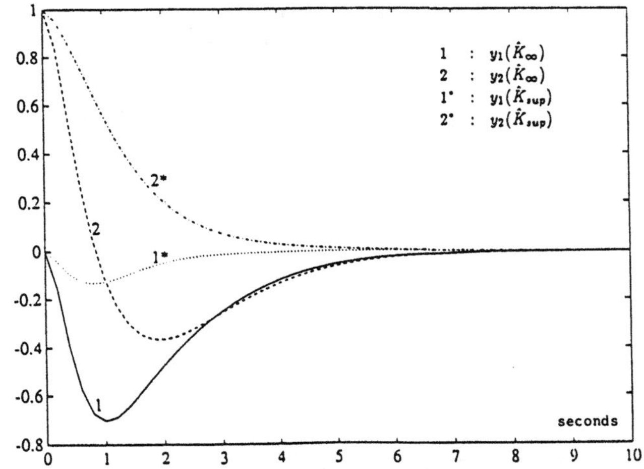


Figure 3: Plant output response to unit step disturbance in output channel 2

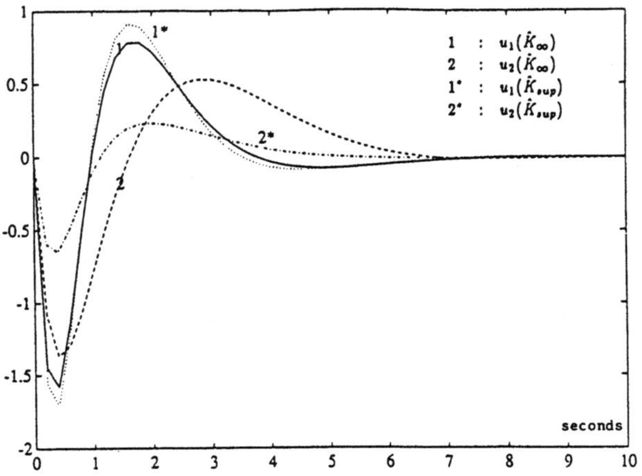


Figure 4: Plant input response to unit step disturbance in output channel 1

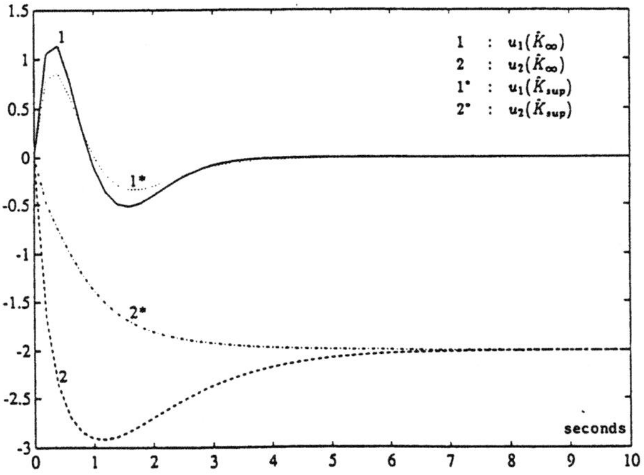


Figure 5: Plant input response to unit step disturbance in output channel 2

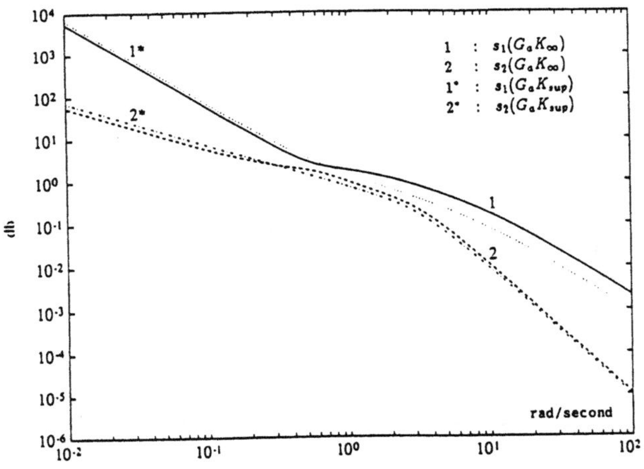


Figure 6: Singular values of the loop gain : $G_a$ K

## $H_\infty$ CONTROL WITH STATE FEEDBACK

*A.A. Stoorvogel*

**Abstract**

This paper considers $H^\infty$–control under the assumption that all the states are available for feedback. It can be shown that in that case we can restrain ourselves to static feedback. This paper gives extensions and a more intuitive explanation of recent results. Under a number of assumptions necessary and sufficient conditions are given for the existence of a stabilizing feedback such that the closed loop system has $H^\infty$ norm less than or equal to some predetermined bound $\gamma$. It is shown that if these assumptions are not met then by disturbing the system in such a way that these assumptions are satisfied we can find results about the existence of these desired feedbacks in more general cases.

## 1. INTRODUCTION

We consider the following linear time–invariant system :

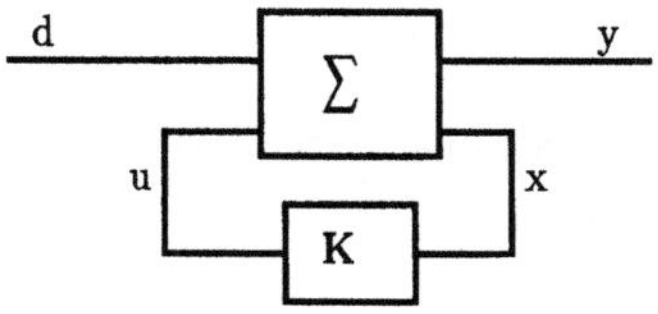


where $\Sigma$ is given by the equations:

$$\Sigma: \quad \begin{aligned} \dot{x} &= Ax + B_1 d + B_2 u \\ y &= Cx + D_1 d + D_2 u \end{aligned} \tag{1.1}$$

The general purpose of $H_\infty$ control is to find a feedback law which stabilizes the system and makes the $H_\infty$ norm as small as possible. We will limit ourselves to the case that all the states are measured. Recently it has been shown that in that case we can restrain ourselves to static feedback.(see e.g. [3]) In this paper we will only require that we can get the $H_\infty$ norm less or equal to some a priori given value $\gamma$, i.e. we would like to find a matrix K such that the state feedback u = Kx internally stabilizes the system and makes the $H_\infty$–norm of the closed loop system smaller than or equal to a <u>predescribed</u>

value $\gamma$. This problem has received a lot of attention ( [2],[3],[4] and [7] ). In these references it is required that the norm becomes *strictly* smaller than $\gamma$. For example in [4] it is shown that, for a special case, the existence of K is equivalent with the existence of an $\varepsilon > 0$ for which a certain algebraic Riccati equation has a solution. It was also shown that, if K exists, a matrix K with the desired properties can be found in terms of the solution of this Riccati equation. Later these results were extended in [2],[3] and [7] to more general cases. The problem however is to test if the Riccati equation has a solution for some unknown $\varepsilon$. In principle one would have to perform such a check for infinitely many values of $\varepsilon$. Recently another paper appeared [1]. Under a number of assumptions this paper shows that you can get rid of the $\varepsilon$ in the Riccati equation. In chapter 3 we also will be looking for cases where we don't need this $\varepsilon$ in the Riccati equation. We will see that for a large class of systems the existence of a stabilizing positive semi-definite solution of an algebraic Riccati equation is a necessary and sufficient condition for the existence of the desired K. We can also find such a K in terms of the solution of this ARE. In our opinion, another unsatisfactory point in the above-mentioned references was that they don't give any intuitive reasoning for the introduction of this $\varepsilon$. In chapter 4 we will, for cases for which we have to introduce an $\varepsilon$, give a much more intuitive introduction of $\varepsilon$. We disturb the system a little in such a way that we can use our results from chapter 3. If this disturbance satisfies some prerequisites then the $H_\infty$ norm depends continuously on the disturbance and hence we can get the norm below our bound by choosing a sufficiently small disturbance. It turns out that the results of [2,3,4] can be reobtained by choosing a specific disturbance.

## 2. PRELIMINARY RESULTS

Let $G(s) = C(sI-A)^{-1}B+D$. Our standing assumption in this chapter is that we have an arbitrary $\gamma$ such that $\gamma^2I-D^TD > 0$. Define

$$A_N := A+B(\gamma^2I-D^TD)^{-1}D^TC, \qquad B_N := B(\gamma^2I-D^TD)^{-1/2},$$
$$C_N := [I+D(\gamma^2I-D^TD)^{-1}D^T]^{1/2}C, \qquad G_N := C_N(sI-A_N)^{-1}B_N.$$

Lemma 2.1 If A has no eigenvalues in $i\mathbb{R}$ then $\|G_N\|_\infty \le 1$ if and only if $\|G\|_\infty \le \gamma$.

Proof: This is a simple modification from [1,lemma 2.1] ■

Lemma 2.2 Assume A is stable and $\| C(sI-A)^{-1}B \|_\infty \le 1$ then there exists a positive semi-definite matrix X such that

$$A^TX + XA + XBB^TX + C^TC = 0$$

and all eigenvalues of $A+BB^TX$ lie in the closed left half plane.

Proof: If (A,B,C) is a minimal realization then this is a known result (see [6]). The general result can be obtained by some convenient basis transformations. ■

Theorem 2.1 Assume $\| D \| < \gamma$ and (C,A) detectable. Then the following two statements are equivalent.

1) A is stable and $\| C(sI-A)^{-1}B+D \|_{\infty} \leq \gamma$

2) $\exists$ $X \geq 0$ which satisfies the ARE

$$A^TX + XA + (B^TX+D^TC)^T(\gamma^2I-D^TD)^{-1}(B^TX+D^TC) + C^TC = 0$$

Proof "$\Rightarrow$" We first show that $A_N$ as defined before is stable. Assume $A_N x = \lambda x$, $\mathrm{Re}\lambda \geq 0$ and $x \neq 0$ then we know $(\lambda I-A)$ is invertible since A is stable and $y := Cx \neq 0$ since otherwise $x = 0$ which would be a contradiction. Define $z := (\gamma^2I-DD^T)^{-1}y \neq 0$, then it can be shown that

$$\gamma^2 z = [\ C(\lambda I-A)^{-1}B+D\ ]D^Tz$$

This yields a contradiction since $\| D^T \| < \gamma$ and $\| C(\lambda I-A)^{-1}B+D \| \leq \gamma$ for all $\lambda$ with $\mathrm{Re}\lambda \geq 0$. Hence $A_N$ is stable

By lemma 2.1 and lemma 2.2 we know that there exists an $X \geq 0$ such that

$$A_N^TX + XA_N + XB_NB_N^TX + C_N^TC_N = 0.$$

If we rewrite this ARE using the definitions of $A_N$, $B_N$ and $C_N$ we get the desired ARE. "$\Leftarrow$" We first show that A is stable. Assume $Ay = \lambda y$, $\mathrm{Re}\lambda \geq 0$ and $y \neq 0$ then by applying $y^T$ to the left and y to the right hand side of the algebraic Riccati equation we find that, since $X \geq 0$ and $\mathrm{Re}\lambda \geq 0$ we must have $Cy = 0$ which contradicts our detectability assumption. Hence A is stable. Define $\underline{D} := \gamma^2I-D^TD$. We can rewrite the ARE as

$$(sI+A^T)X - X(sI-A) + (B^TX+D^TC)^T\underline{D}^{-1}(B^TX+D^TC) + C^TC = 0$$

where $s \in i\mathbb{R}$. We premultiply with $B^T(sI+A^T)^{-1}$ and postmultiply with $(sI-A)^{-1}B$. These inverses exist since A is stable. Then the resulting equation can be rewritten as

$$I-G^{\sim}(s)G(s) = K^{\sim}(s)K(s)$$

where $K(s) := -\underline{D}^{-1/2}(B^TX+D^TC)(sI-A)^{-1}B + \underline{D}^{1/2}$ and hence $\| G \|_{\infty} \leq 1$ ■

## 3. $H^{\infty}$-CONTROL USING STATE FEEDBACK - THE REGULAR CASE

Suppose we have system (1.1). We will try to find a matrix K with the desired properties. Our only restriction will be that we assume that $\| D_1 \| < \gamma$. In this chapter we will investigate the case that $D_2$ is injective. In correspondence with the LQ problem we will call this the regular case.

Definition 3.1 The system zeros of $(A,B_2,C,D_2)$ are defined as all the $\lambda \in \mathbb{C}$ for which the matrix

$$\begin{bmatrix} A-\lambda I & B_2 \\ C & D_2 \end{bmatrix}$$

loses rank, i.e. the rank is smaller than the normrank.

Theorem 3.1 Assume $D_2$ is injective and $(A,B_2,C,D_2)$ has no system zeros on the imaginary axis. Moreover assume that $\| D_1 \| < \gamma$ then there exists a K such that

1) $A+B_1K$ is asymptotically stable,
2) $\| (C_1+D_2K)(sI-A-B_2K)^{-1}B_1 + D_1 \|_\infty \leq \gamma$,

if and only if $\exists$ $X \geq 0$ which satisfies the ARE,

$$(A_H-B_H\Xi_H F_H^T C_H)^T X + X(A_H-B_H\Xi_H F_H^T C_H) + XD_H^T D_H X \\ - XB_H\Xi_H B_H^T X + C_H^T(I-F_H\Xi_H F_H^T)C_H = 0 \quad (3.1)$$

and which is such that the matrix

$$A_H - B_H\Xi_H(B_H^T X + F_H^T C_H)$$

has all its eigenvalues in the closed left half plane. Here

$$A_H := A + B_1(\gamma^2-D_1^T D_1)^{-1}D_1^T C, \qquad B_H := B_2 + B_1(\gamma^2-D_1^T D_1)^{-1}D_1^T D_2,$$
$$C_H := [\, I+D_1(\gamma^2-D_1^T D_1)^{-1}D_1^T \,]^{1/2}C, \qquad D_H := B_1(\gamma^2-D_1^T D_1)^{-1/2},$$
$$F_H := [\, I+D_1(\gamma^2-D_1^T D_1)^{-1}D_1^T \,]^{1/2}D_2, \qquad \Xi_H := [\, D_2^T(\, I+D_1(\gamma^2-D_1^T D_1)^{-1}D_1^T \,)D_2]^{-1}.$$

In that case one possible K is given by

$$K := -\Xi_H(B_H^T X + F_H^T C_H).$$

In order to prove this we need one small lemma:

Lemma 3.2 Let X be a positive semi-definite solution to the following ARE,

$$F^T X + XF + X(G_1-G_2)X + H_1 = 0 \quad (3.2)$$

such that $F-G_2X$ is asymptotically stable, where $G_1 \geq 0$, $G_2 \geq 0$ and $H_1 \geq 0$. Let $H_2$ be such that $H_1 \geq H_2 \geq 0$ then there also exists a positive semi-definite solution to the following ARE,

$$F^T Y + YF + Y(G_1-G_2)Y + H_2 = 0 \quad (3.3)$$

such that $F-G_2Y$ has all its eigenvalues in the *closed* left half plane.

Proof Let $P_1 \geq 0$ and $P_2 \geq 0$ be solutions of the following algebraic Riccati equations:

$$F^T P_1 + P_1 F - P_1 G_2 P_1 + H_1 = 0$$
$$F^T P_2 + P_2 F - P_2 G_2 P_2 + H_2 = 0$$

such that the matrices $F-G_2P_1$ and $F-G_2P_2$ have all their eigenvalues in the closed left half plane. These matrices exist by standard LQ theory. We have $P_1 \leq X$. This can be shown by using the fact that X is the solution of a differential game and $P_1$ is the solution of the LQ problem. For a proof see [5, proof of theorem 6.1]. Since $H_1 \geq H_2 \geq 0$ we have $P_1 \geq P_2$ and hence $P_2 \leq X$. We look at the following differential equation:

$$\dot{K} = F^T K + KF + K(G_1 - G_2)K + H_2 \qquad K(0) = P_2.$$

Using some standard theory we note that there exists a solution to this differential equation for all $t \geq 0$ and $0 \leq K(t) \leq X$. Using [5, proof of theorem 6.1] it is easily seen that K is increasing in t. Therefore $\lim_{t\to\infty} K(t) = Y$ exists and will satisfy our ARE (3.3). Assume $x \neq 0$ is an eigenvector of $A - G_2 Y$ with eigenvalue $\lambda$ such that $\operatorname{Re}\lambda > 0$. Then by applying x to both sides of (3.3) it can be shown that $Yx=0$ which implies $P_2 x = 0$ since $Y \geq P_2$. Therefore $(A - G_2 P_2)x = \lambda x$ which yields a contradiction. ■

<u>Proof of theorem 3.1</u> "⇒" Assume we have such a desired K, i.e. $\exists K \geq 0$ such that $A + B_2 K$ is stable and

$$\| (C + D_2 K)(sI - A - B_2 K)^{-1} B_1 + D_1 \|_\infty \leq \gamma$$

Hence by theorem 2.1 there exists an $X \geq 0$ such that

$$A_K^T X + X A_K + (B_1^T X + D_1^T C_K)^T (\gamma^2 I - D_1^T D_1)^{-1} (B_1^T X + D_1^T C_K) + C_K^T C_K = 0$$

where $A_K := A + B_2 K$ and $C_K := C + D_2 K$. We can rewrite this equation as

$$A_H^T X + X A_H + X D_H D_H^T X + C_H^T C_H + K^T F_H^T F_H K - K^T (F_H^T C_H + B_H^T X) \\ - (F_H^T C_H + B_H^T X)^T K = 0$$

This can then be written as

$$(A_H - B_H \Xi_H F_H^T C_H)^T X + X(A_H - B_H \Xi_H F_H^T C_H) + X D_H D_H^T X - X B_H \Xi_H B_H^T X \\ + C_H (I - F_H \Xi_H F_H^T) C_H + W^T W = 0 \qquad (3.4)$$

where

$$W := \Xi_H^{1/2} F_H^T C_H - \Xi_H^{-1/2} K + \Xi_H^{1/2} B_H^T X$$

It can be shown that if y is an eigenvector of $(A_H - B_H \Xi_H F_H^T C_H) - B_H \Xi_H B_H^T X$ with eigenvalue $\lambda$ where $\operatorname{Re}\lambda \geq 0$ then by applying y to both sides of (3.4) we find $(A + B_2 K)y = \lambda y$ Since $A + B_2 K$ is stable this gives a contradiction. Hence $(A_H - B_H \Xi_H F_H^T C_H) - B_H \Xi_H B_H^T X$ is asymptotically stable. Since $C_H^T (I - F_H \Xi_H F_H^T) C_H \geq 0$ we can use lemma 3.2 which tells us that since (3.4) has a solution also (3.1) has a solution Y such that $A_H - B_H \Xi_H (F_H^T C_H + B_H^T X)$ has all its eigenvalues in the closed half plane.

"⇐" Assume we have a solution of (3.1) satisfying the stability requirement. Then we can rewrite the equation as

$$(A+B_2K)^TX + X(A+B_2K) + (XB_1+(C+D_2K)^TD_1)(\gamma^2I-D_1^TD_1)^{-1}$$
$$\bullet(B_1^TX+D_1^T(C+D_2K)) + (C+D_2K)^T(C+D_2K) = 0$$

where we chose the K as suggested. We know that all eigenvalues of $A+B_2K$ are in the closed left half plane. By theorem 2.1 it remains to be shown that $(C+D_2K,A+B_2K)$ has no unobservable eigenvalues on the imaginary axis. Let $(A+B_2K)y=\lambda y$ and $(C+D_2K)y=0$ where $y \neq 0$ and Re $\lambda = 0$. Applying $y^T$ and y to the left and right hand side of the ARE respectivelygives us $B_1^TXy = 0$. Using the stabilizability of $(A,B_2)$ it is then easily derived that $Xy=0$. Hence y satisfies $[A-B_2\Xi_HF_H^TC_H]y=\lambda y$ and $[C-D_2\Xi_HF_H^TC_H]y=0$. Defining $p:= \Xi_HF_HC_Hy$ gives us

$$\begin{bmatrix} A-\lambda I & B_2 \\ C & D_2 \end{bmatrix}\begin{bmatrix} y \\ p \end{bmatrix} = 0$$

which is a contradiction with our assumption that there were no system zeros on the imaginary axis. ■

## 4. $H^\infty$-CONTROL USING STATE FEEDBACK - THE SINGULAR CASE

Assume we have the same system (1.1) but this time our $D_2$ matrix is not injective or the system $(A,B_2,C,D_2)$ has system zeros on the imaginary axis. In that case we can't apply our results of chapter 3 and we have to do something else. In this chapter we will solve this problem by disturbing the C and the $D_2$ matrices in such a way that they will satisfy the conditions of chapter 3.
Assume we have the following disturbed system

$$\dot{x} = Ax + B_1d + B_2u$$
$$y = C(\varepsilon)x + D_1d + D_2(\varepsilon)u$$

Furthermore assume $C(\varepsilon)$ and $D_2(\varepsilon)$ satisfy the following assumptions :

A1) $C(\varepsilon)$ and $D_2(\varepsilon)$ are continuous at $\varepsilon = 0$

A2) $C(0) = C$ and $D_2(0) = D_2$

A3) $C^T(\varepsilon)C(\varepsilon)$ and $D_2^T(\varepsilon)D_2^T(\varepsilon)$ are increasing in $\varepsilon$

A4) $C^T(\varepsilon)D_2(\varepsilon) = C^TD_2$ and $D_1^T[\ C(\varepsilon) \ \vdots\ D_2(\varepsilon)] = D_1^T[\ C \ \vdots\ D_2\ ]$

Theorem 4.1 Assume $C(\varepsilon)$ and $D_2(\varepsilon)$ satisfy A1–A4 then

$$\lim_{\varepsilon \downarrow 0} \gamma_\varepsilon = \gamma_0$$

where

$$\gamma_\omega := \inf \left\{ \| (C(\omega)+D_2(\omega)K)(sI-A-B_2K)^{-1}B_1+D_1 \|_\infty \mid K \in \mathbb{R}^{m\times n} \text{ is such that } A-B_2K \text{ is a stability matrix} \right\}.$$

Proof For each stabilizing K we have

$$\| (C(\varepsilon)+D_2(\varepsilon)K)(sI-A-B_2K)^{-1}B_1+D_1 \|_\infty$$

is increasing in $\varepsilon$ which can be checked easily using A3–A4. Hence $\gamma_0 \le \gamma_\varepsilon$. Remains to be shown that for all $\delta > 0$ there exists an $\varepsilon_1$ such that $\gamma_\varepsilon < \gamma_0 + \delta$ for all $\varepsilon \in (0,\varepsilon_1]$. By definition of $\gamma_0$ there exists a stabilizing $K_0$ such that

$$\| (C+D_2K_0)(sI-A-B_2K_0)^{-1}B_1+D_1 \|_\infty < \gamma_0 + \delta/2$$

Let $M := \| (sI-A-B_2K_0)^{-1}B_1 \|_\infty$. By A1–A2 there exists an $\varepsilon_1$ such that

$$\|[C-C(\varepsilon)]+[D_2-D_2(\varepsilon)] \| < \delta/(2M) \quad \text{for all } \varepsilon \in (0,\varepsilon_1]$$

Hence

$$\| (C(\varepsilon)+D_2(\varepsilon)K_0)(sI-A-B_2K_0)^{-1}B_1+D_1 \|_\infty < \gamma_0 + \delta$$

for all $\varepsilon \in (0,\varepsilon_1]$. Therefore $\gamma_\varepsilon < \gamma_0 + \delta$ for all $\varepsilon \in (0,\varepsilon_1]$ ■

Assume we have again a $\gamma$ and we have the same goals as in the previous chapter. If $\gamma = \gamma_0$ then we can't use theorem 4.1. Suppose $\gamma > \gamma_0$. Then we disturb C and $D_2$ in such a way that $D_2(\varepsilon)$ is injective, $(A,B_2,C(\varepsilon),D_2(\varepsilon))$ has no purely imaginary system zeros and A1–A4 are satisfied. One possible way of doing this is

$$C(\varepsilon) := \begin{bmatrix} C \\ 0 \\ \varepsilon I \end{bmatrix} \quad \text{and } D_2(\varepsilon) := \begin{bmatrix} D_2 \\ \varepsilon I \\ 0 \end{bmatrix}$$

Then by theorem 4.1 we know there exists a $\varepsilon_1$ such that $\gamma_\varepsilon < \gamma$ for all $\varepsilon \in (0,\varepsilon_1]$. We solve, using the techniques of chapter 3, this regular problem and find a K which satisfies our demands for the perturbed problem. However our perturbation didn't change the set of stabilizing K and increased the $H_\infty$–norm. Hence this K satisfies also our demands for the original singular problem.

Remark In [2,3,4] they don't split the problem up into a singular and a regular case but look at one ARE which should have a solution for some $\varepsilon > 0$. We get this same ARE by choosing an appropriate disturbance. They choose an $\phi$ such that Ker $\phi$ = Im $D_2^T$. We construct a $\psi$ such that $\psi^T\psi = (\phi^T\phi)^+$ where the + denotes the Moore–Penrose inverse. Then we get the same ARE by choosing the following disturbance,

$$C(\varepsilon) := \begin{bmatrix} C \\ 0 \\ \varepsilon I \end{bmatrix} \quad \text{and } D_2(\varepsilon) := \begin{bmatrix} D \\ \varepsilon\psi \\ 0 \end{bmatrix}$$

They guarantee a norm strictly smaller then $\gamma$, we can guarantee the same if our disturbance is such that $C^T(\varepsilon)C(\varepsilon)$ is strictly increasing in $\varepsilon$. This can be shown by noting that :

$$G_\varepsilon^\sim(s)G_\varepsilon(s) = G_0^\sim(s)G_0(s) + F^\sim(s)\left[C^T(\varepsilon)C(\varepsilon)-C^TC\right]F(s) + F^\sim(s)K^T\left[D_2^T(\varepsilon)D_2(\varepsilon)-D_2^TD_2\right]KF(s)$$

The advantage of our approach is that we only have to use the $\varepsilon$ in some special cases.

## 5. CONCLUSION

It appears that using a state space approach to this $H_\infty$ problem offers in most cases a nice and easily verifiable condition. When we have the bad luck that we have either system zeros on the imaginary axis or a $D_2$ matrix which is not injective then we can use a disturbance which makes the system satisfy this conditions. In this paper it is seen that we have more freedom in this disturbance then the specific choice which leads to the results of [2,3,4].

### References


[1] J.C. Doyle, K. Glover, P.P. Khargonekar and B.A. Francis, *State space solutions to standard $H_2$ and $H_\infty$ control problems*, American control conference, Atlanta, 1988.

[2] P.P Khargonekar,I.R. Petersen and M.A. Rotea, *$H^\infty$-optimal control with state feedback*, To appear in IEEE Trans. Automat. Contr.,1988.

[3] P.P Khargonekar,I.R. Petersen and K.Zhou, *Robust stabilization and $H^\infty$ control*, Submitted for publication IEEE Trans. Aut. Contr.,September 1987.

[4] I.R. Petersen, *Disturbance attenuation and $H^\infty$ optimization: A design method based on the algebraic Riccati equation*, IEEE Trans. Aut. Contr.,Vol AC–32,No. 5, May 1987,pp. 427–429.

[5] A.A. Stoorvogel, *The singular zero sum differential game with stability* using $H_\infty$ *control theory*, Submitted for publication Math. Contr. Sign. Syst..

[6] J.C. Willems,*Least squares optimal control and the algebraic Riccati equation*, IEEE Trans. Aut. Contr.,Vol 16, No.6, December 1971, pp. 621–634.

[7] K. Zhou and P.P. Khargonekar, *An algebraic Riccati equation approach to $H^\infty$ optimization*, Syst. Contr. Lett., Vol. 11, No.2,August 1988, pp. 85–91.

# CONVERGENCE ANALYSIS OF SELF-TUNING CONTROLLERS BY BAYESIAN EMBEDDING

P. R. Kumar

Abstract

We analyze adaptive control schemes which use recursive least squares parameter estimates.

## 1. Introduction.

Consider a linear stochastic system

$$y(t+1) = \sum_{i=0}^{p} [a_i y(t-i) + b_i u(t-i-d+1)] + w(t+1) \tag{1}$$

where $\{w(t)\}$ is a mean 0, standard deviation $\sigma^2$, white, Gaussian noise and $u(t)$ is a measurable function of $(y(0),\ldots,y(t))$. The parameter vector $\theta^0 = (a_0, \ldots, a_p, b_0, \ldots, b_p)^T$ is unknown.

In this paper, we consider adaptive control schemes which use the recursive least squares parameter estimates generated by:

$$\hat{\theta}(t+1) = \hat{\theta}(t) + R^{-1}(t)\phi(t)[y(t+1) - \phi^T(t)\hat{\theta}(t)] \; ; \; \hat{\theta}(0) = \overline{\theta} \tag{2}$$

$$R(t+1) = R(t) + \phi(t+1)\phi^T(t+1) \; ; \; R^{-1}(-1) = P_0 = P_0^T > 0 \, . \tag{3}$$

The results of this paper are drawn from [1], to which the reader is referred for more details.

2. Convergence of parameter estimates.

Let us "pretend" that $\theta^0$ is randomly chosen from a Gaussian distribution with mean $\overline{\theta}$ and covariance $\frac{1}{\sigma^2} P_0$, and independent of w. Then $\hat{\theta}(t)$ can be regarded as the conditional mean of $\theta^0$, and is therefore a martingale. By this procedure, which we call Bayesian embedding, we are able to establish the following Theorem providing the convergence of the parameter estimates.

THEOREM 1:

There exists a set $N \subseteq R^n$ of Lebesgue measure 0, so that if $\theta^0 \notin N$, then $\lim_{t \to +\infty} \hat{\theta}(t) = \hat{\theta}(\infty)$ exists almost surely.

This approach has been previously utilized also by Sternby [2] and Rootzen and Sternby [3]. In order to show that $\hat{\theta}(t)$ is indeed the conditional mean, we have found it necessary to weaken the standard conditions under which the Kalman filter is valid, eg. Liptser and Shiryayev [4, Assumptions 1-4, page 62 and Theorem 13.4, or Example 1, page 85] which require a square integrability condition on $\phi(t)$. This extension has been done in Chen, Kumar and van Schuppen [5].

3. The normal equations of least squares.

The recursive least squares estimates satisfy the normal equations:

$$\left[\sum_{t=0}^{N-1} \phi(t)\phi^T(t)\right]\hat{\theta}(N) = \left[\sum_{t=0}^{N-1} \phi(t)y(t+1)\right] + P_0^{-1}[\overline{\theta} - \hat{\theta}(N)] .$$

From this we are able to establish the following fundamental results:

THEOREM 2:

Let $r(n) := \sum_{t=0}^{N} \phi^T(t)\phi(t)$ where $\phi(t) := (y(t),\ldots,y(t-p),u(t),\ldots,u(t-p))^T$.

Then

$$\text{(i)} \quad \lim_{N \to +\infty} \frac{1}{r(N)} \sum_{t=0}^{N} \phi(t)\phi^T(t)[\hat{\theta}(t) - \theta^0] = 0 \ \text{ a.s.} \tag{4}$$

$$\text{(ii)} \quad \lim_{N \to +\infty} \frac{1}{r(N)} \sum_{t=0}^{N} \phi(t)\phi^T(t)[\hat{\theta}(\infty) - \theta^0] = 0 \ \text{ a.s.} \tag{5}$$

This result is the basis of all further results.

4. The stability of indirect adaptive control schemes.

Let us consider a general adaptive control law of the form:

$$R(\hat{\theta}(t);q^{-1})u(t) = S(\hat{\theta}(t);q^{-1})y(t) + T(\hat{\theta}(t);q^{-1})z^*(t) \quad \text{a.s.}$$

and where R, S and T are polynomials whose coefficients are continuous in $\theta$, $\{z^*(t)\}$ is a bounded, deterministic sequence. We shall make the following assumptions:

(i) There exists a polynomial $R'(\theta;q^{-1})$ whose coefficients are continuous in $\theta$, so that

$$R(\theta;q^{-1}) = R'(\theta;q^{-1})\, B(\theta;q^{-1})\,.$$

(ii) All the roots of

$$H(\theta;q^{-1}) := A(\theta,q^{-1})\, R'(\theta;q^{-1}) - q^{-d}\, S(\theta;q^{-1})$$

are inside the open unit disk.

(iii) The system (1), which we rewrite as

$$A(q^{-1})y(t) = q^{-d}B(q^{-1})u(t) + w(t)\,,$$

is strictly minimum phase.

An analysis of the results (4,5) of Theorem 2 allows us to establish that <u>all</u> such indirect adaptive control laws are stable.

<u>THEOREM 3:</u>

$$\limsup_{N\to+\infty} \frac{1}{N} \sum_{t=1}^{N} [y^2(t) + u^2(t)] < +\infty \qquad \text{a.s.}$$

5. The performance of indirect adaptive control schemes.

The fundamental results (4,5) of Theorem 2 allow us to characterize the asymptotic performance of all indirect adaptive control schemes.

THEOREM 4:

(i) $$[A(\hat{\theta}(\infty);q^{-1})-A(q^{-1})]\,R(\hat{\theta}(\infty);q^{-1}) = q^{-d}[B(\hat{\theta}(\infty);q^{-1}) - B(q^{-1})]\,S(\hat{\theta}(\infty);q^{-1}) \quad \text{a.s.}$$

(ii) $$\lim_{N\to+\infty} \frac{1}{N} \sum_{t=0}^{N-1} [H(\hat{\theta}(\infty);q^{-1})y(t) - q^{-d}T(\hat{\theta}(\infty);q^{-1})z^*(t) - R'(\hat{\theta}(\infty);q^{-1})w(t)]^2 = 0 \quad \text{a.s.}$$

(iii) $$\lim_{N\to\infty} \frac{1}{N} \sum_{t=0}^{N-1} [(B(\hat{\theta}(\infty);q^{-1}) - B(q^{-1}))\,T(\hat{\theta}(\infty);q^{-1})z^*(t)]^2 = 0 \quad \text{a.s.}$$

(iv) Suppose $d \geq 2$, and denote

$$A(\hat{\theta}(\infty);q^{-1}) = 1 - \sum_{i=0}^{b} \hat{a}_i(\infty)q^{-i-1} \quad .$$

Then

$$\hat{a}_i(\infty) = a_i \quad \text{for} \quad 0 \leq i \leq d-2 \quad .$$

The result (i) above yields all "self-tuning" type results, while (ii) yields all "optimality" type results. The third result (iii) yields all results based on "persistency of excitation" of $\{z^*(t)\}$. Finally (iv) shows that the leading coefficients of the A-polynomial are correctly estimated for high delay systems.

## 6. Applications.

The above results can be applied in a straightforward fashion to determine the asymptotic behavior of a variety of adaptive control schemes. The specific schemes thus analyzed in [1] are the self-tuning regulator for unit delay systems, an "indirect" self-tuning regulator for systems with delay $d \geq 2$, and pole-zero placement schemes. As an example, we are able, for the first time, to establish the self-tuning property of many of these schemes.

Acknowledgments

The research reported here has been supported by the U.S. Army Office under Contract No. DAAL-03-88-K-0046.

References

[1] P. R. Kumar, "Convergence of Adaptive Control Schemes using Least-Squares Parameter Estimates," University of Illinois, January 1989. To appear in *IEEE Transactions on Automatic Control.*

[2] J. Sternby, "On Consistency for the Method of Least-Squares using Martingale Theory," *IEEE Transactions on Automatic Control*, vol. AC-22, p. 346, 1977.

[3] H. Rootzen and J. Sternby, "Consistency in Least-Squares Estimation: A Bayesian Approach," *Automatica*, vol. 20, no. 4, pp. 471-475, 1984.

[4] R. S. Liptser and A. N. Shiryayev, *Statistics of Random Processes II: Applications*, New York, Springer-Verlag, 1977.

[5] H.-F. Chen, P. R. Kumar and J. H. van Schuppen, "On Kalman Filtering for Conditionally Gaussian Systems with Random Matrices," University of Illinois, February 1989. To appear in *Systems and Control Letters.*

# ON BOUNDED ADAPTIVE CONTROL WITH REDUCED PRIOR KNOWLEDGE

Romeo Ortega , Yu Tang , Laurent Praly

**Abstract**. In this paper we address the problem of insuring boundedness of all signals for adaptive controllers in closed loop with unknown plants. Our main concerns in this paper are: First, that the choice of the controller structure should be free to the designer and the number of adjustable parameters determined by the number the controller needs to achieve acceptable performance. Second, that the robustness measure be given in terms of the best achievable transfer function by the chosen controller structure and not in terms of the distance from the plant to some "ideal plant" set. Using a new parameter update law, conditions for stability are given for plants (possibly nonlinear time varying, but linear in the control) in closed loop with stable controllers which are linear in the adjustable parameters.

## I. INTRODUCTION

In recent years ,there has been considerable interest in designing modifiedparameter update laws for direct adaptive controllers in order to preserve global boundedness in spite of the presence of unmodeled dynamics. Most of the existing results [1]-[3], (see also [6] for a recent survey), are concerned with schemes where the number of adjustable parameters is determined by the assumed order of the plant and the controller structure is dictated by the theory requirements. It is assumed that the plant to be controlled belongs to the family

$$\mathbb{F}_{\mu}:=\{G(p):G(p)=G_0(p)[1+\mu\Delta_m(p)]+\mu\Delta_a(p)\} \tag{1.1}$$

where $p:=d/dt, G(p)\in \mathbb{R}(p)$ is linear time invariant (LTI) stably invertible of known order and high frequency gain and $\Delta m(p), \Delta a(p)\in\mathbb{R}(p)$ are LTI and stable. That is, the plant lies in the neighborhood of a set of "ideal plants" where the unmodeled dynamics are scaled by the factor $\mu>0$. Using various modified parameter update laws (see [5] for a unified treatment) it is shown in [1]-[3] that for an adaptive controller designed for $\mu=0$ there exists a $\mu^*>0$ such that for all $\mu\in[0,\mu^*)$ a class of $G\in\mathbb{F}\mu$ (see [6,7] for details on this class) are globally stabilized in the $L_\infty$ sense, i.e. all signals remain bounded. In the authors' opinion, in spite of its unquestionable theoretical significance, these results do not provide a real design methodology, in the sense of giving a yes or no answer to a specific robust control problem, but rather constitute some kind of "continuity of stability" or well-posedness statements. See [6] for further discussions.

In this paper we are interested in adaptively controlling plants which do not necessarily belong to the perturbation family $\mathbb{F}\mu$. Also, we will not prescribe the controller structure and will only impose on it the conditions of stability and linearity on the adjustable parameters. Our main contribution is the introduction of a new parameter adaptation algorithm for which the following result can be established: Given a plant, possibly nonlinear time-varying (NLTV), but linear in the control signal, in closed loop with a (fairly general) linearly parametrized controller, the overall system is globally $L_\infty$ stable if slow adaptation is enforced, a "good" performance is achievable with the given controller and some prior knowledge on this controller parametrization is available. Unfortunately, the prior knowledge condition mentioned above implies the knowledge of a (fixed) stabilizing controller parametrization. Therefore, our result should be viewed as a preservation of stability, rather than a stabilization condition. In view of the generality of the problem formulation, this situation is not surprising since it is natural that for its solution we have to introduce more information about the plant and that we pursue a less ambitious control objective.

Notation. Standard notation of input-output theory is used throughout the paper, see e.g. [9] for further details. The output of an operator $\mathfrak{M}$ acting on a signal x(t) will be denoted $\{\mathfrak{M}x\}(t)$. The symbols $\|\ \|$, $\|\ \|_\infty$ will be used to denote the sup vector norms on $\mathbb{R}^n$ and the $L_\infty$ norm respectively.

## II. ADAPTIVE CONTROL LAW

The problem we address in the paper is the following: Given a (fairly general) parametrized controller find a parameter update law such that, without assumptions on the structure of the plant or controller, we can define a class of plants for which all closed loop signals are bounded. In this case we will say that the closed loop system is $L_\infty$ stable. In this section we present the proposed underlying controller structure and the parameter update law. The only substantial restrictions imposed on the controller are that it be stable and linear in the adjustable parameters. For ease of exposition we will present a simple output feedback linear compensator where all the coefficients of the numerator are adjusted on line. As will become clear later other controller configurations, nonlinear controllers or the case when only a few parameters are adjusted, may be similarly analyzed.

Consider the parametrized controller $\hat{C}:=C(\hat{\theta}):u(t)\rightarrow y(t)$ defined by

$$u(t)=\{\hat{C}y\}(t)=\hat{\theta}^T(t)\phi(t) \tag{2.1.a}$$

$$\phi(t)=F(p)y(t) \tag{2.1.b}$$

where u(t), y(t) are the plant input and output respectively, $\hat{\theta}(t)\in\mathbb{R}^{n_\theta}$ is a vector of adjustable parameters and $F(p)\in\mathbb{R}^{n_C}(p)$ is a stable LTI operator of order $n_F$ and realization

$$F(p)=(pI-A_F)^{-1}b_F \tag{2.1.c}$$

with $A_F$ diagonalizable. We propose the parameter update law

$$\dot{\hat{\theta}}(t)=-\sigma(t)[\hat{\theta}(t)-\theta_0]-\gamma\phi(t)y(t)/\rho^2(t),\ \gamma>0 \tag{2.2.a}$$

$$\dot{\rho}(t)=-\sigma_0\rho(t)+\delta[1+|y(t)|],\ \delta,\sigma_0>0 \tag{2.2.b}$$

$$\sigma(t)=\delta[1+|y(t)|]/\rho(t) \tag{2.2.c}$$

$$\hat{\theta}(0)=\theta_0,\rho(0)=\rho_0>0 \tag{2.2.d}$$

where $\delta$, $\sigma_0$ and $\rho_0$ are chosen such that

$$\delta\ge\|b_F\|\ \|Q\|\ \|Q^{-1}\| \tag{2.3.a}$$

$$\sigma_0\le \min_i|\mathbb{R}_e\{\lambda_i(A_F)\}| \tag{2.3.b}$$

$$\rho_0\ge \|\phi(0)\|\ \|Q\|\ \|Q^{-1}\| \tag{2.3.c}$$

where $A_F=Q\Lambda_F Q^{-1}$ with $\Lambda_F$ diagonal.

Remark 2.1. Three are the main features of the proposed estimator: the normalization factor $\rho(t)$ [4]; the driving term $\theta_0$, which will incorporate the prior information on the stabilizing parameters; and the inclusion of a time varying leakage $\sigma(t)$.

Remark 2.2. It is important to remark that the estimator (2.2) is not a parameter search procedure as the ones used in universal stabilizers [8]. As is well known the latter controllers are only of theoretical interest, i.e. to determine limits of adaptive stabilizability.

Remark 2.3. The estimator (2.2) is related with both, the fixed $\sigma$-modification of [1], and the $e_1$-modification proposed in [2]. The idea of the latter is to ensure that the update law has a fixed point corresponding to reference behaviour. This is not the case here. However, both modifications behave similarly for large signals. Also, it can be shown [12] that it has a behaviour similar to the fixed $\sigma$-modification close to the equilibrium points.

The lemma below is instrumental for the establishment of the stability results. Its proof is given for the sake of self-containment.

**Lemma 2.1.** Consider the regressor vector $\phi(t)$ defined by (2.1.b) and the normalization factor $\rho(t)$ given by (2.2.b) with (2.3). Under these conditions

$$\bar{\phi}(t):=\frac{\phi(t)}{\rho(t)}\le 1 \tag{2.4}$$

Proof. (2.1)→

$$\|\phi(t)\|\le\|e^{A_Ft}\| \ \|\phi(0)\|+\|b_F\|\int_0^t\|e^{A_F(t-\tau)}\| \ |y(\tau)|d\tau \tag{2.5}$$

In view of (2.3.b)

$$\|e^{A_Ft}\|\le\|Q\| \ \|Q^{-1}\|e^{-\sigma_0 t} \tag{2.6}$$

On the other hand (2.2.b) gives

$$\rho(t)=e^{-\sigma_0 t}\rho(0)+\delta\int_0^t e^{-\sigma_0(t-\tau)}[\,|y(\tau)|+1]d\tau \tag{2.7}$$

Thus, combining (2.5) and (2.7) and taking into account (2.3.a),(2.3.c),(2.6), we complete the proof.

□□□

Remark 2.4. It is clear from the proof of Lemma 2.1 that F in (2.1.c) could be any NLTV operator satisfying

$$\|\phi(t)\|\le c_1\|\phi(0)\|+c_2\int_0^t e^{-\lambda(t-\tau)}|y(\tau)|d\tau \tag{2.8}$$

with $c_{1,2}$,$\lambda$ known positive constants.

## III. NON LINEAR TIME VARYING PLANTS: SMALL GAIN ANALYSIS

In this section we define a class of NLTV plants for which the adaptive controller (2.1)-(2.3) yields an $L_\infty$-stable closed loop system. The main result, obtained from a direct application of the small gain theorem [9], is contained on the proposition below.

**Proposition 3.1.** Consider a causal NLTV plant, linear in the control signal u(t), and described by the mapping $G:L_{\infty e}\rightarrow L_{\infty e}$, that is

$$y(t)=\{Gu\}(t) \tag{3.1}$$

in closed loop with the adaptive control (2.1)-(2.3). Assume the closed loop system is well posed and that there exists a parametrization of the controller $\theta_*\in\mathbb{R}^{n_\theta}$ such that

$$\frac{g_*}{\sigma_0}(\gamma+\delta\|\theta_0-\theta_*\|)<1 \tag{3.2}$$

with $g_*$ the $L_\infty$ gain of $H_*$ and

$$H_*:=H(\theta_*)=(1-G\theta_*^TF)^{-1}G \tag{3.3}$$

Under these conditions, the adaptive system is $L_\infty$ stable, that is y(t), $\theta(t)$ are uniformly bounded, and the bounds are computable.

Proof. Define the vector

$$z(t):=[\hat{\theta}(t)-\theta_0]\rho(t) \tag{3.4}$$

From (2.2) it is easy to see that z(t) satisfies

$$z(t)=\{\frac{\gamma}{p+\sigma_0}\bar{\phi}y\}(t)\ ,\ z(0)=0 \tag{3.5}$$

Also, combining (2.1) and (3.1) and using the linearity in u(t) of G we get

$$y(t)=\{H_*(\hat{\theta}-\theta_*)^T\phi\}(t)$$

$$=\{H_*\bar{\phi}^Tz\}(t)+\{H_*(\theta_0-\theta_*)^T\bar{\phi}\rho\}(t) \tag{3.6}$$

where to get the last identity we have used (2.4) and (3.4). Similarly, we can write (2.2.b) in operator notation as

$$\rho(t)=\{\frac{\delta}{p+\sigma_0}(|y|+1)\}(t) \tag{3.7}$$

In the sequel we will refer to (3.5)-(3.7) as the error model for the adaptive system. Noting that lemma 2.1 insures $\bar{\phi}(t)$ is a bounded signal, and taking $L_\infty$ gains in (3.6) yields

$$\|y(t)\|_\infty \le g_*\{\|z(t)\|_\infty+\|\theta_0-\theta_*\|\ \|\rho(t)\|_\infty\}+k_1$$

$$\le \frac{g_*}{\sigma_0}\left[\gamma+\delta\|\theta_0-\theta_*\|\right]\|y(t)\|_\infty+k_2 \tag{3.8}$$

where $k_i$ are constants and the last inequality is obtained from (3.5) and (3.7). Using condition (3.2) in (3.8) insures y(t) is uniformly bounded (by a computable bound). To prove that $\theta(t)\in L_\infty$ notice from (3.5) that z(t) is bounded if y(t) is bounded, and that $\rho(t)$ is bounded away from zero by construction (2.2.b).

□□□

Remark 3.1. The proposition above shows that the proposed adaptive controller insures global boundedness of all signals if the following conditions are satisfied: There exists a "good" controller parametrization ($C^*=C(\theta^*)$) that yields a closed loop system with small $L_\infty$ gain; ii) Adaptation is slow, i.e., small adaptation gain; iii) Some prior knowledge on $\theta^*$ is available, or the requirement i) may be attained with "low gain" control ,i.e., $\|\theta_*\|$ small.

The result above is derived in a very high level of generality. Its limitation as an adaptive stabilization theorem stems from the fact below.

**Fact 3.1.** Condition (3.2) implies that

$$\theta_0 \in \Theta_s := \{\theta \in R^{n_\theta} : H(\theta) \text{ is } L_\infty \text{ stable}\} \tag{3.9}$$

where

$$H(\theta) := (1 - G\theta^T F)^{-1} G$$

is the operator that describes the closed loop plant. In other words, if (3.2) holds then $C_0 := C(\theta_0)$ stabilizes the plant.

<u>Proof.</u> Consider the implications

$$\hat{\theta} \equiv \theta_0 \Rightarrow z = 0$$

$$\Rightarrow y(t) = \{H_*(\theta_0 - \theta_*)^T \bar{\phi} \frac{\delta}{p + \sigma_0}(|y| + 1)\}(t) \tag{3.10}$$

where the last expression is obtained replacing (3.7) in (3.6). On the other hand

$$(3.2) \Rightarrow g_* \| \theta_0 - \theta_* \| \frac{\delta}{\sigma_0} < 1 \tag{3.11}$$

which, from inspection of (3.10), is a sufficient condition for boundedness of y(t).

□□□

<u>Remark 3.2.</u> Fact 3.1 is tantamount to saying that if the adaptive system satisfies the stability condition of Proposition 3.1 then a stabilizing controller ($C(\theta_0)$, $\theta_0 \in \Theta_s$) is known a priori. Notice however, that the proposition provides only a sufficient condition for stability.

<u>Remark 3.3.</u> It is interesting to note that for all $\theta_0 \in \Theta_s$ we can derive from (3.2) an upperbound on the adaptation gain such that the system preserves stability as

$$\gamma < \sigma_0 / g_0 \tag{3.12}$$

where $g_0$ is the $L_\infty$ gain of $H_0 := H(\theta_0)$.

## IV. NUMERICAL EXAMPLE

As an illustration of the theorem described above we consider the problem of simultaneous stabilization of plants with unknown sign of the high frequency gain which has attracted the attention of researchers for some time, see e.g. [11]. To this end ,we consider the following plants

$$\dot{y}(t) = y(t) + bu(t); \quad b = \pm 1; \quad y(0) = y_0 \tag{4.1}$$

in closed loop with the controller

$$u(t)=\hat{\theta}(t)\phi(t) \tag{4.2.a}$$

$$\dot{\phi}(t)=-10\phi(t)+y(t); \qquad \phi(0)=0 \tag{4.2.b}$$

where the parameter is updated using (2.2) with $\delta=1, \sigma_0=10$ and $\rho(0)=1$.

The stabilizing set for both plants is

$$\Theta_s=\{\theta:\theta b<-10\} \tag{4.3}$$

Notice that there is no single controller gain that will insure simultaneous stabilization.

If, without loss of generality, we restrict $b\theta^* \in [-30.25,-10)$ the closed loop system $H_*$ has real distinct roots, and

$$g_*=-10|b|/(10+b\theta_*) \tag{4.4}$$

In this case,condition (3.2) of Proposition (3.1) for b=1 becomes

$$1+\gamma+1.1\theta_*<\theta_0<-(1+\gamma)+.9\theta_* \tag{4.5}$$

which, as expected, requires $\theta o$ to belong to the stabilizing set. Notice that (4.5) also imposes an upperbound on the adaptation gain, i.e., $\gamma<2.025$. An interval for $\theta o$, similar to (4.5), may be obtained for b=-1. The important point being that the intervals are non-overlapping. This fact shows that Proposition 3.1 does not allows us to find a single controller that will globally stabilize both plants.

## VI CONCLUDING REMARKS

We have studied the problem of defining a class of plants for which stabilization is possible with a given (fairly general) parametrized controller whose parameters are updated with a new law. Two are the main features of our approach: First, we do not impose a particular controller structure or number of adjustable parameters. Second, no assumptions on the plant being close to some "ideal plant" are required for the analysis.

Instrumental for the establishment of our results is the introduction of a new parameter update law. The latter, is related with the $e_1$-modification of [2] in far from equilibrium situations, and essentially reduces to the fixed $\sigma$-modification [17] close to the equilibrium points.

**REFERENCES**

[1] Ioannou, P. and Tsakalis, K., "A robust direct adaptive controller," *IEEE Trans. Automat. Contr.* Vol.AC-31, No.11, Nov. 1986

[2] Narendra, K., and Annaswamy, A. M., "A new adaptive law for robust adaptation without persistent excitation," *IEEE Trans. Automat. Contr.* Vol.AC-32, 1987

[3] Kreisselmeier, G. and Anderson, B. D. O., "Robust model reference control," *IEEE Trans. Automat. Contr.* Vol. AC-31, No.2, Feb. 1986

[4] Praly, L., "Robustness of model reference adaptive control," *Proc. 3rd Yale Workshop Adapt. Syst.*, New Haven, CT, 1983

[5] Ioannou, P., and Sun J., "Theory and design of robust of direct and indirect adaptive control schemes," *Int. J. Contr.*, Vol.47 No.3, 1988

[6] Ortega, R. and Tang Y., "Robustness of adaptive controllers: a survey," *Automatica* (to appear Aug. 1989)

[7] Khargonekar, P. and R.Ortega, "Comments on the robust stability analysis of adaptive controllers using normalizations, *IEEE Trans. Aut. Cont.*, Vol. 34, No. 4, 1989.

[8] Martensson, B., "The order of any stabilizing regulator is sufficient a priori information for adaptive stabilization," *Syst. & Contr. Lett.* Vol.6, 1985

[9] Willems, J. C., *The Analysis of Feedback Systems,* MIT Press, Cambridge Massachusetts, 1971

[10] Hsu, L. and Costa, R. R., "Bursting phenomena in continuous-time adaptive systems with a $\sigma$-modification," *IEEE Trans. Automat. Contr.* Vol. AC-32, No.1, Jan. 1987

[11] Nussbaum, R. D., "Some remarks on a conjecture in parameter adaptive control," *Syst. & Contr. Lett.*, Vol.3, No.5, 1983

[12] Espinosa, G., "Stability of the equilibria in adaptive systems with modified estimators", *M. Sc. Thesis CINVESTAV IPN*, May 1989. (In spanish).

# Indirect Techniques for Adaptive Input Output Linearization of Nonlinear Systems

Andrew Teel Raja Kadiyala Petar Kokotovic Shankar Sastry

**Abstract**

**A technique of indirect adaptive control based on certainty equivalence for input output linearization of nonlinear systems is proven convergent. It does not suffer from the overparameterization drawbacks of the direct adaptive control techniques on the same plant. This paper also contains a semi-indirect adaptive controller which has several attractive features of both the direct and indirect schemes.**

## 1 Introduction

There has been much recent research in the usage of adaptive control techniques for improving the input output linearization by state feedback of nonlinear systems with parametric uncertainty. Techniques of direct adaptive control (with no explicit identification) were proposed and developed in [13, 6, 3, 12], (see also [11]). In this paper we continue a program of investigating indirect adaptive control of nonlinear systems. Nonlinear indirect adaptive control is motivated by the fact that, with exact knowledge of the plant parameters, a nonlinear state feedback law and a suitable set of coordinates can be chosen to produce linear input-output behavior. In the case of parameter uncertainty, intuition suggests that parameter estimates which are converging to their true values can be used to asymptotically linearize the system. This heuristic is known as the *certainty equivalence principle*. Indirect adaptive control differs from direct adaptive control in that it relies on an observation error to update the plant parameters rather than relying on an output error. Indirect adaptive control can be broken down into two parts. First, a parameter identifier is attached to the plant and adjusts the parameter estimates on line. These estimated parameters are then used in the linearizing control law.

Our work is part of a continuing effort in indirect adaptive control of nonlinear systems initiated in [1, 2, 10]. Our results are an extension of those in [2] in that they address in detail, input output linearizable nonlinear systems (rather than the full state linearizable case treated in [1, 2]); they are also more specifically related to adaptive tracking (rather than stabilization as treated in [10]). Also, our assumptions are somewhat different. This paper is a condensed version of a full report [14]. An outline is as follows:

In section 2, we review two identifier structures for nonlinear systems, (they have appeared in [1, 7, 8, 9]). Section 3 gives an outline of an indirect adaptive controller based on certainty equivalence along with a proof of convergence. We also present a semi-indirect adaptive controller which contains attractive features of the direct and indirect schemes. Section 4 contains a simulation comparison of a direct, indirect adaptive and non-adaptive controller methodology. Section 5 gives some conclusions.

## 2 Identifier Structures

Consider the system

(1) $\dot{x} = f(x,\theta^*) + g(x,\theta^*)u$

with $x \in \mathbf{R}^n, u \in \mathbf{R}, \theta^* \in \mathbf{R}^p$ and $f$, $g$ are assumed to be smooth vector fields on $\mathbf{R}^n$. Further let $f(x,\theta^*)$ and $g(x,\theta^*)$ have the form

$$(2)\quad \begin{aligned} f(x,\theta^*) &= \sum_{i=1}^{p} \theta_i^* f_i(x) \\ g(x,\theta^*) &= \sum_{i=1}^{p} \theta_i^* g_i(x) \end{aligned}$$

Here $\theta_i^*$, $i = 1, \ldots, p$, are unknown parameters, which appear linearly, and the smooth vector fields $f_i(x)$, $g_i(x)$ are known. If we formulate the regressor

$$(3)\quad w^T(x,u) = [f_1(x) + g_1(x)u, \ldots, f_p(x) + g_p(x)u]$$

so that $w^T(x,u) \in \mathbf{R}^{n \times p}$ contains all of the nonlinearities of the system, then (1) can be written as

$$(4)\quad \dot{x} = w^T(x,u)\theta^*$$

For a system with multiple inputs, the regressor is formed in an analogous manner and (4) holds except that the notation used to define $w$ is more involved.

## 2.1 Observer-based Identifier

To estimate the unknown parameters, we will use the identifier system

$$(5)\quad \begin{aligned} \dot{\hat{x}} &= A(\hat{x} - x) + w^T(x,u)\hat{\theta} \\ \dot{\hat{\theta}} &= -w(x,u)P(\hat{x} - x) \end{aligned}$$

Here $A \in \mathbf{R}^{n \times n}$ is a Hurwitz matrix and $P \in \mathbf{R}^{n \times n} > 0$ is a solution to the Lyapunov equation

$$(6)\quad A^T P + PA = -Q, \quad Q > 0$$

This identifier is reminiscent of one proposed in [8], [7]. Note that $A = -\sigma I$ is a special case of the identifier. If we define $e_1 = \hat{x} - x$, the observer state error, and $\phi = \hat{\theta} - \theta^*$, the parameter error, and assume $\theta^*$ to be constant but unknown then we have the error system

$$(7)\quad \begin{aligned} \dot{e}_1 &= Ae_1 + w^T(x,u)\phi \\ \dot{\phi} &= -w(x,u)Pe_1 \end{aligned}$$

One should note the similarity of the error equation above with that of the error equation of a full order observer, although all the states are available by assumption.

**Theorem 2.1 Stability of Observer-based Identifier**
*Consider the observer-based identifier of equation (7),*

then
1. $\phi \in L_\infty$,
2. $e_1 \in L_\infty \cap L_2$,
3. If $w(x,u)$ *is bounded,*
   then $\dot{e}_1 \in L_\infty$ *and* $\lim_{t\to\infty} e_1(t) = 0$.

**Remarks:**

1. The proof is a standard Lyapunov argument on the function

   $$(8)\quad V(e_1,\phi) = e_1^T P e_1 + \phi^T \phi$$

2. The condition on the boundedness of $w$ is a stability condition. In particular, if the system is bounded-input bounded-state (bibs) stable with bounded input, then $w$ is bounded. (see [11])

3. Theorem 2.1 makes no statement about parameter convergence. As is standard in the literature one can conclude from (7) that $e_1$ and $\phi$ both converge exponentially to zero if $w$ is sufficiently rich, ie., $\exists \alpha_1, \alpha_2, \delta > 0$ such that

   $$(9)\quad \alpha_1 I \geq \int_s^{s+\delta} w w^T dt \geq \alpha_2 I$$

   This condition is impossible to verify explicitly ahead of time since $w$ is a function of $x$.

## 2.2 Filtered Regressor Identifier

Consider filtered forms of $w$,$x$ given by $W$,$W_0$ and defined by

$$(10)\quad \begin{array}{rcl} \epsilon\dot{W} & = & -W + \epsilon w^T(x,u) \\ \epsilon\dot{W}_0 & = & -W_0 + x \end{array}$$

The state can be reconstructed from the filtered regressor and filtered state as

$$(11)\quad x = W\theta^* + W_0 + [x(0) - W(0)\theta^* - W_0(0)]e^{-\frac{t}{\epsilon}}$$

The equivalence is shown by observing that

$$(12)\quad \begin{array}{rcl} \dot{x} & = & \dot{W}\theta^* + \dot{W}_0 - \frac{1}{\epsilon}[x(0) - W(0)\theta^* - W_0(0)]e^{-\frac{t}{\epsilon}} \\ & = & \frac{1}{\epsilon}[-W + \epsilon w^T(x,u)]\theta^* + \frac{1}{\epsilon}[-W_0 + x] - \frac{1}{\epsilon}[x(0) - W(0)\theta^* - W_0(0)]e^{-\frac{t}{\epsilon}} \\ & = & \frac{1}{\epsilon}[-W + \epsilon w^T(x,u)]\theta^* + \frac{1}{\epsilon}\{-W_0 + W\theta^* + W_0 + [x(0) - W(0)\theta^* - W_0(0)]e^{-\frac{t}{\epsilon}}\} \\ & & -\frac{1}{\epsilon}[x(0) - W(0)\theta^* - W_0(0)]e^{-\frac{t}{\epsilon}} \\ & = & w^T(x,u)\theta^* \end{array}$$

We can form the estimated state as

$$(13)\quad \hat{x} = W\hat{\theta} + W_0$$

and then, if we define $e_2 = \hat{x} - x$, we have

$$(14)\quad e_2 = W\phi - [x(0) - W(0)\theta^* - W_0(0)]e^{-\frac{t}{\epsilon}}.$$

This form of the identifier was proposed in [10], [1].

### 2.2.1 Gradient Algorithm

To estimate the unknown parameters, we can use the *gradient algorithm* which yields the following error system:

$$(15)\quad \dot{\phi} = -gW^T e_2 \qquad g > 0$$

**Theorem 2.2 Stability of Filtered Regressor Identifier Using the Gradient Method**
*Consider the filtered regressor identifier and the gradient algorithm of equation (15),*

then
1. $\phi \in L_\infty$,
2. $e_2 \in L_2$,
3. If $w(x,u)$ *is bounded,*
   then $e_2, \dot{e}_2 \in L_\infty$ *and* $\lim_{t\to\infty} e_2(t) = 0$.

**Remarks:**

1. The proof is a standard Lyapunov argument on the function

   $$(16)\quad V(\phi) = \frac{1}{2}\phi^T\phi$$

2. The condition on the boundedness of $w$ is a stability condition. In particular, if the system is bounded-input bounded-state (bibs) stable with bounded input, then $w$ is bounded. (see [11])

3. Theorem 2.2 makes no statement about parameter convergence. Parameter convergence is implied by $w$ being sufficiently rich (cf. equation (9).)

### 2.2.2 Least-Squares Identifier

Another approach for estimating the parameters is the *least-squares algorithm* which can be used with the filtered regressor identifier but not with the observer-based identifier. This algorithm produces the following error system:

$$(17)\quad \begin{aligned} \dot{\phi} &= -\gamma\Gamma W^T e_2 \\ \dot{\Gamma} &= -\gamma\Gamma W^T W\Gamma \qquad \gamma > 0 \quad \Gamma(0) > 0 \end{aligned}$$

**Theorem 2.3 Stability of Filtered Regressor Identifier Using the Least-Squares Method**
*Consider the filtered regressor identifier and the least-squares algorithm of equation (17),*

then *1.* $\phi \in L_\infty$,

*2.* $e_2 \in L_2$,

*3.* If $w(x,u)$ *is bounded,*
then $e_2, \dot{e}_2 \in L_\infty$ *and* $\lim_{t\to\infty} e_2(t) = 0$.

**Remarks:**

1. The proof is a standard Lyapunov argument on the function

   $$(18)\quad V(\phi) = \phi^T\Gamma^{-1}\phi$$

2. The same remarks as those after Theorem 2.2 concerning parameter convergence hold.

# 3 Indirect Adaptive Control

Nonlinear *indirect* adaptive control is motivated by the fact that, with exact knowledge of the plant parameters, a nonlinear state feedback law and a suitable set of coordinates can be chosen to produce linear input-output behavior. Linear system theory can then be applied to control the linearized portion of the system. In the case of parameter uncertainty, intuition suggests that parameter estimates which are converging to their true values can be used to asymptotically linearize the system. This heuristic is known as the *certainty equivalence principle.*

To fix notation, we review, following [5], the basic linearizing theory. Consider a single-input single-output system

$$(19)\quad \begin{aligned} \dot{x} &= f(x) + g(x)u \\ y &= h(x) \end{aligned}$$

with $x \in \mathbf{R}^n, u \in \mathbf{R}$ and $f$, $g$, $h$ smooth. Differentiating $y$ with respect to time, one obtains

$$(20)\quad \dot{y} = L_f h + L_g h u$$

Here $L_f h$, $L_g h$ stand for the Lie derivatives of $h$ with respect to $f$, $g$ respectively. If $L_g h(x) \neq 0$ $\forall x \in \mathbf{R}^n$ then the control law

$$(21)\quad u = \frac{1}{L_g h}(-L_f h + v)$$

yields the linear system

$$(22)\quad \dot{y} = v.$$

If $L_g h(x) \equiv 0$, one continues to differentiate obtaining

$$(23)\quad y^{(i)} = L_f^i h + L_g L_f^{i-1} h u \quad i = 1, 2, \ldots$$

If there is a fixed integer $\gamma$ such that $\forall x \in \mathbf{R}^n$ $L_g L_f^i h \equiv 0$ for $i = 0, \ldots, \gamma - 2$ and $L_g L_f^{\gamma-1} h(x) \neq 0$ then the control law

$$(24)\quad u = \frac{1}{L_g L_f^{\gamma-1} h(x)}(-L_f^\gamma h(x) + v)$$

yields

(25) $y^{(\gamma)} = v.$

We stress that the linearization conditions hold in all of $\mathbf{R}^n$. Some completeness conditions on vector fields involving $f$,$g$ are sufficient for this (for details see [5] chapter 2).

The integer $\gamma$ is called the ***strong relative degree*** of system (19). We will not consider the case where the relative degree is not defined; namely, where $L_g L_f^{\gamma-1} h(x) = 0$ for some values of $x$.

For a system with a strong relative degree $\gamma$, it is easy to verify that at each $x^0 \in \mathbf{R}^n$ there exists a neighborhood $U^0$ of $x^0$ such that the mapping

$$\Phi : U^0 \longrightarrow \mathbf{R}^n$$

defined as

$$(26)\quad \begin{array}{rcccl} \Phi_1(x) & = & \xi_1 & = & h(x) \\ \Phi_2(x) & = & \xi_2 & = & L_f h(x) \\ & & \vdots & & \\ \Phi_\gamma(x) & = & \xi_\gamma & = & L_f^{\gamma-1} h(x) \end{array}$$

with

$$d\Phi_i(x) g(x) = 0 \quad for \quad i = \gamma+1, \ldots, n$$

is a diffeomorphism onto its image.

If we set $\eta = (\Phi_{\gamma+1}, \ldots, \Phi_n)^T$ it follows that the system may be written in the ***normal form*** ([5]) as

$$(27)\quad \begin{array}{rcl} \dot{\xi}_1 & = & \xi_2 \\ & \vdots & \\ \dot{\xi}_{\gamma-1} & = & \xi_\gamma \\ \dot{\xi}_\gamma & = & b(\xi, \eta) + a(\xi, \eta) u \\ \dot{\eta} & = & q(\xi, \eta) \\ y & = & \xi_1. \end{array}$$

In equation (27), $b(\xi, \eta)$ represents the quantity $L_f^\gamma h(x)$ and $a(\xi, \eta)$ represents $L_g L_f^{\gamma-1} h(x)$. We assume that $x = 0$ is an equilibrium point of the system (ie. $f(0) = 0$) and we assume that $h(0) = 0$. Then the dynamics

(28) $\dot{\eta} = q(0, \eta)$

are referred to as the ***zero-dynamics*** (see [5] section 4.3 for details). The nonlinear system (19) is said to be minimum phase if the zero-dynamics are asymptotically stable.

## 3.1 Non-Adaptive Tracking

We now apply the normal form and the minimum phase property to the tracking problem. We desire to have $y(t)$ track a given $y_M(t)$. We start by choosing $v$ in (24) as

(29) $v = y_M^{(\gamma)} + \alpha_1 (y_M^{(\gamma-1)} - y^{(\gamma-1)}) + \ldots + \alpha_\gamma (y_M - y)$

with $\alpha_1, \ldots, \alpha_\gamma$ chosen so that

(30) $s^\gamma + \alpha_1 s^{\gamma-1} + \ldots + \alpha_\gamma$

is a Hurwitz polynomial. Note that $y^{(i-1)} = \xi_i$. If we define $e_i = y^{(i-1)} - y_M^{(i-1)}$ then we have

$$(31)\quad \begin{array}{rcl} \dot{e} & = & Ae \\ \dot{\eta} & = & q(\xi, \eta) \\ \xi_i & = & e_i + y_M^{(i-1)} \end{array}$$

where $A$ is the companion matrix associated with (30), and hence is a Hurwitz matrix.

It is easy to see that this control results in asymptotic tracking and bounded states $\xi$ provided $y_M, \dot{y}_M, \ldots, y_M^{(\gamma-1)}$ are bounded.

It can be also be shown that $\eta$ remains bounded as well, assuming exponentially stable zero-dynamics and $q(\xi, \eta)$ is Lipschitz in $\xi, \eta$, by using a converse Lyapunov approach. Thus, this control yields bounded tracking. (see [12]).

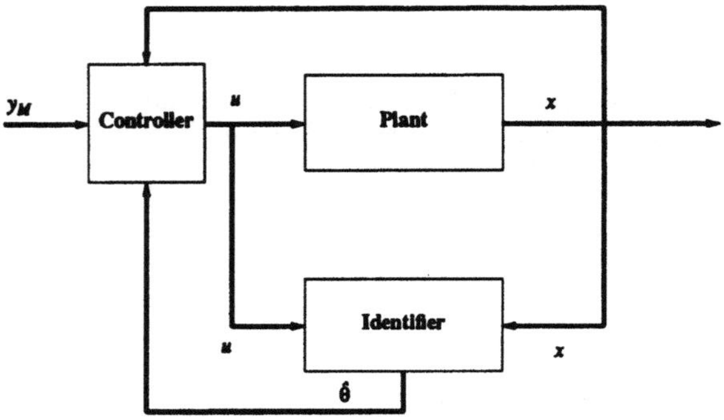


Figure 1: Block Diagram of an Indirect Adaptive Controller

## 3.2 Indirect Adaptive Tracking

In the case of parameter uncertainty, we have the system

$$\begin{aligned} \dot{x} &= f(x,\theta^*) + g(x,\theta^*)u \\ y &= h(x,\theta^*) \end{aligned} \tag{32}$$

with $\theta^* \in \mathbb{R}^p$ the vector of unknown parameters. We will make the following assumptions:

**Assumption 1 Linear Parameter Dependence**

*The vector fields $f(x,\theta^*)$, $g(x,\theta^*)$ and the output function $h(x,\theta^*)$ in the system (32) depend linearly on the unknown parameters as*

$$\begin{aligned} f(x,\theta^*) &= \textstyle\sum_{i=1}^p \theta_i^* f_i(x) \\ g(x,\theta^*) &= \textstyle\sum_{i=1}^p \theta_i^* g_i(x) \\ h(x,\theta^*) &= \textstyle\sum_{i=1}^p \theta_i^* h_i(x) \end{aligned}$$

*where $f_i(x), g_i(x)$ are known smooth vector fields on $\mathbb{R}^n$ and $h_i(x)$ are known smooth scalar functions.*

**Assumption 2 Relative Degree**

*The relative degree of the true system (32) is $\gamma$, and for all $\hat{\theta}$ in a ball around $\theta^*$ and all $x$ in a neighborhood of $x^0$*

$$L_{g(x,\hat{\theta})} L_{f(x,\hat{\theta})}^{\gamma-1} h(x,\hat{\theta})$$

*is bounded away from zero.*

In the discussion that follows we will be using the implicit summation notation (ie. there is a summation over repeated indices) to keep the expressions manageable. For example, we will write $f(x,\theta^*)$ as $\theta_j^* f_j(x)$. Now if we pick the following diffeomorphism

$$\begin{aligned} \Phi(x,\theta^*) &= \begin{bmatrix} h(x,\theta^*) \\ L_{f(x,\theta^*)} h(x,\theta^*) \\ \vdots \\ L_{f(x,\theta^*)}^{\gamma-1} h(x,\theta^*) \\ \Phi_{\gamma+1}(x,\theta^*) \\ \vdots \\ \Phi_n(x,\theta^*) \end{bmatrix} = \begin{bmatrix} \theta_{j_0}^* h_{j_0}(x) \\ \theta_{j_1}^* \theta_{j_0}^* L_{f_{j_1}} h_{j_0}(x) \\ \vdots \\ \theta_{j_{\gamma-1}}^* \cdots \theta_{j_0}^* L_{f_{j_{\gamma-1}}} \cdots L_{f_{j_1}} h_{j_0}(x) \\ \Phi_{\gamma+1}(x,\theta^*) \\ \vdots \\ \Phi_n(x,\theta^*) \end{bmatrix} \\ &= \begin{bmatrix} \Phi_\xi(x,\theta^*) \\ -------- \\ \Phi_\eta(x,\theta^*) \end{bmatrix} = \begin{bmatrix} \xi \\ -- \\ \eta \end{bmatrix} \end{aligned} \tag{33}$$

with

$$L_g\Phi_i = d\Phi_i g(x,\theta^*) = 0 \quad i = \gamma+1,\ldots,n$$

and $\Phi_{\gamma+1}(x,\theta^*),\ldots,\Phi_n(x,\theta^*)$ chosen so that $\Phi(x,\theta^*)$ has a nonsingular jacobian matrix at $x^0$, then we have, in the normal form,

$$(34)\quad \begin{aligned} \dot{\xi}_1 &= \xi_2 \\ &\vdots \\ \dot{\xi}_\gamma &= \theta^*_{j_\gamma}\cdots\theta^*_{j_0}L_{f_{j_\gamma}}\cdots L_{f_{j_1}}h_{j_0}(x) + \theta^*_{j_\gamma}\cdots\theta^*_{j_0}L_{g_{j_\gamma}}L_{f_{j_{\gamma-1}}}\cdots L_{f_{j_1}}h_{j_0}(x)u \\ \dot{\eta} &= q_{\theta^*}(\xi,\eta) \\ y &= \xi_1 \end{aligned}$$

where

$$(35)\quad q_{i\theta^*}(\xi,\eta) = L_{f(x,\theta^*)}\Phi_i(x,\theta^*) \quad \gamma+1 \le i \le n.$$

We assume that $x = 0$ is an equilibrium point of the system (32) (ie. $f(0,\theta^*) = 0$) and we assume $h(0,\theta^*) = 0$. Then the dynamics

$$(36)\quad \dot{\eta} = q_{\theta^*}(0,\eta)$$

are referred to as the zero-dynamics. The nonlinear system (32) is said to be minimum phase if the zero-dynamics are asymptotically stable. We will now impose the following assumption:

**Assumption 3 Exponentially Stable Zero Dynamics**
*The equilibrium point $\eta = 0$ of the zero-dynamics of the true system (32) is exponentially stable.*

Now let us consider our choice for the control law. The certainty equivalence principle suggests that we pick the appropriate linearizing control law but with the unknown parameters replaced by their estimates. We choose

$$(37)\quad u = \frac{1}{\hat{\theta}_{j_\gamma}\cdots\hat{\theta}_{j_0}L_{g_{j_\gamma}}L_{f_{j_{\gamma-1}}}\cdots L_{f_{j_1}}h_{j_0}(x)}[\hat{\theta}_{j_\gamma}\cdots\hat{\theta}_{j_0}L_{f_{j_\gamma}}\cdots L_{f_{j_1}}h_{j_0}(x) + \hat{v}]$$

To achieve tracking we pick $\hat{v}$ in the form of (29). However, we do not have exact expressions for the derivatives of $y$ which involve unknown parameters. Instead we will use estimates of the derivatives of $y$ obtained from the parameter estimates:

$$(38)\quad \hat{v} = y_M^{(\gamma)} + \alpha_1(y_M^{(\gamma-1)} - \hat{y}^{(\gamma-1)}) + \ldots + \alpha_\gamma(y_M - \hat{y})$$

where

$$(39)\quad \hat{y}^{(i)} = \hat{\theta}_{j_i}\cdots\hat{\theta}_{j_0}L_{f_{j_i}}\cdots L_{f_{j_1}}h_{j_0}(x)$$

Now let us return to the normal form. Observe that $\dot{\xi}_\gamma$ can be written as

$$(40)\quad \begin{aligned} \dot{\xi}_\gamma &= \theta^*_{j_\gamma}\cdots\theta^*_{j_0}L_{f_{j_\gamma}}\cdots L_{f_{j_1}}h_{j_0}(x) + \theta^*_{j_\gamma}\cdots\theta^*_{j_0}L_{g_{j_\gamma}}L_{f_{j_{\gamma-1}}}\cdots L_{f_{j_1}}h_{j_0}(x)u \\ &- [\hat{\theta}_{j_\gamma}\cdots\hat{\theta}_{j_0}L_{f_{j_\gamma}}\cdots L_{f_{j_1}}h_{j_0}(x) + \hat{\theta}_{j_\gamma}\cdots\hat{\theta}_{j_0}L_{g_{j_\gamma}}L_{f_{j_{\gamma-1}}}\cdots L_{f_{j_1}}h_{j_0}(x)u] \\ &+ [\hat{\theta}_{j_\gamma}\cdots\hat{\theta}_{j_0}L_{f_{j_\gamma}}\cdots L_{f_{j_1}}h_{j_0}(x) + \hat{\theta}_{j_\gamma}\cdots\hat{\theta}_{j_0}L_{g_{j_\gamma}}L_{f_{j_{\gamma-1}}}\cdots L_{f_{j_1}}h_{j_0}(x)u] \end{aligned}$$

If we define the (large dimensional) vector of all multilinear parameter product errors,

$$(41)\quad \chi = (\hat{\theta}_{j_\gamma}\cdots\hat{\theta}_{j_0}) - (\theta^*_{j_\gamma}\cdots\theta^*_{j_0})$$

then

$$(42)\quad \begin{aligned} \dot{\xi}_\gamma &= \hat{\theta}_{j_\gamma}\cdots\hat{\theta}_{j_0}L_{f_{j_\gamma}}\cdots L_{f_{j_1}}h_{j_0}(x) + \hat{\theta}_{j_\gamma}\cdots\hat{\theta}_{j_0}L_{g_{j_\gamma}}L_{f_{j_{\gamma-1}}}\cdots L_{f_{j_1}}h_{j_0}(x)u \\ &\quad + \bar{z}^T(x,u)\chi \end{aligned}$$

Note that if $\hat{\theta} - \theta^* \equiv \phi \to 0$ as $t \to \infty$ then $\chi \to 0$ as $t \to \infty$.
Substituting the certainty equivalence control law, we have

$$(43)\quad \dot{\xi}_\gamma = \hat{v} + \bar{z}^T(x,u)\chi$$

Now notice that $\hat{v}$ can be written as

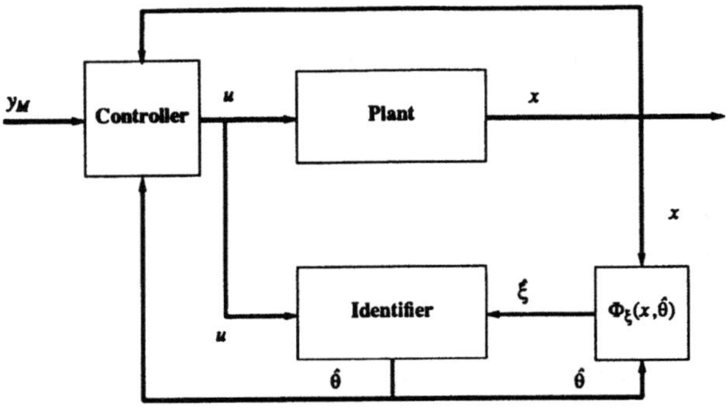


Figure 2: Block Diagram of a Semi-Indirect Adaptive Controller

$$\begin{aligned}\hat v &= y_M^{(\gamma)} + \alpha_1(y_M^{(\gamma-1)}) - y^{(\gamma-1)}) + \alpha_1(y^{(\gamma-1)} - \hat y^{(\gamma-1)}) \\ &\quad + \ldots + \alpha_\gamma(y_M - y) + \alpha_\gamma(y - \hat y)\end{aligned} \tag{44}$$

which can be seen as the exact tracking law plus an offset which is a function of the parameter error. Therefore, in the closed loop we have

$$\begin{aligned}\dot e &= Ae + z^T(x,u)\chi \\ \dot\eta &= q(\xi,\eta) \\ \xi_i &= e_i + y_M^{(i-1)}\end{aligned} \tag{45}$$

where $A$ is a Hurwitz matrix.

We will now state the following bounded tracking result under parameter uncertainty:

**Theorem 3.1 Convergence of Indirect Adaptive Controller When Identifier Input Is Sufficiently Rich**

*Consider the plant of equation (32) and the control objective of tracking the trajectory $y_M$.*

If (A1) *Assumption 1 holds (Linear Parameter Dependence),*

(A2) *Assumption 2 holds (Relative Degree),*

(A3) *Assumption 3 holds (Exponentially Stable Zero Dynamics),*

(A4) $|\chi| \to 0$ *as* $t \to \infty$,

(A5) $z^T(x,u)$ *is "cone bounded" in $x$ and uniform in $u$,*
*ie.* $|z^T(x,u)| \le \ell_{\bar z}|x| \; \forall u \in \mathbf{R}$,

(A6) $A$ *is a Hurwitz matrix,*

(A7) $q(\xi,\eta)$ *is globally Lipschitz in* $\xi,\eta$,

(A8) $y_M, \dot y_M, \ldots, y_M^{(\gamma-1)}$ *are bounded*

then *the control law given by (37) and (38) results in bounded tracking for the system (32). (ie., $x \in \mathbf{R}^n$ is bounded and $y(t) \to y_M(t)$.)*

**Proof:** See [14] □.

The drawback with this result is that it needs the convergence of the identifier for its proof of asymptotic tracking. In turn, this requires the presence of sufficient richness which is not explicit in terms of conditions on the input. This is in contrast to the direct adaptive controller ([12]) where parameter convergence is not needed for stability and asymptotic tracking.

## 3.3 Semi-Indirect Adaptive Tracking

In this section we give a modified scheme which combines attractive features of the direct and indirect schemes; as in direct adaptive control, parameter convergence is not necessary to achieve asymptotic tracking; as in indirect adaptive control, it is not necessary to overparameterize the system. The scheme uses an observer-based identifier that is similar to the one described in section 2.1 but here the states are estimated in the coordinates of the diffeomorphism. Consequently, exact knowledge of the diffeomorphism is necessary. This is made possible by using an estimated diffeomorphism that is a function of the time-varying parameter estimate (see figure 2). These results are an extension of those found in [2].

Consider the system (32) and allow assumption 3 to hold. We will modify assumption 1 so that $h(x,\theta^*)$ is no longer required to be a linear function of the parameters.

**Assumption 1 A Linear Parameter Dependence in $f$ and $g$**

*The vector fields $f(x,\theta^*)$ and $g(x,\theta^*)$ in the system (32) depend linearly on the unknown parameters as*

$$\begin{aligned} f(x,\theta^*) &= \textstyle\sum_{i=1}^{p} \theta_i^* f_i(x) \\ g(x,\theta^*) &= \textstyle\sum_{i=1}^{p} \theta_i^* g_i(x) \end{aligned}$$

*where $f_i(x), g_i(x)$ are known smooth vector fields on $\mathbf{R}^n$. The output function $h(x,\theta^*)$ is not required to have this structure.*

We will also modify assumption 2 as follows:

**Assumption 2 A Constant Relative Degree**

*For all $\hat\theta$ in a ball around $\theta^*$ and for all $x$ in a neighborhood of $x^0$,*

$$L_{g(x,\hat\theta)}h(x,\hat\theta) = L_{g(x,\hat\theta)}L_{f(x,\hat\theta)}h(x,\hat\theta) = \ldots = L_{g(x,\hat\theta)}L^{\gamma-2}_{f(x,\hat\theta)}h(x,\hat\theta) = 0$$

*and*

$$L_{g(x,\hat\theta)}L^{\gamma-1}_{f(x,\hat\theta)}h(x,\hat\theta)$$

*is bounded away from zero.*

This assumption is reasonable in the adaptive case because the structure of the system is known. The relative degree will drop only in very special cases. This assumption can be relaxed if parameter convergence is assumed. This trade-off will be discussed in more detail later.

For the development that follows, also consider the parametrized model

$$\begin{aligned} \dot x &= f(x,\theta) + g(x,\theta)u \\ y &= h(x,\theta) \end{aligned} \tag{46}$$

where $\theta \in \mathbf{R}^p$ is *fixed* and *known*. From linearization theory, if we pick the following diffeomorphism for the system (46)

$$\Phi(x,\theta) = \begin{bmatrix} h(x,\theta) \\ L_{f(x,\theta)}h(x,\theta) \\ \vdots \\ L^{\gamma-1}_{f(x,\theta)}h(x,\theta) \\ \Phi_{\gamma+1}(x,\theta) \\ \vdots \\ \Phi_n(x,\theta) \end{bmatrix} = \begin{bmatrix} \Phi_\xi(x,\theta) \\ -------- \\ \Phi_\eta(x,\theta) \end{bmatrix} = \begin{bmatrix} \xi \\ -- \\ \eta \end{bmatrix} \tag{47}$$

with

$$L_g\Phi_i = d\Phi_i g(x,\theta) = 0 \quad i = \gamma+1,\ldots,n$$

and $\Phi_{\gamma+1}(x,\theta),\ldots,\Phi_n(x,\theta)$ chosen so that $\Phi(x,\theta)$ has a nonsingular jacobian matrix at $x^0$, and if we choose the following control law

$$u = \frac{1}{L_{g(x,\theta)}L^{\gamma-1}_{f(x,\theta)}h(x,\theta)}[-L^{\gamma}_{f(x,\theta)}h(x,\theta) + v] \tag{48}$$

then we have the resulting closed loop system

$$\begin{aligned} \dot\xi_1 &= \xi_2 \\ &\vdots \\ \dot\xi_{\gamma-1} &= \xi_\gamma \\ \dot\xi_\gamma &= v \\ \dot\eta &= q_\theta(\xi,\eta) \\ y &= \xi_1 \end{aligned} \tag{49}$$

where

$$q_{i\theta}(\xi,\eta) = L_{f(x,\theta)}\Phi_i(x,\theta) \quad \gamma+1 \le i \le n$$

We can achieve bounded tracking ($y(t) \rightarrow y_M(t)$) for the system (46) in the same way as described in section 3.1.

Now consider the actual plant given in (32). We will choose, for this system, the diffeomorphism given in (47) but now $x$ is the actual state of the plant. We will replace $\theta$ by $\theta^*$ in each of the $\Phi_i(x,\theta)$ $\gamma+1 \le i \le n$. For $\Phi_i(x,\theta)$ $1 \le i \le \gamma$, $\theta$ will be replaced by $\hat{\theta}$, the time varying parameter estimate. Observe that, under these conditions, the $\xi$ states are no longer related simply by a chain of integrators. The chain of integrators structure is perturbed by the time varying nature of $\hat{\theta}$ and the fact that the time derivatives of $\xi$ are taken along the trajectories of the plant states which are a function of $\theta^*$. Consider the following two functions of $x$:

$$\begin{aligned} \hat{\xi} &= \Phi_\xi(x,\hat{\theta}) \\ \eta &= \Phi_\eta(x,\theta^*). \end{aligned} \tag{50}$$

This transformation is the same functional form as (47) but different in that $\hat{\xi}$ is evaluated along the estimates of $\theta$. Taking the time derivative along the trajectories of (32) we have

$$\begin{aligned} \dot{\hat{\xi}} &= \frac{\partial\Phi_\xi(x,\hat{\theta})}{\partial x}\dot{x} + \frac{\partial\Phi_\xi(x,\hat{\theta})}{\partial\theta}\dot{\hat{\theta}} \\ &= \frac{\partial\Phi_\xi(x,\hat{\theta})}{\partial x}[f(x,\theta^*) + g(x,\theta^*)u] + \frac{\partial\Phi_\xi(x,\hat{\theta})}{\partial\theta}\dot{\hat{\theta}} \\ &= \begin{bmatrix} L_{f(x,\theta^*)}h(x,\hat{\theta}) \\ \vdots \\ L_{f(x,\theta^*)}L^{\gamma-1}_{f(x,\hat{\theta})}h(x,\hat{\theta}) \end{bmatrix} \\ &\quad + \begin{bmatrix} L_{g(x,\theta^*)}h(x,\hat{\theta}) \\ \vdots \\ L_{g(x,\theta^*)}L^{\gamma-1}_{f(x,\hat{\theta})}h(x,\hat{\theta}) \end{bmatrix} u + \frac{\partial\Phi_\xi(x,\hat{\theta})}{\partial\theta}\dot{\hat{\theta}} \\ \dot{\eta} &= \frac{\partial\Phi_\eta(x,\theta^*)}{\partial x}\dot{x} \\ &= \begin{bmatrix} L_{f(x,\theta^*)}\Phi_{\gamma+1}(x,\theta^*) \\ \vdots \\ L_{f(x,\theta^*)}\Phi_n(x,\theta^*) \end{bmatrix} \\ &= q_{\theta^*}(\xi^*,\eta) \end{aligned} \tag{51}$$

where

$$\xi^* = \Phi_\xi(x,\theta^*).$$

The vector of tracking errors is defined as

$$e_i = \xi_i - y_M^{(i-1)} \quad 1 \le i \le \gamma \tag{52}$$

and thus, the derivative of the tracking error is

$$\begin{aligned} \dot{e} &= \begin{bmatrix} L_{f(x,\theta^*)}h(x,\hat{\theta}) \\ \vdots \\ L_{f(x,\theta^*)}L^{\gamma-1}_{f(x,\hat{\theta})}h(x,\hat{\theta}) \end{bmatrix} + \begin{bmatrix} L_{g(x,\theta^*)}h(x,\hat{\theta}) \\ \vdots \\ L_{g(x,\theta^*)}L^{\gamma-1}_{f(x,\hat{\theta})}h(x,\hat{\theta}) \end{bmatrix} u \\ &\quad + \frac{\partial\Phi_\xi(x,\hat{\theta})}{\partial\theta}\dot{\hat{\theta}} - \begin{bmatrix} \dot{y}_M \\ \vdots \\ y_M^{(\gamma)} \end{bmatrix} \\ &= \begin{bmatrix} L_{f(x,\theta^*)}h(x,\hat{\theta}) \\ \vdots \\ L_{f(x,\theta^*)}L^{\gamma-1}_{f(x,\hat{\theta})}h(x,\hat{\theta}) \end{bmatrix} + \begin{bmatrix} L_{g(x,\theta^*)}h(x,\hat{\theta}) \\ \vdots \\ L_{g(x,\theta^*)}L^{\gamma-1}_{f(x,\hat{\theta})}h(x,\hat{\theta}) \end{bmatrix} u \end{aligned} \tag{53}$$

$$
-\left(\begin{bmatrix} L_{f(x,\hat\theta)}h(x,\hat\theta) \\ \vdots \\ L_{f(x,\hat\theta)}L^{\gamma-1}_{f(x,\hat\theta)}h(x,\hat\theta)\end{bmatrix} + \begin{bmatrix} L_{g(x,\hat\theta)}h(x,\hat\theta) \\ \vdots \\ L_{g(x,\hat\theta)}L^{\gamma-1}_{f(x,\hat\theta)}h(x,\hat\theta)\end{bmatrix} u\right)
$$

$$
+\left(\begin{bmatrix} L_{f(x,\hat\theta)}h(x,\hat\theta) \\ \vdots \\ L_{f(x,\hat\theta)}L^{\gamma-1}_{f(x,\hat\theta)}h(x,\hat\theta)\end{bmatrix} + \begin{bmatrix} L_{g(x,\hat\theta)}h(x,\hat\theta) \\ \vdots \\ L_{g(x,\hat\theta)}L^{\gamma-1}_{f(x,\hat\theta)}h(x,\hat\theta)\end{bmatrix} u\right)
$$

$$
+\frac{\partial\Phi_\xi(x,\hat\theta)}{\partial\theta}\dot{\hat\theta} - \begin{bmatrix} \dot y_M \\ \vdots \\ y_M^{(\gamma)}\end{bmatrix}
$$

Observe that, from the structure of $f(x,\cdot)$,$g(x,\cdot)$,

$$
(54)\quad \begin{aligned}
L_{f(x,\theta^*)}L^i_{f(x,\hat\theta)}h(x,\hat\theta) &= \textstyle\sum_{j=1}^p \theta_j^* L_{f_j(x)}L^i_{f(x,\hat\theta)}h(x,\hat\theta)\\
L_{f(x,\hat\theta)}L^i_{f(x,\hat\theta)}h(x,\hat\theta) &= \textstyle\sum_{j=1}^p \hat\theta_j L_{f_j(x)}L^i_{f(x,\hat\theta)}h(x,\hat\theta)\\
L_{g(x,\theta^*)}L^{\gamma-1}_{f(x,\hat\theta)}h(x,\hat\theta) &= \textstyle\sum_{j=1}^p \theta_j^* L_{g_j(x)}L^{\gamma-1}_{f(x,\hat\theta)}h(x,\hat\theta)\\
L_{g(x,\hat\theta)}L^{\gamma-1}_{f(x,\hat\theta)}h(x,\hat\theta) &= \textstyle\sum_{j=1}^p \hat\theta_j L_{g_j(x)}L^{\gamma-1}_{f(x,\hat\theta)}h(x,\hat\theta)
\end{aligned}
$$

Also recall that $u$ defined in equation (48) produces an exponentially stable tracking system for $\hat\theta$ fixed. Therefore, we pick $u$ according to (48) with the expressions for the derivatives of $y$ in the tracking law (29) determined assuming $\hat\theta$ fixed. (ie. $y^{(i-1)} = \xi_i$). Then, using assumption 2A and simplifying we have

$$
(55)\quad \begin{aligned}
\dot e &= \begin{bmatrix} \sum_{j=1}^p(\theta_j^* - \hat\theta_j)L_{f_j(x)}h(x,\hat\theta) \\ \vdots \\ \sum_{j=1}^p(\theta_j^* - \hat\theta_j)L_{f_j(x)}L^{\gamma-1}_{f(x,\hat\theta)}h(x,\hat\theta)\end{bmatrix}\\
&+ \begin{bmatrix} \sum_{j=1}^p(\theta_j^* - \hat\theta_j)L_{g_j(x)}h(x,\hat\theta) \\ \vdots \\ \sum_{j=1}^p(\theta_j^* - \hat\theta_j)L_{g_j(x)}L^{\gamma-1}_{f(x,\hat\theta)}h(x,\hat\theta)\end{bmatrix} u + Ae + \tfrac{\partial\Phi_\xi(x,\hat\theta)}{\partial\theta}\dot{\hat\theta}
\end{aligned}
$$

Define a new variable $\bar\xi \in \mathbf{R}^\gamma$ with dynamics given by

$$
(56)\quad \begin{aligned}
\dot{\bar\xi} &= \tfrac{\partial\Phi_\xi(x,\hat\theta)}{\partial x}[f(x,\hat\theta) + g(x,\hat\theta)u] + \tfrac{\partial\Phi_\xi(x,\hat\theta)}{\partial\theta}\dot{\hat\theta} + \Omega(\bar\xi - \hat\xi)\\
\bar\xi(0) &= \hat\xi(0)
\end{aligned}
$$

where $\Omega$ is a Hurwitz matrix. This equation resembles (51) with two differences: (1) $\dot x$ is replaced by $f(x,\hat\theta) + g(x,\hat\theta)u$ and (2) the additional term $\Omega(\bar\xi - \hat\xi)$ appears. Define

(57) $\varepsilon = \bar\xi - \hat\xi$.

Then

$$
(58)\quad \begin{aligned}
\dot\varepsilon &= \Omega\varepsilon + \begin{bmatrix} \sum_{j=1}^p(\hat\theta_j - \theta_j^*)L_{f_j(x)}h(x,\hat\theta) \\ \vdots \\ \sum_{j=1}^p(\hat\theta_j - \theta_j^*)L_{f_j(x)}L^{\gamma-1}_{f(x,\hat\theta)}h(x,\hat\theta)\end{bmatrix}\\
&+ \begin{bmatrix} \sum_{j=1}^p(\hat\theta_j - \theta_j^*)L_{g_j(x)}h(x,\hat\theta) \\ \vdots \\ \sum_{j=1}^p(\hat\theta_j - \theta_j^*)L_{g_j(x)}L^{\gamma-1}_{f(x,\hat\theta)}h(x,\hat\theta)\end{bmatrix} u
\end{aligned}
$$

We now specify the parameter update law

(59) $\dot{\hat\theta} = g(x, u, \hat\theta, t)$.

Define

$$
(60) \quad \begin{aligned} M(x,\hat{\theta},u) &= \begin{bmatrix} L_{f_1(x)}h(x,\hat{\theta}) & \cdots & L_{f_p(x)}h(x,\hat{\theta}) \\ \vdots & & \vdots \\ L_{f_1(x)}L_{f(x,\hat{\theta})}^{\gamma-1}h(x,\hat{\theta}) & \cdots & L_{f_p(x)}L_{f(x,\hat{\theta})}^{\gamma-1}h(x,\hat{\theta}) \end{bmatrix} \\ &+ \begin{bmatrix} L_{g_1(x)}h(x,\hat{\theta})u & \cdots & L_{g_p(x)}h(x,\hat{\theta})u \\ \vdots & & \vdots \\ L_{g_1(x)}L_{f(x,\hat{\theta})}^{\gamma-1}h(x,\hat{\theta})u & \cdots & L_{g_p(x)}L_{f(x,\hat{\theta})}^{\gamma-1}h(x,\hat{\theta})u \end{bmatrix} \end{aligned}
$$

and

(61) $\phi = \hat{\theta} - \theta^*$.

Then we have

$$
(62) \quad \begin{aligned} \dot{\varepsilon} &= \Omega\varepsilon + M\phi \\ \dot{\phi} &= g(x,u,\hat{\theta},t). \end{aligned}
$$

Using the Lyapunov function candidate

(63) $V(\varepsilon,\phi) = \varepsilon^T P_0 \varepsilon + \phi^T\phi \qquad \Omega^T P_0 + P_0\Omega = -I$

and taking the time derivative along the trajectories of (62) leads to choosing

(64) $g(x,u,\hat{\theta},t) = -M^T P_0 \varepsilon$

for the parameter update law. In this case, since

(65) $\dot{V} = -\varepsilon^T\varepsilon$

we can conclude that, $\forall t \geq 0$,

$$
(66) \quad \begin{aligned} |\varepsilon(t)| &\leq \rho|\phi(0)| \quad \rho = \sqrt{\lambda_{min}^{-1}(P_0)} \\ |\phi(t)| &\leq |\phi(0)| \end{aligned}
$$

and hence $\varepsilon$ is a bounded $L_2$ function.

To study the stability of the tracking error system (55) we will define

(67) $\zeta = e + \varepsilon$

Then the tracking error $e$ can be seen as the output of a linear, time-varying filter given by

$$
(68) \quad \begin{aligned} \dot{\zeta} &= A\zeta + [(\Omega - A) - \tfrac{\partial\Phi_\xi(x,\hat{\theta})}{\partial\theta}M^T P_0]\varepsilon \\ e &= \zeta - \varepsilon \end{aligned}
$$

We will now state the following bounded tracking result under parametric uncertainty:

**Theorem 3.2 Convergence of Semi-Indirect Adaptive Controller** *Consider the plant of equation (32) and the control objective of tracking the trajectory $y_M$.*

If (A1) *Assumption 1A holds (Linear parameter dependence in f,g),*

(A2) *Assumption 2A holds (Constant relative degree),*

(A3) *Assumption 3 holds (Exponentially stable zero dynamics),*

(A4) *$q_{\theta^*}(\xi^*,\eta)$ is globally Lipschitz in $\xi^*,\eta$,*

(A5) *$\Phi_\xi(x,\theta)$ is globally Lipschitz in $\theta$ and uniform in $x$,*
*ie. $|\Phi_\xi(x,\theta^*) - \Phi_\xi(x,\hat{\theta})| \leq \ell_\phi|\phi| \ \forall x \in \mathbf{R}^n$,*

(A6) *$A$ is a Hurwitz matrix,*

(A7) *$\frac{\partial\Phi_\xi(x,\hat{\theta})}{\partial\theta}M^T$ is "cone bounded" in $x$ and uniform in $u,\hat{\theta}$,*
*ie. $|\frac{\partial\Phi_\xi(x,\hat{\theta})}{\partial\theta}M^T| \leq \ell_{\tilde{N}}|x| \ \forall u \in \mathbf{R}, \forall\hat{\theta} \in \mathbf{R}^p$,*

(A8) *$y_M, \dot{y}_M, \ldots, y_M^{(\gamma-1)}$ are bounded,*

(A9) *$|\phi(0)|$ bounded as a function of specified Lipschitz constants*

then *the control law $u$ given in (48) results in bounded tracking for the system (32). (ie. $x \in \mathbf{R}^n$ is bounded and $y(t) \to y_M(t)$ as $t \to \infty$).*

**Proof:** See [14] □.

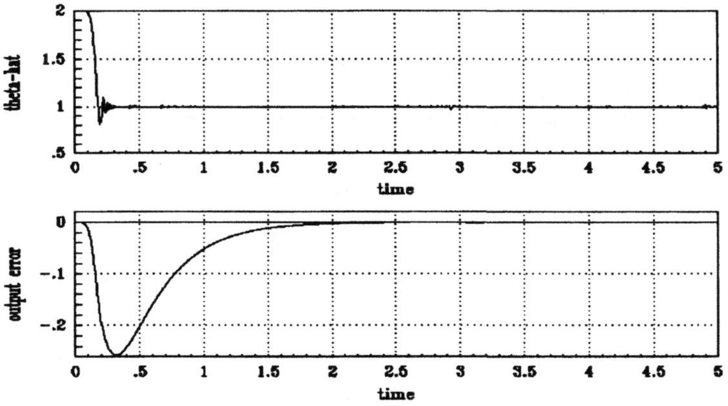


Figure 3: Indirect Adaptive Controller - Observer Based

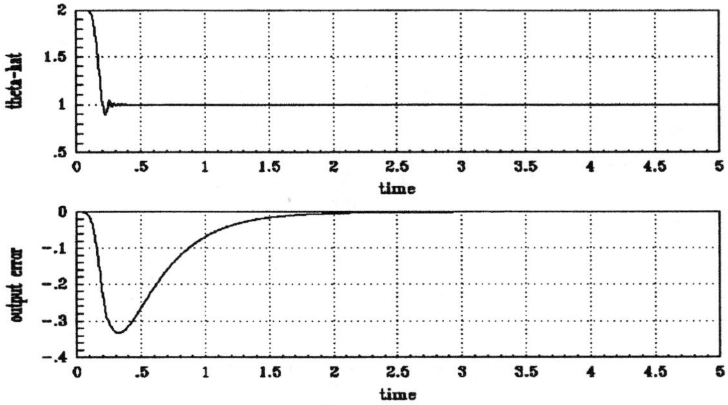


Figure 4: Semi-indirect Adaptive Controller

# 4 Closed Loop Simulations

## 4.1 Comparison of Methods

We will qualitatively compare five nonlinear control schemes, namely direct, indirect, and semi-indirect adaptive control, non-adaptive nonlinear control and sliding mode control. The system we choose to simulate is:

$$\begin{aligned} \dot{x}_1 &= x_2 + \theta\psi(x_1, x_2) \\ \dot{x}_2 &= u \\ y &= x_1 \\ \psi(x_1, x_2) &= x_1[10 + \sin(x_1)] \end{aligned} \tag{69}$$

This plant is easily linearized with

$$u = -\frac{\partial\psi}{\partial x_1}[\theta x_2 + \theta^2\psi(x_1, x_2)] + v \tag{70}$$

and output tracking is achieved by

$$v = \ddot{y}_M + \alpha_1(\dot{y}_M - \dot{y}) + \alpha_2(y_M - y) \tag{71}$$

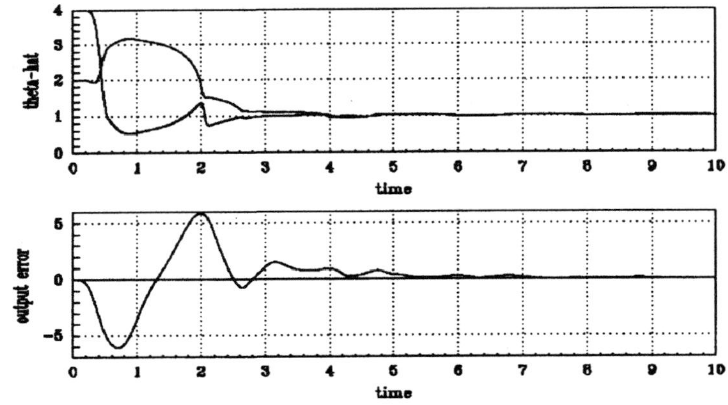


Figure 5: Direct Adaptive Controller

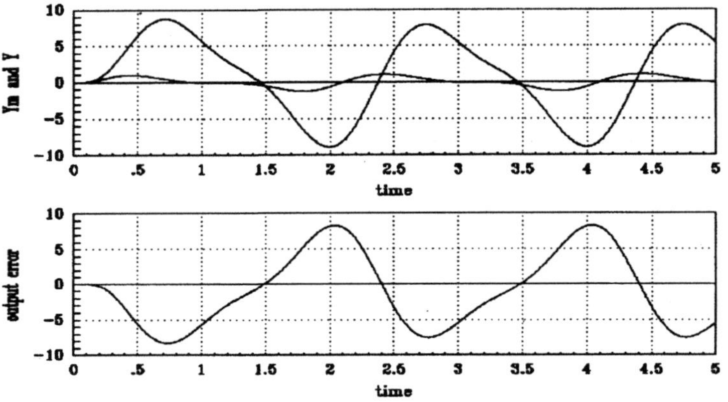


Figure 6: Non-Adaptive Controller

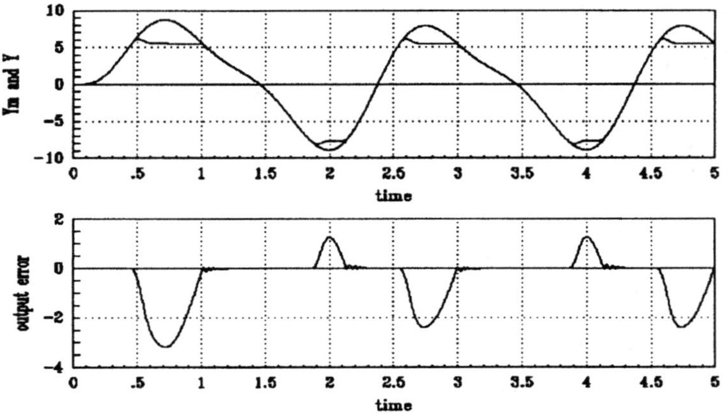


Figure 7: Sliding Mode Controller

except in the case of sliding mode, where the $\alpha_2(y_M - y)$ term is replaced by $k\text{sgn}(y_M - y)$. We picked $\alpha_1 = 30$, $\alpha_2 = 200$, and $k = 2000$ to provide good nominal tracking.

The equation for the semi-indirect parameter update is:

$$(72)\quad \dot{\hat{\theta}} = \psi(x_1, x_2)g_{11}(\hat{\xi}_1 - \bar{\xi}_1) + \psi(x_1, x_2)\frac{\partial\psi}{\partial x_1}g_{12}(\hat{\xi}_2 - \bar{\xi}_2)$$

with

$$(73)\quad \begin{aligned} \dot{\bar{\xi}}_1 &= x_2 + \psi(x_1, x_2)\hat{\theta} - g_{21}(\bar{\xi}_1 - \hat{\xi}_1) \\ \dot{\bar{\xi}}_2 &= \tfrac{\partial\psi}{\partial x_1}\hat{\theta}(x_2 + \psi(x_1, x_2)\hat{\theta}) + \psi(x_1, x_2)\dot{\hat{\theta}} - g_{22}(\bar{\xi}_2 - \hat{\xi}_2) + u \end{aligned}$$

where the constants $g_{ij}$ are gains and from equation (50)

$$(74)\quad \begin{aligned} \hat{\xi}_1 &= x_1 \\ \hat{\xi}_2 &= x_2 + \hat{\theta}\psi(x_1, x_2) \end{aligned}$$

The true value was $\theta^* = 1$ and $\hat{\theta}$ was initially at 2. For the direct adaptive controller a second parameter had to be added, $\theta_2 = \theta^2$, with an initial value of 4. The reference signal was picked to be $10sin(\pi t) + 5sin(2\pi t)$ to provide adequate excitation.

We picked $g = 500$ and $\sigma = 50$ for the indirect adaptive controller with the observer based identifier. For the semi-indirect scheme we $g_{ij} = 50$, but the update gain was scaled back to 1 since our error in the transformed space was smaller for the semi-indirect. The update gains for the direct controller were set to 1000 and 2000 for the first and second components of the regressor.

## 4.2 Simulation Results

The indirect scheme, with the observer based identifier, and the semi-indirect controller performed quite well compared with the other methods, as can be seen in figure 3 and figure 4. The parameter $\theta$ converged to the correct value in less the one second and the output error, $y - y_M$ was driven to zero. The identifier was quite robust to choices of update gains and estimator gains, $\sigma$. Virtually all reasonable values yielded convergence in less than one second. The indirect scheme was also able to handle larger perturbations in $\dot{x}_1$, such as $\psi(x_1, x_2) = {x_1}^2$ for large values of $x_1$. With this $\psi(x_1, x_2)$, the non-adaptive controller became unstable, but for the indirect scheme the identifier was able to converge quickly enough to stabilize the system. It does seem that in most cases the excitation provided by system instability drives the parameters to their true values, thus allowing the controller to stabilize the plant.

$$(75)\quad \frac{d\hat{\theta}}{dt} = -g\psi(x_1, x_2)(x_1 - \hat{x}_1)$$

so the estimation error, $x_1 - \hat{x}_1$, is driven to zero.

The direct scheme did not converge nearly as fast as the indirect, as shown in figure 5 – note the different time scale. The parameters were approaching their true values around six seconds, and the output error was driven to zero, which is what the direct method guarantees without any claims on excitation. Any hopes of speeding up the convergence would be by increasing the update gain or by increasing the amplitude of the reference signal. This would increase the elements in the regressor, and would cause the identifier to be ill-conditioned. In fact the update gain had to be reduced by a factor proportional to the square of the increase in the reference signal amplitude. The identifier in the direct scheme also has more states than the observer based identifier for the indirect. These extra states, six in all, come from filtering the regressor for the generation of the augmented output error used to drive the parameter updates and also from the additional parameter, $\theta^2$, which needs to be identified. The adaptive schemes are compared in figure 8.

The non-adaptive scheme, figure 6, performed as well as could be expected. The tracking gains could have been increased in hopes of swamping out the perturbation caused by $\psi(x_1, x_2)$, but in anything other than a noiseless environment this would be ill advised.

The sliding mode method steered the output error to zero, but when the perturbation was large, $\psi(x_1, x_2)$ was at its maximum, the system could not swamp it out as quickly. The gain $k$ was set at 2000. Larger gains caused considerable chattering in the regions were $\psi(x_1, x_2)$ was not at its maximum and would also send the numerical integrator into fits. The results are shown in figure 7.

| Comparison of Adaptive Methods | | | |
|---|---|---|---|
| Criterion | Scheme | | |
| | Indirect | Semi Indirect | Direct |
| Parameter Convergence | Fastest | Fast | Slow |
| Sensitivity to Adaptation Gains | Not Very | Slightly | Very |
| Ease of Implementation | Easy | Moderate | Difficult |
| Needs Overparametrization | No | No | Yes |
| Needs Constant Relative Degree to Prove Tracking | No | Yes | No |
| Needs Parameter Convergence to Prove Tracking | Yes | No | No |

Figure 8: Method Comparison

### 4.3 Non-Constant Relative Degree

We investigated the semi-indirect control scheme further by simulating a system which does not have a constant relative degree. Clearly the constant relative degree assumption is sufficient for asymptotic tracking, but, as will be seen, it is not a necessary condition.

The system we picked was a simple third order plant described by

$$(76)\quad \begin{aligned} \dot{x}_1 &= x_2 + \theta x_3 \\ \dot{x}_2 &= x_3 \\ \dot{x}_3 &= u \\ y &= x_1 \end{aligned}$$

We let the initial $\hat{\theta}$ be 0.1, and had $\theta^* = 0$. Hence, the relative degree would decrease for non-zero $\theta$, and the actual relative degree would be different from the initial relative degree of the estimated system. The linearizing control law was applied to the system with the same type of tracking law to close the loop as above, namely

$$(77)\quad v = y_M{}^{(3)} + \alpha_1(\ddot{y}_M - \ddot{y}) + \alpha_2(\dot{y}_M - \dot{y}) + \alpha_3(y_M - y)$$

where $\alpha_1 = 9$, $\alpha_2 = 26$, $\alpha_3 = 24$, and the input was picked to be $6[sin(2\pi t) + sin(0.25\pi t)]$. The results, shown in figure 9, reveals that the closed loop system was able to track the input, thus showing it is not necessary to have the relative degree fixed. In fact, it turns out, as previously stated, that if the relative degree were changing then we must have parameter convergence for the semi-indirect method to asymptotically track an input. Thus, if we assume constant relative degree for the semi-indirect method, then we do not need to have parameter convergence to achieve asymptotic tracking, but if we did not want to assume constant relative degree then we would need to have the parameters converge. The later is, interestingly enough, the same assumption necessary to show tracking in the indirect case. The semi-indirect scheme thus allows us two scenarios. If we are certain of the structure of our plant and can guarantee that the relative degree will not change in the neighborhood of interest, then we do not need to have strict requirements on the richness of the input. On the other hand, if we are not sure of the structure of our plant or have parasitic effects which may easily change the relative degree, then we must have a rich input to assure parameter convergence,thus giving us asymptotic tracking. It should be noted that in all the simulations that have been run (numerous but certainly not exhaustive) a system has yet to be seen where the parameters do not converge in the closed loop with just about any non-zero input.

## 5 Conclusion

In this paper, we have presented convergence results for two nonlinear adaptive control schemes. We presented an output tracking result using indirect adaptive control. This approach was based on certainty equivalence for input output linearization of nonlinear systems. Examples of identification schemes based on observation errors were also presented. The form of the identifier did not need to be specified for the convergence result and overparameterization was not necessary. However, the result was based on an assumption of identifier convergence. Simulation results were presented for this indirect adaptive control scheme using a familiar induction motor model. Simulation results

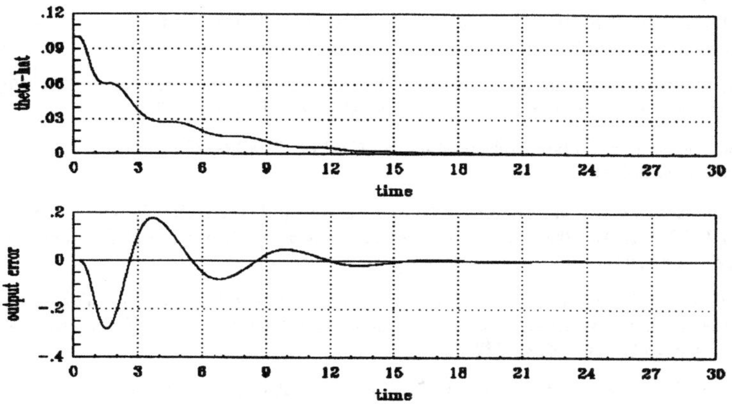


Figure 9: Semi-indirect Adaptive Controller with a Non-Constant Relative Degree Plant

were presented to compare this scheme with a direct adaptive scheme, a non-adaptive control scheme and a sliding mode scheme.

We also presented an output tracking result using a semi-indirect adaptive control scheme. The result was an extension of that found in [2]. This approach was similarly based on certainty equivalence for input output linearization of nonlinear systems. The result did not require overparameterization and did not assume parameter convergence. In contrast to the indirect scheme, this semi-indirect tracking result was dependent on the identification scheme used. Simulation results for the scheme were also presented.

**Acknowledgements**

This research was supported in part by the Army under grant ARO DAAL-88-K0572, NASA under grant NAG2-243, and NSF under grant ECS 87-15811.

# References

[1] G. Bastin and G. Campion. Indirect adaptive control of linearly parametrized nonlinear systems. To be presented IFAC Symposium, Glasgow, UK, 19-21 April 1989.

[2] G. Campion and G. Bastin. *Indirect Adaptive State Feedback Control of Linearly Parametrized Nonlinear Systems.* Technical Report AP 89.09, Laboratoire D'Automatique, Dynamique et Analyse des Systèmes, Université Catholique de Louvain, 1989.

[3] G. Georgiou and D. Normand Cyrot. De'couplage et linearisation adaptative d'un moteur à induction. Prepublication Report, CNRS-ESE, Paris, 1989.

[4] John Hauser, Shankar Sastry, and George Meyer. *Nonlinear Controller Design for Flight Control Systems.* Technical Report UCB/ERL M88/76, Electronics Research Laboratory, University of California, Berkeley, 94720, 1988.

[5] Alberto Isidori. *Nonlinear Control Systems: An Introduction.* Springer-Verlag, 1989.

[6] I. Kanellakopoulos, P. V. Kokotovic, and R. Marino. Robustness of adaptive nonlinear control under an extended matching condition. In *Nonlinear Control Systems Design, Preprints of the IFAC Symposium, Capri, Italy,* June 1989.

[7] G. Kreisselmeier. Adaptive observers with exponential rate of convergence. *IEEE Transactions on Automatic Control,* 22:2–8, 1977.

[8] P. Kudva and K.S. Narendra. Synthesis of an adaptive observer using Lyapunov's direct method. *Int. Journal of Control,* 18:1201–1210, 1973.

[9] G. Luders and K.S. Narendra. An adaptive observer and identifier for a linear system. *IEEE Transactions on Automatic Control*, 18:496–499, 1973.

[10] Jean-Baptiste Pomet and Laurent Praly. Indirect adaptive nonlinear control. In *27th IEEE Conference on Decision and Control*, pages 2414–2415, December 1988.

[11] Shankar Sastry and Marc Bodson. *Adaptive Control: Stability, Convergence, and Robustness*. Prentice Hall, 1989.

[12] S.S. Sastry and Alberto Isidori. *Adaptive Control of Linearizable Systems*. Technical Report UCB/ERL M87/53, Electronics Research Laboratory, University of California, Berkeley, 94720, June 1987.

[13] David G. Taylor, Petar V. Kokotovic, Riccardo Marino, and Ioannis Kanellakopoulos. Adaptive regulation of nonlinear systems with unmodeled dynamics. *IEEE Transactions on Automatic Control*, 34:405–412, 1989.

[14] Andrew Teel, Raja Kadiyala, Petar Kokotovic, and Shankar Sastry. *Indirect Techniques for Adaptive Input Output Linearization of Nonlinear Systems*. Technical Report UCB/ERL M89/92, Electronics Research Laboratory, University of California, Berkeley, 94720, August 1989.

# INTERPOLATION APROACH TO THE $H_\infty$ FILTER PROBLEM

D. Ho, U. Shaked, M.J. Grimble and A. Elsayed

Abstract

The $H_\infty$ optimization problem is solved by the equalizing solutions technique which requires a special type of Diophantine equations to be solved. This type of Diophantine equation is difficult to solve numerically in the multivariable case. The application of the directional interpolation approach to the $H_\infty$ filtering design problem is considered in this paper. This approach provides an explicit solution to the problem without the need for approximation techniques.

## 1. Introduction

The interpolation approach is one of the important methods developed to solve the $H_\infty$ minimization problem in control theory. The basic principle is to find some inner matrices to satisfy given interpolation conditions (Francis, Helton and Zames, 1984 [12]. These matrices are the extension of the single-input, single-output and all pass functions.

Two main approaches have been suggested for solving the inners interpolation problem. The first is based on the Pick and Nevanlinna theory. This approach was recently extended by Kimura (1987 [2]). The second approach is based on Hankel-norm techniques (Glover 1984) [1]) which is efficient in computations, but lacks structural simplicity when compared with the first approach. The calculations in the first approach are complicated, even using the directional interpolation method to simplify the procedure [2]. Shaked [11] introduced an explicit method to solve the problem without using an iterative algorithm. This approach uses the interpolation directions in a solution to the $H_\infty$ optimization problem, in state space form. The results are obtained by a simple substitution of the problem parameters.

### 1.1 *$H_\infty$ filtering problem*

The $H_\infty$ filtering problem was motivated by Grimble [8]. A theoretical analysis of the problem has been developed in [9] for scalar discrete time systems (via solving two diophantine equations). Their approaches are related to the solution of $\ell_2$ problem via the Kwakernaak [5] lemma. The results have also been extended to the multivariable case [10]. However, the problem is difficult to solve numerically, since it involves a special type of diophantine equation. In this paper, the directional interpolation method [11] will be applied to this problem so that the Diophantine equation can be replaced by a set of parameters. The $H_\infty$ filter is obtained by a simple substitution of the parameters. The notation and variables of this papers are the same as those in [10] and [11], and $\bar{X}$ is defined to be complex conjugate of X.

## 1.2 *Multivariable $H_\infty$ Filtering*

Consider the discrete-time multivariable, linear, time-invariant system shown in Fig. 1. The signals in the system may be detailed as

| | | | |
|---|---|---|---|
| Output | : | $y(t) = W(z^{-1})\xi(t) \qquad t = 0, 1, 2,..$ | (1.1) |
| Observations | : | $z(t) = M(z^{-1})W(z^{-1})\xi(t) + W_n(z^{-1})(\omega(t)+v(t)$ | (1.2) |
| Estimated Output | : | $\hat{y}(t+k \mid t) = H_f(z^{-1})z(t) \qquad k \geq 0$ | (1.3) |
| Prediction error | : | $\tilde{y}(t+k \mid t) = y(t+k) - \hat{y}(t+k \mid t)$ | (1.4) |

where $W = A_d^{-1}C_d$, $M = A_m^{-1}B_m z^{-k_m}$, $W_n = A_n^{-1}C_n$ and $W_p = A_p^{-1}B_p$ also $W \varepsilon R^{rxm}(z^{-1})$, $M \varepsilon R^{rxr}(z^{-1})$, $W_n \varepsilon R^{rxq}(z^{-1})$ and $W_p \varepsilon R^{rxr}(z^{-1})$.

The white driving noise sources $\{\xi(t)\}$, $\{w(t)\}$ and $\{v(t)\}$ are assumed to be zero-mean and independent. The covariances of the $\{\xi(t)\}$, $\{w(t)\}$ are taken to be identity and $\text{cov}[\,v(t), v(\tau)\,] = R\delta_{t\tau}$, $R \geq 0$.

It is required to find the $H_\infty$ estimator that minimises the cost function.

$$J_k^\infty = \sup_{|z|=1} (P_o(z^{-1})\Phi_{\tilde{y}\tilde{y}}(z^{-1})) \tag{1.5}$$

where $P_o$ is a dynamic weighting term and $\Phi_{\tilde{y}\tilde{y}}$ is the power spectral density matrix of the error signal $\tilde{y}(t)$

Define the left coprime factorization

$$A_f^{-1}[\tilde{C}_n, \tilde{C}_d] = [W_n, z^{k_m}MW] \tag{1.6}$$

and the generalized spectral factor $Y = A_f^{-1}D_f$ where $D_f$, $D_{fo}$ satisfy

$$D_f D_f^* = \tilde{C}_d\tilde{C}_d^* + \tilde{C}_n\tilde{C}_n^* + A_f R A_f^* = D_{fo}^* D_{fo} \tag{1.7}$$

where $D_f^t(z^{-1}) = D_f^*(z)$

Define $g = \max\{\deg(D_f), k + k_m + \deg(C_d)\}$ where $D_f(o)$ is full rank and $D_f(z^{-1}) \varepsilon R^{rxr}$ is strictly Schur. Let the following left coprime polynomial matrices be defined as

$$A_2^{-1}B_2 = B_p A_d^{-1} \tag{1.8}$$

$$A_3^{-1}B_3 = B_p B_m^{-1} A_m A_n^{-1} \tag{1.9}$$

$$\tilde{A}_2^{-1}\tilde{B}_2 = F_{1s}W_{pn}(A_d S_n)^{-1} \tag{1.10}$$

$$\tilde{A}_3^{-1}\tilde{B}_3 = F_{1s}W_{pn}(A_nMS_n)^{-1} \tag{1.11}$$

where $F_{1s}$ and $W_{pn}$ are defined in (1.16) and (1.17), respectively.
The theorem in [10] is presented below for reference purposes:

<u>*Theorem 1:*</u> *Multivariable $_\infty$ Filtering (k=0)*

*Consider the system shown in Fig. 1 and assume that the cost index (1.5) is to be minimized. The spectral factor $D_f$ is defined by (1.7). The optimal $H_\infty$ filter is computed from the solution* ($G_1$, $S_1$, $F_1$), *with* $F_1$ of minimal degree, of the equations:

$$G_1D_f^*z^{-g} + \tilde{A}_2F_1D_{fo} = \tilde{B}_2C_d\tilde{C}_d^*z^{k+k_m-g} \tag{1.12}$$

$$S_1D_f^*z^{-g} - \tilde{A}_3F_1D_{fo}z^{-k-k_m} = \tilde{B}_3(C_n\tilde{C}_n^* + A_nRA_f^*)z^{-k_m-g} \tag{1.13}$$

*where $\tilde{A}_2, \tilde{B}_2, \tilde{A}_3, \tilde{B}_3$ satisfy the following left coprime factorizations:*

$$\tilde{A}^{-1}\tilde{B}_2 = F_{1s}W_{pn}S_n^{-1}A_d^{-1} \tag{1.14}$$

$$\tilde{A}_3^{-1}\tilde{B}_3 = F_{1s}W_{pn}S_n^{-1}M^{-1}A_n^{-1} \tag{1.15}$$

*and the strictly Hurwitz spectral factor $F_{ls}$ satisfies*

$$F_{ls}F^*_{ls} = F_lF_l^* \tag{1.16}$$

*The scalar g is the smallest positive integer to make these equations polynomial in* $z^{-1}$. *The right coprime Hurwitz polynomial matrices $S_n$ and $W_{pn}$ satisfy.*

$$W_{pn}S_n^{-1} = S^{-1}W_{po} \tag{1.17}$$

*here* $S \,\varepsilon\, R^{rxr}(z^{-1})$ *is obtained by spectral factorization using:*

$$SS^* = \lambda^2I_r - A_{po}^{-1}B_{po}A_d^{-1}C_d(I_r - \tilde{C}_d^*D_f^{*-1}D_f^{-1}\tilde{C}_d)C_d^*A_d^{*-1}B_{po}^*A_{po}^{*-1} \tag{1.18}$$

*$H_\infty$ filter transfer function matrix:*

$$H_f = W_{po}^{-1}SF_{1s}^{-1}\tilde{A}_2^{-1}G_1D_f^{-1}A_f \tag{1.19}$$

*Implied equation:*

$$\tilde{A}_2^{-1}G_1z^{-k-k_m} + \tilde{A}_3^{-1}S_1 = F_{1s}S^{-1}W_{po}B_m^{-1}A_mA_f^{-1}D_f \tag{1.20}$$

*Optimum function:*

$$X = \lambda^2I_r \tag{1.21}$$

<u>Proof</u>: [10] •
•

Conclusions which may be drawn from this theorem are that the $H_\infty$ estimator $H_f$ in (1.19) can be expressed in terms of the unknown transfer function matrices $S(\lambda)$, $F_{1s}$ and $G_1$. Notice that $F_{1s}$ and G are the solution of the special type of Diophantine equations (1.12) to (1.15). By substituting (1.14), (1.15) to (1.12), (1.13), respectively (see Section 3), $F_{1s}$ will appear on the right hand sides of (1.12) and (1.13). It is not a standard Diophatine equations problem and $F_{1s}$ is also required to satisfy the condition in (1.16). The solution in the SISO case can be easily obtained, however, it is a difficult problem in the multivariable case. There is no exact numerical solution to the problem of finding $F_{1s}$ and $F_1$. The solution $H_f$ also depends on the transfer function matrix $S(\lambda)$ that satisfies (1.18). An alternative way to find $H_f$ is by using the interpolation approach which will be discussed in Section 3. An explicit method to find the spectral factor of $S(\lambda)$ will also be introduced in Section 5.

## 2. Interpolation Approach

The directional interpolation approach using inner matrix properties is discussed in this section. The required inner matrices are defined as:

$$U^*(s)U(s) = I_r \text{ where } U^*(s) = U^t(-s). \tag{2.1}$$

The directional interpolation problem is to find an inner matrix U and the minimum $|\lambda|$ such that

$$U(\alpha_i)\,\underline{d}_i = \lambda^{-1}\,\underline{w}_i \quad ; \; Re(\alpha_i) > 0, \qquad i = 1,\ldots,p. \tag{2.2}$$

This inner matrix U has a special structure which solves the $H_\infty$ optimization problem. In fact the motivation of this work came from the equalizing solutions of the $H_\infty$ work.

### 2.1 *$H_\infty$ equalizing solution*

It is required to find an $M_1 \in H^\infty_{mxr}$ such that:

$$M_1T_1 - T_2 = \lambda U \tag{2.3}$$

where $T_1 \in H^\infty_{r\,xr}$, $T_2 \in H^\infty_{mxr}$, U is inner in $H^\infty_{mxr}$ and min $|\lambda|$ $\{\alpha_i, i=1,..p\}$ are the distinct right half plane zeros of $T_1$. The $\alpha_i$ are the corresponding input zero-directions such that:

$$T_1(\alpha_i)\underline{d}_i = 0 \tag{2.4}$$

$$\underline{w}_i \triangleq -T_2(\alpha_i)\underline{d}_i \tag{2.5}$$

In (2.3) $M_1$, U and $\lambda$ are unknowns. By multiplying $\underline{d}_i$ on both sides of (2.3)

$$M_1T_1(\alpha_i)\underline{d}_i - T_2(\alpha_i)\underline{d}_i = \lambda U(\alpha_i)\underline{d}_i \tag{2.6}$$

Hence (2.6) becomes the same form as the interpolation problem in (2.2) for $M_1T_1(\alpha_i)\underline{d}_i = 0$. Thus $\lambda$ and U can be obtained first via (2.2), then $M_1$ can be calculated from (2.3).

The matrix U(s) in (2.2) has to satisfy a special structure as in (2.1). The following lemma and theorem establish a direct relationship to the filtering result discussed in the next section.

Lemma 1:

Define

$$F_\lambda = \{f_{ij}\} \text{ where } f_{ij} = (\bar{\alpha}_i + \alpha_j)^{-1}[\underline{\bar{d}}_i^t \underline{d}_j - \lambda^{-2}\underline{\bar{w}}_i^t \underline{w}_j] \tag{2.7}$$

or $F_\lambda$ can be found by

$$\bar{D}^t D - \lambda^{-2}\bar{W}^t W = \bar{\Lambda} F_\lambda + F_\lambda \Lambda_\alpha$$

where $D = [\underline{d}_1, \underline{d}_2, ..., \underline{d}_p]$, $W = [\underline{w}_1, ..., \underline{w}_p]$ and $\Lambda_\alpha = \text{diag}\{\alpha_i, i = 1, ..., p\}$ (2.8)

Hence, U(s) solves the interpolation problem only if $F_\lambda \geq 0$.

Proof: [ 11 ].

Note: For $F_\lambda \geq$ means $F_\lambda = (F^{1/2})^t (F^{1/2})$ (2.9)

where $F^{1/2} \triangleq [\underline{f}_1, \underline{f}_2, ..., \underline{f}_p]$

*Theorem 2*

*For $F_\lambda \geq 0$, the $H_\infty$ interpolation problem can be solved by*

$$U(s) = [I_m - \tilde{C}(sF_\lambda + \tilde{A})^{-1}\tilde{C}^t]\theta \tag{2.10}$$

$$\text{if } \tilde{A} + \tilde{A}^t = \tilde{C}^t \tilde{C} \tag{2.11}$$

$$\text{where } C \triangleq \theta D - \lambda^{-1} W \quad , \tilde{A} \triangleq \tilde{C}^t \theta D - F_\lambda \Lambda_\alpha, \tag{2.12}$$

$\theta \ \varepsilon \ R_{mxr}$ is an arbitrary matrix subject to $\theta^t \theta = I_r$ (m > r).

Proof: [ 11 ] To prove U(s) is an inner matrix satisfying (2.1), the sufficient condition (2.11) is required.

Assuming a special structure for U in (2.10), $F_\lambda$ is identical to the Pick-Nevanlinna theory [ 11 ]. As $\lambda$ decreases towards $\lambda_m$, one of the poles of U(s) approaches $\infty$ where $\det\{F_{\lambda m}\} = 0$. Therefore $\lambda_m$ is the minimum and U(s) is thus the optimal solution. Notice that C, $F_\lambda$ and $\tilde{A}$ may be complex, however, U(s) is a real-rational matrix. In the next section the $H_\infty$ filtering equation can be written in the form of (2.3) and $H_f$ is written in terms of parameters $T_1$, $T_2$ and U(s).

## 3. Application of the Interpolation Approach to $H_\infty$ Filtering

In this section the interpolation approach is applied to the $H_\infty$ filtering calculation. $F_1$ and $F_{1s}$ in (1.16) will be replaced by $UU^*$ in solving the problem. Multiply (1.12) from the left by $A_2^{-1}$ and substitute (1.14):

$$\tilde{A}_2^{-1} G_1 D_f z^{-g} + F_1 D_{fo} = F_{1s} W_{pn} S_n^{-1}(\lambda) A_d^{-1} \tilde{C}_d \tilde{C}_d^* z^{k+k_m-g} \tag{3.1}$$

Multiply both sides of the last equation by $F_{1s}^{-1}$ and $D_{fo}^{-1}$ from the left and the right, respectively,

$$F_{1s}^{-1}\tilde{A}_2^{-1}G_1D_f^*D_{fo}^{-1}z^{-g} + F_{1s}^{-1}F_1 = W_{pn}S_n^{-1}(\lambda)A_d^{-1}\tilde{C}_d\tilde{C}_d^*D_{fo}^{-1}z^{k+k_m-g} \tag{3.2}$$

Define the following parameters

$$U = -F_{1s}^{-1}F_1 \tag{3.3}$$

$$M_1 = F_{1s}^{-1}\tilde{A}_2^{-1}G_1 \tag{3.4}$$

$$T_1 = D_f^*D_{fo}^{-1}Z^{-g} \tag{3.5}$$

$$T_2 = W_{pn}S_n^{-1}(\lambda)A_d^{-1}\tilde{C}_d\tilde{C}_d^*D_{fo}^{-1}z^{k+k_m-g} \tag{3.6}$$

Hence (3.2) can be written as

$$M_1T_1 - T_2 = U \tag{3.7}$$

The problem is to find U and M, for a given $\lambda$. The $T_1$ and $T_2$ are known transfer function matrices.

In order to show that this problem is equivalent to the $H_\infty$ directional interpolational problem, the matrix U has to be shown to be an inner matrix.

Use the identity in (1.16),

$$F_{1s}F_{1s}^* = F_1F_1^* \tag{3.8}$$

By multiplying both sides by $F_{1s}^{-1}$ and $(F_{1s}^*)^{-1}$, from the left and the right, respectively

$$(F_{1s}^{-1}F_1)(F_1^*F_{1s}^{*-1}) = UU^* = I \tag{3.9}$$

The matrix U can therefore be obtained in the form of Theorem 2.

$$U = I - C(sF_\lambda + \tilde{A})^{-1}C^t \tag{3.10}$$

The necessary condition for U to exist is $F \geq 0$, if $\lambda$ is sufficiently large. Also

$$U^*U = (F_{1s}^{-1}F_1)^*(F_{1s}^{-1}F_1) = F_1^*(F_{1s}F_1^*)^{-1}F_1 = F_1^*(F_1F_1^*)^{-1}F_1 = I$$

The matrix U can therefore be obtained in the form of (2.10)

$$U = I - \tilde{C}(sF_\lambda + \tilde{A})^{-1}\tilde{C}^t$$

C, $F_\lambda$ and $\bar{A}$ are defined in (2.7) and (2.12). The necessary condition for U to exist is $F_\lambda \geq o$, if $\lambda$ is sufficiently large.

When $\lambda$ decreases to a value $\lambda_m$, F will become nearly singular. For $\lambda = \lambda_m$, det(F) will be zero and the minimum $\lambda$ will be $\lambda_m$.

Hence $M_1$ can also be obtained from (3.4) and (3.7) as:

$$M_1 = F_{1s}^{-1}\tilde{A}_2^{-1}G_1 = (U + T_2)T_1^{-1} \tag{3.11}$$

3.1 *Filter equation*

The transfer-function matrix of the filter $H_f$ is given from (1.19), hence from (3.11) $H_f$ can be rewritten as;

$$H_f = W_{po}^{-1}S(\lambda)M_1D_f^{-1}A_f \tag{3.12}$$

or $$H_f = W_{po}^{-1}S(\lambda)(U + T_2)T_1^{-1}D_f^{-1}A_f \tag{3.13}$$

$$= W_{po}^{-1}S(\lambda)(U + T_2)D_{fo}D_f^{*-1}D_f^{-1}A_fz^g \qquad [\text{from } (3.5)] \tag{3.14}$$

From (1.7) $$(D_f^*)^{-1}D_f^{-1} = D_{fo}^{-1}(D_{fo}^*)^{-1} \tag{3.15}$$

$H_f$ can be written as

$$H_f = W_{po}^{-1}S(\lambda)(U + T_2)(D_{fo}^*)^{-1}A_fz^g \tag{3.16}$$

It is not essential to calculate $G_1$, $\tilde{A}_2$, $\tilde{B}_2$, F and $F_{1s}$, all that is required is U. The inner U can be obtained from (2.10) and (2.11). The $H_f$ has been shown to be stable in [10]. Expression (3.16) is a simplified version of (1.19) and the characteristic of $H_f$ will not change. A detailed procedure of computing the matrix U is shown in the next section.

4. Computation Procedures

1. Calculate the right half plane zeros $\{\alpha_i, i = 1,...,p\}$ and the input directions input directions $\{\underline{d}_i, i=1,..,p\}$ of $T_1$ from (3.5) where p is the McMillan deg of $D_{fo}^{-1}$ and $D = [\underline{d}_1,..,\underline{d}_p]$ and $\Lambda_\alpha = \text{diag}\{\alpha_i,..,i=1,p\}$.

2. Set a sufficiently large initial value of $\lambda$.

3. Calculate $T_2$ from (3.6) and $w_i = -T_2(\alpha_i)d_i$ for all i=1,..,p. Set $W = [\underline{w}_1,..,\underline{w}_p]$

4. Find

$$F = \{f_{ij}\} \quad f_{ij} = \frac{\bar{\underline{d}}_i^t\underline{d}_j - \bar{\underline{w}}_i^t\underline{w}_j}{\bar{s}_i + s_j}$$

where $$s_i = \frac{\alpha_i - 1}{\alpha_i + 1}$$

or find F from (2.3) using the Sylvestor equation.

5. Check whether F is > 0, if Yes reduce $\lambda$ and go to step 4. If F is indefinite, increase $\lambda$ and go to step 4. If the smallest eigenvalue of F is positive and less than a tolerance, then go to 4.

6. Set $\lambda_m = \lambda$. Calculate U as follows

$$U(z) = I - \tilde{C}\left(\frac{z-1}{z+1} F_\lambda + \tilde{A}\right)^{-1}\tilde{C}^t$$

$$\tilde{A} = C^t D - F_\lambda \Lambda_\alpha \ , \tilde{C} = \theta D - \lambda^{-1} W$$

7. $H_f$ is obtained by (3.16) for $\lambda = \lambda_m$.

Remarks:

(1) The interpolation approach discussed in Section 2 is in the continuous time (s-domain), bilinear-transformations from s to z domains are used in step 4 and 6.

(2) The calculation of $T_2$ requires the spectral factor $S(\lambda)$ in (1.18).

At present, the spectral factorization is not available in z-domain, hence $SS^*(\lambda)$ has to be converted from discrete time using bilinear transformation, hence, ill conditioned transfer function matrix may arise. Further investigations are required to improve the spectral factorization of transfer-function matrix in order to obtain a robust numerical $H_\infty$ filtering solution.

## 6. Conclusions

The $H_\infty$ filtering problem considered in this paper represents a new class of filtering problem motivated by the need to keep the estimation error spectrum small over a range of frequencies. The $H_\infty$ approach used here is based on the polynomial approach in the related $H_\infty$ robust control problem. The $H_\infty$ filtering problem in multivariable discrete time systems is difficult to sovle numerically. Using the directional interpolation method [ 11 ], the special type diophantine equations can be replaced by a set of parameters and the filter can be computed without any approximation technique.

## References


1. Glover, K. 'All optimal Hankel-norm approximations of linear multivariable system and their $L_\infty$-error bounds, Int. J. Contr., Vol. 39, 1115-1119, 1984.
2. H. Kimura, *Directional interpolation approach to $H_\infty$-optimization and robust stabilization*, IEEE Trans. Auto Contr., Vol. 32, pp. 1085-1093, 1987.
3. M.J. Grimble, *$H_\infty$ and LQG robust design methods for uncertain linear systems*, IFAC Workshop on Model Error Concepts and Compensation, Boston, 1985.
4. M.J. Grimble, *Optimal $H_\infty$ robustness and the relationship to LQG design problems*, Int. J. Control, Vol. 43, No. 2, pp. 351-372, 1986.
5. H. Kwakernaak, *Robustness optimization of linear feedback systems*, in Proc. 22nd IEEE Conf. Decision Contr., San Antonio, Texas, 1983.

6. H. Kwakernaak, *Minimax frequency domain performance and robustness optimization of linear feedback systems*, IEEE Trans. Automat. Contr., Vol. AC-30, pp. 994-1004, 1985.
7. G. Zames and B.A. Francis, *Feedback, minimax sensitivity and optimal robustness*, IEEE Trans. Automat. Contr., Vol. AC-28, pp. 585-601, 1983.
8. M.J. Grimble, *Minimax design of optimal linear filters*, Presented at the MTNS Conference, Arizona, 1987.
9. M.J. Grimble and A. Elsayed, *$H_\infty$ design of optimal digital linear filters*, IEEE Colloquium on Digital Signal Processing, London 1987.
10. A. Elsayed, M.J. Grimble and D. Ho., *On $H_\infty$ filtering in multivariable discrete-time systems*, IEEE International Symposium on Circuits and Systems, Helsinki, 1988.
11. U. Shaked, *The explicit structure of inner matrices and its application in $H_\infty$ optimization*, submitted to IEEE Trans. Auto. Contr.,
12. B.A. Francis, J.W. Helton and G. Zames *$H_\infty$ optimal feedback controllers for linear controllers for linear mutlivariable systems*, IEEE Trans. Automat. Contr., Vol. AC-29 888-900, 1984.
13. U. Shaked, *Explicit expression for the special factor of Para-hermitian matrices*, Internal Report. (submitted).

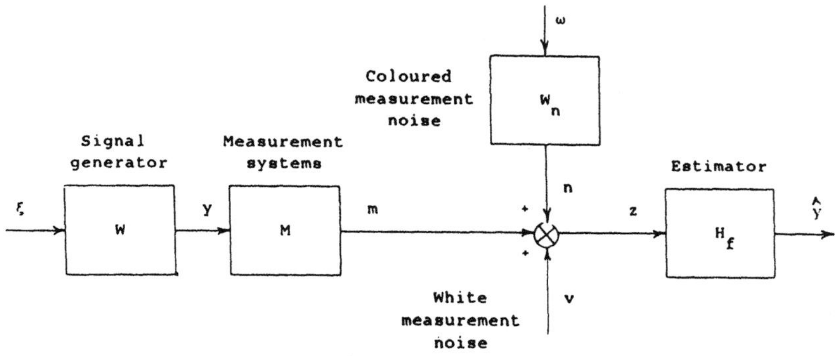


Figure 1. Signal and noise models.

# Stochastic Disturbance Decoupling

Hiroshi Inaba† and Naohisa Otsuka†

**Abstract.** *For a linear control system having white Gaussian processes as its system and measurement disturbances, we first formulate a stochastic version of the usual disturbance decoupling problem. More precisely, contructing a Kalman-Bucy filter to generate the best estimate of the state, we consider the problem of finding a feedback of the state estimate so as to minimize the corresponding output in the mean square sense. Then we prove necessary and sufficient conditions for its solvability.*

**Key Words.** disturbance decoupling, stochastic system, multivariable linear system.

## I. Introduction.

In the framework of the so-called geometric approach of Wonham[3], various disturbance decoupling problems (DDP) for linear control systems have been extensively investigated in the recent literature(See e.g.,[3] – [6]). The DDP using state feedback with or without stability has been studied by Wonham[3]. More general DDP's are those using measurement feedback, and this type of problems has also recieved a great deal of interest in the recent papers[4] – [6].

However, in the previous investigations it is assumed that all the disturbances are deterministic and no measurement disturbances are present. The first author[9] of the present paper has already formulated and briefly discussed a more realistic case where a measurement disturbance is present and all the disturbances involved are white Gaussian processes. In this paper, we will study this stochastic disturbance decoupling problem (SDDP) in more detail, and obtain more general results than those in [9].

---

† Department of Information Sciences, Tokyo Denki University, Hatoyama-Machi, Hiki-Gun, Saitama 350 – 03, Japan.

In Section II, we will first formulate our SDDP. It is then shown that the influence of stochastic disturbances on the output can be decomposed into a sum of two terms one of which gives the lowest bound of the influence and the other of which is only the term that may be acturally (a) minimized to zero or (b) reduced arbitrarily close to zero. Accordingly, we propose the two problems, SDDPZ and SDDPA, corresponding to the cases (a) and (b), respectively. We then give preliminaries in Section III for investigating those problems. Finally, in Section IV it is first shown that the SDDPZ is equivalent to the usual DDP and a necessary and sufficient condition for its solvability is proved. It is then shown that the SDDPA is equivalent in a sense to an almost disturbance decoupling problem of Willems[8], and a necessary and sufficient condition for its solvability is obtained.

## II. Formulation of the Stochastic Disturbance Decoupling Problem.

The stochastic linear control system to be considered is described as

(2.1)
$$\Sigma : \begin{cases} \dfrac{d}{dt}x(t) = Ax(t) + Bu(t) + Gw(t), \qquad x(t_0) = x_0 \\ y(t) = Cx(t) + v(t) \\ z(t) = Dx(t) \end{cases}$$

where $A \in \mathbf{R}^{n \times n}$, $B \in \mathbf{R}^{n \times r}$, $G \in \mathbf{R}^{n \times q}$, $C \in \mathbf{R}^{m \times n}$ and $D \in \mathbf{R}^{p \times n}$ are real constant matrices, $w(t)$ and $v(t)$ are independent white Gaussian processes with zero means and covariance intensity matrices $W > 0$ and $V > 0$, i.e.,

$$\mathrm{E}[w(t)w^{\mathrm{T}}(s)] = W\delta(t-s) \text{ and } \mathrm{E}[v(t)v^{\mathrm{T}}(s)] = V\delta(t-s),$$

and $x_0$ is a Gaussian random vector. The $y(t)$ and $z(t)$ represent its measurement and output processes, respectively.

Suppose that the control input $u(t)$ at time $t$ is constructed as a certain functional $f$ of the past measurement data $Y(t;t_0) := \{y(\tau)\,;\, t_0 < \tau \leqq t\}$, i.e.,

(2.2)
$$u(t) = f(t, Y(t;t_0)).$$

Then, roughly speaking, our stochastic disturbance decoupling problem is to find a functional $f$ which minimizes in a certain sense the influence of the disturbances $w(t)$ and $v(t)$ on the output $z(t)$. In this investigation, its admissible functionals are restricted to be linear, and in particular only the following special form will be considered:

(2.3)
$$u(t) = -F\hat{x}(t)$$

where $\hat{x}(t)$ is the best estimate of the state $x(t)$ in the least mean square sense based on the given data $Y(t;t_0)$, and $F$ is a matrix in $\mathbf{R}^{r \times n}$. Of course, $\hat{x}(t)$ equals the conditional expectation $\mathrm{E}[x(t)|Y(t;t_0)]$, and it can be obtained by a Kalman-Bucy filter[1].

To formulate our problem more precisely, we make the following assumptions.

(2.4) **Assumption.** System $\Sigma$ of (2.1) is assumed to satisfy the following conditions.

(i) $(A,G)$ is stabilizable.

(ii) $(A,C)$ is detectable.

(iii) $(A,B)$ is stabilizable. □

Let us denote the state estimation error by $\tilde{x}(t)$, and its variance matrix by $\tilde{Q}(t)$, i.e.,

(2.5) $$\tilde{x}(t) := x(t) - \hat{x}(t),$$

(2.6) $$\tilde{Q}(t) := \mathrm{E}[\tilde{x}(t)\tilde{x}^{\mathrm{T}}(t)].$$

It is well known ( See, e.g., [2,p.367] ) that under Assumption (2.4, i, ii) the estimate $\hat{x}(t)$ satisfies the following steady-state Kalman-Bucy filter :

(2.7) $$\frac{d}{dt}\hat{x}(t) = (A - KC)\hat{x}(t) + Ky(t) + Bu(t).$$

Here, $K$ is the steady-state gain matrix given by

(2.8) $$K = \tilde{Q}C^{\mathrm{T}}V^{-1}$$

where $\tilde{Q}$ is a nonnegative-definite symmetric matrix defined by either

$$\tilde{Q} := \lim_{t_0 \to -\infty} \tilde{Q}(t) = \lim_{t \to \infty} \tilde{Q}(t)$$

or the unique nonnegative-definite solution of the algebraic Riccati equation

(2.9) $$A\tilde{Q} + \tilde{Q}A^{\mathrm{T}} + GWG^{\mathrm{T}} - \tilde{Q}C^{\mathrm{T}}V^{-1}C\tilde{Q} = 0.$$

Now, substituting (2.3) into (2.1) and (2.7), and arranging terms with (2.5), we can easily obtain the following composite system

(2.10) $$\Sigma_F : \begin{cases} \dfrac{d}{dt}\begin{bmatrix} \hat{x}(t) \\ \tilde{x}(t) \end{bmatrix} = \begin{bmatrix} A - BF & KC \\ 0 & A - KC \end{bmatrix}\begin{bmatrix} \hat{x}(t) \\ \tilde{x}(t) \end{bmatrix} + \begin{bmatrix} 0 & K \\ G & -K \end{bmatrix}\begin{bmatrix} w(t) \\ v(t) \end{bmatrix} \\ z(t) = D\hat{x}(t) + D\tilde{x}(t). \end{cases}$$

We note that by virtue of Assumption (2.4, i, ii) the matrix $(A-KC)$ is stable, and furthermore that Assumption (2.4, iii ) ensures the existence of an $F$ such that $A-BF$ is a stable matrix. So we denote by $\mathbf{F}_s$ the set of all those $F$'s such that $A-BF$ are stable, i.e.,

$$\mathbf{F}_s := \{ F \in \mathbf{R}^{r\times n} \mid A - BF \text{ is stable} \}.$$

For every $F \in \mathbf{F}s$, System $\Sigma_F$ of (2.10) is asymptotically stable, and hence the steady-state solutions for $\hat{x}(t)$ and $\tilde{x}(t)$ exist. For such steady-state solutions, it is meaningful to define

(2.11) $$J(F) := \mathrm{E}\,\|\,z(t)\,\|^2, \qquad F \in \mathbf{F}_s$$

because $z(t)$ is a stationary process and hence $\mathrm{E}\,\|\,z(t)\,\|^2$ is independent of $t$.

It is ready to state formally our stochastic disturbance decoupling problem.

(2.12) **Stochastic Disturbance Decoupling Problem**(SDDP). Suppose that System $\Sigma$ of (2.1) satisfies Assumption (2.4), and construct the composite system $\Sigma_F$ of (2.10). Then, find (if possible) an $F \in \mathbf{F}_s$ which minimizes $J(F)$. □

It is easily seen that the steady-state solutions $\hat{x}(t)$ and $\tilde{x}(t)$ of (2.10) have zero means. Moreover, it is obvious that $\hat{x}(t)$ and $\tilde{x}(t)$ are statistically independent because they are Gaussian and orthogonal. Hence, we easily see from (2.10) that

(2.13) $$J(F) = \mathrm{E}\,\|\,Dx(t)\,\|^2 = \mathrm{E}\,\|\,D\hat{x}(t)\,\|^2 + \mathrm{E}\,\|\,D\tilde{x}(t)\,\|^2$$

$$= \mathrm{Tr}\,D\hat{Q}D^{\mathrm{T}} + \mathrm{Tr}\,D\tilde{Q}D^{\mathrm{T}}$$

where

(2.14) $$\hat{Q} := \mathrm{E}[\hat{x}(t)\hat{x}(t)^{\mathrm{T}}] \text{ and } \tilde{Q} := \mathrm{E}[\tilde{x}(t)\tilde{x}(t)^{\mathrm{T}}].$$

However, since $\tilde{x}(t)$ is independent of $F$ and so is $\tilde{Q}$, it follows from (2.13) that

(2.15) $$J(F) \geqq \mathrm{Tr}\,D\tilde{Q}D^{\mathrm{T}} \qquad \text{for all } F \in \mathbf{F}_s.$$

Based on the above argument we thus propose the following two problems.

(2.16) **Problem**(SDDPZ). Let System(2.1) be given, Assumption(2.4) be satisfied, and the composite system(2.10) be constructed. Then, the SDDPZ is to find (if possible) an $F \in \mathbf{F}_s$ such that $\mathrm{Tr}D\hat{Q}D^{\mathrm{T}} = 0$, or equivalently $J(F) = \mathrm{Tr}D\tilde{Q}D^{\mathrm{T}}$. □

(2.17) **Problem**(SDDPA). Let the same hypotheses as those of (2.16) be given. Then, the SDDPA is to find (if possible) an $F \in \mathbf{F}_s$ for each $\varepsilon > 0$ such that $\mathrm{Tr}D\hat{Q}D^{\mathrm{T}} < \varepsilon$, or equivalently $J(F) < \varepsilon + \mathrm{Tr}D\tilde{Q}D^{\mathrm{T}}$. □

It is easily seen that if the SDDPZ is solvable then so are the SDDPA and the SDDP. However, the solvability of the SDDPA does not necessarily imply that of the SDDP, and hence when this is the case we say that SDDP (2.12) is solvable in the *almost sense*.

## III. Preliminaries.

Recall that the steady-state solutions $\hat{x}(t)$ and $\tilde{x}(t)$ of (2.10) have zero means, and are statistically independent. It is not difficult to calculate their variance matrices $\hat{Q}$ and $\tilde{Q}$ defined by (2.14). In fact, we can show [2,p.104] that for each $F \in \mathbf{F}_s$, $\hat{\mathrm{Q}}$ is given by the unique solution of

(3.1) $$(A-BF)\hat{Q}+\hat{Q}(A-BF)^{\mathrm{T}}+KVK^{\mathrm{T}}=0.$$

Moreover, we can easily obtain the following expression for $\hat{Q}$:

(3.2) $$\hat{Q}=\int_0^{\infty} e^{(A-BF)t}KVK^{\mathrm{T}}e^{(A-BF)^{\mathrm{T}}t}dt.$$

Similarly, the error variance matrix $\tilde{Q}$ which is given by the unique solution of (2.9) can be easily shown to have the expression

(3.3) $$\tilde{Q}=\int_0^{\infty} e^{(A-KC)t}(GWG^{\mathrm{T}}+KVK^{\mathrm{T}})e^{(A-KC)^{\mathrm{T}}t}dt.$$

We now introduce some notations used in the following discussion. First, we use notation $<A|\mathrm{Im}B>$ for the reachable subspace of $(A,B)$, i.e.,

(3.4) $$<A|\mathrm{Im}B>:=\mathrm{Im}B+A\mathrm{Im}B+\cdots+A^{n-1}\mathrm{Im}B=\bigcup_{t\geq 0}\mathrm{Im}\,e^{At}B$$

where $\mathrm{Im}B$ stands for the image of $B$. Further, the norm $\|M\|$ of a matrix $M$ is defined to be the maximum singular value of $M$, i.e., $\|M\|:=[\lambda_{\max}(M^{\mathrm{T}}M)]^{1/2}=[\lambda_{\max}(MM^{\mathrm{T}})]^{1/2}$ where $\lambda_{\max}(\cdot)$ indicates the maximum eigenvalue.

In the next section, we use the following lemmas to solve Problems(2.16) and (2.17).

(3.5) **Lemma.** Let $P\in\mathbf{R}^{l\times l}$ be positive-definitė and symmetric, $S\in\mathbf{R}^{q\times q}$ be stabe, $U\in\mathbf{R}^{q\times l}$ and $H\in\mathbf{R}^{p\times q}$. Then,

$$\int_0^{\infty} He^{St}UPU^{\mathrm{T}}e^{S^{\mathrm{T}}t}H^{\mathrm{T}}dt=0 \Leftrightarrow He^{St}U=0 \ \ \textit{for all}\ t\geqq 0 \Leftrightarrow <S|\mathrm{Im}\,U>\subset \mathrm{Ker}H.$$

**Proof.** We first note that since $S$ is a stable matrix

$$\int_0^{\infty}\|e^{St}\|dt<\infty \ \text{ and } \int_0^{\infty}\|e^{St}\|^2dt<\infty.$$

We now have

$$\int_0^{\infty} He^{St}UPU^{\mathrm{T}}e^{S^{\mathrm{T}}t}H^{\mathrm{T}}dt=0$$

$$\Leftrightarrow \int_0^{\infty}\|\xi He^{St}UP^{1/2}\|^2dt=\int_0^{\infty}\xi He^{St}UPU^{\mathrm{T}}e^{S^{\mathrm{T}}t}H^{\mathrm{T}}\xi^{\mathrm{T}}dt=0 \ \ \textit{for all}\ \xi\in\mathbf{R}^{p}$$

$$\Leftrightarrow He^{St}U=0 \ \text{ for all } t\geqq 0 \ (\text{since } P^{1/2}>0)$$

$$\Leftrightarrow \mathrm{Im}\,e^{St}U\subset \mathrm{Ker}\,H \ \text{ for all } t\geqq 0$$

$$\Leftrightarrow \quad <S|\mathrm{Im}\, U> = \bigcup_{t\geq 0} \mathrm{Im}\, e^{St}U \subset \mathrm{Ker}\, H. \quad \square$$

(3.6) **Lemma.** Let $P, S, U$ and $H$ be those of (3.5), and define

$$R := \int_0^\infty e^{St}UPU^{\mathrm{T}}e^{S^{\mathrm{T}}t}dt \quad \text{and} \quad \mu := \mathrm{rank} HRH^{\mathrm{T}}$$

Then,

$$\int_0^\infty \|He^{St}UP^{1/2}\|^2 dt \leqq \mathrm{Tr}\, HRH^{\mathrm{T}} \leqq \mu \int_0^\infty \|He^{St}UP^{1/2}\|^2 dt.$$

**Proof.** First note that since $HRH^{\mathrm{T}} \geqq 0$ and rank $HRH^{\mathrm{T}} = \mu$, the symmetric matrix $HRH^{\mathrm{T}}$ has only $\mu$ positive eigenvalues $\lambda_1 \geqq \lambda_2 \geqq \cdots \geqq \lambda_\mu > 0$. Now we have

$$\begin{aligned}
\int_0^\infty \|He^{St}UP^{1/2}\|^2 dt &= \int_0^\infty \|He^{St}UPU^{\mathrm{T}}e^{S^{\mathrm{T}}t}H^{\mathrm{T}}\| dt \\
&\leqq \int_0^\infty \mathrm{Tr}\, He^{St}UPU^{\mathrm{T}}e^{S^{\mathrm{T}}t}H^{\mathrm{T}} dt \\
&= \mathrm{Tr}\, HRH^{\mathrm{T}} = \sum_{i=1}^{\mu} \lambda_i \\
&\leqq \mu\lambda_1 = \mu \|HRH^{\mathrm{T}}\| \\
&\leqq \mu \int_0^\infty \|He^{St}UPU^{\mathrm{T}}e^{S^{\mathrm{T}}t}H^{\mathrm{T}}\| dt \\
&= \mu \int_0^\infty \|He^{St}UP^{1/2}\|^2 dt. \quad \square
\end{aligned}$$

We now introduce various notions of invariant subspaces and of reachability subspaces.

(3.7) **Definition.** Let $(A,B)$ be those given in (2.1).

(i) A subspace $\mathbf{V} \subset \mathbf{R}^n$ is said to be *feedback $(A,B)$-invariant* if there exists an $F \in \mathbf{R}^{r\times n}$ such that $(A-BF)\mathbf{V} \subset \mathbf{V}$.

(ii) A subspace $\mathbf{R} \subset \mathbf{R}^n$ is called a *reachability subspace* for $(A,B)$ if there exist an $F \in \mathbf{R}^{r\times n}$ and a subspace $\mathbf{B}_1 \subset \mathbf{B} := \mathrm{Im} B$ such that

$$\mathbf{R} = < A-BF \,|\, \mathbf{B}_1 > := \mathbf{B}_1 + (A-BF)\mathbf{B}_1 + \cdots + (A-BF)^{n-1}\mathbf{B}_1.$$

(iii) A subspace $\mathbf{R}_a \subset \mathbf{R}^n$ is called an *almost reachability subspace* for $(A,B)$ if there exist an $F \in \mathbf{R}^{r\times n}$ and a sequence $\mathbf{B}_1 \supset \mathbf{B}_2 \supset \cdots \supset \mathbf{B}_n$ of subspaces of $\mathbf{B}$ such that

$$\mathbf{R}_a = \mathbf{B}_1 + (A-BF)\mathbf{B}_2 + \cdots + (A-BF)^{n-1}\mathbf{B}_n. \quad \square$$

The notion of almost reachability subspaces was first introduced by Willems [7],[8]. Definition(3.7,iii) is not the one originally given by him, but an equivalent one.

Let $\mathbf{K} \subset \mathbf{R}^n$ be a subspace. We denote by $\mathbf{V}(\mathbf{K})$ the family of feedback $(A,B)$-invariant subspaces contained in $\mathbf{K}$, i.e.,

(3.8) $$\mathbf{V}(\mathbf{K}) := \{\phi \subset \mathbf{K} : \exists F \in \mathbf{R}^{n\times n}, (A-BF)\phi \subset \phi\}$$

and write $\mathbf{F}(\mathbf{K})$ for the set of all those $F$'s. Moreover, we introduce a family of subspaces by

(3.9) $$\mathbf{V}_s(\mathbf{K}) := \{\phi \in \mathbf{V}(\mathbf{K}) : \exists F \in \mathbf{F}(\phi), \mathrm{Re}\sigma[(A-BF)|_\phi] < 0\}$$

where $(A-BF)|_\phi$ means the restriction of $A-BF$ to $\phi$, and $\sigma[\cdot]$ denotes the set of the eigenvalues. We furthrmore denote by $\mathbf{R}(\mathbf{K})$ and $\mathbf{R}_a(\mathbf{K})$, respectively, the families of reachability subspaces and of almost reachability subspaces contained in $\mathbf{K}$, i.e.,

(3.10) $$\mathbf{R}(\mathbf{K}) := \{\mathbf{R} \subset \mathbf{K} : \mathbf{R} \text{ is a reachability subspace}\}, \text{ and}$$

(3.11) $$\mathbf{R}_a(\mathbf{K}) := \{\mathbf{R}_a \subset \mathbf{K} : \mathbf{R}_a \text{ is an almost reachability subspace}\}.$$

Then the following lemmas hold.

(3.12) **Lemma**[8,p.237]. The families $\mathbf{V}(\mathbf{K})$, $\mathbf{R}(\mathbf{K})$ and $\mathbf{R}_a(\mathbf{K})$ are all closed under subspace addition (i.e.,$\mathbf{V}_1, \mathbf{V}_2 \in \mathbf{V}(\mathbf{K}) \Rightarrow \mathbf{V}_1 + \mathbf{V}_2 \in \mathbf{V}(\mathbf{K})$, etc.). Consequently, each family has a unique supremal element, i.e.,

$$\sup\mathbf{V}(\mathbf{K}) =: \mathbf{V}_\mathbf{K}^* \in \mathbf{V}(\mathbf{K}),\ \sup\mathbf{R}(\mathbf{K}) =: \mathbf{R}_\mathbf{K}^* \in \mathbf{R}(\mathbf{K}), \text{ and } \sup\mathbf{R}_a(\mathbf{K}) =: \mathbf{R}_a{}^*{}_{,\mathbf{K}} \in \mathbf{R}_a(\mathbf{K}). \quad \square$$

(3.13) **Lemma**[3,pp.114-115]. The family $\mathbf{V}_s(\mathbf{K})$ has a unique supremal element, i.e.,

$$\sup\mathbf{V}_s(\mathbf{K}) =: \mathbf{V}_s{}^*{}_{,\mathbf{K}}(\mathbf{K}) \in \mathbf{V}_s(\mathbf{K}) \quad \square$$

## IV. Solvability for the Stochastic Disturbance Decoupling Problem.

Before discussing the solvability for the stochastic disturbance decoupling problem, let us briefly consider the usual deterministic disturbance decoupling problem. The system to be studied is given by

(4.1) $$S : \begin{cases} \dfrac{d}{dt}x(t) = Ax(t) + Bu(t) + G\xi(t) \\ z(t) = Dx(t) \end{cases}$$

and the problem is to find a state feedback

$$u(t) = -Fx(t)$$

such that in the closed loop system

$$(4.2) \qquad S_F: \begin{cases} \dfrac{d}{dt}x(t) = (A - BF)x(t) + G\xi(t) \\ z(t) = Dx(t) \end{cases}$$

there is no influence of the deterministic disturbance $\xi(t)$ on the output $z(t)$ and the system $S_F$ is stable. This problem can be formally stated as follows.

(4.3) **Disturbance Decoupling Problem with Stability(DDPS)**[3,p.113]. Find (if possible) an $F \in \mathbf{F}_s$ such that

$$<A - BF|\mathrm{Im}\, G> \subset \mathrm{Ker}\, D. \quad \square$$

For this problem the following holds.

(4.4) **Lemma**[3,p.116]. DDPS (4.3) is solvable if and only if

(i) $\mathrm{Im} G \subset \mathbf{V}_s{}^*{}_{,\mathrm{Ker}\, D}$, and

(ii) $(A,B)$ of (4.1) is stabilizable. $\square$

Now we are ready to prove our first main theorem.

(4.5) **Theorem.** Let System (2.1) be given, and satisfy Assumption (2.4). Then, SDDPZ (2.16) is solvable if and only if

$$\mathrm{Im}\, K \subset \mathbf{V}_s{}^*{}_{,\mathrm{Ker}\, D}$$

where $K$ is the steady-state Kalman-Bucy filter gain given by (2.8), and $\mathbf{V}_s{}^*{}_{,\mathrm{Ker}\, D}$ is the maximal member of the family $\mathbf{V}_s(\mathrm{Ker}\, D)$.

**Proof.** By definition, SDDPZ (2.16) is solvable if and only if

$$\mathrm{Tr}\, D\hat{Q}D^{\mathrm{T}} = 0.$$

Applying Lemma (3.5), and using (3.2), we easily obtain

$$\mathrm{Tr}\, D\hat{Q}D^{\mathrm{T}} = 0 \quad \Leftrightarrow \quad {}^{\exists}F \in \mathbf{F}_s, \ <A - BF|\mathrm{Im}\, K> \subset \mathrm{Ker}\, D$$

Thus, SDDPZ (2.16) has been converted to a DDPS(4.3) with $G$ replaced by $K$. Since by Assumption (2.4, iii) $(A,B)$ is stabilizable, we can use Lemma (4.4) to complete our proof. $\square$

Now, we quote a lemma from Willems [8] which will play an essential step to prove our second theorem.

(4.6) **Lemma**[8,p.248]. Assume that $(A,B)$ of (2.1) is stabilizable. Then, for every $\varepsilon > 0$, there exists an $F \in \mathbf{F}_s$ such that

$$\int_0^\infty \| D e^{(A - BF)t} G \|^2 dt < \varepsilon$$

if and only if

$$\mathrm{Im}G \subset \mathbf{V}_{b^*,\mathrm{Ker}\,D} := \mathbf{R}_{b,^*\mathrm{Ker}\,D} + \mathbf{V}_{s,^*\mathrm{Ker}\,D}$$

where

$$\mathbf{R}_{b^*,\mathrm{Ker}\,D} := A\mathbf{R}_{a,^*\mathrm{Ker}\,D} + \mathrm{Im}B$$

$\mathbf{R}_{a^*,\mathrm{Ker}\,D} := \sup\mathbf{R}_a(\mathrm{Ker}D)$ and $\mathbf{V}_{s,^*\mathrm{Ker}\,D} := \sup\mathbf{V}_s(\mathrm{Ker}D)$. □

Next, we will prove our second main theorem.

(4.7) **Theorem.** Let System (2.1) be given, and satisfy Assumption (2.4). Then, SDDPA (2.17) of stochastic disturbance decoupling in the almost sense is solvable, i.e., for every $\varepsilon > 0$ there exists an $F \in \mathbf{F}_s$ such that $\mathrm{Tr}D\hat{Q}D^{\mathrm{T}} < \varepsilon$ if and only if

$$\mathrm{Im}K \subset \mathbf{V}_{b,^*\mathrm{Ker}\,D} := \mathbf{R}_{b,^*\mathrm{Ker}\,D} + \mathbf{V}_{s,^*\mathrm{Ker}\,D}.$$

**Proof.** We first recall that

$$\hat{Q} = \int_0^\infty e^{(A-BF)t} KVK^T e^{(A-BF)^{\mathrm{T}}t}\, dt, \qquad F \in \mathbf{F}_s$$

So by Lemma (3.6) it is easily seen that for some $\mu > 0$

$$\int_0^\infty \|De^{(A-BF)t}KV^{1/2}\|^2 dt \leqq \mathrm{Tr}\,D\hat{Q}D^{\mathrm{T}} \leqq \mu \int_0^\infty \|De^{(A-BF)t}KV^{1/2}\|^2 dt, \qquad F \in \mathbf{F}_s. \quad (1)$$

Since $V^{1/2} > 0$ we have $\| V^{1/2} \| > 0$ and hence

$$\|D\, e^{(A-BF)t}KV^{1/2}\|^2 \geq \frac{1}{\|V^{-1/2}\|^2} \|D\, e^{(A-BF)t}K\|^2.$$

Therefore, it follows from (1) that

$$\frac{1}{\|V^{-1/2}\|^2}\int_0^\infty \|De^{(A-BF)t}K\|^2 dt \leqq \mathrm{Tr}\,DQD^{\mathrm{T}} \leqq \mu\, \|V^{1/2}\|^2 \int_0^\infty \|De^{(A-BF)t}K\|^2 dt.$$

The above inequalities imply that

$$^\forall\varepsilon > 0,\ ^\exists F \in \mathbf{F}_s,\quad \mathrm{Tr}\,D\hat{Q}D^{\mathrm{T}} < \varepsilon \qquad (2)$$

$$\Leftrightarrow\ ^\forall\varepsilon > 0,\ ^\exists F \in \mathbf{F}_s,\ \int_0^\infty \|D\, e^{(A-BF)t}K\|^2 dt < \varepsilon. \qquad (3)$$

Hence, we have shown that SDDPA (2.17) is solvable if and only if (3) is satisfied. Now, Lemma (4.6) with $G$ replaced by $K$ can be directly used to complete our proof of the theorem. □

## V. Concluding Remarks.

A stochastic disturbance decoupling problem (SDDP) was formulated for linear multivariable systems which are corrupted by white Gaussian processes as external disturbances and measurement disturbances. It was first shown that the SDDP can be converted to one of the two problems : (a) a usual deterministic disturbance decoupling problem and (b) an almost disturbance decoupling problem of Willems[8]. Then, for each problem a necessary and sufficient condition for its solvability was proved, and in particular it was turned out that the stead-state Kalman-Bucy filter gain matrix for the state plays an essential role in the SDDP.

## References.

[1] R. E. Kalman and R. S. Bucy, New results in linear filtering and prediction theory, J.Basic Engr., Trans. ASME, Ser. D, vol.83, pp.95 – 108, 1961.

[2] H. Kwakernaak and P. Sivan, Linear Optimal Control Systems, Wiley-Interscience, 1972.

[3] W. M. Wonham, Linear Multivaliable Control, a Geometric Approach, 2nd ed., Springer, 1979.

[4] H. Akashi and H. Imai, Disturbance localization and output deadheat through an observer in discrete-time linear multivariable systems, IEEE Trans. Automat. Contr., AC-24, pp.621 – 627, 1979.

[5] J. M. Schumacher, Compensator synthesis using (C,A,B)-pairs, IEEE Trans. Automat. Contr., AC-25, pp.1133 – 1138, 1980.

[6] J. C. Willems and C. Commault, Disturbance decoupling by measurement feedback with stability or pole placement, SIAM J. Contr. Optimiz., vol. 19,pp.490 – 504, 1981.

[7] J. C. Willems, "Almost A ( modB )-invariant subspaces," Asterisque, vol. 75-76, pp.239 – 248, 1980.

[8] J. C. Willems, Almost invariant subspaces : an approach to high gain feedback design – part 1 : almost controlled invariant subspaces, IEEE Trans. Autmat. Contr., AC-26, pp.235 – 252, 1981.

[9] H. Inaba, A note on stochastic disturbance decoupling, Proc. JAACE Symp. Stochastic Systems Theory and Its Applications, pp.237-240, Oct. 1986, Tokyo.

# DISCRETE–TIME FILTERING FOR LINEAR SYSTEMS IN CORRELATED NOISE WITH NON–GAUSSIAN INITIAL CONDITIONS: FORMULAS AND ASYMPTOTICS

Richard B. Sowers and Armand M. Makowski

## ABSTRACT

We consider the one–step prediction problem for discrete–time linear systems in correlated plant and observation noises, and non–Gaussian initial conditions. Explicit representations are obtained for the MMSE and LMMSE (or Kalman) estimates of the state given past observations, as well as for the expected square of their difference. These formulae are obtained with the help of the Girsanov transformation for Gaussian white noise sequences, and display explicitly the dependence of the quantities of interest on the initial distribution. With the help of these formulae, we completely characterize the asymptotic behavior of the error sequence in the scalar time–invariant case.

## I. INTRODUCTION

We consider the one–step prediction problem associated with the stochastic discrete–time linear dynamical system

$$\begin{aligned} X^\circ_{t+1} &= A_t X^\circ_t + W^\circ_{t+1} \\ X^\circ_0 &= \xi \\ Y_t &= H_t X^\circ_t + V^\circ_{t+1} \end{aligned} \qquad t = 0, 1, \ldots \quad (1.1)$$

defined on some probability triple $(\Omega, \mathcal{F}, P)$ which carries the $\mathbb{R}^n$–valued plant process $\{X^\circ_t,\ t = 0, 1, \ldots\}$ and the $\mathbb{R}^k$–valued observation process $\{Y_t,\ t = 0, 1, \ldots\}$. Here, for all $t = 0, 1, \ldots$, the matrices $A_t$ and $H_t$ are of dimension $n \times n$ and $k \times n$, respectively. Throughout we make the following assumptions (A.1)–(A.3), where

**(A.1):** The process $\{(W^\circ_{t+1}, V^\circ_{t+1}),\ t = 0, 1, \ldots\}$ is a zero–mean Gaussian White Noise (GWN) sequence with covariance structure $\{\Gamma_{t+1},\ t = 0, 1, \ldots\}$ given by

$$\Gamma_{t+1} := \text{Cov}\begin{pmatrix} W^\circ_{t+1} \\ V^\circ_{t+1} \end{pmatrix} = \begin{pmatrix} \Sigma^w_{t+1} & \Sigma^{wv}_{t+1} \\ \Sigma^{vw}_{t+1} & \Sigma^v_{t+1} \end{pmatrix}; \qquad t = 0, 1, \ldots \quad (1.2)$$

**(A.2):** For all $t = 0, 1, \ldots$, the covariance matrix $\Sigma^{v}_{t+1}$ is positive definite; and

**(A.3):** The initial condition $\xi$ has distribution $F$ with finite first and second moments $\mu$ and $\Delta$, respectively, and is independent of the process $\{(W^{\circ}_{t+1}, V^{\circ}_{t+1}),\ t = 0, 1, \ldots\}$. No other ***a priori*** assumptions are enforced on $F$.

The (one–step) prediction problem associated with (1.1) is defined as the problem of evaluating the conditional expectation

$$E[\phi(X^{\circ}_{t+1}) |\, Y_0, \ldots, Y_t] \qquad\qquad t = 0, 1, \ldots (1.3)$$

for all bounded Borel mappings $\phi : I\!R^n \to \mathcal{C}$, with $\mathcal{C}$ denoting set of the complex numbers. In this paper, we solve the prediction problem (1.3) associated with (1.1)–(1.2). For each $t = 0, 1, \ldots$, once the conditional distribution of $X^{\circ}_{t+1}$ given $\{Y_0, \ldots, Y_t\}$ is available, it is possible to construct the MMSE estimate $\hat{X}_{t+1} := E[X^{\circ}_{t+1} |\, Y_0, \ldots, Y_t]$ of $X^{\circ}_{t+1}$ on the basis of $\{Y_0, \ldots, Y_t\}$. In general, $\hat{X}_{t+1}$ is a *non–linear* function of $\{Y_0, \ldots, Y_t\}$, in contrast with the LLMSE (or Kalman) estimate $\hat{X}^{K}_{t+1}$ of $X^{\circ}_{t+1}$ computed on the basis of $\{Y_0, \ldots, Y_t\}$, which is by definition linear in these quantities. We shall find representations for both $\{\hat{X}_t,\ t = 0, 1, \ldots\}$ and $\{\hat{X}^{K}_t,\ t = 0, 1, \ldots\}$, and then form the mean square error $\epsilon_t := E[\|\hat{X}_t - \hat{X}^{K}_t\|^2]$ for $t = 1, 2, \ldots$.

When the plant and observation noises are *uncorrelated*, and the observation noise sequence $\{V_t,\ t = 0, 1, \ldots\}$ is standard (i.e., $\Sigma^{wv}_{t+1} = 0$ and $\Sigma^{v}_{t+1} = I_n$ for all $t = 0, 1, \ldots$), the prediction problem posed above is the discrete–time counterpart of the situation investigated in [3]. In Section II we state the main results for the nonlinear prediction problem, and outline the proofs in Section III, thus indicating how the technique of [3] extends to the correlated noise situation without major difficulties. We then use these results in Section IV to obtain representations for $\{\hat{X}_t,\ t = 1, 2, \ldots\}$, $\{\hat{X}^{K}_t,\ t = 1, 2, \ldots\}$, and $\{\epsilon_t,\ t = 1, 2, \ldots\}$. These expressions explicitly display the dependence of the initial distribution $F$, and form the basis for the large time asymptotic analysis carried out in [6] on the error terms $\{\epsilon_t,\ t = 1, 2, \ldots\}$. In Section V we consider this asymptotic behavior in the scalar time–invariant case, and give a complete characterization of these asymptotics in terms of the plant gain $a$, the observation gain $h$, and the noise covariance matrix $\Gamma$. Many details have been omitted for the sake of brevity; additional information and material can be found in the thesis [4] and in [5].

## II. THE FILTERING PROBLEM

### II.1. The notation

A word on the notation: For any positive integers $n$ and $m$, we denote the space of $n \times m$ real matrices by $\mathcal{M}_{n\times m}$ and the cone of $n \times n$ symmetric positive-definite matrices by $\mathcal{Q}_n$. As in [3], for every $\Sigma$ in $\mathcal{Q}_{2n}$, let $X_\Sigma$ and $B_\Sigma$ denote generic $\mathbb{R}^n$–valued random variables (RVs) such that $(X_\Sigma, B_\Sigma)$ is a $\mathbb{R}^{2n}$–valued zero–mean Gaussian RV with covariance matrix $\Sigma$. For every bounded Borel mapping $\phi : \mathbb{R}^n \to \mathcal{C}$, we define the mappings $\mathcal{T}\phi : \mathbb{R}^n \times \mathbb{R}^n \times \mathcal{Q}_{2n} \to \mathcal{C}$ and $\mathcal{U}\phi : \mathbb{R}^n \times \mathbb{R}^n \times \mathcal{Q}_n \times \mathcal{M}_{n\times n} \times \mathcal{Q}_{2n} \to \mathcal{C}$ by

$$\mathcal{T}\phi[x, b; \Sigma] := \mathcal{E}\left[\phi(x + X_\Sigma)\exp[b' B_\Sigma]\right] \tag{1.4}$$

and

$$\mathcal{U}\phi[x, b; \Lambda, \Psi; \Sigma] := \mathcal{E}\left[\mathcal{T}\phi[x + \Psi\xi, \xi; \Sigma]\exp[b'\xi - \frac{1}{2}\xi'\Lambda\xi]\right] \tag{1.5}$$

with $\mathcal{E}$ denoting integration with respect to the Gaussian distribution of the RV $(X_\Sigma, B_\Sigma)$.

Throughout, let $I_n$ denote the unit matrix in $\mathcal{M}_{n\times n}$, and let $O_n$ denote the zero element in $\mathcal{M}_{n\times n}$, i.e., the $n \times n$ matrix whose elements are all zero. Elements of $\mathbb{R}^n$ are always interpreted as column vectors, and transposition is denoted by $'$. Finally, let $\Psi(\cdot,\cdot)$ be the state transition matrix given by

$$\begin{aligned} &\Psi(t,t) = I_n \\ &\Psi(s+1,t) = [A_s - \Sigma^{wv}_{s+1}(\Sigma^v_{s+1})^{-1}H_s]\Psi(s,t). \qquad s = t, t+1, \ldots \\ &\qquad\qquad\qquad\qquad\qquad\qquad\qquad\qquad\qquad t = 0, 1, \ldots \end{aligned} \tag{1.6}$$

## II.2. The main results

We define the $\mathcal{Q}_n$–valued sequence $\{P_t,\ t = 0, 1, \ldots\}$ by the recursions

$$\begin{aligned} &P_{t+1} = A_t P_t A_t' - [A_t P_t H_t' + \Sigma^{wv}_{t+1}][H_t P_t H_t' + \Sigma^v_{t+1}]^{-1}[A_t P_t H_t' + \Sigma^{wv}_{t+1}]' + \Sigma^w_{t+1} \\ &P_0 = O_n \\ &\qquad\qquad\qquad\qquad\qquad\qquad t = 0, 1, \ldots \end{aligned} \tag{2.1}$$

and for convenience, we introduce the $\mathcal{Q}_k$–valued sequence $\{J_t,\ t = 0, 1, \ldots\}$, where

$$J_t := H_t P_t H_t' + \Sigma^v_{t+1}. \qquad t = 0, 1, \ldots \tag{2.2}$$

The deterministic sequences $\{Q_t,\ t = 0, 1, \ldots\}$ and $\{R_t,\ t = 0, 1, \ldots\}$ take values in $\mathcal{M}_{n\times n}$ and $\mathcal{Q}_n$, respectively, and are defined recursively by

$$\begin{aligned} &Q_{t+1} = A_t Q_t - [A_t P_t H_t' + \Sigma^{wv}_{t+1}]J_t^{-1}H_t\big(Q_t + \Psi(t,0)\big) + \Sigma^{wv}_{t+1}(\Sigma^v_{t+1})^{-1}H_t\Psi(t,0) \\ &R_{t+1} = R_t - \big(Q_t + \Psi(t,0)\big)' H_t' J_t^{-1} H_t \big(Q_t + \Psi(t,0)\big) + \Psi'(t,0)H_t' H_t \Psi(t,0) \\ &\qquad\qquad\qquad\qquad\qquad\qquad t = 0, 1, \ldots \end{aligned} \tag{2.3}$$

with initial conditions $Q_0 = R_0 = O_n$. From these sequences, we form the $\mathcal{Q}_{2n}$–valued sequence $\{\Sigma_t,\ t = 0, 1, \ldots\}$ by setting

$$\Sigma_t = \begin{pmatrix} P_t & Q_t \\ Q_t' & R_t \end{pmatrix}. \qquad t = 0, 1, \ldots \tag{2.4}$$

We also generate the $\mathbb{R}^n$–valued processes $\{\bar{X}_t, t = 0, 1, \ldots\}$ and $\{\bar{B}_t, t = 0, 1, \ldots\}$ via the recursive relations

$$\begin{aligned} \bar{X}_{t+1} &= [A_t - [A_t P_t H_t' + \Sigma_{t+1}^{wv}] J_t^{-1} H_t] \bar{X}_t + [A_t P_t H_t' + \Sigma_{t+1}^{wv}] J_t^{-1} Y_t \\ \bar{B}_{t+1} &= \bar{B}_t - (Q_t + \Psi(t,0))' H_t' J_t^{-1} H_t \bar{X}_t + (Q_t + \Psi(t,0))' H_t' J_t^{-1} Y_t \end{aligned}$$

$$t = 0, 1, \ldots \quad (2.5)$$

with initial values $\bar{X}_0 = \bar{B}_0 = 0$.

The solution to the prediction problem associated with (1.1) can now be given. On $\Omega$ define the filtration $\{\mathcal{Y}_t,\ t = 0, 1, \ldots\}$ generated by the observations $\{Y_t,\ t = 0, 1, \ldots\}$, i.e.,

$$\mathcal{Y}_t := \sigma\{Y_0, Y_1, \ldots, Y_t\}. \qquad t = 0, 1, \ldots \quad (2.6)$$

**Theorem 1.** *For any bounded Borel mapping $\phi : \mathbb{R}^n \to \mathbb{R}$ and any $t = 0, 1, \ldots$, the relationship*

$$E[\phi(X_{t+1}^{\circ}) | \mathcal{Y}_t] = \frac{\mathcal{U}\phi[\bar{X}_{t+1}, \bar{B}_{t+1}; M_{t+1}, \Psi(t+1, 0); \Sigma_{t+1}]}{\mathcal{U}\mathbb{1}[\bar{X}_{t+1}, \bar{B}_{t+1}; M_{t+1}, \Psi(t+1, 0); \Sigma_{t+1}]} \quad P - a.s. \qquad (2.7)$$

*holds true, where $\mathbb{1}$ denotes the constant mapping $\mathbb{R}^n \to \mathbb{R} : x \to 1$ and the $\mathcal{Q}_n$–valued sequence $\{M_t,\ t = 0, 1, \ldots\}$ is defined recursively by*

$$M_{t+1} = M_t + \Psi(t,0)' H_t' (\Sigma_{t+1})^{-1} H_t \Psi(t,0) \qquad t = 0, 1, \ldots \quad (2.8)$$

*with $M_0 = O_n$*

The probabilistic interpretation as covariance matrices for $\{M_t,\ t = 0, 1, \ldots\}$ is available in [4, 5]. We readily see that the structure of the predictor in the general situation is not markedly different from that for the uncorrelated case [3]. The noise correlation is encoded in the *universal* sufficient statistics [3] that parametrize the predictor, but does *not* affect the *form* of the statistics bearing functionals.

## III. PROOFS

Only the structure of the proof is outlined as details are available in the thesis [4] and in [5]. The approach used here extends the one introduced in [3], and is again based on finding a probability measure $\bar{P}$, absolutely continuous with respect to the original measure $P$ on $\mathcal{F}$, under which the statistical calculations are readily performed. Here, as explained in [4], the *arbitrary* covariance structure of the plant and observation noise sequences leads to the use of a Girsanov transformation on the *joint* $\mathbb{R}^{n+k}$–valued sequence $\{(W_{t+1}^{\circ}, V_{t+1}^{\circ}),\ t = 0, 1, \ldots\}$ (and not merely on the observation noise sequence $\{(V^\circ_{t+1}),\ t=0,1,\ldots\}$ as in the uncorrelated case [3, 4]). To that end, define the filtration $\{\mathcal{F}_t,\ t=0,1,\ldots\}$ by

$$\mathcal{F}_{t+1} := \mathcal{F}_0 \vee \sigma\{W^\circ_{t+1}, V^\circ_{s+1},\ s=0,1,\ldots t\} \qquad t=0,1,\ldots \quad (3.1)$$

with $\mathcal{F}_0 := \sigma\{\xi\}$, and the $\mathbb{R}^{n+k}$–valued sequence $\{(W_{t+1}, V_{t+1}),\ t=0,1,\ldots\}$ by

$$\begin{pmatrix} W_{t+1} \\ V_{t+1} \end{pmatrix} = \begin{pmatrix} W^\circ_{t+1} \\ V^\circ_{t+1} \end{pmatrix} - \begin{pmatrix} \Sigma^w_{t+1} & \Sigma^{wv}_{t+1} \\ \Sigma^{vw}_{t+1} & \Sigma^v_{t+1} \end{pmatrix} \begin{pmatrix} \varphi^w_t \\ \varphi^v_t \end{pmatrix} \qquad t=0,1,\ldots \quad (3.2)$$

where $\{\varphi^w_t,\ t=0,1,\ldots\}$ and $\{\varphi^v_t,\ t=0,1,\ldots\}$ are $\mathcal{F}_t$–adapted sequences taking values in $\mathbb{R}^n$ and $\mathbb{R}^k$, respectively, yet to be specified. Recalling the Girsanov transformation [2], we see that if for *any* two such sequences $\{\varphi^w_t,\ t=0,1,\ldots\}$ and $\{\varphi^v_t,\ t=0,1,\ldots\}$, we define $\{(W_{t+1}, V_{t+1}),\ t=0,1,\ldots\}$ by (3.2), then for each $T=0,1,\ldots$ we can find a probability measure $\bar{P}$ on $(\Omega,\mathcal{F})$ satisfying (B) where

**(B):** *The probability measure $\bar{P}$ is mutually absolutely continuous with $P$ on $\mathcal{F}$ and agrees with $P$ on $\mathcal{F}_0$. Furthermore, $\{(W_{t+1}, V_{t+1}),\ t=0,1,\ldots,T\}$ is a zero–mean $(\mathcal{F}_t,\bar{P})$ GWN sequence with the same covariance structure under $\bar{P}$ as the covariance structure under $P$ of the original noise sequence $\{(W^\circ_{t+1}, V^\circ_{t+1}),\ t=0,1,\ldots,T\}$.*

Following [3, 4], we decompose the plant process $\{X^\circ_t,\ t=0,1,\ldots\}$ as

$$X^\circ_t = X_t + Z_t \qquad t=0,1,\ldots \quad (3.3)$$

where the $\mathbb{R}^n$–valued $\{X_t,\ t=0,1,\ldots\}$ carries the randomness due to the plant noise process $\{W^\circ_{t+1},\ t=0,1,\ldots\}$, and where the $\mathbb{R}^n$–valued $\{Z_t,\ t=0,1,\ldots\}$ contains only the randomness due to the initial condition $\xi$. It is argued in [3, 4] that the sequences $\{\varphi^w_t,\ t=0,1,\ldots\}$ and $\{\varphi^v_t,\ t=0,1,\ldots\}$ in (3.2) must necessarily have the form

$$\varphi^w_t = \varphi_t \quad \text{and} \quad \varphi^v_t = -(\Sigma^v_{t+1})^{-1}[\Sigma^{vw}_{t+1}\varphi_t + H_t Z_t] \qquad t=0,1,\ldots \quad (3.4)$$

for some unspecified $\mathcal{F}_t$–adapted sequence $\{\varphi_t,\ t=0,1,\ldots\}$ taking values in $\mathbb{R}^n$. Injecting (3.4) into (3.2), we obtain

$$W_{t+1} = W^\circ_{t+1} + \Sigma^{wv}_{t+1}(\Sigma^v_{t+1})^{-1}H_t Z_t[\Sigma^w_{t+1} - \Sigma^{wv}_{t+1}(\Sigma^v_{t+1})^{-1}\Sigma^{vw}_{t+1}]\varphi_t \qquad t=0,1,\ldots \quad (3.5)$$

and the Girsanov theorem gives the appropriate probability measure $\bar{P}$ which satisfies **(B)**, via

$$\frac{d\bar{P}}{dP} = \exp\left[\sum_{s=0}^{T}\left[\varphi'_s[W^\circ_{s+1} - \Sigma^{wv}_{s+1}(\Sigma^v_{s+1})^{-1}V^\circ_{s+1}] - Z'_s H'_s(\Sigma^v_{s+1})^{-1}V^\circ_{s+1}\right]\right.$$
$$\left. + \frac{1}{2}\sum_{s=0}^{T}\left[\varphi'_s[\Sigma^w_{s+1} - \Sigma^{wv}_{s+1}(\Sigma^v_{s+1})^{-1}\Sigma^{vw}_{s+1}]\varphi_s + Z'_s H'_s(\Sigma^v_{s+1})^{-1}H_s Z_s\right]\right]. \quad (3.6)$$

In order to complete the description of the probability measure (3.6), we must specify $\{X_t,\ t=0,1,\ldots\}$, $\{Z_t,\ t=0,1,\ldots\}$, and $\{\varphi_t,\ t=0,1,\ldots\}$. To that end we rewrite the evolution of $\{X_t^\circ,\ t=0,1,\ldots\}$ in terms of $\{X_t,\ t=0,1,\ldots\}$, $\{Z_t,\ t=0,1,\ldots\}$ and $\{W_{t+1},\ t=0,1,\ldots\}$. Since we wish to use the properties of $\bar{P}$, it is more natural to write this evolution in terms of $\{W_{t+1},\ t=0,1,\ldots\}$ rather than in terms of $\{W_{t+1}^\circ,\ t=0,1,\ldots\}$, and this leads to

$$\begin{aligned} X_{t+1}+Z_{t+1} &= A_tX_t^\circ + W_{t+1}^\circ \\ &= A_t(X_t+Z_t) + W_{t+1} - \Sigma_{t+1}^{wv}(\Sigma_{t+1}^{v})^{-1}H_tZ_t \\ &\quad + [\Sigma_{t+1}^{w} - \Sigma_{t+1}^{wv}(\Sigma_{t+1}^{v})^{-1}\Sigma_{t+1}^{vw}]\varphi_t \qquad\qquad t=0,1,\ldots \qquad (3.7)\\ &= A_tX_t + [A_t - \Sigma_{t+1}^{wv}(\Sigma_{t+1}^{v})^{-1}H_t]Z_t + W_{t+1} \\ &\quad + [\Sigma_{t+1}^{w} - \Sigma_{t+1}^{wv}(\Sigma_{t+1}^{v})^{-1}\Sigma_{t+1}^{vw}]\varphi_t. \end{aligned}$$

This suggests a separation of the dynamics in the form

$$\begin{aligned} X_{t+1} &= A_tX_t + W_{t+1} + [\Sigma_{t+1}^{w} - \Sigma_{t+1}^{wv}(\Sigma_{t+1}^{v})^{-1}\Sigma_{t+1}^{vw}]\varphi_t - \pi_t \\ Z_{t+1} &= [A_t - \Sigma_{t+1}^{wv}(\Sigma_{t+1}^{v})^{-1}H_t]Z_t + \pi_t \end{aligned} \qquad t=0,1,\ldots \qquad (3.8)$$

with initial conditions $X_0=\zeta$ and $Z_0=\xi-\zeta$ where $\zeta$ and $\{\pi_t,\ t=0,1,\ldots\}$ are $I\!R^n$–valued RVs yet to be specified. At this point, we simply assume

$$\varphi_t=0, \quad \pi_t=0 \quad \text{and} \quad \zeta=0 \qquad\qquad t=0,1,\ldots \qquad (3.9)$$

and summarize the relevant quantities under this constraint (3.9).

- **The effect of the initial condition**

$$\begin{aligned} Z_{t+1} &= [A_t - \Sigma_{t+1}^{wv}(\Sigma_{t+1}^{v})^{-1}H_t]Z_t \\ Z_0 &= \xi, \end{aligned} \qquad t=0,1,\ldots \qquad (3.10)$$

so that $Z_t=\Psi(t,0)\xi$ for $t=0,1,\ldots$.

- **The noise processes**

$$\begin{aligned} \begin{pmatrix} W_{t+1} \\ V_{t+1} \end{pmatrix} &= \begin{pmatrix} W_{t+1}^\circ \\ V_{t+1}^\circ \end{pmatrix} - \begin{pmatrix} \Sigma_{t+1}^{w} & \Sigma_{t+1}^{wv} \\ \Sigma_{t+1}^{vw} & \Sigma_{t+1}^{v} \end{pmatrix} \begin{pmatrix} 0 \\ -(\Sigma_{t+1}^{v})^{-1}H_tZ_t \end{pmatrix} \\ &= \begin{pmatrix} W_{t+1}^\circ + \Sigma_{t+1}^{vw}(\Sigma_{t+1}^{v})^{-1}H_tZ_t \\ V_{t+1}^\circ + H_tZ_t \end{pmatrix}. \end{aligned} \qquad t=0,1,\ldots \qquad (3.11)$$

- **The auxiliary system**

$$\begin{aligned} X_{t+1} &= A_tX_t + W_{t+1} \\ X_0 &= 0 \\ Y_t &= H_tX_t + V_{t+1}. \end{aligned} \qquad t=0,1,\ldots \qquad (3.12)$$

- **The change of measure**

$$\frac{d\bar{P}}{dP} = \exp\Big[-\sum_{s=0}^{T} Z_s' H_s' (\Sigma_{s+1}^v)^{-1} V_{s+1}^\circ + \frac{1}{2}\sum_{s=0}^{T} Z_s' H_s' (\Sigma_{s+1}^v)^{-1} H_s Z_s\Big]. \qquad (3.13)$$

The properties of our decomposition and change of measure are summarized in

**Proposition 1.** *Let the filtration $\{\mathcal{F}_t,\ t = 0,1,\ldots\}$ be given by (3.1). If the sequences $\{X_t,\ t = 0,1,\ldots\}$, $\{Z_t,\ t = 0,1,\ldots\}$ and $\{(W_{t+1}, V_{t+1}),\ t = 0,1,\ldots\}$ are defined by (3.10)–(3.12) and if the probability measure $\bar{P}$ is defined by (3.13), then $\bar{P}$ and $P$ are mutually absolutely continuous on $\mathcal{F}$, agree on $\mathcal{F}_0$, and the process $\{(W_{t+1}, V_{t+1}),\ t = 0,1,\ldots,T\}$ is a zero-mean $(\mathcal{F}_t, \bar{P})$ GWN sequence with covariance structure $\{\Gamma_{t+1},\ t = 0,1,\ldots,T\}$.*

Motivated by the form of (3.13), we define the $I\!R$–valued sequence $\{L_t,\ t = 0,1,\ldots\}$ by

$$L_{t+1} := \exp\Big[-\sum_{s=0}^{t} Z_s' H_s' (\Sigma_{s+1}^v)^{-1} V_{s+1}^\circ + \frac{1}{2}\sum_{s=0}^{t} Z_s' H_s' (\Sigma_{s+1}^v)^{-1} H_s Z_s\Big] \qquad t = 0,1,\ldots \qquad (3.14)$$

with $L_0 = 1$, and observe that $d\bar{P}/dP = L_{T+1}$. The arguments of [3] and [4] can now be applied *in extenso* to yield the results (2.1)–(2.8) over the finite horizon.

The final step now consists in extending these results from the finite horizon $t = 0,1,\ldots,T$ to the infinite horizon $t = 0,1,\ldots$. To that end, note the following: The dynamics of the sequences $\{(\bar{X}_t, \bar{B}_t),\ t = 0,1,\ldots,T+1\}$ and $\{\Sigma_t,\ t = 0,1,\ldots,T+1\}$ are *independent* of $T$. Moreover, although the transformed measure $\bar{P}$ used in the derivation depends *a priori* on $T$, the definitions of the mappings $\mathcal{T}\phi$ and $\mathcal{U}\phi$ are independent of $T$. These remarks are sufficient to yield Theorem 1 from the finite–horizon results of this section. ∎

Following on the comments made at the end of the proof, we could have displayed explicitly the dependence of the transformed measure $\bar{P}$ on the parameter $T$, say through the notation $\bar{P}_{T+1}$. Although $\bar{P}_{T+1} = \bar{P}_T$ on the $\sigma$-field $\mathcal{F}_T$ for all $T = 0,1,\ldots$, and the probability measure $\bar{P}_{T+1}$ is mutually absolutely continuous with respect to $P$, it is *not* true in general [4] that the *projective system* $\{\bar{P}_T,\ T = 0,1,\ldots\}$ has a limit $\bar{P}$ which is absolutely continuous with respect to $P$ on the $\sigma$-field $\vee_T \mathcal{F}_T$, i.e., there does not exist necessarily a probability measure $\bar{P}$ on $\vee_T \mathcal{F}_T$ such that $\bar{P}$ is absolutely continuous with respect to $P$, and $\bar{P}_T = \bar{P}$ on the $\sigma$-field $\mathcal{F}_T$ for all $T = 0,1,\ldots$. Although this could *a priori* complicate matters for the infinite-horizon situation, we shall not concern ourselves with this difficulty in what follows. Indeed, in the remainder of this paper, only statements for finite $t$ will be made and the notation $\bar{P}$ (and $\bar{E}$) will be used throughout with the understanding that $\bar{P} = \bar{P}_{T+1}$ for some $t < T$. As should be clear from earlier comments, the exact choice of $T$ is irrelevant.

## IV. REPRESENTATIONS FOR $\{\epsilon_t,\ t=0,1,\ldots\}$.

Using Theorem 1, we now develop formulae for $\{\hat{X}_t,\ t=0,1,\ldots\}$, $\{\hat{X}^K_t,\ t=0,1,\ldots\}$ and $\{\epsilon_t,\ t=0,1,\ldots\}$. We do this under the additional assumption (A.4), where

**(A.4):** The covariance matrix $\Delta$ is positive-definite.

To state these representation results, we find it convenient to introduce the auxiliary quantities $\{Q^*_t,\ t=0,1,\ldots\}$ and $\{R^*_t,\ t=0,1,\ldots\}$ in $\mathcal{M}_{n\times n}$ and $\mathcal{Q}_n$, respectively, by setting

$$Q^*_t := Q_t + \Psi(t,0) \quad \text{and} \quad R^*_t := M_t - R_t. \qquad t=0,1,\ldots \quad (4.1)$$

With this notation, we have

**Theorem 2.** *For all $t=0,1,\ldots$, the representations*

$$\hat{X}_{t+1} = \bar{X}_{t+1} + Q^*_{t+1}\frac{\int_{\mathbb{R}^n} z\exp\left[z'\bar{B}_{t+1} - \frac{1}{2}z'R^*_{t+1}z\right]dF(z)}{\int_{\mathbb{R}^n} \exp\left[z'\bar{B}_{t+1} - \frac{1}{2}z'R^*_{t+1}z\right]dF(z)} \qquad (4.2)$$

*and*

$$\hat{X}^K_{t+1} = \bar{X}_{t+1} + Q^*_{t+1}\left[R^*_{t+1} + \Delta^{-1}\right]^{-1}\left[\bar{B}_{t+1} + \Delta^{-1}\mu\right] \qquad (4.3)$$

*hold $P$-a.s.*

Before discussing a proof of this result, several points are in order:

**(i):** The expression (4.3) provides a non-standard representation for the Kalman filter associated with system (1.1). This representation is notable in that it explicitly displays the effects of the mean $\mu$ and covariance $\Delta$ of the initial condition $\xi$; the only dependence of the filtering formulae on $\mu$ and $\Delta$ is through the affine mapping $x \mapsto [R^*_{t+1} + \Delta^{-1}]^{-1}[x + \Delta^{-1}\mu]$.

**(ii):** We readily see from (1.6), (2.3) and (2.8) that

$$\begin{aligned} Q^*_{t+1} &= \left[A_t - [A_tP_tH'_t + \Sigma^{wv}_{t+1}]J_t^{-1}H_t\right]Q^*_t \\ R^*_{t+1} &= R^*_t + Q^{*\prime}_t H'_t J_t^{-1} H_t Q^*_t \end{aligned} \qquad t=0,1,\ldots \quad (4.4)$$

with initial conditions $Q^*_0 = I_n$ and $R^*_0 = O_n$. Note also that the dynamics (2.5) then simplifies into

$$\begin{aligned} \bar{B}_{t+1} &= \bar{B}_t - Q^{*\prime}_t H'_t J_t^{-1} H_t \bar{X}_t + Q^{*\prime}_t H'_t J_t^{-1} Y_t \\ \bar{B}_0 &= 0. \end{aligned} \qquad t=0,1,\ldots \quad (4.5)$$

The following two technical lemmas will be useful in the forthcoming discussion. The proofs are available in [4] and are omitted here in the interest of brevity.

**Lemma 1.** *For $t = 0, 1, \ldots$, $\bar{B}_{t+1}$ is a zero-mean Gaussian RV with covariance $R^*_{t+1}$ under $\bar{P}$.*

**Lemma 2.** *For any $t = 0, 1, \ldots$ and any $I\!R$-valued, nonnegative $\mathcal{Y}_t \vee \sigma\{\xi\}$-measurable RV $X$, the relation*

$$E[X] = \bar{E}\left[X \exp[\xi' \bar{B}_{t+1} - \frac{1}{2}\xi' R^*_{t+1}\xi]\right] \tag{4.6}$$

*holds true.*

**A proof of Theorem 2.** The first step consists in finding a representation for the conditional characteristic function $E\left[\exp[i\theta' X^\circ_{t+1}] \middle| \mathcal{Y}_t\right]$. Under the enforced moment assumptions on $\xi$, an expression for the conditional mean is recovered by differentiating this characteristic function with respect to $\theta$ and then setting $\theta = 0$. Finally, by substituting a Gaussian distribution for $F$ in this representation for $\hat{X}_{t+1}$, we obtain a formula for $\hat{X}^K_{t+1}$. Details are available in [4].

■

Theorem 2 now leads us to a simple representation of the errors $\{\epsilon_{t+1},\ t = 0, 1, \ldots\}$. In what follows, for each $\Lambda$ in $\mathcal{Q}_n$, $G_\Lambda$ denotes a normal distribution with zero mean and covariance $\Lambda$.

**Theorem 3.** *The representation*

$$\epsilon_{t+1} = \int_{I\!R^n} \frac{\left\| Q^*_{t+1} \int_{I\!R^n} \{z - [R^*_{t+1} + \Delta^{-1}]^{-1}[b + \Delta^{-1}\mu]\} \exp[z'b - \frac{1}{2}z' R^*_{t+1} z] dF(z) \right\|^2}{\int_{I\!R^n} \exp[z'b - \frac{1}{2}z' R^*_{t+1} z] dF(z)} \, dG_{R^*_{t+1}}(b) \tag{4.7}$$

*holds true for all $t = 0, 1, \ldots$*

**Proof.** We observe directly from Theorem 2 that

$$Q^*_{t+1} \cdot \frac{\int_{I\!R^n} \{z - [R^*_{t+1} + \Delta^{-1}]^{-1}[\bar{B}_{t+1} + \Delta^{-1}\mu]\} \exp[z'\bar{B}_{t+1} - \frac{1}{2}z' R^*_{t+1} z] dF(z)}{\int_{I\!R^n} \exp[z'\bar{B}_{t+1} - \frac{1}{2}z' R^*_{t+1} z] dF(z)} = \hat{X}_{t+1} - \hat{X}^K_{t+1}$$

for all $t = 0, 1, \ldots$; therefore upon changing to the measure $\bar{P}$ and using Lemma 2, we find that $\epsilon_{t+1}$ is given by the expectation (under $\bar{P}$) of the ratio

$$\frac{\left\| Q^*_{t+1} \int_{I\!R^n} \{z - [R^*_{t+1} + \Delta^{-1}]^{-1}[\bar{B}_{t+1} + \Delta^{-1}\mu]\} \exp[z'\bar{B}_{t+1} - \frac{1}{2}z' R^*_{t+1} z] dF(z) \right\|^2}{\int_{I\!R^n} \exp[z'\bar{B}_{t+1} - \frac{1}{2}z' R^*_{t+1} z] dF(z)}$$

We now obtain (4.7) by a simple application of Lemma 1 on this last expression.

■

## V. ASYMPTOTICS – THE SCALAR CASE

We now use the representation result of Theorem 3 to investigate the asymptotic behavior of the sequence $\{\epsilon_t,\ t = 1, 2, \ldots\}$ in the scalar case, i.e., $n = k = 1$, when the system dynamics are time invariant.

Let us first fix some notation. In accordance with common usage, we use lower case letters to denote scalar quantities so that $X_t^\circ = x_t^\circ$, $A_t = a$, $H_t = h$, etc. Let $\mathcal{D}$ denote the collection of square-integrable distribution functions on $I\!R$, and let $\mathcal{D}_0$ denote the zero–mean elements of $\mathcal{D}$. For $r \geq 0$, let $G_r$ denote the zero–mean Gaussian distribution with variance $r$.

We characterize the asymptotic behavior of $\{\epsilon_t,\ t = 1, 2, \ldots\}$ in terms of the auxiliary quantities

$$\bar{a} := a - \frac{\sigma^{wv} h}{\sigma^v} \quad \text{and} \quad \bar{c} := \sigma^w - \frac{(\sigma^{wv})^2}{\sigma^v}. \tag{5.1}$$

As a final remark before presenting the asymptotic analysis, observe that there is no loss of generality in considering only initial distributions in $\mathcal{D}_0$. Indeed, since both the true and wide–sense conditional expectation operators are linear and since the state $x_t^\circ$ is affine in the initial condition $\xi$, we may subtract out $E[\xi]$ when forming the difference $\epsilon_t = \hat{x}_t - \hat{x}_t^K$. For $F$ in $\mathcal{D}_0$, (4.7) reduces to

$$\epsilon_t = (q_t^*)^2 I_F(r_t^*) \qquad t = 1, 2, \ldots \tag{5.2}$$

where

$$I_F(r) := \int_{I\!R} \frac{\left| \int_{I\!R} \left\{ z - \frac{b}{r+1/\Delta} \right\} \exp[zb - \frac{1}{2}z^2 r] dF(z) \right|^2}{\int_{I\!R} \exp[zb - \frac{1}{2}z^2 r] dF(z)} dG_r(b), \quad r \geq 0. \tag{5.3}$$

Note that for any non–Gaussian $F$ in $\mathcal{D}_0$, $I_F$ is positive definite in that $I_F(r) = 0$ if and only if $r = 0$ [4].

Moreover, if $h = 0$ or $\bar{a} = 0$, then the system dynamics imply $r_t^* = 0$ for all $t = 0, 1, \ldots$, whence $\epsilon_t = 0$ for all $t = 1, 2, \ldots$ and all distributions $F$ in $\mathcal{D}_0$. Thus only the cases $h \neq 0$ and $\bar{a} \neq 0$ are of interest. The main result of this section is

**Theorem 4.** *We have the following convergence results when $n = k = 1$ and when $h \neq 0$ and $\bar{a} \neq 0$.*

1. *If $\bar{c} \neq 0$, or if $|\bar{a}| \leq 1$ and $\bar{c} = 0$, then $\lim_t \epsilon_t = 0$ for any distribution $F$ in $\mathcal{D}$, whereas if $\bar{c} = 0$ and $|\bar{a}| > 1$, then the asymptotic behavior of $\epsilon_t$ depends nontrivially upon $F$ in $\mathcal{D}$.*

*Moreover we also have the following estimates:*

2. *If $\bar{c} \neq 0$, or if $\bar{c} = 0$ and $|\bar{a}| < 1$, $\lim_t \epsilon_t = 0$ at an exponential rate independent of $F$ for $F$ in $\mathcal{D}$ non–Gaussian whereas if $\bar{c} = 0$ and $|\bar{a}| = 1$, then the rate depends non-trivially upon $F$.*

The proof of Theorem 4 is given in Propositions 2–4 below by considering all possible cases.

**Proposition 2.** *Assume $h \neq 0$ and $\bar{a} \neq 0$.*

1. *If $\bar{c} \neq 0$, $\lim_t \epsilon_t = 0$ for all distributions $F$ in $\mathcal{D}$, with rate*

$$\lim_t \frac{1}{t} \ln \epsilon_t = 2 \ln \left| \bar{a} \left( \frac{\sigma^v}{h^2 p_\infty + \sigma^v} \right) \right| < 0 \tag{5.4}$$

*for all non-Gaussian distributions $F$ in $\mathcal{D}$, where $p_\infty := \lim_t p_t$.*

2. *If $\bar{c} = 0$ and $|\bar{a}| < 1$, then $\lim_t \epsilon_t = 0$ for all distributions $F$ in $\mathcal{D}$ with rate*

$$\lim_t \frac{1}{t} \ln \epsilon = 2 \ln |\bar{a}| < 0 \tag{5.5}$$

*for all non-Gaussian $F$ in $\mathcal{D}$.*

**Proof.** If $\bar{c} \neq 0$, then the pair $(\bar{a}, \bar{c})$ is controllable and $p_\infty$ is well-defined, finite and positive and by standard results [1, Theorem 5.1 and Appendix 1] we conclude that

$$\left| a - \frac{a p_\infty h + \sigma^v}{h^2 p_\infty + \sigma^v} h \right| = \left| \frac{\bar{a} \sigma^v}{h^2 p_\infty + \sigma^v} \right| < 1. \tag{5.6}$$

It is not difficult to see from (5.3) that

$$\lim_t \frac{1}{t} \ln(q_t^*)^2 = 2 \ln \left| \frac{\bar{a} \sigma^v}{h^2 p_\infty + \sigma^v} \right| < 0, \tag{5.7}$$

and that $r_\infty^*$ thus must be finite and positive. It then follows from standard arguments that

$$0 < \liminf_t \, I_F(r_t^*) \leq \limsup_t \, I_F(r_t^*) < \infty, \tag{5.8}$$

and this, together with (5.7), is sufficient to prove claim 1 for $F$ in $\mathcal{D}_0$. The proof of claim 2 is similar. When $0 < |\bar{a}| < 1$ and $\bar{c} = 0$, the dynamics of $q_t^*$ yield $\lim_t t^{-1} \ln(q_t^*)^2 = 2 \ln |\bar{a}| < 0$. The dynamics of $r_t^*$ then require that $0 < r_\infty^* < \infty$, and that again (5.8) hold. The combination of (5.2) and (5.8) prove claim 2 for $F$ in $\mathcal{D}_0$.

■

The dependencies given in Theorem 4 when $\bar{c} = 0$ and $|\bar{a}| \geq 1$ are now illustrated through some simple examples. First, however, we verify a general result.

**Proposition 3.** *For any distribution $F$ in $\mathcal{D}_0$, $\limsup_t \, t I_F(t) < \infty$ so that $\lim_t \, I_F(t) = 0$.*

**Proof.** Since the functional $I_F$ is *independent* of the system dynamics $(a, h, \Gamma)$, we may assume for the purpose of argumentation that our system is

$$x_t^\circ = \xi, \quad y_t = \xi + v_{t+1}^\circ. \qquad t = 0, 1, \ldots \quad (5.9)$$

Here $a = h = \sigma^v = 1$ and $\sigma^w = 0$ so that $\bar{c} = 0$ and $|\bar{a}| = 1$. Consequently $q_t^* = 1$ and $r_t^* = t$ for all $t = 1, 2, \ldots$, whence $\epsilon_t = I_F(t)$ for $t = 0, 1, \ldots$. For all $t = 0, 1, \ldots$, define the linear estimate $\check{x}_t$ of $x_t^\circ$ on the basis of $\{y_0, \ldots, y_t\}$ to be

$$\check{x}_{t+1} := \frac{1}{t+1} \sum_{s=0}^{t} y_s \qquad t = 0, 1, \ldots \quad (5.10)$$

with $\check{x}_0 := 0$. Since $\check{x}_t$ is a *linear* estimator, $\hat{x}_t^K$ the LLMSE estimator and $\hat{x}_t$ the MMSE estimator, we conclude that $E[|\hat{x}_t - \hat{x}_t^K|^2] \leq 4E[|\check{x}_t - x_t^\circ|^2]$ by a straightforward application of the triangle inequality. From (5.10), we verify that $I_F(t) = \epsilon_t \leq 4/(t+1)$, and the claim is immediate. ∎

We now consider the following two distributions $F_1$ and $F_2$ in $\mathcal{D}_0$.

**Distribution $\mathbf{F_1}$.** Distribution $F_1$ admits a density with respect to Lebesgue measure $\lambda$ on $I\!R$ given by

$$\frac{dF_1}{d\lambda}(z) = \sum_{i=1}^{n} \alpha_i \frac{1}{\sqrt{2\pi\rho^2}} \exp\left[-\frac{1}{2}\frac{(z-\mu_i)^2}{\rho^2}\right] \qquad z \in I\!R \quad (5.11)$$

where $\rho > 0$, $0 < \alpha_i \leq 1$ for $i = 1, 2, \ldots, n$, $\sum_{i=1}^n \alpha_i = 1$, and $\sum_{i=1}^n \alpha_i \mu_i = 0$. We exclude the case where $F_1$ is actually Gaussian.

**Distribution $\mathbf{F_2}$.** Under $F_2$, the RV $\xi$ takes on a finite number of values $z_1 < z_2 \ldots < z_n$ with probabilities $p_1, p_2, \ldots, p_n$ respectively and $\sum_{i=1}^n p_i z_i = 0$.

The following two facts are proved in [4].

**Fact 1.** *We have* $\lim_t (\rho^2 t + 1)^2 I_{F_1}(t) = K$ *for some* $K > 0$.

**Fact 2.** *We also have* $\lim_t t I_{F_2}(t) = 1$.

We now can prove the rest of Theorem 4.

**Proposition 4.** *Assume* $h \neq 0$ *and* $\bar{c} \neq 0$.

1. *If* $|\bar{a}| = 1$, *then* $\lim_t \epsilon_t = 0$ *for any distribution* $F$ *in* $\mathcal{D}$, *the rate of convergence depending nontrivially upon* $F$ *for* $F$ *non-Gaussian,*

2. *If* $|\bar{a}| > 1$, *then* $\limsup_t \epsilon_t < \infty$ *for all distributions* $F$ *in* $\mathcal{D}$, *with the asymptotic behavior depending nontrivially upon* $F$ *for* $F$ *not Gaussian.*

**Proof.** *Claim 1.* Under the enforced assumptions, we have $\epsilon_t = (1)^t I_F(t)$ for all $t = 0, 1, \ldots$ and all $F$ in $\mathcal{D}_0$. By Proposition 2, $\lim_t \epsilon_t = 0$; however, if $F = F_1$, $\lim_t (\ln \epsilon_t / \ln t) = -2$, whereas if $F = F_2$, $\lim_t (\ln \epsilon_t / \ln t) = -1$. *Claim*

*2.* It is easy to verify that under the hypotheses on $(a, h, \Gamma)$, $\lim_t r_t^* = \infty$ but $\lim_t (q_t^*)^2/r_t^* = \sigma^v(\bar{a}^2 - 1)/h^2$. For $F$ in $\mathcal{D}_0$, then

$$\epsilon_t = \frac{(q_t^*)^2}{r_t^*}(r_t^* I_F(r_t^*)). \qquad t = 1, 2, \ldots \quad (5.12)$$

Again applying Proposition 2, we get $\limsup_t \epsilon_t < \infty$ for all $F$ in $\mathcal{D}_0$. However, if $F = F_1$, $\lim_t \epsilon_t = 0$, whereas if $F = F_2$, then $\lim_t \epsilon_t = 1$.

■

## AFFILIATIONS

The work of R. B. Sowers was supported through an ONR Graduate Fellowship while completing the requirements towards the M.S. degree. The work of A. M. Makowski was supported partially through the Engineering Research Centers Programs NSF CDR 88–03012 and partially through NSF grant ECS 83–51836.

## REFERENCES

[1] P. Caines and D. Mayne, "On the discrete–time Ricatti equation of optimal control," *Inter. J. of Control* **12**, pp. 785–794 (1970) and **14**, pp. 205–207 (1971).

[2] G. DiMasi and W. Runggaldier, "On measure transformations for combined filtering and parameter estimation in discrete time," *Systems & Control Letters* **2**, pp. 57–62 (1982).

[3] A. M. Makowski, "Filtering formulae for partially observed linear systems with non–Gaussian initial conditions," *Stochastics* **16**, pp. 1–24 (1986).

[4] R. B. Sowers, *New Results in Discrete–Time Nonlinear Filtering*, M.S. Thesis, Electrical Engineering Department, University of Maryland, College Park (MD), August 1988.

[5] R.B. Sowers and A.M. Makowski, "Discrete-time filtering for linear systems in correlated noise with non–Gaussian initial conditions," *Proceedings of the Conference on Information Sciences and Systems*, The Johns Hopkins University, Baltimore (MD), March 1989.

[6] R. B. Sowers and A.M. Makowski, "Filtering for discrete–time linear systems with non–Gaussian initial conditions: Asymptotic behavior of the difference between the MMSE and LMSE estimates," submitted to *IEEE Transactions on Automatic Control* (1989).

# BOUNDARY FEEDBACK STABILIZATION OF DISTRIBUTED PARAMETER SYSTEMS

C. I. Byrnes and D. S. Gilliam

**Abstract**

**In this paper we present a root locus approach to boundary feedback stabilization for a special class of distributed parameter systems. The techniques are particularly well suited for problems arising from initial boundary value problems for partial differential equations with inputs and outputs occurring as point actuators and sensors located on the boundary of the spatial domain.**

In an earlier paper [3], the authors showed how one could interpret quite general boundary conditions in terms of a root locus theory and outlined a program for giving a rigorous justification of root locus methods for certain classes of distributed parameter systems. As indicated in [3], in spite of the intuitive appeal of this method, some care must be taken in this analysis, as illustrated by the example of F. Rellich [8]. For the systems considered in [3] and regular boundary conditions in [1], we show in the present paper, that, if the boundary input and output satisfy certain general assumptions, then the pathologies observed in Rellich's example are avoided. Further, employing simple proportional error feedback laws we present the basis for a rigorous root locus analysis for the boundary feedback stabilization of a certain class of distributed parameter systems.

In the classic work [1], G.D. Birkhoff analyzed the spectral properties of boundary conditions for ordinary differential operators on a finite interval. Under the assumption that the boundary conditions are "regular" it was shown that the resulting operators always possess a discrete spectrum consisting of eigenvalues of finite multiplicity with all but a finite number being simple. Further, an asymptotic representation for the eigenvalues and eigenvectors was obtained as well as an expansion of Fourier type in terms of these eigenvectors. In his dissertation entitled "A comparison of the series of Fourier and Birkhoff", M.H. Stone [9] generalized the work of Birkhoff eliminating the requirement that the coefficients of the operator be smooth and showing that in a very definite sense the expansions of Fourier and Birkhoff are equivalent. A good historical discussion and more complete set of references can be found in [4].

In this paper we exploit the work in [1,4,7,9] in order to develop our root locus approach to boundary feedback stabilization for distributed systems whose spatial part is governed by an ordinary differential operator on a finite interval with regular boundary conditions.

In what follows, we consider a distributed control system

$$\begin{aligned} \dot{w}(x,t) &= Aw(x,t), \\ Bw(t) &= u(t), \\ w(x,0) &= f(x) \in L^2(0,1), \\ y(t) &= Cw(t) \end{aligned} \tag{1}$$

where $A$ is an even order ordinary differential operator of the form

$$A = i^{(n-2)} \frac{d}{dx}^{n} + \sum_{j=2}^{n} p_j(x) \frac{d}{dx}^{n-j}, \quad i = \sqrt{-1} \tag{2}$$

acting in the state space $L^2(0,1)$. The functions $\{p_j(x)\}_{j=2}^{n}$ are assumed $C^\infty[0,1]$. The domain of the unbounded operator $A$ is defined, in terms of $n$ linearly independent, linear, homogeneous boundary conditions

$$\begin{aligned} W_i(w) &\equiv A_i(w) + B_i(w) = 0 \\ A_i(w) &\equiv \alpha_i w^{(m_i)}(0) + \sum_{j=0}^{m_i-1} \alpha_{ij} w^{(j)}(0), \\ B_i(w) &\equiv \beta_i w^{(m_i)}(1) + \sum_{j=0}^{m_i-1} \beta_{ij} w^{(j)}(1), \; i = 1, \cdots, n \end{aligned} \tag{3}$$

The dense domain of $A$ is denoted by

$$D(A) = \{f \in L^2(0,1) \; : \; f \in H^{(n)}(0,1), \; W_i(f) = 0, \; i = 1, \; 2, \; \cdots, n\}$$

where $H^{(n)}$ is the ususal Sobolev space.

The adjoint $A^*$ of $A$ is defined analogously, although, in what follows direct mention of detailed properties of the adjoint system will not be required. We forego a complete description for the sake of brevity.

The input $u(t)$ is assumed to occur through the boundary as in [3] and without loss of generality it will be assumed that it appears in $W_1$, i.e.,

$$Bw(t) \equiv W_1(w)(t) = u(t) \tag{4}$$

It is further assumed that the output sensor has the form

$$y(t) = C(w)(t) = w^{(j_0)}(0,t) \tag{5}$$

The principal result of this work is the analysis of the spectru of the closed loop system obtained from (1)-(3) via a scalar feedback law of the form

$$u(t) = -ky(t) \tag{6}$$

As in [3] we define the closed loop system by introducing the spatial operator $A_k$ as the operator $A$ subject to perturbed boundary conditions obtained from the feedback law (6), ie,

$$\tilde{W}_1(w) = \tilde{A}_1(w) + B_1(w) = 0$$

with

$$\tilde{A}_1(w) \equiv \alpha_1 w^{(m_1)}(0) + \sum_{j=0, j\neq j_0}^{m_1-1} \alpha_{1j} w^{(j)}(0) + (\alpha_{1j_0} + k) w^{(j_0)}(0) \tag{7}$$

and for $i = 2,\ 3,\ \cdots,\ n$ from (3)

$$\tilde{W}_i(w) \equiv W_i(w) = 0. \tag{8}$$

The domain of $A_k$ is thus given by

$$D(A_k) = \{f \in L^2(0,1)\ :\ f \in H^{(n)}(0,1),\ \tilde{W}_i(f) = 0,\ i = 1,\ 2,\ \cdots, n\}$$

and the resulting closed loop system has the form

$$\begin{aligned} \dot{w}(x,t) &= A_k w(x,t) \\ w(x,0) &= f(x) \\ \tilde{W}_i(w) &= 0, \quad i = 1,\ 2,\ \cdots,\ n. \end{aligned} \tag{9}$$

Following [1], the boundary conditions in (3) are assumed to be normalized subject to the following conditions. The orders $m_j$ $(m_1 \leq (n-1))$ of the boundary conditions form a non-increasing sequence with no three successive $m_j$'s equal and each $W_i$ is of exact order $m_i$, i.e., either $\alpha_i \neq 0$ or $\beta_i \neq 0$. Further it is assumed that the set of boundary conditions $W_i$ is regular in the sense of Birkhoff [1] (see Assumption 1 below) which guarantees that $A$ is a discrete spectral operator ([4]) in $L^2(0,1)$.

The essential ingredient of the work in [1] is an explicit asymptotic representation for a basis of solutions of the eigenvalue equation

$$Af = \lambda f \tag{10}$$

In order to present the results found in [1,4] it is necessary to introduce some notation. Let $n = 2\nu$ and let $\omega_j$, $j = 0, 1, \cdots, n-1$ denote the $n$th roots of unity, enumerated so that $\omega_0 = 1 = -\omega_\nu$, the imaginary part of $\omega_j$ is positive for $0 < j < \nu$ and negative for $\nu < j < 2\nu$. Let $\lambda$ be an arbitrary complex number and $z = z(-\lambda)$ denote the $n$th root of $-\lambda$ which lies in the sector

$$S = \{z \in \mathbb{C}\ :\ -\pi/n < \arg(z) \leq \pi/n\}.$$

Define

$$\begin{aligned} \sigma_j(x,z) &= e^{iz\omega_j x}, \quad 0 \le k \le \nu, \\ \sigma_j(x,z) &= e^{iz\omega_j(x-1)}, \quad \nu < k < 2\nu, \end{aligned} \tag{11}$$

With this notation we can transform the eigenvalue equation (10) to the form examined in [4], namely,

$$Af = -z^n f \tag{12}$$

Thus the eigenvalues for (10) are related to (12) by

$$\lambda = -z^n. \tag{13}$$

The result of [1] regarding the asymptotic representation for a basis of solutions for (12) can now be stated as

**Theorem 1** *For each positive integer $p, q = 0, 1, \cdots$, (12) has $n$ linearly independent solutions $f_j$ satisfying*

$$f_j^{(p)}(x) = (i\omega_j z)^p \sigma_j(x,z) \left\{ 1 + \sum_{\ell=1}^{q} \frac{A_{\ell p}(x)}{(i\omega_j z)^\ell} + \frac{E_{jp}(x,z)}{z^{q+1}} \right\} \tag{14}$$

*$j = 1, \cdots, n$ where $A_{\ell p}(x)$ is continuous together with its derivatives of all orders and $E_{jp}(x,z)$ is analytic in $z$ and bounded for all $z \in S$ and $x \in (0,1)$.*

The nonzero spectrum of $A$ consists of the zeroes of the determinant $\det(W_i(f_j))$. For $z \in S$ and our choice of ordering for the $n$th roots of unity $\omega_j$, it is easily established that the terms $\exp(i\omega_j z)$ in this determinant are exponentially small for $j = 1, \cdots, \nu$ and the terms $\exp(-i\omega_j z)$ are exponentially small for $j = \nu + 1, \cdots, 2\nu - 1$. Also for $z \in S$, as in [4], it can be shown there are positive constants $t$ and $s$ so that the determinant being zero implies that either $|z| < t$ or $|\mathrm{Im}(z)| < s$. In particular the asymptotic behaviour of the eigenvalues depends only on the highest order derivative terms in each boundary condition.

Thus we define

$$g(z) \equiv \prod_{j=1}^{n} (iz)^{-m_j} (\det(W_i(f_j))) \tag{15}$$

Hence for large $|z|$ and any $q$, there are constants $c_{0j}$, $c_{1j}$, $c_{2j}$, $d_0$, $d_1$, $d_2$ for which $g(z)$ can be expressed as

$$g(z) = \left( \Theta_0 + \sum_{j=1}^{q-1} \frac{c_{0j}}{z^j} + \frac{d_0}{z^q} \right)$$

$$+ \quad e^{iz}\left(\Theta_1+\sum_{j=1}^{q-1}\frac{c_{1j}}{z^j}+\frac{d_1}{z^q}\right) \tag{16}$$

$$+ \quad e^{-iz}\left(\Theta_2+\sum_{j=1}^{q-1}\frac{c_{2j}}{z^j}+\frac{d_2}{z^q}\right)$$

where

$$\Theta_0+\Theta_1 s+\Theta_2 s^{-1} = \tag{17}$$

$$\begin{vmatrix} a_1 & \omega_1^{m_1}\alpha_1 & \cdots & b_1 & \omega_{\nu+1}^{m_1}\beta_1 & \cdots & \omega_{2\nu-1}^{m_1}\beta_1 \\ a_2 & \omega_1^{m_2}\alpha_2 & \cdots & b_2 & \omega_{\nu+2}^{m_2}\beta_2 & \cdots & \omega_{2\nu-1}^{m_2}\beta_2 \\ \vdots & \vdots & \vdots & \vdots & \vdots & \vdots & \vdots \\ a_n & \omega_1^{m_n}\alpha_n & \cdots & b_n & \omega_{\nu+1}^{m_n}\beta_n & \cdots & \omega_{2\nu-1}^{m_n}\beta_n \end{vmatrix}$$

$$a_1=(\alpha_1+s\beta_1),\ \cdots,\ a_n=(\alpha_n+s\beta_n)$$

$$b_1=(-1)^{m_1}(\alpha_1+\beta_1/s),\ \cdots,\ b_n=(-1)^{m_n}(\alpha_n+\beta_n/s)$$

Following Birkhoff [1] we impose the following regularity hypotheses on the boundary conditions.

**Assumption 1** *The constants $\Theta_1$ and $\Theta_2$ in (17) are not zero and the constant $\Theta_0 \neq \pm 1$.*

It was shown in [7] that if the boundary conditions are separated, i.e., in $W_i$ the terms $B_i \equiv 0$ for $i = 1, \cdots, \nu$ and the terms $A_i \equiv 0$ for $i = \nu+1, \cdots, 2\nu$, then the regularity condition in Assumption 1 is always satisfied. Further, in the separated case it can be shown [7,4] that the constant $\Theta_0$ is always zero so that the asymptotic behavior of the zeroes discussed below is extremely simple. The case of separated boundary conditions and feedback law (6) corresponds to a co-located actuator and sensor which is the situation that occurs most often in applications.

The zeroes of $g(z)$ in $S$ are asymptotic to the zeroes of

$$\Theta_0+\Theta_1 e^{iz}+\Theta_2 e^{-iz} \tag{18}$$

and the zeros of this expression are easily obtained from elementary trigonometry.

Recalling the relationship between $z$ and $\lambda$ in (12) one obtains the formulas of Birkhoff [1] concerning the asymptotic distribution of the eigenvalues.

The eigenvalues $\lambda$ of $A$ with boundary conditions $\{W_i\}$ are in general simple (multiplicity one) and form a pair of infinite sequences $\lambda_{I\ell}$, $\lambda_{II\ell}$ $\ell = 1,\ 2,\ \cdots$ such that

$$\lambda_{I\ell} \quad = \quad -(2\pi\ell)^n\left(1+\sum_{j=1}^{q-1}\frac{g_{Ij}}{\ell^j}+\frac{E_{I\ell}}{\ell^q}\right) \tag{19}$$

$$\lambda_{II\ell} = -(2\pi\ell)^n \left(1 + \sum_{j=1}^{q-1} \frac{g_{IIj}}{\ell^j} + \frac{E_{II\ell}}{\ell^q}\right)$$

where $g_{Ij}$ and $g_{IIj}$ are constants and $|E_{I\ell}| < M$, $|E_{II\ell}| < M$.

The following assumption guarantees that as the gain $k$ in (6) varies over a compact set, only the terms $g_{Ij}$, $g_{IIj}$, $E_{I\ell}$, $E_{II\ell}$ in (19) are changed and these expressions vary continuously in $k$. Therefore the asymptotic distribution of the eigenvalues is unchanged. Another consequence is that $k$ does not effect the regularity condition in Assumption 1. Thus the closed loop system obtained for nonzero $k$ remains a complete, discrete spectral system.

**Assumption 2** *The order of the derivative $j_0$ in the output (5) is strictly less than the order $m_1$ of the boundary condition $W_1$ for the input.*

From (7), (8), (14)-(16), we find that the spectrum of $A_k$ is given as the zeroes of

$$\Delta_k(z) = \det(W_i(f_j)) + k\det(\tilde{W}_i(f_j)) \tag{20}$$

and as in (15) we consider

$$f(z,k) = \prod_{j=1}^{n} (iz)^{-m_j} \Delta_k(z) \equiv g(z) + kh(z) \tag{21}$$

where $g(z)$ is given in (15).

Assumption 2 implies that the zeroes of $f(z,k)$ have the same asymptotic behaviour as $g(z)$ for each $k$ and is determined by $\Theta_0$, $\Theta_1$, $\Theta_2$.

The main result can now be stated as

**Theorem 2** *If the system (1)-(3) satisfies Assumption 1, the output (5) satisfies Assumption 2 and the closed loop system (9) is exponentially stable for $k = k_0$, then there exists an $\epsilon > 0$ such that the system remains exponentially stable for all $k \in (k_0 - \epsilon, k_0 + \epsilon)$.*

Complete details of the proof can be found in a forthcoming paper by the authors. This paper also includes several examples including stabilization of a nondamped cantilever beam.

From Assumptions 1, 2 and the results of Birkhoff regarding the aymptotic distribution of the eigenvalues of the open and closed loop systems we see that for $k \in K$ a compact neighborhood of $k_0$, the zeros $z_j(k)$ of $f(z,k)$ all have the same asymptotic behavior. Hence for a large constant $R > 0$ there exists an $N$ such that for all $k \in K$ the roots $\{z_j(k)\}_{j=N+1}^{\infty}$ (here we do not distinguish between the two asymptotic types) satisfy $\mathrm{Re}(z_j(k)) < -R$ and are simple. Assume that the first finitely many roots of $f(z,k_0)$ are enumerated as $\{z_j(k_0)\}_{j=1}^{N_0}$ each of finite multiplicity $\mu_j$ and all lie in the compact ball of radius $R$ about 0 in the $z$-plane.

If $(z_0, k_0)$ is one of the finitely many zeros $z_j(k_0) \in S_R(0)$ of multiplicity $\mu_0$ (if $\mu_0 = 1$ the situation is simpler and can be handled using the implicit function theorem) define $w = z - z_0$ and $\alpha = k - k_0$, then the function

$$F(w, \alpha) = f(w + z_0, \alpha + k_0)$$

is analytic and has a zero at $(w, \alpha) = (0, 0)$ of multiplicity $\mu_0$. By the Weierstrass preparation theorem [5] there exists unique functions $h$, $W$ such that

$$F(w, \alpha) = h(w, \alpha) W(w, \alpha)$$

where $h$ does not vanish in a neighborhood of $(0, 0)$ and $W$ is a Weierstrass polynomial

$$W(w, \alpha) = w^{\mu_0} + \sum_{j=0}^{\mu_0 - 1} a_j(\alpha) w^j$$

with the functions $a_j$ analytic and vanishing at $\alpha = 0$. Now according to [6] Lemma 1.2, page 275, $W$ has the unique factorization

$$W(w, \alpha) = \prod_{j=0}^{\mu_0} (w - \phi_j(\alpha))$$

where

$$\phi_j(\alpha) = \sum_{k=1}^{\infty} b_{jk} \alpha^{(k/\mu_0)}$$

and each $\phi_j$ is a continuous but not necessarily $C^1$ function of $\alpha$. In fact, for each $j$ there exists an $\epsilon > 0$ such that $\phi_j$ is $C^1$ for $\alpha$ in $(-\epsilon, 0)$ and $(0, \epsilon)$.

Thus there exist finitely many neighborhoods $U_j$ about $z_j(k_0)$ and $V_j$ about $k_0$ and continuous functions $\psi_{ij} : V_j \mapsto U_j$ $i = 1, \cdots, \mu_j$ such that $\psi_{ij}(k_0) = z_j$ for all $i$ and such that if $(z, k) \in U_j \times V_j$ is a zero of $f$ then $z = \psi_{ij}(k)$ for some $i$ and $f(\psi_{ij}(k), k) = 0$ for all $k \in V_j$. Further if $(z, k) \notin \bigcup(U_j \times V_j)$ then there is a ball about $(z, k)$ on which $f \neq 0$ and hence by a simple compactness argument we see that all the finitely many roots of $f(z, k)$ correspond to the roots of $f(z, k_0)$ in a continuous fashion for some neighborhood of $k_0$. Therefore $\epsilon$ can be chosen so that for $k \in S_\epsilon(k_0)$ the eigenvalues of $A_k$ remain bounded away from the imaginary axis in the left half plane and the closed loop system is exponentially stable.

**Acknowledgments:** The work of the first author is supported in part by a grant from AFOSR and NSF. The second author is supported in part by a grant from AFOSR and Texas Advanced Research Program grant # 0219-44-5195.

## References


[1] G.D. Birkhoff, "Boundary value and expansion problems of ordinary linear differential equations", Trans. Amer. Math. Soc., Vol. 9, (1908), 373-395.

[2] C.I. Byrnes, "Adaptive stabilization of infinite dimensional linear systems," Proc. of IEEE Conf. on Dec. and Control, Los Angeles, 1987.

[3] C.I. Byrnes, D.S. Gilliam, "Asymptotic properties of root locus for distributed parameter systems," Proc. of IEEE Conf. on Dec. and Control, Austin, 1988.

[4] N.Dunford, J. Schwartz, "Linear Operators, Vols. I, II, III," Interscience, N.Y. (1963).

[5] L. Hörmander, "An introduction to complex analysis in several variables," Van Nostrand, New Jersey, (1966).

[6] L. Hörmander, "Linear partial differential operators, I," Springer-Verlag, New York, (1969).

[7] H.P. Kramer, "Perturbation of differential operators," Pacific J. Math., Vol. 7, (1957), 1405-1435.

[8] F. Rellich, "Perturbation Theory of Eigenvalue Problems," Gordon and Breach, New York, (1969).

[9] M.H. Stone, "A comparison of the series of Fourier and Birkhoff," Transactions AMS, (1926), 695-761.

# OPTIMAL NONLINEAR FEEDBACK SYSTEM DESIGN FOR A GENERAL TRACKING PROBLEM

Guanrong Chen and Rui J.P.de Figueiredo

**Abstract** *An optimal feedback design strategy for a tracking problem of general nonlinear systems is posed and solved in a Banach space setting in the time domain. An existence theorem is established, a convergent recursive algorithm for solving the problem is given, and a simple example is included for the purpose of illustration.*

## 1. Introduction.

Motivated by the elegant and attractive approach of $H^\infty$ optimization to feedback control designs for MIMO linear systems, we have recently developed a new approach to the optimal feedback control strategies for *nonlinear** systems (cf. Chen and de Figueiredo [4,5] and de Figueiredo and Chen [8]). We first formulated the optimal disturbance rejection problem in [8] and then the robust stabilization problem in [4] for general MIMO nonlinear systems in a Banach space setting in the time domain. Based on some nonlinear operator-theoretic techniques, we have been able to establish existence theorems and convergent recursive algorithms for solving these problems.

On the other hand, some significant results on optimization theory for nonlinear systems from an entirely different approach have been obtained by Ball and Helton [1] and Foias and Tannenbaum [9].

Differing from the nonlinear interpolation technique of Ball, Foias, Helton, and Tannenbaum (see also [2]), our approach is to design a nonlinear compensator stabilizing the closed-loop system while minimizing the operator norm of a nonlinear operator which reflects the uniform bound of certain unknown error. Our problem is formulated in a Banach space setting in the time domain, and is based on the theory of nonlinear Lipschitz operators. The main advantage of our approach is that typical optimization problems in the nonlinear systems control theory and engineering can be formulated very precisely. Moreover, the procedure for obtaining an optimal solution is relatively simple.

In this paper, we will further investigate a tracking problem formulated in a manner similar to the optimal disturbance rejection problem investigated in de Figueiredo and Chen [8] for general MIMO nonlinear control systems. We will

* Here and throughout the paper, by "nonlinear" we mean "not necessarily linear"

establish an existence theorem and a convergent recursive computational scheme for the problem, and give a simple example for the purpose of illustration.

## 2. Statement of the problem and preliminary results.

Consider a general MIMO nonlinear closed-loop system defined in a Banach space setting:

$$\begin{cases} e = r - y \\ y = P(C(e)) + W_2 d \\ r = W_1 u\,, \end{cases} \tag{1}$$

the configuration of which is depicted in Figure 1, where $r$ denotes the external reference input signal modeled as the output of a linear filter $W_1$ driven by a source $u$, $e$ the error signal between the reference $r$ and the system output $y$ which is required to check the given reference $r$, $\tilde{d}$ the possible disturbance modeled as the output of a linear filter $W_2$ driven by an external source $d$, where $W_2$ may be zero in case no disturbance is considered, and $P$ and $C$ are respectively the plant and compensator operators.

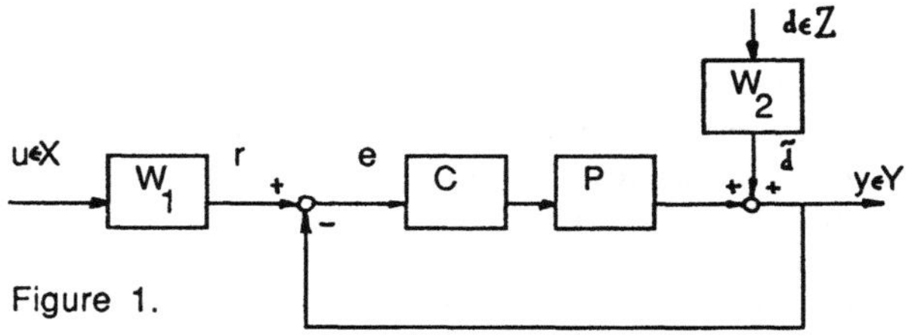


Figure 1.

Let $X, Y$, and $Z$ be three Banach spaces of $l$-,$p$-, and $q$-tuples of real-valued functions defined on the time domain [0,T] where $T \leq \infty$, such that $u \in X$, $d \in Z$, and $y, r, e \in Y$. Assume that $W_1 : X \to Y$ and $W_2 : Z \to Y$ are bounded linear operators such that $R(W_1)$, $R(W_2) \subset D(C)$; and $P : R(C) \to Y$ and $C : Y \to D(P)$ are bounded nonlinear operators* such that $R(C) \subset D(P)$ and $R(P) \subset D(C)$. Here and throughout the paper, we use the notation $D(A) =$ the domain of the operator $A$ and $R(A) =$ the range of $A$. Moreover, we require that all these operators as well as their compositions through feedback be *causal*.

Let us introduce the admissible class of source inputs and disturbances:

$$U = \{\, u \in X : \; \|u\|_X \leq M_u < \infty \,\}; \tag{2a}$$

and

$$N = \{\, d \in Z : \; \|d\|_Z \leq M_d < \infty \,\}. \tag{2b}$$

Because of the feedback in the nonlinear system, we need to require that the plant output $y$ as well as the reference input $r$ be both restricted to a bounded set $\Omega$ of $Y$ defined by

$$\Omega_0 = \{\, y \in Y : \; \|y\|_Y \leq M_y < \infty \}. \tag{3a}$$

---

* This does not imply the stability of the system because of the closed-loop configuration

This assumption makes sense based on practical considerations. For example, the choice of $M_y$ could be dictated by the allowable maximum dynamic range or maximum power (when the $L^2$ space is considered) for the output.

$$\Omega = \{ \, e = r - y : \quad r, y \in \Omega_0 \, \}, \tag{3b}$$

and assume that $D(C) \subset \Omega$.

Now, given $P$, $W_1$, $W_2$, $U$, $N$, and $\Omega$, our *objective* is: *to design a compensator $C$ belonging to certain admissible class $S$ so as to stabilize the system while minimizing the error $e$ between the system output $y$ and the reference signal $r$ .*

In order to pose the problem precisely, we need to introduce an admissible class $S$ of nonlinear compensator operators. First, let $Lip(\Omega, \Omega)$ be the family of Lipschitz nonlinear operators from from $\Omega$ into $Y$, where the operator norm of a Lipschitz operator $T \in Lip(\Omega, \Omega)$ is defined by (cf. Martin [10]).

$$|||T||| = \|T(0)\|_Y + \sup_{\substack{e_1, e_2 \in D(T) \\ e_1 \neq e_2}} \frac{\|T(e_1) - T(e_2)\|_Y}{\|e_1 - e_2\|_Y} . \tag{4}$$

Let

$$S = \{ \, C(\cdot) \in Lip(\Omega, \Omega) : \qquad P(C(\cdot)) \in Lip(\Omega, \Omega) \, \} . \tag{5}$$

be the admissible class of nonlinear compensator operators. Since the domain of admissible $C$ (which is also the domain of $P(C(\cdot))$) is the closed bounded set $\Omega \subset Y$, this assumption is not too restrictive. A number of nonlinear operators such as the polynomic and exponential operators satisfy this condition•on $\Omega$. The following three lemmas have been established in de Figueiredo and Chen [8]:

**Lemma 1.** *$S$ is an (infinite-dimensional) Banach space.*

This lemma shows that the admissible class of compensators is a very large family.

**Lemma 2.** *Given $P$ and $S$. Then, the subset $\Sigma$ of $S$ defined below consists of compensators $C$ that stabilize the system* (1)*:*

$$\Sigma = \{ \, C \in S : \; |||P(C(\cdot))||| < 1 \} . \tag{6}$$

**Lemma 3.** *Under the condition stated in* (6)*, the nonlinear operator*

$$I(\cdot) + P(C(\cdot)) : \quad \Omega \to \Omega$$

*is invertible, and its inverse, denoted by $(I + P(C))^{-1}$, is also a Lipschitz operator and satisfies*

$$|||(I + P(C))^{-1}||| \le \beta(C) + \frac{1}{1 - |||P(C)|||} \tag{7}$$

*where the constant $\beta(C) = \|(I + P(C))^{-1}(0)\|_Y$.*

Based on Lemmas 2 and 3, the tracking problem under consideration can be formulated as follows: First, it follows from (1) and Lemma 3 that

$$e = (I + P(C))^{-1}(r - W_2 d). \tag{8}$$

Since in (8) $r$ and $d$ are independent variables where $W_2$ may be zero when no disturbance is considered as mentioned above, and both of them are uniformly bounded, we will minimize the operator norm $\|(I + P(C))^{-1}\|$ and pose the problem as follows:

$$\min_{C \in \Sigma} \{ \|(I + P(C))^{-1}\| \}. \tag{OPT}$$

### 3. An existence theorem.

The following existence result can be established by imitating the proof of Theorem 1 in de Figueiredo and Chen [8]:

**Theorem 1.** *The objective functional*

$$f(C) := \|(I + P(C))^{-1}\| \tag{9}$$

*is continuous on $\Sigma$ defined by (6). Consequently, (OPT) always has a solution under one of the following conditions:*

(i) *Let $S_c$ be a compact subset of $S$ and let*

$$\Sigma_C = \{ C \in S_c : \quad \|P(C)\| < 1 \}. \tag{10}$$

*Then, (OPT) has a solution in $\Sigma_C$.*

(ii) *Let $S_c$ be a bounded set in any finite − dimensional subspace of $S$ and let*

$$\Sigma_C = \{ C \in S_c : \quad \|P(C)\| \leq 1 - \epsilon \}, \tag{11}$$

*where $\epsilon > 0$ arbitrarily small. Then (OPT) has a solution in $\Sigma_C$.*

(iii) *Let $S_c$ be any finite − dimensional subspace of $S$ and suppose that the given plant operator $P$ has the growing property that it maps an unbounded set to an unbounded set. Let*

$$\Sigma_C = \{ C \in S_c : \quad \|P(C)\| \leq 1 - \epsilon \}, \tag{12}$$

*where $\epsilon > 0$ arbitrarily small. Then (OPT) has a solution in $\Sigma_C$.*

Since the polynomic operators of finite degree constitutes a dense set in the Banach space $S$ of Lipschitz operators, the following result is immediate. The result states that the infimum of the objective functional $f(C)$ can be approached as close as possible. Hence, from the application point of view, this result is important and useful.

**Corollary 1.** *Let*

$$\delta = \inf_{C \in \Sigma} |||(I + P(C))^{-1}|||$$

*and*

$$\delta_k = \inf_{C_k \in \Sigma_{C_k}} |||(I + P(C))^{-1}|||\,,$$

*where* $\Sigma$ *is defined by* (6) *and* $\Sigma_{C_k}$ *is defined by* (11) *or* (12) *in the* $k$*-dimensional subspace of* $S_c$. *Then, we have* $\delta_k \to \delta$ *as* $k \to \infty$.

We remark that the family of analytic systems with finite order Volterra series expansions (cf. de Figueiredo [7]) is also a dense set in the admissible class $S$. Moreover, a great deal of plant operators $P$ such as polynomic and exponential systems have the growing property stated in part (iii) although they are of the Lipschitz type when restricted on the bounded set $\Omega$.

We also remark that there are many nontrivial examples of Lipschitz operators the totality of which consists of a compact set in the infinite dimensional space $S$. One example may be found in de Figueiredo and Chen [8] which consists of nonlinear systems with fading memory (cf. Boyd and Chua [3]).

## 4. A convergent recursive algorithm for solving *OPT*.

In this section, we establish a convergent recursive scheme for solving (OPT) under the conditions posed in Theorem 1. The algorithm is essentially based on the Neumann-type expansion formula for the Lipschitz operator $(I + PC)^{-1}$; namely: let $A_0 = I$, and $A_n = I - P(C(A_{n-1}))$, then

$$(I + PC)^{-1}(e) = \lim_{n \to \infty} A_n(e)$$

for all $e \in \Omega$, (cf. Martin [10]). Recall that

$$\Sigma_C = \{\, C \in S_c :\ |||P(C)||| \le 1 - \epsilon \,\}.$$

Define

$$\Sigma_n = \left\{\, C \in S_c :\ |||P(C)||| \le 1 - \epsilon - \frac{\epsilon}{n+1} \,\right\},$$

and

$$\begin{aligned}
&A_0 = I\,,\\
&A_1(C_1) = I - P(C_1)\,,\\
&A_n(C_n; C_{n-1}, \ldots, C_1) = I - P(C_n(A_{n-1}(C_{n-1}; C_{n-2}, \ldots, C_1)))\,,
\end{aligned}$$

$n = 2, 3, \ldots$. Moreover, let

$$f_n(C_n; C_{n-1}, \ldots, C_1) = |||A_n(C_n; C_{n-1}, \ldots, C_1)|||$$

and

$$f(C) = \| (I + P(C))^{-1} \| .$$

Then, a generalized recursive nonlinear programming algorithm for solving (OPT) can be established as follows. For each $n, n = 1, 2, \ldots$ , solve successively the following minimization:

$$\min_{C_n \in \Sigma_n} \{ \| I - P(C_n(A_{n-1}(C_{n-1}^*; C_{n-2}^*, \ldots, C_1^*))) \| \} . \qquad (GRNP)$$

The verification of the convergence of the scheme, namely: $C_n^* \to C^*$ as $n \to \infty$, can be found in de Figueiredo and Chen [8], which is based on a result of Daniel [6].

Note that for each $n$, the objective functional in (GRNP) is continuous in $C_n$, and hence can be solved by the nonlinear programming discussed below. Moreover, it may be reduced to a standard min-max problem in some simple cases such as in the example shown in the next section.

In order to solve $(GRNP)$, we propose the following generalized nonlinear programming scheme. For a general form, consider the $(OPT)$ posed in Section 2 and set

$$\begin{cases} \min_{C \in \Sigma} \phi(C) \\ g(C) \leq 1 - \epsilon \end{cases} \qquad (GNP)$$

where $\phi(C) = I - P(C)$, namely:

$$\phi(C) = \|PC(0)\|_Y + \sup_{\substack{e_1, e_2 \in \Omega \\ e_1 \neq e_2}} \left\{ \frac{\|(e_1 - e_2) + PC(e_1) - PC(e_2)\|_Y}{\|e_1 - e_2\|_Y} \right\}, \qquad (13)$$

and

$$g(C) = \|PC(0)\|_Y + \sup_{\substack{e_1, e_2 \in \Omega \\ e_1 \neq e_2}} \frac{\|PC(e_1) - PC(e_2)\|_Y}{\|e_1 - e_2\|_Y}, \qquad (14)$$

We close this section by a remark that in practice, $\Sigma_{C_k}$ may be chosen to be the $k$-dimensional Fock space of nonlinear compensators (cf., Chen and de Figueiredo [5]), or simply the space of $k$th order polynomic compensators of the form

$$C(\cdot)(t) = \sum_{i=0}^{k} f_i(t)(\cdot)^i .$$

In such cases, on one hand the feedback is realizable, and on the other hand (GNP) becomes a nonlinear programming problem since the variables of the objective functional $\phi(C) = \phi(\sum_{i=0}^{k} f_i(t)(\cdot)^i)$ are real-valued functions. It differs from the conventional nonlinear programming only in the supremum setting, which is a computational but not conceptual issue. In some simple cases, such as in the example shown below, this difficulty can be overcome.

**5. An example.**

Consider the nonlinear $P$ given by

$$P(\cdot) = (\cdot)^2$$

and let the compensator $C$ belong to the class of quadratic systems of the form

$$C(e)(t) = c_1(t)e(t) + c_2(t)e^2(t),$$

in which we let the leading term $c_0(t) = 0$ so that $P(C(0)) = 0$ for simplicity. Moreover, let $X = Y = Z = L_2$ and use the notation $\|\cdot\| = \|\cdot\|_2$.

Then, we have

$$\begin{aligned} g(C) &= \sup_{\substack{e_1,e_2\in\Omega \\ e_1\neq e_2}} \frac{\|(c_1e_1 + c_2e_1^2)^2 - (c_1e_2 + c_2e_2^2)^2\|}{\|e_1 - e_2\|} \\ &\leq \sup_{\substack{e_1,e_2\in\Omega \\ e_1\neq e_2}} \|c_1(e_1+e_2) + c_2(e_1^2 + e_2^2)\| \left\| \frac{c_1(e_1 - e_2) + c_2(e_1^2 - e_2^2)}{\|e_1 - e_2\|} \right\| \\ &\leq (2M_y\|c_1\| + 2M_y^2\|c_2\|)(\|c_1\| + 2M_y\|c_2\|) \\ &:= a(c_1, c_2). \end{aligned}$$

Set

$$a(c_1, c_2) \leq 1 - \epsilon.$$

Then, we have

$$\Sigma_C = \{\, c_1, c_2 \in L_2 : \; a(c_1, c_2) \leq 1 - \epsilon \,\}.$$

Secondly, note that $\phi(c_1, c_2, e_1, e_2)$ is a non-negative real functional of $c_1, c_2, e_1$ and $e_2$, and is given by

$$\begin{aligned} \phi(c_1, c_2, e_1, e_2) = \Bigg\| &\frac{e_1 - e_2}{\|e_1 - e_2\|} + (c_1(e_1 + e_2) + \\ &c_2(e_1^2 + e_2^2))(c_1\frac{e_1 - e_2}{\|e_1 - e_2\|} + c_2\frac{e_1^2 - e_2^2}{\|e_1 - e_2\|}) \Bigg\| \end{aligned}$$

We also have

$$\Omega = \{\, e \in L_2 : \quad \|e\| \leq M_y \,\}.$$

Hence, the (GNP) is finally reduced to the following:

$$\min_{c_1,c_2\in\Sigma_0} \; \max_{e_1,e_2\in\Omega} \phi(c_1, c_2, e_1, e_2). \tag{15}$$

Problem (15) is now a min-max optimization and hence can be solved by standard or modified techniques available in the literature.

This work was supported in part by the ONR Contract N00014-88-K-0127 and the State of Texas grant TATP 2892

REFERENCES

[1] J.A. Ball and J.W. Helton, "Sensitivity bandwidth optimization for nonlinear feedback systems," Tech Report, Dept. of Math., Univ. of California at San Diego, 1988.

[2] J.A. Ball, C. Foias, J.W. Helton, and A. Tannenbaum, "On a local nonlinear commutant lifting theorem," Indiana J. Mathematics, 36 (1987), 693-709.

[3] S. Boyd and L. Chua, "Fading memory and the problem of approximating nonlinear operators with Volterra series," IEEE Trans. Circ. & Sys.,32 (1985), 1150-1161.

[4] G. Chen and Rui J.P. de Figueiredo, "On robust stabilization of nonlinear control systems," Contr. Sys. Lett., 1989 in press.

[5] G. Chen and Rui J.P. de Figueiredo, "An RKHS of bounded nonlinear operators for nonlinear systems control," TREE 8817, Dept. of Elec. Comp. Eng., Rice Univ., 1988, submitted.

[6] J.W. Daniel, *The Approximate Minimization of Functionals*, Englewood Cliffs: Prentice-Hall, 1971.

[7] Rui J.P. de Figueiredo, "A generalized Fock space framework for nonlinear system and signal analysis," IEEE Trans. Circ. & Sys., 30 (1983), 637-647.

[8] Rui J.P. de Figueiredo and G. Chen, "Optimal disturbance rejection for nonlinear control systems, TREE 8805, Dept. of Elec. Comp. Eng., Rice Univ., 1988, to appear in IEEE Trans. Auto. Contr.

[9] C. Foias and A. Tannenbaum, "Weighted optimization theory for nonlinear systems," 1988, submitted.

[10] R.H. Martin, *Nonlinear Operators and Differential Equations in Banach Spaces* Wiley, New York, 1976.

# STABILITY THEORY FOR DIFFERENTIAL/ALGEBRAIC SYSTEMS WITH APPLICATION TO POWER SYSTEMS

DAVID J. HILL
IVEN M.Y. MAREELS

KEYWORDS: Differential/algebraic systems, Lyapunov methods, Krasovskii and Lur'e type Lyapunov functions, power system stability.

ABSTRACT Some results on Lyapunov stability for a class of differential/algebraic systems are presented. The stability results are applied to power systems.

## 1. INTRODUCTION

Many physical problems yield mathematical descriptions which are a mixture of ordinary differential equations and algebraic equations. Power systems and electric circuits fall into this category. For circuit theory it is reasonable to infer dynamical behaviour from a singularly perturbed system which allows for parasitic elements [7,12]. However, in many applications such as some power systems problems there is no obvious physical interpretation for the perturbation. Thus the differential algebraic (DA) model should be dealt with directly.

Unfortunately DA systems in their full generality have some peculiar properties which make their mathematical analysis rather intricate and technical [13,9]. As a consequence rather little is known about their solutions except in the linear case and some special nonlinear situations.

In this report, we study DA systems under some additional smoothness and regularity conditions, simplifying the technical details considerably, yet obtaining results which are relevant for a large class of engineering problems. Our attention is focused especially on Lyapunov type stability results. It is our aim to present rigourously what is meant by Lyapunov stability in the DA system context and to demonstrate some basic results in this area.

This work is motivated by recent stability analyses for power systems using a DA system description [10, 5, 2,14]. Imprecise statements have been made by referring to stability concepts pertaining to ordinary differential equation (ODE) descriptions. The aim here is to give a more solid basis for such discussions.

All proofs are omitted; for complete details see [6]. The structure of the paper is as follows. Section 2 gives the basic DA problem description. Sections 3 and 4 develop the stability results. Section 5 discusses the application to power systems. Section 6 gives some comments on generalizations and conclusions.

## 2. PROBLEM DESCRIPTION

We consider DA systems in the format:

$$\dot{x} = f(x, y) \tag{2.1.1}$$
$$0 = g(x, y) \tag{2.1.2}$$

with some compatible initial conditions, $(x_0, y_0)$ i.e.

$$0 = g(x_0, y_0) \tag{2.1.3}$$

where

$$f : R^n \times R^m \to R^n \quad ; (x, y) \mapsto f(x, y) \tag{2.1.4}$$
$$g : R^n \times R^m \to R^m \quad ; (x, y) \mapsto g(x, y) \tag{2.1.5}$$

Simplifying, we assume throughout that f and g are twice continuously differentiable in some open connected set $\Omega \subset R^n \times R^m$;

$$f, g \in C^2(\Omega) \tag{A1}$$

and that the Jacobian of g with respect to y has constant full rank on $\Omega$ :

$$rank\ (D_2 g(x, y)) = m \qquad \forall (x, y) \in \Omega \tag{A2}$$

We use the following notations:

$x(t, x_0, y_0)$, $y(t, x_0, y_0)$ are solutions of (2.1) as a function of time and initial conditions

$B_\epsilon = \{\ (x, y) \in R^n \times R^m : \ \ \|(x, y)\| < \epsilon\ \}$;

$G = \{\ (x, y) \in R^n \times R^m : \ \ g(x, y) = 0\ \}; \Omega_G = \Omega \cap G$ ;

$\overline{\Omega}$ = *closure of* $\Omega$ *in* $R^n \times R^m$ ; $\delta(\Omega)$ = *boundary of* $\Omega$ *in* $R^n \times R^m$

*int* $(\Omega)$ = *interior of* $\Omega$ *in* $R^n \times R^m$ ; $K = \{\ a : R_+ \to R_+$ , continuous, strictly increasing, $a(0) = 0\ \}$; $\dot{V}_{(n.m)}$ = derivative of the function V with respect to time along the solution of the system with equations (n.m)

## 3. LYAPUNOV STABILITY RESULTS

In this section, we consider stability properties of equilibria of the DA system (2.1). We assume existence and uniqueness of solutions for the system (2.1). Conditions for this are developed in [6].

It follows from the implicit function theorem and Assumption A2 that given $(\bar{x}, \bar{y}) \in \Omega_G$ there is some neighbourhood $U \subset R^n$ of $\bar{x}$ and a unique twice differentiable function

$$u : R^n \to R^m \quad ; \ x \mapsto u(x), \quad u \in C^2(U)$$

such that

$$0 = g(x,u(x)) \;\; \forall\, x \in U \;\; \text{and} \;\; (U \times u(U))_G \subset \Omega_G$$

with Jacobian

$$(Du)(x) = -(D_2 g)^{-1}(x, u(x)) \cdot (D_1 g)(x_1, u(x)) \quad \forall x \in U$$

The following immediate result describes the reduced system (RS) for (2.1)

**Lemma 3.1** In the neighbourhood $(U \times u(U))_G \subset \Omega_G$, the system (2.1) reduces to

$$\dot{x} = f(x,u(x))\,;\, y=u(x) \tag{3.1}$$ ❑

We assume that the system (2.1) has a unique (isolated) equilibrium in $\Omega$, which we regard to be the origin, without loss of generality:

$$\text{In } \Omega,\; f(x,y) = 0 \text{ and } g(x,y) = 0 \quad \text{iff} \quad (x,y) = 0 \tag{A3}$$

When discussing stability in the DA system context it should be clear that we only consider stability with respect to perturbations which satisfy the algebraic constraints. When using the RS representation, this feature has been accounted for.

We now present the formal definitions of stability of the trivial solution (x(t,0,0), y(t,0,0)) ≡ (0,0) of the DA system (2.1).

**Definition 1.** The trivial solution of (2.1) is called stable if given $\epsilon > 0$, there exists a $\delta > 0$ such that for all $(x_0, y_0) \in \Omega_G \cap B_\delta$ then $(x(t, x_0, y_0), y(t, x_0, y_0)) \in \Omega_G \cap B_\epsilon$, $\forall t \in R_+$.

**Definition 2.** The trivial solution of (2.1) is called asymptotically stable if it is stable and there exists $\eta > 0$ such that for all $(x_0, y_0) \in \Omega_G \cap B_\eta$ then

$$\lim_{t \uparrow \infty} \| (x(t, x_0, y_0), y(t, x_0, y_0)) \| = 0$$

It is straightforward to derive versions of the basic Lyapunov stability arguments for DA systems. The following results serve the needs of the sequel.

**Theorem 2.** Let $\Omega' \subset \Omega$ be open connected and contain the origin. Suppose there exists a $C^1(\Omega')$ function $V : \Omega' \to R_+$ such that V is positive definite and has negative semi definite derivative on $\Omega'_G$, i.e.

$V(x, y) \geq a(\|(x, y)\|)$ and $\dot{V}_{(2.1)} \leq 0$ on $\Omega'_G$ for some $a \in K$. Then the trivial solution (x(t,0,0),y(t,0,0)) ≡ (0,0) of the DA system (2.1) is stable. ❑

The corresponding result for asymptotic stability follows. For later convenience, we add an estimate of the domain of attraction.

**Theorem 3.** Under the conditions of Theorem 2., suppose that V has negative definite derivative on $\Omega'_G$, i.e. $\dot{V}_{(2.1)} \leq -c(\|(x, y)\|)$ on $\Omega'_G$ for some $c \in K$.

De-

fine $\alpha = \sup_{\gamma \in R_+} \{\gamma : G \cap B_\gamma \subset \Omega'' \cap \Omega'_G\}$; $V_\alpha^{-1} = \{(x, y) \in \Omega'_G : V(x, y) < a(\alpha)\}$.

The trivial solution $(x(t,0,0), y(t,0,0)) \equiv (0, 0)$ of the DA system (2.1) is asymptotically stable with domain of attraction containing $V_a^{-1}(\subset \Omega'_G \subset \Omega_G)$. ❑

As for time-invariant ODEs the condition on $\dot{V}_{(2.1)}$ in Theorem 3 can be relaxed to provide a counterpart of the LaSalle invariance principle [4]. Let

$$\dot{V}_{(2.1)}(x,y) \leq 0 \quad \text{on} \quad \Omega'_G$$

$$S = \{(x,y) \in \Omega'' \cap \Omega'_G \; : \; \dot{V}_{(2.1)}(x,y) \equiv 0\}$$

Then it can be proved that trajectories approach the largest invariant set in S. Under Assumption A3, this is (0, 0). The conclusion of Theorem 3 remains unchanged.

**Remarks.**

(3.2) If $V$ is required to be decresent, i.e. $V(x, y) < b(\|(x,y)\|)$ on $\Omega'_G$ for some $b \in K$, then the conditions of Theorems 2, 3 ensure uniform stability and asymptotic stability respectively.

We can now rephrase a classic result due to Krasovskii in the context of DA systems:

**Theorem 4.** Assume that (A1), (A2) and (A3) hold, and furthermore that

$$A(x,y) := (D_1 f)(x,y) - (D_2 f)(x,y)(D_2 g)^{-1}(x,y)(D_1 g)(x,y)$$

satisfies

$$\epsilon I + A^T(x,y) + A(x,y) \leq 0$$

where $\Omega'$ is an open connected subset of $\Omega$ containing the origin. Under the assumptions (A1) to (A4) the trivial solution $(x(t, 0, 0), y(t, 0, 0) = (0, 0)$ of the DA system (2.1) is asymptotically stable. ❑

**Corollary 1.** Under the assumptions (A1)–(A3) and if on some open set $\Omega'_G$ containing the origin there holds:

$$(D_2 f)(x,y) = (D_1 g)^T(x,y) \tag{C1}$$

$$(D_2 g)(x,y) = (D_2 g)^T(x,y) > 0 \tag{C2}$$

$$(D_1 f)(x,y) + (D_1 f)^T(x,y) \geq \epsilon I \tag{C3}$$

then the trivial solution $(0,0)$ of the DA system (2.1) is asymptotically stable with domain of attraction $\Omega_G \cup \Omega_G'$. ❑

**Remarks.**

(3.3) Notice that the proposed Lyapunov function in general will not be positive definite in $\Omega$.

(3.4) In the event of multiple isolated equilibria $(x_i, y_i)$ the above argument demonstrates that provided conditions (A1) to (A4) are met in appropriate sets $\Omega'_{iG}$, $\Omega_{iG}$ (containing $(x_i, y_i)$ ) then the <u>same</u> Lyapunov function $V = f^T(x,y)f(x,y)$ can be used to demonstrate uniform asymptotic stability of each such equilibrium. This observation is essential when addressing set stability.

(3.5) The stability condition of Theorem 4 is illuminated by closer connection with the local ODE description (3.1). Defining;

$$h : R^n \to R^n \; ; \quad x \mapsto h(x) = f(x, u(x)) \, , \quad h \in C^2(U)$$

we see that, by virtue of assumptions (A3), $h(x) = 0$ in $U$ iff $x = 0$, hence $h^T(x)h(x)$ is a positive definite function in $U$. Furthermore the Jacobian of $h$ can in $U$ be expressed as:

$$(Dh)(x) = (D_1f)(x, u(x)) - (D_2f)(x, u(x))(D_2g)^{-1}(x, u(x))(D_1g)(x, u(x))$$

and therefore satisfies by assumption (A4):

$$\epsilon I + (Dh)^T(x) + (Dh)(x) \leq 0$$

on some $U' \subset U$, $0 \in U' \subset U$, and for some $\epsilon > 0$

which establishes uniform asymptotic stability for the trivial solution of (3.1). This is equivalent to Theorem 4.

We now note that if condition (C3) is strengthened to make $D_1f(x,y)$ symmetric then (C1)-(C3) guarantee the existence of a scalar function $V(x,y)$ such that :

$$D_1V(x, y) = f(x, y) \qquad (3.2.1)$$

$$D_2V(x, y) = g(x, y) \qquad (3.2.2)$$

holds on $\Omega'_G$. Further $V$ is positive definite in some neighbourhood of the origin with negative definite derivative in $\Omega'_G$. This observation makes it natural to proceed to study the DA version of gradient systems [8]. The following result observes that conditions (3.2) can be relaxed in a useful way.

**Theorem 5.** Suppose the assumptions (A1)-(A3) hold and there exists a C1 positive definite function $V: R^n \times R^m \to R$ such that on the open set $\Omega'_G$ containing the origin:

$$D_1V(x, y) = -H_1(x, y)f(x, y) \qquad (D1)$$

$$D_2V(x, y) = -H_2(x, y)f(x, y) \qquad (D2)$$

with $H_1$ positive semi-definite. Then the trivial solution of DA system (2.1) is stable.

**Remark:**

(3.6) There are opportunities to generalize conditions (D1), D2). For instance, suppose (D2) is replaced by

$$D_2V(x, y) = -g_1(x, y) \qquad (D3)$$

where $g(x,y) = g_1(x,y) + g_2(y)$. Then a result can be built around the Lyapunov function

$$V_1(x,y) = V(x,y) + \int_0^y g_2^T(\mu)d\mu$$

if $g_2$ gives positivity of the integral term.

Obviously, the choice $H_1 = I$ restores the gradient system case. Another interesting case is

$$H_1 = \begin{pmatrix} 0 & -1 \\ 1 & 0 \end{pmatrix}$$

which corresponds to the Hamiltonian system:

$$\dot{x}_1 = D_2H(x_1,x_2,y)\,;\dot{x}_2 = -D_1H(x_1,x_2,y)\,;0 = D_3H(x_1,x_2,y) \tag{3.3}$$

**Corollary 2.** Suppose $H: R^{n_1} \times R^{n_2} \times R^m \to R$ is $C1$, s.t. assumptions (A1)–(A2) are satisfied in $\Omega'_G$ containing the origin. Then the trivial solution of system (3.3) is stable. ❑

## 4. LUR'E - POSTNIKOV TYPE RESULTS FOR DA SYSTEMS.

In this section we consider the DA system's equivalent of the Lur'e problem [1] which is particularly relevant for the power system application discussed briefly in Section 6. Consider the DA system in the format

$$\dot{x} = Ax - Bh(Cx,y)\,;0 = g(Cx,y) \tag{4.1}$$

where $A$, $B$ and $C$ are real matrices of dimensions respectively nxn, nxp, pxn, $B,C$ have full column rank and row rank respectively. $h$, $g$ are $C2$ functions on some open connected set $\Lambda \subset R^p \times R^m$:

$$\Lambda = \{(u,y) \in R^m : (u,y) = (Cx,y),\quad (x,y) \in \Omega \subset R^n \times R^m\}$$

where $\Omega$ is an open connected set in $R^n \times R^m$ containing $(0,0)$. Thus we have

$$h: R^p \times R^m \to R^p \qquad h \in C^2(\Lambda) \tag{L1}$$

$$g: R^p \times R^m \to R^m \qquad g \in C^2(\Lambda)$$

We assume also that

$$rank\ (D_2g\ (u,y)) = m \qquad \forall (u,y) \in \Lambda \tag{L2}$$

and that

$$h(0,0) = 0 \text{ and } g(0,0) = 0 \tag{L3}$$

We have the following result:

**Theorem 6.** Assume that the transfer function

$$(N + Qs)C(sI - A)^{-1}B \tag{L4}$$

is positive real, (for some real matrices $N = N^T \geq 0, Q = Q^T \geq 0$); that $(A,B)$ is reachable and $(C,A)$ observable, and that there is no pole zero cancellation between $(N + Qs)$ and $C(sI - A)^{-1}B$. Assume further that

$$h^T(Cx,y)NCx \geq 0 \quad \text{on } \Omega_G$$
$$h^T(Cx,y)NCx = 0 \quad \text{on } \Omega_G \text{ only if } (Cx,y) = 0 \tag{L5}$$

Assume that there exists a scalar function W(x,y) in C2(Ω) such that:

$$W : R^n \times R^m \to R \; ; \; (x,y) \mapsto W(x,y), \quad W \in C^2(\Omega)$$

$$\left.\begin{array}{l} D_1W(x,y) = C^TQh(x,y) \quad \text{on } \Omega_G \\ D_2W(x,y) = 0 \quad \text{on } \Omega_G \\ W \text{ is nonnegative definite} \quad \text{on } \Omega_G \end{array}\right\} \tag{L6}$$

Under the assumptions (L1–L6) the trivial solution of the DA system (4.1) is asymptotically stable. ❑

## 5. APPLICATION TO POWER SYSTEMS

DA models for power systems arise from the inclusion of nonlinear load models [10, 5, 2, 14]. The functions $f$ and $g$ correspond to the real and reactive power equations in the usual load flow. From [2] we simply quote the model:

$$\dot{\gamma}_g = -S[P_g(a_g, a_l, V) - P_M \tag{5.1.1}$$

$$\dot{a}_g = \gamma_g \tag{5.1.2}$$

$$0 = P_l(a_g, a_l, V) + P_d \tag{5.1.3}$$

$$0 = [V]^{-1}[Q_b(a_g, a_l, V) + Q_d(V) \tag{5.1.4}$$

α is the vector of load angles relative to a reference bus; γ is the vector of relative generator frequencies, V is the amplitude of the bus voltages, $P$ and $Q$ refer to real and reactive powers. Subscripts g, l refer to generator, load buses in the network respectively. $P_M$ is the mechanical input power. Finally $S = T_gM_g^{-1}T_g^T$ where $M_g$ is a diagonal matrix of inertia constants and $T_g$ a matrix with elements 1 or −1 used in forming relative angles.

The powers $P_g$, $P_l$ and $Q_b$ are given by $P_b = (P_l, P_g)$ and

$$P_{b_i}(a,V) = \sum_{j=1}^{n} V_iV_jB_{ij}\sin(a_i - a_j)$$

$$Q_{b_i}(a,V) = -\sum_{j=1}^{n} V_iV_jB_{ij}\cos(a_i - a_j)$$

In arriving at the DA system (5.1) several assumptions and variable transformations are used; refer to [2] for the details.

From [2] we see that under 'normal' operating conditions the DA model satisfies assumptions (A1) to (A3). The equilibrium $(\gamma_g^s, a^s, V^s)$ is given by

$$\gamma_g = 0 \tag{5.2.1}$$

$$P_b(a,V) - \tilde{P} = 0 \tag{5.2.2}$$

$$[V]^{-1}[Q_b(a,V) + Q_d(a,V)] = 0 \tag{5.2.3}$$

where $\tilde{P} = (-P_d, P_M)$. Equations (5.2) are the load flow equations.

In the formulation of a stability result, it is convenient to use

$$B(a, V) = -\frac{1}{2}\sum_{i=1}^{n}\sum_{j=1}^{n} V_i V_j B_{ij}\cos(a_i - a_j)$$

and then

$$H(\gamma_g, a, V) = \frac{1}{2}\gamma_g^T S^{-1}\gamma_g + B(a, V)$$

It is easy to verify that DA system (5.1) can be rewritten as

$$\dot{\gamma}_g = -SD_2H(\gamma_g, a_g, (a_l, V)) + SP_M \tag{5.3.1}$$

$$\dot{a}_g = SD_1H(\gamma_g, a_g, (a_l, V)) \tag{5.3.2}$$

$$0 = D_3H(\gamma_g, a_g, (a_l, V)) + \begin{bmatrix} P_b \\ -[V]^{-1}Q_d(V) \end{bmatrix} \tag{5.3.3}$$

Clearly, system (5.3) is a modification to the Hamiltonian form (3.4).

Motivated by Remark (3.6), we consider

$$W(\gamma_g, a, V) = H(\gamma_g, a, V) - B(a^s; V^s) + P_d^T a_l - P_M^T a_g + \int_{V^s}^{V} Q_d^T(\mu)[\mu]^{-1} d\mu \tag{5.4}$$

From [2], we can show that $W$ is positive definite in a neighbourhood containing the equilibrium point. Also we can verify that $\dot{V}_{(5.1)} = 0$ everywhere. The following result can be stated.

**Theorem 7.** Under normal operating conditions [2], the corresponding equilibrium point $(\gamma_g^s, a^s, V^s)$ satisfying (5.2) is stable (in the sense of Definition 1). ❑

## 6. CONCLUSIONS

This report has studied differential algebraic systems (DA) in the most simple situation where the algebraic constraints can be solved for the auxiliary variables as a function of the states in some neighbourhood. This enables arguments for ordinary differential equations (ODEs) to be employed once some care is taken of neighbourhoods of validity for models and stability conditions to hold. Precise Lyapunov type stability concepts and results are briefly introduced and applied to some specific situations ending of generalized gradient and Hamiltonian systems. An interesting application is seen to be in power system stability.

As indicated throughout, there is much scope for generalisations. Obviously, the smoothness restrictions on $V$ can be relaxed along lines used in ODE results [4,11]. Relaxing the requirement that $D_2g$ has full rank allows for non-unique solutions, but stability analysis can still be carried out [11]. However, such extensions will bring all the peculiar behaviour of DA systems to the problem forefront [13], [9]. A further issue suggested by [5] is to study connections with a singularly perturbed ODE embedding the DA system. Of course it is desirable to obtain methods for stabilising nonlinear DA systems via feedback.

## 7. REFERENCES


[1] B.D.O. Anderson, "Stability of control systems with multiple nonlinearities", J. of the Franklin Institute, Vol. 282, No.3, September 1966, pp.155 - 160.

[2] C.N. Chong and D.J. Hill, "Energy function stability analysis of undamped power systems with nonlinear loads", to appear IEEE Trans. Circuits and Systems.

[3] J. Dieudonne, Foundation of Modern Analysis, Academic Press, New York, 1969.

[4] J.K. Hale, Ordinary Differential Equations, Wiley-Interscience, New York, 1969.

[5] D.J. Hill and C.N. Chong, "Lyapunov functions of Lur'e -Postnikov form for structure preserving models of power systems", to appear Automatica.

[6] D.J. Hill and I.M.Y. Mareels,"Stability Theory for Differential/Algebraic Systems with Application to Power Systems,EE8836, University of Newcastle,internal report

[7] M. Hasler and J.Neirynck,Nonlinear Circuits,Artech House Inc, Norwood,MA,1986

[8] M.W. Hirsch and S. Smale, Differential Equations, Dynamical Systems and Linear Algebra, Academic Press, New York, 1974.

[9] S.-E. Mattson, "On differential/algebraic systems", Department of Automatic Control, Lund Institute of Technology, Report TFRT-7327, September 1986.

[10] N. Narasimhamurthi and M.T. Musavi, "A generalized energy function for transient stability analysis of power systems", IEEE Trans. Circuits and Systems, Vol. CAS - 31, No.7, July 1984, pp.637 - 645.

[11] N. Rouche, P. Habets and M. Laloy, Stability Theory by Liapunov's Direct Method, Springer-Verlag, Applied Mathematical Sciences, Vol.22, 1977.

[12] S.S.Sastry and C.A.Desoer,"Jump Behaviour of Circuits and Systems",IEEE CAS, Vol CAS28, No12, Dec1981, pp1109-1124.

[13] F. Takens, "Constrained equations; a study of implicit differential equations and their discontinuous solutions", in Structural Stability, the Theory of Catastrophes and Applications in the Sciences, Ed. P. Hilton, Lecture Notes in Mathematics Vol.525, Springer-Verlag, 1976.

[14] Th. Van Cutsem and M. Ribbens-Pavella, "Structure preserving direct methods for transient stability analysis of power systems", Proc. 24th Conference on Decision and Control, Ft. Landerdale, Florida, December 1985.

# FEEDBACK EQUIVALENCE OF PLANAR SYSTEMS AND STABILIZABILITY

B. Jakubczyk and W. Respondek

Abstract

We consider local feedback equivalence and local weak feedback equivalence of control systems. The later equivalence is up to local coordinate changes in the state space, local feedback transformations, and state dependent changes of the time scale. We show that, under such equivalence, there are only five nonequivalent local canonical forms (some with parameters) for generic control-affine systems in the plane. For a more general class of planar systems, excluding only a class of infinite codimension, we propose a general normal form. A subclass of such systems, including all systems of codimension 3, can be brought to canonical forms. We examine stabilizability of each of these canonical forms under smooth feedback. As stabilizability under smooth feedback is invariant under weak feedback equivalence, this solves the stabilizability problem for the above mentioned class of systems.

## 1. Introduction.

Feedback transformations and feedback equivalence proved to be very useful in analysis and design of nonlinear control systems. However, the feedback classification problem as well as the feedback equivalence problem are not well understood, in general (compare a discussion given in [11]). In particular, functional invariants should appear in normal forms, as the results of [11] (Proposition 3.12), and [15] show. These invariants have not been constucted, so far, except of some regular (nonsingular) cases when the method of Cartan can be applied [9]. Most of the cases interesting for engineering are singular, however. There are only restricted classes of systems which are locally classified under feedback, like well known feedback linearizable systems (the Brunovsky classification). Some feedback classification results were recently given for quadratic systems [3] and planar systems with no singularities [10], and there is hope for further progress.

For systems with sufficiently many controls a generic local classification under weak feedback equivalence, and under mild feedback equivalence, was proposed in [11]. We continue this analysis here for control-affine systems in the plane, with scalar control. We show that, generically, there are five nonequivalent local canonical forms, under feedback equivalence. Three of them have no parameters, one has a scalar invariant, and one has a functional invariant. We also present a general normal form with functional parameters, for which genericity is not required. These functional parameters are also invariants in a sense which will be explained elsewhere.

We use our feedback analysis of planar systems (Section 2) to study the problem of feedback stabilization in Section 3. The nonlinear stabilization problem is a crutial one in theory and applications of nonlinear control systems and has recently attracted a lot of attention (see e.g. [1], [2], [4], [6], [12], [13], [14]), and various methods like center manifold, high-gain feedback, Lyapunov functions, zero dynamics, and others have been applied. We solve completely the problem of stabilization under smooth feedback for the class of systems described by the following condition: there exists a $k \leq 3$ such that $g$ and $\mathrm{ad}_g^k f$ are linearly independent, locally.

While working on this paper we received a very interesting study of Dayawansa, Martin, and Knowles [8] concerning stabilization of planar systems. Our stabilization results follow from their's (in fact, they prove much more). However, we are additionally able to give an invariant characterization of the cases considered and our analysis illustrates usefulness of our normal forms.

## 2. Feedback equivalence and normal forms.

We shall consider the following class of planar control systems

$$\Sigma: \quad \dot{z} = f(z) + ug(z), \quad z \in \mathbb{R}^2, \ u \in \mathbb{R},$$

where the vector fields $f$ and $g$ are assumed to be of class $C^\infty$ (for most problems cosidered here finite differentiability of an appropriate order is sufficient, also). We shall say that a system of this form is *locally feedback equivalent* to a system in the form

$$\Sigma': \quad \dot{z}' = f'(z') + u'g(z'), \quad z' \in \mathbb{R}^2, \ u' \in \mathbb{R}.$$

if there are local transformations of class $C^\infty$,

$$z' = \phi(z), \quad u' = \alpha(z) + \beta(z)u, \tag{2.1}$$

which transform the second system into the first (where $\alpha(z)$ and $\beta \neq 0$ are smooth functions). We shall say that these systems are *locally weakly feedback equivalent* if they are equivalent up to the above transformations and a nondegenerate smooth change of time scale

$$dt' = \gamma(z)dt, \quad \gamma(z) > 0.$$

In other words, two systems represented by pairs of vector fields $(f, g)$ and $(f', g')$ are locally weakly feedback equivalent if the systems represented by the pairs $(\gamma f, \gamma g)$ and $(f', g')$ are locally feedback equivalent, where $\gamma$ is a suitable positive function of the state. Finally, if the function $\gamma$ above satisfies the additional constraint $L_g\gamma = 0$, then we say that the two systems are *locally mildly feedback equivalent* [11]. It is easy to see that weak feedback equivalence preserves stabilizability under smooth feedback.

We shall consider our systems $\Sigma$ around points $z_0$ at which the vector field $g$ does not vanish, and we shall transform them to normal forms at the origin, i.e. we shall take $\phi(z_0) = 0$. We denote $z = (x, y)$.

The following proposition is a reformulation of Theorem 3.4 in [11], and its proof is analogous to the proof of this theorem.

*Proposition 2.1.* Any system $\Sigma$ generic in the $C^\infty$ Whitney topology is locally feedback equivalent, around any point at which $g$ does not vanish, to one of the following systems:

$$(a) \qquad \dot{x} = y + 1, \quad \dot{y} = u,$$

$$(b) \qquad \dot{x} = y, \quad \dot{y} = u,$$

$$(c) \qquad \dot{x} = y^2 \pm 1, \quad \dot{y} = u,$$

$$(d) \qquad \dot{x} = y^2 + \lambda x, \quad \dot{y} = u, \quad \lambda \in \mathbb{R}, \ \lambda \neq 0,$$

$$(e) \qquad \dot{x} = y^3 + xy + a(x), \quad \dot{y} = u,$$

where $a$ is a smooth function and $a(0) \neq 0$.

For any such system the following sets are smooth curves in $\mathbb{R}^2$:

$$S_1 = \{ \ z \mid f(z) \text{ and } g(z) \text{ are linearly dependent } \},$$

$$S_2 = \{ \ z \mid [f, g](z) \text{ and } g(z) \text{ are linearly dependent } \}.$$

The feedback transformations (2.1) map these curves into corresponding ones defined by the equivalent system, according to the formula

$$\phi(S_1) = S_1', \quad \phi(S_2) = S_2'.$$

Canonical form (a) appears at points outside both curves. Points in $S_1$, at which $S_1$ is transversal to $g$ lead to canonical form (b). Similar points in $S_2$ lead to canonical form (c). Finally, isolated points at which the curve $S_1$ is tangent to the trajectories of $g$ give canonical form (d), and isolated points of tangency of $S_2$ to these trajectories gives canonical form (e). The points of type (d) (of tangency of $S_1$ and $g$) coincide with the intersection points of $S_1$ and $S_2$.

Above, and further, we use the Lie bracket of vector fields $[f, g] = \frac{\partial g}{\partial x} f - \frac{\partial f}{\partial x} g$. We shall also apply the usual notation $\mathrm{ad}_g f = [g, f]$, and $\mathrm{ad}_g^k f = \mathrm{ad}_g \mathrm{ad}_g^{k-1} f$ for the iterated Lie bracket.

*Remark 2.2.* The function $a$ can not be removed from the canonical form (e) as a functional (differential) invariant appears in this case. Informally, this invariant can be described as follows. There is a feedback $u = \alpha(z)$ which makes the curve $S_2$ invariant under the flow of the system $\Sigma$ represented now by the vector field $f + \alpha g$. This flow, restricted to $S_2$, defines a unique vector field $F$ on $S_2$. Locally, around the points of tangency of $S_2$ to $g$, there is also a second vector field on $S_2$. Namely, if $g$ is made horizontal (of the form $\partial/\partial y$) and the x-coordinate of the curve $S_2$ reaches a local maximum (or minimum) at a point $z_0$, then the trajectories of $g$ intersect $S_2$ twice and so they define a local transformation (involution) of $S_2$ around this point. The image of $F$ under this transformation gives another vector field on $S_2$. It is well known that two vector fields on $\mathbb{R}$ define a functional invariant. A detailed discussion of this and similar cases will be given elsewhere.

For weak feedback equivalence the canonical forms (a), (c), and (e) are equivalent to the form $\dot{x} = 1$, $\dot{y} = u$, so only three nonequivalent cases appear, and the parameter in (d) disappears.

For nongeneric systems we have the following result.

*Theorem 2.3* If $g(z_0) \neq 0$ and the system $\Sigma$ satisfies the accessibility condition at $z_0$ (equivalently, the Lie algebra generated by the vector fields $f$ and $g$ is of full rank at $z_0$), then either $f(z_0)$ and $g(z_0)$ are linearly independent, and $\Sigma$ is locally weekly feedback equivalent at $z_0$ to the system:

$$\dot{x} = 1, \qquad \dot{y} = u,$$

or $f(z_0)$ and $g(z_0)$ are linearly dependent, and then $\Sigma$ is locally feedback equivalent around $z_0$ to the following system at the origin:

$$\dot{x} = y^{k+1} + y^{k-1}a_{k-1}(x) + \cdots + ya_1(x) + a_0(x), \qquad \dot{y} = u,$$

where $a_0, \ldots, a_{k-1}$ are smooth functions of $x$, $a_0(0) = \cdots = a_{k-1}(0) = 0$, and $k \geq 0$. The number $k$ is the minimal number $r$ such that $\mathrm{ad}_g^{r+1} f(z_0)$ is linearly independent of $g(z_0)$.

If all the derivatives of $a_0$ at zero vanish up to order $q - 1$ and the derivative of order $q$ does not vanish, then we may take $a_0 = \epsilon x^q + cx^{2q-1}$, $\epsilon = \pm 1$, $c \in \mathbb{R}$, in the above normal form.

If we replace feedback equivalence with weak feedback equivalence and, for a fixed $i, j$, $0 \leq i, j \leq k - 1$, the first nonvanishing (at zero) derivatives of $a_i$ and $a_j$ are, respectively, of order $p$ and $q$, then, in the above normal form, we may take

$$a_i = \epsilon x^p, \quad a_j = \delta x^q, \quad \text{provided that} \quad p(k - j + 1) \neq q(k - i + 1).$$

*Proof.* In the first step we choose local coordinates so that the vector field $g = \partial/\partial y$. Then, after applying a feedback we can assume that $f = f^1\partial/\partial x$. In the case when $f(z_0)$ and $g(z_0)$ are linearly independent we may apply the change of time scale $\gamma = (f^1)^{-1}$ which, together with a feedback transformation, gives the first normal form of our theorem.

In the other case, we may assume that $f^1(0) = 0$. Then, also, $f(z_0) = 0$ and from the accessibility rank condition at $z_0$ it follows that there is a $k$ such that $(\mathrm{ad}_g^{k+1} f)(z_0)$ is linearly independent of $g(z_0)$. Let us fix such a minimal $k$ (note that $k$ is invariant under feedback equivalence). Then $f = f^1(x,y)\partial/\partial x$, where all the partial derivatives with respect to $y$ of $f^1$ vanish at $z_0$ up to order $k$, and the derivative of order $k+1$ does not vanish. Preserving this form of the system, we see that we still have freedom of using coordinate changes of the form $\tilde{x} = \phi(x)$, $\tilde{y} = \psi(x,y)$. Thus, Mather's theorem on versal unfoldings of functions of one variable (in our case $y$) is applicable here (cf. [5]) and we can transform the component $f^1$ to the new form

$$f^1(\tilde{x},\tilde{y}) = \phi'(\phi^{-1}(\tilde{x}))(\tilde{y}^{k+1} + \tilde{y}^{k-1}a_{k-1}(\tilde{x}) + \cdots + \tilde{y}a_1(\tilde{x}) + a_0(\tilde{x})).$$

Changing coordinates, again, for $x = \tilde{x}$, $y = (\phi'(\phi^{-1}(\tilde{x}))^{1/(k+1)}\tilde{y}$, and applying a feedback transformation, we obtain our system in the desired normal form.

If the function $a_0$ has the first nonvanishing derivative at 0 of order $q$, then there exists a change of coordinate $\tilde{x} = \phi(x)$ which transforms the vector field $a_0\partial/\partial x$ to the form $(\epsilon x^q + cx^{2q-1})\partial/\partial x$. Applying the change of coordinates of the form $\tilde{x} = \phi(x)$, $\tilde{y} = y(\phi'(\phi^{-1}(\tilde{x})))^{1/(k+1)}$ and an appropriate feedback, we get $a_0 = \epsilon x^q + cx^{2q-1}$.

Finally, if $p(k-j+1) \neq q(k-i+1)$, then we can simplify the functions $a_i(x) = A_i(x)x^p$, and $a_j(x) = A_j(x)x^q$, $A_i(0) \neq 0 \neq A_j(0)$, as follows. It is enough to find transformations $\tilde{x} = \phi(x) = \Phi(x)x$, $\tilde{y} = Q(x)y$, and a time scale $\gamma$ such that the coefficients of $f^1$ at $y^{k+1}$, $y^i$, and $y^j$ take the desired form, that is

$$\phi'\gamma Q^{-k-1} = 1, \quad \phi'\gamma Q^{-i}A_i = \Phi^p\delta, \quad \phi'\gamma Q^{-j}A_j = \Phi^q\epsilon,$$

where $\delta = \mathrm{sgn}(A_i(0))$ and $\epsilon = \mathrm{sgn}(A_j(0))$. Finding $\phi'\gamma = Q^{k+1}$ from the first equation, and substituting it into the other two equations gives that $Q^{k-i+1}A_i = \Phi^p\delta$, $Q^{k-j+1}A_j = \Phi^q\epsilon$. Thus, eliminating $\Phi$, we get

$$Q = (\delta^q\epsilon^{-p}A_i^{-q}A_j^p)^s, \quad s = (q(k-i+1) - p(k-j+1))^{-1}.$$

Putting $Q$ into the above equations we find easily $\Phi$, $\phi'$, and $\gamma$, which satisfy these equations. ■

*Remark 2.4.* It can be shown that in the analytic case, and under proper assumptions, the functions $a_1, \ldots, a_{k-1}$ in the normal form in our theorem are invariants of feedback equivalence, modulo the action of a finite group. We plan to discuss this in detail in a future study.

The following corollary follows easily from the above theorem.

*Corollary 2.5.* Assume that $f(0) = 0$, $g(0) \neq 0$, and denote by $k$ the minimal number such that $g$, and $\mathrm{ad}_g^k f$ are linearly independent at the origin. Then, all systems $\Sigma$ with $k \leq 3$ (excluding a class of infinite codimension) are locally weakly feedback equivalent around the origin to one of the following systems:

(*i*) $\quad \dot{x} = y, \quad \dot{y} = u, \quad$ if $k$ is equal to 1,

(*ii*) $\quad \dot{x} = y^2 + \epsilon x^q, \quad \dot{y} = u, \quad \epsilon = \pm 1, \quad$ if $k$ is equal to 2, or, if $k = 3$, to one of the systems

$$(iii)' \qquad \dot{x} = y^3 + \delta y x^p + \epsilon x^q, \quad \dot{y} = u, \quad \delta, \epsilon = \pm 1, \quad p < q, \quad \text{if } 3p \neq 2q,$$

$$(iii)'' \\ \dot{x} = y^3 + a(x)y + \epsilon x^q, \quad \dot{y} = u, \quad \text{where } a(x) = A(x)x^p, \quad A(0) \neq 0, \quad \text{if } 3p = 2q,$$

$$(iii)''' \qquad \dot{x} = y^3 + \epsilon x^q, \quad \dot{y} = u.$$

Note that the case $q \leq p$ in (iii)' can be reduced to (iii)"' by taking $\delta y x^p + \epsilon x^q = x^q(\epsilon + \delta y x^{p-q}) = x^q Q$, applying the time scale $\gamma = Q^{-1}$, and introducing the new coordinate $\tilde{y} = yQ^{-1/3}$ and an appropriate feedback.

*Remark 2.6.* In the above corollary some combinations of signs of $\delta$ and $\epsilon$ give equivalent systems. To eliminate this, the following combinations of signs should be taken only (they cover all the nonequivalent cases). In the canonical forms (iii)' and (iii)" one should take:

$$\text{if} \quad q = 2r, \quad \text{then} \quad (\delta, \epsilon) = (+,+) \text{ or } (-,+),$$

$$\text{if} \quad q = 2r+1, \ p = 2\ell + 1, \quad \text{then} \quad (\delta, \epsilon) = (+,+) \text{ or } (+,-),$$

$$\text{if} \quad q = 2r+1, \ p = 2\ell, \quad \text{then} \quad (\delta, \epsilon) = \text{any combination}.$$

In the canonical form (ii) both signs of $\epsilon$ should appear, as well as in (iii)"' with $q$ odd (in the latter case $\epsilon = +$, when $q$ is even).

## 3. Stabilization.

In this section we study the problem of feedback stabilization of $\Sigma$, i.e. we seek for a feedback control $u = \alpha(z)$ such that the closed loop system

$$\dot{z} = f(z) + g(z)\alpha(z), \tag{3.1}$$

is locally asymptotically stable at an equilibrium $z_0 = (x_0, y_0)$. We will solve completely the stabilization problem for a subclass of $\Sigma$ described by Corollary 2.5, i.e. in the case when there exists a $k \leq 3$ such that $g$ and $\text{ad}_g^k f$ are linearly independent at $z_0$. For every such system, listed in Corollary 2.5, we determine whether or not it can be stabilized by a (smooth) feedback and, in the case of affirmative answer, we give an explicit formula for the stabilizing feedback.

We want to emphasize that, given a system in the form $\dot{x} = f_1(x, y)$, $\dot{y} = u$, stabilizability results presented below follow also from the deep study of Dayawansa et al. [8]. However, in the studied cases we provide an invariant characterization of stabilizability which is independent of particular coordinate representation and of the chosen feedback. For completness we give proofs which are simple in the considered cases.

We shall use the concept of center manifold [7] which can be briefly summarized as follows (cf. also [1], [2], [6]) . Consider the dynamical system

$$\dot{x} = Ax + f(x, y), \tag{3.2a}$$

$$\dot{y} = By + g(x, y), \tag{3.2b}$$

where $x \in \mathbb{R}^n$, $y \in \mathbb{R}^m$, and $A$ and $B$ are constant matrices such that all the eigenvalues of $A$ have zero real parts, while all the eigenvalues of $B$ have negative real parts. The functions $f$ and $g$ are $C^r$, $r \geq 2$, with $f(0,0) = g(0,0) = 0$ and $f'(0,0) = g'(0,0) = 0$, where $f'$ denotes the Jacobian matrix of $f$. A manifold $H = \{ (x, y) \mid y = h(x) \}$ is called a center manifold if $h$ is smooth, $h(0) = h'(0) = 0$ and, moreover, $H$ is a local invariant manifold of (3.2).

*Theorem 3.1* (a) There exists a local center manifold $y = h(x)$, $|x| < \delta$, where $h$ is of class $C^r$.

(b) Assume that the zero solution of the system

$$\dot{w} = Aw + f(w, h(w)), \tag{3.3}$$

which governs the flow on the center manifold is asymptotically stable. Then the zero solution of (3.2) is asymptotically stable and, moreover, there exists a solution $w(t)$ of (3.3) such that

$$x(t) = w(t) + O(e^{-\gamma t}), \tag{3.4a}$$

$$y(t) = h(w(t)) + O(e^{-\gamma t}), \tag{3.4b}$$

where $x(0)$ and $y(0)$ are sufficiently small and $\gamma > 0$ is a constant.

Using this result we prove the following simple stabilization result for control systems of the form

$$\dot{x} = f(x, y), \quad \dot{y} = u, \tag{3.5}$$

where $x, y \in \mathbb{R}$ and $f(x_0, y_0) = 0$, $f'(x_0, y_0) = 0$ (more elaborate results of this type are proved in [2], [14]).

*Lemma 3.2.* Suppose there exists a function $g \in C^r$, $r \geq 2$, such that the dynamical system $\dot{x} = f(x, g(x))$ is asymptotically stable at $x_0$. Then there exists a feedback $u = \alpha(x, y)$ of class $C^{r-1}$ such that the system

$$\dot{x} = f(x, y), \quad \dot{y} = \alpha(x, y) \tag{3.6}$$

is asymptotically stable at $(x_0, y_0)$.

*Proof.* Let us introduce the new coordinates $\tilde{x} = x$, $\tilde{y} = y - g(x)$. In these coordinates (denoted again by $x$ and $y$) our system (3.5) takes the form

$$\dot{x} = f(x, y + g(x)), \quad \dot{y} = u - \frac{\partial g}{\partial x} f(x, g(x)). \tag{3.7}$$

Put

$$u = \alpha(x, y) = \frac{\partial g}{\partial x} f(x; g(x)) - y, \tag{3.8}$$

then the second equation of (3.7) becomes $\dot{y} = -y$. Therefore, the center manifold $H$ is given by $y = 0$, with the flow on $H$ governed by

$$\dot{x} = f(x, g(x)). \tag{3.9}$$

From the previous theorem we conclude that (3.8) asymptotically stabilizes (3.5). ■

Let $k$ denote the smallest integer such that $g(0)$ and $\mathrm{ad}_g^k f(0)$ are linearly independent. One can check by a direct computation that $k$ is independent of the chosen weak feedback $(\alpha, \beta, \gamma)$. For $k = 2$ we define another invariant of mild feedback (i.e. when $L_g\gamma = 0$) as follows. There exists $\alpha$ such that the curve $S_2 = \{z \mid \det(\mathrm{ad}_g f(z), g(z)) = 0\}$ is invariant under the flow of the smooth vector field $f + \alpha g$. This vector field clearly vanishes at $(0,0)$ and its order, understood as the order of zero of the function $(f + \alpha g)(t)$, where $t$ parametrizes $S_2$, gives a mild feedback invariant, denoted by $q$. Observe that we also get another invariant $\epsilon$, namely the sign of the q-th derivative at $t = 0$ of $f + \alpha g$.

To illustrate the above definition consider the system $\dot{x} = y^2 + \epsilon x^q$, $\dot{y} = u$ (compare Corollary 2.5). We have $S_2 = \{(x, y) | y = 0\}$. Therefore, $\alpha = 0$ and we get that $f + \alpha g$ restricted to $S_2$ is $\epsilon x^q$.

We are now in a position to state our stabilizability result. Let $k$ and the pair $(q, \epsilon)$, when $k = 2$, be as above.

*Proposition 3.3.* Assume that $k \leq 3$. Then $\Sigma$ is stabilizable by a smooth feedback if and only if $k = 1$, or $k = 2$ and $\epsilon = -1$, or $k = 3$ and the linear approximation has no unstable uncontrollable modes.

*Proof.* We use the classification obtained in Corollary 2.5.

(i) It is a standard result from linear control theory that the system (i) is asymptotically stabilizable by a linear feedback $u = k_1 x + k_2 y$, for suitable constants $k_1$, $k_2$.

(ii) Case $q = 2r$, $\epsilon = 1$. It is obvious that this system can not be stabilized by any feedback, since $\dot{x} > 0$ for $x \neq 0$.

Case $q = 2r$, $\epsilon = -1$. We see that in this case Lemma 3.2 holds for $g(x) = -x^r + x^{r+1}$. Indeed, $(-x^r + x^{r+1})^2 - x^{2r} = -x^{2r+1} + x^{2r+2}$, thus $\dot{x} = g(x) - x^{2r}$ is asymptotically stable, and so is our system (ii).

Case $q = 2r + 1$, $\epsilon = 1$. Observe that, if $x_0 > 0$, then $\dot{x} > 0$ and therefore $x(t) > 0$. Then (ii) is not stabilizable by any feedback.

Case $q = 2r + 1$, $\epsilon = -1$. In this case our lemma can be applied for $g(x) = 0$ and thus a stabilizing feedback is simply given by $u = \alpha(x, y) = -y$ (one can also observe that $V = x^2 + y^2$ is a Lyapunov function).

(iii)', (iii)" Both cases, with $q \geq 3$ and $p \geq 2$, can be dealt together (observe that if $3p = 2q$, i.e. when we are in (iii)", then automatically $q \geq 3$ and $p \geq 2$). It is always possible to choose $g(x) = -Kx$, $K > 0$, with $K$ big enough, to ensure that $f(x, g(x)) = (-Kx)^3 - K\alpha(x)x + \epsilon x^q = -Mx^3 + o(x^3)$, where $M > 0$, thus satisfying the assumptions of Lemma 3.2. Therefore, (iii)' and (iii)" are asymptotically stabilizable by the smooth feedback computed via (3.8).

Case $q = 3$, $p = 1$. According to Remark 2.6 we can always put $\delta = 1$. Assume $\epsilon = 1$, then $g(x) = -2x^2$ satisfies the assumptions of Lemma 3.2 since

$(-2x^2)^3 - 2x^3 + x^3 = -x^3 + o(x^3)$. In the case $\epsilon = -1$ the feedback $u = -y$ does the job since $g(x) = 0$ satisfies the assumptions of Lemma 3.2.

Case $q = 2$, $p = 1$. By Remark 2.6 we can consider only the case $\epsilon = 1$. If $\delta = 1$, then $g(x) = -x$ satisfies the assumptions of Lemma 3.2 and the flow on the center manifold is governed by $\dot{x} = -x^3$. If $\delta = -1$, then we put $g(x) = x + 2x^2$ and we obtain $f(x, g(x)) = -x^3 + o(x)$. Thus, in both cases the system is stabilizable.

(iii)"' Case $q \geq 3$. We take $g(x) = -2x$ which satisfies the assumptions of Lemma 3.2 (compare the case $q \geq 3$, $p \geq 2$ of (iii)').

Case $q = 2$. It is enough to consider only $\epsilon = 1$ (see Remark 2.6). To prove that the feedback

$$u = -(2x + 2xy) - (y^3 + x^2) \tag{3.10}$$

asymptotically stabilizes the system we shall consider the Lyapunov function

$$V = x^2 + \frac{1}{4}y^4 + x^2 y.$$

We have that $\dot{V} = -(y^3 + x^2)^2$. Since $V$ is positive definite and $\{(x, y) \mid y^3 + x^2 = 0\}$ is not an invariant set then, by LaSalle's theorem, the system fed by (3.10) is asymptotically stable. Observe the following interestintg property of the studied system. Consider the system $\dot{x} = x^2 + v^3$ evolving on $\mathbb{R}$, with control $v$ appearing nonlinearly. Clearly, this system can not be stabilized by any $C^1$ feedback $v = v(x)$. On the other hand, its dynamic extension $\dot{x} = x^2 + v^3$, $\dot{v} = u$ can be stabilized, as we have just shown. For another example of this phenomenon see [2].

Case $q = 1$. If $\epsilon = -1$, then the system is clearly stabilizable by $u = -y$. The case $\epsilon = 1$ was considered in a pioneering paper by Kawski [12], where it is shown that the system can be stabilized by a Hölder-continuous feedback (clearly, any $C^1$ feedback can not stabilize the system since its linear approximation has an uncontrollable mode associated with eigenvalue 1. ■

Observe the following interestintg property of the case (iii)"', $q = 2$. Consider the system $\dot{x} = x^2 + v^3$ evolving on $\mathbb{R}$, with control $v$ appearing nonlinearly. Clearly, this system can not be stabilized by any $C^1$ feedback $v = v(x)$. On the other hand, its dynamic extension $\dot{x} = x^2 + v^3$, $\dot{v} = u$ can be stabilized, as we have just shown. For another example of this phenomenon see [2]. Observe the following interestintg property of the studied system. Consider the system $\dot{x} = x^2 + v^3$ evolving on $\mathbb{R}$, with control $v$ appearing nonlinearly. Clearly, this system can not be stabilized by any $C^1$ feedback $v = v(x)$. On the other hand, its dynamic extension $\dot{x} = x^2 + v^3$, $\dot{v} = u$ can be stabilized, as we have just shown. For another example of this phenomenon see [2].

To summarize, we see that (ii) is stabilizable if and only if $\epsilon = -1$, and asymptotic stability is achieved via $C^\infty$ feedback in this case. Systems (iii)', (iii)" are always stabilizable via $C^\infty$ feedback which guarantees existence of a center manifold with "cubic" stability on it. A similar situation holds for (iii)"' in the case $q \geq 3$ (i.e. stability due to existence of a center manifold with "cubic" stability on it). In the case $q = 2$ any center manifold can not be obtained, although existence of a stabilizing feedback can be shown using Lyapunov functions. The case $q = 1$, $\epsilon = 1$ is the most involved one. As shown by Kawski, continuous stabilizing feedback exists, although any $C^1$ feedback can not do the job. We want to emphasize that this case is of codimension 3 i.e. it appears generically when systems with one parameter are considered.

References

[1] D. Aeyels, "Stabilization of a class of nonlinear systems by smooth feedback control", Systems & Control Letters 5 (1985), 289-294.

[2] W.M. Boothby, R. Marino, "Feedback stabilization of planar nonlinear systems", Systems & Control Letters 12 (1989), 87-92.

[3] B. Bonnard, "Quadratic control systems", preprint.

[4] R.W. Brockett, "Asymptotic stability and feedback stabilization", in Differential Geometric Control Theory, R.W. Brockett, R.S. Millmann and H.J. Sussmann eds., Progress in Mathematics 27, Birkhäuser, Boston 1983, 181-191.

[5] T. Bröcker, L. Lander, Differential Germs and Catastrophes, London Math. Soc. Lecture Note Series 17, Cambridge University Press, Cambridge 1975.

[6] C.I. Byrnes and A. Isidori, "The analysis and design of nonlinear feedback systems, I,II: Zero Dynamics and Global Normal Forms, preprint.

[7] J. Carr, Applications of Center Manifold Theory, Springer, New York 1981.

[8] W.P. Dayawansa, C.F. Martin, and G. Knowles, Asymptotic stabilization of a class of smooth two dimensional systems, submitted.

[9] R.B. Gardner, "Differential geometric methods interfacing control theory", in Differential Geometric Control Theory, R.W. Brockett, R.S. Millmann and H.J. Sussmann eds., Progress in Mathematics 27, Birkhäuser, Boston 1983, 181-191.

[10] R.B. Gardner and W.F. Shadwick, "Feedback equivalence of control systems", System & Control Letters 8 (1987), 463-465.

[11] B. Jakubczyk, "Equivalence and invariants of nonlinear control systems", in Nonlinear Controllability and Optimal Control, H.J. Sussmann ed., Marcel Decker (to appear).

[12] M. Kawski, "Stabilization of nonlinear systems in the plane", Systems & Control Letters 12 (1989), 169-175.

[13] R. Marino, High gain stabilization and partial feedback linearization, Proceedings 25-th CDC, Athens 1986, 209-213.

[14] H.J. Sussmann and E.D. Sontag, "Further comments on the stabilizability of the angular velocity of a rigid body", Systems & Control Letters 12 (1989), 213-217.

[15] K. Tchoń, On normal forms of affine systems under feedback, Proc. Conference, Nantes 1988, Lecture Notes in Control and Information Sciences, Springer 1989.

# TOPOLOGICAL DYNAMICS OF DISCRETE-TIME SYSTEMS

J. Tsinias - S. Kotsios -N. Kalouptsidis.

**Abstract**

In this paper necessary and sufficient Lyapunov-like conditions are presented for the absolutely asymptotic stability of discrete-time control systems in the vicinity of a compact set.

## 1. Stability analysis.

We consider discrete-time control systems

$$x(n+1)=f(x(n),u(n)), n\in \mathbb{N}=\{0,1,2,...\}$$
$$x \in X, \quad u \in \Omega\subset \mathbb{R}^{\ell}$$

on a locally compact metric space X with distance $d(\cdot,\cdot)$, where the inputs u( ) take values on a connected subset $\Omega$ of $\mathbb{R}^{\ell}$. We assume that the mapping $f: X \times \Omega \to X$ is continuous. The reachable map of the system of x at time n is the multivalued map (m.m.)R defined as follows

$$R(n,x)= \{y \in X : \text{there are } u(1),...,u(n) \in \Omega \text{ such that } y = f(... f(f(x,u(1)), u(2)),..., u(n))\}$$

A m.m. $F:(n,x) \to F(n,x)$ is called lower semicontinuous (L.S.C) on $S\subset X$ if for each $(n,x) \in \mathbb{N} \times S$, $y \in F(n,x)$ and sequence $x_i \to x$, there is a sequence $y_i \to y$ with $y_i \in F(n,x)$, $\forall i=1,2,....$ F is called upper semicontinuous (U.S.C) on $S\subset X$, if for each $(n,x) \in \mathbb{N} \times S$; $x_i \to x$ and $y_i \in R(n,x_i)$ there is a sequence $w_i \in R(n,x)$ such that $d(y_i,w_i) \to 0$.

As in the time-continuous case we can show that for each $n \in \mathbb{N}$ and $x\in X$, R(n,x) is connected and R is L.S.C. on X. If further the map $X \ni x \to f(x,u)$ satisfies a Lipschitz condition uniformly on u, then R is U.S.C. on X.

We shall say that the m.m. $\Gamma_1 = DR$:

$$DR(n,x)=\{ y \in X: \text{there exist sequences } x_i \to x,\ y_i\to y \text{ with } y_i \in R(n,x_i)\}$$

is the prolongation (of R) of order 1. The transitized map of order 2 is the m.m. $R_2 = \mathcal{T}\Gamma_1$:

$\Gamma_1(n,x)=\{y \in X: \exists\ n_1,n_2,\ldots,n_k \in \mathbb{N}$ and $x_2,\ldots,x_{k-1} \in X$
such that $\sum_{i=1}^{k} n_i=n$ and $y \in \Gamma_1(n_k,x_k),\ x_k \in \Gamma_1(n_{k-1},x_{k-1}),\ldots,$
$x_2 \in \Gamma_1(n_1,x)\}$

Example 1.1. Consider the planar system

$$\begin{bmatrix} x_1 \\ x_2 \end{bmatrix}(n+1) = \begin{bmatrix} \frac{1}{2}\, x_1 \sin(x_2 u) \\ x_1 x_2 \end{bmatrix}(n), \quad n \in \mathbb{N}, \quad x=(x_1,x_2)' \in \mathbb{R}^2, \ u \in U=\mathbb{R}$$

We can easily evaluate that for any $n \in \mathbb{N}$, $R(n;(x_1,0)')= 0 \in \mathbb{R}^2$, whereas $R_2(n;(x_1,0)')=[-2^{-n}x_1,\ 2^{-n}x_1] \times \{0\}$.

As in the continuous - time case [1-4] higher order prolongational and transitized maps are defined by induction on the set of the ordinal numbers. Let $\alpha$ be an ordinal number and suppose that $\Gamma_\beta$ and $R_\beta$ have been defined for all $\beta<\alpha$. Then $R_\alpha = \cup\{\mathcal{T}\Gamma_\beta,\ \beta<\alpha\}$ and $\Gamma_\alpha=DR_\alpha$ are the transitized and the prolongational maps of order a respectively. Similar to the continuous-time case we have the following result.

Proposition 1.2. For any ordinal $\alpha$

(i) $\Gamma_\beta \subset R_\alpha \subset \Gamma_\alpha,\ \forall\ \beta < \alpha$

(ii) $R_\alpha$ is transitive and $\Gamma_\alpha$ is cluster, namely $\mathcal{T} R_\alpha = R_\alpha$ and $D\Gamma_\alpha = \Gamma_\alpha$

(iii) $\Gamma_\gamma = \cup\{\Gamma_\beta,\ \beta<\gamma\} = R_\gamma$, where $\gamma$ is the first uncountable ordinal number.

In the sequel we denote $\Gamma \overset{def}{===} \Gamma_\gamma = R_\gamma$. Obviously $\Gamma$ is the smallest transitive and cluster m.m. which contains R. Furthermore,if we assume that R is U.S.C. on X ,then $\Gamma = clR$.

Let M be a compact subset of X and $\alpha$ an ordinal number. M is called $\alpha$-stable if for any neighborhood U of M there is a neighborhood W of M with $R_\alpha(n,W) \subset U,\ \forall\ n \in \mathbb{N}$. M is called $\alpha$-uniform attractor if for any x near M and for any $\varepsilon>0$ there exist a $\delta>0$ and an integer N such that

$$R_\alpha(n,S(x,\delta)) \subset S(M,\varepsilon),\ \forall\ n \geq N$$

where $S(x,\delta)=\{y \in X: d(x,y)<\delta\}$ and $S(M,\varepsilon)=\{y \in X: d(y,M)<\varepsilon\}$. M is called absolutely asymptotically stable (A.A.S.) if it is $\gamma$-stable and a $\gamma$-uniform attractor. M is called $\alpha$-positively invariant ($\alpha$-p.i.) if

$R_\alpha(n,M) \subset M, \quad \forall\, n \in \mathbb{N}.$

Similar to the continuous-time case [3,4] we can establish the following proposition.

Proposition 1.3. Suppose that M is a compact subset of X which is A.A.S. and $\Gamma$ is L.S.C. on X. Then

(i) There exists a fundamental system of neighborhoods $W_i$ of M that are $\gamma$-p.i., namely $\Gamma(n,W_i) \subset W_i$, $\forall\, n \in \mathbb{N}$, $i=1,2,\ldots$

(ii) $\Gamma$ satisfies the semigroup property:

$$\Gamma(n_1,\Gamma(n_2,x)) = \Gamma(n_1+n_2,x), \quad \forall\, n_1,n_2 \in \mathbb{N}$$

and for any x near M.

Let M be a compact subset of X. A Lyapunov function of M is a continuous real function $\phi$ defined on a neighborhood W of M such that

(i) $\phi(x)=0$ for $x \in M$ and $\phi(x)>0$ for $x \in W-M$.

(ii) $\phi(f(x,u)) \leq \phi(x)$, for any $u \in \Omega$ and $x$, $f(x,u) \in W$.

A strong Lyapunov function $\Phi$ of M is a Lyapunov function such that there exists a strictly increasing continuous real function $c\colon \mathbb{R}^+ \to \mathbb{R}^+$ with $c(0)=0$ and

$$\Phi(f(x,u)) \leq \Phi(x) - c(d(x,M)), \quad \forall\, u \in \Omega, \quad x \in W$$

Theorem 1.4. Let M be a compact subset of X, that has a Lyapunov function $\phi$ defined on $S(M,\delta)$, $\delta>0$. Then for any ordinal $\alpha$

(i) There exists an open neighborhood $O \supset M$ which is $\alpha$-p.i. and a positive $\lambda$ with $S[M,\lambda] \subset O$ so that $\phi(y) \leq \phi(x)$ for any $y$, $x \in O$, $y \in R_\alpha(n,x)$, $n \in \mathbb{N}$.

(ii) M is a $\alpha$-stable and $\alpha$-p.i.

(iii) There is a fundamental system of $\alpha$-p.i. open neighborhoods $O_n$ of M and a fundamental system of $\alpha$-p.i. closed neighborhoods $W_n$ of M and a sequence $\{\lambda_n\}$, $\lambda_n>0$ with $S[M,\lambda_n] \subset O_n$ and $S[M,\lambda_n] \subset W_n$.

Proof: Let $u \in \Omega$. By continuity of f there is a neighborhood V of M such that $f(V,u) \subset S(M,\delta)$. Let $0<\varepsilon<\delta$ such that $S(M,\varepsilon) \subset V$ and $m=\inf\{\phi(x), d(x,M)>\varepsilon, x \in S(M,\delta)\}$. Obviously $m>0$. Let $O=\{x \in S(M,\varepsilon) : \phi(x)<\mu<m\} \subset V$. Then O is 1-p.i. Indeed, if this was not true, $R(\ell,x) \not\subset O$ for some $x \in O$ and $\ell \in \mathbb{N}$.

Let k be the first time for which $R(k,x)\not\subset O$ and so $R(m,x)\subset O$, $\forall m<k$. We distinguish two cases. The first is $R(k,x)\cap O\neq\phi$. By the connectedness of $R(k,x)$ there is $y\in R(k,x)\cap\partial O$ and $z\in O$ such that $y\in R(1,z)$. If $z\in M$ we have $\phi(y)\leq\phi(z)$, $\phi(z)=0$ and so $\phi(y)=0$ which implies $y\in M\subset O$, a contradiction. If $z\in O-M$ then $\phi(y)\leq\phi(z)<\mu<m$. But $m>\varepsilon$, thus $y\in O$, a contradiction. The other case is $R(k,x)\subset X-O$ and $R(k-1,x)\subset O\subset V$. Let $z\in R(k-1,x)$. Then $y=f(z,u)\in R(1,R(k-1,x))=R(k,x)$ and $y\in S(M,\delta)\subset O$ and so $R(k,x)\cap O\neq\phi$, which is a contradiction. The rest of the proof follows similar to the continuous-time case [4].

Theorem 1.5. Let M be a compact subset of X.

(i) If M has a strong Lyapunov function, then it is A.A.S.

(ii) Conversely, if $\Gamma$ is L.S.C. (or R is U.S.C.) on a neighborhood of M and M is A.A.S., then M has a strong Lyapunov function.

Outline of the proof (i) If M has a strong Lyapunov function $\Phi$ then by Theorem 1.4 M is $\gamma$-stable. To establish that M is a $\gamma$-uniform attractor we proceed as in Theorem 19 [4]. In particular we consider a $\gamma$-p.i. neighborhood O of M and for any $\varepsilon>0$ such that $S(M,\varepsilon)\subset O$ we consider the $\gamma$-p.i. neighborhood P of M defined as $P=\{x\in S(M,\varepsilon/2):\Phi(x)\leq m/3\}$, where $m=\inf\{\Phi(x): d(M,x)=\varepsilon/2\}$. Let $\upsilon>0$ such that $S(M,\upsilon)\subset P$. Using an induction argument on the set of the ordinal numbers we can show that

$$\Phi(y)\leq\Phi(x)-nc(\upsilon)$$

for all $y\in\Gamma(n,x)$, $y, x\in G \overset{\text{def.}}{=} O-P$. To complete the proof if suffices to prove that for each $x\in O$ there is $N\in\mathbb{N}$ and $\delta>0$ so that $\Gamma(n, S(x,\delta))\subset P\subset S(M,\varepsilon)$, $\forall n\geq N$. Indeed, if this is not true there are two possibilities. First there are $x_o\in G$, sequences $x_i\in G$ with $x_i\to x_o$, $n_i\in\mathbb{N}$ with $n_i\to+\infty$ and $y_i\in\Gamma(n_i,x_i)$ with $y_i\in P$. Since $y_i\in G$, we have $\Phi(y_i)\leq\Phi(x_i)-n_ic(\upsilon)$, $\forall i$, a contradiction since $c(\upsilon)>0$, $n_i\to+\infty$ and $\Phi(y_i)\geq 0$. The other possibility is $x_o\in P$. In this case we define $P'=\{x\in S[M,\varepsilon/2]:\Phi(x)<m/2\}$, which is a $\gamma$-p.i. neighborhood of M and $P\subset P'\subset S(M,\varepsilon)$. Thus there is $\delta>0$ such that $S(x_o,\delta)\subset P'$ and $\Gamma(n,S(x,\gamma))\subset\Gamma(n,P')\subset S(M,\varepsilon)$ for each $n\geq\mathbb{N}=0$.

(ii) Assume now that $\Gamma$ is L.S.C. and M is A.A.S. We define

$$\ell(n,x) = \sup\{ d(y,M),\ y \in \Gamma(n,x) \},$$

$$\phi(x) = \sup \{\ell(n,x),\ n \in \mathbb{N}\} \quad \text{and}$$

$$\upsilon(n,x) = \sup \{ \phi(y),\ y \in \Gamma(n,x) \},\ n \in \mathbb{N}.$$

As in the continuous - time case we can show that since $\Gamma$ is cluster and L.S.C., then for any $n \in \mathbb{N}$ the functions $\ell(n,x)$ and $\upsilon(n,x)$ are continuous for every x belonging to a neighborhood O of M and $\phi$ is a Lyapunov function of M. Furthermore we can establish the following properties

$$\upsilon(0,x) > 0, \quad \forall x \in O-M,$$

$$\upsilon(n,x) = 0, \quad \forall n \in \mathbb{N} \text{ and } x \in M,$$

$$\upsilon(n,y) \leq \upsilon(n+m,x),\ \forall n,\ m \in \mathbb{N},\ y \in \Gamma(m,x),\quad y,x \in O$$

and there exists a positive strictly increasing real function $\alpha : \mathbb{R}^+ \to \mathbb{R}^+$ such that $\alpha(0)=0$ and

$$\alpha(d(x,M)) \leq \phi(x),\ \forall x \in O.$$

Finally A.A. stability implies the existence of a strictly increasing real function $G: \mathbb{R}^+ \to \mathbb{R}^+$, $G(0) = 0$ such that the function

$$\Phi(x) \overset{\text{def.}}{=\!=} \sum_{n=0}^{\infty} G(\upsilon(n,x))$$

exists and is continuous and positive definite locally around M. Furthermore for each $y=f(x,u)$, $u \in \Omega$ we have

$$\Phi(y) = \sum_{n=0}^{\infty} G(\upsilon(n,y)) \leq \sum_{n=0}^{\infty} G(\upsilon(n+1,x)) = \sum_{n=1}^{\infty} G(\upsilon(n,x)) =$$

$$\Phi(x) - G(\upsilon(0,x)) \leq \Phi(x) - G(\phi(x)) \leq \Phi(x) - G(\alpha(d(x,M)))$$

Therefore $\Phi$ is a strong Lyapunov function with $c=G\circ\alpha$.

<u>Example 1.6.</u> We can easily establish that the map $\Phi(x)=x_1^2+x_2^2$, $x=(x_1,x_2)' \in \mathbb{R}^2$ is a strong Lyapunov function of $0 \in \mathbb{R}^2$ for the planar control system of the example 1.1. Therefore $0 \in \mathbb{R}^2$ is an A.A.S. equilibrium.

2. Applications.

The theory we developed is applicable to several problems in control theory as the stabilization in the presence of disturbances, observer design etc. In this section we shall state without proof a theorem that provides a sufficient Lyapunov condition for the output feedback stabilization of discrete-time systems in the presence of disturbances:

$$x(n+1) = f(x(n), u(n), w(n))$$
$$y(n) = h(x(n)) \quad , \quad n \in \mathbb{N}$$
$$x \in X = \mathbb{R}^n, \quad y \in \mathbb{R}^k, \ u \in \Omega \subset \mathbb{R}^\ell, \ w \in W \subset \mathbb{R}^\ell,$$

where u represents the input by means of which we can stabilize the system, w is the disturbance and y is the output of the system. The mappings $f(\cdot,\cdot,\cdot)$ and h are continuous with $f(0,0,w) = 0$, $\forall w \in W$ and $h(0)=0$. Further h is Lipschitz continuous and an open mapping. We assume that

(A1) There exist continuous mappings F and G and a positive constant $k>0$ with

$$||f(x,u,w)-F(x,w)-G(x,w)u \ || \leq k \ ||u||^2$$

for any (x,u) near zero and $w \in W$. ($||\ \ ||$ denotes the usual Euclidean norm).

(A2) There is a continuous real function $V: X \to \mathbb{R}$ such that

(i) V is convex and Lipschitz continuous on a neighborhood of zero.

(ii) V is positive definite, that is $V(0)=0$ and $V(x)>0$ for $x \neq 0$ near zero.

(iii) There are positive real constants $0<\ell<1$ and $M>0$ such that for any y near $0 \in \mathbb{R}^k$ there is $u \in \Omega$ with,

$$||u|| < M \ || \ y \ ||$$

and

$$V(F(x,w) + G(x,w)u) \leq V(x) - \ell ||x||$$

for any $0 \neq x \in h^{-1}(y)$ near zero and for all $w \in W$.

(A3) W is compact and $\Omega$ contains $0 \in \mathbb{R}^\ell$.

<u>Theorem 2.1</u>. Under the assumptions (A1), (A2) and (A3) exists a Lipshitz continuous output feedback law $\Phi: \mathbb{R}^k \to \Omega$ such that V is a strong Lyapunov function of $0 \in X$ for the

resulting system

$$x(n+1) = f(x(n), \quad \Phi(h(x(n))), w(n))$$

and so $0 \in X$ is an A.A.S. equilibrium.

The proof of the previous theorem as well as further results concerning the stabilization problem of discrete-time systems will presented in [ 5 ].

## Reference

[1] N. Kalouptsidis, D. Elliot, Stability Analysis of the orbits of Control Systems, Math. Systems Theory, 15, 323-342, 1982.

[2] N. Kalouptsidis, Prolongations and Lyapunov functions in Control systems, Math. Systems theory, 16, 233-249, 1983.

[3] J. Tsinias, N. Kalouptsidis, A. Bacciotti, Lyapunov functions and stability of Dynamical Polysystems. Math. Systems Theory, 19, 333-354, 1987.

[4] J. Tsinias, N. Kalouptsidis, Prolongations and stability analysis via Lyapunov functions of Dynamical Polysystems. Math. Systems Theory, 20, 215-233, 1987.

[5] J. Tsinias, N. Kalouptsidis, Output feedback stabilization of discrete-time systems, submitted.

# ANOTHER APPROACH TO THE LOCAL DISTURBANCE DECOUPLING PROBLEM WITH STABILITY FOR NONLINEAR SYSTEMS.

Leo van der Wegen

**Abstract**
The local disturbance decoupling problem with stability is the problem of finding a regular state feedback such that for the feedback system the outputs are decoupled from the disturbances and the equilibrium $x_e$ is locally exponentially stabilized. A crucial role in the solution of this problem is played by a distribution that is in some sense the smallest locally controlled invariant distribution containing the disturbance vector fields. Essential in order that the problem is solvable is that the drift dynamics restricted to the leaf of this distribution through $x_e$ is locally exponentially stabilizable. Conditions for solvability of the local disturbance decoupling problem with stability for a nonlinear system are given in terms of conditions for the linearization of such a system around $x_e$.

## 1. Introduction

In [6] and [7] the local disturbance decoupling problem with stability (LDDPS) is considered for the class of affine nonlinear control systems having $x = 0$ as an equilibrium of the drift dynamics. The LDDPS is the problem of finding locally around $x = 0$ a regular state feedback such that after feedback the disturbances do not influence the outputs and $x = 0$ is a locally exponentially stable equilibrium of the modified drift dynamics. Loosely speaking, the LDDPS is solvable if and only if the disturbance vector fields are contained in $\Delta_s^*$, the largest locally controlled invariant, exponentially stabilizable distribution contained in the kernel of the output mapping. This result very much resembles the solution of the DDPS for linear systems. However, a drawback of this method is that the distribution $\Delta_s^*$ is hard to calculate analytically.

Therefore, we propose another method here to solve the LDDPS. We define the smallest locally controlled invariant, involutive distribution $(D_e)_*$ contained in the kernel of the output mapping h that contains the

disturbance vector fields e and the largest local controllability distribution $\mathcal{S}^*$ in ker dh.
Then, under some technical conditions, the LDDPS is solvable if the dynamics of the system restricted to the leaf $L_0$ of $(D_e)_*$ through $x = 0$ can be stabilized exponentially. This is shown is section 2. In section 3 we give (under technical conditions) necessary and sufficient conditions for the solvability of the LDDPS for a nonlinear system in terms of solvability of the DDPS for its linearization around $x = 0$ and the distribution $(D_e)_*$.

## 2. The local disturbance decoupling problem with stability

Consider the analytic control system

$$(2.1)\qquad \begin{cases} \dot{x} = f(x) + g(x)u + e(x)d := f(x) + \sum_{i=1}^{m} g_i(x)u_i + \sum_{i=1}^{r} e_i(x)d_i \\ y = h(x)\ , \quad y \in \mathbb{R}^{\ell}\ , \quad x \in \mathbb{R}^n \end{cases}$$

with $f(0) = 0$, $h(0) = 0$ and $\dim sp\{g_1,\ldots,g_m\} = m$ on a neighborhood $\Omega_0$ of $x = 0$. We give the exact problem formulation now.
The local disturbance decoupling problem with stability (LDDPS) for (2.1) is defined as follows:
Find -if possible- a locally defined feedback

$$(2.2)\qquad u = \alpha(x) + \beta(x)v\ , \quad \alpha(0) = 0\ , \quad \beta(x) \text{ invertible on } \Omega_0$$

and an involutive locally controlled invariant distribution $\Delta$ contained in ker dh, such that

$$(2.3)\qquad [f+g\alpha,\Delta] \subset \Delta,\ [(g\beta)_i,\Delta] \subset \Delta,\ i = 1,\ldots,m,\ e_j \in \Delta,\ j = 1,\ldots,r$$

and $x = 0$ is a locally exponentially stable equilibrium of the drift dynamics $\dot{x} = (f+g\alpha)(x)$, i.e. $\sigma(\frac{\partial(f+g\alpha)}{\partial x}(0)) \subset \mathbb{C}^-$, the open left-half plane.

**Notation**

By $\tilde{f}$ and $\tilde{g}_i$, $i = 1,\ldots,m$ we will denote the vector fields $f + g\alpha$ and $(g\beta)_i = \sum_{j=1}^{m} g_j\beta_{ij}$. Moreover, the second and third condition in (2.3) will be abbreviated as $[\tilde{g},\Delta] \subset \Delta$ and $e \in \Delta$.

Let $\Delta^*$ denote the largest locally controlled invariant distribution in ker dh (which is by definition involutive) and assume that

(A1) $e \in \Delta^*$

(A2) $\Delta^*$ constant dimensional on $\Omega_0$.

Note that (A1) implies that the local disturbance decoupling problem (LDDP) is solvable. Let $\mathcal{S}^*$ denote the largest regular local controllability distribution contained in ker dh (see [4]). Suppose that the feedback (2.2) is such that $[\tilde{f},\Delta^*] \subset \Delta^*$ and $[\tilde{g},\Delta^*] \subset \Delta^*$ and that (A1) and (A2) hold. We define a distribution $D_e^{\tilde{f},\tilde{g}}$ now by the following algorithm.

**Algorithm 2.1**

1. $D_0 := sp\{e_i,\ i=1,\ldots,r\} + \mathcal{S}^*$
2. $D_{k+1} := D_k + [\tilde{f},D_k] + \sum_{i=1}^{m} [\tilde{g}_i,D_k]$

Assume that

(A3) the distributions $D_k$ have constant dimension on $\Omega_0$ for all k.

Then the algorithm stops after $k^* < n$ steps. Let $D_e^{\tilde{f},\tilde{g}}$ denote the involutive closure of $D_{k^*}$. Suppose that

(A4) $D_e^{\tilde{f},\tilde{g}}$ is constant dimensional.

Note that $D_e^{\tilde{f},\tilde{g}}$ is contained in $\Delta^*$ and locally controlled invariant. The following theorem holds.

**Theorem 2.1**

Assume that (A1) - (A4) hold. Then $D_e^{\tilde{f},\tilde{g}}$ is the smallest constant dimensional involutive, locally controlled invariant distribution contained in the kernel of the output mapping h that contains $\mathcal{S}^*$ and the disturbance vector fields e.

**Proof**

Since $D_e^{\tilde{f},\tilde{g}}$ and $\Delta^*$ are involutive and constant dimensional, there exists a coordinate transformation $z = z(x)$ such that $D = sp\{\frac{\partial}{\partial z_1}\}$, $\Delta^* = sp\{\frac{\partial}{\partial z_1},\frac{\partial}{\partial z_2}\}$. In these coordinates system (2.1,2) has the following form:

$$(2.4)\quad \begin{cases} \dot z_1 = \tilde f^1(z_1,z_2,z_3) + \tilde g^{11}(z_1,z_2,z_3)v_1 + \tilde g^{21}(z_1,z_2,z_3)v_2 \\ \qquad\qquad + e^1(z_1,z_2,z_3)d \\ \dot z_2 = \tilde f^2(z_2,z_3) \qquad + \tilde g^{22}(z_2,z_3)v_2 \\ \dot z_3 = \tilde f^3(z_3) \qquad + \tilde g^{23}(z_3)v_2 \\ y = h(z_3) \end{cases}$$

where $G = sp\{g_1,\ldots,g_m\}$, $G \cap \Delta^* = sp\{\tilde g^1\} = sp\{\tilde g_1,\ldots,\tilde g_s\}$ and $\Delta^* \cap sp\{\tilde g^2\} = 0$. Then $\dim sp\{\tilde g_1^{23},\ldots,\tilde g_{m-s}^{23}\} = m-s$. Hence any additional feedback $v = \tilde\alpha(z) + \tilde\beta(z)w$ with $\tilde\alpha(0) = 0$ and $\tilde\beta(z)$ invertible on $\Omega_0$ applied to (2.4) such that $\Delta^*$ is invariant under $\tilde f + \tilde g\tilde\alpha$ and $\tilde g\tilde\beta$ fulfills the condition $v_2 = \tilde\alpha_2(z_3) + \tilde\beta_{22}(z_3)w_2$. This implies that application of such a v to (2.4) does not influence the structure of the equations (2.4). In other words, the distribution $D_e^{\tilde f,\tilde g}$ is invariant under $\tilde f + \tilde g\tilde\alpha$ and $\tilde g\tilde\beta$ and contains e and $\mathcal{S}^*$. Hence if the smallest involutive distribution with these properties ($D_e^{\tilde f+\tilde g\tilde\alpha,\tilde g\tilde\beta}$) is constant dimensional, then it is contained in $D_e^{\tilde f,\tilde g}$. By converting arguments it follows that $D_e^{\tilde f,\tilde g}$ equals $D_e^{\tilde f+\tilde g\tilde\alpha,\tilde g\tilde\beta}$. So the distribution $D_e^{\tilde f,\tilde g}$ is independent of the feedback (2.2) as long as (2.2) is such that $\Delta^*$ is invariant under $\tilde f$ and $\tilde g$. Suppose now that $\Delta$ is a constant dimensional locally controlled invariant involutive distribution in ker dh containing e and $\mathcal{S}^*$. Assume that $\Delta$ is invariant under $\tilde f$ and $\tilde g$, but $\Delta^*$ is not. Since $\Delta \subset \Delta^*$ and both distributions are constant dimensional and involutive there exists a coordinate transformation $z = z(x)$ such that $\Delta = sp\{\frac{\partial}{\partial z_1}\}$, $\Delta^* = sp\{\frac{\partial}{\partial z_1},\frac{\partial}{\partial z_2}\}$. Moreover, there exists a feedback $v = \tilde\alpha(z) + \tilde\beta(z)w$ with $\tilde\alpha(0) = 0$, $\tilde\beta(z)$ invertible on $\Omega_0$ such that $\Delta$ and $\Delta^*$ are invariant under $\tilde{\tilde f} := \tilde f + \tilde g\tilde\alpha$ and $\tilde{\tilde g} := \tilde g\tilde\beta$ and $\frac{\partial\tilde{\tilde f}}{\partial z_1} = \frac{\partial\tilde f}{\partial z_1}$ and $\frac{\partial\tilde{\tilde g}}{\partial z_1} = \frac{\partial\tilde g}{\partial z_1}$ (see [5]). Since $\Delta$ is invariant under $\tilde{\tilde f}$ and $\tilde{\tilde g}$ and contains e and $\mathcal{S}^*$, it is obvious that if $D_e^{\tilde{\tilde f},\tilde{\tilde g}}$ is constant dimensional, then it is contained in $\Delta$. □

**Corollary 2.1**

Under the assumptions (A1) - (A4) there exists a uniquely defined smallest constant dimensional, involutive, locally controlled invariant distribution in ker dh that contains e and $\mathcal{S}^*$. It will be denoted by $(D_e)_*$.

The solution of the LDDPS is given by:

**Theorem 2.2**

Assume that (A1) - (A4) hold. Suppose that the drift dynamics on $L_0$, the leaf of $(D_e)_*$ through $x = 0$, can be stabilized exponentially and that the linearization of the dynamics of the system (2.1) modulo $(D_e)_*$ is stabilizable. Then the LDDPS is solvable. On the other hand, if the LDDPS for (2.1) is solvable by making a constant dimensional distribution $\Delta$ invariant, then the drift dynamics on the leaf of $\Delta$ through $x = 0$ can be stabilized exponentially and the linearization of (2.1) modulo $\Delta$ is stabilizable.

The proof of this theorem is straightforward and will be omitted. In case $\mathcal{S}^*$ equals the zero distribution, the drift dynamics on $L_0$ is fixed (see [3]). In order that the LDDPS is solvable, this dynamics has to be exponentially stable. In case $\mathcal{S}^* \neq 0$ the dynamics on $\tilde{L}_0$, the leaf of $\mathcal{S}^*$ through $x = 0$ can be stabilized exponentially if the linearization of the system restricted to $\mathcal{S}^*$ is stabilizable.

**Remark**

Our results generalize similar ones developed in [1] for linear systems in order to evade the explicit calculation of $\mathcal{V}_s^*$, the largest stabilizable controlled invariant subspace in the kernel of the output mapping (see [8]).

## 3. Disturbance decoupling with stability for a nonlinear system and for its linearization around an equilibrium

Consider the analytic control system (2.1) and its linearization around the equilibrium $x = 0$

$$(3.1)\qquad \begin{cases} \dot{z} = Az + Bv + Ed\ , & A = \dfrac{\partial f}{\partial x}(0)\ ,\ B = g(0)\ ,\ E = e(0) \\ y = Cz & C = \dfrac{\partial h}{\partial x}(0) \end{cases}$$

Define characteristic numbers $\rho_i(x)$ , $i = 1,\ldots,\ell$ for (2.1) as follows: $\rho_i(x)$ is the smallest integer such that $L_g L_f^k h(x) = 0$, $k < \rho(x)$ and $L_g L_f^{\rho(x)} h(x) \neq 0$. Assume that

(A5) the characteristic numbers $\rho_i(x)$, $i = 1,\dots,\ell$ are finite and constant and equal to $\rho_i$ on $\Omega_0$ and the decoupling matrix A(x) has full row rank on $\Omega_0$.

Here the decoupling matrix A(x) is defined by $(A(x))_{ij} := L_{g_j} L_f^{\rho_i} h_i(x)$ (see [4]). Under this assumption the characteristic numbers of (3.1) are equal to those of (2.1) (see [2]). If, moreover, $[f,\Delta^*] \subset \Delta^*$ and $\Delta^*(0)$ is identified with a subspace of $\mathbb{R}^n$, then $\Delta^*(0) = \mathcal{V}^*$ and $A\mathcal{V}^* \subset \mathcal{V}^*$, where $\mathcal{V}^*$ denotes the largest controlled invariant subspace in ker C. Suppose now that the feedback

(3.2) $\quad u = \alpha(x) + \beta(x)w\ ,\quad \alpha(0) = 0,\quad \beta(x)$ invertible on $\Omega_0$

and the locally controlled invariant involutive constant dimensional distribution $\Delta$ solve the LDDPS for (2.1). Then the disturbance decoupling problem with stability (DDPS) for (3.1) is solved by the linearization $v = Fz + Gw$ of the feedback (3.2). It follows from $\sigma((A+BF)|_{\Delta(0)}) \subset \mathbb{C}^-$ and $(D_e)_*(0) \subset \Delta(0)$ that $(D_e)_*(0) \subset \mathcal{V}_s^*$ (the largest stabilizable controlled invariant subspace in ker C). A partial converse of this assertion can be proven.

**Theorem 3.1**

Consider system (2.1) and its linearization (3.1). Suppose that (A1) - (A5) hold. Assume that

(A6) im $E \subset \mathcal{V}_s^*$ ,

(A7) the linearization of the dynamics of (2.1) modulo $(D_e)_*$ is stabilizable,

(A8) $\mathcal{S}^*$ is constant dimensional on $\Omega_0$ and the linearization of the dynamics on $\mathcal{S}^*$ is stabilizable.

The the LDDPS for (2.1) is solvable if $(D_e)_*(0) \subset \mathcal{V}_s^*$.

**Proof**

First, consider the case $m = \ell$. Then $\mathcal{S}^* = 0$. Suppose that (3.2) is a feedback that leaves $\Delta^*$ and therefore also $(D_e)_*$ invariant. Let $L_0$ denote the leaf of $(D_e)_*$ through $x = 0$. It follows from $(D_e)_*(0) \subset \mathcal{V}_s^*$ that the linearization $(A+BF)_{(D_e)_*(0)}$ of $(f+g\alpha)|_{L_0}$ fulfills $\sigma((A+BF)|_{(D_e)_*(0)}) \subset \mathbb{C}^-$. Consequently, the drift dynamics of (2.1, 3.2) restricted to $L_0$ is exponentially stable. the LDDPS is solvable now by (A7).

Now consider the general case, It can easily be seen that $\dim \mathcal{S}^* \geq \dim \mathcal{R}^*$ and $\mathcal{R}^* \subset \mathcal{S}^*(0)$, where $\mathcal{R}^*$ denotes the largest controllability subspace in ker C (see [8]). Without loss of generality the drift dynamics of (2.1) on the leaf $\tilde{L}_0$ of $\mathcal{S}^*$ through $x = 0$ can be stabilized exponentially by a feedback (3.2) that leaves $\Delta^*$ and thus $(D_e)_*$ and $\mathcal{S}^*$ invariant (see [5]). The rest of the proof goes along the lines of the $m = \ell$ case. □

**Remark**

The results in this section show that under technical conditions the LDDPS for the static decouplable nonlinear system (2.1) is solvable if and only if the following three conditions hold (i) the LDDP for (2.1) is solvable, (ii) the DDPS for its linearization (3.1) is solvable and (iii) $(D_e)_*(0) \subset \mathcal{V}_s^*$.

**Example**

Consider the system (2.1) with

$$(3.4) \qquad f(x) = \begin{pmatrix} 2x_1 + x_2 + x_1x_2 + x_1x_3 \\ -x_2 \\ -x_3 + x_4x_5^2 \\ 3x_4 + 3x_5 \\ 4x_5 \end{pmatrix}, \quad g(x) = \begin{pmatrix} 1 & 0 \\ 0 & 0 \\ 1 & 0 \\ 0 & 0 \\ 0 & 1 \end{pmatrix}, \quad h(x) = x_5,$$

$$e(x) = \begin{pmatrix} 1 + x_3^2 \\ 1 \\ 0 \\ 0 \\ 0 \end{pmatrix}$$

Since $\langle dh, g_2\rangle(x) = 1$, it follows that $\rho = 0$ and $A(x)$ has full row rank. $\Delta^* = \mathrm{sp}\{\frac{\partial}{\partial x_1}, \ldots, \frac{\partial}{\partial x_4}\}$ is invariant under $f$ and $g$ and contains $e$. Furthermore, $\mathcal{S}^* = \mathrm{sp}\{\frac{\partial}{\partial x_1}, \frac{\partial}{\partial x_3}\}$ and $(D_e)_* = \mathrm{sp}\{\frac{\partial}{\partial x_1}, \ldots, \frac{\partial}{\partial x_3}\}$ are constant dimensional. The linearization of (2.1, 3.4) around $x = 0$ is given by (3.1) with

$$(3.5) \qquad A = \begin{pmatrix} 2 & 1 & 0 & 0 & 0 \\ 0 & -1 & 0 & 0 & 0 \\ 0 & 0 & -1 & 0 & 0 \\ 0 & 0 & 0 & 3 & 3 \\ 0 & 0 & 0 & 0 & 4 \end{pmatrix}, \quad B = \begin{pmatrix} 1 & 0 \\ 0 & 0 \\ 1 & 0 \\ 0 & 0 \\ 0 & 1 \end{pmatrix}, \quad C = (0 \;\; 0 \;\; 0 \;\; 0 \;\; 1), \quad E = \begin{pmatrix} 1 \\ 1 \\ 0 \\ 0 \\ 0 \end{pmatrix}$$

Note that $\mathcal{V}_s^* = \mathrm{sp}\{e_1, e_2, e_3\}$ and that (3.1,5) is stabilizable, hence the DDPS for (3.1) is solvable. By applying the feedback

(3.6) $u_1 = -10x_1 + v_1, \; u_2 = v_2$

the system (2.1, 3.4, 3.6) is such that the drift dynamics $(f+g\alpha)|_{L_0}$ is locally exponentially stable. Since the linearization of this system modulo $(D_e)_*$ is controllable, the LDDPS for (2.1, 3.4) is solvable.

**Acknowledgements**

The author wants to thank Dr. Henk Nijmeijer for his suggestions on the contents and organization of the paper and Dr. Arjan van der Schaft for bringing a preliminary version of [1] under his attention.

**References**

[1] G. Basile, G. Marro, *Self-bounded controlled invariants versus stabilizability,* **Journal of Optimization Theory and Appl.** 48 (1986), 245-263.

[2] L.C.J.M. Gras, H. Nijmeijer, *Decoupling in nonlinear systems: from linearity to nonlinearity,* **IEE Proc. Pt. D.,** 136 (1989), 53-62.

[3] A, Isidori, J.W. Grizzle, *Fixed modes and nonlinear noninteracting control with stability,* **IEEE Trans. Autom. Control** 33 (1988), 907-914.

[4] A. Isidori, **Nonlinear Control Systems: An Introduction (Lecture notes in Control and Information Sciences 72)**, Springer, New York, 1985.

[5] H. Nijmeijer, *The triangular decoupling problem for nonlinear control systems,* **Nonlinear Analysis, Theory, Methods and Appl.** 8 (1984), 273-279.

[6] L.L.M. van der Wegen, H. Nijmeijer, *The local disturbance decoupling problem with stability for nonlinear systems,* **Systems and Control Letters** 12 (1989), 139-149.

[7] L.L.M. van der Wegen, *On the use of stable distributions in design problems,* **preprints of the IFAC Symposium on Nonlinear Control System Design** (1989)

[8] W.M. Wonham, **Linear Multivariable Control: A Geometric Approach,** Springer, New York, 1979.

# STABILIZATION OF NONLINEAR SYSTEMS AND COPRIME FACTORIZATION

M.S. VERMA and THO PHAM

In this paper, the problem of constructing a right coprime factorization of a system is considered. The approach is based on the concept of right-coprimeness introduced recently and on a stabilizing state feedback. Since our approach does not require construction of solutions to a Bezout identity, the construction of a stabilizing output injection is not needed. Some relationships between the existence of stabilizing output feedbacks and solutions to Bezout identities for (time-varying) linear systems are discussed and a development of left-coprime factorization of a nonlinear system is presented. We also prove that the existence of two coprime factorizations will imply the existence of the other two when the plant and the controller are both linear and the unity feedback system is finite-gain stable.

## I. Introduction

One of the main problems in the theory of coprime fractional representations for nonlinear systems is how to construct these factorizations. The approach employed in this paper follows from the work of Verma [6]. In [6], the concept of right-coprimeness for a system is seen as the set of all stable input-output pairs, i.e., the graph of a system. In the following section, we consider the problem of constructing a right-coprime factorization (r.c.f.) for a system. We first construct an r.c.f. for a finite-dimensional linear time-invariant (FDLTI) system in a state-variable description by finding its stabilizing state feedback. Then this approach is extended to nonlinear systems. In section 3, we then develop a framework of left-coprime factorization (l.c.f.) for nonlinear systems similar to r.c.f. developed in [6] and discuss the relationships between coprimeness and stabilizing output feedback of a (time-varying) linear system. We also prove that the existence of two coprime factorizations will imply the existence of the other two when the plant and the controller are both linear and the unity feedback system is finite-gain stable. Finally, we present some concluding remarks in section 4.

II. RCF Construction of A System

As developed in [6], we view stable systems as those that take a class of inputs designated as stable into a class of outputs also designated as stable and unstable systems as those that may take only some, but not all, stable inputs into stable outputs. Let U be a space of inputs and V be a space of outputs, both assumed to be linear vector spaces, and let G: U→V be an input-output map which describes a physical system of interest. Let Us U and Vs V be subspaces of U and V representing stable inputs and outputs, respectively. We say G is stable if GUs Vs. Given a mapping G, which is not necessarily stable, we denote

$$Do(G) = \{u \in Us: \; Gu \in Vs\}$$
$$Ra(G) = \{Gu: \; u \in Do(G)\}$$
$$Graph(G) = \{[Gu \;\; u]': \; u \in Do(G)\}.$$

Depending on the context, the space $L^p$ will consist of vector-valued or scalar-valued p-integrable functions defined on $IR_+$, $1 \le p \le \infty$. In our framework, U and V can be thought as $L^{pe}$ and Us and Vs can be thought as $L^p$. A mapping G: U→V is said to be bounded on its domain if there exists a real constant $0 < k \in IR_+$ such that $\|Gu\| \le k\|u\|$ for all $u \in Do(G)$ and $\| . \|$ denotes some appropriate norm. $\|G\|$ is defined as the infimum of all such k's. G is finite-gain stable if G is stable and $\|G\|$ is finite. In order to construct an r.c.f. for a nonlinear system based on its graph, we first construct an r.c.f. of a causal FDLTI system. We then later extend this approach to a class of time- varying nonlinear systems considered by Desoer and Kabuli [1] with some additional assumptions.

Theorem 1 If a causual FDLTI system realized in a state-variable description is stabilizable and detectable, then its r.c.f. can be obtained from its graph and stabilizing state-feedback.

Proof : Let G(s) be the mxn proper rational transfer martrix of a causal FDLTI system with the following stabilizable and detectable state-variable description

$$\dot{x} = Ax + Bu, \quad x(0) = 0 \qquad (2.1)$$
$$y = Cx + Du \qquad (2.2)$$

Let F be a constant matrix such that all the eigenvalues of A + BF are in the left-half plane. Let

$$u = Fx + v \qquad (2.3)$$

be a state feedback to the system (2.1) - (2.2). Then the resulting system is

$$\dot{x} = (A + BF)x + Bv \qquad (2.4)$$
$$y = (C + DF)x + Dv \qquad (2.5)$$

Let D(s) be the transfer matrix of the system (2.3) - (2.4) with input v and output u and N(s) be the transfer matrix of the state-variable realization (2.4) - (2.5) with input v and output y. The validity of the fractional representation $G(s) = N(s)\, D^{-1}(s)$ is obvious. Let $U = L^{2e}$, $V = L^{2e}$, $U_s = L^2$, and $V_s = L^2$. Also let G(s) be viewed as a mapping (denoted G) from U to V in terms of a convolution integral representation [3,10]. Note that this mapping can also be represented by (2.1) - (2.2) since the initial condition on the state x is zero. Similarly, N(s) and D(s) are viewed as the mappings N: $U \rightarrow V$ and D: $U \rightarrow U$, respectively. From (2.3), it follows that if $v \notin Us$, then either u or x must be unstable since F is a constant matrix. If u belongs to Us and x is unstable, then we have y in V - Vs by (2.2) and the detectability of (A,C). Hence when $v \notin Us$, either u or y is unstable. From [6], we know that $G = ND^{-1}$ is an r.c.f. over the space of stable maps. By construction, N,D, and $D^{-1}$ are causal, N and D are finite-gain (f.g.) stable . Since N(s) and D(s) do not have common zeros on the jw-axis, we have a r.c.f. over the space of causal and f.g. stable maps [6]. Q.E.D.

Remark: Note that the right-coprimeness of N(s) and D(s) is proved in [5, 9] in terms of solutions to a Bezout identity. However, the construction of solutions to the Bezout identity requires the construction of a stabilizing output injection matrix for G(s). This operation is in addition to that of finding the state feedback matrix F and amount to obtaining a left coprime factorization of G(s).

Since the idea of constructing r.c.f. for G(s) is based on its state-variable description, this approach of constructing r.c.f. may extend to nonlinear systems. Starting from a state variable description

$$\dot{x} = f(x,u,t),\ x(o) = x_o \qquad (2.6)$$
$$y = g(x,u,t) \qquad (2.7)$$

of a nonlinear system [8], it may be possible to utilize the state-feedback configuration and construct an r.c.f. for the corresponding input-output map by finding a stabilizing state-feedback $u=k(x,v,t)$: $L^{pe} \times L^{pe} \times IR_+ \rightarrow L^{pe}$ for the system. Before proceeding to construct an r.c.f. for a nonlinear system, we need to obtain a notion of detectability in terms of its input and output. Motivated by the proof of Theorem 1, we introduce the following definition of input-output detectablity. P is said to be input-output detectable if $u \in Us$ and

$y \in Ys$ imply that x is stable and there exists a constant $\beta>0$ such that $\| x \| \leq \beta \| \begin{bmatrix} y \\ u \end{bmatrix} \|$.
From [1], a construction of a r.f. of a class of non-linear systems specified by the following state-space description,

$$x = A(t)x+f(t,x)+B(t)u, \ x(o) = o \quad (2.11)$$
$$y = h(t,x,u) \quad (2.12)$$

has been obtained under some suitable assumptions on the functions A(t), B(t), f(t,x), h(t,x,u) and the construction of a stabilizing state feedback u(t) = K(t)x(t) + v(t). In the case when h (t,x,u) = x, the factorization $P=ND^{-1}$ of the system in (2.11) - (2.12) is shown to be right-coprime by constructing solutions U and V to the Bezout identity

$$UN + VD = I. \quad (2.13)$$

It is shown that the choice V = I and U: $y \rightarrow v$, $v(t) := -K(t)y(t)$ satisfies (2.13). Using the approach suggested here, we prove the right-coprimeness of the factorization provided in [1] for the class of nonlinear system (2.11) - (2.12) even when $h(t,x,u) \neq x$.

Theorem 2 If the mapping P of the system (2.11) - (2.12) is input-output detectable, then P has a r.c.f. over the space of f.g. stable map.

Proof: Assume that $Hvx : v(t) \rightarrow x(t)$ is stable. Then from an assumption in [1] for the system (2.11) - (2.12), i.e., for any causal stable map $Hx : L_{\infty e} \rightarrow L_{\infty e}$, $H_X : u(t) \rightarrow x(t)$, the causal map $Hy : u(t) \rightarrow y(t)$ defined by $y(t) = h(t, (H_X u)(t), u(t))$ is a stable map where $h: IR_+ \times IR^n \times IR^{ni} \rightarrow IR^{no}$, $Hvy: v(t) \rightarrow y(t)$ is stable. From the state feedback configuration, $u = Kx + v = (KHvx + I)v$ where I is the identity map. From [1], it is known that there exists an $\beta >0$ such that $\sup \| K(t) \| \leq \beta$ for all t in $IR_+$. Hence the map $Hvu : v \rightarrow u$ is stable. From [1], it has been shown that $Hvu^{-1}$ exists and is causal. Hence, $P \triangleq Hvy\, Hvu^{-1}$ is a right factorization. From [6], we know that P is right-coprime if and only if $v \in U - Us$ implies $u \in U - Us$ or $y \in Y - Ys$. Assume that $v \in U - Us$. Then if $u \in U - Us$, we are done. If $u \in Us$, then we need to show that $y \in Y - Ys$. But from the mapping $Hvu : v \rightarrow u$, we know that x is unstable and hence $y \in Y - Ys$ since P is input-output detectable. Therefore, P as constructed is an r.c.f. over the space of stable maps. Now, also from the state feedback configuration, we have $v = u - Kx$. So $\|v\| \leq \|u\| + \|Kx\| \leq \|u\| + \beta\|x\|$. Since P is input-output detectable, there exists an $k > 0$ such that $\|x\| \leq k\| \begin{bmatrix} y \\ u \end{bmatrix} \|$. So $\|v\| \leq (1 + \beta k)\| \begin{bmatrix} y \\ u \end{bmatrix} \|$ since $\|u\| \leq \| \begin{bmatrix} y \\ u \end{bmatrix} \|$. Hence, P has an r.c.f. over the space of f.g. stable map. Q.E.D.

Remark: It may also be possible to construct r.c.f. of other classes of nonlinear systems in an analogous manner. This is so because the existence of stabilizing state-feedbacks may be known for a larger class of nonlinear systems described by (2.6) - (2.7). The theorem below is currently being applied to constructing r.c.f. for the class of feedback linearizable nonlinear system [7].

Theorem 3 Consider the system described by (2.6) - (2.7) with the following assumptions:

a) The function f is such that $x \in L^{pe}$ is uniquely determined for all $u \in L^{pe}$ and $x_0 \in IR^q$.
b) The pair f, g is input-output detectable.
c) The mapping from v to u defined by $u = K(x,v,t)$ is invertible for all $t \in IR_+$, $x \in IR^q$, i.e., $v = K'(x,u,t)$
d) The mappings from (v,x) to u and (u,x) to v determined by K and K′ respectively are causal and finite-gain stable.
e) Let P be the I/0 map from $L^{pe}$ to $L^{pe}$ with $x(0) = x_0 \in IR^q$ fixed ( P: u →y). Let N: v → y, P: v → u be the I/0 maps associated with

$$\dot{x} = f(x, K(x,v,t), t),\ x(o) = x_0$$
$$y = g(x, K(x,v,t), t)$$
$$u = K(x,v,t)$$

where K is a stabilizing feedback,i.e., N and D are finite-gain stable. Then $P = ND^{-1}$ is an r.c.f. of P.

Proof: By construction, N and D are causal. From c), $D^{-1} : L^{pe} \to L^{pe}$ is the (causal) map associated with $\dot{x} = f(x,u,t)$, $v = K'(x,u,t)$ with $x(o)=x_0$ from u to v. Hence $y=Pu = PDv$. But $y = Nv$ for all $v \in L^{pe}$. So $PD = N$. Since N and D are f.g. stable and causal, D is invertible and $PD=N$, $P=ND^{-1}$ is a r.f. Note that P and $D^{-1}$ are causal. Let $v \in L^{pe}$, $u \in L^p$, $y \in L^p$. Then b) implies that $x \in L^p$. Also d) implies that $v \in L^p$. From [6], graph $(P) = \begin{bmatrix} N \\ D \end{bmatrix} L^p$. Also from c) and d), we have $\| x \| \le \beta \| \begin{bmatrix} y \\ u \end{bmatrix} \|$ and $\| v \| \le \gamma \| \begin{bmatrix} x \\ u \end{bmatrix} \|$, respectively. So

$$\begin{aligned} \|v\| &\le \gamma(\| x \| + \| u \|) \\ &\le \gamma(\beta \| \begin{bmatrix} y \\ u \end{bmatrix} \| + \|u\|) \\ &\le \gamma(\beta + 1) \| \begin{bmatrix} y \\ u \end{bmatrix} \| . \end{aligned}$$

So $\| \begin{bmatrix} y \\ u \end{bmatrix} \| = \| \begin{bmatrix} N \\ D \end{bmatrix} v \| \ge (1/\gamma(\beta + 1)) \|v\|$ and $\alpha = (1/\gamma(\beta + 1)) > 0$. Therefore, $P = ND^{-1}$ is a r.c.f. over the space of causal f.g. stable maps. Q.E.D.

III. <u>On the development of l.c.f. of a nonlinear system and the relationships between stablizing output feedbacks and Bezout identity solutions</u>

For the purpose of understanding the role of l.c.f. and constructing them, let $G = \tilde{D}^{-1}\tilde{N}$ be a left fractional (l.f.) representation of G over the space of causal stable maps, i.e., if $G: U \rightarrow Y$ is a mapping, then $D : Y \rightarrow Y$, $N : U \rightarrow Y$ are causal stable maps with D invertible such that $\tilde{D}G = \tilde{N}$. Equivalently,

$$\tilde{D}G - \tilde{N} = [\tilde{D} - \tilde{N}] \begin{bmatrix} G \\ I \end{bmatrix} = 0$$

on U. Consider the mapping $[\tilde{D} - \tilde{N}] : YxU \rightarrow Y$ and its kernel. It follows that

$$\ker [\tilde{D} - \tilde{N}] \supset \begin{bmatrix} G \\ I \end{bmatrix} U.$$

Conversely, let $(y,u) \in YxU$ be such that

$$\begin{bmatrix} y \\ u \end{bmatrix} \in \ker [\tilde{D} - \tilde{N}],$$

i.e., $\tilde{D}y - \tilde{N}u = 0$. Since $\tilde{D}$ is invertible, this implies that $y = \tilde{D}^{-1}\tilde{N}u = Gu$, i.e.,

$$\begin{bmatrix} y \\ u \end{bmatrix} \in \begin{bmatrix} G \\ I \end{bmatrix} U$$

Hence, we have that

$$\ker [\tilde{D} - \tilde{N}] = \begin{bmatrix} G \\ I \end{bmatrix} U$$

In a similar manner, we can show that

$$\begin{bmatrix} G \\ I \end{bmatrix} Do\,(G) = \ker [\tilde{D} - \tilde{N}] \cap [V_S \times U_S]$$

In an intuitive manner, the role of l.c.f. can be seen in mapping all input-output pairs of a system to the null element without first transforming an unstable input-output pair to a stable one. Thus, a distinction is made between input-output pairs of a system which are stable, i.e., they belong to the graph of the system, and those both of whose elements are unstable. In other words, the system has no input-output pair $(y,u) \in (Y - Y_S) \times (U - U_S)$ such that $\tilde{D}y = \tilde{N}u \in Y_S$.

Definition 1: $G = D_l^{-1}N_l$ is a left-coprime factorization (l.c.f.) over the space of causal stable maps if it is a l.f. over the space of causal stable maps, and $(\mathrm{Im}G \cap V - V_s) = S$ where $S = \{ v \in V - V_S: D_l v \in V_S\}$.

Definition 2: $G = D_l^{-1}N_l$ is an l.c.f.over the space of causal f.g. stable maps if

a. it is an l.c.f. over the space of causal stable maps

b. it is an l.c.f. over the space of causal f.g. stable maps,i.e., it is an l.f. over the space of causal f.g. stable maps and $D_l^{-1}$ is causal if and only if G is causal.

Similar to [6], we obtain here a simple characterization of an l.c.f. of G.

Lemma 1: Assume that $G = D_l^{-1}N_l$ is a l.f. over the space of stable maps. Let us denote

$$Y1 = \{ y \in (Y - Y_S): y = Gu, u \in Us\}$$
$$Y2 = \{ y \in (Y - Y_S): y = D_l^{-1}z, z \in Y_S\}$$

G is an l.c.f. over the space of stable maps if and only if given $z = D_l y = N_l u$, $y \in (Y - Y_S)$, $u \in (U - U_S)$ imply that $z \in (Y - Y_S)$.

Sufficiency. Let $y \in (Y - Y_S)$ such that $y = Gu$ and $u \in Us$. Since $G = D_l^{-1}N_l$, we have $y = D_l^{-1}N_l u$ or $y = D_l^{-1}z$ since $z = N_l u$. But $z \in Ys$ since $N_l$ is a stable map and $u \in Us$. Hence,we have Y1 Y2. Now if $y = D_l z \in (Y - Y_S)$ and $z \in Ys$, we have $y = D_l^{-1}N_l u = Gu$ since $z = N_l u$. Since $z \in Ys$ and $y \in (Y - Y_S)$, we must have $u \in Us$ by assumption. So Y2 Y1. Therefore, Y1 = Y2 and G is a l.c.f. over the space of stable maps.

Necessity. Assume that G is an l.c.f. over the space of stable maps. Let $z = D_l y = N_l u$ and $y \in (Y - Y_S)$. If $z \in Ys$ then $u \in Us$ by the left-coprime property of G. Hence if $u \in (U - U_S)$, we must have $z \in (Y - Y_S)$. Q.E.D.

From [6], it is known that if the plant P and the controller C have r.c.f. $P = ND^{-1}$ and $C = XY^{-1}$ over the space of causal and f.g. stable maps, then the unity feedback system is well-posed, i.e., a well-defined causal f.g. stable map, if the following mapping

$$M = \begin{bmatrix} Y & N \\ -X & D \end{bmatrix}$$

is unimodular. The same characterization of stability of the unity feedback system is obtained when P and C are both causal FDLTI systems [9]. Note that in our framework the characterization involves r.c.f. of <u>both</u> P and C, unlike [9] where stability of the unity feedback system is characterized in terms of an r.c.f. of P and a l.c.f. of C and vice-versa in the case when P and C are both linear. It is known from [9] that the unity feedback system is stable if the following double Bezout identity

$$\begin{bmatrix} \tilde{D} & -\tilde{N} \\ \tilde{X} & \tilde{Y} \end{bmatrix} \begin{bmatrix} Y & N \\ -X & D \end{bmatrix} = \begin{bmatrix} I & O \\ O & I \end{bmatrix}$$

can be solved for the doubly coprime factorization $P = ND^{-1} = \tilde{D}^{-1}\tilde{N}$ over the ring of proper and stable rational matrices, i.e., the space of causal and f.g. stable maps in our framework. Note that the inverse of M is given by

$$M^{-1} = \begin{bmatrix} \tilde{D} & -\tilde{N} \\ \tilde{X} & \tilde{Y} \end{bmatrix}$$

Hence if the inverse of M can be explicitly computed, then a partitioning of it yields solutions to the Bezout identity corresponding to an r.c.f. of P as well as a l.c.f. of P. A major difficulty in extending this approach to nonlinear systems is that $M^{-1}$ may not be block partitioned in this case. However, it would be interesting to see if under some conditions on the nonlinear plant P or the controller C, $M^{-1}$ could be block partitioned and this approach can be extended to nonlinear systems.

<u>Lemma 2:</u> Suppose that C is linear and $C = Y_l^{-1} X_l$ is a l.c.f. If the unity feedback system is f.g. stable, then P has an r.c.f.

<u>Proof</u>: Write $P = P(I+CP)^{-1}[(I+CP)^{-1}]^{-1} = ND^{-1}$ with $N \triangleq P(I + CP)^{-1}$, $D \triangleq (I + CP)^{-1}$. Since the unity feedback system is f.g. stable, we have N and D causal and f.g. stable. Hence $P = ND^{-1}$ is a r.f. From [6], it follows that P is an r.c.f. if and ony if Do (C) $\supset$ Ra (P) and the mapping [C I]: Graph(P) $\to L^p$ is bounded. If either of these conditions is not satisfied, then let T: $L^{pe} \to L^{pe}$ be a causal, f.g. stable and invertible operator such that

a. $T^{-1}$ is causal
b. $T[(I + CP). \text{Do}(P)] = L^p$
c. $T[C\ I]$: Graph(P) $\to L^p$ is bounded.

Then $\hat{P} = N_1 D_1^{-1}$ is an r.c.f. with $N_1 \triangleq NT^{-1} = P(I + CP)T^{-1}$ and $D_1 \triangleq DT^{-1} = (I + CP)^{-1} T^{-1}$. It is clear that $[X_1 \ \ Y_1] (L^{pe} \times L^{pe}) = L^{pe}$. We claim that $[X_1 \ \ Y_1] : L^p \times L^p \rightarrow L^p$ is onto. Let $z \in L^p$. Then there exist $y \in L^{pe}$ and $u \in L^{pe}$ such that $z = Y_1 y + X_1 u$. So $Y_1^{-1} z = y + Cu$. From Lemma 1, if $Y_1^{-1} z \in L^{pe} - L^p$, then there exists $u1 \in L^p$ such that $Y_1^{-1} z = Cu1$. Let $y = 0$ and $u = u1$, then $z = Y_1 y + X_1 u$. If $Y_1^{-1} z \in L^p$ let $y = Y_1^{-1} z$ and $u = 0$, then claim is established. So we have $L^p = [X_1 \ \ Y_1] (L^p \times L^p) = [X_1 \ \ Y_1] \begin{bmatrix} I & P \\ -C & I \end{bmatrix}$ Do (C) x Do (P) = $[0 \ \ X_1 P + Y_1]$Do (C) x Do (P) = $(X_1 P + Y_1)$ Do (P) = $Y_1 (I + CP)$ Do(P).
HenceT=$Y_1$ will satisfy the conditions a-c.
So $P = N_1 D_1^{-1} = P(Y_1 + X_1 P)^{-1} \ [(Y_1 + X_1 P)^{-1}]^{-1}$ is an r.c.f over the space of causal f.g. stable maps. Q.E.D.

In a similar way, we can show that when C is linear and $C = ND^{-1}$ is a r.c.f., P has an l.c.f. if the unity feedback system is f.g. stable.

Remarks

1. If C has both r.c.f. and l.c.f., then parameterization of all stabilizing controllers is possible. This can be useful since C may sometimes have simpler structure than P.
2. When P and C are both linear and the unity feedback system is f.g. stable, then the existence of any two coprime factorizations will imply the existence of the remaining two coprime factorizations.

## IV. Concluding Remarks

From the framework developed in [6], we construct an r.c.f. for a causal FDLTI system which is stabilizable and detectable by using a stabilizing state feedback. We then extend this approach to construct an r.c.f. for a class of time-varying nonlinear systems considered by Desoer and Kabuli (see [1]). Using the approach developed in this paper, it may be possible to construct an r.c.f. of other classes of nonlinear systems since the existence of their stabilizing state-feedbacks may be known. We also introduce a new notion of detectability and assume the state-space realization of nonlinear system to satisfy this property. Since our approach does not require construction of solutions to a Bezout identity, the construction of a stabilizing output injection is not needed. Some relationships between the existence of stabilizing output feedbacks and solutions to Bezout identity for (time-varying) linear systems are discussed and a development of l.c.f. of a nonlinear system is presented. We also prove that the existence of two coprime factorizations will imply the existence of the other two when the plant and the controller are both linear and the unity feedback system is f.g. stable.

This research was supported by NSF Grant no. ECS - 8810578.

References


[1] C.A. Desoer and M.G. Kabuli, "Right Factorizations of a Class of Time-Varying Nonlinear Systems," IEEE Trans. Automatic Control 33, 1988, 755-757.

[2] C.A. Desoer and M.G. Kabuli, "Stabilization and Robustness of Nonlinear Unity-Feedback System: Factorization Approach," Int. J. Control, 47, 1988, 1133-1148.

[3] C.A. Desoer and M. Vidyasagar, Feedback Systems: Input-Output Properties, Academic Press, 1975.

[4] V. Manousiouthakis, "On Time-Varying Control," Proc. American Control Conf. 3, 1987, 2050-2053.

[5] C.N. Nett, C.A. Jacobson and M.J. Balas, "A Connection Between State-Space and Doubly Coprime Fractional Representation," IEEE Trans. Automatic Control 29, 1984, 831-832.

[6] M.S. Verma, "Coprime Fractional Representations and Stability of Nonlinear Feedback Systems," Int. J. Control, 48, 1988, 897-918.

[7] M.S. Verma and L.R. Hunt, "Coprime Factorizations of Linearizable Nonlinear Systems," to be submitted.

[8] M. Vidyasagar, Nonlinear Systems Analysis, Prentice-Hall, Englewood Cliffs, NJ, 1978.

[9] M. Vidyasagar, Control System Synthesis: A Factorization Approach, MIT Press, Cambridge, Massachusetts, 1985.

[10] J.C. Willems, Analysis of Feed-back Systems, MIT Press, Cambridge, Massachusetts, 1971.

# Identification of Linear Systems by Prony's Method

G. Ammar D. Cheng W. Dayawansa
C. Martin

**Abstract**

In this paper a variation of Prony's method is used in a problem of system identification. It is shown that methods developed for Gaussian quadrature can be used in the problem of exponential interpolation which is equivalent to identification for single input/single output linear systems.

## 1 Introduction.

Consider the following interpolation problem: Given $2n$ numbers $\eta_j$, $j = 0, 1, \ldots, 2n-1$, determine distinct numbers $\{\lambda_k\}_{k=1}^n$ and corresponding $\{w_k\}_{k=1}^n$, $k = 1, 2, \ldots, n$ such that

$$\eta_j = \sum_{k=1}^{n} w_k \, e^{\lambda_j k} \quad (j = 0, 1, \ldots, 2n-1). \tag{1.1}$$

We refer to this problem as the *exponential interpolation problem.* Methods for solving this problem date to the work of Prony in 1795. [3]. Also, see [5]. We will say that the input data set $\{\eta_j\}$ is *regular* if this problem has a solution.

Exponential interpolation is closely related with system identification problems for linear dynamical systems. Specifically, consider the linear single-input \ single-output linear system

$$\dot{x}(t) = A\ x(t) + b\ u(t),\ y(t) = c\ x(t);\ x(0) = x_0 \tag{1.2}$$

where the input $u(t)$ and the output $y(t)$ are scalar functions, $x(t) \in \mathbf{R}^n$ for each $t$, and $A \in \mathbf{R}^{n\times n}, b \in \mathbf{R}^{n\times 1}$ and $c \in \mathbf{R}^{1\times n}$. Then the output is given by

$$y(t) = ce^{At}x_0 + \int_0^t ce^{A(t-s)}bu(s)ds. \tag{1.3}$$

If the state matrix $A$ has distinct eigenvalues $\lambda_1, \ldots, \lambda_n$, then there exists scalars $w_1, \ldots, w_n$ such that the free response ($u(t) \equiv 0$) of the system is given by

$$y_f(t) = ce^{At}x_0 = \sum_{j=1}^{n} w_j e^{\lambda_j t} \tag{1.4}$$

Observe that this representation is invariant under change of basis in the state space. If we seek to determine the representation (1.4) from observations $y(j)$, $j = 0, 1, \ldots 2n-1$, we see that this is equivalent with an exponential interpolation problem.

In this paper we describe a numerical procedure for solving the exponential interpolation problem for regular input data. The method is closely related with Prony's method, and is based on ideas for orthogonal polynomials extended to the case of an indefinite bilinear form. We then consider the use of this method in system identification problems.

If the output data set $\{y_j\}$ is irregular, then the representation (1) is invalid, and more general techniques must be used. See [4] for consideration along this line. Also see [2] for a study of general partial realization problems.

## 2 Exponential Interpolation.

We assume that the observations $\{y_j\}_{j=0}^{2n-1}$ form a regular data system, so that the exponential interpolation problem has a solution.

Define the $n$-vectors

$$w = \begin{bmatrix} w_1 \\ w_2 \\ \vdots \\ w_n \end{bmatrix}, \quad y_s = \begin{bmatrix} \eta_s \\ \eta_{s+1} \\ \vdots \\ \eta_{s+n-1} \end{bmatrix}$$

and also define

$$V_s = \begin{bmatrix} e^{\lambda_1 s} & e^{\lambda_2 s} & \dots & e^{\lambda_n s} \\ e^{\lambda_1(s+1)} & e^{\lambda_2(s+1)} & \dots & e^{\lambda_n(s+1)} \\ e^{\lambda_1(s+2)} & e^{\lambda_2(s+2)} & \dots & e^{\lambda_n(s+2)} \\ \vdots & \vdots & & \vdots \\ e^{\lambda_1(s+n-1)} & e^{\lambda_2(s+n-1)} & \dots & e^{\lambda_n(s+n-1)} \end{bmatrix}$$

for $s = 0, 1, \dots, n$. Then (1) implies that for each $s, V_s w = y_s$. Consequently, $y_{s+1} = V_{s+1} V_s^{-1} y_s$. This relationship implies that $V_{s+1} V_s$ is a Frobenious matrix,

$$V_{s+1}\ V_s^{-1} = \begin{bmatrix} 0 & 1 & & \\ & 0 & \ddots & \\ & & \ddots & 1 \\ \alpha_0 & \alpha_1 & \dots & \alpha_{n-1} \end{bmatrix}.$$

Furthermore, this Frobenious matrix is equal to $V_1 V_0^{-1}$, and is therefore independent of $s$. Thus, if (1.1) is satisfied then the $\eta_j$ satisfy a linear constant-coefficient difference equation of order $n$. Consequently, the quantities $\{e^{\lambda_j}\}_{j=1}^{n}$ are the zeros of the polynomial

$$p_n(z) = z^n - \alpha_{n-1} z^{n-1} - \dots - \alpha_1 z - \alpha_0. \tag{2.5}$$

The vector of coefficients $a = [\alpha_j]_{j=1}^{n-1}$ satisfies the Hankel system of equations

$$H_n a = y_n, \tag{2.6}$$

where $H_n = [\eta_{j+k}]_{j,k=0}^{n-1}$. These observations form the basis of Prony's method: Solve the Hankel equations 2. 6, determine the zeros of $p_n(z)$, and solve the Vandermonde system $V_0 w = y_0$. This analysis also shows that the exponential interpolation problem has a solution provided the Hankel matrix $H_n$ is nonsingular and that the zeros of the polynomial $p_n(z)$ are distinct.

It is well established that the exponential interpolation problem can be severely ill conditioned, so any numerical method for the problem will yield inaccurate results for some input data sets. Much of the problem of condition can be attributed to the fact that the Hankel matrix $H_n$ is often nearly singular. However, the above description leaves open the method for obtaining the zeros of $p_n$. It is well known that since small changes in a polynomial's coefficients can lead to large changes in its zeros, so the power basis is often not the best way to represent the polynomial. Furthermore, one must specify the procedure for finding the zeros. We could find the eigenvalues of the Frobeneous matrix, which can also be a very ill-conditioned problem.

We describe below a procedure that is analogous with standard procedures for orthogonal polynomials on the real line. In particular, we construct a tridiagonal matrix whose eigenvalues are the zeros of 2. 5.

Note that the exponential interpolation reduces to a classical moment problem when $\lambda_j \in \mathbf{R}$ and $w_k > 0$. The techniques of constructing a discrete measure from given moments can be extended to our case: namely, when the Hankel matrix is formed by moments of an indefinite bilinear form, using algorithms described by Gragg [1].

If all the Hankel matrices $H_k, k = 1, \ldots, n$ are nonsingular, then there exists a unit lower triangular matrix $L_n$ such that $L_n H_n L_N^* =$ diag $[\sigma_j]_{j=0}^{n-1}$. Then the $k$th row of $L_n$ contains the coefficients of the monic *Lanczos polynomial* $p_k(z)$ of degree $k$ determined by $H_n$. the Lanczos polynomials are orthogonal polynomials when the bilinear form determined by $H_n$ is an inner product. Moreover, the Lanczos polynomials satisfy a three-term recurrence relation:

$$p_{k+1}(z) = (z - \alpha_{k+1})p_k(z) - \beta_k^2 p_{k-1}(z).$$

Thus, the Lanczos polynomials are the characteristic polynomials of the leading sections of the tridiagonal matrix

$$J_n = \begin{bmatrix} \alpha_1 & 1 & 0 & \cdots & 0 \\ \beta_1^2 & \alpha_2 & 1 & \ddots & \vdots \\ 0 & \beta_2^2 & \alpha_3 & \ddots & 0 \\ \vdots & \ddots & \ddots & \ddots & 1 \\ 0 & \cdots & 0 & \beta_{n-1}^2 & \alpha_n \end{bmatrix}.$$

If the Hankel matrix $H_n$ is real, then the $\alpha_k$ and $\beta_k^2$ are real, but $\beta_k$ may be real or imaginary. The recursion coefficients $\alpha_k$, $\beta_k^2$ as well as the matrix $L_n = [l_{j,k}]_{j,k=0}^{n-1}$ can be generated from the following algorithm [1].

**Algorithm 1.**

**input:** $\{\alpha_j\}_{j=0}^{2n-1}$ such that the Hankel matrices $H_m = [\alpha_{j-k}]_{j,k=0}^{m-1}$ is nonsingular for each $1 \leq m \leq n$.

**output:** the unit lower triangular matrix $L_n = [\mathcal{L}_{j,k}]_{j,k=0}^{n-1}$ and the recursion coefficients $\{a_j\}, \{b_j\}$.

$\sigma_{-1} = 1,\ , \tau_0 = 0,\ \ell_{00} = 1$
for $n = 0,\ 1, \cdots, N-1$

$$
\begin{array}{|l}
\sigma_n = \sum_{j=0}^{n} \ell_{nj}\alpha_{n+j}, \\
\tau_{n+1} = \left(\sum_{j=0}^{n} \ell_{nj}\alpha_{n+j+1}\right)/\sigma_n \\
b_n^2 = \sigma_n/\sigma_{n-1}. \\
a_{n+1} = \tau_{n+1} - \tau_n, \\
\ell_{n-1\,n} = \ell_{n\,-1} = 0, \\
\ell_{n+1\,n+1} = 1, \\
\text{for } j = 0,\ 1, \cdots, n \\
|\ell_{n+1\,j} = \ell_{n\,j-1} - a_{n+1}\ell_{nj} - b_n^2\ell_{n-1\,j}
\end{array}
$$

This algorithm allows us to construct the tridiagonal matrix $T_n$; standard software can then be used to calculate the eigenvalues and eigenvectors of $T_n$. It can then be shown that the weight vector $w$ is the componentwise product of the first row of the eignevector matrix $Z$ and the first column of $Z^{-1}$.

Consider again the dynamical system 1. 2, and assume that we know the output at $t = j, j = 0, 1, \ldots, N$. We ask for $e^{\lambda_j}$ and $w_j$ such that

$$ce^{At}b = \sum_{j=1}^{n} w_j e^{\lambda_j t}. \tag{2.7}$$

Assume that $x(0) = 0$, and choose $u(t) = e^{\alpha t}$. Further assume that $\alpha \neq \lambda_j$, for each $j$. Then from 1. 3 we obtain

$$y(t) = \sum_{j=1}^{n} \tilde{w}_j (e^{\alpha t} - e^{\lambda_j t})$$

where $\tilde{w}_j = \frac{w_j}{\alpha - \lambda_j}$. One way to identify the representation (2.5) from $\{y(j)\}$ is by solving the standard exponential interpolation problem

$$y(j) = \sum_{k=0}^{n} \tilde{w}_k e_{\lambda_k j} \tag{2.8}$$

If this problem is solved exactly, then $e^{\alpha t}$ will appear as one of the computed eigenvalues $e^{\lambda_0}$, and moreover, the corresponding coefficient will satisfy $w_0 = -\sum_{j=1}^{n} \tilde{w}_j$. These conditions can therefore be checked using the computed answer to (2.6); if the conditions are not satisfied to a certain degree of accuracy, that would indicate that one should not expect the other computed $\lambda_j$ to be accurate either. (just an indication; not a proof.)

Another approach to this problem is to first identify the eigenvalues $\{e^{\lambda_j}\}_{j=1}^{n}$ from the free response as described in the introduction. Then from equation (2.8) we have

$$y_s = (W_s - V_s)\tilde{w} \tag{2.9}$$

where

$$W_s = \begin{bmatrix} e^{\alpha s} & e^{\alpha s} & \cdots & e^{\alpha s} \\ e^{\alpha(s+1)} & e^{\alpha(s+1} & \cdots & e^{\alpha(s+1)} \\ e^{\alpha(s+2)} & e^{\alpha(s+2)} & \cdots & e^{\alpha(s+2)} \\ \vdots & \vdots & & \vdots \\ e^{\alpha(s+n-1)} & e^{\alpha(s+n-1)} & \cdots & e^{\alpha(s+n-1)} \end{bmatrix}$$

The equations (2.9) can be solved by solving a rank-one modification of a Vandermonde system, which can be achieved using well-known methods using $O(n^2)$ operations.

We have seen that problems of system identification are closely related with exponential interpolation problems. More general techniques must be used when the system matrix has multiple eigenvalues, [5].

The first author was supported in part by NSF Grant # DMS-8704196. The third author was supported in part by NSF Grant # ECS-8802483 and the last author was supported in part by NSA Grant # MDA904-85-H0009 and NASA Grant # NAG2-82.

## References

[1] Gragg, W., *Matrix interpretations and applications of the continued fraction algorithm,* Rocky Mountain Journal of Mathematics, Vol. 4, 1974, pp. 213-225.

[2] Gragg, W. and Lindquist, A., *On the partial realization problem,* Linear Algebra And Its Applications, Vol. 50, 1983, pp. 277-319.

[3] de Prony, R., *Essai expérimental et analytique,* J. École Polytech. (Paris), Vol. 1, 1795, pp. 24-76.

[4] Rissanen, J., *Recursive identification of linear sequences,* SIAM J. Cont., Vol. 9, 1971, pp. 420-430.

[5] Ammar, G., Dayawansa, W., and Martin, C., *Exponential interpolation: theory and numerical algorithms,* preprint.

# On observability of chaotic systems: an example

Giovanni B. Di Masi
Andrea Gombani

**Abstract:** the concept of observability of a special chaotic system, namely the dyadic map, is studied here in case the observation is not exact. The usual concept of observable subspace does not distinguish among the behaviour of different models. It turns out that a suitable extension of this concept can be obtained using the idea of Hausdorff dimension. It is shown that this dimension increases as the observation error becomes smaller, and is equal to one only if the system is observable.

## §1. Introduction

The study of nonlinear dynamical systems has recently attracted the attention of an increasing group of scientists involved in theoretical as well as in applicative fields. In particular there has been a growing interest for chaotic systems since it has been recognized that chaotic and random behaviour of solutions of deterministic systems is an inherent feature of many physical and engineering phenomena.

Since a possible characterization of chaos is that, under a suitable observation mechanism, the output of the system behaves as a purely nondeterministic process, it is of interest to study the observability properties of such systems. Results in this direction can be found in [1],[4],[8].

In this paper we examine the observability properties of a simple chaotic system described by the dyadic map, whose dynamic behaviour can be effectively characterized in terms of symbolic dynamics.

It turns out that a natural extension of the concept of dimension of the observability space for linear systems can be given in terms of Hausdorff dimension of the observable set. The tool of Hausdorff dimension has been used in investigations on chaotic systems in connection with the study of the dimension of strange attractors [6].

## §2. The problem

Let I be the unit interval. By *chaotic system* it is usually meant a map f: I $\rightarrow$ I with the following properties [5]:

1. f has sensitive dependence on the initial conditions, i.e. there exists a $\delta > 0$ s.t. for each $x,y \in I$ there exists $n \in \mathbf{N}$, s.t. $|f^n(x) - f^n(y)| > \delta$.

2. periodic points are dense in I.

3. f is topologically transitive, i.e. for any pair of open sets $U,V \subset I$, there exist $k > 0$ s.t. $f^k(U) \cap V \neq \emptyset$.

It is fairly straightforward to check that the dyadic (figure 1)

$$f(x) = \begin{cases} 2x & \text{for } 0 \leq x < 1/2 \\ 2x - 1 & \text{for } 1/2 \leq x \leq 1 \end{cases} \tag{1}$$

satisfies the above requirements.

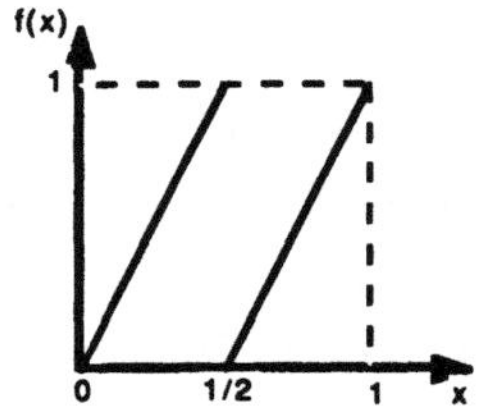


figure 1

There is another way to describe this model which is often used, called *symbolic dynamics* (see [5], [6]). By this term it is meant a representation of f in terms of a shift on a set of binary sequences. Define the set of binary sequences

$$\Sigma_2 := \{s = (s_0 s_1 s_2 \ldots) \mid s_j = 0 \text{ or } 1\}$$

endowed with the following metric:

$$d(s,t) = \sum_{i=0}^{\infty} \frac{|s_i - t_i|}{2^i}$$

On $\Sigma_2$ define a shift as follows

$$U(s_0 s_1 s_2 \ldots) = (s_1 s_2 s_3 \ldots)$$

Denoting by $\pi$ the map from I onto $\Sigma_2$ which associates to a real number its nonterminating binary representation, it is clear that the following diagram commutes:

$$\begin{array}{ccc} I & \xrightarrow{\ f\ } & I \\ \pi \big\downarrow & & \big\downarrow \pi \\ \Sigma & \xrightarrow{\ U\ } & \Sigma \end{array}$$

By *dynamical system* we mean here a mathematical model evolving in time whose trajectories $\{x(t), y(t)\}_{t\in \mathbf{N}}$ (behaviours, see [7]) admit a representation through a pair of functions (F,H) on a suitable domain

$$\begin{cases} x(t+1) & = F(x(t)) \\ y(t) & = H(x(t)) \end{cases} \tag{2}$$

The variables x and y are called state and observation of the system. The above system is said to be *observable at time t* if there exists an injection form the range of x(t) into the cartesian product of the observation up to time t $\{y(1), y(2),\ldots, y(t)\}$. The system is *observable* if there exists a $t_0$ (possibly infinite) such that the system is observable at each $t \geq t_0$. Since F is a deterministic function, in this context, the system is observable as soon as the initial condition can be determined exactly. More generally, even for an unobservable system, we shall say that an initial condition *x is observable* if it is uniquely determined by the observation of the corresponding trajectory.

Consider now a particular example of (2), where F is the dyadic map (1) and H is the following two state observation function

$$h_0(x) = \begin{cases} 0 & \text{for } 0 \leq x < 1/2 \\ 1 & \text{for } 1/2 \leq x \leq 1 \end{cases} \tag{3}$$

We get the following dynamical system

$$(S_0) \quad \begin{cases} x(t+1) = f(x(t)) \\ y(t) = h_0(x(t)) \end{cases} \tag{4}$$

It is easy to see that this model is observable over an infinite interval of time, i.e. the initial condition (and hence the whole trajectory) is determined uniquely by the infinite observation of the system (to this end take the binary representation of an initial state $x_0$: this will coincide with the history of the observations). One reason why this model is so interesting is that, in spite of its complete observability, any observation over a finite time interval is indistinguishable from the outcome of a coin tossing (see[6]). Another reason is that it introduces the *symbolic dynamics* in very natural way. In fact, the history of the observations of $f^n(x)$ under h is precisely $\pi(x)$. Observability here depends on the fact that the inverse images under $h_0$ of the states 0 and 1 coincide exactly with the two intervals [0,1/2), [1/2,1) (these are called Markov partitions of f, see [6]). For this reason we call this observation exact. The problem we want to consider now is the following: suppose our observation function does not distinguish exactly between the two intervals, but contains some error $\varepsilon > 0$ as follows:

$$h_\varepsilon(x) = \begin{cases} 0 & \text{for } 0 \le x < 1/2 + \varepsilon \\ 1 & \text{for } 1/2 + \varepsilon \le x \le 1 \end{cases} \tag{5}$$

It can be seen that, for example, the initial conditions x and x + 1/2 are indistinguishable for $x \in [0,\varepsilon)$. We give a precise characterization of this concept in theorem 1 below. However, if $\varepsilon < \eta$, also $h_\varepsilon$ yields a better observation than $h_\eta$ in the sense that more points are distinguishable.The questions we try to answer in this paper are the following:

Does there exist a way to define observability of $S_\varepsilon$ such that:

a) this definition generalizes the usual observability concept for dynamical systems.
b) the function which measures observability of $S_\varepsilon$ is decreasing in $\varepsilon$.

It should be noted that the first thing one would try, namely the measure of the observable set, fails, as shown in Proposition 1 below.

We will study the special case $\varepsilon = 2^{-n}$, and we write, by abuse of notation, with $S_n$, $h_n$, instead of $S_{2^{-n}}$, $h_{2^{-n}}$. We define $\Omega_n$ to be the set of observable points of $S_n$. By the notation $\underbrace{0...0}_{k}$ we mean a sequence of exactly k zeros.

**Theorem 1**: *$\Omega_n$ is the set of points of I whose binary representation has the following properties:*

*a) the sequence $01\underbrace{0...0}_{k}1$ never occurs for $k > n-2$*
*b) the sequence $11\underbrace{0...0}_{k}1$ never occurs for $k > n-1$.*

For the proof we need two lemmas. Define by $Y_n$ the set of trajectories of $S_n$:

$$Y_n := \{s \in \Sigma_2 : s = \{h_n(f^t(x))\}_{t \in N} \text{ for some } x \in [0,1)\}$$

**Lemma 1**: *the set $Y_n$ consists exactly of the points of $\Sigma_2$ in which the sequence $\{1\underbrace{0...0}_{n-1}1\}$ never appears.*

**Proof:** if $\pi(x)$ never has not more than n-2 zeros in a row, then $h(f^t(x)) = h_n(f^t(x))$, and the number of zeros is preserved. If $\pi(x)$ has more than n-1 zeros, the 1 preceeding the zeros is set to zero by $h_n$, and so the output sequence will have at least n zeros. Therefore the sequence $1\underbrace{0...0}_{n-1}1$ can never occur as the output of $S_n$.

**Lemma 2:** *the sequences of $Y_n$ generated by observable points are those which have at most n consecutive zeros.*

**Proof:** if a sequence $s = \{h_n(f^t(x))\}_{t\in N}$ has less than n consecutive zeros, then in view of Lemma 1 it has at most n-2 zeros and thus $h_n(f^t(x)) = h(f^t(x))$. If s has exactly n zeros, than it is seen by inspection that it can only be generated by a sequence $\{11\underbrace{0...0}_{n-1}1\}$. If s has r consecutive zeros, $r > n$, then both the element

$$t_1 = \{11\underbrace{0...0}_{r-1}1\}$$
$$t_2 = \{101\underbrace{0...0}_{r-2}1\}$$

yield the same output and thus the trajectory does not determine uniquely the initial point. ■

**Proof of theorem 1:** to characterize the observable points of [0,1), observe first that a point x for which $\pi(x)$ has a subsequence of n or more consecutive zeros is not observable. In fact, in this case, $s = \{h_n(f^t(x))\}_{t\in N}$ will have at least n+1 zeros, and in view of Lemma 2 this trajectory is generated by more than one point. If there are less than n-1 consecutive zeros, then $h_n(f^t(x)) = h(f^t(x))$ for all t, and the point is observable. If a subsequence with exactly n-1 zeros occurs, there are two possibilities:

a) the subsequence is of the form $01\underbrace{0...0}_{n-1}1$ and it is thus indistinguishable either from $1\underbrace{0...0}_{n}1$ (if there are less than n-1 zeros before the preceeding 1) or from $\underbrace{0...0}_{n+1}1$ (if there are more than n-2 zeros before the preceeding 1).

b) the subsequence is of the form $11\underbrace{0...0}_{n-1}1$. Then the image of the subsequence is $1\underbrace{0...0}_{n}1$, which in view of lemma 2 comes from an observable point. ■

**Corollary:** $\Omega_n \subset \Omega_{n+1}$ . $\qquad 11\underbrace{0...0}_{n-1}1$

**Proposition 1:** *Let $\Omega_n$ be the set of observable points for $S_n$ Then,*

$$\mu(\Omega_n) = 0 \tag{7}$$

*where $\mu$ denotes the Lebesgue measure.*

**Proof:** from theorem 1, $\Omega_n$ is the set of points x such that in $\pi(x)$ some sequences never occur. In view of the Borel-Cantelli lemma the measure of this set is zero. ■

This proposition says that the system $S_0$ is very special with respect to the Lebesgue measure, because it is the only one whose observable set has measure one.

It turns out that a reasonable tool to characterize the magnitude of the observable set is the Hausdorff dimension, as we show below.

## §3. Main result

We are going now to define the Hausdorff dimension of a metric space X. The diameter of a set U of X, is defined as

$$\text{diam}(U) = \sup\{d(x,y) : x,y \in U\}$$

Given $\delta > 0$ we denote by $\mathcal{U}_\delta$ a cover of X such that diam (U) < $\delta$ for all U in $\mathcal{U}_\delta$.

**Definition:** *the Hausdorff dimension of a metric space X is*

$$HD(X) = \inf\left\{\alpha\colon \forall \varepsilon > 0, \exists \text{ a cover } \mathcal{U}_\delta \text{ of } X \text{ s.t. } \sum_{U \in \mathcal{U}_\delta} (\text{diam}\,(U)^\alpha) < \varepsilon\right\} \quad (8)$$

The Hausdorff dimension has several interesting properties (see [3]) :

$$HD(X) \leq HD(X') \quad \text{if} \quad X \subset X' \quad (9a)$$

$$HD(X) = n \quad \text{for } X \subset \mathbb{R}^n \text{ if } \mu(X) > 0 \quad (9b)$$

$$HD(\cup_n X_n) = \sup HD(X_n) \quad (9c)$$

The Hausdorff dimension is equal to the usual dimension in the case of a linear space or of a smooth manifold. As a consequence, we have the following:

**Proposition 2:** *let S be a linear dynamical system with observable space of dimension n. Then also the Hausdorff dimension of this space is n.*

So the Hausdorff dimension is equivalent to the usual one in all classical cases. In general, though, it is a rather difficult object to compute whenever it does not coincide with the usual notion of dimension. Its interest for our application lies in the fact that $S_n$ is not a classical case, but $HD(\Omega_n)$ is still quite easy to compute. Denote, as above, by $\Omega_n$ the observable set of $S_n$

We are now ready to compute the Hausdorff dimension of $\Omega_n$.

**Theorem 2:** $HD(\Omega_n) = \dfrac{2^{n+1} - 6}{2^{n+1} - 3}$

**Proof**: we need first the following result (see [3], Theorem 14.1). Let $u_k(x)$ be the subinterval of I of the form $\left[\frac{j}{2^k}, \frac{j+1}{2^k}\right)$ containing x. Let $\nu$ be a probability measures on I such that $\nu(X) = 1$, and let $\mu$ denote the Lebesgue measure. If

$$X \subset \left\{ x \; : \; \lim_{k \to \infty} \frac{\log \nu(u_k(x))}{\log \mu(u_k(x))} = \delta \right\} \tag{10}$$

then HD (X) = δ. The problem thus becomes to construct the measure ν on the set $\Omega_n$. A standard procedure (see [3], p.143), is to use the measure induced by a Markov chain whose trajectories belong almost surely to $\Omega_n$. Denoting by $p_{ij}$ the transition probabilities and by $p_i$ the invariant measure, it is easily seen that

$$\lim_{k \to \infty} \frac{\log \nu(u_k(x))}{\log \mu(u_k(x))} = -\frac{1}{\log 2} \sum_{i,j=1}^{n} p_i p_{ij} \log p_{ij} \tag{11}$$

In our case, $\Omega_n$ is the set of all numbers whose binary expansion never has the sequences $01\underbrace{0...0}_{k}1$ for k >n-2 and the sequence $11\underbrace{0...0}_{k}1$ for k >n-1. We now construct the Markov chain z(t) as follows: if the first two digits of x(t) are 01 followed by i-1 zeros, set z(t) = i for i = {1,..., n-1}. If x(t) terminates with exactly i zeros preceeded by 11 then set z(t) = n+i. it is easy to see that z(t) has transition probabilities

$$[p_{ij}] = \begin{cases} 1/2 & \text{for } j=i+1, i\neq n-1,\ 2n-1 \\ & \text{for } i=j=n \text{ and } i=1,\ j=n \\ & \text{for } i=2,\dots,n-2,n+1,\dots,2n-2 \text{ and } j=1 \\ 1 & \text{for } i=n-1,\ i=2n-1, j=1 \text{ and } j=1 \\ 0 & \text{otherwise} \end{cases} \tag{12}$$

The invariant measure for $[p_{ij}]$ is seen to be

$$p_1 = p_n = \frac{2^{n-1}}{2^{n+1}-3}$$

$$p_i = \frac{2^{n-i}}{2^{n+1}-3} \qquad i = 2,\dots, n-1$$

$$p_i = \frac{2^{2n-i-1}}{2^{n+1}-3} \qquad i = n+1,\dots, 2n-1$$

a simple substitution in (11) yields the result.

We still need to justify the choice of (12). It is easy to see that, when we condition on $\{x \in \Omega_n\}$, the probability measure induced by the Lebesgue measure is exactly the measure induced by the Markov chain (12). To see that this conditional probability is indeed the one with support $\Omega_n$, we refer the reader to the original paper of Billingsley [2]. ■

Another and perheaps more natural way to look at the observability problem is the one concerned with the set of possible output trajectories, $Y_n$. This set of binary strings can be inbedded in [0,1) in an obvious way, and we can thus define, with abuse of notation, the Hausdorff dimension of $Y_n$. In a fashion completely similar to that of theorem 2 we can prove the following

***Theorem 3****: the Hausdorff dimension of* $Y_n$ *is* $\frac{2^{n+1}}{2^{n+1}+1}$

We would like to remark that the dimensions computed in theorems 2 and 3 converge to 1 as n goes to infinity, yielding thus that consistency which was seeked in the beginning of the paper.

## §4. Conclusions.

We have presented an example where the definition of dimension of the observable subspace of a dynamical system is extended to the case of noninteger numbers. We conjecture that this procedure can be generalized to a system of the form (2) whenever the function F admits a Markov partition on its domain and H takes only finite values. This problem is currently being investigated by the authors.

**Acknowledgments**: it is a pleasure to thank Professor Tom Taylor for drawing our attention on this subject, and for the useful discussions we had together.

This research took parly place while the second author was visiting the Division of Systems and Decision Sciences, IIASA, Laxenburg, Austria.

## References

[1] Aeyels D., Generic Observability of Differentiable Systems, *SIAM J. Control and Optimization,* Vol 19, 1981, 595-603.

[2] Billingsley P., Hausdorff dimension in probability theory II, *Illinois Journal of Mathematics,* 5,1961, 291-298.

[3] Billingsley P., *Ergodic theory and information,* Wiley, 1965.

[4] Drager L., Martin C., Global observability of a class of nonlinear discrete time systems, *Systems and Control Letters,* 6, 1985,65-68.

[5] Devaney R.L., *Chaotic Dynamical systems,* Benjamin/Cummings 1986.

[6] Guckenheimer J., Holmes P., *Nonlinear Oscillations, Dynamical Systems, and Bifurcation of Vector Fields,* Springer-Verlag, 1983.

[7] Willems J.C., From time series to linear systems, Part I. Approximate modelling. *Automatica,* Vol 22, N.5, 1986, 561-580.

[8] Taylor T.J.S., On observation of chaotic systems and randomness, Proc. MNTS 1987.

# INTERPOLATING UNIQUELY WITH ONLY A FINITE CLASS OF POLYNOMIALS

A. DeStefano
P. Kostelec
D.I. Wallace

**Abstract**

This paper draws from theorems in transcendental number theory to answer questions about interpolation with a finite class of multivariable polynomials. In particular we describe sets of data points at which a unique polynomial from a particular class gives us a good approximation of the output.

In this paper, we consider the problem of approximating a function using a class of simpler functions. Our point of view is that we wish to recover a function $f$ from partial information about it given at a set of data points in the domain of $f$.

Our motivation comes from studying observability of dynamical systems. In particular, we are interested in the question: Does sampling preserve observability? By sampling, we mean obtaining measurements of output from experiments conducted at a finite set of data points. Using the output obtained, we would like to recover the particular solution to the dynamical system (or equivalently to recover the initial data). If this can be done, we say that sampling ***preserves observability*** or that the system is ***discretely observable.***

The case of a linear dynamical system in one variable is considered by Martin and Smith and the problem of whether the dynamical system is discretely observable is shown to be equivalent to an interpolation problem. We also take this point of view.

Much work has been done in approximation theory on finding methods of interpolation and theorems on existence and uniqueness of solutions to interpolation problems for particular classes of functions. Classical polynomial interpolation in one variable of a finite set of data points has a unique solution (Davis, [2]). However, we have no such theorem for multivariable polynomial interpolation. There has been work done on which data points can be chosen to give a unique solution (Chui and Lai, [1]).

Here we are not so much interested in how to obtain a solution but rather if we have a solution we would like to be able to determine whether it is unique. We will restrict our class of interpolating functions to a finite set of polynomials.

By a solution to an interpolation problem we really mean a "good" approximation, since with only a finite set of polynomials from which to choose we cannot hope for an exact solution. We now state the problem in more precise terms.

Let

$$\begin{gathered} x_{11}, \cdots, x_{1s} \\ x_{21}, \cdots, x_{2s} \\ \vdots \\ x_{r1}, \cdots, x_{rs} \end{gathered}$$

be $r$ data points in $\mathbf{R}^s$.

Let $y_1, \cdots, y_r$ be the output obtained from some experimental measure at the data points. There are two problems to consider.

1. If we choose a finite class of polynomials $\mathcal{P}$, can we find $\epsilon > 0$ such that there exists a unique solution $P \in \mathcal{P}$ such that

$$|P(x_{i1} \ldots, x_{is}) - y_i| < \epsilon$$

$$\text{for } i = 1, 2, \ldots, r,$$

and on what does $\epsilon$ depend?

2. If we choose a finite class of polynomials $\mathcal{P}$ and an $\epsilon > 0$, at which data points should we test so that

$$|P(x_{i1}, \ldots, x_{is}) - y_i| < \epsilon$$

$$\text{for } i = 1, 2, \ldots, r\ ?$$

We should note that in particular applications all measurements are rational numbers. Also, since the rational numbers are dense in $\mathbf{R}$, we will limit our finite class of polynomials to have rational coefficients. This is justified since for practical purposes, say using a computer, all numbers are rational.

We have stated the general problem and will next illustrate how some results of A. O. Gelfond help us to attack this problem. But prior to stating our first lemma, whose proof appears in Gelfond, [3], a few definitions are in order.

Let $\zeta$ be a number in an algebraic field $K$ of degree $n$ over the rationals, and let $\omega_1, \ldots, \omega_n$ denote the basis for the ring of integers in this field. Now the number $\zeta$ can be represented in an infinite number of ways, in the form $\frac{p_1\omega_1 + \cdots + p_n\omega_n}{q_1\omega_1 + \cdots + q_n\omega_n}$, where $p_1, \ldots, p_n, q_1, \ldots, q_n$ are rational integers with greatest common divisor 1. Let $\phi(p_1, \ldots, q_n) = \max[\,|p_1|, \ldots, |q_n|\,]$. Then,

**Definition 1.** The *measure* of $\zeta$, denoted by $m(\zeta)$, is equal to $\min \phi(p_1, \ldots, q_n)$, where the minimum is taken over all possible representations of $\zeta$.

So, for example, if $\zeta = \frac{a}{b}$ is a rational number, and $a$ and $b$ are relatively prime, then the measure of $\zeta$ is $\max[\,|a|, |b|\,]$.

**Definition 2.** Let $P(x_1, \ldots, x_s)$ be a polynomial, in $s$ variables, with rational integral coefficients. Then the *height of* $P(x_1, \ldots, x_s)$, denoted by $H(P(x_1, \ldots, x_s))$, is the maximum absolute value of its coefficients.

With these two definitions in hand, we can now proceed to the statement of Lemma 1.

**Lemma 1.** (Gelfond): Suppose $K$ is a field of degree $\nu$ over the rationals, and suppose further that $P(x_1, \ldots, x_s)$ is a polynomial with integral coefficients, of degree $n_i$ in $x_i$, $1 \le i \le s$ and $n = \sum_{i=1}^{s} n_i$. Now fix a basis for the ring of integers in $K$, and pick algebraic numbers $\alpha_1, \ldots, \alpha_s$ from $K$. Let $q_i = m(\alpha_i)$, for $i = 1, \ldots, s$. Then either

$$P(\alpha_1, \ldots, \alpha_s) = 0$$

or

$$|P(\alpha_1, \ldots, \alpha_s)| > H^{-\nu+1} \prod_{i=1}^{s} q_i^{-\nu n_i} e^{-\gamma n},$$

where $H =$ the height of $P(x_1, \ldots, x_s)$, and $\gamma$ is a constant which depends only on $K$ and the basis chosen.

Notice that, once a basis for the ring of integers in $K$ is fixed, an estimate as to the size of the constant $\gamma$ in Lemma 1 can be made. Elementary, though tedious, manipulations will yield that $n\gamma < n\nu s(\log \omega) + \nu s \log n$, and $\gamma < \nu s(\log \omega + 1)$, where $\nu$ is the degree of the number field, and $\omega = max_{1 \le i \le s}\{\ |\omega_i|\ \}$.

So what does Lemma 1 mean? Basically, it states that either by a stroke of luck $P(\alpha_1, \ldots, \alpha_s) = 0$, or else $|P(\alpha_1, \ldots, \alpha_s)|$ can be no smaller than a particular value, determined by the height of the polynomial, the integral basis chosen, the degree of the field, and the measures of the algebraic numbers chosen.

An obvious corollary can be obtained from Lemma 1:

**Corollary 1.** Let $P(x_1, \ldots, x_s)$ be as in Lemma 1, except that this polynomial has coefficients which are rational numbers with denominator $D$. $H$ is still the height of $P(x_1, \ldots, x_s)$. Then, assuming the hypothesis of Lemma 1, either

$$P(\alpha_1, \ldots, \alpha_s) = 0$$

or

$$|P(\alpha_1, \ldots, \alpha_s)| > D^{-\nu} H^{-\nu+1} \prod_{i=1}^{s} q_i^{-\nu n_i} e^{-\gamma n}.$$

Now we will apply these results to an interpolation problem. For another example of similar flavor, see Martin and Wallace, [5].

Let $\mathcal{P}$ be a finite set of polynomials over $\mathbb{Z}$ (or $\mathbb{Q}$). We will use Gelfond's lemma to see where two elements can approximate the same output data. To be precise, suppose $\mathcal{P}$ consists of polynomials in $s$ variables, $x_1, \ldots, x_s$ and let $y_1, \ldots y_r$ be a set of output data. We wish to fit a polynomial $P(x_1, \ldots, x_s)$ to the data so that

$$P(x_{i1}, \ldots, x_{is}) = y_i \quad , \forall_i .$$

Of course, since $\mathcal{P}$ is a finite set we can only hope to approximate the data so that for some $\epsilon > 0$,

$$|P(x_{i1}, \ldots, x_{is}) - y_i| < \epsilon/2 \quad , \forall_i .$$

Begging the question for the moment of whether we can, in fact, find such a $P \in \mathcal{P}$ for all possible output vectors, it is natural to ask how we can avoid finding two polynomials, $P \neq Q$, both of which lie in an $\epsilon/2$ ball around the output data. Obviously this is equivalent to asking when

$$|P - Q| = |R(x_{i1}, \ldots, x_{is}) - 0| < \epsilon.$$

In other words, if we look at the family $\mathcal{P}'$ (henceforth referred to as the "derived family of $\mathcal{P}$") of differences between members of $\mathcal{P}$, we are asking when the elements of $\mathcal{P}'$ are bounded away from zero at the test points. To implement the corollary to Lemma 1, we will choose our $x_{ij}$ to lie in some number field of degree $\nu$ over $\mathbb{Q}$. The right hand side of the inequality in Lemma 1 will play the role of $\epsilon$. To get a lower bound on the quantity

$$D^{-\nu} H^{-\nu+1} \left[ \prod_{i=1}^{s} q_i^{-\nu n_i} \right] e^{-\gamma n},$$

we must bound the quantities $H, q_i, \nu, n_i$ from above. By the note after the lemma,

$$\gamma < \nu s(\log \omega + 1)$$

where $\nu$ is the degree of the number field, and $\omega$ is the largest absolute value taken over all the basis elements in some fixed integral basis. Also

$$n\gamma < n\nu s(\log \omega) + \nu s(\log n)$$

which was an intermediate (and better) estimate. We then get the following corollary to Lemma 1.

**Corollary 2.** Let $\mathcal{P}$ be a family of polynomials in $s$ variables of total degree $\leq n$ whose coefficients are rational numbers with denominator $D$ and whose heights are bounded by $H/2$. Let $x_{i1}, \ldots, x_{is}, 1 \leq i \leq r$ be sample points with outputs $y_i$, such that the $x_{ij}$ are chosen from some number field $K$ of degree $\nu$ so that the measure of $x_{ij} \leq q$ with respect to some fixed integral basis. Then for any $R \in \mathcal{P}'$ for any $i$, either

$$R(x_{i1}, \ldots, x_{is}) = 0$$

or

$$|R(x_{i1}, \ldots, x_{is})| \geq D^{-\nu} H^{-\nu+1} q^{-\nu n} e^{-\gamma\nu},$$

for

$$\gamma < \nu s(\log \omega + 1).$$

Next we want to pick $x_{ij}$ to rule out the possibility that $R = 0$. Having bounded the degree of the polynomials in $\mathcal{P}$, let us also bound the degree of $x_{is}$ by $n_j$. Then we must pick $s$ different number fields with the $j^{th}$ number field of degree $\geq n_j$. Furthermore, we want the elements $x_{ij}$ (for $j$ fixed) to always be in an extension of degree $\geq n_j$ over the other fields. (For example, we throw out rational entries). Then we can guarantee that $R \neq 0$.

**Corollary 3.** Same hypothesis as in Corollary 2 except that, in addition, the $j^{th}$ coordinate of the sample point, $x_{ij}$, shall be chosen from some field $K_j$ and shall have degree at least $n_j$ over the remaining fields. Then

$$|R(x_{i1}, \ldots, x_{is})| \geq D^{-\nu} H^{-\nu+1} q^{-\nu n} e^{-\gamma\nu}.$$

**Example.** Suppose $\mathcal{P}$ consists of polynomials of the form

$$P(x_1, x_2) = ax_1x_2 + bx_1 + cx_2 + d.$$

Suppose we take $H \leq 1$ and $D = 2$, so that $a, b, c, d$ are of the form $\pm\frac{1}{2}, \pm 1$ or $0$. There are $5^4$ such polynomials and they are bounded near the origin, which constrains our outputs, $y_i$, a lot. Now suppose we choose $x_1$ from $\mathbf{Q}(\sqrt{2})$ and $x_2$ from $\mathbf{Q}(\sqrt{3}) - \mathbf{Q}$. Suppose also they have measure $\leq q$ and we have chosen the obvious bases so that $\omega \leq \sqrt{3}$, where $\omega$ is defined as before. Any difference of two of these polynomials has height $\pm 2$, and Corollary 3 yields

$$|P - Q| \geq 2^{-4} q^{-2\cdot 2} e^{-\gamma\cdot 2}, \ \gamma < 2 \cdot 2 \cdot (\log \sqrt{3} + 1).$$

Setting $q = 2$ we have

$$|P - Q| \geq 2^{-8} e^{-8(\log \sqrt{3}+1)}.$$

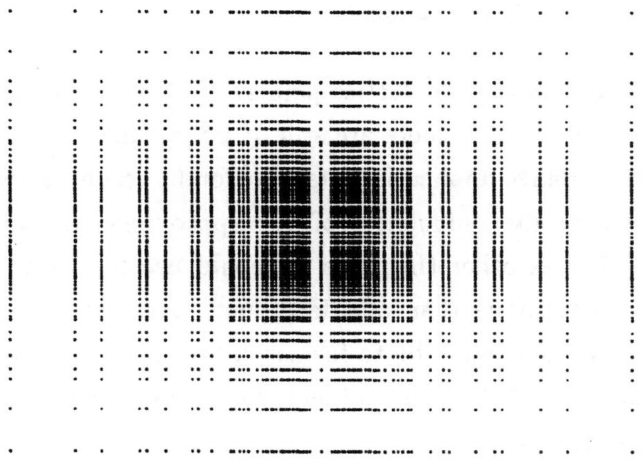

FIGURE 1. Subset of $\mathbf{Q}(\sqrt{2}) \times \mathbf{Q}(\sqrt{3})$

If this seems small, remember that we have chosen on the order of $5^8$ points of which we will take a subset. Now obviously it is not going to be possible to approximate outputs on any subset of $5^8$ points by only $5^4$ polynomials, so we have chosen our set of sample points too big.

Let us instead choose $x_{i1}$ to be rational and $x_{i2}$ from $\mathbf{Q}(\sqrt{2}) - \mathbf{Q}$. Then the total degree of our field is $\nu = 2$ and the largest integer in the basis is $\sqrt{2}$. We then get (for $q = 2$)

$$\begin{aligned}\gamma &< 2 \cdot 2 \cdot (\log\sqrt{2} + 1),\\ |R| = |P - Q| &\geq 2^{-2} \cdot 1 \cdot 2^{-4} e^{-\gamma \cdot 2}\\ &> 2^{-8} e^{-8(\log\sqrt{2}+1)}\\ &= 2^{-8} e^{-8} 2^{-4}.\end{aligned}$$

Now, with these choices we have about $5^2 \cdot 5^3$ lattice points.

Setting $q = 1$ we get

$$\begin{aligned}|R| = |P - Q| &\geq 2^{-2} 1^{-1} 1^{-4} e^{-4(\log\sqrt{2}+1)}\\ &\geq 2^{-2} \cdot 2^{-2} e^{-4}\\ &\geq 2^{-4} e^{-4}\\ &\geq 5^{-4} = .0016\end{aligned}$$

whereas there are about $2 \cdot 3^3$ lattice points, all of which are on one of three lines $x_{i1} - 1, 0$ or $-1$, and there are still $5^4$ polynomials to use.

As we have applied it so far, Gelfond's results allow us to make claims concerning the uniqueness of our approximating function at one data point. He has other theorems we could use to study simultaneous approximations for more than one data point. Also we must consider the question of how big $\mathcal{P}$ must be to furnish a "good" approximation for any data set of size $n$. All these are subjects for future research.

References

[1] Chui, Charles and Lai, Hang-Chin, "Vandermonde determinant and Lagrange interpolation in $\mathbf{R}^s$", *Lecture Notes in Pure and Applied Mathematics*, **107** 23-35, 1987.

[2] Davis, P., *Interpolation and Approximation*, Blaisdell Publishing Company, New York, 1963.

[3] Gelfond, A. O., *Transcendental and Algebraic Numbers*, Dover Publications, New York, 1960.

[4] Martin, Clyde and Smith, Jennifer, "Approximation, Interpolation and Sampling", *Contemporary Mathematics*, **68**, 227-252, 1987.

[5] Martin, C. F. and Wallace, D. I. , "Observability and Transcendental Number Theory", to appear.

# SINC APPROXIMATION METHOD FOR COEFFICIENT IDENTIFICATION IN PARABOLIC SYSTEMS

John Lund

Abstract

A parabolic partial differential equation is discretized using sinc expansion in both the spatial and temporal domains. The resulting Sinc-Galerkin scheme is illustrated in the solution of a (singular) forward problem and a parameter identification problem.

## 1. Introduction and Summary

A fully Galerkin scheme for the numerical solution of the parameter identification problem

$$
\begin{aligned}
&L(p)u \equiv \frac{\partial u}{\partial t} - \frac{\partial}{\partial x}\left(p(x)\,\frac{\partial u}{\partial x}\right) = r(x,t)\ ;\ x \in (0,1)\ ,\ t > 0 \\
&u(0,t) = u(1,t) = 0\ ,\ t > 0 \\
&u(x,0) = 0\ \ ,\ \ x \in (0,1)
\end{aligned}
\tag{1.1}
$$

was developed in [4]. Here parameter identification will mean that with given observations of $u$ and assuming a known input $r$ then one is to determine the system parameter $p$. The scheme in [4] is based on tensor product approximations by splines (in the spatial domain) and by sincs (in the temporal domain). In the case of singular problems a recovery scheme based on spline approximation could be enhanced by switching to a spatial base that is more amenable to the approximation of singular functions. This is a propitious property of sinc interpolation — its persistent exponential convergence in the presence of singularities.

If the solution of the forward problem (1.1) is denoted by

$$
u \equiv L(p)^{-1} r \tag{1.2}
$$

and point evaluations of this solution at $p_i$ in $(0,1) \times (0,\infty)$ are given by

$$[K_m u]_i = u(p_i) \tag{1.3}$$

then the solution of the inverse problem is given by

$$F(p) = K_m \big(L(p)\big)^{-1} r \quad . \tag{1.4}$$

If the solution has been sampled at $m$ data points $d^T = (d_1, d_2, \ldots, d_m)$ then the recovery of $p$ is obtained by solving the minimization problem

$$\min_p \left\{ \frac{1}{2} \|F(p) - d\|^2 \right\} \tag{1.5}$$

where the two-norm minimization is over an appropriate parameter space.

The present method of discretization of the forward problem is somewhat atypical in that the approximation is fully Galerkin (in contrast to a Galerkin spatial expansion followed by a time-stepping ordinary differential equation solver). The evaluation of the system (1.1) as well as a closely allied partial differential equation is required in the numerical optimization problem (1.5) (the former for the evaluation of $F$ and the latter for its Jacobian). The sinc basis used in the present development provides a very accurate forward solver in both problems. Whereas this technique requires an additional dimension of complexity it bypasses the ordinary differential equation stage of the solution phase in the time-stepping methods. A precise measurement of the comparative efficiency between these methods remains to be studied.

In Section 2 the full discretization of (1.1) is carried out for arbitrary spatial and temporal basis functions. The solution method of the forward problem (1.1) via partial temporal diagonalization of the system is recorded. The final form of the discrete system is then in block diagonal form which may be solved by block Gauss Elimination methods (vector architectures [2]). A brief review of the sinc basis is given in the next section with some detail for the sinc-spatial approximation of the term

$$S(p)u \equiv -\frac{d}{dx}\left(p(x)\frac{du}{dx}\right) \quad . \tag{1.6}$$

The emphasis here is on this discretization since it proceeds somewhat differently than the procedures in [2], [3] or [5]. That is, the procedure is carried out in a fashion which avoids the differentiation of the parameter

$p$. As this discretization has never heretofor received numerical testing the example closing Section 3 is used to illustrate the method's viability in the approximation of a forward problem with a singular solution. For the same problem, this forward solution method is used for the recovery of the parameter $p$ in (1.1) by the method outlined in [4] which is based on the minimization problem in (1.5).

The results of the computations in Section 3 illustrate the numerical potential of this fully Sinc-Galerkin methodology for the identification problem. A number of issues remain besides the previously mentioned numerical comparison. Perhaps the most important issue is the formulation of an analytic framework within which is housed the convergence criteria of the approximate parameter $p$ to the true parameter. Indeed, a framework proceeding along the lines of the very nice development found in [1] would analytically support the numerics of the present paper.

## 2. Discretization of the Forward Problem

In a fully Galerkin approach to (1.1) with basis elements $X_k(x)$ and $T_\ell(t)$ an assumed solution of (1.1) takes the form with $q = x$ or $t$

$$\hat{u}(x,t) = \sum_{k=-M_x}^{N_x} \sum_{\ell=-M_t}^{N_t} u_{k\ell} X_k(x) T_\ell(t) \quad , \quad m_q = M_q N_q \quad . \tag{2.1}$$

The coefficients $\{u_{k\ell}\}_{k,\ell}$ are obtained by orthogonalizing the residual

$$R(p) \equiv L(p)\hat{u} - r \tag{2.2}$$

with respect to the basis elements $X_j(x)T_i(t)$ in the inner product

$$< f, g > \equiv \int_0^\infty \int_0^1 f(x,t) g(x,t) v(x) w(t) dx dt \tag{2.3}$$

where the weight function $v(x)w(t)$ will be spelled out in the development. Replacing $f$ by (2.2) and $g$ by $X_i T_j$ in (2.3) leads to the $m_x \times m_t$ matrix equation

$$C_x U B^T + A(p) U C_t^T = R \tag{2.4}$$

where $U = [u_{ij}]_{m_x \times m_t}$ and

$$[C_x]_{ik} = \int_0^1 X_i(x) X_k(x) v(x) dx \quad , \quad -M_x \leq i \, , \, k \leq N_x \quad , \tag{2.5}$$

$$(2.6) \qquad [C_t]_{j\ell} = \int_0^\infty T_j(t) T_\ell(t) w(t) dt \ , \quad -M_t \le j \ , \ \ell \le N_t \ ,$$

$$(2.7) \qquad \begin{gathered} [R]_{ij} = \int_0^\infty \int_0^1 r(x,t) X_i(x) T_j(t) v(x) w(t) dx\, dt \\ -M_x \le i \le N_x \ , \ -M_t \le j \le N_t \ , \end{gathered}$$

$$(2.8) \qquad [B]_{j\ell} = \int_0^\infty T_j(t) \frac{dT_\ell}{dt}(t) w(t) dt \ , \quad -M_t \le j \ , \ \ell \le N_t \ ,$$

and

$$(2.9) \qquad \begin{gathered} [A(p)]_{ik} = -\int_0^1 X_i(x) \left\{ \frac{d}{dx} \left[ p(x) \frac{d}{dx} X_k(x) \right] \right\} v(x) dx \\ -M_x \le i \ , \ k \le N_x \ . \end{gathered}$$

Multiplying (2.4) on the right by $(C_t^T)^{-1}$ and assuming an eigen-decomposition of $B^T(C_t^T)^{-1}$, i.e.

$$(2.10) \qquad B^T(C_t^T)^{-1} = Z_t \Lambda_t Z_t^{-1} \ , \quad \Lambda_t = \mathrm{diag}(\lambda_1, \ldots, \lambda_{m_t})$$

yields the equivalent matrix system

$$(2.11) \qquad C_x V \Lambda_t + A(p) V = G$$

where

$$(2.12) \qquad V \equiv U Z_t \quad \text{and} \quad G = R(C_t^T)^{-1} Z_t \ .$$

The system (2.11) is solved via solving the $m_t$ systems

$$(2.13) \qquad \{\lambda_j C_x + A(p)\} v^{(j)} = g^{(j)} \ , \quad j = 1, 2, \ldots, m_t$$

where $v^{(j)}(g^{(j)})$ is the $j^{\text{th}}$ column of the matrix $V(G)$. If the $m_x m_t \times 1$ vector obtained from $V$ by "stacking" the columns of $V$ one upon another is denoted by $co(V) = (v^{(1)}, v^{(2)}, \ldots, v^{(m_t)})^T$ then the system (2.11) may be written in the form

$$(2.14) \qquad \{(\Lambda_t \otimes C_x) + (I \otimes A(p))\} co(V) = co(G) \ .$$

This Kronecker sum form of the system is in the same form as the system in [2] where a vector block Gauss Elimination procedure is implemented.

## 3. The Fully Sinc-Galerkin Method

The expansion functions for (2.1) are derived from the sinc functions which are defined on $\mathbb{R}^1$ by

$$S_k(\xi) \equiv \frac{\sin(\pi(\xi - kh)/h)}{\pi(\xi - kh)/h} \quad , \quad h > 0 \tag{3.1}$$

where $k$ is an integer. In order that $S_k$ be defined on $(0,1)$ and $(0,\infty)$ the variable $\xi$ in (3.1) is defined by

$$\xi \equiv \phi(x) = \ell n\left(\frac{x}{1-x}\right) \quad , \quad x \in (0,1) \tag{3.2}$$

and

$$\xi \equiv \psi(t) = \ell n(t) \quad , \quad t > 0 \quad , \tag{3.3}$$

respectively. These two maps were originally used in [6] for the discretization of (1.1) ($p = 1$) and received substantial numerical testing in [3]. As shown in the latter, if the weight function $v(x)w(t)$ in (2.3) is defined by

$$v(x) = (\phi'(x))^{-1/2} \quad , \quad w(t) = (\psi'(t))^{1/2} \tag{3.4}$$

then the solution of (1.1) may be computed with exponential accuracy even in the case of singular solutions $u$. The procedure in [6] may be applied to the term

$$S(p)u \equiv -\frac{d}{dx}\left(p(x)\,\frac{du}{dx}\right) \tag{3.5}$$

upon carrying out the differentiations and applying the discretizations in [6] to $p\,\frac{d^2u}{dx^2}$ and $\frac{dp}{dx}\,\frac{du}{dx}$. This procedure is not desirable in the solution of problem (1.1) with singular solutions $u$. One approach involving collocation may be found in [5] but the following procedure does not directly differentiate the parameter $p$ in (1.1).

To develop the discretization of (3.5) the notation is conveniently compacted by the introduction of the matrices

$$\mathcal{A} \equiv \frac{-1}{h^2}\,I^{(2)} + \frac{1}{4}\,I \tag{3.6}$$

and

$$\mathcal{B} \equiv \frac{1}{h}\,I^{(1)} + D\left(\frac{1-2x_i}{4}\right) \tag{3.7}$$

where the matrices $I^{(j)}$ ($j = 1$ or $2$) have $ik^{\text{th}}$ entries given by

$$h \frac{d}{d\xi}\left(S_k(\xi)\right)\Big|_{\xi=ih} = \begin{cases} 0 & i = k \\ \frac{(-1)^{i-k}}{i-k} & i \neq k \end{cases} \tag{3.8}$$

and

$$h^2 \frac{d^2}{d\xi^2}\left(S_k(\xi)\right)\Big|_{\xi=ih} = \begin{cases} -\pi^2/3 & i = k \\ \frac{(-1)^{i-k}}{(i-k)^2} & i \neq k \end{cases} \quad . \tag{3.9}$$

The non-superscripted $I$ is the identity matrix and $D(g(\xi_i))$ is a diagonal matrix with $ii^{\text{th}}$ entry given by $g(\xi_i)$ where $\xi_i = x_i = \phi^{-1}(ih)$ ($t_i = \psi^{-1}(ih)$) in the case of spatial (temporal) nodes.

If $p$ in (3.5) has the assumed form

$$p_a(x) = \sum_{\ell=-M_p}^{N_p} p_\ell S_\ell \circ \phi(x) \tag{3.10}$$

then the approximate solution of $\mathcal{S}(p)u$ is obtained through integration by parts as follows:

$$\begin{aligned} &\int_0^1 (p_a u')' S_k \circ \phi(x) \frac{dx}{\sqrt{\phi'(x)}} \\ &\qquad = -\sum_\ell p_\ell \int_0^1 S_\ell \circ \phi(x) u'(x) \left[S_k \circ \phi \frac{1}{\sqrt{\phi'}}\right]' dx \\ &\qquad = \sum_\ell p_\ell \int_0^1 u(x) \left\{S_\ell \circ \phi(x) \left[S_k \circ \phi \frac{1}{\sqrt{\phi'}}\right]'\right\}' dx \quad . \end{aligned} \tag{3.11}$$

Now assume an approximate solution to the one-dimensional problem $\mathcal{S}(p)u = r$ with homogeneous boundary conditions $u(0) = u(1) = 0$ of the form

$$\hat{u}(x) = \sum_{k=-M_x}^{N_x} u_k S_k \circ \phi(x) \quad . \tag{3.12}$$

Replacing $u$ by $\hat{u}$ in (3.11) followed by an application of the sinc quadrature formula in [6] results in the approximation

$$A(p)\vec{u} = D\left(r/(\phi')^{3/2}\right) \tag{3.13}$$

for the coefficients $\vec{u} = (u_{-M_x}, \ldots, u_{N_x})^T$ in (3.12) where

$$A(p_a) \equiv \mathcal{A} D(\sqrt{\phi'}) D(p_a) + \frac{1}{h} \mathcal{B} D(\sqrt{\phi'}) D(I^{(1)} p_a) \quad . \tag{3.14}$$

The matrices $\mathcal{A}$, $\mathcal{B}$ and $D(\cdot)$ are as defined in the previous paragraph.

Returning to the discretization of (1.1) a short computation using the weight (3.4) in (2.8) yields

$$B = D(\sqrt{\psi'})\left[\frac{1}{h}\, I^{(1)} - \frac{1}{2}\, I\right] D(\sqrt{\psi'}) \quad . \tag{3.15}$$

Hence the solution of the forward problem (1.1) is obtained from (2.14) with $A(p_a)$ given by (3.12) after back substitution using (2.12).

This scheme is implemented on the problem (1.1) with $p(x) = 1 + x(1-x)$ and true solution $u(x,t) = \big(x(1-x)\big)^{1/2} t e^{-t}$. The approximate solution (2.1) has $M_x = N_x = M_t = P$ and $N_t$ as displayed. The sample $m$ in (1.3) is taken from nine equispaced nodes in $(0,1)$ with six points in $(0,\infty)$ spaced one half unit apart beginning at $t = 1/2$. The two columns headed $\|E\|_s$ and $\|E\|_u$ are the errors $\max\limits_{i,j} |u(x_i,t_j) - \hat{u}(x_i,t_j)|$ on the sinc grid $(x_i,t_j) = \big(\phi^{-1}(ih), \psi^{-1}(jh)\big)$, $-P \le i \le P$, $-P \le j \le N_t$, and the uniform grid $(x_i,t_j) = (i/100, j/5) : 1 \le i,j \le 100$, respectively. The mesh $h = \pi/\sqrt{P}$ is taken the same in space and time.

TABLE 3.1

| | | Forward | | Parameter |
|---|---|---|---|---|
| $P$ | $N_t$ | $\|E\|_s$ | $\|E\|_u$ | $\|p_a\|_u$ |
| 2 | 1 | .75 − 2 | .31 − 1 | .21 − 0 |
| 4 | 2 | .43 − 2 | .14 − 1 | .67 − 1 |
| 8 | 4 | .15 − 2 | .92 − 2 | .37 − 1 |

$.aa - d \equiv .aa \times 10^{-d}$

The column headed $\|p_a\|_u$ is the error in approximating the true parameter $p$ with the approximate (3.10) with $N_p = M_p = P$. The measurement is based on the maximum deviation of the $p_\ell$ in (3.10) with the true value of $p(x_\ell)$. The approximate parameter is obtained from (1.5) where $F(p_a)$ is obtained from solving (2.14) for each iterate of a quasi-Newton method. Graphs of the approximates corresponding to $N_p = 2$ ($\times\times\times$), 4 (ooo) and 8 (+++) are displayed in Figure 3.1 below.

FIGURE 3.1

Graphs of $p_a(x)$ versus $p(x)$

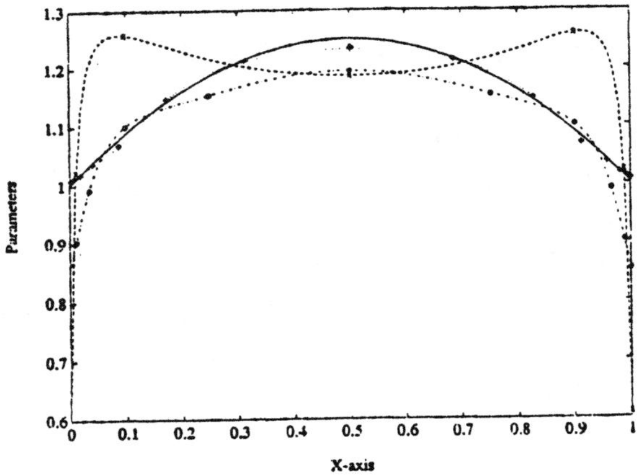


It is a pleasure to thank my friend and colleague Ralph Smith for his painstaking computations; and Rene' Tritz, whose expert typing is, as always, appreciated.

References

[1] H.T. Banks, P.D. Lamm, "Estimation of Variable Coefficients in Parabolic Systems," *IEEE Trans. on Aut. Control* 30 (1985), 386–398.

[2] K.L. Bowers, J. Lund, K.M. McArthur, "Symmetrization of the Sinc-Galerkin Method with Block Techniques for Elliptic Equations," *IMA J. of Num. Anal.* 9 (1989), 29–46.

[3] D.L. Lewis, J. Lund, K.L. Bowers, "The Space-Time Sinc-Galerkin Method for Parabolic Problems," *Inter. J. Num. Meth. Eng.* 24 (1987), 1629–1644.

[4] J. Lund, C.R. Vogel, "A Fully Galerkin Method for the Numerical Solution of Inverse Problems in Parabolic Partial Differential Equations," to appear in *Inverse Problems* (1989).

[5] K.M. McArthur, "A Collocative Variation of the Sinc-Galerkin Method for Second Order Boundary Value Problems," *Progress in Systems and Control Theory* 1 (1989).

[6] F. Stenger, "A Sinc-Galerkin Method of Solution of Boundary Value Problems," *Math. Comp.* 33 (1979), 85–109.

# Observability and Harish-Chandra Modules

JOSEPH A. WOLF

**Abstract.**

In an earlier note [10] we interpreted some questions of discrete observability of finite linear systems $dx/dt = Ax$ in terms of finite dimensional group representation theory. The main result said that a certain sort of observability can be cast into the language of group representation theory. Then, discrete observability comes down to whether the representation in question is cocyclic (dual to a cyclic representation) with the observation set up as a cocyclic vector (cyclic for the dual representation). Here we describe a setting in the representation theory of semisimple Lie groups where analogous results hold for infinite linear systems.

## 1. The Representation-Theoretic Interpretation of Observability.

In this section we recall the principal results of [10] connecting discrete observability and group representation theory.

**1.1. Definition.** Let $\pi$ be a representation of a group $G$ on a vector space $V$ of dimension $n < \infty$ . Fix a vector $x_0 \in V$, a (co)vector $c'$ in the linear dual space $V'$ of $V$, and a subset $S = \{g_1, ..., g_n\} \subset G$. The triple $(\pi, c', S)$ is **discretely observable** if we can always solve for $x_0$ in the system of equations

$$c' \cdot \pi(g_i)x_0 = e_i, \qquad 1 \leqq i \leqq n \tag{1.2}$$

Discrete observability of $(\pi, c', S)$ is equivalent to nonsingularity of the matrix

$$M = M(\pi, c', S) = \begin{pmatrix} c' \cdot \pi(g_1) \\ \vdots \\ c' \cdot \pi(g_n) \end{pmatrix} . \tag{1.3}$$

The notion of discrete observability for a linear system $dx/dt = Ax$ with constant coefficients, corresponds to the case of a 1-parameter linear group, where $G$ is the additive group of real numbers, $A$ is an $n \times n$ matrix,

$\pi(t) = exp(tA)$, and $g_i = t_i$ for some real numbers $t_1, \ldots, t_n$, so that $\pi(g_i) = exp(t_i A)$. See [6].

This interpretation has a useful formulation [10]:

**1.4. Theorem.** Let $\pi'$ denote the dual of $\pi$ , representation of $G$ on the linear dual $V'$ of $V$. Let $H$ denote the subgroup of $G$ generated by $S$. If $(\pi, c', S)$ is discretely observable then $c'$ is a cyclic vector for $\pi'|_H$ .

In particular, in Theorem 1.4, $c'$ is a cyclic vector for $\pi'$, so $\pi'$ is a cyclic representation, i.e. $\pi$ is a **cocyclic representation.**

**1.5. Corollary.** There exist $c' \in V'$ and $S \subset G$ such that $(\pi, c', S)$ is discretely observable, if and only if the representation $\pi$ is cocyclic.

In order to be able to use this result, we proved [10]

**1.6. Theorem.** Let $\pi$ represent a group $G$ on a finite dimensional vector space over a field $\mathbf{F}$ . Then $\pi$ is cocyclic if and only if every $\mathbf{F}$–irreducible summand of the maximal semisimple subrepresentation of $\pi$ has multiplicity bounded by its $\mathbf{F}$–degree.

## 2. Harish-Chandra's K-Multiplicity Theorem.

In this section we describe certain results from the representation theory of semisimple[1] Lie groups. These results give a multiplicity bound much like that in Theorem 1.6.

Let $G$ be a connected semisimple Lie group with finite center. Every compact subgroup of $G$ is contained in a maximal compact subgroup, and any two maximal compact subgroups are conjugate. Now fix a maximal compact subgroup $K \subset G$; because of the conjugacy it doesn't matter which one we use.

**2.1. Definitions.** Let $\pi$ be a representation of $K$ on a complex vector space $V$. A vector $v \in V$ is called **$K$-finite** if $\pi(K) \cdot v$ is contained in a finite dimensional subspace of $V$. A subspace $U \subset V$ is called **$K$-isotypic** if it is $\pi(K)$-invariant, if the resulting action of $K$ on $U$ is a direct sum of copies of some irreducible representation of $K$, and if $U$ is not properly contained in a larger subspace of $V$ with those properties. If $\psi$ is the irreducible representation of $K$ in question, then $U$ is called the **$\psi$-isotypic component** of $V$, and the representation of $K$ on $U$ is called the **$\psi$-isotypic component** of $\pi$.

Let $\mathfrak{g}_0$ denote the (real) Lie algebra of $G$ and $\mathfrak{g}$ its complexification.

[1] The results of this section are true in somewhat greater generality than the setting described here. See the Appendix.

Similarly $\mathfrak{k}_0$ will be the (subalgebra of $\mathfrak{g}_0$ that is the) real Lie algebra of $K$ and $\mathfrak{k}$ is the complexification of $\mathfrak{k}_0$.

**2.2. Definition.** A **$(\mathfrak{g}, K)$-module** is a complex vector space $V$ that is simultaneously a $\mathfrak{g}$-module and a $K$-module, say through representations

$$\pi : \mathfrak{g} \longrightarrow End(V) \quad \text{and} \quad \pi : K \longrightarrow End(V)$$

in such a way that (i) every vector $v \in V$ is $K$-finite, (ii) the differential of $\pi$ as a representation of $K$ coincides with the $\mathfrak{k}$-restriction of $\pi$ as a representation of $\mathfrak{g}$, and (iii) if $k \in K$ and $\xi \in \mathfrak{g}$ then $\pi[Ad(k)\xi] = \pi(k) \cdot \pi(\xi) \cdot \pi(k)^{-1}$.

**2.3. Definitions.** By **Harish-Chandra module** for $(\mathfrak{g}, K)$ we mean a $(\mathfrak{g}, K)$-module in which the $K$-isotypic subspaces are finite dimensional. A Harish-Chandra $(\mathfrak{g}, K)$-module $V$ is **irreducible** if it is irreducible as a $\mathfrak{g}$-module, **indecomposable** if it is indecomposable as a $\mathfrak{g}$-module, **cyclic** if it is cyclic as a $\mathfrak{g}$-module, etc.

The point of these definitions is a celebrated series of foundational results of Harish-Chandra, a few of which can be summarized as follows.

**2.4. Theorem.** Let $\pi$ be an irreducible unitary representation of $G$, say on the Hilbert space $V_\pi$, and let $V$ be the space of all $K$-finite vectors in $V_\pi$. Then $V$ dense in $V_\pi$ and $V$ is an irreducible Harish-Chandra module for $(\mathfrak{g}, K)$.

**2.5. Theorem[2].** Let $V$ be an irreducible Harish-Chandra module for $(\mathfrak{g}, K)$. Let $\pi$ denote the representation of $K$ on $V$. If $\psi$ is any irreducible representation of $K$ and if $U$ is the $\psi$-isotypic component of $V$, then $\dim(U) \leqq \deg(\psi)^2$, that is, the multiplicity of $\psi$ in $\pi$ is bounded by the degree of $\psi$.

One needs somewhat more than plain topological irreducibility of a continuous representation $\pi$ of $G$, say on a complete locally convex topological vector space (or even a Banach space) $V_\pi$, for the sort of result just described. The appropriate general notion is that of topologically completely irreducible (TCI) representation. One proves that $\pi$ is TCI if and only if the space $V$ of all $K$-finite vectors in $V_\pi$ is an irreducible $(\mathfrak{g}, K)$ Harish-Chandra module and is dense in $V_\pi$. See [7] or [9]. In the context of semisimple groups it is usually more convenient to use the notion of admissible representation: $\pi$ is **admissible** if $V$ is dense in $V_\pi$ and $V$ is a $(\mathfrak{g}, K)$ Harish-Chandra module. One can prove that every $(\mathfrak{g}, K)$ Harish-Chandra module is the space of all $K$-finite vectors for an admissible representation of $G$.

---

[2]This is due to Harish-Chandra for linear groups as an easy consequence of his Subquotient Theorem [1]. For non-linear groups Harish-Chandra proved $\dim(U) \leqq c_\pi \cdot \deg(\psi)^2$, for some integer $c_\pi \geqq 1$. That is not quite good enough for our purposes. Later Lepowsky gave an algebraic argument [5] for Theorem 2.5, and more recently Casselman proved a Submodule Theorem [1] which strengthens the Subquotient Theorem so that Theorem 2.5 follows easily.

The connection between unitary representations, Harish-Chandra modules, and discrete observability, is given by comparing the multiplicity statements in Theorems 1.6 and 2.5. One concludes, for example,

**2.6. Theorem.** Let $V$ be an irreducible Harish-Chandra module for $(\mathfrak{g}, K)$, let $W$ be any finite dimensional $K$-invariant subspace, and let $\phi$ denote the representation of $K$ on $W$. Then the representation $\phi$ is cocyclic. In other words, there exist $c' \in W'$ and $S \subset K$ such that $(\phi, c', S)$ is discretely observable.

## 3. Approximate Observability.

Let $V$ be an irreducible $(\mathfrak{g}, K)$ Harish-Chandra module. Write $\widehat{K}$ for the unitary dual of $K$, i.e. the (set of equivalence classes of) irreducible unitary representations. Given a Cartan subalgebra $\mathfrak{t}_0 \subset \mathfrak{k}_0$ and a root ordering, $\psi \in \widehat{K}$ is specified by its highest weight $\nu \in \sqrt{-1}\mathfrak{t}_0^*$, which we abbreviate by $\psi = \psi_\nu$. Given $m \geqq 0$ we have the finite set

$$\widehat{K}_m = \{\psi_\nu \in \widehat{K} \mid ||\nu|| \leqq m\}$$

of representations of $K$. For each $\psi_\nu \in \widehat{K}$ let $V[\nu]$ denote the $\psi_\nu$-isotypic subspace of $V$. Then $m \geqq 0$ specifies a finite dimensional $K$-invariant subspace

$$V_m = \sum_{\psi_\nu \in \widehat{K}_m} V[\nu].$$

We are going to obtain a variation on Theorem 2.6 for $V$ by applying that theorem to the $V_m$ as $m \to \infty$.

We start by realizing $V$ as the underlying Harish-Chandra module of a TCI Banach representation $\pi$ of G on a Hilbert space $V_\pi$, in such a way that $\pi|_K$ is unitary. This is a standard procedure, using Casselman's Submodule Theorem [1] (which strengthens Harish-Chandra's Subquotient Theorem [2]) to locate $V$ as a submodule of the Harish-Chandra module underlying a nonunitary principal series[3] representation of $G$. Let $\pi'$ denote the dual representation. Its representation space is $V_{\pi'} = V'_\pi$, and the subspace $V'$

[3]The "principal series" or "unitary principal series" of $G$ consists of the representations of the form $Ind_P^G(\mu \otimes \alpha)$ where $P = MAN$ is a minimal parabolic subgroup of $G$, where $A$ is the vector group part of a maximally noncompact Cartan subgroup of $G$ and $\alpha$ is a unitary character on $A$, where $\mu$ is an irreducible representation of the centralizer $M$ of $A$ in $K$, and where $N$ is a certain nilpotent normal subgroup of $P$. Since $M$ is compact, $\mu$ is finite dimensional and may be assumed to be unitary. Implicitly $\mu \otimes \alpha$ is extended from $MA$ to $P = MAN$ by triviality on $N$. The "nonunitary principal series" is obtained by dropping the requirement that $\alpha$ be unitary, i.e. by taking $\alpha$ to be any 1-dimensional complex representation of $A$. In any case, $Ind\,_P^G(\mu \otimes \alpha)|_K = Ind_M^K(\mu)$ and thus is unitary.

of $K$-finite vectors is the Harish-Chandra module dual to $V$. The finite dimensional subspace $(V')_m$ is naturally identified with the dual $(V_m)'$ of $V_m$, so we simply denote it by $V'_m$.

The cardinality of $\widehat{K}_m$ is bounded by a polynomial $p(m)$ because highest weights $\nu$ are confined to a lattice in $\sqrt{-1}\mathfrak{t}_0^*$. So it is easy to see

**3.1. Lemma.** Choose cyclic vectors $c'_\nu \in V'[\nu]$, for every $\psi_\nu \in \widehat{K}$. Then the $c'_\nu$ can be rescaled so that $\sum c'_\nu$ converges absolutely in $V'_\pi$.

With this in mind, we define

**3.2. Definition.** Let $\pi$ by a TCI Banach representation of $G$ such that the space $V$ of $K$-finite vectors in $V_\pi$ is a $(\mathfrak{g}, K)$ Harish-Chandra module. A vector $c \in V_\pi$ is **approximately cyclic** for $K$ if $c = \sum c_\nu$, absolutely convergent in $V_\pi$, where each $c_\nu$ is a $K$-cyclic vector in $V[\nu]$. A vector $c' \in V'_\pi$ is **approximately cocyclic** for $K$ if $c' = \sum c'_\nu$, absolutely convergent in $V'_\pi$, where each $c'_\nu$ is a $K$-cyclic vector in $V'[\nu]$.

**3.3. Definition.** Let $\pi$ be a TCI Banach representation of $G$. Fix $c' \in V'_\pi$. Then $(\pi, c')$ is **approximately discretely observable** for $K$ just when $c' = \lim c'_m$ absolutely convergent with $c'_m \in V'_m$, and we have an increasing sequence of subsets $S_m \subset K$ with cardinality $|S_m| = dim V'_m$, so that we can always solve the system of equations

$$c' \cdot \pi(g_i)x_0 = e_i, \qquad 1 \leqq i \leqq n$$

for $x_m \in V_m$.

The idea of Definition 3.3 is that, in a clearly measured way, one can come as close as desired to observability – at the price of sufficiently many observations. Now Theorem 2.6 and Lemma 3.1 combine to yield

**3.4. Theorem.** Let $\pi$ be a TCI Banach representation of $G$. Then $\pi'$ is approximately cocyclic. Let $c' \in V'_\pi$ be an approximately cocyclic vector. Then $(\pi, c')$ is approximately discretely observable.

## Appendix. K-Multiplicities for General Semisimple Groups.

In this Appendix we indicate how the results of §2 extend to a class of reductive Lie groups that contains all connected semisimple groups and all groups of Harish-Chandra class.

The **general semisimple groups** studied in [3], [4] and [8] are the reductive Lie groups $G$ (i.e. $\mathfrak{g} = \mathfrak{s} \oplus \mathfrak{z}$ with $\mathfrak{s}$ semisimple and $\mathfrak{z}$ commutative) that satisfy the conditions

(A.1) $G$ has a normal abelian subgroup $Z$ which centralizes the identity component $G^0$ of $G$ and such that $Z \cdot G^0$ has finite index in $G$, and

(A.2) if $x \in G$ then conjugation $Ad(x)$ is an **inner** automorphism on the complexified Lie algebra $\mathfrak{g}$.

This is a convenient class in which to do representation theory.

Fix a general semisimple group $G$. There is no loss of generality in expanding $Z$ to $Z \cdot Z_{G^0}$ where $Z_{G^0}$ is the center of $G^0$.

Let $Z_G(G^0)$ denote the centralizer of $G^0$ in $G$. Denote $G^\dagger = Z_G(G^0) \cdot G^0$. Many arguments for a general semisimple group $G$ go from $G^0$ to $G^\dagger$ to $G$.

The analog of maximal compact subgroup for $G^0$ is just the full inverse image $K^0$ of a maximal compact subgroup in the connected linear semisimple Lie group $G^0/Z_{G^0}$. The analog of maximal compact subgroup for $G^\dagger$ is just $K^\dagger = Z_G(G^0) \cdot K^0$, which in fact is the full inverse image of a maximal compact subgroup in $G^0/Z_{G^0} = G^\dagger/Z_G(G^0)$. The analog of a maximal compact subgroup $K$ for $G$ can be equivalently defined as the $G$-normalizer of $K^0$, the $G$-normalizer of $K^\dagger$, or the full inverse image of a maximal compact subgroup in $G/Z$ or in $G/Z_G(G^0)$. We refer to these groups $K$, $K^\dagger$ and $K^0$ respectively as **maximal compactly embedded subgroups** of $G$, $G^\dagger$ and $G^0$. If $Z$ is compact, they are just the maximal compact subgroups.

By **Cartan involution** of $G$ we mean an involutive automorphism whose fixed point set is a maximal compactly embedded subgroup. All the standard results hold: every maximal compactly embedded subgroup of $G$ is the fixed point set of a unique Cartan involution, and every Cartan involution of $\mathfrak{g}_0$ extends uniquely to a Cartan involution of $G$. See [8].

A technique developed in [8] reduces the proofs of Theorems 2.4 and 2.5 for connected reductive Lie groups $G^0$ to the case where $Z_{G^0}$ is compact, and there one can use Harish-Chandra's arguments without change.

Passage from $G^0$ to $G^\dagger$ is based on two straightforward facts.

(A.3) The irreducible representations of $G^\dagger$ are just the $\pi^\dagger = \xi \otimes \pi^0$ where $\xi$ is an irreducible, necessarily finite dimensional, representation of $Z_G(G^0)$, where $\pi^0$ is an ireducible representation of $G^0$, and where $\xi$ and $\pi^0$ agree on $Z_{G^0}$.

(A.4) The irreducible subrepresentations of $\pi^\dagger|_{K^\dagger}$ are just the $\psi^\dagger = \xi \otimes \psi^0$ where $\xi$ is the irreducible finite dimensional representation of $Z_G(G^0)$ mentioned above, and where $\psi^0$ is an irreducible representation of $\pi^0|_{K^0}$.

In Theorem 2.4 now $V_{\pi^\dagger} = E_\xi \otimes V_{\pi^0}$. Since the representation space $E_\xi$ of $\xi$ is finite dimensional, the spaces of $K^\dagger$-finite and $K^0$-finite vectors are related by $V^\dagger = E_\xi \otimes V^0$. The validity of the assertion passes directly from $G^0$ to $G^\dagger$. In Theorem 2.5 the Harish-Chandra modules are related by $V^\dagger = E_\xi \otimes V^0$, so again the result for $(\mathfrak{g}, K^0)$ Harish-Chandra modules implies the result for $(\mathfrak{g}, K^\dagger)$ Harish-Chandra modules.

Passage from $G^\dagger$ to $G$ uses a variation on the classical Schur's Lemma.

(A.5) If $\pi^\dagger$ is an irreducible unitary representation of $G^\dagger$ then the induced representation $Ind_{G^\dagger}^G(\pi^\dagger)$ is a finite sum of irreducible unitary representations of $G$. If $\pi$ is an irreducible unitary representation of $G$ then $\pi|_{G^\dagger}$ is a finite sum of irreducible unitary representations of $G^\dagger$. The multiplicity of $\pi$ in $Ind_{G^\dagger}^G(\pi^\dagger)$ is equal to the multiplicity of $\pi^\dagger$ in $\pi|_{G^\dagger}$.

Let $\pi$ be an irreducible unitary representation of $G$, say on a Hilbert space $V_\pi$, and let $V$ be the space of $K$-finite vectors. Realize $\pi$ as a subrepresentation of $Ind_{G^\dagger}^G(\pi^\dagger)$ for some irreducible unitary representation $\pi^\dagger$ of $G^\dagger$. The representation space of $Ind_{G^\dagger}^G(\pi^\dagger)$ is the space

$$Ind_{G^\dagger}^G(V_{\pi^\dagger}) = [L^2(G) \otimes V_{\pi^\dagger}]^{G^\dagger}$$

of $G^\dagger$-fixed vectors, where $G^\dagger$ acts on $L^2(G)$ by right translation and on $V_{\pi^\dagger}$ by $\pi^\dagger$. $G$ acts on $Ind_{G^\dagger}^G(V_{\pi^\dagger})$ by left translation on the $L^2(G)$ factor. The subspace of $K$-finite vectors is

$$Ind_{G^\dagger}^G(V^\dagger) = [L^2(G)'' \otimes V^\dagger]^{G^\dagger}$$

where $L^2(G)''$ consists of the elements of $L^2(G)$ that are $K$-finite on the left and the right. If we assume Theorem 2.4 for the representation $\pi^\dagger$ then it follows that the space $Ind_{G^\dagger}^G(V^\dagger)$ of $K$-finite vectors for $Ind_{G^\dagger}^G(V_{\pi^\dagger})$ is dense and is a Harish-Chandra module, i.e. that Theorem 2.4 holds for $\pi$.

The restriction of $\xi$ to $Z_{G^0}$ is a multiple of a unitary character $\zeta$. The left regular representations of the groups $K^0$, $K^\dagger$ and $K$ relative to $\zeta$ are

$$\lambda^0 = Ind_{Z_{G^0}}^{K^0}(\zeta), \quad \lambda^\dagger = Ind_{Z_{G^0}}^{K^\dagger}(\zeta), \quad \lambda = Ind_{Z_{G^0}}^{K}(\zeta).$$

Induction by stages says that $\lambda = Ind_{K^\dagger}^K(\lambda^\dagger)$. Theorem 2.5 for the $(\mathfrak{g}, K^\dagger)$ Harish-Chandra module $V^\dagger$ just says that the representation $\pi^\dagger$ of $K^\dagger$ is equivalent to a subrepresentation of $\lambda^\dagger$. It follows that the induced representation of $K$ is equivalent to a subrepresentation of $\lambda$. In other words, Theorem 2.5 follows for the $(\mathfrak{g}, K)$ Harish-Chandra module $V$.

Theorems 2.6 and 3.4 now hold for irreducible Harish-Chandra $(\mathfrak{g}, K)$-modules and TCI Banach representations $\pi$ of $G$, where $G$ is a general semisimple group and $K$ is a maximal compactly embedded subgroup.

Research partially supported by NSF Grant DMS-88-05816.

**References.**

[1] W. Casselman, *Jacquet Modules for Real Reductive Groups*, in "Proceedings of the International Congress of Mathematicians", Helsinki, 1978, pp. 557-563.

[2] Harish-Chandra, *Representations of Semisimple Lie Groups II*, Trans. Amer. Math. Soc., vol. 76 (1954), pp. 26-65.

[3] R. A. Herb and J. A. Wolf, *The Plancherel Theorem for General Semisimple Groups*, Compositio Mathematica, vol. 57 (1986), pp. 271-355.

[4] R. A. Herb and J. A. Wolf, *Rapidly Decreasing Functions on General Semismple Groups*, Compositio Mathematica, vol. 58 (1986), pp. 73-110.

[5] J. Lepowsky, *Algebraic Results on Representations of Semisimple Lie Groups*, Trans. Amer. Math. Soc., vol. 176 (1973), pp. 1-44.

[6] C. Martin and D. I. Wallace, *Observability and Transcendental Number Theory*, to appear.

[7] G. Warner, "Harmonic Analysis on Semi-Simple Lie Groups", I, Springer Verlag, 1972.

[8] J. A. Wolf, "Unitary Representations on Partially Holomorphic Cohomology Spaces" (The Action of a Real Semisimple Lie Group on a Complex Flag Manifold, II), Memoirs Amer. Math. Soc., Number 138, 1974.

[9] J. A. Wolf, *Foundations of Representation Theory for Semisimple Lie Groups*, in "Harmonic Analysis and Representations of Semisimple Lie Groups" ed. Wolf, Cahen and DeWilde, in series "Mathematical Physics and Applied Mathematics", vol 5 (1980), Reidel, Dordrecht, pp. 69-130.

[10] J. A. Wolf, *Observability and Group Representation Theory*, "Proceedings of the Conference on Computation and Control" ed. Bowers and Lund, in series "Progress in Systems and Control Theory", vol. 2 (1989), Birkhauser, Boston, to appear.

## MODELLING AND NONLINEAR CONTROL OF AN OVERHEAD CRANE

**B. d'Andréa-Novel, J. Lévine**

**Abstract:** In this work, we study the following positionning problem: we consider a platform moving along the horizontal axis, equipped with a winch around which a cable is enrolled, ended by the load. Large and fast movements are considered. Stabilization of a reference trajectory is studied via static state feedback and dynamic state feedback.
We show that the system can be linearized by dynamic feedback and the performances of the nonlinear dynamic controller are studied.

# 1 Introduction

An overhead crane, made of a platform moving along the horizontal axis, equipped with a winch around which a cable is enrolled, ended by the load, is considered (see Figure 1).

The control of overhead cranes is usually operated manually. However, improvements in accurate positionning for manipulations in hostile environment and in productivity gains (high speed, multiple tasks) are strongly needed and motivate the development of automatic control algorithms.

More precisely, we aim at driving the system along fast trajectories with stability and at reaching the endpoint with relatively high accuracy. Moreover, robustness of the control law is needed.

Section 2 is devoted to the modelization of the overhead crane. We assume that the cable and the load mass can be considered as a rigid pendulum. Thus, the overhead crane can be seen as a controlled mechanical system, with 3 degrees of freedom (2 prismatic, 1 rotational) and 2 actuators, the number of actuators being smaller than the number of degrees of freedom. It results that the classical control techniques (see [16,7,12,13]) do not allow a full state linearization by static feedback.

In fact, it can be deduced from [1] where the particular structure of mechanical systems is exploited, that full state linearization (see [16,12]) is not possible by static feedback and that output functions can be obtained to partially linearize by feedback the system with stability around an equilibrium point (see [4,2,3]). These results are recalled in Section 3.

We show in Section 4 that the system can be linearized by dynamic feedback linearization and simulation results are displayed in Section 5.

## 2 Modelling of the overhead crane

The cable ended by the load mass is considered as a rigid pendulum, with variable length (see Figure 1). Denoting :

$M$ : the platform mass, $m$ : the load mass,

$x_p$ : the platform abscissa, $L$ : the length of the cable,

$\theta$ : the angle of the cable with the vertical axis,

$R$ : the radius of the winch, $J$ : the inertia moment of the winch,

$u_1$ : the external force applied by the motor to the platform,

$u_2$ : the control of the winch,

$X$, $Z$ : the cartesian coordinates of the load mass in an absolute system coordinates,

writing the dynamic equations of the load mass, the platform and the winch:

$$\begin{cases} m\ddot{X} &= -Tsin\theta \\ m\ddot{Z} &= -Tcos\theta + mg \\ M\ddot{x}_p &= u_1 + Tsin\theta \\ \frac{J}{R^2}\ddot{L} &= -\frac{u_2}{R} + T \end{cases} \tag{1}$$

with

$$\begin{cases} X = x_p + Lsin\theta \\ Z = Lcos\theta \end{cases} \tag{2}$$

differentiating (2) twice and eliminating the tension $T$, we obtain the nonlinear state space representation of the system:

$$\begin{cases} \dot{x} = f(x) + \sum_{i=1}^{m=2} u_i g_i(x) \\ x = (x_p\,,\ L\,,\ \theta\,,\ \dot{x}_p\,,\ \dot{L}\,,\ \dot{\theta})' \end{cases} \tag{3}$$

The state vector $x$ is supposed to be measured. The open-loop vector field $f$ and the control vector fields $g_i$ are of the following form:

$$\left\{\begin{array}{l} f = \begin{pmatrix} \dot{x}_p \\ \dot{L} \\ \dot{\theta} \\ \dfrac{J sin\theta(g cos\theta + L\dot{\theta}^2)}{d(\theta)} \\ \dfrac{MR^2(g cos\theta + L\dot{\theta}^2)}{d(\theta)} \\ \dfrac{-2d(\theta)\dot{\theta}\dot{L} - g sin\theta(J+J_0) - JL sin\theta cos\theta\dot{\theta}^2}{L d(\theta)} \end{pmatrix} \\ g_1 = \begin{pmatrix} 0 \\ 0 \\ 0 \\ \dfrac{J_0}{M d(\theta)} \\ -\dfrac{R^2 sin\theta}{d(\theta)} \\ -\dfrac{J_0 cos\theta}{M L d(\theta)} \end{pmatrix} , \quad g_2 = \begin{pmatrix} 0 \\ 0 \\ 0 \\ \dfrac{R sin\theta}{d(\theta)} \\ -R\dfrac{\frac{M}{m} + sin^2\theta}{d(\theta)} \\ -R\dfrac{sin\theta cos\theta}{d(\theta)} \end{pmatrix} \\ J_0 = M(\frac{J}{m} + R^2) \ , \ d(\theta) = J_0 + J sin^2\theta \end{array}\right. \tag{4}$$

Remark that $d(\theta)$ is never vanishing.

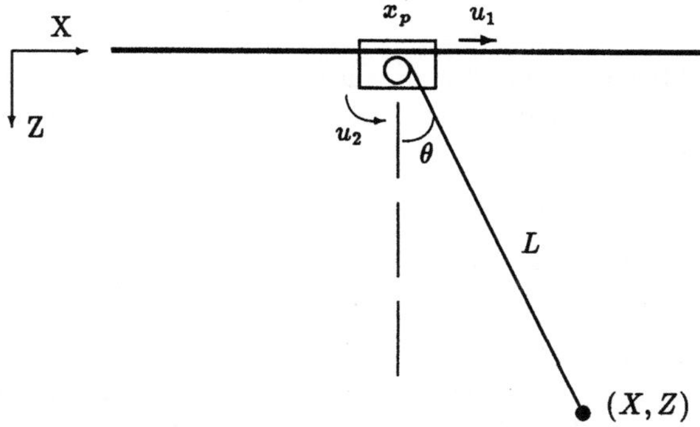


FIGURE 1 : The overhead crane

# 3 Static feedback approaches

We recall that system (3)-(4) is a mechanical system with 2 motors and 3 degrees of freedom and that it has been shown in [1] that these systems are not linearizable by a static state feedback law of the form:

$$u(x) = \alpha(x) + \beta(x)v \tag{5}$$

$\beta$ being a non singular $m \times m$ matrix, $m$ being the number of inputs and $v$ an auxiliary input.

Nevertheless, since the tangent linearization of (3)-(4) at an equilibrium point $x_e$ is controllable, one can show (see [4,3]) that there exist a feedback control of the form (5) and "output" functions $h_1$, $h_2$ such that, denoting $y = h(x)$:

1) the input-output mapping $v \longrightarrow y$ is linear,

2) the closed-loop system (3)-(4) with (5) is stable at $x_e$.

According to the input-output approach, only 4 dimensions are linearized by feedback and a 2-dimensional submanifold becomes unobservable by feedback.

Locally around the equilibrium $x_e$, this approach is equivalent to the construction of a linear stabilizing state feedback law as shown in [4,3].

This approach can be extended (see [19,1,2]) to follow slowly varying reference trajectories of the form:

$$x_r = (x_{p_r}(t)\,,\ L_r(t)\,,\ 0\,,\ \dot{x}_{p_r}(t)\,,\ \dot{L}_r(t)\,,\ 0)' \tag{6}$$

with $L_r$ belonging to a sufficiently small interval $[L_1, L_2]$. Note that in (6) the position $x_{p_r}$ can be arbitrarily chosen, since the linear approximation of (3)-(4) does not depend on it.

# 4 Full linearization via dynamic feedback control

We propose now to use "dynamic" state feedback laws to accomplish full linearization. Consequently, with such a control law, the stabilization of any admissible fast trajectory is guaranteed.

Dynamic compensators of the form:

$$\begin{cases} u = \alpha(x,w) + \beta(x,w)v \\ \dot{w} = a(x,w) + b(x,w) \end{cases} \tag{7}$$

have been initially considered by Singh [18] for the nonlinear input-output invertibility problem. The nonlinear decoupling problem by dynamic feedback [11,8,9, 14,17], the model matching problem [10] and the input-output linearization [15] have also received successfull answers.

However, in our case, no priviledged outputs are to be considered and the state space approach appears to be more natural.

In [5,6] the extended state space linearization is considered, namely we aim at finding a dynamic compensator (7) and an extended state space diffeomorphism:

$$\xi = \phi(x,w) \tag{8}$$

such that the transformed system is linear and controllable:

$$\dot{\xi} = F\xi + Gv \tag{9}$$

Sufficient conditions are obtained in [6] for the particular class of dynamic compensators made of elementary chains of integrators, followed by a static feedback on the extended space. In our situation, this result does not apply.

Nevertheless, we show that there exists a dynamic compensator, made of a static feedback first, followed by a double chain of integrators on the first input channel and a diffeomorphism which achieve full linearization.

**Theorem 1**
*System (3)-(4) is dynamic feedback linearizable using the following dynamic compensator:*

$$\left\{ \begin{array}{l} u(x) = \alpha(x) + \beta(x)w \\ \alpha(x) = \begin{pmatrix} -mgsin\theta cos\theta \\ mRgcos\theta - \frac{J}{R^2}L\dot{\theta}^2 \end{pmatrix} \quad \beta(x) = \begin{pmatrix} msin\theta & M \\ -\frac{mR^2+J}{R} & \frac{Jsin\theta}{R} \end{pmatrix} \\ \ddot{w}_1 = v_1 \\ w_2 = v_2 \end{array} \right. \tag{10}$$

*and the extended state space diffeomorphism:*

$$\xi = (X\,,\ \frac{dX}{dt}\,,\ \frac{d^2X}{dt^2}\,,\ \frac{d^3X}{dt^3}\,,\ Z\,,\ \frac{dZ}{dt}\,,\ \frac{d^2Z}{dt^2}\,,\ \frac{d^3Z}{dt^3})' \tag{11}$$

*where X and Z are given by (2).*

**Proof :**

The reader can check that the closed-loop extended system with state variables $\tilde{x} = (x\,,\ w_1\,,\ \dot{w}_1)'$ is static feedback linearizable (see [16,12]) and that it can be expressed as a linear controllable system in coordinates (11). □

**Remark :** The diffeomorphism (11) is singular when:

$$cos\theta(gcos\theta - w_1) = 0 \tag{12}$$

# 5 Simulation results

A first simulation consists in following the trajectory:

$$\dot{x}_p = 3\,m/s\,,\ \dot{\theta} = 0\,rd/s\,,\ \dot{L} = -0.2\,m/s \tag{13}$$

with the initial conditions:

$$\left\{ \begin{array}{l} x_p(0) = 0\,m\,,\ \theta(0) = 0\,rd\,,\ L(0) = 2.1\,m \\ \dot{x}_p(0) = 0\,m/s\,,\ \dot{\theta}(0) = 0\,rd/s\,,\ \dot{L}(0) = 0\,m/s \end{array} \right. \tag{14}$$

The dynamic feedback law smoothly stabilizes the system whereas static feedback laws lead to oscillations on $\theta$ and $x_p$.

To illustrate once more the fact that static feedback laws usually ensure only local stability, we have taken the following initial conditions:

$$x_p(0) = 0\,m\,,\ \theta(0) = 0.72\,rd\,,\ L(0) = 2.1\,m \tag{15}$$

and we want the system to reach:

$$x_p(\infty) = 0\,m\,,\ \theta(\infty) = 0\,rd\,,\ L(\infty) = 0\,m \tag{16}$$

By dynamic feedback the system's state reaches the equilibrium with stability whereas static feedback produces a divergence.

**Acknowledgements:**
This work was partially supported by EDF-DER Chatou grant.
The authors are indebted to F. Boustany for his efficient contribution concerning the computational aspects and to B. Charlet and P. Martin for useful discussions concerning dynamic feedback linearization.

# References

[1] B. d'Andréa-Novel. *Commande non linéaire des robots. Editions Hermès, 1988.*

[2] B. d'Andréa, J. Lévine. *Synthesis of nonlinear state feedback for the stabilization of a class of manipulators. Information and Decision Technologies 14, 1988, 145-168.*

[3] B. d'Andréa, L. Praly. *About finite nonlinear zeros for decouplable systems. Systems and Control Letters 10, 1988, 103-109.*

[4] C.I. Byrnes, A. Isidori. *Local stabilization of minimum phase nonlinear systems. Systems and Control Letters 11, 1988, 9-17.*

[5] B. Charlet, J. Lévine, R. Marino. *On dynamic feedback linearization. Systems and Control Letters, to appear.*

[6] B. Charlet, J. Lévine, R. Marino. *Sufficient conditions for dynamic feedback linearization. submitted.*

[7] D. Claude, M. Fliess, A. Isidori. *Immersion directe et par bouclage d'un système non linéaire dans un linéaire. C.R.A.S. t. 296 I, 1983, 237-240.*

[8] J. Descusse, C.H. Moog. *Decoupling with dynamic compensation for strong invertible affine nonlinear systems. Int. J. Control, Vol. 42 N. 6, 1985, 1387-1398.*

[9] J. Descusse, C.H. Moog. *Dynamic decoupling for right invertible nonlinear systems. Systems and Control Letters, Vol. 8 N. 4, 1987, 345-349.*

[10] M.D. Di Benedetto, A. Isidori. *The matching of nonlinear models via dynamic state feedback. SIAM J. Control and Optimization, Vol. 24, 1986, 1063-1075.*

[11] M. Fliess. *A new approach to the noninteracting control problem in nonlinear system theory. Proc. 23th Allerton Conf., 1985, 123-129.*

[12] L.R. Hunt, R. Su, G. Meyer. *Design for multi-input nonlinear systems. Diff. Geometric Control Theory, R.W. Brockett, R.S. Millman, H.J. Sussmann, Eds. Birkhauser, 1983, 268-298.*

[13] A. Isidori. *Nonlinear Control Systems: An introduction. Lecture Notes in Control and Information Sciences, Springer-Verlag, Berlin, New-York, 1985.*

[14] A. Isidori. *Control of nonlinear systems via dynamic state feedback. Conf. Algebraic and Geometric Methods in Nonlinear Control Theory, Paris, 1985.*

[15] A. Isidori, C.H. Moog, A. de Luca. *A sufficient condition for full linearization via dynamic state feedback Proc. 25th. IEEE CDC, Athens, 1986.*

[16] B. Jakubczyk, W. Respondek. *On linearization of control systems. Bull. Acad. Pol. Sci., Ser. Sci. Math., 28, 1980, 517-522.*

[17] H. Nijmeijer, W. Respondek. *Dynamic input-output decoupling of nonlinear control systems. IEEE Trans. on Automatic Control, AC-33, 1988, 1065-1070.*

[18] S. Singh. *Decoupling of invertible nonlinear systems with state feedback and precompensation. IEEE Trans. on Automatic and Control, AC-25, 1980, 1237-1239.*

[19] M. Vidyasagar. *Nonlinear Systems Analysis. Prentice Hall, 1978.*

# ON ADAPTIVE LINEARIZING CONTROL OF OMNIDIRECTIONAL MOBILE ROBOTS

G. CAMPION, G. BASTIN.


## *ABSTRACT*

We discuss the design of linearizing control algorithms for a particular class of non linear systems : "omnidirectional mobile robots". We analyse 2 kinds of control : input state linearization and input-output linearization. We discuss also the design of adaptive linearizing control to cope with lack of knowledge of the mass distribution of the payload.


## INTRODUCTION

Several prototypes of "omnidirectional mobile" robots have been recently described in the literature. Two typical examples are the 4-wheels URANUS robot of the Robotics Institute of Carnegie Mellon University (Muir and Neuman, 1987) and the UCL 3-wheels robot of the University of Louvain (Campion 1988). Such mobile robots are called "omnidirectional" because they exhibit *perfect* mobility in the horizontal plane. Thanks to special constructive features (which are described hereafter), they can achieve simultaneous motions along the three degrees of freedom in the plane (longitudinal, transverse and rotational) and hence perform unusual motions like cute angle turns, straight motion with simultaneous rotation, or straight motion in any direction without preliminary reorientation. Our purpose in this paper is to discuss the design of adaptive linearizing control algorithms for these systems.

## 1. ROBOT DESCRIPTION

Omnidirectional mobile robots are constituted by a rigid trolley equipped with a set of N wheels whose orientation is fixed with respect to the trolley, and which are driven by DC motors. The perfect mobility of the robot results from a particular feature of the wheels : their tread is not a tyre, like in a conventional wheel, but is formed by a set of *free* rollers. The rotation axes of these rollers have also a fixed orientation with respect to the wheels.

The global position of the robot in the plane is described by the vector q(t):

$$q(t) \equiv [\, x(t), y(t), \theta(t) \,]^T \tag{1.1}$$

where $(x(t), y(t))$ are the coordinates of an arbitrary reference point of the frame and $\theta(t)$ its angular orientation. The rotation angle of each wheel (index $i = 1,...,N$) is denoted $\varphi_i(t)$ and we define the N-vector $\varphi(t)$ :

$$\varphi(t) \equiv [\ \varphi_1(t) \ldots \varphi_N(t)\ ] \tag{1.2}$$

The configuration of each wheel is described by the following 5 constants (fig.1): $r_i$ the radius of the wheel, $l_i$ the distance between the wheel and the reference point, $\alpha_i$ and $\beta_i$ the angles which characterize the position of the wheel with respect to the trolley, $\gamma_i$ the position of the roller which is in contact with the ground.As a matter of illustration, the numerical values of these constants for the URANUS and UCL robots are shown in the following table (see also fig. 2 and 3).

| | URANUS | | | | UCL | | |
|---|---|---|---|---|---|---|---|
| Wheel | 1 | 2 | 3 | 4 | 1 | 2 | 3 |
| $\alpha$ | 45° | 135° | 225° | 315° | 60° | 180° | 240° |
| $\beta$ | -45° | 45° | -45° | 45° | 0 | 0 | 0 |
| $\gamma$ | 45° | -45° | 45° | 45° | 0 | 0 | 0 |

## 2. KINEMATICAL CONSTRAINTS

The (3+N) generalized coordinates [q(t), $\varphi$(t) ] describing the robot motion are obviously not independent. For each wheel there is a kinematical constraint expressing the pure rolling condition for the roller in contact with the ground. These N constraints are expressed as follows :

$$A_1\ R(\theta)\dot{q} + A_2\ \dot{\varphi} = 0 \tag{2.1}$$

where $A_1$ is a (Nx3) constant matrix:

$$A_1 \equiv \begin{bmatrix} -\sin(\alpha_1+\beta_1+\gamma_1) & -\cos(\alpha_1+\beta_1+\gamma_1) & l_1\cos(\beta_1+\gamma_1) \\ \cdot & \cdot & \cdot \\ \cdot & \cdot & \cdot \\ \cdot & \cdot & \cdot \\ -\sin(\alpha_N+\beta_N+\gamma_N) & -\cos(\alpha_N+\beta_N+\gamma_N) & l_N\cos(\beta_N+\gamma_N) \end{bmatrix} \tag{2.2}$$

R($\theta$) and $A_2$ are respectively as (3x3) orthogonal matrix and a (NxN) diagonal matrix :

$$R(\theta) = \begin{bmatrix} \cos\theta & \sin\theta & 0 \\ -\sin\theta & \cos\theta & 0 \\ 0 & 0 & 1 \end{bmatrix} \qquad A_2 = \begin{bmatrix} r_1\cos\gamma_1 & & 0 \\ & \ddots & \\ 0 & & r_N\cos\gamma_N \end{bmatrix} \tag{2.3}$$

Notice that $A_1$ and $A_2$ are constant matrices depending only on the wheels configuration. We introduce the following assumptions.

**H 1** *The wheel characteristics are such that $A_1$ and $A_2$ are full rank.*

**H2** *The wheel characteristics are such that :*

$$\sum_{i=1}^{N} \sin(\alpha_i + \beta_i + \gamma_i) = \sum_{i=1}^{N} \cos(\alpha_i + \beta_i + \gamma_i) = 0$$

Notice that these assumptions are satisfied for both URANUS and UCL robots. Under assumption $H_1$ we may assume that the matrix $A_1$ has been arranged in such a way that the three first rows are linearly independent. Then we introduce the following partitions of $A_1$, $\varphi$, $A_2$:

$$A_1 = \begin{bmatrix} A_{1a} \\ A_{1b} \end{bmatrix} \quad \varphi = \begin{bmatrix} \varphi_a \\ \varphi_b \end{bmatrix} \quad A_2 = \begin{bmatrix} A_{2a} & 0 \\ 0 & A_{2b} \end{bmatrix} \qquad (2.4)$$

where $A_{1a}$ is a (3x3) non singular matrix. It follows then from (2.1) that:

$$A_{1a} R(\theta) \dot{q} + A_{2a}\dot{\varphi}_a = 0 \text{ and } A_{1b} R(\theta) \dot{q} + A_{2b}\dot{\varphi}_b = 0 \qquad (2.5)$$

which imply that:

$$\dot{\varphi}_b = A_{2b}^{-1} A_{1b} A_{1a}^{-1} A_{2a} \dot{\varphi}_a \equiv C\dot{\varphi}_a \qquad (2.6.a)$$

Without loss of generality we may suppose that $\varphi(0) = 0$. Hence:

$$\varphi_b(t) = C\varphi_a(t) \qquad (2.6.b)$$

The constraints (2.1) are not completely integrable : there exists no analytical relationship between q(t) and $\varphi(t)$ . In other words $\varphi(t)$ does not depend only on q(t) but on the full history $(q(\tau), \tau \leq t)$ :

$$\varphi(t) = -A_2^{-1} A_1 \int_0^T R\,[\theta(\tau)]\, \dot{q}(\tau)d\tau \qquad (2.7)$$

But under assumption $H_2$, the constraints are *partially* integrable under the form

$$\theta(t) - \theta(0) = [\sum_{i=1}^{N} l_i \cos(\beta_i + \gamma_i)]^{-1} \sum_{i=1}^{N} [r_i\, \varphi_i(t) \cos\gamma_i\,] \qquad (2.9)$$

This means that the angular orientation $\theta(t)$ of the robot can be computed at time t from the measurement of the wheels rotation angles $\varphi_i(t)$.

## 3. ROBOT DYNAMICS.

The motion equations are now derived via Lagrange formalism. Defining $T(q,\dot{q},\dot{\varphi})$ as the kinetic energy, $f_2(\dot{\varphi})$ as the N vector of the friction torques, and u as the N vector of the torques provided by the motors, the robot dynamics is written:

$$\frac{d}{dt}\left(\frac{\partial T}{\partial \dot{q}}\right) - \frac{\partial T}{\partial q} = R^T(\theta)\, A_1^T \lambda \qquad (3.1.a)$$

$$\frac{d}{dt}\left(\frac{\partial T}{\partial \dot{\varphi}}\right) = f_2(\dot{\varphi}) + A_2\,\lambda + u \qquad (3.1.b)$$

where $\lambda$ is the N vector of Lagrange multipliers associated with the constraints (2.1). These equations can also be expressed as follows:

$$M_1\, R(\theta)\,\ddot{q} + v_1\dot{\theta}^2 = A_1^T\,\lambda \qquad (3.2.a)$$

$$M_2\,\ddot{\varphi} - f_2\,(\dot{\varphi}) = A_2^T\,\lambda + u \qquad (3.2.b)$$

where $M_1$, $M_2$ and $V_1$ are respectively a (3x3)-matrix , a (NxN)-matrix and a 3-vector which are constant and depend on the mass distribution in the robot. By elimination of $\lambda, \dot{\varphi}, \ddot{\varphi}$ between (3.2), (2.1) and the time derivative of (2.1), we obtain:

$$M\, R(\theta)\ddot{q} + f(q,\dot{q}) = -A_1^T\,A_2^{-1}\,u \qquad (3.4)$$

where:

$$M = M_1 + A_1^T\,A_2^{-1}\,M_2\,A_2^{-1}A_1 \qquad (3.5.a)$$

$$f(\theta,\dot{q}) = v_1\,\dot{\theta}^2 + A_1^T\,A_2^{-1}\,M_2\,A_2^{-1}A_1\left(\dot{\theta}\frac{\partial R}{\partial \theta}\right)\dot{q} + A_1^T\,A_2^{-1}\,f_2\,[A_2^{-1}\,A_1\,R(\theta)\,\dot{q}] \qquad (3.5.b)$$

The dynamical equations (3.4, 3.5) together with the constraint (2.1) constitute the complete dynamical description of the robot motion.

## 4. LINEARIZINGCONTROL

### 4.1. Input-state linearization.

We first consider the case where full state $(q,\dot{q})$ measurement is available through an appropriate remote sensing device. In that case, it is immediate that any state feedback control law $u(q,\dot{q})$ which satisfies:

$$A_1^T\,A_2^{-1}u = -\,MR(\theta)\,v - f(\theta,\dot{q}) \qquad (4.1)$$

ensures that $\ddot{q} = v$, for any v, and hence achieves input-state linearization of (3.4). The existence of the solution of (4.1) is guaranteed by assumption $H_1$. For N = 3, the solution is unique. For N > 3, there is an infinite number of solutions, among which we can select the one which minimizes the input energy || u ||:

$$u(q,\dot{q}) = A_2(A_1^T)^+[-MR(\theta)v - f(\theta,\dot{q})]$$

If we suppose that the goal is to track a desired trajectory $q_d(t)$, then the choice:

$$v = \ddot{q}_d + K_1(\dot{q}_d - \dot{q}) + K_2(q_d - q) \tag{4.2}$$

ensures the following linear dynamics for the tracking error $\tilde{q} = q_d - q$:

$$\ddot{\tilde{q}} + K_1\dot{\tilde{q}} + K_2\tilde{q} = 0 \tag{4.3}$$

where $K_1$, $K_2$ are design matrices at the user's choice.

**4.2. Input-output linearization.**

The control strategy of section 4.1 is made implementable only if an accurate remote sensing system is available. Such systems are complex and expensive. It is therefore clearly of interest to look for alternative control strategies which require only low cost measurement devices. Assume that the wheels angular positions and velocities $(\varphi, \dot{\varphi})$ are the only available measurements and consider $\varphi_a(t)$ as the system output (recall that $\varphi_b(t)$ is proportional to $\varphi_a(t)$ (2.6b)). We have:

$$\dot{\varphi}_a = -A_{2a}^{-1} A_{1a} R(\theta)\dot{q} \tag{4.4}$$

Then from (4.4) and using (3.4), it is easy to check that any control u satisfying:

$$A_1^T A_2^{-1} u = M A_{1a} A_{2a}^{-1} v + M\left(\dot{\theta}\frac{\partial R}{\partial \theta}\right)\dot{q} - f(\theta,\dot{q}) \tag{4.5}$$

ensures that $\ddot{\varphi}_a = v$, for any v, and achieves input output linearization. This control always exists under assumption H1. It is unique for N = 3 and can be selected in order to minimize ||u||, for N>3. In practice, the implementation of the control law defined by (4.5) requires the on line knowledge of $(\theta,\dot{q})$. They can actually be observed from the measurements $(\varphi,\dot{\varphi})$ as follows: $\theta(t)$ is obtained from (2.9), and $\dot{q}(t)$ from the kinematical constraint (2.1).

We define a desired reference trajectory $\varphi_d(t)$ which corresponds to $q_d(t)$ according to (2.1):

$$\varphi_d(t) = -A_2^{-1} A_1 \int_0^t R\,[\theta_d(\tau)]\,\dot{q}_d(\tau)\,d\tau \qquad (4.6)$$

and the induced partition: $\varphi_d(t) = [\varphi_{da}(t), \varphi_{db}(t)]^T$. Then the choice:

$$v = \ddot{\varphi}_{da} + K_1(\dot{\varphi}_{da} - \dot{\varphi}_a) + K_2(\varphi_{da} - \varphi_a) \qquad (4.8)$$

involves the following linear dynamics for the outpout error $\tilde{\varphi}_a$:

$$\ddot{\tilde{\varphi}}_a + K_1\dot{\tilde{\varphi}}_a + K_2\tilde{\varphi}_a = 0 \qquad (4.9)$$

With this control law the global internal stability of the closed loop follows from assumption $H_2$. Indeed it can be easily checked that the following transformation is a diffeomorphism :

$$z = [z_1\ z_2\ z_3]^T = W(\dot{x}, \dot{y}, \theta, \dot{\theta}, \varphi_a) \qquad (4.10)$$

$$z_1 = \varphi_a \qquad z_2 = \dot{\varphi}_a = -A_{2a}^{-1} A_{1a} R(\theta)\,\dot{q} \qquad z_3 = \theta$$

The linearizing control law (4.5), (4.8) ensures that $(\varphi_a, \dot{\varphi}_a)$ and hence, from (4.7), that $(\varphi_b, \dot{\varphi}_b)$ are bounded. This implies, using (2.9), the boundedness of $\theta$ and hence that:

$$\lim_{t\to\infty} \tilde{\varphi}(t) = 0 \qquad \lim_{t\to\infty} (\theta_d(t) - \theta(t)) = 0 \qquad \lim_{t\to\infty} (\dot{q}_d(t) - \dot{q}(t)) = 0$$

This establishes the global internal stability of the closed loop. Notice however that the convergence to zero of the tracking error ($q_d$ - q) is not guaranteed. But simulation results have shown that the linearizing control strategy decreases significantly this tracking error, compared with less sophisticated methods (like PID control for instance).

## 5. ADAPTIVE CONTROL

The feedback linearizing control laws of section 4 require a perfect knowledge of the robot model and mainly of its inertial parameters. However, if the robot is devoted to the transport of various loads which can be heavier than the robot itself, it is clear that the presumption of a perfect knowledge of the load related inertial parameters is completely unrealistic. It is therefore of interest to examine the performance of adaptive control algorithms designed to cope with this parametric uncertainty. We consider an adaptive version of the input-output linearizing control law of section 4.2 derived from the indirect adaptive control algorithm of Middleton and Goodwin (1986). We present simulation results relative to the UCL robot. The robot without load is assumed to be perfectly known but the inertia characteristics of the payload (mass, position of the mass center, vertical inertia moment) are supposed to be unknown. The reference trajectory is a circle with maximum speed 1m/sec. The actual payload mass is 140 kg (for a robot of 40 kg). Fig. 4 compares 3 trajectories :

- the ideal desired trajectory;
- the trajectory obtained with a fixed linearizing controller designed fo the robot without load;
- the trajectory obtained with the adaptive controller.

The improvement due to the parameter adaptation is obvious since the asymptotic tracking error ($q_d$ - q) is clearly significantly reduced.

## AKNOWLEDGMENTS.

The results presented in this paper have been obtained within the framework of the Belgian Programme on Concerted Research Actions and on Interuniversity Attraction Poles initiated by the Belgian State, Prime Minister's Office, Science Policy Programming. The scientific responsibility rests with its authors.

## REFERENCES

Campion G., Bastin G., Rolin D., Raucent B (1988) "External linearization control for an omnidirectional mobile robot" Proc IFAC-SYROCO-1988 Karlsruhe pp 71.1 - 71.6.

Middleton R., Goodwin G.C. (1986) "Adaptive computed torque control for rigid link manupulators" IEEE 25th CDC - Athens Dec.1986 pp 68-73

Muir P.F., Neuman C.P. (1987) "Kinematic modeling for feedback control of an omnidirectional wheeled mobile robot" Proc IEEE Symp. on Robotics and Automation Ap1987 - pp 1772-1778

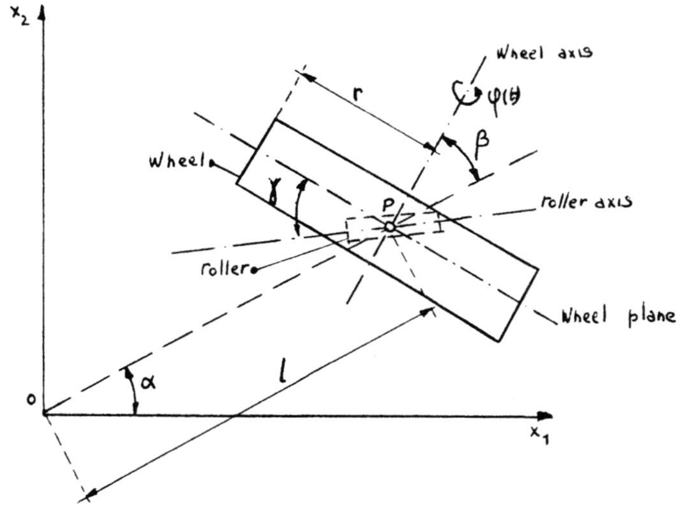


Fig. 1. Wheel characteristics.

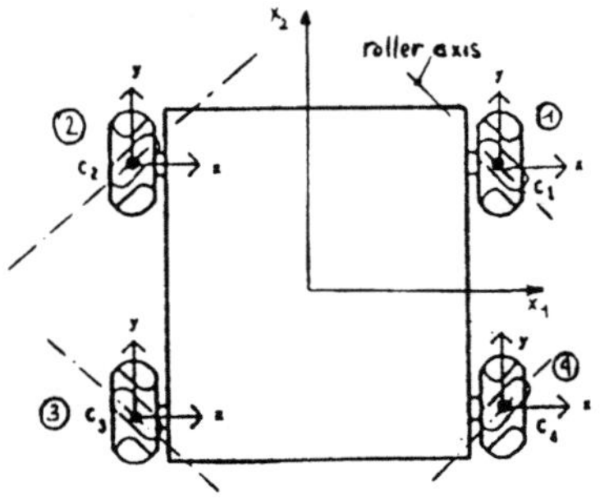


Fig. 2. Uranus robot.

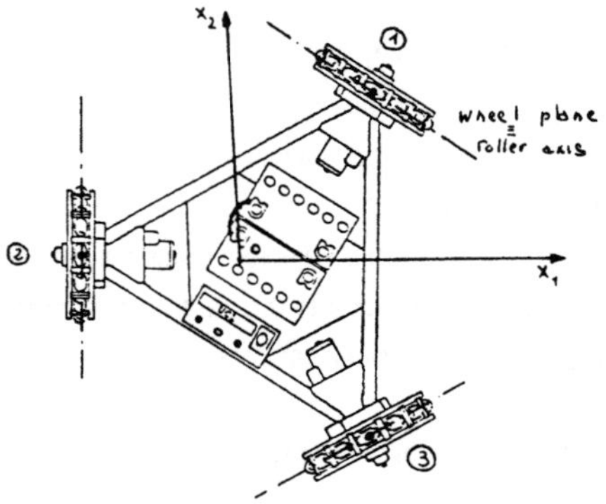


Fig. 3. UCL robot.

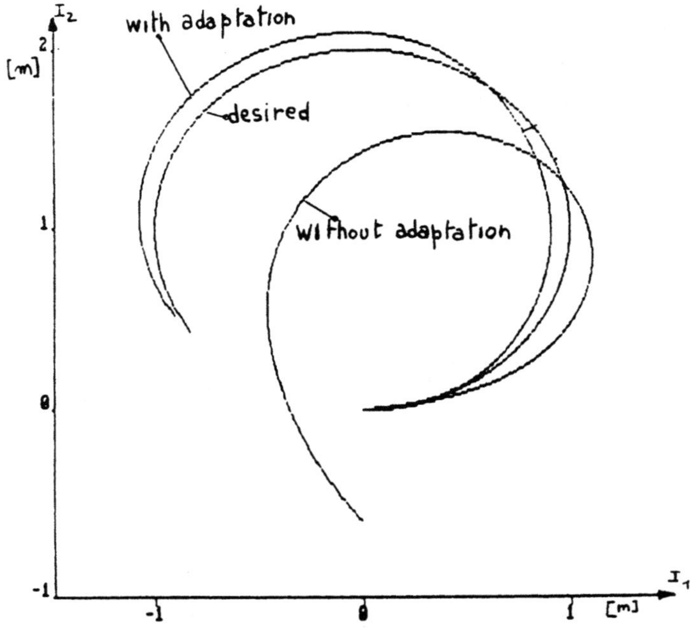


Fig. 4. Control trajectories.

# Robot Control via Nonlinear Observer

*C. Canudas de Wit* , K.J . Åström** and N. Fixot**

***Abstract.*** High precision measurements of joint displacements are available on robot manipulators. In contrast, the velocity measurements obtained through tachometers are in many cases contaminated by noise. It is therefore economically and technically interesting to investigate the possibility of stabilize the robot dynamics using only direct available measurements such as the angular positions. This paper deals with the problem of trajectory tracking in robot motion via nonlinear observers. Local conditions for exponential stability of the closed-loop system are given.

## 1. Introduction

High precision measurements of joint displacements are available on robot manipulators. In contrast, the velocity measurements obtained through tachometers are in many cases contaminated by noise. It is therefore economically and technically interesting to investigate the possibility of stabilize the robot dynamics using only direct available measurements such as the angular positions.

Control of *elastic* robots via nonlinear observers were analyzed before based on the so-called pseudo-linearization technique [3] [4] [7]. The pseudo-linearization consists in finding a state-space change of coordinates such that, in the transformed state-space, the system admits a linear tangent model independently to the operating point [5]. A linear Luenberger observer can thus be implemented. A drawback of this approach is that it requires an important amount of calculations and that it only gives approximated closed-loop stability results (under the hypothesis that the linear tangent model dominate the other nonlinearities).

Another approach for designing speed observers consists in defining a hypersurface which is a function of the observation errors and in determining the conditions that make this hypersurface attractive. Once the error trajectories have reached this hypersurface, a switching error based action makes the observation errors "slide" to zero. This type of observer is called a sliding observer and belongs to a special class of variable structure nonlinear systems [6]. In robot manipulators, global conditions for the invariance of the hypersurface cannot always be guaranteed. Instead, local conditions are easily derived and can be arbitrarily established by increasing the observer gains. These gains need not always remain large, but they can exponentially decrease in time. A sliding observer for robot manipulators with time-varying gains was studied by Canudas de Wit and Slotine [2]. Closed-loop results are not yet available, however.

In this paper we study another observer type, namely "smooth nonlinear observer" (i.e. an observer with "smooth" or "differentiable" gains) together with a nonlinear control law which is only a function of estimated velocities and measurement positions. The main difference with respect to the sliding approach is that the switching gains are replaced by differentiable nonlinear functions which are adjusted in order to render the closed-loop system exponentially stable. This approach also differs from the linearization techniques in the sense that local convergence is obtained without the need of a nonlinear change of coordinates. Instead, the nonlinear structure and some of the physical properties of the robot model are exploited. The stability analysis of the augmented system which results from the combination of the smooth observer dynamics and the control law, is performed locally (via standard Lyapunov functions) around a constant or time-varying bounded vector instead of being performed around an attractive manifold, or through diffeomorphic transformations.

# 2. Problem statement

This section describes the robot model dynamics resulting from the Lagrange equations. This model can be expressed in an observable state-space representation and possesses some important properties which are useful for designing nonlinear state-space observers.

## 2.1 Model Description

Consider a rigid robot having $n$ revolute joints expressed as:

$$H(q)\ddot{q} + C(q,\dot{q})\dot{q} + \tau_g(q) = \tau \qquad (2.1)$$

where $q, \dot{q}, \ddot{q} \in \mathcal{R}^n$ are respectively the link displacements, velocity and acceleration. H(q) is the $nxn$ inertia matrix, $C(q, \dot{q})\dot{q}$ the Coriolis and centripetal forces, $\tau_g(q)$ the gravity components and $\tau$ the applied motor torque. Friction is not included in model (2.1) in order to simplify the discussion.

Introducing the state vectors $x_1 = q$ and $x_2 = \dot{q}$, model (2.1) has the following state-space representation :

$$\dot{x}_1(t) = x_2(t) \qquad (2.2.a)$$

$$\dot{x}_2(t) = \beta(x_1,x_2) + u(t) \qquad (2.2.b)$$

$$y(t) = x_1(t) \qquad (2.2.c)$$

where $\beta(x_1, x_2)$ and u(t) are given as,

$$\beta(x_1, x_2) = - H(x_1)^{-1} [C(x_1, x_2)\, x_2 + \tau_g(x_1)] \qquad (2.3)$$

$$u(t) = H(x_1)^{-1}\, \tau \qquad (2.4)$$

Note that the new "input" u(t) is measurable, since the measurable "output" y(t) is equal to $x_1(t)$ and hence u(t) can be computed from (2.4). This state-space representation is locally observable, see **[2]** for further details.

## 2.2. Model Properties

Certain properties inherent to robot dynamics are useful in designing nonlinear observers. These properties are summarized hereafter :

*(i)* $H(x_1) = H(x_1)^T > 0 \quad \rightarrow \quad H(x_1)^{-1}$ exists

*(ii)* $\sigma_0 > \|H(x_1)\| > 0$ and $\sigma'_0 > \|H(x_1)^{-1}\| > 0$

*(iii)* $\tau_g\,(q) \leq \sigma_1$

*(iv)* The $i^{th}$ element of vector $C(x_1,x_2)x_2$ is $x_2^T N_i(x_1) x_2$, where the $N_i(x_1)$ matrices are symmetric and composed of bounded elements. ($\|N_i(x_1)\| < \infty$, $\forall\, i = 1..\, n$).

*(v)* At least one possible parametrisation for $C(x_1, x_2)$ exists such that any vector $\zeta$ with bounded norm satisfies :

$$\zeta^T \left[ \frac{d}{dt} \{H(x_1)\} - 2\, C(x_1, x_2) \right] \zeta = 0$$

*(vi)* $\|C(x_1, x_2)\, x_2\| \le \sigma_2 \|x_2\|^2$

*(vii)* $C\,(x_1,x_2)\, x_2 \;=\; C\,(x_1,\zeta)\, \zeta + \pi_0\, (x_1,\zeta)\, \tilde{\zeta} + o^2(\tilde{\zeta})$

*(viii)* $C\,(x_1,x_2)\, \zeta \;=\; C\,(x_1,\zeta)\, \zeta + \pi_1\, (x_1,\zeta)\, \tilde{\zeta}$

where $\tilde{\zeta} = x_2 - \zeta$ and

$$\pi_0(x_1, \zeta) = \frac{\partial}{\partial x_2} \{C(x_1, x_2) x_2\}_{x_2 = \zeta}$$

$$\pi_1\, (x_1, \zeta) = \frac{\partial}{\partial\, x_2} \{C(x_1, x_2)\, \zeta\}_{x_2 = \zeta}$$

*(ix)* $\|\pi_0\, (x_1, \zeta)\| \le \sigma_3 \|\zeta\|$ ; $\sigma_3 = 2n\, \{\sup \|N_i(x_1)\|\}$ $\forall$ i, $x_1$

*(x)* $\|\pi_1\, (x_1, \zeta)\| \le \sigma_4 \|\zeta\|$ ; $\sigma_4 = n\, \{\sup \|N_i(x_1)|\}$ $\forall$ i, $x_1$

*(xi)* $\|o^2(\tilde{\zeta})\| \le \sigma_5 \|\tilde{\zeta}|^2$

where $\|.\|$ denotes any norm and the $\sigma_i$'s are positive bounded non-zero constants. Properties*(vi)* -*(xi)* hold for all $\zeta$ and $x_1 \in \mathbb{R}^n$ with $\|\zeta\| < \infty$.

# 3. Control design

## 3.1 Nonlinear observer

The observer design is carried out on the basis of the nonlinear state-space structure (2.2), with additional correcting error terms introduced to ensure good observer tracking properties. The structure is the following :

$$\dot{\hat{x}}_1(t) = -\Gamma_1(t)\, \tilde{x}_1(t) + \hat{x}_2(t) \quad (3.1.a)$$

$$\dot{\hat{x}}_2(t) = -\Gamma_2(t)\, \tilde{x}_1(t) + \hat{\beta}(t) + u(t) \quad (3.1.b)$$

with,

$$\tilde{x}_1(t) = \hat{x}_1(t) - x_1(t)$$

$$\tilde{x}_2(t) = \hat{x}_2(t) - x_2(t)$$

where $\Gamma_1(t)$ and $\Gamma_2(t)$ are the *nxn* design matrices. They may be linear or nonlinear functions of the observer states or simply constant matrices. The nonlinear vector $\hat{\beta}(t)$, which may also depend on the system states and on the estimated state vector, is introduced to compensate for the nonlinear effects of $\beta(x_1, x_2)$. It does not necessarily have the same structure as $\beta(x_1, x_2)$ but it will closely resemble it. Its choice, as well as the selection of the observer gain matrices, will be discussed later. Since the position $x_1(t)$ is accessible from direct measurement, the observer structure (3.1) only contains terms in $x_1(t)$ and not on its estimate $x_1(t)$.

The error system is thus obtained by subtracting the system (2.2) from the observer (3.1),

$$\dot{\tilde{x}}_1(t) = -\Gamma_1(t)\, \tilde{x}_1(t) + \tilde{x}_2(t) \quad (3.2.a)$$

$$\dot{\tilde{x}}_2(t) = -\Gamma_2(t)\, \tilde{x}_1(t) + \Delta\beta(t) \quad (3.2.b)$$

where,

$$\Delta\beta(t) = \hat{\beta}(t) - \beta(x_1, x_2)$$

The observer design then consists in finding suitable functions $\Gamma_1$, $\Gamma_2$ and $\hat{\beta}$, such that the error system (3.2) has an asymptotically stable solution. Using property *(vii)*, the vector $\beta(x_1, x_2)$ can be rewritten as a function of the time-varying bounded vector $\zeta = \dot{q}_r$ :

$$\beta(x_1,x_2) = -H(x_1)^{-1}\, [C(x_1,\zeta)\zeta + \pi_0(x_1,\zeta)(x_2-\zeta) + \tau_g(x_1) + 0^2(\tilde{\zeta})] \quad (3.3)$$

where $\tilde{\zeta} = x_2 - q_r$, which suggests defining $\hat{\beta}(t)$ as,

$$\hat{\beta}(t) = \beta(x_1,\hat{x}_2,\zeta) = -H(x_1)^{-1}\,[C(x_1,\zeta)\zeta + \pi_0(x_1,\zeta)(\hat{x}_2-\zeta) + \tau_g(x_1)] \tag{3.4}$$

giving :

$$\Delta\beta = \hat{\beta} - \beta = -H(x_1)^{-1}\,[\pi_0(x_1,\zeta)\,\tilde{x}_2 - o^2(\tilde{\zeta})]$$

with this choice, the error system becomes:

$$\dot{\tilde{x}}_1(t) = -\Gamma_1(t)\,\tilde{x}_1(t) + \tilde{x}_2(t) \tag{3.5.a}$$

$$\dot{\tilde{x}}_2(t) = -\Gamma_2(t)\,\tilde{x}_1(t) - H(x_1)^{-1}\,\pi_0(x_1,\zeta)\,\tilde{x}_2(t) + H(x_1)^{-1}\,o^2(\tilde{\zeta}) \tag{3.5.b}$$

Letting the observer gains $\Gamma_1(t)$ and $\Gamma_2(t)$ be defined as:

$$\Gamma_1(t) = \Gamma_1^* + \Gamma_1^*(t)$$

$$\Gamma_2(t) = \Gamma_2^* + \Gamma_2^*(t)$$

where $\Gamma^*_1$ and $\Gamma^*_2$ are constant matrices and $\Gamma^*_1(t)$ and $\Gamma^*_2(t)$ are time varying matrices, possibly also state dependent, the system (3.5) can be rewritten as:

$$\dot{\tilde{x}}(t) = [A_1 + B_1(t)]\,\tilde{x}(t) + f_1(\tilde{\zeta}) \tag{3.6}$$

with,

$$A_1 = \begin{bmatrix} -\Gamma_1^* & I \\ -\Gamma_2^* & 0 \end{bmatrix} \qquad B_1(t) = \begin{bmatrix} -\Gamma_1^*(t) & 0 \\ -\Gamma_2^*(t) & -H(x_1)^{-1}\pi_0(x_1,\zeta) \end{bmatrix}$$

$$f_1(\tilde{\zeta}) = \left[\,0\,,\quad H(x_1)^{-1}o^2(\tilde{\zeta})\right]^T$$

with $o^2(\tilde{\zeta})$ bounded by the square norm of $\tilde{\zeta}$ as indicated by property *(xi)* .

## 3.2 Control law

Let $q_r(t)$ be the twice-differenciable reference trajectory with bounded second order derivatives, and consider the following control law:

$$\tau = H(x_1)[\ddot{q}_r - K_v(t)\,(\hat{x}_2-\dot{q}_r) - K_p^*\,(x_1-q_r)] + C(x_1,\dot{q}_r)\dot{q}_r + \tau_g(x_1) \tag{3.7}$$

with,

$$K_v(t) = K_v^* - H(x_1)^{-1}\,\pi_0(x_1,\dot{q}_r) \tag{3.8}$$

where $K_v^*$ and $K_p^*$ are constant matrices to be defined later. Introducing the tracking error e(t) as :

$$e(t) = q(t) - q_r(t)$$

The closed loop system is given as,

$$H(x_1)\,[\ddot{e} + K_v(t)\,(\hat{x}_2 - \dot{q}_r) + K_p^*\,e] = C(x_1,\dot{q}_r)\dot{q}_r - C(x_1,x_2)x_2$$

$$= \pi_0\,(x_1,\dot{q}_r)\dot{e} - o^2(\dot{e}) \tag{3.9}$$

where the last equality is obtained by using property *(vii)* , with $\zeta = \dot{q}_r$ .With the time varying gain (3.8), system (3.9) can be rewritten as:

$$\ddot{e} + K_v^*\,\dot{e} + K_p^*\,e = -K_v(t)\,\tilde{x}_2 - H(x_1)^{-1}\,o^2(\dot{e}) \tag{3.10}$$

or in the following state space form, (with $w_1 = e$, $w_2 = \dot{e}$ and $w^T = [w_1^T, w_2^T]$ )

$$\dot{w}(t) = A_2\,w(t) + [\,A_3 + B_2(t)\,]\,\tilde{x}(t) + f_2(w) \tag{3.11}$$

where

$$A_2 = \begin{bmatrix} 0 & I \\ -K_p^* & -K_v^* \end{bmatrix} \qquad A_3 = \begin{bmatrix} 0 & 0 \\ 0 & -K_v^* \end{bmatrix}$$

$$B_2(t) = \begin{bmatrix} 0 & 0 \\ 0 & H(x_1)^{-1}\pi_0(x_1,\dot{q}_r) \end{bmatrix}$$

$$f_2(w) = \begin{bmatrix} 0 & -H(x_1)^{-1}o^2(w) \end{bmatrix}^T$$

The system equation (3.6) and (3.11) yields

$$\begin{Bmatrix} \dot{\tilde{x}} \\ \dot{w} \end{Bmatrix} = \begin{bmatrix} A_1 & 0 \\ A_3 & A_2 \end{bmatrix} \begin{Bmatrix} \tilde{x} \\ w \end{Bmatrix} + \begin{bmatrix} B_1(t) & 0 \\ B_2(t) & 0 \end{bmatrix} \begin{Bmatrix} \tilde{x} \\ w \end{Bmatrix} + \begin{Bmatrix} f_1(\tilde{x}) \\ f_2(w) \end{Bmatrix} \tag{3.12}$$

which can be rewritten as

$$\dot{z}(t) = [\,A + B(t)\,]\,z(t) + f(z) \tag{3.13}$$

where $z^T = [x^T, w^T]$ and where we have used $B(t) = B(x_1,\dot{q}_r)$. The matrix B(t) is smooth and bounded for any $x_1$ and for any bounded and differenciable $\dot{q}_r$.The vector f(z) verifies f(0) = 0. The following lemma gives local conditions for the stability of system (3.13).

## 3.3 Closed loop stability

***Lemma 1*** : Consider the system (3.13) under the following assumption :

*(i)* All the eigenvalues of A have a negative real part :

$$\| e^{At} \| \leq a_1 e^{-\alpha t} \; ; \; \| e^{A^T t} \| \leq a_2 e^{-\alpha t}$$

*(ii)* $\| B(t) \| \leq b_0 < \infty$ for all $t \geq 0$.

*(iii)* $\| f(z) \| \leq \rho_0 \| z \|^2$

then the norm of z(t ) verifies:

$$\|z(t)\|^2 \leq \varepsilon_0 \|z(0)\|^2 e^{-\varepsilon_1 (t)} \qquad \forall \, t \geq 0$$

$$\varepsilon_1(t) = \int_0^t \left( \frac{1}{p_0} - 2\, b_0 - 2\, \rho_0 \|z(\tau)\| \right) d\tau$$

where $\varepsilon_0 \geq 1$ and $p_0$ is an upper bound on the symmetric and positive definite matrix P obtained from the solution of the following equation :

$$A^T P + P A = - I \quad , \qquad \|P\| \leq p_0$$

Furthermore, if the constants $b_0$, $\varepsilon_0$ and $p_0$ verify the following inequality,

$$(iv) \qquad 1 - 2 b_0 p_0 \left( 1 - \frac{1}{\sqrt{\varepsilon_0}} \right) > 0 \tag{3.14}$$

then the sets $\Omega_0(z,r_0)$ and $\Omega_{\varepsilon 0}(z, r_0)$ and a positive constant $r_0$ exist,

$$\Omega_{\varepsilon 0}(z, r_0) = \{ \, z : \|z\|^2 < r_0^2 / \varepsilon_0 \, \} \qquad , \; \varepsilon_0 \geq 1$$

$$\Omega_0(z,r_0) = \{ \, z : \|z\|^2 < r_0^2 \, \}$$

with $r_0$ given as,

$$r_0 = \frac{1 - 2b_0p_0}{2\rho_0p_0}$$

such that any trajectory z(t) starting in $\Omega_{\varepsilon 0}(z, r_0)$ remains within $\Omega_0(z, r_0) \supseteq \Omega_{\varepsilon 0}(z, r_0)$ and asymptotically tends to zero with a rate given by $\varepsilon_1(t) > 0$.

***Proof.*** Define $V = z^T Pz$ with $P = P^T > 0$ and $p_0 \geq \|P\|$. Using assumptions *(ii)* and *(iii)*, we can easily see that the time derivative of V(t) is bounded as follows :

$$\begin{aligned}\dot{V} &= z^T[A^TP + PA]\,z + z^T\,[B^TP + PB]\,z + 2\,z^TP\,f(z)\\ &\leq -\|z\|^2\,[1 - 2\,b_0p_0 - 2\,\rho_0p_0\,\|z\|] = -\|z\|^2\varepsilon(z)\end{aligned}$$

where P is the solution of $A^TP + P\,A = -\,I$, which due to hypothesis *(i)*, exists and is uniquely defined as :

$$P = \int_0^{\infty} e^{At}\,e^{A^Tt}\,dt$$

Since V(t) is upperbounded as $V(t) \leq \|z(t)\|^2\,p_0$, we obtain,

$$\frac{\dot{V}(z)}{V(z)} \leq -\frac{\|z\|^2\,\varepsilon(z)}{\|z\|^2\,p_0} = -\frac{\varepsilon(z)}{p_0}$$

integrating from 0 to t, we get

$$V(z(t)) = V(z(0))\,e^{\frac{-1}{p_0}\int_0^t \varepsilon\,(z(\tau))\,d\tau} = V(z(0))\,e^{-\varepsilon_1(t)}$$

using lower and upper bounds on V the above expression gives

$$\|z(t)\|^2 \leq \varepsilon_0\,\|z(0)\|^2\,e^{-\varepsilon_1(t)} \quad ; \quad \varepsilon_0 = \frac{\lambda_{sup}(P)}{\lambda_{min}(P)}$$

Condition *(iv)* implies that $1 - 2\,b_0\,p_0 > 0$, so that $r_0$ is positive and therefore $\varepsilon(z)$ and hence $\varepsilon_1(t)$ remain positive if z verifies the following

$$\|z(t)\| \leq \frac{1 - 2b_0p_0}{2\rho_0p_0} = r_0 \quad \Leftrightarrow \quad \|z(t)\|^2 \leq r_0^2$$

that is as long as z remains in $\Omega_0$. Any initial condition $z(0) \in \Omega_{\varepsilon 0}$ implies that the norm of z is bounded as follows:

$$\|z(t)\|^2 \leq \varepsilon_0 \|z(0)\|^2 e^{-\varepsilon_1(t)} \leq \frac{\varepsilon_0 r_0^2}{\varepsilon_0} e^{-\varepsilon_1(t)} = r_0^2 e^{-\varepsilon_1(t)}$$

therefore z remains in $\Omega_0$ and asymptotically decrease to zero. The maximum archivable rate is obtained as $\|z\|$ tends to zero, i.e.

$$\lim_{\|z\| \to 0} e^{-\varepsilon_1(t)} = e^{-(1 - 2 b_0 p_0)t}$$

Note that $\varepsilon_1(t)$ remains positive as long as the inequality (3.14) is verified. On the other hand, since A is an asymptotically stable matrix, positive constants $a_1$, $a_2$ and $\alpha$ exist such that,

$$\|e^{At}\| \leq a_1 e^{-\alpha t} \quad \text{and} \quad \|e^{A^T t}\| \leq a_2 e^{-\alpha t}$$

where $\alpha \leq \mathrm{Re}\{\lambda_{max}[A]\}$. Including these bounds on the solution of P, we obtain a bounded $p_0$, i.e.

$$\|P\| \leq a_1 a_2 \int_0^{\infty} e^{-2\alpha t} \, dt = \frac{a_1 a_2}{2\alpha} = p_0$$

∇∇∇

To increase the size of $\Omega_{\varepsilon 0}$ and $\Omega_0$, it is advisable to try to diminish $b_0$ by reducing the norm of B(t ). The observer gains, $\Gamma_1(t)$, $\Gamma_2(t)$, and the controller gains, $K_p^*$ and $K_v^*$, are then designed seeking to minimize $\|B(t)\|$ and render A asymptotically stable.

***Lemma 2.*** (Asymptotic stability of A). Let, $\Gamma_1^* = \gamma_1 I$, $\Gamma_2^* = \gamma_2 I$, $K_p^* = k_p I$ and $K_v^* = k_v I$, then the eigenvalues of A are given by the solution of the following second order equations :

$$\det(\lambda I - A) = \det(\lambda I - A_1)\det(\lambda I - A_2) = (\lambda^2 + \gamma_1\lambda + \gamma_2)^n(\lambda^2 + k_v\lambda + k_p)^n$$

***Proof.*** Using block determinant properties we obtain,

$$\det(\lambda I - A_1) = \begin{vmatrix} (\lambda + \gamma_1)I & -I \\ \gamma_2 I & \lambda I \end{vmatrix} = |\lambda I| \, |(\lambda + \gamma_1)I - (-I)(\lambda I)^{-1}(\gamma_2 I)|$$

$$= |\lambda I| \; |(\lambda + \gamma_1 + \frac{\gamma_2}{\lambda}) I| = \lambda^n (\lambda + \gamma_1 + \frac{\gamma_2}{\lambda})^n = (\lambda^2 + \gamma_1 \lambda + \gamma_2)^n$$

$$\det(\lambda I - A_2) = \begin{vmatrix} \lambda I & -I \\ k_p I & (\lambda + k_v) I \end{vmatrix} = |\lambda I| \, |(\lambda + k_v) I + \frac{k_p}{\lambda} I|$$

$$= \lambda^n \left( \lambda + k_v + \frac{k_p}{\lambda} \right) = (\lambda^2 + k_v \lambda + k_p)^n$$

$\nabla\nabla\nabla$

Simple setting of the eigenvalues of A can thus be performed by tuning the positive scalar $\gamma_1$, $\gamma_2$, $k_v$ and $k_p$. Although this choice fulfills condition *(i )* ,it does not necessarily gives the smallest norm for $p_0$.Increasing arbitrarily $\alpha$ does not necessarily mean that $p_0$ can be decreased.

***Lemma 3.*** If $\Gamma_1{}^*(t) = \Gamma_2{}^*(t) = 0$, then an upperbound on the norm of B(t) is minimized.

***Proof*** ( Outline ). Note that the norm $\|B(t)\|$ is bounded as follows:

$$\|B(t)\| \le \| B_1(t) \| + \| B_2(t) \|$$

and that $B_2(t)$ does not depend on $\Gamma_1{}^*(t)$ or $\Gamma_2{}^*(t)$. The norm of $B_1(t)$ can be expressed in terms of the eigenvalues of $B_1(t)^T B_1(t)$, which are all real. Using Gersgorin's Theorem, one can proove that the smallest upperbound on $\lambda_{sup}\{B_1{}^T B_1\}$ is then obtained for $\Gamma_1{}^*(t) = \Gamma_2{}^*(t) = 0$.

$\nabla\nabla\nabla$

***Corollary***. (Determination of $b_0$). If $\Gamma_1{}^*(t) = \Gamma_2{}^*(t) = 0$ and $\dot{q}_r(t)$ is a bounded vector, ($\|\dot{q}_r(t)\| < \zeta_0$, $\forall\, t > 0$), a bounded positive constant $b_0$ exists such that :

$$\|B(t)\| \le \|B_1(t)\| + \|B_2(t)\| = b_0 \qquad \forall\; t > 0$$

***Proof.*** With $\Gamma_1{}^*(t) = \Gamma_2{}^*(t) = 0$, we have $B_1(t) = B_2(t)$ and hence

$$\|B(t)\| \le 2\,\|B_2(t)\| = 2\,\|H(x_1)^{-1}\,\pi_0(x_1,\dot{q}_r)\| \le 2\,\|H(x_1)^{-1}\|\,\|\pi_0(x_1,\dot{q}_r)\|$$
$$\le 2\,\sigma'_0\,\sigma_3\,\|\dot{q}_r\| \le 2\,\sigma'_0\,\sigma_3\,\zeta_0 = b_0$$

where the upperbounds $\sigma'_0$, $\sigma_3$ and $\zeta_0$ are obtained from properties *(ii)* and *(ix)*, and from the boundedness of $\dot{q}_r(t)$.

∇∇∇

The constant $b_0$ depends on the norm of $\| \dot{q}_r \|$ and essentially, on the eigenvalues of the inertia matrix, hence $\sigma'_0$ and $\sigma_3$. This upperbound cannot be modified by the observer gains. It will be established by the mechanical characteristics of the robot and by a nominal velocity, $\dot{q}_r$, around which the observer stability is studied. The following theorem resumes the previous results.

***Theorem 1.*** The following "smooth" observer :

$$\dot{\hat{x}}_1(t) = -\Gamma_1(t)\, \tilde{x}_1(t) + \hat{x}_2(t) \tag{3.15.a}$$

$$\dot{\hat{x}}_2(t) = -\Gamma_2(t)\, \tilde{x}_1(t) \; -H(x_1)^{-1}\, [C(x_1,\dot{q}_r)\dot{q}_r + \pi_0(x_1,\dot{q}_r)(\hat{x}_2-\dot{q}_r) + \tau_g(x_1)] + u(t) \tag{3.15.b}$$

together with the control law (2.4)-(3.7)-(3.8) is locally stable ( in the sense of Lemma1) provided that the eigenvalues of $A_1$ and $A_2$ are tuned so that Condition (3.14) is verified for any z(0) belonging to $\Omega_{\varepsilon 0}(z)$.

# 4. Conclusion

In this paper we have proposed a modified computed-torque law which together with the nonlinear observer yields a locally exponentially stable closed-loop system. These results can be interpreted as a local stabilisation of a robot manipulator with a nonlinear control law based only on position. These results are important in industrial robot applications because they help to reduce the cost and dimension of the sensors. The analysis technique used for the smooth observer design yields quite conservative results leading, in general, to over-dimensioned observer gains. In practice, one can take smaller gains without affecting the observer's properties.

# 5. References

[1] ***Canudas de Wit, C. (1988).*** "Adaptive Control for partially known Systems: Theory and Applications", Elsevier Publishers, Studies in Automation and Control, V. 7

[2] ***Canudas de Wit, C. and Slotine, J.J.E. (1989).*** "Sliding Observers in Robot Manipulators", IFAC Symposium on Nonlinear Control Systems Design, Capri, Italy,June 1989.

[3] ***Nicosia, S., Tomei, P. and Tornambe, A. (1986).*** "Feedback control of elastic robots by Pseudo-Linearisation Techniques", 25th Conference on Decision and Control, Athens, Greece, 1986, pp 397-40

[4] ***Nicosia, S., Tomei, P. and Tornambe, A. (1987).*** "Approximate Asymptotic Observers for a class of Nonlinear Systems",Proc 26th Conference on Decision and Control, Los Angeles, CA, Dec. 1987.

[5] ***Reboulet, C. and Champetier, C. (1984).*** "A new method for linearizing non-linear systems: the pseudolinearization", Int. Journal of Control, 1984, vol 40, n° 4, pp 631-638.

[6] ***Slotine, J.J.E., Hedrick, J.K. and Misawa, E.A. (1987).*** "Sliding Observers for Nonlinear Systems", ASME-Journal of Dynamics Systems, Measurements and Control, Vol. 109 n° 3.

[7] ***Tornambe, A. and Tomei, P. (1986).*** "A new approach to observe and control a one link flexible robot arm", Proc. 8th, IASTED Int. Symp. Robotics and Artificial Intelligence, Toulouse 1986.

# Modelling and Simulation of Flexible Beams using Cubic Splines and Zero-Order Holds.

Simon Dancose, Jorge Angeles

**Abstract**

The modelling of linearly elastic beams using cubic splines is introduced in this paper. This technique makes it possible to model beams of any shape more efficiently than with existing methods. In fact, cubic splines are shown to be more generally applicable than the assumed-mode approach. Moreover, they are computationally less costly than finite-element methods, and hence, are more suitable for real-time applications. As well, a zero-order-hold scheme is implemented for the simulation of beams undergoing large rigid-body motions. Finally, an example is included to illustrate the effectiveness of the techniques proposed here. In this example, a stepped beam free at one end and clamped at a rotating hub, is simulated under deterministic initial disturbances.

## 1. Introduction

Present day industrial robots are easy to control since they are designed to be rigid and slow. The consequence is a low weight-carrying capacity which increases costs and decreases the speed of operation. To improve this ratio, lightweight robots with flexible links are required. The best example of an existing flexible manipulator is the Space Shuttle Remote Manipulator System, developed and built by Spar Aerospace Inc.for NASA: it moves relatively slowly so as to minimize the elastic deformations. It is expected that a variety of new manipulators will be developed as part of the Space Station project under NASA direction, and these will have to be much faster so as to be used as aids for space-structures assembly and for many specialized tasks such as in-orbit satellite repair and refueling (Lacombe and Berger [7]).

Two of the main reasons discouraging the flexible robot arm design are: the arising complex mathematical models, which are costly to handle, and difficult to control. Some of the better known techniques available to derive the governing equations of motion for an elastic system include: finite element analysis (Laurenson [8]; Bayo [2], Menq and Chen [10]); normal mode analysis (Meirovitch [9]; Cannon and Schmitz [4]; Hastings and Book [13]); transfer functions(Skaar and Tucker [14]) and lumping techniques (Book [3]). Although these methods have been successfully applied in some simple examples, more complicated mechanical systems, like multi-link manipulators, call for more efficient modelling and solution techniques. Here lies the motivation behind the spline-based method of solution described in this paper. The idea of using cubic splines comes from their natural association with elastic beams under static conditions, that is, both have a vanishing fourth derivative. Furthermore, the zero-order hold is introduced for the time-discretisation of the finite-dimensional continuous-time model derived from the use of cubic splines. It is shown that, by suitably exploiting the structure of the model at hand, the zero-order hold can be derived very efficiently from the viewpoints of economy and accuracy of computations.

## 2. Space-Discretisation Using Cubic Splines

The kinetic and potential energy, $T$ and $V$, respectively, of the flexible beam rotating about one of its ends, as shown in Fig 1, are the following:

$$T = \frac{1}{2}\int_a^l \rho(x)s(x)\dot{\mathbf{r}}\cdot\dot{\mathbf{r}}\,dx + \frac{1}{2}I_h\dot{\theta}^2 \quad , \quad V = \frac{1}{2}\int_a^l EI(x)\left[u''(x,t)\right]^2 dx \tag{1a \& b}$$

where

| | |
|---|---|
| $a$ | : length of hub from center of rotation to beam [m] |
| $l$ | : length [m] |
| $\rho(x)$ | : mass density [kg/m$^3$] |
| $s(x)$ | : cross-sectional area [m$^2$] |
| $EI(x)$ | : flexural rigidity [kg m$^3$/s$^2$] |
| $u(x,t)$ | : deflection of beam from its neutral axis [m] |
| $u''(x,t)$ | : curvature of beam [m$^{-1}$] |
| $I_h$ | : moment of inertia of the hub [kg m$^2$] |
| $\theta$ | : rotation of hub [rad] |

The following assumptions are made:

*i*) The deflection $u(x,t)$ is small ($\leq 0.1\,l$) and any extension is neglected.

*ii*) As a consequence of *i*, the Euler-Bernoulli model is used, for which rotary inertia and shear deformation effects are ignored.

*iii*) Beam is long and slender, i.e., the cross-sectional dimensions are smaller than 10% of the length

*iv*) Hub is rigid, and hence, its kinetic energy, $T_h$, can be written as $I_h\dot{\theta}^2/2$.

From these assumptions the following expressions are derived for the position vector $\mathbf{r}$ and the velocity $\dot{\mathbf{r}}$ of an arbitrary point on the beam:

$$\mathbf{r} = x\,\mathbf{i} + u\,\mathbf{j} \quad , \quad \dot{\mathbf{r}} = -u\dot{\theta}\,\mathbf{i} + (\dot{u} + x\dot{\theta})\mathbf{j} \tag{2a \& b}$$

If $u(x,t)$ is approximated by cubic spline functions, then we have:

$$u_k(x,t) = A_k(x-x_k)^3 + B_k(x-x_k)^2 + C_k(x-x_k) + D_k \tag{3}$$

for $x_k \leq x \leq x_{k+1}$, where $x_k$ is the abscissa of the $k$th supporting point (SP) of the spline (Späth [15]), and

$$A_k = \frac{1}{6\Delta x_k}(u''_{k+1} - u''_k) \tag{4a}$$

$$B_k = \frac{1}{2}u''_k \tag{4b}$$

$$C_k = (u_{k+1} - u_k)/\Delta x_k - \frac{1}{6}\Delta x_k(u''_{k+1} + 2u''_k) \tag{4c}$$

$$D_k = u_k \tag{4d}$$

where

$$\Delta x_k = x_{k+1} - x_k \tag{5}$$

Note that Timoshenko beams can also be modelled with spline functions. In this case, however, a second spline function would be needed to account for the rotation of the cross section of the beam.

The beam can be modelled as a *cantilever beam* rotating about its clamped end where the boundary conditions are the following:

$$\text{at } x = a, \quad \begin{cases} u = 0, \\ \frac{du}{dx} = 0. \end{cases} \tag{6a \& b}$$

These mean that, at the clamped end, both the displacement and the slope of the beam vanish. Moreover, at the free end, both the moment and the shear stress exerted on the beam vanish. Hence,

$$\text{at } x = l, \quad \begin{cases} \frac{d^2u}{dx^2} = 0, \\ \frac{d^3u}{dx^3} = 0. \end{cases} \tag{7a \& b}$$

For $n$ supporting points, the continuity and smoothness conditions at the SP yield the following relation:

$$\mathbf{A}_s\mathbf{u}'' = 6\mathbf{C}_s\mathbf{u} \tag{8}$$

where $\mathbf{A}_s$ and $\mathbf{C}_s$ are the following $n' \times n'$ matrices:

$$\mathbf{A}_s = \begin{bmatrix} 2\alpha_1 & \alpha_1 & 0 & & 0 \\ \alpha_1 & 2\alpha'_1 & \alpha_2 & \cdots & 0 \\ 0 & \ddots & & & \vdots \\ 0 & \cdots & \alpha_{n-3} & 2\alpha'_{n-3} & \alpha_{n-2} \\ 0 & \cdots & 0 & \alpha_{n-2} & 2\alpha'_{n-2} \end{bmatrix} \tag{9}$$

and

$$\mathbf{C}_s = \begin{bmatrix} \beta_1 & 0 & 0 & \cdots & & 0 \\ -\beta'_1 & \beta_2 & 0 & & & \\ \beta_2 & -\beta'_2 & \beta_3 & 0 & & \\ 0 & & \ddots & & & \vdots \\ 0 & \cdots & 0 & \beta_{n-2} & -\beta'_{n-2} & \beta_{n'} \end{bmatrix} \tag{10}$$

with

$$\alpha_i = \Delta x_i, \quad \beta_i = 1/\Delta x_i, \quad \text{for } i = 1, \ldots, n', \quad \text{where} \quad n' = n - 1$$

$$\alpha_i' \equiv \alpha_i + \alpha_{i+1}, \text{ for } i = 1, \ldots n-3;\ 2\alpha_{n-2}' = 2\alpha_{n-2} + 3\alpha_{n-1};$$

$$\beta_i' = \beta_i + \beta_{i+1} \quad \text{for } i = 1, \ldots, n-2$$

and $\mathbf{u}$ and $\mathbf{u}''$ are the following $n'$-dimensional vectors:

$$\mathbf{u} = [u_2, \ldots, u_n]^T, \quad \mathbf{u}'' = [u_1'', \ldots, u_{n-2}'', 0]^T \qquad (11a \ \& \ b)$$

where $\mathbf{u}$ is the vector of time-varying displacement and $\mathbf{u}''$ is the vector of time-varying curvature at the SPs, and hence, these vectors contain neither $u_1$ nor $u_{n-1}''$. Thus, eq.(8) leads to

$$\mathbf{u} = \mathbf{N}\mathbf{u}'', \quad \text{where} \quad \mathbf{N} = \frac{1}{6}\mathbf{C}_s^{-1}\mathbf{A}_s \qquad (12a \ \& \ b)$$

which gives a very useful linear relationship between displacement and curvature. Using the above relationship, and after some manipulation, the following energy expressions are obtained:

$$T = \frac{1}{2}\left[\dot{\theta}^2\left(I_a + \mathbf{u}^T\mathbf{M}'\mathbf{u}\right) + \dot{\mathbf{u}}^T\mathbf{M}'\dot{\mathbf{u}} + 2\dot{\theta}\vec{\gamma}^T\mathbf{N}^{-1}\dot{\mathbf{u}}\right], \quad V = \frac{1}{2}\mathbf{u}^T\mathbf{K}'\mathbf{u} \qquad (13a \ \& \ b)$$

Now, eqs.(13$a$ & $b$) are rewritten in the forms $T = T(\theta, \mathbf{u}'', \dot{\mathbf{u}}'')$ and $V = V(\theta, \mathbf{u}'')$, which are more convenient for experimental purposes, i.e., $\mathbf{u}''$ can be measured directly using straingauges, while $\mathbf{u}$ is more difficult to measure, since it requires the use of vision systems that are not capable of operating at the high frequencies of a beam (Piedboeuf and Hurteau [12]). The energy expressions now take on the form

$$T = \frac{1}{2}\left[\dot{\theta}^2\left(I_a + \mathbf{u}''^T\mathbf{M}''\mathbf{u}''\right) + \dot{\mathbf{u}}''^T\mathbf{M}''\dot{\mathbf{u}}'' + 2\dot{\theta}\vec{\gamma}^T\dot{\mathbf{u}}''\right], \quad V = \frac{1}{2}\mathbf{u}''^T\mathbf{K}''\mathbf{u}'' \qquad (14a \ \& \ b)$$

where

$$I_a = I_b + I_h \qquad (15)$$

and $I_b$ is the moment of inertia of the unflexed beam. Moreover, $\mathbf{M}''$, $\mathbf{K}''$, and $\vec{\gamma}$ are all constant coefficients dependent upon the beam's configuration. It is underlined that $\mathbf{M}''$ and $\mathbf{K}''$ are $n' \times n'$ symmetric positive definite matrices, and so are $\mathbf{M}' = \mathbf{N}^{-T}\mathbf{M}''\mathbf{N}^{-1}$ and $\mathbf{K}' = \mathbf{N}^{-T}\mathbf{K}''\mathbf{N}^{-1}$

Now, the term $\mathbf{u}''^T\mathbf{M}''\mathbf{u}''$, is assumed to be negligible as compared to $I_a$, which is verified later in the simulation. Then, $T$ and $V$ can be represented as

$$T = \frac{1}{2}\dot{\mathbf{q}}^T\tilde{\mathbf{M}}\dot{\mathbf{q}}, \quad V = \frac{1}{2}\mathbf{q}^T\tilde{\mathbf{K}}\mathbf{q} \qquad (16a \ \& \ b)$$

where

$$\tilde{\mathbf{M}} = \begin{bmatrix} \mathbf{M}'' & \vec{\gamma} \\ \vec{\gamma}^T & I_a \end{bmatrix}, \quad \tilde{\mathbf{K}} = \begin{bmatrix} \mathbf{K}'' & \mathbf{0} \\ \mathbf{0}^T & 0 \end{bmatrix}, \quad \text{and} \quad \mathbf{q} = \begin{bmatrix} \mathbf{u}''^T & \theta \end{bmatrix}^T \qquad (17c - e)$$

Next, the dynamical model is derived using the Euler-Lagrange equations, thus obtaining the following:

$$I_a\ddot{\theta} + \vec{\gamma}^T \ddot{\mathbf{u}}'' = \tau(t) \tag{18a}$$

$$\vec{\gamma}\ddot{\theta} + \mathbf{M}''\ddot{\mathbf{u}}'' + \mathbf{K}''\mathbf{u}'' = \mathbf{0} \tag{18b}$$

in which $\mathbf{0}$ is the $n'$-dimensional zero vector and $\tau(t)$ is the applied torque. Moreover, eq.(18$a$) is a scalar equation and eq.(18$b$) represents a set of $n'$ scalar equations. In compact form, the foregoing equations take on the form:

$$\tilde{\mathbf{M}}\ddot{\mathbf{q}} + \tilde{\mathbf{K}}\mathbf{q} = \begin{bmatrix}\mathbf{0}^T & \tau(t)\end{bmatrix}^T \tag{19}$$

These equations can be reduced by substituting eq.(18$a$) into eq.(18$b$), thus yielding the following set of $n'$ equations, which are also independent of $\theta$:

$$\mathbf{M}\ddot{\mathbf{u}}'' + \mathbf{K}\mathbf{u}'' = \vec{\phi}(t) \tag{20a}$$

where

$$\mathbf{M} = \mathbf{M}'' - \frac{1}{I_a}\vec{\gamma}\vec{\gamma}^T, \quad \mathbf{K} = \mathbf{K}'', \quad \text{and} \quad \vec{\phi}(t) = -\frac{1}{I_a}\vec{\gamma}\tau(t) \tag{20b – d}$$

The time response of the system described by eq.(20$a$) will be calculated using the method discussed in the following section.

Once eq.(20$a$) is integrated, the hub motion is obtained from eq.(18$a$) by simple quadrature as follows:

$$\theta(t) = \frac{1}{I_a}\left\{\int_0^t \left[\int_0^{t'} \tau(\sigma)d\sigma\right] dt' - \vec{\gamma}^T\mathbf{u}''(t)\right\} + C_1 t + C_2 \tag{21}$$

where constants $C_1$ and $C_2$ are computed from initial conditions.

## 3. Zero-Order-Hold Time Discretisation of the Finite-Dimensional Dynamical Model of the Vibrating Beam

A linear stationary—i.e. time-invariant—dynamical system has the standard state-variable representation that follows (Kailath [6]):

$$\dot{\mathbf{x}}(t) = \mathbf{A}\mathbf{x}(t) + \mathbf{B}\mathbf{u}(t), \quad \mathbf{x}(0) = \mathbf{x}_0 \tag{22a}$$

$$\mathbf{y}(t) = \mathbf{C}\mathbf{x}(t) \tag{22b}$$

where $\mathbf{x}$, $\mathbf{u}$ and $\mathbf{y}$ are $2n'$-, $p$- and $m$- dimensional vectors of state, control and output variables respectively. Moreover $\mathbf{A}$, $\mathbf{B}$ and $\mathbf{C}$ are, correspondingly, $2n' \times 2n'$, $2n' \times p$ and $m \times 2n'$ constant matrices.

The discrete-time version of the same system, when sampled with a zero-order hold (Åström and Wittenmark [1]), at time intervals of amplitude $h$, is:

$$\mathbf{x}(k+1) = \mathbf{F}\mathbf{x}(k) + \mathbf{G}\mathbf{u}(k), \quad \mathbf{x}(0) = \mathbf{x}_0 \tag{23a}$$

$$\mathbf{y}(k) = \mathbf{H}\mathbf{x}(k) \tag{23b}$$

where **F** and **G** are $2n' \times 2n'$ and $2n' \times p$ constant matrices defined as:

$$\mathbf{F} = e^{\mathbf{A}h}, \quad \mathbf{G} = \int_0^h e^{\mathbf{A}\tau} d\tau \mathbf{B}, \quad \mathbf{H} = \mathbf{CF} \tag{24a - c}$$

and clearly.

$$\mathbf{x}(k) = \mathbf{x}(t_k), \quad \mathbf{u}(k) = \mathbf{u}(t_k), \quad t_k = kh \tag{24d}$$

Therefore, the task of simulating eq.(22) becomes a problem of evaluating matrices **F** and **G**. Methods for computing these matrices are available (Åström and Wittenmark [1]; Moler and Van Loan [11]; Fulmer [5]; Ward [16]), and hence, these need no further discussion. However, the said methods rely on the computation of the exponential of a $2n' \times 2n'$ matrix **A**. Presented next is an approach allowing the computation of the exponential via matrix functions of an $n' \times n'$ positive definite matrix, which is much faster to compute and provides a highly reliable means to assess accuracy.

First, the Cholesky decomposition of **M**, namely, $\mathbf{M} = \mathbf{N}^T\mathbf{N}$ which is possible because **M** is positive definite, is computed. Next, eq.(20$a$) is pre-multiplied by $\mathbf{N}^{-T}$ where **N** is lower triangular. For reasons that will be explained later, the decoupled system, eq.(20$a$), is used instead of eq.(19), i.e.,

$$\mathbf{N}^{-T}\mathbf{N}^T\mathbf{N}\ddot{\mathbf{u}}'' + \mathbf{N}^{-T}\mathbf{K}\mathbf{u}'' = \mathbf{N}^{-T}\vec{\phi}(t) \tag{25}$$

Letting $\mathbf{v} = \mathbf{N}\mathbf{u}''$, the following equation is obtained

$$\ddot{\mathbf{v}} + \mathbf{W}\mathbf{v} = \mathbf{N}^{-T}\vec{\phi}(t) \tag{26a}$$

where

$$\mathbf{W} = \mathbf{N}^{-T}\mathbf{K}\mathbf{N}^{-1} \tag{26b}$$

in which **W** is also positive definite. Thus, by letting $\mathbf{x} = \begin{bmatrix}\mathbf{v}^T & \dot{\mathbf{v}}^T\end{bmatrix}^T$ and the scalar input $u(t) = \tau(t)$, matrix **A** turns out to be

$$\mathbf{A} = \begin{bmatrix} \mathbf{O} & \mathbf{1} \\ -\mathbf{W} & \mathbf{O} \end{bmatrix} \tag{27a}$$

where **O** and **1** are the $n' \times n'$ zero and identity matrices, respectively, and

$$\mathbf{B} = \begin{bmatrix} \mathbf{0} \\ \mathbf{N}^{-1}\mathbf{w} \end{bmatrix}, \quad \mathbf{w} = \begin{bmatrix} 0 \\ 1 \end{bmatrix} \tag{27b}$$

where **0** is the $(2n-1)$-dimensional vector.

Since it is desired to measure the curvature at the SPs, the output becomes $\mathbf{y} = \mathbf{u}''$. Putting this in terms of **x** yields:

$$\mathbf{C} = \begin{bmatrix} \mathbf{N}^{-1} & \mathbf{O} \end{bmatrix} \tag{27c}$$

This system can easily be shown to be observable; therefore, just by measuring $\mathbf{u}''$ and $\theta$, the state of the system can be determined. The system is also controllable, which implies that any state can be achieved by choosing the right input $u$.

Because of the simple structure of $\mathbf{A}$, the exponential matrix $\mathbf{F}$ can be calculated in a simplified form. In fact, the $k$th power of $\mathbf{A}$ takes on the following form:

$$\mathbf{A}^k = \begin{bmatrix} (-1)^{k/2}\mathbf{\Omega}^k & \mathbf{O} \\ \mathbf{O} & (-1)^{k/2}\mathbf{\Omega}^k \end{bmatrix} \tag{28a}$$

for $k$ even, and

$$\mathbf{A}^k = \begin{bmatrix} \mathbf{O} & (-1)^{\frac{k-1}{2}}\mathbf{\Omega}^{k-1} \\ (-1)^{\frac{k+1}{2}}\mathbf{\Omega}^{k+1} & \mathbf{O} \end{bmatrix} \tag{28b}$$

for $k$ odd, where $\mathbf{\Omega}$ is the positive definite square root of $\mathbf{W}$, i.e.,

$$\mathbf{\Omega} = \mathbf{W}^{1/2} > \mathbf{O} \tag{28c}$$

From this, it can readily be shown that:

$$\mathbf{F} = \begin{bmatrix} \cos\mathbf{\Omega}h & \mathbf{\Omega}^{-1}\sin\mathbf{\Omega}h \\ -\mathbf{\Omega}\sin\mathbf{\Omega}h & \cos\mathbf{\Omega}h \end{bmatrix} \tag{29}$$

which requires simply the calculation of the *cos* and *sin* functions of the $n' \times n'$ matrix $\mathbf{\Omega}h$.

To calculate $\mathbf{G}$, we exploit the fact that $\mathbf{A}$ is invertible and hence, eq.(24$b$) leads to

$$\mathbf{G} = \mathbf{A}^{-1}\left(e^{\mathbf{A}h} - \mathbf{1}\right)\mathbf{B} \tag{30a}$$

where $\mathbf{1}$ is the identity $n' \times n'$ matrix and

$$\mathbf{A}^{-1} = \begin{bmatrix} \mathbf{O} & -\mathbf{W}^{-1} \\ \mathbf{1} & \mathbf{O} \end{bmatrix} \tag{30b}$$

Thus

$$\mathbf{G} = \begin{bmatrix} -\mathbf{W}^{-1}(\cos\mathbf{\Omega}h - \mathbf{1}) \\ (\sin\mathbf{\Omega}h)\mathbf{\Omega}^{-1} \end{bmatrix}\mathbf{N}^{-1}\mathbf{w} \tag{31}$$

Finally, $\mathbf{H}$ is found to be:

$$\mathbf{H} = \begin{bmatrix} \mathbf{N}^{-1}\cos\mathbf{\Omega}h & \mathbf{N}^{-1}\mathbf{\Omega}^{-1}\sin\mathbf{\Omega}h \end{bmatrix} \tag{32}$$

By looking at eqs.(29–31) it is clear why eq.(20) was used instead of eq.(19). Indeed, by using eq.(20) matrix $\mathbf{A}$ is nonsingular thus making it possible to calculate $\mathbf{W}^{-1}$ and $\mathbf{\Omega}^{-1}$. The singularity of $\mathbf{A}$, that would have resulted from eq.(19), is caused by the singularity of $\mathbf{K}_c$—no elasticity in the motor shaft.

All the required matrix functions can be calculated by placing **W** in its diagonal form:

$$\mathbf{W} = \mathbf{S}\mathbf{\Lambda}_{\mathbf{W}}\mathbf{S}^{-1}$$

where the columns of **S** are the—mutually orthonormal—proper vectors of **W** and **Λ** is a diagonal matrix composed of the—real positive—proper values of **W**.

## 4. Example

The simulation of the non-uniform beam shown in Fig.2 and described in table 1 was performed. The initial conditions and input values are given in Table 2. The calculated natural frequencies of the beam are stored in vector $\mathbf{w}_n$ which is displayed next. The calculated matrices $\mathbf{M}''$ and $\mathbf{K}''$, and the vector $\vec{\gamma}$ are also given next. Figures 3 to 4 are the time histories of the beam.

Table 1

| | Hub | Section 1 | Section 2 |
|---|---|---|---|
| $n$ (No of SPs) | 0 | 5 | 4 |
| $L$ [m] | $a = 0.1$ | 0.5 | 0.4 |
| $\rho$ [kg/m$^3$] | 2712 | 2712 | 2712 |
| $E$ [GPa] | — | 71.0 | 71.0 |
| $b$ [m] | — | 0.1 | 0.1 |
| $d$ [mm] | — | 2.0 | 1.0 |

$$I_h = 4.26 \times 10^{-2} \quad \text{kgm}^2$$

The initial conditions are:

$$\theta(0) = -0.0056 \quad \text{rad}$$
$$\dot{\theta}(0) = 0 \quad \text{rad/s}$$

Table 2

| SP | $x$[m] | $u$[m] | $\dot{u}$[m/s] |
|---|---|---|---|
| 1 | 0.100 | 0.000 | 0 |
| 2 | 0.225 | 0.001 | 0 |
| 3 | 0.350 | 0.002 | 0 |
| 4 | 0.475 | 0.004 | 0 |
| 5 | 0.600 | 0.008 | 0 |
| 6 | 0.733 | 0.016 | 0 |
| 7 | 0.867 | 0.032 | 0 |
| 8 | 1.000 | 0.064 | 0 |

$$\mathbf{M}'' = \begin{bmatrix} 0.287 & 0.481 & 0.348 & 0.228 & 0.134 & 0.066 & 0.022 \\ 0.482 & 0.816 & 0.598 & 0.397 & 0.237 & 0.117 & 0.040 \\ 0.349 & 0.599 & 0.452 & 0.309 & 0.189 & 0.095 & 0.033 \\ 0.230 & 0.400 & 0.311 & 0.221 & 0.141 & 0.073 & 0.026 \\ 0.136 & 0.240 & 0.191 & 0.142 & 0.096 & 0.052 & 0.019 \\ 0.067 & 0.118 & 0.096 & 0.073 & 0.052 & 0.031 & 0.012 \\ 0.022 & 0.040 & 0.033 & 0.026 & 0.019 & 0.012 & 0.005 \end{bmatrix} \times 10^{-3}\text{kg m}^4$$

$$\mathbf{K}'' = \begin{bmatrix} 0.197222 & 0.098611 & 0 & 0 & 0 & 0 & 0 \\ 0.098611 & 0.394444 & 0.098611 & 0 & 0 & 0 & 0 \\ 0 & 0.098611 & 0.394444 & 0.098611 & 0 & 0 & 0 \\ 0 & 0 & 0.098611 & 0.394444 & 0.098611 & 0 & 0 \\ 0 & 0 & 0 & 0.098611 & 0.223519 & 0.013148 & 0 \\ 0 & 0 & 0 & 0 & 0.013148 & 0.052593 & 0.013148 \\ 0 & 0 & 0 & 0 & 0 & 0.013148 & 0.052593 \end{bmatrix} \text{N m}^3$$

$$I_a = 0.152346 \quad \text{kg m}^2$$

$$\vec{\gamma} = \begin{bmatrix} 5.3993 \\ 8.9375 \\ 6.3445 \\ 4.1223 \\ 2.4528 \\ 1.2107 \\ 0.4048 \end{bmatrix} \times 10^{-3}\text{kg m}^3\text{s}, \quad \mathbf{w}_n = \begin{bmatrix} 4.0121 \\ 9.5265 \\ 26.8127 \\ 45.7768 \\ 74.9687 \\ 102.205 \\ 111.893 \end{bmatrix} \text{Hz}$$

Furthermore, as stated in Section 2, $I_a >> \mathbf{u}^T\mathbf{M}'\mathbf{u}$ since for the worst case scenario, where $\mathbf{u}$ is taken at its maximum allowable value—to remain within the linearly elastic model from assumption *ii* in Section 2—, the following ratio is obtained

$$I_a/(\bar{\mathbf{u}}^T\mathbf{M}'\bar{\mathbf{u}}) = 0.00535$$

where $\bar{\mathbf{u}}$ is the vector of the maximum allowable displacements.

## 5. Conclusions

The space discretisation of a beam under rigid-body motions was achieved using cubic splines. This procedure allows the direct use of either displacement or curvature as the generalized coordinates, which permits the use of strain-gauges to measure the curvature, thus yielding the displacement from eq.(12$a$). The use of strain-gauges is much better than vision feedback since it is much more accurate and much faster. This spline model also makes it possible to simulate beams with shape discontinuities such as the one in the example which would be extremely tedious to solve using assumed modes.

The time-discretisation of the continuous-time finite-dimensional model of the beam was performed with zero-order holds. The procedure followed in this paper performs very rapidly and accurately and seems suitable for the modelling, simulation and control of mechanical systems composed of multiple flexible bodies. This improvement is introduced by the fact that some matrix functions are performed on $n' \times n'$-dimensional positive definite matrices rather than taking matrix functions on $2n' \times 2n'$-dimensional non-symmetric matrices.

## 6. References

[1] Åström K.J. and Wittenmark B.. 1984. *Computer Controlled Systems: Theory and Design*. Prentice-Hall. Englewood Cliffs. N.J..

[2] Bayo, E.. 1987. "A Finite-Element Approach to Control the End-Point Motion of a Single-Link Flexible Robot." *Journal of Robotic Systems*, Vol.4, No.1. pp 63-75.

[3] Book, W.J.. 1979. "Analysis of Massless Elastic Chains with Servo Controlled Joints," *Journal of Dynamic Systems, Measurement, and Control*. Trans. of ASME. Vol.101, pp. 187-192.

[4] Cannon, R. and Schmitz E.. "Initial Experiments on the End-Point Control of a Flexible One-Link Robot". *The International Journal of Robotics Research*. The MIT press. Cambridge, Massachusetts. Vol.3, No.3, Fall 1984.

[5] Fulmer, E.P., 1975. Computation of the Matrix Exponential, *Amer. Math. Monthly*, 82, pp. 156-159.

[6] Kailath, T., 1980. *Linear systems*, Prentice-Hall, N.J.

[7] Lacombe, J.L. and Berger, G., 1982. "In-space assembly and maintenance of unmanned spacecraft," *Proceedings of the International Federation of Automatic Control in Space*, Noordwijkerhout, The Netherlands.

[8] Laurenson, R.M., 1976. "Modal Analysis of Rotating Flexible Structures," *AIAA 14th Aerospace Sciences Meeting*, Washington, D.C., Vol.14, No.10, pp. 1444-1450.

[9] Meirovitch, L.. *Elements of Vibration Analysis*, McGraw-Hill, 1975.

[10] Menq, C.-H., and Chen, J.-S., 1988. "Dynamic modeling and payload-adaptive control of a flexible manipulator," *IEEE Robotics and Automation*, pp. 488-493.

[11] Moler, C. and Van Loan, C., 1978. Ninteen Dubious Ways to Compute the Exponential of a Matrix, *SIAM Reviews*, Vol.20, No.4, pp.801-836.

[12] Piedboeuf, J.C., and Hurteau, R., 1987. "Modelling and Identification of Parameters of a Flexible One-Link Arm Using Visual Feedback," *Proc. IEEE Conference on Robotics and Automation*, p.1555.

[13] Sakawa, Y. , Matsuno F. and Fukushima S., 1985. "Modeling and Feedback Control of Flexible Arm," *Journal of Robotics Systems*, John Wiley & Sons, Inc. Vol.2, pp. 453-472.

[14] Skaar, S.B. and Tucker, D., 1986. "Point Control of a One-Link Flexible Manipulator," *ASME Journal of Applied Mechanics*, Vol.53, pp.23-27.

[15] Späth, H., 1978, *Splines Algorithmen zur Konstruktion glatter Kurven und Flachen*, Second Edition, R. Oldenbourg Verlag, Munich-Vienna.

[16] Ward, R.C., 1977. "Numerical computation of the matrix exponential with accuracy estimate", *SIAM J. of Numerical Analysis*, pp.600-610.

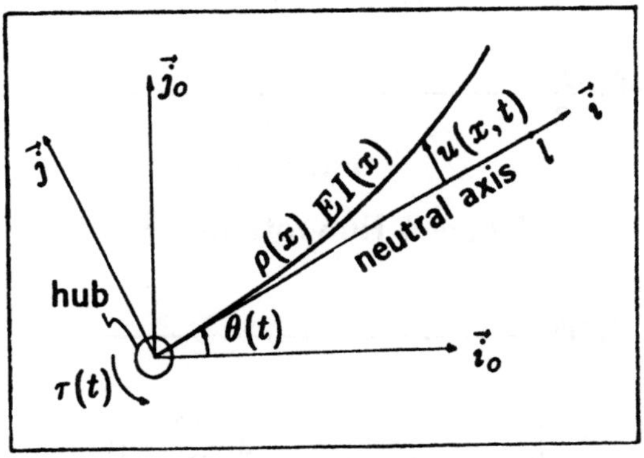


Figure 1 Rotating Flexible Beam

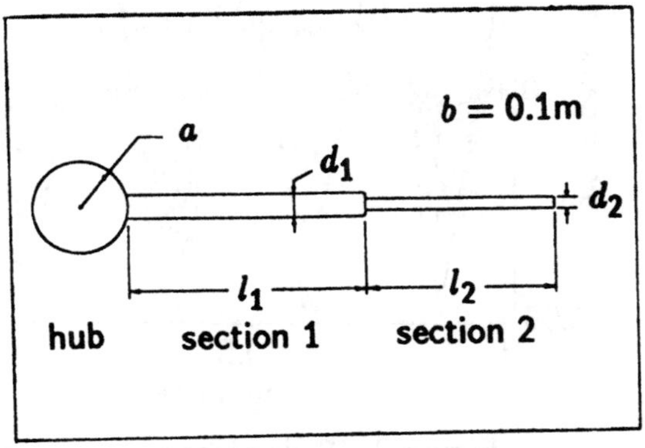


Figure 2 Beam Used in Example

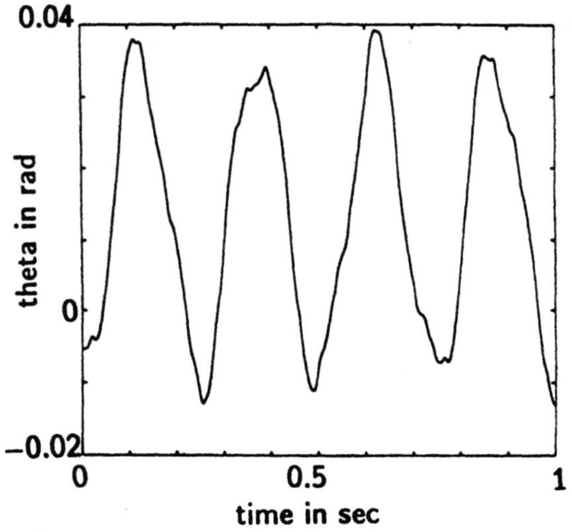


Figure 4 Hub rotation vs time

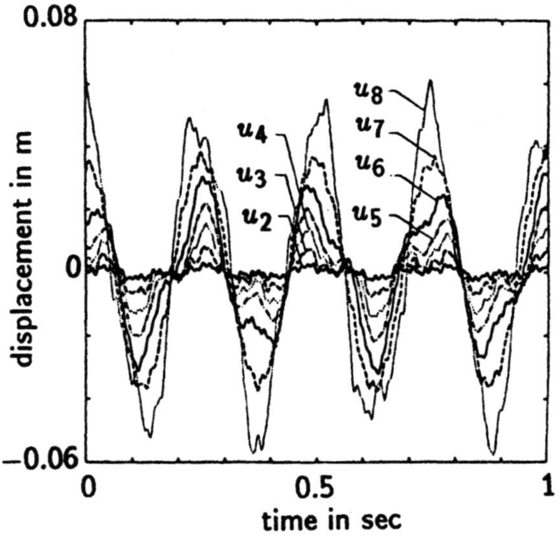


Figure 3 SP displacement vs time

# Towards a differential topological classification of robot manipulators

K. Tchoń

**Abstract**

The paper is concerned with a classification of models of rigid manipulators via state feedback and local diffeomorphisms of the state and output spaces. Using some methods of singularity theory, finite lists of normal forms are provided for generic manipulators with the number of degrees of freedom $n \geq 2$ , $n \neq 8$ . If $n = 8$ , the classification is proved to be infinite.

# 1 Introduction

This study has originally been motivated by the output tracking problem in robot manipulators. The problem amounts to finding driving forces or torques which, applied at the joints of the manipulator, would make its effector to travel along a prescribed trajectory in the workspace.

The output tracking problem has a well-known solution in the case when the manipulator's kinematics are non-singular, [6]. In [11] a singular tracking problem in non-redundant manipulators is dealt with and solved under assumption that singularities in the internal coordinates of the manipulator are either crossed transversally, if of codimension 1, or avoided, if of higher codimensions. In general, however, when solving the tracking problem or the inverse kinematic problem, one has to deal with complex singularities of kinematics, [3],[5], [8],[9],[12],[15],[18]. Hence, a useful preliminary step toward solving the problem may be to classify possible singularities of kinematics of the manipulator. This paper presents such a classification for "almost all" manipulators. We believe that the classification might be interesting on its own, so we have concentrated here just on the classification deferring some robotic applications to a separate paper , [18].

More specifically, we shall consider affine control systems with outputs which describe both dynamics and kinematics of the manipulator. Since the dynamics are typically non-singular, by appropriate static feedback and the state space coordinates change, we reduce the classification problem of manipulators to classifying singularities of their kinematics via pairs of local diffeomorphisms. Having formulated the problem in this way, we are able to apply some machinery of singularity theory to produce lists of normal forms of the manipulators in generic cases. The lists are finite for all numbers of degrees of freedom $n \geq 2$, except $n = 8$. From the systemic point of view this paper may be thought of as a development of certain ideas stated in [10].

The paper is composed as follows. In section 2 we introduce some basic concepts, in particular describe the transformations under which our classification is being made. Section 3 contains the main result (Theorem 3.1). Proof of the theorem is sketched in Section 4. The paper is concluded with Section 5, where a classification is provided of manipulators whose kinematic transformations are restricted to either a change of the effector position or of the orientation.

# 2 Basic concepts

We shall consider the Lagrange model of dynamics of a rigid, n-degree of freedom manipulator consisting of the 5th order kinematic pairs. The model can be expressed in the following form, [4],[17],

$$\begin{aligned} \dot{x} &= \theta \\ \dot{\theta} &= f(x,\theta) + g(x)u \\ y &= h(x) \end{aligned} \tag{1}$$

Hereabove $x, \theta$ stand for, respectively, internal coordinates and velocities (positions and velocities at the joints), u is a vector of controls (driving forces or torques at the joints), y results from the kinematic transformation accomplished by the manipulator. It will be assumed that $x, \theta$ belong to an open neighbourhood of $0 \in R^n$ or just to $R^n$. The outputs $y$ are elements of the Euclidean group $SE(3) \cong SO(3) \times R^3$, and descibe positions and orientations of the effector w.r.t. a fixed base coordinate frame. All the maps displayed in (1) are required to be smooth, i.e. of class $C^\infty$. Matrix $g(x)$ is invertible everywhere, [17].

Given a system (1) one can apply the feedback $u = -g(x)^{-1}f(x,\theta) + g(x)^{-1}v$ and transform (1) to the form $\mu$:

$$\begin{aligned} \dot{x} &= \theta \\ \dot{\theta} &= u \\ y &= h(x) \end{aligned} \tag{2}$$

which will serve us in what follows as a basic reference model. Denote by $M$ the class of systems (2). Then it is clear that $M$ can be identified with the set of smooth maps $C^\infty(R^n, SE(3))$, hence topologized by furnishing the last set with the weak $C^\infty$ topology, [7],[13]. We recall here that the weak $C^\infty$ topology is a topology of uniform convergence on compact subsets of $R^n$, of maps along with their partial derivatives of all orders. It is well-known that $\mathcal{M} \cong C^\infty(R^n, SE(3))$ is a Baire space, i.e. residual subsets (countable intersections of open-dense subsets of $\mathcal{M}$ ) are dense in $\mathcal{M}$ . Further on a property possessed by systems $\mu$ belonging to a residual subset of $\mathcal{M}$ will be called *generic*. If the residual set is clear from the context, its elements will often be referred to as generic.

In this paper we are considering the problem of classification of systems (2). More specifically, we want to find a finite list of *simple normal forms* to which one could locally transform systems (2) from a *residual subset* of $\mathcal{M}$, using some *natural transformations*. The transformations admitted are of the following type:

i. Local diffeomorphic change of $(x, \theta)$ coordinates.

ii. Static state feedback.

iii. Local diffeomorphism of output coordinates.

In particular, if $\varphi$ is a local diffeomorphism of internal coordinates and $\omega$ is a local diffeomorphism of SE(3) then the transformations are defined as follows:

$$\begin{array}{lcl} (x,\theta) & \mapsto & (\varphi(x), \frac{\partial\varphi(x)}{\partial x}\theta), \\ u & \mapsto & \frac{\partial^2\varphi(x)}{\partial x^2}(\theta,\theta) + \frac{\partial\varphi(x)}{\partial x}u, \\ y & \mapsto & \omega(y). \end{array} \tag{3}$$

Clearly, in (3) $\frac{\partial\varphi(x)}{\partial x}$ stands for the Jacobi matrix of $\varphi$ at x , while the ith coordinate of $\frac{\partial^2\varphi(x)}{\partial x^2}(\theta,\theta)$ equals $\theta^T \frac{\partial^2\varphi_i(x)}{\partial x^2}\theta$.

Given (3) one easily deduces the following result which deserves to be stated separately.

**Proposition 2.1** *The action of transformations (3) on a system (2) is equivalent to* $h \mapsto \omega \circ h \circ \varphi^{-1}$.

In view of the above, the problem of classification of systems (2) has been reduced to the *classification of smooth maps* under the so-called RL-equivalence, [1],[13]. Indeed, we shall prove that a finite classification of generic systems exists for all $n \geq 2$ , except $n = 8$ .

# 3 Main result

In this section we are going to state a theorem on the generic classification of systems (2). The theorem establishes a collection of lists of normal forms, specified by the number of degrees of freedom n , to which a generic system (2) can be transformed via transformations (3). Having fixed n , let $\mathcal{M}_n$ denote the space of systems (2) with n degrees of freedom. We observe that, since dynamics of (2) are not affected by the transformations (see Proposition 2.1), the normal forms are completely defined just by their output maps.

**Theorem 3.1** *For every $n = 2, 3, \ldots$ , except $n = 8$ , there exists a residual subset $\mathcal{R}_n \subset M_n$ such that any $\mu \in \mathcal{R}_n$ is locally transformed via (S) to a normal form from the lists given below. If $n = 8$ , the classification by normal forms is infinite. The finite lists of normal forms are described by the following output maps $h(x)$:*

- *a.* $n = 2$ :
- $(x_1, x_2, 0, 0, 0, 0)$
- *b.* $n = 3$ :
- $(x_1, x_2, x_3, 0, 0, 0)$
- *c.* $n = 4$ :
  - *c1.* $(x_1, x_2, x_3, x_4, 0, 0)$
  - *c2.* $(x_1^2, x_2x_1, x_3x_1, x_2, x_3, x_4)$
- *d.* $n = 5$
  - *d1.* $(x_1, x_2, x_3, x_4, x_5, 0)$
  - *d2.* $(x_1^2, x_2x_1, x_2, x_3, x_4, x_5)$
  - *d3.* $(x_1^3 + x_2x_1, x_3x_1^2 + x_4x_1, x_2, x_3, x_4, x_5)$
- *e.* $n = 6$
  - *e1.* $(x_1, x_2, x_3, x_4, x_5, x_6)$
  - *e2.* $(x_1^2, x_2, x_3, x_4, x_5, x_6)$
  - *e3.* $(x_1^3 + x_2x_1, x_2, x_3, x_4, x_5, x_6)$
  - *e4.* $(x_1^4 + x_2x_1 + x_3x_1^2, x_2, x_3, x_4, x_5, x_6)$
  - *e5.* $(x_1^5 + x_2x_1 + x_3x_1^2 + x_4x_1^3, x_2, x_3, x_4, x_5, x_6)$
  - *e6.* $(x_1^6 + x_2x_1 + x_3x_1^2 + x_4x_1^3 + x_5x_1^4, x_2, x_3, x_4, x_5, x_6)$
  - *e7.* $(x_1^7 + x_2x_1 + x_3x_1^2 + x_4x_1^3 + x_5x_1^4 + x_6x_1^5, x_2, x_3, x_4, x_5, x_6)$
  - *e8.* $(x_1^2 \pm x_2^2 + x_3x_1 + x_4x_2, x_1x_2, x_3, x_4, x_5, x_6)$
  - *e9.* $(x_1^2 + x_2^3 + x_3x_1 + x_4x_2 + x_5x_2^2, x_1x_2, x_3, x_4, x_5, x_6)$
  - *e10.* $(x_1^3 + x_2^3 + x_3x_1 + x_4x_2 + x_5x_1^2 + x_6x_2^2, x_1x_2, x_3, x_4, x_5, x_6)$
  - *e11.* $(x_1^3 + x_3x_1 + x_4x_2 + x_5x_1x_2, x_1^2 + x_2^2 + x_6x_1, x_3, x_4, x_5, x_6)$
- *f.* $n = 7$ :
  - *f1.* $(x_2, x_3, x_4, x_5, x_6, x_7)$
  - *f2.* $(\pm x_1^3 + x_2^2, x_3, x_4, x_5, x_6, x_7)$

- *f3.* $(x_1^2 + x_2^3 + x_3x_2, x_3, x_4, x_5, x_6, x_7)$
- *f4.* $(\pm x_1^2 + x_2^4 + x_3x_2 + x_4x_2^2, x_3, x_4, x_5, x_6, x_7)$
- *f5.* $(x_1^2 + x_2^5 + x_3x_2 + x_4x_2^2 + x_5x_2^3, x_3, x_4, x_5, x_6, x_7)$
- *f6.* $(\pm x_1^2 + x_2^6 + x_3x_2 + x_4x_2^2 + x_5x_2^3 + x_6x_2^4, x_3, x_4, x_5, x_6, x_7)$
- *f7.* $(x_1^2+x_2^7+x_3x_2+x_4x_2^2+x_5x_2^3+x_6x_2^4+x_7x_2^5, x_3, x_4, x_5, x_6, x_7)$
- *f8.* $(x_1^2x_2 \pm x_2^3 + x_3x_1 + x_4x_2 + x_5x_1^2, x_3, x_4, x_5, x_6, x_7)$
- *f9.* $(x_1^2x_2 + x_2^4 + x_3x_1 + x_4x_2 + x_5x_1^2 + x_6x_2^2, x_3, x_4, x_5, x_6, x_7)$
- *f10.* $(x_1^2x_2 \pm x_2^5 + x_3x_1 + x_4x_2 + x_5x_1^2 + x_6x_2^2 + x_7x_2^3, x_3, x_4, x_5, x_6, x_7)$
- *f11.* $(x_1^3 + x_2^4 + x_3x_1 + x_4x_2 + x_5x_2^2 + x_6x_1x_2 + x_7x_1x_2^2, x_3, x_4, x_5, x_6, x_7)$
- *f12.* $(\pm x_1^2 + x_2^2 + x_3^2, x_1x_3 + x_4x_1 + x_5x_2 + x_6x_3, x_4, x_5, x_6, x_7)$
- *f13.* $(x_1x_2 - x_2x_3 + x_4x_3, x_1x_2 - x_1x_3 + x_5x_1 + x_6x_2, x_4, x_5, x_6, x_7)$
- *f14.* $(x_1x_2 - x_1x_3 + x_4x_1 + x_5x_2, x_2^2 + x_3^2 + x_1^3 + x_6x_1 + x_7x_1^2, x_4, x_5, x_6, x_7)$
- *f15.* $(x_1x_2 - x_1x_3 + x_4x_1 + x_5x_2, x_2x_3 + x_1^3 + x_6x_1 + x_7x_1^2, x_4, x_5, x_6, x_7)$

***Now let $n \geq 9$. Define $q_{n-k} = \pm x_1^2 \pm x_2^2 \ldots \pm x_{n-k}^2, r_{n-4} = (x_{n-4}, x_{n-3}, \ldots, x_n)$. Then there are the following forms:***

- ***h.*** $n \geq 9$ :
  - *h1.* $(x_{n-5}, r_{n-4})$
  - *h2.* $(q_{n-6} + x_{n-5}^2, r_{n-4})$
  - *h3.* $(q_{n-6} + x_{n-5}^3 + x_{n-4}x_{n-5}, r_{n-4})$
  - *h4.* $(q_{n-6} + x_{n-5}^4 + x_{n-4}x_{n-5} + x_{n-3}x_{n-5}^2, r_{n-4})$
  - *h5.* $(q_{n-6} + x_{n-5}^5 + x_{n-4}x_{n-5} + x_{n-3}x_{n-5}^2 + x_{n-2}x_{n-5}^3, r_{n-4})$
  - *h6.* $(q_{n-6} + x_{n-5}^6 + x_{n-4}x_{n-5} + x_{n-3}x_{n-5}^2 + x_{n-2}x_{n-5}^3 + x_{n-1}x_{n-5}^4, r_{n-4})$
  - *h7.* $(q_{n-6} + x_{n-5}^7 + x_{n-4}x_{n-5} + x_{n-3}x_{n-5}^2 + x_{n-2}x_{n-5}^3 + x_{n-1}x_{n-5}^4 + x_nx_{n-5}^5, r_{n-4})$
  - *h8.* $(q_{n-7} + x_{n-6}^2x_{n-5} \pm x_{n-5}^3 + x_{n-4}x_{n-6} + x_{n-3}x_{n-5} + x_{n-2}x_{n-6}^2, r_{n-4})$
  - *h9.* $(q_{n-7} + x_{n-6}^2x_{n-5} + x_{n-5}^4 + x_{n-4}x_{n-6} + x_{n-3}x_{n-5} + x_{n-2}x_{n-6}^2 + x_{n-1}x_{n-5}^2, r_{n-4})$
  - *h10.* $(q_{n-7} + x_{n-6}^2x_{n-5} \pm x_{n-5}^5 + x_{n-4}x_{n-6} + x_{n-3}x_{n-5} + x_{n-2}x_{n-6}^2 + x_{n-1}x_{n-5}^2 + x_nx_{n-5}^3, r_{n-4})$

– *h11.* $(q_{n-7}+x_{n-6}^3+x_{n-5}^4+x_{n-4}x_{n-6}+x_{n-3}x_{n-5}+x_{n-2}x_{n-5}^2+ x_{n-1}x_{n-6}x_{n-5}+x_nx_{n-6}x_{n-5}^2,\ r_{n-4})$

A sketch proof of the theorem is deferred to the next section. To conclude this section we observe that, since the spaces $M_n$ are Baire, the theorem results in the following.

**Corollary 3.1** *Let $n \geq 2$, $n \neq 8$. Then, for any $\mu \in M_n$, there exist transformations (3) which applied to $\mu$ make it arbitrarily close to a normal form from among the listed above (locally, in the sense of the weak $C^\infty$ topology). Furthermore, if $\mu \in R_n$, the transformations make $\mu$ to locally coincide with such a normal form.*

# 4 Proof of the main result

As a matter of fact, proof of Theorem 3.1, although rather long and technically involved, is just a compilation of known results and techniques of singularity theory. Hence, we are going to sketch here main ideas of the proof, and to give references to the literature, such that enable the reader with some basic acquaintance with singularity theory to reconstruct all the details.

First we note that, since the normal forms introduced are local, we can assume w.l.o.g. that output maps of (2) take their values in a real space $R^m$. Furthermore, the Euclidean group $SE(3)$ is six-dimensional, so $m = 6$. Thus, we shall deduce our result from the existing knowledge on classification of generic map-germs $h : R^n \longrightarrow R^6$ w.r.t. local diffeomorphisms in the source and target spaces. The equivalence of map-germs defined by the diffeomorphisms is called RL-equivalence, [1]. It is well-known, [1],[13], that for infinitesimally stable map-germs the classification by RL-equivalence reduces to the classification of so-called genotypes of the map-germs by the contact equivalence. Therefore, our strategy will follow that of [13], i.e. we shall classify the genotypes, and then construct the normal forms via stabilization by unfolding. Genericity of our normal forms results from the fact that, except for $n = 8$, we are working within the so-called nice range of dimensions, where the stability is generic, [14].

To be more specific, we shall consider infinitesimally stable map-germs from $(R^{p+s}, 0)$ to $(R^{p+t}, 0)$, where $p + s = n$, $p + t = 6$, whose rank at 0 equals $p$. By a transversality argument generic map-germs must satisfy the condition $p + s \geq st$, referred to in the sequel as the transversality condition. The stable map-germs are treated as stable p-parameter unfoldings of elements belonging to contact orbits of codimension $\leq p + s$ of map-germs $(R^s, 0) \mapsto (R^t, 0)$ whose rank at $0 \in R^s$ is equal to 0. Such maps-germs are called genotypes of the unfoldings. Hence, our primary task is to find canonical forms of contact orbits of genotypes, up to codimension $p+s$ . First, let us look at non-singular cases. They are represented by $s=0$ , if $n \leq 6$ , or $t=0$ , if $n \geq 6$ . Consequently, we arrive at normal forms of immersions (a,b,c1,d1,e1) or submersions (e1,f1,h1) under the RL-equivalence. It is easily checked that for $n=2$ or $n=3$ the immersions are the only cases met generically. Next, we let $n=4$. It follows that $s \leq t$ , so by the transversality condition $s=1$, $t=3$, $p=3$. In effect we need to classify genotypes $R \mapsto R^3$ . Equivalently, we can deduce the normal form c2 from a theorem of Morin, [16] . Similar argument leads to the normal forms d2 and d3 for $n=5$ . These are just the Morin canonical forms of Thom-Boardman singularities $S_{1,0}$ ,$S_{1,1,0}$ , [1]. Now consider the case $n=6$ . By the transversality condition we need to find canonical forms for contact orbits $K(5,1,1)$ and $K(4,2,2)$ in terminology of [13]. As before, the stable unfoldings of genotypes belonging to $K(5,1,1)$ follow immediately from the theorem of Morin for the singularities $S_{1,0}$, $S_{1_2,0}, \ldots, S_{1_6,0}$ . In this way we obtain normal forms e2-e7 . The contact orbits $K(4,2,2)$ have been characterized by Arnold, [2], by the following canonical forms: $(x_1^2 \pm x_2^2, x_1x_2)$, $(x_1^2+x_2^3, x_1x_2)$, $(x_1^3+x_2^3, x_1x_2)$, $(x_1^3, x_1^2+x_2^2)$. Normal forms e8-e11 are then produced by the unfolding of the above genotypes. The case of $n=7$ reduces, by the transversality condition, to classifying the contact orbits $K(5,2,1)$ and $K(4,3,2)$ . Indeed, $K(5,2,1)$ are classified by singularities of functions $R^2 \mapsto R$ of codimension $\leq 5$ and corank 1 or 2 . But this is just elementary catastrophe theory, [13]. By the unfolding of the genotypes of corank 1 (cuspoids), we obtain the normal forms f2-f7 , while f8-f11 result, by unfolding, from genotypes $D_4^{\pm}$ ,$D_5^{\pm}$, $D_6^{\pm}$, $E_6$ of Arnold, [1]. Next, the orbits $K(4,3,2)$ can be classified by singularities of maps $R^3 \mapsto R^2$ , of codim $\leq 4$ . There are five of them: three of codim 3 : $(\pm x_1^2+x_2^2+x_3^2, x_1x_3)$, $(x_1x_2-x_2x_3, x_1x_2-x_1x_3)$, [10], and two of codim 4 : $(x_1x_2-x_1x_3, x_2^2+x_3^2+x_1^3)$, $(x_1x_2-x_1x_3, x_2x_3+x_1^3)$, [14]. From these genotypes we deduce normal forms f12-f15 by unfolding. Observe that in fact the forms f2-f7 coincide with those found by Morin, [16].

In a sense the case of $n=7$ has been the hardest one. For, if $n \geq 9$ then, by the transversality condition, we only need to consider contact orbits $K(5, n-5, 1)$ , i.e. to classify singularities of functions $R^{n-5} \mapsto R$ of codim$\leq$ 5 and corank 1 or 2 .But, up to some Morse terms, the classification is the same as for $n=7$ (f2-f11).

Eventually we have arrived at the case of $n=8$ . The problem of producing generic normal forms in this case amounts, by the transversality condition, to classifying contact orbits $K(5,3,1)$ and $K(4,4,2)$ . Clearly, there is a normal form of the submersion type, and a finite list of normal forms obtained by the unfolding of singularities of functions $R^3 \mapsto R$ of codim$\leq$ 5 and corank 1 or 2 . So $K(5,3,1)$ unfolds to a finite collection of normal forms. Complications appear when analyzing $K(4,4,2)$ . Namely, it has been proved in [13] that $K(4,4,2)$ contains infinite number of orbits. Consequently, by a theorem of Mather, cf.[14], Th.1.2, p.217 (the "only if" part), there would be infinitely many non-equivalent 4-parameter unfoldings of genotypes representing the orbits. This finishes the proof.

# 5 Conclusion

We want to conclude this study with a natural by-product of Sections 3,4 , namely with an analog of Theorem 3.1 for systems (2) whose output map $h$ takes values either in $R^3$ or in $SO(3)$. The robotic interpretation of this particular $h$ is clear: it means that one simply pays attention either to the position or orientation of the effector. Especially the first case is often met in applications.

Using the same methods as those employed in the proof of Theorem 3.1 , we are able to give a complete, finite classification of systems (2) with $h : R^n \mapsto R^3$ or $SO(3)$, under suitably adjusted transformations (3). The result is as follows.

**Theorem 5.1** *Let $M_n$ denote the topological space of systems (2) with $h(x) \in R^3$ or $SO(3)$, acted on by transformations (3) with $\omega$ being a local diffeomorphism of $R^3$ or $SO(3)$ . Then, for every $n \geq 2$ , there exists a residual subset $\mathcal{R}_n \subset M_n$ such that any $\mu \in \mathcal{R}_n$ can be transformed locally to a normal form from the finite lists displayed below. The lists are determined by the form of $h(x)$.*

- *a.* $n = 2$ :
  - *a1.* $(x_1, x_2, 0)$
  - *a2.* $(x_1^2, x_2x_1, x_2)$
- *b.* $n = 3$ :
  - *b1.* $(x_1, x_2, x_3)$
  - *b2.* $(x_1^2, x_2, x_3)$
  - *b3.* $(x_1^3 + x_2x_1, x_2, x_3)$
  - *b4.* $(x_1^4 + x_2x_1 + x_3x_1^2, x_2, x_3)$
- *c.* $n \geq 4$ :
  - *c1.* $(x_{n-2}, x_{n-1}, x_n)$
  - *c2.* $(\pm x_1^2 ... \pm x_{n-3}^2 \pm x_{n-2}^2, x_{n-1}, x_n)$
  - *c3.* $(\pm x_1^2 ... \pm x_{n-3}^2 + x_{n-2}^3 + x_{n-1}x_{n-2}, x_{n-1}, x_n)$
  - *c4.* $(\pm x_1^2 ... \pm x_{n-3}^2 + x_{n-2}^4 + x_{n-1}x_{n-2} + x_nx_{n-2}^2, x_{n-1}, x_n)$

# 6 References


- [1] V. I. Arnold et al., Singularities of Differentiable Maps. Birkhäuser, Boston, 1985.
- [2] V.I.Arnold, Singularities of smooth maps. Usp. Mat. Nauk 23(1968) ,pp.3-44 (in Russian).
- [3] D.R.Baker, C.W.Wampler II, On the inverse kinematics of redundant manipulators. Int. J. Rob. Res., 7(1988), pp.3-21.
- [4] R.W.Brockett, Robotic manipulators an the product of exponentials formula. In: Math. Theory of Networks and Systems, Springer-Verlag, Berlin, 1984, pp.120-129.
- [5] J.H.Davis, R.M.Hirschorn, Tracing control of a flexible robot arm. IEEE Trans. Aut. Contr., 33(1988), pp.238-248.
- [6] E.Freund, Fast nonlinear control with arbitrary pole placement for industrial robots and manipulators. Robotics Res., 1(1982), pp.65-78.
- [7] M.W.Hirsch, Differential Topology. Springer-Verlag, Berlin, 1976.
- [8] R.M.Hirschorn, J.H.Davis, Output tracking for nonlinear systems with singular points. SIAM J. Contr. Opt., 25(1987), pp.547-557.
- [9] R.M.Hirschorn, J.H.Davis, Global output tracking for nonlinear systems. SIAM J. Contr. Opt., 26(1988), pp.1321-1330.
- [10] B.Jakubczyk, K.Tchoń, Singularities and normal forms of observed dynamics. Math.of Control, Signals and Systems,2(1989), pp.19-31.
- [11] A.Kasiński, K.Tchoń, Internal trajectory calculation for manipulators with locally degenerate kinematics. Proc. 2nd Nat. Conf. on Robotics, Technical Univ. of Wroclaw Publ., vol.2(1988), pp.126-134 (in Polish).
- [12] F.Lamnabhi-Lagarrigue, P.E.Crouch, I.Ighneiwa, Tracking through singularities. To appear in Proc. of Colloque International Automatique Nonlineaire, Nantes 1988.
- [13] J.Martinet, Singularities of Smooth Functions and Maps. Cambridge Univ. Press, Cambridge, 1982.
- [14] J.N.Mather, Stability of $C^\infty$ mappings: VI. The nice dimension. In: Liverpool Singularities I, ed. by C.T.C.Wall. Springer-Verlag, Berlin, 1971, pp. 207-253.

- **[15] N.H.McClamroch, D.Wang, Feedback stabilization and tracking of constrained robots. IEEE Trans. Aut. Control, 33, No5 (1988), pp.419-426.**

- **[16] B.Morin, Formes canoniques des singularitws d'une application differentiable. Compt. Rendus Acad. Sci. Paris 260(1965), pp.5662-5665,6503-6506.**

- **[17] R.P.Paul, Robot Manipulators: Mathematics, Programming, and Control. MIT Press, Cambridge, 1981.**

- **[18] K.Tchoń, Differential topology of the inverse kinematic problem for redundant robot manipulators. Submitted for publication.**

# AN ADAPTIVE PD CONTROL ALGORITHM FOR ROBOTS

P. Tomei

*Abstract.* This paper deals with the dynamic control of robots, whose parameters are not exactly known. The traditional PD control law is made adaptive with respect to the gravity parameters. Simulation tests are included to show the effectiveness of the proposed control law.

## 1. Introduction and robot model.

We refer to the so-called *point to point* control of robot manipulators. As well known [1], control laws based on the feedback from the state variables of joint angles and their derivatives have been shown to be globally asymptotically stable. Moreover, such control algorithms are robust with respect to uncertainties on the inertia parameters; that is, even if the inertia parameters are not known the global asymptotic stability is ensured.

Conversely, uncertainties on the gravity parameters (such as the payload) may lead to undesired steady state errors.

In this paper we show how an adaptive PD control law can be designed. The proposed controller ensures the asymptotic stability even if the inertia and gravity parameters are unknown, provided that upper and lower bounds of the inertia matrix are available.

Following the Lagrangian formulation the dynamics of a robot is described by

$$B(q)\ddot{q} + C(q,\dot{q})\dot{q} + e(q) + F\dot{q} = u \tag{1}$$

where $q$ is the vector of generalized coordinates, $B(q)$ is the symmetric positive definite inertia matrix that is bounded for any $q$, $C(q,\dot{q})$ takes into account the Coriolis and centrifugal forces and is linear with respect to $\dot{q}$ and bounded with respect to $q$, $F$ is the symmetric positive semidefinite matrix of the viscous friction coefficients, $u$ is the vector of the applied torques and $e(q)$ is the vector of the gravity forces.

Two simplifying properties should be noted about the dynamic model (1). First, as remarked by many authors [1,2], matrices $B(q)$ and $C(q,\dot{q})$, given a suitable definition of $C$, are related by the fact that matrix $\dot{B} - 2C$ is skew-symmetric. This implies that

$$\dot{B} = C + C^T \tag{2}$$

The other important property is that matrices $B$ and $C$ and vector $e$ are linear in terms of robot and load parameters [3].

2. Adaptive PD controller.

As known, referring to point to point control, an independent local PD feedback at each joint ensures the global asymptotic stability, provided that the gravity terms are compensated. This convergence is stated by the following theorem, that we recall for the sake of comprehension.

*Theorem* 1 [1]. Given a control law

$$u = e(q) - K_p(q - q_0) - K_v\dot{q} \tag{3}$$

where $K_p$ and $K_v$ are symmetric positive definite matrices, the equilibrium point $q = q_0, \dot{q} = 0$, of system (1),(3) is globally asymptotically stable.

*Proof.* Select as candidate Lyapunov function

$$v(q - q_0, \dot{q}) = \frac{1}{2}\dot{q}^T B\dot{q} + \frac{1}{2}(q - q_0)^T K_p(q - q_0)$$

We obtain

$$\dot{v}(q - q_0, \dot{q}) = \frac{1}{2}\dot{q}^T \dot{B}\dot{q} + \dot{q}^T(-C\dot{q} - F\dot{q} - K_p(q - q_0) - K_v\dot{q}) + \dot{q}^T K_p(q - q_0) \tag{4}$$

Recalling that $\dot{B} - 2C$ is skew-symmetric, from (4) we have

$$\dot{v} = -\dot{q}^T(K_v + F)\dot{q}$$

and therefore $\dot{v}$ is negative semidefinite. A direct application of the Lasalle theorem [4, p.108] gives the thesis. △

When the gravity vector $e(q)$ is not perfectly known, it cannot be exactly compensated. In this case the PD control law will be

$$u = \hat{e}(q) - K_p(q - q_0) - K_v\dot{q} \tag{5}$$

where $\hat{e}(q)$ is the available estimate of $e(q)$.

Observing (1) and (5) we note that, in general, $q = q_0, \dot{q} = 0$ is no more an equilibrium point of (1),(5). In fact, supposing that the following matrix is nonsingular

$$K_p + \frac{\partial(e - \hat{e})}{\partial q} \tag{6}$$

the equilibrium point becomes $q = \hat{q}_0, \dot{q} = 0$, where $\hat{q}_0$ is given by the solution of the algebraic equation

$$K_p(\hat{q}_0 - q_0) + e(\hat{q}_0) - \hat{e}(\hat{q}_0) = 0 \tag{7}$$

Note that the new equilibrium point can be made arbitrarily close to the previous one by increasing the proportional gain matrix $K_p$. However, the control law (5) does not ensure the asymptotic stability of the new equilibrium point [1].

A possible way to overcome the previous difficulty may be to make the PD algorithm adaptive. An intermediate step toward this solution is to find a strict global Lyapunov function which allows us to prove the stability of the PD controller (with exact gravity compensation) without the use of the Lasalle theorem.

In the sequel we adopt as a norm of a $n \times 1$ vector $x$

$$\|x\| = \sqrt{\sum_{i=1}^{n} x_i^2}$$

and as a norm of a matrix $A$ the corresponding induced norm

$$\|A\| = \sqrt{\max_{\text{eigenvalue}} A^T A}$$

Moreover, we indicate by $A_M$ and $A_m$, respectively, the maximum and the minimum eigenvalues of a symmetric positive definite bounded matrix $A(x)$, for any $x$.

*Alternative proof of Theorem 1.* As suggested by a work of Koditschek on the adaptive control for attitude tracking [5], we select as candidate Lyapunov function

$$v(q-q_0,\dot{q}) = \beta\left[\gamma\left(\frac{1}{2}\dot{q}^T B\dot{q} + \frac{1}{2}(q-q_0)^T K_p(q-q_0)\right) + \frac{2\dot{q}^T B(q-q_0)}{1+2(q-q_0)^T(q-q_0)}\right] \tag{8}$$

where $\beta$ and $\gamma$ are positive constants. Equation (8) implies

$$v \geq \beta\left[\gamma(\frac{1}{2}B_m\|\dot{q}\|^2 + \frac{1}{2}K_{p_m}\|q-q_0\|^2) - 2B_M\frac{\|\dot{q}\|\|q-q_0\|}{1+2\|q-q_0\|^2}\right] \tag{9}$$

From (9) we obtain that if

$$\gamma > \frac{2B_M}{\sqrt{B_m K_{p_m}}} \tag{10}$$

then $v$ is a positive definite radially unbounded function. The time derivative of (8) is given by

$$\begin{aligned}\dot{v} = \beta\Bigg[&\gamma\left(\frac{1}{2}\dot{q}^T\dot{B}\dot{q} + \dot{q}^T(-C\dot{q} - F\dot{q} - K_p(q-q_0) - K_v\dot{q}) + \dot{q}^T K_p(q-q_0)\right)\\ &+ 2\frac{\dot{q}^T\dot{B}(q-q_0) + \dot{q}^T B\dot{q} - (\dot{q}^T C^T + (q-q_0)^T K_p + \dot{q}^T K_v + \dot{q}^T F)(q-q_0)}{1+2(q-q_0)^T(q-q_0)}\\ &- 8\frac{\dot{q}^T B(q-q_0)\dot{q}^T(q-q_0)}{[1+2(q-q_0)^T(q-q_0)]^2}\Bigg]\end{aligned} \tag{11}$$

From (11), recalling (3) and the skew-simmetry of $\dot{B} - 2C$, we obtain

$$\begin{aligned}\dot{v} = \beta\Bigg[&-\gamma\dot{q}^T(K_v + F)\dot{q} - 2\frac{(q-q_0)^T K_p(q-q_0)}{1+2(q-q_0)^T(q-q_0)}\\ &+ 2\frac{\dot{q}^T B\dot{q}}{1+2(q-q_0)^T(q-q_0)} + 2\frac{\dot{q}^T C(q-q_0)}{1+2(q-q_0)^T(q-q_0)}\\ &+ 2\frac{\dot{q}^T(K_v + F)(q-q_0)}{1+2(q-q_0)^T(q-q_0)} - 8\frac{\dot{q}^T B(q-q_0)\dot{q}^T(q-q_0)}{[1+2(q-q_0)^T(q-q_0)]^2}\Bigg]\end{aligned} \tag{12}$$

It can be easily verified that

$$2\frac{\dot{q}^T C(q-q_0)}{1+2(q-q_0)^T(q-q_0)} \leq \frac{\alpha_C}{\sqrt{2}}\|\dot{q}\|^2$$
$$2\frac{\dot{q}^T B\dot{q}}{1+2(q-q_0)^T(q-q_0)} \leq 2B_M\|\dot{q}\|^2$$
$$8\frac{\dot{q}^T B(q-q_0)\dot{q}^T(q-q_0)}{[1+2(q-q_0)^T(q-q_0)]^2} \leq 2B_M\|\dot{q}\|^2$$

where $\alpha_C$ is such that

$$\|C(q,\dot{q})\| \leq \alpha_C\|\dot{q}\| \tag{13}$$

The constant $\alpha_C$ surely exists, since $C$ is linear in $\dot{q}$ and bounded in $q$. Hence,

$$\dot{v} \leq \beta\Bigg[-\gamma(K_{v_m}+F_m)\|\dot{q}\|^2 - 2K_{p_m}\frac{\|q-q_0\|^2}{1+2\|q-q_0\|^2} + (4B_M+\frac{\alpha_C}{\sqrt{2}})\|\dot{q}\|^2 + 2\frac{(K_{v_M}+F_M)\|\dot{q}\|\|q-q_0\|}{1+2\|q-q_0\|^2}\Bigg] \tag{14}$$

From (14) it follows that if

$$\gamma > \frac{1}{K_{v_m}}\left[\frac{(K_{v_M}+F_M)^2}{2K_{p_m}} + 4B_M + \frac{\alpha_C}{\sqrt{2}}\right] \tag{15}$$

$\dot{v}$ is negative definite. Therefore, taking $\gamma$ as a positive constant that simultaneously satisfies (10) and (15) the global asymptotic stability of the equilibrium point $q = q_0, \dot{q} = 0$ is proved, since the function (8) becomes a strict global Lyapunov function. $\triangle$

At this point we are ready to give the main result of this paper. Since the gravity vector e(q) is linear in terms of robot parameters, it can be expressed as

$$e(q) = E(q)p \tag{16}$$

where $p$ is the parameter vector and $E(q)$ is a known matrix. Consider the control law

$$u = -K_p(q-q_0) - K_v\dot{q} + E(q)\hat{p} \tag{17}$$

with the parameter adaptation dynamics

$$\dot{\hat{p}} = -\beta E^T(q)\left[\gamma\dot{q} + \frac{2(q-q_0)}{1+2(q-q_0)^T(q-q_0)}\right] \tag{18}$$

where $\gamma$ is such that

$$\gamma > \max\left\{\frac{2B_M}{\sqrt{B_m K_{p_m}}}, \frac{1}{K_{v_m}}\left[\frac{(K_{v_M}+F_M)^2}{2K_{p_m}} + 4B_M + \frac{\alpha_C}{\sqrt{2}}\right]\right\} \tag{19}$$

and $\beta$ is a positive constant.

Even if the inertia matrix is supposed unknown, we assume known upper and lower bounds of its eigenvalues. Moreover, we assume known the constant $\alpha_C$ defined in (13).

*Theorem* 2. Consider system (1). The control law (17),(18) is such that

$$\lim_{t\to\infty} \left\| \begin{bmatrix} q - q_0 \\ \dot{q} \end{bmatrix} \right\| = 0$$

*Proof.* Choose as candidate Lyapunov function

$$\begin{aligned} v(q-q_0,\dot{q},\hat{p}-p) = \beta\Big[&\gamma(\frac{1}{2}\dot{q}^T B\dot{q} + \frac{1}{2}(q-q_0)^T K_p (q-q_0)) \\ &+ \frac{2\dot{q}^T B(q-q_0)}{1+2(q-q_0)^T(q-q_0)}\Big] + \frac{1}{2}(\hat{p}-p)^T(\hat{p}-p) \end{aligned}$$

The time derivative of $v$, recalling the alternative proof of Theorem 1 and condition (19), satisfies the inequality

$$\begin{aligned} \dot{v} \le \beta\Big[&-a\|\dot{q}\|^2 - b\|q-q_0\|^2 \\ &+ (\gamma\dot{q}^T E(q) + \frac{2(q-q_0)^T E(q)}{1+2(q-q_0)^T(q-q_0)})\Big](\hat{p}-p) + \dot{\hat{p}}^T(\hat{p}-p) \end{aligned} \tag{20}$$

where $a$ and $b$ are positive constants. Substituting (18) into (20) we obtain

$$\dot{v} \le -a\|\dot{q}\|^2 - b\|q-q_0\|^2$$

By applying the Lasalle theorem [4, p.108] the thesis is proved. △

*Remark.* An alternative way to eliminate the steady state errors caused by imperfect gravity compensation is that of adding an integral action to the PD controller [1]. However, the resulting PID controller needs as many integrators as the number of the links. On the other hand, the number of integrators required by the adaptive PD controller is equal to the number of unknown parameters. Hence, in the usual case in which the only unknown parameter is the payload, one integrator suffices to implement the adaptive controller.

## 3. Simulation results.

The adaptive PD controller proposed in Section 2 has been tested by simulation on a three revolute jointed robot, having links 0.5 $m$ long. Viscous frictional forces were neglected.

The non-zero entries of the inertia matrix $B$ and of the gravity vector $e$, that completely characterize the robot model (1), are given by

$$\begin{aligned} B_{11} &= a_1 + a_2\cos^2 q_2 + a_3\cos^2(q_2+q_3) + a_4\cos q_2\cos(q_2+q_3) \\ B_{22} &= a_5 + a_4\cos q_3 \\ B_{23} &= B_{32} = a_6 + a_7\cos q_3 \\ B_{33} &= a_8 \\ e_2 &= b_1\cos q_2 + b_2\cos(q_2+q_3) \\ e_3 &= b_2\cos(q_2+q_3) \end{aligned} \tag{21}$$

In Table 1 are reported the values of the parameters $a_i$ and $b_i$ referred to payloads $m_p$ of 0 and 5 $kg$.

The aim of the simulation tests was to compare the traditional PD control law (5) with the adaptive algorithm (17),(18), assuming a nominal payload of 5 $kg$ and actual payloads of 0 and 5 $kg$.

As displayed by (21), vector $e$ is linear in terms of parameters $b_1$ and $b_2$. However, since actually only one parameter (the payload) is considered to be unknown, we can obtain an adaptation dynamics of order 1, instead of order 2. Indeed, observe that $b_1$ and $b_2$ can be expressed as

$$\begin{aligned} b_1 &= 189.1708 + 4.9008\, m_p \\ b_2 &= 52.9286 + 4.9008\, m_p \end{aligned} \tag{22}$$

that lead to the following expression for $e$

$$e(q) = e_A(q) + e_B(q)\, m_p \tag{23}$$

in which $e_A$ and $e_B$ are known vectors.
Consequently, the control law (17),(18), for this specific application, can be modified as follows

$$\begin{aligned} u &= -K_p(q - q_0) - K_v \dot{q} + e_A(q) + e_B(q)\hat{m}_p \\ \dot{\hat{m}}_p &= -\beta e_B^T(q) \left[ \gamma \dot{q} + \frac{2(q - q_0)}{1 + 2(q - q_0)^T (q - q_0)} \right] \end{aligned} \tag{24}$$

The problem that has been considered is that of regulation about a reference position. In the first set of simulation runs the PD control law (5) was used, assuming an available estimate of the payload equal to 5 $kg$, $K_p$ = diag[10000] and $K_v$ = diag[3000]. Figures 1 and 2 illustrate the results of simulation. In these figures is reported the time history of the distance $e_{ABS}$ between the reference and the actual position of the end-effector, in the Cartesian space. As one can see, when the available estimate of $m_p$ is different from the actual value we have a steady state error.

Analogous simulations have been carried out for the adaptive PD algorithm (24). Matrices $K_p$ and $K_v$ were set as above and the adaptation gains were chosen as $\gamma = 0.1$, $\beta = 100$. The initial value of the payload estimate was $\hat{m}_p(0) = 5\, kg$.

In figures 3 and 4 are shown the time histories of the distance $e_{ABS}$ and of the payload estimate $\hat{m}_p$, relative to actual payloads of 5 and 0 $kg$. Note that, since $e_B(q_0) \neq 0$, the payload estimate tends toward the true value.

The PD controller yields slightly better dynamic performances when the payload value is known. On the other hand, the adaptive PD controller ensures that the error goes to zero even if the payload is not exactly known.

## 4. Conclusions.

In practice, the robot parameters are never exactly known. Therefore, the use of robust control laws is preferable. The simple PD controller is robust with respect to the inertia parameters. Unfortunately, it is not robust with respect to the gravity parameters that have to be compensated.

In this paper an adaptive version of the PD controller has been proposed that avoids the need of exact knowledge of the gravity parameters. The proposed adaptive PD controller ensures global convergence, provided that upper and lower bounds of the inertia matrix are available. Simulation tests show that such a controller gives a satisfactory dynamic behavior.

*Acknowledgement.* This work was supported by MPI funds.

References

[1] S. Arimoto and F. Miyazaki, "Stability and robustness of PID feedback control for robot manipulators of sensory capability", Robotics Research, eds. M. Brady and R.P. Paul, Cambridge: MIT Press, 1984.
[2] D. Koditschek, "Natural motion of robot arms", 23rd IEEE Conf. Decision Control, Las Vegas, 1984.
[3] P. Khosla and T. Kanade, "Parameter identification of robot dynamics", 24th IEEE Conf. Decision Control, Fort Lauderdale, 1985.
[4] W. Hahn, "Stability of motion", Berlin: Springer-Verlag, 1967.
[5] D. Koditschek, "Application of a new Lyapunov function to global adaptive attitude tracking", 27th Conf. Decision Control, Austin, 1988.

| $m_p = 0\,kg$ | $m_p = 5\,kg$ |
|---|---|
| $a_1 = 23.380$ | $a_1 = 23.380$ |
| $a_2 = 9.2063$ | $a_2 = 10.456$ |
| $a_3 = 2.4515$ | $a_3 = 3.7015$ |
| $a_4 = 5.4000$ | $a_4 = 7.9000$ |
| $a_5 = 82.399$ | $a_5 = 84.899$ |
| $a_6 = 2.6274$ | $a_6 = 3.8774$ |
| $a_7 = 2.7000$ | $a_7 = 3.9500$ |
| $a_8 = 25.779$ | $a_8 = 27.027$ |
| $b_1 = 189.17$ | $b_1 = 213.67$ |
| $b_2 = 52.928$ | $b_2 = 77.432$ |

TABLE 1 Robot parameters.

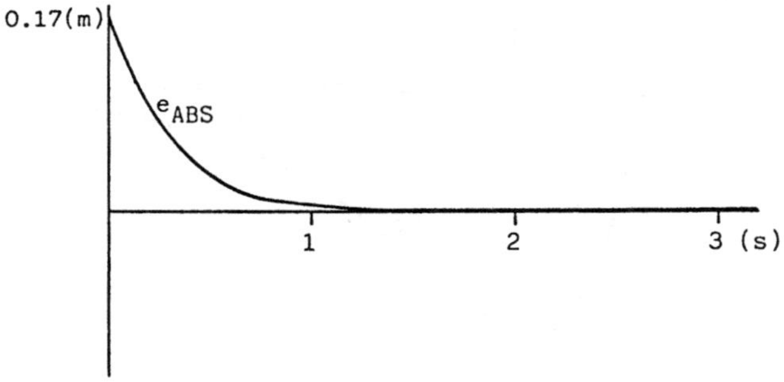


FIGURE 1 PD control law ($m_p = 5\,kg$).

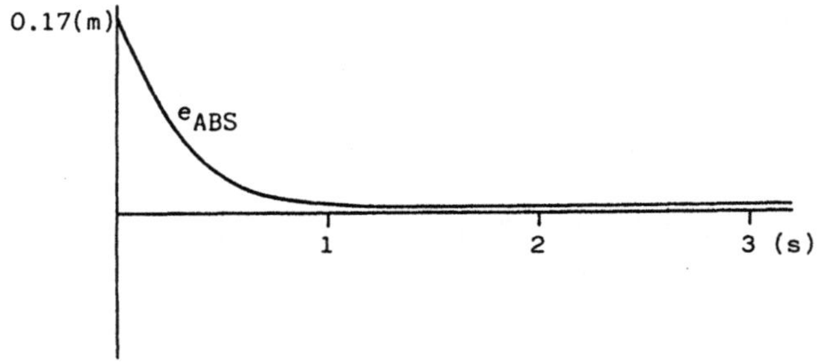


FIGURE 2 PD control law ($m_p = 0\,kg$).

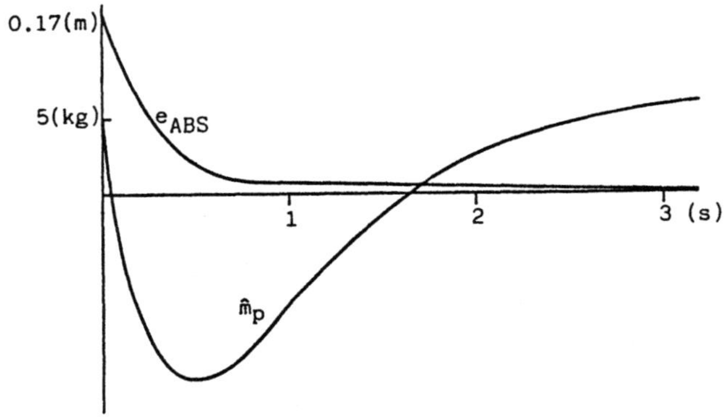


FIGURE 3 Adaptive PD control law ($m_p = 5\,kg$).

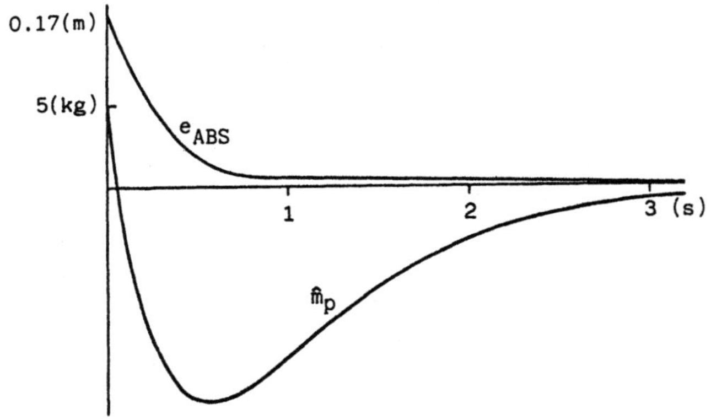


FIGURE 4 Adaptive PD control law ($m_p = 0\,kg$).

## COMPARISON OF ROBUSTNESS MEASURES FOR A FLEXIBLE STRUCTURE

Jan Bontsema and Taco van der Vaart

**Abstract**

We investigate the applicability of the theory of robust stabilization with respect to additive, stable perturbations of a normalized left–coprime factorization to controller design of a flexible beam with uncertain parameters.

### 1. Introduction.

The problem of robust stabilization with respect to additive stable perturbations of a normalized left–coprime factorization has been considered by several authors [11], [9], [10], [4] and [7]. In [9] and [10], it was shown that for the rational case this problem has an elegant explicit solution in terms of Riccati equations and in [4] this was extended to a class of infinite–dimensional systems. In [7] it was shown that the problem of robustness optimization for normalized coprime factor perturbations is equivalent to robust optimization in the gap metric. This covered both finite– and infinite–dimensional systems.

Of course these theories only consider unstructured perturbations, whereas in flexible structures (and other applications) one usually has to take structured perturbations into account. Here we consider a prototype example of a p.d.e. model of a damped beam in which we suppose that the damping and the stiffness coefficient are unknown. This p.d.e. model retains some essential characteristics, typical of large flexible structures, such as uncertain damping, and point actuators and sensors, while at the same time it is possible to obtain a rigorous mathematical formulation in both time and frequency domain [1]. In particular, it belongs to the class of infinite–dimensional systems discussed in [4] (see [1]). In the absence of a theory for robustness optimization under structured perturbations for infinite–dimensional systems, we decided to investigate how unstructured theories of robustness optimization would work on this prototype model. In [1] the theory of robustness optimization under additive perturbations from [5] was applied and here we apply the theory of robustness optimization with respect to additive stable perturbations of a normalized left–coprime factorization. Although there are countless numerical examples in the literature demonstrating that various controller theories produce good controllers for flexible systems [10], they do not provide much insight into the effect of uncertainty in the damping or other parameters. It is hoped that this parameter study will help in this direction.

The first part of investigation was the dependence of the maximal robustness margin on the damping and stiffness parameters. Then taking a fixed pair of parameter values to define the nominal model we designed the controller which (nearly) achieves the maximal robustness margin. Then we mapped the parameter region this controller actually stabilized, which was larger than the region predicted by the theory. The region predicted by the theory can be calculated in terms of the $T$–gap between the nominal plant and the perturbed plant [7].

For the infinite–dimensional theory of robustness optimization with respect to additive stable perturbations of a normalized left–coprime factorization we refer to the paper [4] in this volume. In section 2 we summarize the relevant results on the relationships to the $T$–gap metric from [7].

An important question is how one can best apply the infinite–dimensional theories to this p.d.e. model and this is discussed in section 3 together with various approximation questions. In section 4 we give our numerical results and in section 5 we given some concluding remarks.

## 2. Optimal robustness and the gap metric

Here we summarize results from [7] which we need in the sequel. For simplicity we suppose that $P_1$ and $P_2$ are in the Pritchard–Salamon class defined in [4], although the results in [7] apply to more general plants. The directed $T$–gap is defined by

$$\vec{\delta}_T(P_1,P_2) = \inf_{Q \in H_\infty} \| [\tilde{M}_1,\tilde{N}_1] - Q[\tilde{M}_2,\tilde{N}_2] \|_\infty \tag{2.1}$$

and the $T$–gap by

$$\delta_T(P_1,P_2) = \max\{\vec{\delta}_T(P_1,P_2), \vec{\delta}_T(P_2,P_1)\} \tag{2.2}$$

where $P_i = \tilde{M}_i^{-1}\tilde{N}_i$ is a normalized left–coprime factorization of $P_i$, $i = 1,2$.

In general $\vec{\delta}_T(P_1,P_2) \neq \vec{\delta}_T(P_2,P_1)$, but if $\delta_T(P_1,P_2) < 1$, then they are both equal.

There exists the following relationship between the Glover–McFarlane class of perturbations $\mathcal{G}_\varepsilon$ of [4], eqn. (4.1) and the directed $T$–gap ball:

$B_T(P,\varepsilon)=\{P_1:\vec{\delta}_T(P,P_1)<\varepsilon\}=\{P_1=(\tilde{M}+\Delta_M)^{-1}(\tilde{N}+\Delta_N):P=\tilde{M}^{-1}\tilde{N}$ is a normalized coprime factorization, $\Delta_M$, $\Delta_N\in H_\infty$ and $\|[\Delta_M,\Delta_N]\|_\infty<\varepsilon\}$ (2.3)

Notice that the set in (2.3) is larger than the set $\mathcal{G}_\varepsilon$ in [4], eqn. (4.1) where $\Delta_m$, $\Delta_n$ are restricted to the smaller class $M(\hat{A}_-(0))$. The main result in [7] is that a controller $K\in H_\infty$ stabilizes all $P_1$ with $\vec{\delta}_T(P,P_1)<\varepsilon$ if and only if $K$ stabilizes all $P_1$ with $\delta_T(P,P_1)<\varepsilon$, if and only if $K$ stabilizes all $P_1=(\tilde{M}+\Delta_M)^{-1}(\tilde{N}+\Delta_N)$ where $\Delta_M$, $\Delta_N\in H^\infty$ satisfy $\|[\Delta_m,\Delta_n]\|_\infty<\varepsilon$. The advantage of the gap metric is that it can be calculated (at least for. finite-dimensional plants) as a 2-block $H^\infty$-optimization problem [6], [3].

## 3. Flexible beam model and approximations

Consider the following model for a flexible Euler-Bernoulli beam with viscous damping: ([1])

$$w_{tt}(x,t)+\alpha_1 Aw_t(x,t)+\alpha_2 Aw(x,t)=\frac{1}{\rho a}(\delta(0)u_1(t)-\delta_x(0)u_2(t)) \qquad (3.1a)$$

$$y_1(t)=w(0,t);\ \ y_2(t)\ =\ w_x(0,t) \qquad (3.1b)$$

where $w(x,t)$ is the deflection of the beam, $-1\le x\le 1$, $\alpha_1$ and $\alpha_2$ are the damping and stiffness coefficient, $\rho a$ is the mass per unit length and $A$ is the stiffness operator $A=\dfrac{d^4}{dx^4}$ with domain

$$D(A)=\{f\in H^4(-1,1)\,|\,w_{xx}(\pm 1,t)=0;\ w_{xxx}(\pm 1,t)=0\}$$

It is easy to show that the transfer matrix for this model is given by

$$G(s)=\frac{1}{\rho a}\begin{bmatrix}\dfrac{1}{2s^2} & 0\\ & \dfrac{3}{2s^2}\\ 0 & \end{bmatrix}+\frac{1}{\rho a}\sum_{i=3}^{\infty}\begin{bmatrix}v_i^2(0) & 0\\ 0 & v_{ix}^2(0)\end{bmatrix}/\,(s^2+\alpha_1\lambda_i^4 s+\alpha_2\lambda_i^4) \qquad (3.1c)$$

where $\lambda_i$ and $v_i(x)$ are the eigenvalues and eigenfunctions of the stiffness operator $A$. For more details see [1] or [2].

It has been shown in [1] that this beam model has a Pritchard-Salamon state-space realization, and so we may apply the theory of [4] directly. However, this would involve solving infinite-dimensional Riccati equations, which is very time consuming at best and in fact there are no known convergence results for the Riccati equations of our example. They have so-called "unbounded" $B$ and $C$-operators and very little work has been done on the numerical approximation of solutions of such Riccati equations.

So we shall take an approximation approach using the known properties of the transfer function (3.1c). In [1] it is shown that $G$ is the sum of an finite-dimensional part $G_f$, which contains the unstable modes and an infinite-dimensional stable part $G_s$:

$$G = G_f + G_s \tag{3.2}$$

where $G_s \in \mathcal{M}(\hat{\mathcal{A}}_-(0))$. Furthermore $G_s$ is nuclear, which means that it is easy to approximate by a finite-dimensional system and a bound on the error can be calculated ([8]). For our investigations we take the 8th order approximation

$$G_f = \frac{1}{\rho a}\begin{bmatrix} \frac{1}{2s^2} & 0 \\ 0 & \frac{3}{2s^2} \end{bmatrix} + \frac{1}{\rho a}\sum_{i=3}^{4}\begin{bmatrix} v_i^2(0) & 0 \\ 0 & v_{ix}^2(0) \end{bmatrix} / (s^2 + \alpha_1\lambda_i^4 s + \alpha_2\lambda_i^4) \tag{3.3}$$

and for the nominal values of $\rho a = 42.7$, $\alpha_1 = 3.89 \times 10^{-4}$ and $\alpha_2 = 1.129$ we have for the error:

$$\|G_s\|_\infty = \mu = 0.02 \tag{3.4}$$

For other values of $\alpha_1$ and $\alpha_2$ in the neighbourhood of the nominal values the error is of the same order of magnitude.
$G_f$ is rational and we can apply the theory of [4] or [7] to $G_f$ allowing for infinite-dimensional perturbations. If $G_f$ has the normalized coprime factorization

$$G_f = \tilde{M}_f^{-1}\tilde{N}_f \tag{3.5}$$

then

$$G = G_f + G_s = \tilde{M}_f^{-1}(\tilde{N}_f + \Delta_N^f) \tag{3.6}$$

where

$$\|\Delta_N^f\|_\infty = \|\tilde{M}_f G_s\|_\infty \leq \|G_s\|_\infty < \mu \tag{3.7}$$

From (3.7), we may conclude that

$$\vec{\delta}_T(G_f, G) < \mu \tag{3.8}$$

Reversing the roles of $G$ and $G_f$, we can write

$$G_f = G - G_s = \tilde{M}^{-1}(\tilde{N} + \Delta_N) \tag{3.9}$$

where

$$\|\Delta_N\|_\infty = \|\tilde{M} G_s\|_\infty \leq \|G_s\|_\infty < \mu \tag{3.10}$$

and so we can apply the same argument above to conclude that $\vec{\delta}_T(G,G_f) < \mu < 1$ and hence they are equal

$$\vec{\delta}_T(G,G_f) = \vec{\delta}_T(G_f,G) \tag{3.11}$$

Suppose now that we apply the finite–dimensional theory of [10] on $G_f$ to obtain a controller $K_f$ with robustness margin $\varepsilon$. Then from the results quoted in section 2, we see that $K_f$ is a robust controller for $G$ with a robustness margin of at least $\varepsilon - \mu$. In other words, replacing $G$ by $G_f$ in our calculations incurs an error of at most $\mu$ and we have chosen $\mu$ to be negligible compared to the robustness margins of $G_f$ for our range of parameter values. This justifies using $G_f$ in our calculations.

## 4. Numerical results.

We first have considered the dependence of the maximal robustness margin $\varepsilon_{max}$ (see [10]) both on the order of approximation of the infinite–dimensional system as on the parameters $\alpha_1$ and $\alpha_2$.
If the order of approximation is $n=2$, 4 or 6 (cf. eqn. (3.3)) then $\varepsilon_{max}=0.3827$ and for $n \geq 8$, $\varepsilon_{max}=0.3828$, where $\alpha_1$ and $\alpha_2$ are equal to the nominal values (cf. eqn. (3.3)).
The dependence of $\varepsilon_{max}$ on the parameters $\alpha_1$ and $\alpha_2$ is shown in table 4.1 (the order of approximation here is taken to be $n=8$).

| $\alpha_2$ \ $\alpha_1$ | $3.89\times10^{-2}$ | $3.89\times10^{-3}$ | $3.89\times10^{-4}$ | $3.89\times10^{-5}$ | $3.89\times10^{-6}$ |
|---|---|---|---|---|---|
| 112.9 | 0.3827 | 0.3827 | 0.3827 | 0.3827 | 0.3827 |
| 11.29 | 0.3827 | 0.3827 | 0.3827 | 0.3827 | 0.3827 |
| 1.129 | 0.3828 | 0.3828 | 0.3828 | 0.3828 | 0.3828 |
| 0.1129 | 0.3838 | 0.3838 | 0.3837 | 0.3836 | 0.3836 |
| 0.01129 | 0.3929 | 0.3935 | 0.3912 | 0.3896 | 0.3894 |

Table 4.1
The dependence of $\varepsilon_{max}$ on $\alpha_1$ and $\alpha_2$.

For the system of order 8 we choose as nominal values $\alpha_1 = 3.89 \times 10^{-4}$ and $\alpha_2 = 1.129$ (cf. eqn. (3.3)). Then $\alpha_1$ and $\alpha_2$ are varied and we calculate the distance between the perturbed plant and the nominal one. If the distance (the directed $T$–gap, see section 2) between nominal and perturbed plant is smaller than $\varepsilon_{max}$ then both plants are guaranteed to be stabilized by a maximally robust controller, designed for the nominal plant (see [10]). In order to calculate the directed $T$-gap (eqn. 2.1) we have to solve a 2-block $H^{\infty}$-problem which can be a numerically hard problem.

For this reason we used the following result for the directed $T$-gap:

Let $G_1$ and $G_2$ have normalized left and right coprime factorizations: $G_1 = \tilde{M}_1^{-1}\tilde{N}_1 = N_1 M_1^{-1}$, $G_2 = \tilde{M}_2^{-1}\tilde{N}_2 = N_2 M_2^{-1}$.

Define $\tilde{R}_1 = \tilde{M}_1 \tilde{M}_2^* + \tilde{N}_1 \tilde{N}_2^*$, $R_2 = \tilde{M}_1 N_2 - \tilde{N}_1 M_2$ then ([3]):

$$\max(\|R_2\|_\infty, \tilde{H}_{R1}) \le \vec{\delta}_T(G_1, G_2) \le (\|R_2\|_\infty^2 + \tilde{H}_{R1}^2)^{\frac{1}{2}} \tag{4.1}$$

where $\tilde{H}_{R1} = \inf_{Q \in H^\infty} \|\tilde{R}_1 - Q\|_\infty$.

Both the upper and lower bounds for the $T$-gap between nominal and perturbed plant, for different values of $\alpha_1$ and $\alpha_2$ are given in table 4.2.

| $\alpha_2$ \ $\alpha_1$ | $3.89\times10^{-2}$ | $3.89\times10^{-3}$ | $3.89\times10^{-4}$ | $3.89\times10^{-5}$ | $3.89\times10^{-6}$ |
|---|---|---|---|---|---|
| 112.9 | 0.2115<br>0.2115 | 0.2115<br>0.2115 | 0.2115<br>0.2115 | 0.2113<br>0.2116 | 0.9098<br>0.9549 |
| 11.29 | 0.2114<br>0.2114 | 0.2113<br>0.2114 | 0.2114<br>0.2114 | 0.5649<br>0.5716 | 0.9895<br>1.0781 |
| 0.129 | 0.2092<br>0.2092 | 0.1901<br>0.1901 | $\approx 0$<br>$\approx 0$ | 0.7985<br>0.8249 | 0.967<br>1.0405 |
| 0.1129 | 0.2114<br>0.2114 | 0.2114<br>0.2114 | 0.5644<br>0.5711 | 0.9892<br>1.0778 | 1.0004<br>1.1148 |
| 0.01129 | 0.2114<br>0.2117 | 0.2124<br>0.2132 | 0.9073<br>0.9525 | 0.9990<br>1.1070 | 0.9999<br>1.1171 |

Table 4.2.

The upper and lower bounds for the $T$-gap between nominal and perturbed plant.

So we can find the region in the $(\alpha_1,\alpha_2)$ plane were a maximally robust controller, designed for the nominal plant, is guaranteed to stabilize the perturbed plant. In order to see how conservative the guaranteed region is we also calculated the actual region where the central maximally robust controller ($K_{max}$) stabilizes the perturbed plant. The same was done for a suboptimal controller ($K_{subopt}$) (see [10]) with robustness margin 0.33. The results are shown in fig. 4.1.

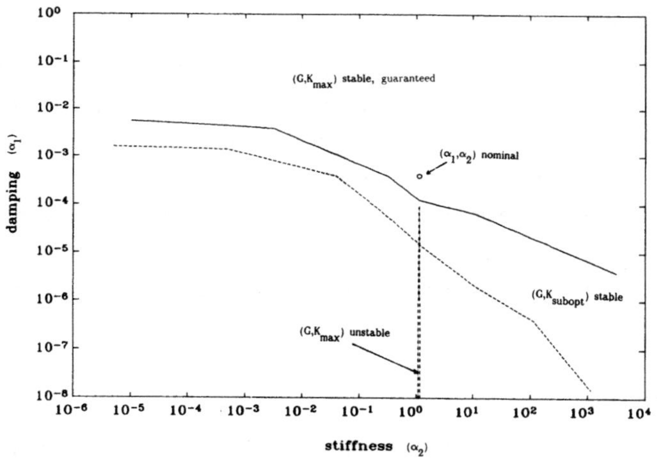


FIGURE 4.1

A priori guaranteed and actual stability regions in the $(\alpha_1,\alpha_2)$–plane.

In [1] robust stabilization w.r.t. additive perturbations on the transfer function was considered and there the objective was to design a controller with an overall decay rate of 0.0065 (i.e. real parts of the closed loop poles left of $-0.0065$). It turned out that the guaranteed stability region in the $(\alpha_1,\alpha_2)$ plane was very small and that the prediction was rather conservative. If the method of [10] is applied to the transfer function $G^{\beta}(s)=G(s-\beta)$ then the controller $K(s)=K^{\beta}(s+\beta)$ will move all the closed loop poles left of $-\beta$. In fig. 4.2 the stability region (with $\beta$=0.0065) is shown near the nominal values of $\alpha_1$ and $\alpha_2$. The actual stability region is much larger.

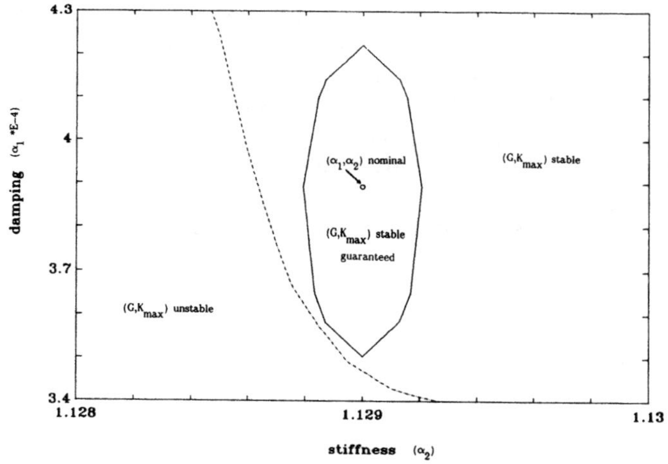


FIGURE 4.2

A priori guaranteed and actual stability region, with $\beta$=0.0065.

## 5. Conclusions.

The theory of robust stabilization with respect to additive, stable perturbations of a normalized left coprime factorization proposed in [8], [10] seems to be a useful method for designing controllers, even when the perturbations are structured. Here we have applied it to stabilize a flexible structure model with parameter uncertainty and it gave better results than the robustness theory with respect to additive perturbations proposed in [5]. For both methods the robustness margin decreases as the desired overal decay rate increases.

## 6. References.


[1] Bontsema, J., Dynamic stabilization of large flexible space structures, Ph.D. thesis University of Groningen, 1989.

[2] Bontsema, J., Curtain, R.F. and Schumacher, J.M., Robust control of flexible systems: a case study, Automatica, vol.24, p.177–186, 1988.

[3] Chu, C.C., Doyle, J.C. and Lee, E.B., The general distance in $H_\infty$-optimal control theory, Int. J. Control, vol.44, p.565–596, 1986.

[4] Curtain, R.F., Robust stabilization for infinite-dimensional linear systems using normalized coprime factorizations (this volume).

[5] Curtain, R.F. and Glover, K., Robust stabilization of infinite-dimensional systems by finite-dimensional controllers, Systems and Control Letters, vol.7, pp.41–47, 1986.

[6] Georgiou, T.T., On the computation of the gap metric, System and Control Letters, vol.11, pp.253–257, 1989.

[7] Georgiou, T.T. and Smith, M.C., Optimal robustness in the gap metric, manuscript, 1989.

[8] Glover, K., Curtain, R.F. and Partington, J.R., Realization and approximation of linear infinite-dimensional systems with error bounds, SIAM J. Control and Optim., vol. 26, pp. 863–899, 1989.

[9] Glover, K. and McFarlane, D., Robust stabilization of normalized coprime factors: An explicit $H_\infty$-solution, Proc. of the ACC, pp.842–847, 1988.

[10] McFarlane, D. and Glover, K., Robust controller design using normalized coprime factor plant descriptions, manuscript, 1989.

[11] Vidyasagar, M. and Kimura, H., Robust controllers for uncertain linear multivariable systems, Automatica, vol.22, pp.85–94, 1986.

# ROBUST STABILIZATION FOR INFINITE-DIMENSIONAL LINEAR SYSTEMS USING NORMALIZED COPRIME FACTORIZATIONS

Ruth F. Curtain

**Abstract**

The problem of robustly stabilizing a linear system subject to $H_\infty$-bounded perturbations in the numerator and the denominator of its normalized left coprime factorizations is considered for a class of infinite-dimensional systems. This class has possibly unbounded, finite-rank input and output operators which includes many delay and distributed systems. The optimal stability margin is expressed in terms of the solutions of the control and filter algebraic Riccati equations.

## 1. Introduction.

This paper extends the results of Glover and McFarlane [GF1], [GF2] on robust stabilizability of normalized coprime factors to a large class of infinite-dimensional systems, the Pritchard-Salamon class. The problem of robust stability of closed-loop systems has received much attention in the literature and for a discussion of and references to the finite-dimensional literature we refer the reader to [GF1], [GF2]. In section 2 the Prichard-Salamon class of systems is introduced and its special properties are listed and discussed. It is a state-space class of systems and its relationships to the Callier-Desoer class of transfer matrices is discussed in section 3, where concepts and properties relevant to control synthesis are detailed. The main result of section 4 is that for the class of Pritchard-Salamon systems, $(C,A,B)$, such that $(A,B)$ is exponentially stabilizable and $(C,A)$ is exponentially detectable it is possible to extend the arguments in [GF1], [GF2] in a natural way. The details of the arguments may be found in [C2].

The exponentially stabilizable and detectable Prichard-Salamon class is not the most general class of infinite-dimensional linear systems (see [W3]), but it does include many delay and p.d.e. systems with unbounded inputs and outputs (see [PS], [C1]).

## 2. The Pritchard Salamon class $(C,A,B)$.

**Definition 2.1.** We suppose that there exist three separable Hilbert spaces $V$, $X$ and $W$ with continuous, dense injections satisfying

(2.1)(i) $\quad \mathcal{Z}_1 = D_V(A) \hookrightarrow W \hookrightarrow X \hookrightarrow V$

(2.1)(ii) $\quad \mathcal{Z}_2^* = D_{W^*}(A^*) \hookrightarrow V^*$

and we suppose that $A$ generates a $C_0$–semigroup on $V$, $X$ and $W$, which we denote by the same symbol, $S(t)$. The input and output maps are $B\in\mathcal{L}(U,V), C\in\mathcal{L}(W,Y)$, where $U=\mathbb{R}^q$ and $Y=\mathbb{R}^p$ are finite–dimensional spaces. Furthermore we suppose that $B$ induces a smooth reachability map with respect to $W$ on $[0,t_1]$ for all finite $t_i$, i.e. for some $\beta>0$ and all $u\in L_2[0,t_1]$

$$(2.2)\qquad \|\int_0^{t_1} S(t_1-s)Bu(s)ds\|_W \le \beta\|u\|_{L_2[0,t_1]}$$

and $C$ induces a smooth observability map which extends to $V$ on $[0,t_1]$ for all finite $t_1$, i.e. for some $\gamma>0$ and all $x\in W$

$$(2.3)\qquad \|CS(\cdot)x\|_{L_2[0,t_1]}\le\gamma\|x\|_V.$$

Under the above assumptions we call $(C,A,B)$ *a Prichard–Salamon system.*

**Remarks.**

**R1:** In fact in Weiss [1] and [2] it is shown that without loss of generality we may always take $\mathcal{Z}=D_V(A)=W$ and $V=D(A^*)^*$ and so (2.1)(i) and (2.1)(ii) are automatically satisfied. However, sometimes it proves convenient to allow the choice of $V$ and $W$ depend on the example in question, as for example in [C1].

**R2:** The role of $X$ is to some extent artificial and its only purpose is to define a dual system $(B^*,A^*,C^*)$. To do this we identify the duals of $U$, Y and $X$. Using $X$ as a pivot space we take $V^*$ to be the space of all linear functionals on $V$ which are continuous with respect to the topology on $X$. $W^*$ is defined analogously and so we have $V^*\hookrightarrow X\hookrightarrow W^*$. Then $S^*(t)\in\mathcal{L}(V^*)\cap\mathcal{L}(W^*)$ generates a $C_0$–semigroup on both $V^*$ and $W^*$ and $B^*\in\mathcal{L}(V^*,U)$ satisfies a (2.3) condition with respect to $S^*$ if and only if $B$ satisfies a (2.2) condition with resepct to $S$. This and a similar statement can be found in [PS]. It is possible to choose $X=V$ or $W$, thus eliminating $X$, but this formulation is not symmetric.

**R3:** $B$ is an admissible control operator for $A$ under the weaker assumption that the controllability map be bounded on $X$ ([W1]) and so (2.2) requires a much smoother reachability map with its range in the smaller space, $W$.

**R4:** $C$ is an admissible observation operator for $A$ under the weaker assumption that the observability map be bounded on $X$ ([W2]) and so (2.3) requires a smoother observability map defined on the larger space, $V$.

**R5:** These assumptions still allow for large classes of p.d.e. and delay systems with unbounded $B$ and $C$ operators and these technical assumptions can be readily verified (see [PS], [C1])

**Important properties of Pritchard Salamon class**

**P1:** The Pritchard–Salamon class is closed under perturbations. If $F\in\mathcal{L}(V,U)$, then $A+BF$ generates a $C_0$–semigroup on $W$, $X$ and $V$ and if $H\in\mathcal{L}(Y,W)$, $A+HC$ generates a $C_0$–semigroup on $W$, $X$ and $V$. ([PS]).

**P2:** The Pritchard–Salamon class has important frequency domain properties which were established in [C1]. For all $s$ in $\mathrm{Re}\, s > w$, the growth bound of $S(t)$, $(sI-A)^{-1}B \in \mathcal{L}(U,V) \cap \mathcal{L}(U,W)$ and $C(sI-A)^{-1} \in \mathcal{L}(W,Y)$ has a bounded extension to $\mathcal{L}(V,Y)$. So the *transfer function* $C(sI-A)^{-1}B$ is a well-defined analytic function in $\mathrm{Re}\, s > w$ and it is norm-bounded there.

**P3:** We say that $(A,B)$ is *exponentially stabilizable* if there exists an $F \in \mathcal{L}(V,U)$ such that $A+BF$ generates an exponentially stable $C_0$–semigroup (on $V$, $X$ and $W$ according to P1). In this case $g(s) = K(sI - A - BF)^{-1}B$ is in the Hardy space $\boldsymbol{H}_2$ and $|g(s)| \to 0$ as $|s| \to \infty$. (i.e. as $\rho \to \infty$ in $\{s \in \mathbb{C}^+ : |s| \geq \rho\}$).

**P4:** We say that $(C,A)$ is *exponentially detectable* if there exists an $H \in \mathcal{L}(Y,W)$ such that $A+HC$ generates an exponentially stable $C_0$–semgroup on $V$, $X$ and $W$. In this case $h(s) = C(sI - A - HC)^{-1}L$ for any $L \in \mathcal{L}(U,V)$ and $|h(s)| \to 0$ as $|s| \to \infty$.

**P5:** If $(C,A)$ is exponentially detectable and $(A,B)$ is exponentially stabilizable, then the *Control Algebraic Riccati Equation* (CARE) (2.4) has a unique non–negative definite self–adjoint solution $Q \in \mathcal{L}(V,V^*)$, and the *Filter Algebraic Riccati Equation* (FARE) (2.5) has a unique non–negative definite self–adjoint solution $P \in \mathcal{L}(W^*,W)$ such that

$$A^*Qz + QAz - QBB^*Qz + C^*Cz = 0 \quad \text{for } z \in \mathcal{Z} \tag{2.4}$$

$$APz + PA^*z - PC^*CPz + BB^*z = 0 \quad \text{for } z \in \mathcal{Z}_2. \tag{2.5}$$

Furthermore both $A - BB^*Q$ and $A - PC^*C$ generate exponentially stable semigroups on $V$, $X$ and $W$. $PQ \in \mathcal{L}(V,W) \cap \mathcal{L}(W) \cap \mathcal{L}(V)$ and $QP \in \mathcal{L}(W^*,V^*) \cap \mathcal{L}(V^*) \cap \mathcal{L}(W^*)$.

## 3. The Callier-Desoer class of transfer functions.

It is known that controller synthesis can be generally formulated over the Callier–Desoer class [CD1–3] of transfer fucntions (see Section 8.2 in [V]).

**Definition 3.1:** Let $A$ comprise all distributions $f(\cdot)$ with support in $(0,\infty)$ of the form

$$f(t) = f_a(t) + \sum_{n+1}^{\infty} f_i \delta(t - t_i) \tag{4.1}$$

where $\delta(\cdot)$ is the delta distribution, $t_0 = 0$ and $t_i$ are positive real numbers; $t_0 \leq t_1 \leq t_2 \leq, \dots$; $f_i$ are real numbers and $f_a$ is a measurable function in $L_1(0,\infty)$ and $\sum_{i=0}^{\infty} |f_i| < \infty$.

We say that $f \in A_-(\alpha)$, if and only if for some $\alpha_1 < \alpha$, $f$ has the decomposition (3.1) where $e^{-\alpha_1 \cdot} f_a(\cdot) \in L_1(0,\infty)$, and $\sum_{i=0}^{\infty} e^{-\alpha_1 t_i} |f_i| < \infty$.

$\hat{A}_-(\alpha)$ denotes the set of Laplace transforms of elements in $A_-(\alpha)$.

$\hat{A}_-^{\infty}(\alpha)$ denotes the subset of $\hat{A}_-(\alpha)$ consisting of those $\hat{f}$ which are bounded away from zero at infinity in $\mathbb{C}_\alpha = \{Re s \geq \alpha\}$.

$\hat{\mathcal{B}}(\alpha)$ is the following commutative algebra of fractions: $\hat{\mathcal{B}}(\alpha) := [\hat{A}_-(\alpha)][\hat{A}_-^{\infty}(\alpha)]^{-1}$.

$M(\hat{\mathcal{B}}(\alpha))$ and $M(\hat{\mathcal{A}}_-(\alpha))$ will denote the class of matrix-valued transfer functions whose elements are in $\hat{\mathcal{B}}(\alpha)$ and $\hat{\mathcal{A}}_-(\alpha)$ respectively. We do not distinguish between the sizes of the matrices.

**Properties of the Callier-Desoer class**

**CD1:** Elements $G \in M(\hat{\mathcal{B}}(\alpha))$ can be decomposed: $G = G_a + G_f$ where $G_a \in M(\hat{\mathcal{A}}_-(\alpha))$ and $G_f$ is a rational transfer function with all its poles in $\operatorname{Re} s \geq \alpha$.

**CD2:** $G \in M(\hat{\mathcal{B}}(\alpha))$ always has a *left coprime* factorization over $M(\hat{\mathcal{A}}_-(\alpha))$: $G = \tilde{M}^{-1}\tilde{N}$, where $\tilde{M}$, $\tilde{N} \in M(\hat{\mathcal{A}}_-(\alpha))$, $\det(\tilde{M}) \in M(\hat{\mathcal{A}}_-^{\infty}(\alpha))$ and there exist $X$, $Y \in M(\hat{\mathcal{A}}_-(\alpha))$ such that $\tilde{N}Y - \tilde{M}X = I$. (i.e. $\tilde{M}$ and $\tilde{N}$ are left coprime over $M(\hat{\mathcal{A}}_-(\alpha))$. An analogous statement holds for *right coprime*. (In the sequel we shall only consider coprime factorizations over $M(\hat{\mathcal{A}}_-(0))$ and so we shall omit the qualification "over $M(\hat{\mathcal{A}}_-(0)$").

**CD3:** If the Pritchard–Salamon system $(C,A,B)$ is such that $(A,B)$ is exponentially stabilizable and $(C,A)$ is exponentially detectable, then its transfer function $G(s) = C(sI-A)^{-1}B$ is in $M(\hat{\mathcal{B}}(0))$. [C1].

**Extra properties of the Pritchard-Salamon class**

**P6:** A *doubly coprime factorization* is a pair of left and right coprime factorizations $G = NM^{-1} = \tilde{M}^{-1}\tilde{N}$, where

$$\begin{bmatrix} \tilde{X} & -\tilde{Y} \\ -\tilde{N} & \tilde{M} \end{bmatrix} \begin{bmatrix} M & Y \\ N & X \end{bmatrix} = \begin{bmatrix} I & 0 \\ 0 & I \end{bmatrix} \quad \text{and } X,\ Y,\ \tilde{X},\ \tilde{Y} \in M(\hat{\mathcal{A}}_-(0)). \tag{3.2}$$

If $(C,A,B)$ is a Prichard–Salamon system such that $(A,B)$ is exponentially stabilizable and exponentially detectable, then $G(s) = D + C(sI-A)^{-1}B$ with $D \in \mathcal{L}(U,Y)$ has the doubly coprime factorization given by

$$\begin{cases} M(s) = I + F(sI-A_F)^{-1}B; & \tilde{M}(s) = I + C(sI-A_H)^{-1}H \\ N(s) = D + C_F(sI-A_F)^{-1}B; & \tilde{N}(s) = D + C(sI-A_H)^{-1}B_H \\ X(s) = I - C_F(sI-A_F)^{-1}H; & \tilde{X}(s) = I - F(sI-A_H)^{-1}B_H \\ Y(s) = -F(sI-A_F)^{-1}H; & \tilde{Y}(s) = -F(sI-A_H)^{-1}H \end{cases} \tag{3.3}$$

where $F \in \mathcal{L}(V,U)$ is chosen such that $A_F = A + BF$ generates an exponentially stable semigroup and $H \in \mathcal{L}(Y,W)$ is chosen such that $A_H = A + HC$ generates an exponentially stable semigroup; $C_F = C + DF$; $B_H = B + HD$.

**P7:** If $(C,A,B)$ is a Pritchard–Salamon system with $(A,B)$ exponentially stabilizable and $(C,A)$ exponentially detectable then $G(s) = C(sI-A)^{-1}B$ has *normalized right and left coprime factorizations*: $G = MN^{-1} = \tilde{M}^{-1}\tilde{N}$ such that

(3.4) $\tilde{N}(j\omega)\tilde{N}^t(-j\omega)+\tilde{M}(j\omega)\tilde{M}^t(-j\omega)=I$

(3.5) $N^t(-j\omega)N(j\omega)+M^t(-j\omega)M(j\omega)=I.$

These normalized factorizations may be obtained by choosing $F=-B^*Q$ and $H=-PC^*$ in (3.3) where $Q$ and $P$ are the solutions of the CARE, (2.4) and the FARE, (2.5).

**Remarks.**

**R6:** CD1 and CD2 are well-known properties which can be found in [CD1-3]. CD3 was proved for bounded $B$ and $C$ in [NBJ1] and for the Pritchard-Salamon class in [C1]. P6 was proved for general rings in [KS] and for bounded $B$ and $C$ in [NJB] and for the Pritchard-Salamon class in [C2]. P7 was proved for bounded $B$ and $C$ in [CW] and for the Pritchard-Salamon class in [C2].

**R7:** Notice that the extension of all these formulas to the Pritchard-Salamon class depends essentially on the special properties P1-P5. This allows us to manipulate expressions as in (3.3) just as if they were all bounded operators (or even matrices). This is of course not true for more general infinite-dimensional systems for which the expressions in (3.3) may not be well-defined, (see [W3]).

## 4. Robust stabilization for normalized coprime factors.

This is a brief outline of the main ideas from [C2] underlying the extension of the finite-dimensional theory of robust stabilization for normalized coprime factors in [GF1,2] to infinite-dimensional plants $G(s)=C(sI-A)^{-1}B$, where $(C,A,B)$ is a Pritchard-Salamon system such that $(A,B)$ is exponentially stabilizable and $(C,A)$ is exponentially detectable. The robust stabilization problem is that of finding a feedback controller $K\in M(\hat{B}(0))$ which stabilizes not only the nominal plant $G$, but also the family of perturbed plants defined by

(4.1) $$\mathcal{G}_\varepsilon=\{G_\Delta=(\tilde{M}+\Delta_M)^{-1}(\tilde{N}+\Delta_N) \text{ such that } \Delta_M,\ \Delta_N\in M(\hat{A}_-(0)) \text{ and } \|[\Delta_M,\Delta_N]\|_\infty<\varepsilon\}$$

where $\tilde{M}^{-1}\tilde{N}$ is a normalized left-coprime factorization of $G\in M(\hat{B}(0))$.

By *stability* of the feedback system $(G,K)$ of figure 4.1 with $G$, $K\in M(\hat{B}(0))$ we mean that $\det(I-GK)\in M(\hat{A}_-^\infty(0))$ and $S=(I-GK)^{-1}$, $KS$, $SG$, $I-KSG\in M(\hat{A}_-(0))$.

We say that $(\tilde{M},\tilde{N},K,\varepsilon)$ is ***robustly stable*** if $(G_\Delta,K)$ is stable for all $G_\Delta\in\mathcal{G}_\varepsilon$. Given $\varepsilon>0$, if there exists a $K$ such that $(\tilde{M},\tilde{N},K,\varepsilon)$ is robustly stable we say that $(G,\varepsilon)$ is ***robustly stabilizable with robustness margin*** $\varepsilon$.

Using the properties of Pritchard-Salamon systems stated in section 2 and 3 we can extend all the finite-dimensional arguments in [GF1,2] in a natural way to reduce our problem to the following minimum distance one:

(4.2) $$\inf_{J\in M(\hat{A}_-(0))} \left\|\begin{bmatrix} R+J \\ I \end{bmatrix}\right\|_\infty.$$

where $R$ is antistable and $R^+(s):=R^t(-s)\in M(\hat{A}_-(0))$ has the realisation

(4.3) $$R^{+}(s)=C(I+PQ)(sI-A+BB^{*}Q)^{-1}B$$

where $P$ and $Q$ are the solution of the FARE, (2.5), and CARE, (2.4).. This realization is a Pritchard–Salamon system and it is exponentially stabilizable and detectable. Furthermore the controllability and observability gramians for this realization of $R^{+}$ are given by

(4.4) $$\begin{cases} Q_0=Q+QPQ\in\mathcal{L}(V\ ,V^{*}) \\ \\ P_0=(\ I\ +PQ\ )^{-1}P\in\mathcal{L}(W^{*},W) \end{cases}$$

(4.5) $$PQ=P_0Q_0\in\mathcal{L}(V,W)\cap\mathcal{L}(W)\cap\mathcal{L}(V)$$

and the singular values of $R^{+}$ equal the square roots of the non-zero eigenvalues of $PQ=P_0Q_0$.

*At this stage we are confronted with a difficulty which is peculiar to the infinite-dimensional case.* While the minimization problem (4.2) always has a solution over the larger space, $\boldsymbol{H}^{\infty}$, ([BH]), namely

(4.6) $$\inf_{J\in\boldsymbol{H}^{\infty}}\left\|\begin{bmatrix} R\ +\ J \\ I \end{bmatrix}\right\|_{\infty}=[1+\lambda_{max}(PQ)]^{\frac{1}{2}}$$

it is not known if the minimizing $J$ is in the smaller space $\mathcal{M}(\hat{A}_{-}(0))$ as required by (4.2). The recent results in [CR] show that the *relaxed distance problem* does have a solution over $\mathcal{M}(\hat{A}_{-}(0))$:

(4.7) $$\inf_{J\in\mathcal{M}(\hat{A}_{-}(0))}\left\|\begin{bmatrix} R\ +\ J \\ I \end{bmatrix}\right\|_{\infty}\leq\sigma>[1+\lambda_{max}(PQ)]^{\frac{1}{2}}$$

and it seems likely that (4.2) will too (cf. [BR]). However, at present our conclusions are a little weaker than those in the finite-dimensional case, namely:
$(G,\varepsilon)$ is robustly stabilizable with robustness margin $\varepsilon$ if $\varepsilon<[1+\lambda_{max}(PQ)]^{-\frac{1}{2}}$ and it will not be robustly stabilizable if $\varepsilon>[1+\lambda_{max}(PQ)]^{-\frac{1}{2}}$.

Finally we remark that it is possible to give explicit formulas for the robust controller analogous to [GF1,2] using the parametrization in [CR]. These are in general infinite-dimensional controllers which depend on the solutions of infinite-dimensional Riccati equations. So we have obtained a nice generalization of the robustness theory of [GF1,2] to the exponentially stabilizable and detectable Pritchard Salamon class of infinite-dimensional linear sytems.

Unfortunately infinite-dimensional Riccati equations are difficult to solve and numerical schemes depend strongly on the special type of system (e.g. delay, parabolic or hyperbolic p.d.e.). Consequently the above theory is not recommended as a practical way of designing robust controllers. A better approach is to consider the infinite-dimensional plant $G$ as the sum of a finite-dimensional approximation $G_f$ and a stable error term $\Delta$, $(G=G_f+\Delta)$, and to apply the above theory to $G_f$.

Optimally robust controllers for $G_f$ with robustness margin $\varepsilon$ will stabilize $G$ if $\|\Delta\|_\infty < \varepsilon$. For applications of coprime robust controllers for delay and pde systems see [KP] and [BV] respectively.

We remark that recently in [GS] another approach was taken to this problem using the gap metric. They obtain a generalization of the theory for a larger class of infinite-dimensional systems, but they do not obtain explicit solutions in terms of Riccati equations. The solution is expressed in terms of normalized coprime factorizations in frequency domain form.

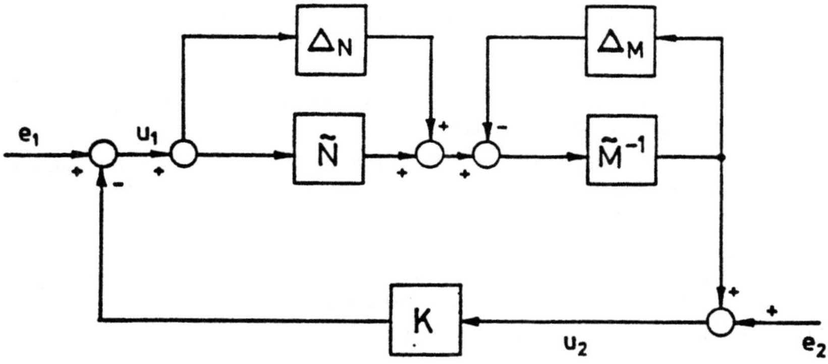


**Fig. 4.1 Left comprime factor perturbations**

## References

[BH] Ball, J.A. and Helton, J.W., "A Beurling-Lax theorem for the Lie group $U(m,n)$ which contains most classical interpolation theory", J. Operator Theory, **8**, p. 107–142, 1983.

[BR] Ball, J.A. and Ran, A., Optimal Hankel Norm Model Reductions and Wiener-Hopf Factorization II: the Non-Canonical Case, Integral Equations and Operator Theory (1987), **10**, p.416–436.

[BV] Bontsema, J. and Vaart van de, T., Comparison of Robust Measures for a Flexible Structure. (this volume)

[CD1] Callier, F.M. and Desoer, C.A., "Simplifications and New Connections on an Algebra of Transfer Functions of Distributed Linear Time-Invariant Systems", IEEE Tran. Circuits of Systems, CAS–27, p. 320–323, 1980.

[CD2] Callier, F.M. and Desoer, C.A., "An Algebra of Transfer Functions for Distributed linear Time-Invariant Systems", IEEE Trans. Circuits and Systems **25**, p. 651–663, 1978.

[CD3] Callier. F.M. and Desoer, C.A., "Stabilization, Tracking and Distributed Rejection in Multivariable Convolution Systems". Ann.

Soc. Sci. Bruxelles, **94**, p. 7–51, 1982.

[CW] Callier, F.M. and Winkin, J., "On Spectral Factorization and LQ–optimal Regulation of Multivariable Distributed Systems", Report 88/9, University of Namen, Belgium.

[C1] Curtain, R.F., "Equivalence of Input–Output Stability and Exponential Stability for Infinite–Dimensional Systems", Math. System Theory, 1988, **21**, p.19–48.

[C2] Curtain. R.F., Robust Stabilizability of Normalized Coprime Factors: The Infinite–Dimensional Case, 1988, Report TW–291, University of Groningen, NL.

[CR] Curtain, R.F. and Ran, A., "Explicit Formulas for Hankel Norm Approximations of Infinite–dimensional Systems", Integral Equations & Operator Theory (to appear).

[GF1] Glover, K. and McFarlane, D., "Robust Stabilization of Normalized Coprime Factors: An Explicit $H_\infty$–solution", to appear in the Proc. of the ACC, 1988.

[GF2] Glover, K. and McFarlane, D., "Robust Stabilization of Normalized Coprime Factor Plant Descriptions with $H_\infty$–bounded Uncertainty", submitted to IEEE Trans. on Automatic Control, 1988.

[GP] Glover, K. and Partington, J.R., Robust Stabilization of Delay Systems. (this volume)

[GS] Georgiou, T.T. and Smith, M.C., Optimal Robustness in the Gap Metric, (manuscript 1989).

[KS] Khargonekar, P.P. and Sontag, E., On the Relation between Stable Matrix Fraction Factorizations and Regulable Realizations of Linear Systems over Rings, IEEE Trans. AC–**27**, 1982, p.627–638.

[NJB1] Nett, C.N., Jacobson, C.A. and Balas, M.J., "Fractional Representation Theory: Robustness with Applications to Finite–Dimensional Control of a Class of Linear Distributed Systems", Proc. IEEE Conf. on Decision and Control, p. 269–280, 1983.

[NJB2] Nett, C.N., Jacobson, C.A., Balas, M.J., "A Connection Between State Space and Doubly Co–prime Fractional Representations", Trans. IEEE, Vol. AC–21, 9, p. 831–832, 1984.

[PS] Pritchard, A.J. and Salamon, D., "The Linear Quadratic Optimal Control Problem for Infinite Dimensional Systems with Unbounded Input and Output Operators", SIAM J. Control and Optimiz, **25**, p. 121–144, 1987.

[V] Vidyasagar, M., "Control System Synthesis: A Coprime Factorization Approach", MIT Press, 1985.

[W1] Weiss, G., "Admissibility of Unbounded Control Operator", SIAM J. Control and Optim. **27**, 1989 (to appear).

[W2] Weiss, G., Admissible Observation Operators for Linear Semigroups, Israel J. Math., **65**, p. 17–43, 1989.

[W3] Weiss, G., The Representation of Regular Linear Systems on Hilbert Spaces, Proceedings of the Conference on Distributed Parameter Systems, Vorau, Austria, July 1988, to appear.

# STANDARD PROBLEM FOR DISTRIBUTED SYSTEMS

Ciprian Foias Allen Tannenbaum

January 2, 1990

**Abstract**

In this note we study the standard $H^\infty$ design (four block) problem for a large class of distributed plants.

## 1 Introduction

In recent years, largely because of the activity in the area of $H^\infty$ design theory, there has evolved a major interest in the employment of operator theoretic methods in systems and control. In particular, there has been a great deal of research in the uses of interpolation and dilation techniques in this context. The point of the present paper is to show how these methods may be used to solve a very general case of the **standard** or **four block problem** in $H^\infty$ design valid for a large class of distributed, i.e., infinite dimensional systems.

The motivations for studying the $H^\infty$ optimization in systems theory lie in the most natural problems of control engineering such as robust stabilization, sensitivity minimization, and model matching. It can be shown that, in the sense of $H^\infty$ optimality, these problems are equivalent, and can be formulated as one *standard problem* [6]. More precisely, consider the feedback diagram in Figure 1. In this configuration $w$, $u$, $y$, and $z$ are vector-valued signals with $w$ the exogenous input representing the disturbances, measurement noises etc., $u$ the command signal, $z$ the output to be controlled, and $y$ the measured output. $G$ represents a combination of the plant and the weights in the control system. The standard $H^\infty$ problem is to find a stabilizing controller $K$ such that the $H^\infty$ norm of the transfer function from $w$ to $z$ is minimized.

Now it is quite well-known that an optimal solution of the standard problem can be reduced to finding the singular values and vectors of a certain operator (the so-called **four block operator**) which will be defined below. For details we refer the reader to [2-7]. Depending on the specific problem considered, the corresponding four block operator can be simplified to a 2-block or a 1-block operator.

Besides appearing in the most general $H^\infty$ synthesis problems, the four block operators also have a number of intriguing mathematical properties in the sense that they are natural extensions of both the Hankel and Toeplitz operators. For this reason they fit into the skew Toeplitz framework developed in [1]. For the full details of our arguments and details about the skew Toeplitz theory applied to this problem we refer the reader to [3]. Here we will just consider the four block problem for single input / single output systems. We must emphasize however that the theory of [3] gives the solution of the standard four block problem in the general multivariable setting. See also the monograph of Francis [6], and the references therein for more details about the engineering aspects of this research area.

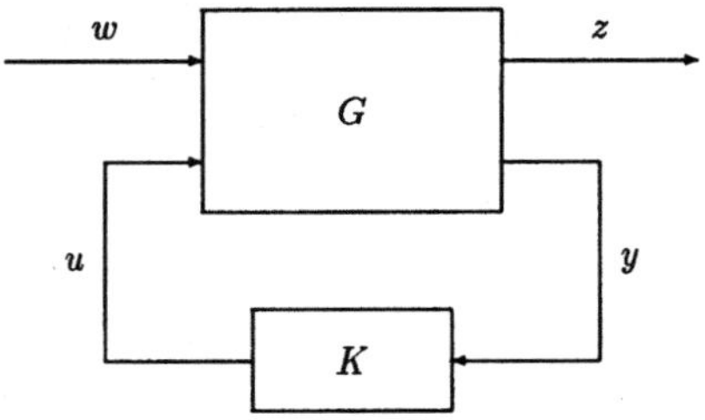


Figure 1.

More precisely, invoking the Youla parametrization and employing standard manipulations involving inner-outer factorizations, for a large class of distributed systems we may reduce the standard problem mentioned above to the following mathematical one. Let $w, f, g, h, \in H^\infty$, where $w, f, g, h$ are rational and $m$ is nonconstant inner. (All of our Hardy spaces will be defined on the unit disc $D$ in the standard way.) Set

$$\mu := \inf\{\| \begin{bmatrix} w - mq & f \\ g & h \end{bmatrix} \|_\infty : q \in H^\infty\}. \tag{1}$$

Then we want to give an algorithm for calculating the quantity $\mu$, and for finding the corresponding $q_{opt} \in H^\infty$, i.e., $q_{opt}$ is such that

$$\mu := \| \begin{bmatrix} w - mq_{opt} & f \\ g & h \end{bmatrix} \|_\infty .$$

Note that for $f = g = h = 0$, this reduces to the classical Nehari problem.

Following [2] and [3], we will identify $\mu$ as the norm of a certain "four block operator" (see Section 2 for the precise definition), and then in Sections 3 and 4 give a determinantal formula for its computation. The techniques given here are based on the previous work in [1], [3], [5].

## 2 The Four Block Operator

We will now define the *four block operator* which will be the major mathematical object of study in this paper. We will not give complete proofs for the various results in this section, and so for all the details we refer the reader to [2] and [3]. We use the notation of the Introduction. Moreover, we let $H(m) := H^2 \ominus mH^2$, $L(m) := L^2 \ominus mH^2$, and we let $P_{H(m)} : H^2 \to H(m)$, $P_{L(m)} : L^2 \to L(m)$ denote the corresponding orthogonal projections. Let $S : H^2 \to H^2$ denote unilateral shift, $T : H(m) \to H(m)$ the compression of $S$, and let $U : L^2 \to L^2$ denote bilateral shift, with $T(m) : L(m) \to L(m)$ the compression of $U$. Then for $w, f, g, h \in H^\infty$ rational, we set

$$A := \begin{bmatrix} P_{L(m)}w(S) & P_{L(m)}f(U) \\ g(S) & h(U) \end{bmatrix}.$$

Note that

$$A = \begin{bmatrix} w(T)P_{H(m)} & f(T(m))P_L(m) \\ g(S) & h(U) \end{bmatrix}$$

(Clearly $A : H^2 \oplus L^2 \to L(m) \oplus L^2$.)

**Proposition 1** *Notation as above. Then* $\|A\| = \mu$.

**Proof.** Use the commutant lifting theorem. (See [2] and [3].) □

Thus in order to solve the four block problem we are required to compute the norm of the operator $A$. This we will show how to do in the next two sections.

In order carry out this program, we will first need to identify the essential norm of $A$ (denoted by $\|A\|_e$). Recall that the essential norm of an operator may be defined as its distance from the space of compact operators. (For details see [3].) We are using the standard notation from operator theory as, for example, given in [7]. In particular $\sigma_e$ will denote the essential spectrum, and $A(\overline{D})$ will stand for the set of analytic functions on $D$ which are continuous on the closed disc $\overline{D}$. We can now state the following result whose proof we refer the reader to [3]:

**Theorem 1** *Notation as above. Let* $w, f, g, h \in A(\overline{D})$, *and set*

$$\alpha := \max\{\|\begin{bmatrix} w(\zeta) & f(\zeta) \\ g(\zeta) & h(\zeta) \end{bmatrix}\| : \zeta \in \sigma_e(T)\} \quad (2)$$

$$\beta := \max\{\|\begin{bmatrix} 0 & 0 \\ g(\zeta) & h(\zeta) \end{bmatrix}\| : \zeta \in \partial D\} \quad (3)$$

$$\gamma := \sup\{\|\begin{bmatrix} f(\zeta) \\ h(\zeta) \end{bmatrix}\| : \zeta \in \partial D\}. \quad (4)$$

*Then*

$$\|A\|_e = \max(\alpha, \beta, \gamma). \quad (5)$$

# 3 Singular System

In this section, we will study the invertibility of certain skew Toeplitz operators as considered in [1], [3] which occur as basic elements in our procedure for computing the norm and singular values of the four block operator. We will show that the calculation of the singular values of the four block operator A amounts to inverting two ordinary Toeplitz operators, and essentially inverting an associated skew Toeplitz operator. The Fredholm conditions on the invertibility of the skew Toeplitz operator (which is essentially invertible), and the coupling between the various systems (expressed as **matching conditions**; see [3]) constitutes a certain linear system of equations called the **singular system** which allows one to determine the invertibility of $A$. The computations for writing down the singular system while straightforward are a bit involved, so we will just give the main idea here of what is involved referring the reader to [1] and [3] for all of the details.

Using the notation of Section 2, we let $\rho > \max(\alpha, \beta, \gamma)$. Note that when $\|A\| > \|A\|_e$, $\|A\|^2$ is an eigenvalue of $AA^*$. By slight abuse of notation, $\zeta$ will denote a complex variable as well as an element of $\partial D$ (the unit circle). The context will always make the meaning clear. Of course, if $\zeta \in \partial D$, then $\overline{\zeta} = 1/\zeta$.

As above, we take $w, f, g, h$ to be rational, and so we can express $w = a/q$, $f = b/q$, $g = c/q$, $h = d/q$, where $a, b, c, d, q$ are polynomials of degree $\leq n$. Then we have that

$$A := \begin{bmatrix} P_{L(m)}(\frac{a}{q})(S) & P_{L(m)}(\frac{b}{q})(U) \\ (\frac{c}{q})(S) & (\frac{d}{q})(U) \end{bmatrix}.$$

Now $\rho^2$ is an eigenvalue of $AA^*$ if and only if

$$\begin{bmatrix} \rho^2 q(T(m))q(T(m))^* & 0 \\ 0 & \rho^2 q(U)q(U)^* \end{bmatrix} \begin{bmatrix} u \\ v \end{bmatrix} - \tag{6}$$

$$\begin{bmatrix} P_{L(m)}a(S) & P_{L(m)}b(U) \\ c(S) & d(U) \end{bmatrix} \begin{bmatrix} a(S)^*P & Pc(U)^* \\ b(U)^*P_L(m) & d(U)^* \end{bmatrix} \begin{bmatrix} u \\ v \end{bmatrix} = 0,$$

for some non-zero

$$\begin{bmatrix} u \\ v \end{bmatrix} \in L(m) \oplus L^2$$

where $P : L^2 \to H^2$ denotes orthogonal projection.

Set

$$u_+ := Pu, \quad u_- := (I - P)u$$

and

$$v_+ := Pv, \quad v_- := (I - P)v, \quad v_{++} := (I - P_{H(m)})v.$$

Then we can write (6) equivalently as

$$\begin{bmatrix} \rho^2 q(T(m))q(T(m))^* - b(T(m))b(T(m))^* & -b(T(m))P_{L(m)}d(U)^* \\ -d(U)b(T(m))^* & \rho^2 q(U)q(U)^* - d(U)d(U)^* \end{bmatrix} \begin{bmatrix} u \\ v \end{bmatrix} \tag{7}$$

$$-\begin{bmatrix} a(T)a(T)^* & a(T)P_{H(m)}c(S)^* \\ c(S)a(T)^* & c(S)c(S)^* \end{bmatrix} \begin{bmatrix} u_+ \\ v_+ \end{bmatrix} = 0.$$

Let $V := U^*|L^2 \ominus H^2$. Further for $q(\zeta) = \sum q_i\zeta^i$, set $q_*(\zeta^{-1}) := \sum \overline{q}_i\zeta^{-1}$. (Similar definitions apply to the other polynomials given below.) If we apply $(I-P)$ to both rows of (7), we see that the basic block operator applied to

$$\begin{bmatrix} u_- \\ v_- \end{bmatrix}$$

is

$$C_- \begin{bmatrix} u_- \\ v_- \end{bmatrix} := \tag{8}$$

$$\begin{bmatrix} \rho^2 q(V^*)q_*(V) - b(V^*)b_*(V) & -b(V^*)d_*(V) \\ -d(V^*)b_*(V) & \rho^2 q(V^*)q_*(V) - d(V^*)d_*(V) \end{bmatrix} \cdot$$

$$\begin{bmatrix} u_- \\ v_- \end{bmatrix}$$

Next applying $(I - P_{H(m)})$ to both rows of (7), we see that the basic operator applied to $v_{++}$ is

$$C_{++}\overline{m}v_{++} := P\{(\rho^2|q|^2 - |c|^2 - |d|^2)\overline{m}v_{++}\}. \tag{9}$$

Finally, applying $P_{H(m)}$ to (7), we derive that the basic operator applied to

$$\begin{bmatrix} u_+ \\ P_{H(m)}v_+ \end{bmatrix}$$

is

$$C_+ \begin{bmatrix} u_+ \\ P_{H(m)}v_+ \end{bmatrix} := \tag{10}$$

$$\begin{bmatrix} \rho^2 q(T)q(T)^* - b(T)b(T)^* - a(T)a(T)^* & -b(T)P_{H(m)}d(S)^* \\ -d(T)b(T)^* & \rho^2 q(T)q(T)^* - d(T)d(T)^* - c(T)c(T)^* \end{bmatrix} \cdot$$

$$\begin{bmatrix} u_+ \\ P_{H(m)}v_+ \end{bmatrix}.$$

The operators $C_-, C_{++}, C_+$ are all skew Toeplitz (see [1] for the precise definition). We will now show how to invert $C_-$ and $C_{++}$ under the assumption $\rho > \|A\|_e$. The essential inversion of $C_+$ can be handled exactly as in [1], [5]. See [3] for the details.

We start with $C_-$ . Namely, let $f_-, g_- \in L^2 \ominus H^2$, and consider the equation

$$C_-\begin{bmatrix} u_- \\ v_- \end{bmatrix} = \begin{bmatrix} f_- \\ g_- \end{bmatrix}. \tag{11}$$

But (11) is equivalent to

$$\begin{bmatrix} \rho^2|q|^2 - |b|^2 & -b\overline{d} \\ -d\overline{b} & \rho^2|q|^2 - |d|^2 \end{bmatrix}\begin{bmatrix} u_- \\ v_- \end{bmatrix} = \begin{bmatrix} f_- \\ g_- \end{bmatrix} + \sum_{j=0}^{n-1} \zeta^j \begin{bmatrix} x_j \\ y_j \end{bmatrix} \tag{12}$$

where

$$\begin{bmatrix} x_j \\ y_j \end{bmatrix} \in \mathbf{C}^2$$

$(0 \leq j \leq n-1)$ are to be determined.

Put $q_o(\zeta^{-1}) := \zeta^{-n}q(\zeta)$. If we multiply (12) by $\zeta^{-n}$, we get that (with all the polynomials in $\zeta^{-1}$)

$$\begin{bmatrix} \rho^2 q_o q_* - b_o b_* & -b_o d_* \\ -d_o b_* & \rho^2 q_o q_* - d_o d_* \end{bmatrix}\begin{bmatrix} u_- \\ v_- \end{bmatrix} = \begin{bmatrix} \zeta^{-n} f_- \\ \zeta^{-n} g_- \end{bmatrix} + \sum_{j=0}^{n-1} \zeta^{j-n} \begin{bmatrix} x_j \\ y_j \end{bmatrix}. \tag{13}$$

Now by definition, $\rho > \|A\|_e$, and so we see that $\rho^2|q|^2 - |b|^2 - |d|^2 > 0$, and hence we can write

$$\det\left\{\begin{bmatrix} \rho^2 q\overline{q} & 0 \\ 0 & \rho^2 q\overline{q} \end{bmatrix} - \begin{bmatrix} \overline{b} \\ \overline{d} \end{bmatrix}[b \quad d]\right\} = \overline{\Delta}\Delta. \tag{14}$$

Thus we see that

$$\zeta^{-n}\det\left\{\begin{bmatrix} \rho^2 q\overline{q} & 0 \\ 0 & \rho^2 q\overline{q} \end{bmatrix}\begin{bmatrix} \overline{b} \\ \overline{d} \end{bmatrix}[bd]\right\} = \Delta_o(\zeta^{-1})\Delta_*(\zeta^{-1}).$$

We now make the following *assumption of genericity*

$$\Delta_o\Delta_* \quad \text{has distinct roots all of which are non-zero.} \tag{15}$$

In [3] it is shown how to remove (15), but for the sake of simplicity of our exposition this assumption will remain in force in what follows. Then it is easy to see that $\Delta_o\Delta_*$ has $2n$ distinct roots $\overline{z}_1, \cdots, \overline{z}_{2n}$ in $D$, and $2n$ distinct zeros $1/z_1, \cdots, 1/z_{2n}$ in the complement of $\overline{D}$. Set

$$\hat{D}(\zeta^{-1}) := \begin{bmatrix} \rho^2 q_o q_* - b_o b_* & -b_o d_* \\ -d_o b_* & \rho^2 q_o q_* - d_o d_* \end{bmatrix}$$

and let $\hat{D}^{ad}(\zeta^{-1})$ denote the algebraic adjoint of $\hat{D}(\zeta^{-1})$. Then if we apply $\hat{D}^{ad}(\zeta^{-1})$ to both sides of (13), we get

$$\Delta_o\Delta_*\begin{bmatrix} u_- \\ v_- \end{bmatrix} = \hat{D}^{ad}(\zeta^{-1})\begin{bmatrix} \zeta^{-n} f_- \\ \zeta^{-n} g_- \end{bmatrix} + \hat{D}^{ad}(\zeta^{-1})\sum_{j=0}^{n-1} \zeta^{j-n}\begin{bmatrix} x_j \\ y_j \end{bmatrix}. \tag{16}$$

Consequently plugging the $1/z_k$ into the last expression, we derive that

$$\widehat{D}^{ad}(z_k)\begin{bmatrix} z_k^n f_-(z_k) \\ z_k^n g_-(z_k) \end{bmatrix} + \widehat{D}^{ad}(z_k)\sum_{j=0}^{n-1} z_k^{-j+n}\begin{bmatrix} x_j \\ y_j \end{bmatrix} = 0 \tag{17}$$

for $k = 1, \cdots, 2n$.

Next note that

$$\Delta_o\Delta_* = \rho^2 q_o q_*(\rho^2 q_o q_* - b_o b_* - d_o d_*).$$

Let $1/z_1, \cdots, 1/z_n$ be such that

$$q_o q_*(z_k) = 0$$

for $1 \leq k \leq n$, and $1/z_{n+1}, \cdots, 1/z_{2n}$ be such that

$$(\rho^2 q_o q_* - b_o b_* - d_o d_*)(z_{n+k}) = 0$$

for $1 \leq k \leq n$. We can now state the following (the proofs of the following results can all be found in [3]):

**Proposition 2** *With the above notation, and under assumption (15), we have that the $x_j, y_j \in \mathbf{C}$, $0 \leq j \leq n-1$, are uniquely defined by*

$$\begin{bmatrix} x_o \\ \cdot \\ \cdot \\ \cdot \\ x_{n-1} \\ y_o \\ \cdot \\ \cdot \\ \cdot \\ y_{n-1} \end{bmatrix} = E^{-1}\hat{E}\begin{bmatrix} z_1^n f_-(z_1) \\ z_1^n g_-(z_1) \\ \cdot \\ \cdot \\ z_{2n}^n f_-(z_{2n}) \\ z_{2n}^n g_-(z_{2n}) \end{bmatrix}$$

*where*

$$\hat{E} := \begin{bmatrix} E_1 & 0_{(n,2n)} \\ 0_{(n,2n)} & E_2 \end{bmatrix}$$

*for*

$$E_1 := \begin{bmatrix} d_o(z_1) & -b_o(z_1) & 0 & 0 & \cdot & 0 & 0 \\ 0 & 0 & d_o(z_2) & -b_o(z_2) & \cdot & 0 & 0 \\ 0 & 0 & 0 & 0 & \cdot & 0 & 0 \\ \cdot & \cdot & \cdot & \cdot & \cdot & \cdot & \cdot \\ 0 & 0 & \cdot & 0 & \cdot & d_o(z_n) & -b_o(z_n) \end{bmatrix},$$

$$E_2 := \begin{bmatrix} b_*(z_{n+1}) & d_*(z_{n+1}) & 0 & 0 & \cdot & 0 & 0 \\ 0 & 0 & b_*(z_{n+2}) & d_*(z_{n+2}) & \cdot & 0 & 0 \\ 0 & 0 & 0 & 0 & \cdot & 0 & 0 \\ \cdot & \cdot & \cdot & \cdot & \cdot & \cdot & \cdot \\ 0 & 0 & 0 & 0 & \cdot & b_*(z_{2n}) & d_*(z_{2n}) \end{bmatrix},$$

*($0_{(n,2n)}$ denotes the $n \times 2n$ matrix all of whose entries are 0), and where*

$$E := \begin{bmatrix} diag(d_o(z_1), \cdots, d_o(z_n))V_1 & diag(-b_o(z_1), \cdots, -b_o(z_n))V_1 \\ diag(b_*(z_{n+1}), \cdots, b_*(z_{2n}))V_2 & diag(d_*(z_{n+1}), \cdots, d_*(z_{2n}))V_2 \end{bmatrix}$$

*for*

$$V_1 := \begin{bmatrix} 1 & z_1 & \cdot & \cdot & \cdot & z_1^{n-1} \\ \cdot & \cdot & \cdot & \cdot & \cdot & \cdot \\ \cdot & \cdot & \cdot & \cdot & \cdot & \cdot \\ \cdot & \cdot & \cdot & \cdot & \cdot & \cdot \\ 1 & z_n & \cdot & \cdot & \cdot & z_n^{n-1} \end{bmatrix}, V_2 := \begin{bmatrix} 1 & z_{n+1} & \cdot & \cdot & \cdot & z_{n+1}^{n-1} \\ \cdot & \cdot & \cdot & \cdot & \cdot & \cdot \\ \cdot & \cdot & \cdot & \cdot & \cdot & \cdot \\ \cdot & \cdot & \cdot & \cdot & \cdot & \cdot \\ 1 & z_{2n} & \cdot & \cdot & \cdot & z_{2n}^{n-1} \end{bmatrix}.$$

*(Note that $diag(a_1, \cdots, a_N)$ denotes the $N \times N$ diagonal matrix with entries $a_1, \cdots, a_N$ on the diagonal.)*

Let $\hat{E}^{-1}E =: [e_{ij}]$ for $1 \leq i \leq 2n$, $1 \leq j \leq 4n$. Then we can now state (again see [3] for the details):

**Corollary 1** *With the above notation, we have that*

$$\begin{bmatrix} u_- \\ v_- \end{bmatrix} = \frac{1}{\Delta_o \Delta_*}\{\hat{D}^{ad}(\zeta^{-1})\begin{bmatrix} \zeta^{-n} f_- \\ \zeta^{-n} g_- \end{bmatrix} + \tag{18}$$

$$\hat{D}^{ad}(\zeta^{-1})\sum_{k=0}^{n-1} \zeta^{k-n} \begin{bmatrix} \sum_{j=1}^{2n} [e_{k+1,2j-1} z_j^n f_-(z_j) + e_{k+1,2j} z_j^n g_-(z_j)] \\ \sum_{j=1}^{2n} [e_{n+k+1,2j-1} z_j^n f_-(z_j) + e_{n+k+1,2j} z_j^n g_-(z_j)] \end{bmatrix}\}.$$

This gives the way to invert $C_-$. Now we consider the inverse of $C_{++}$. Since $\rho > \beta$, we have of course that $C_{++}$ is invertible. For $p(\zeta)$ a polynomial of degree $\leq n$, we let $\tilde{p}(\zeta) := \zeta^n \overline{p(\zeta)}$. Set

$$\lambda(\zeta) := (\rho^2 \tilde{q} q - \tilde{c} c - \tilde{d} d).$$

We now make our second assumption of genericity that

$$\lambda(\zeta) \text{ has distinct nonzero roots all of which are nonzero.} \tag{19}$$

Again in [3], it is shown how to remove assumption (19). However with this assumption, we see that $\lambda(\zeta)$ has $n$ roots $\zeta_1, \cdots \zeta_n \in D$, and $n$ roots $1/\bar{\zeta}_1, \cdots, 1/\bar{\zeta}_n$ which are in the complement of $\overline{D}$. We then have

**Proposition 3** *With assumption (19), if*

$$C_{++}(\bar{m}v_{++}) = f$$

*for* $f \in H^2$, *then*

$$v_{++} := m\begin{pmatrix} \zeta^n f - \sum_{j=1}^{n} \zeta_j \zeta^{n-j} \\ -(\rho^2 \tilde{q} q - \tilde{c} c - d\tilde{d}) \end{pmatrix}$$

*where*

$$\begin{bmatrix} \eta_1 \\ \cdot \\ \cdot \\ \cdot \\ \eta_n \end{bmatrix} = R_1^{-1} \begin{bmatrix} \zeta_1^{n-1} f(\zeta_1) \\ \cdot \\ \cdot \\ \cdot \\ \zeta_n^{n-1} f(\zeta_n) \end{bmatrix}, \quad R_1 := \begin{bmatrix} \zeta_1^{n-1} & \cdot & \cdot & \cdot & 1 \\ \cdot & \cdot & \cdot & \cdot & \cdot \\ \cdot & \cdot & \cdot & \cdot & \cdot \\ \cdot & \cdot & \cdot & \cdot & \cdot \\ \zeta_n^{n-1} & \cdot & \cdot & \cdot & 1 \end{bmatrix}.$$

Based on the above inversion formulae and the essential inversion of the skew Toeplitz operator $C_+$ ([1], [3], [5]), the singular system can be constructed and the following theorem can be proved (even without the above assumptions of genericity; see [3] for all the details as well as for the precise definition of the matrix $M(\rho)$ given below):

**Theorem 2** *There exists an explicitly computable* $5n \times 5n$ *Hermitian matrix* $M(\rho)$ *such that* $\overline{\rho} > \max\{\alpha, \beta, \gamma\}$ *is a singular value of the four block operator* $A$ *if and only if*

$$\det M(\overline{\rho}) = 0.$$

# 4 On Optimal Compensators

The above procedure also gives a way of computing the optimal compensator in a given four block problem [4]. Indeed, from the above determinantal formula one can compute the Schmidt pair $\psi, \eta$ corresponding to the singular value $s := \|A\|$ when $s > \|A\|_e$. See [3], [4]. We will indicate how one derives the optimal interpolant (and thus the opimal compensator) from these Schmidt vectors. In order to do this, notice

$$A\psi = s\eta.$$

Thus, there exists $q_{opt} \in H^\infty$ with

$$(w - q_{opt})\psi_1 + f\psi_2 = s\eta_1$$

$$g\psi_1 + h\psi_2 = s_k\eta_2$$

where

$$\psi = \begin{bmatrix} \psi_1 \\ \psi_2 \end{bmatrix}, \quad \eta = \begin{bmatrix} \eta_1 \\ \eta_2 \end{bmatrix}.$$

One can show (see [4]), that $\psi_1 \neq 0$, so that

$$q_{opt} = w - \frac{s\eta_1 - f\psi_2}{\psi_1}.$$

Note from $q_{opt}$, using the Youla parametrization, we can derive the corresponding optimal controller in a given systems design problem. See also [4] for an extension of the theory of Adamjan-Arov-Krein (valid for the Hankel operator) to the singular values of the four block operator and their relationship to more general interpolation and distance problems.

We conclude this paper by noting that computer programs have been written at the University of Minnesota, and at Honeywell SRC in Minneapolis to carry out the above procedure. So far our computational experience has been very encouraging.

This research was supported in part by grants from the Research Fund of Indiana University, Department of Energy DE-FG02-86ER25020, NSF (ECS-8704047), NSF (DMS-8811084), and from the Air Force Office of Scientific Research AFOSR-90-0024.

## References

[1] H. Bercovici, C. Foias, and A. Tannenbaum, "On skew Toeplitz operators. I," *Operator Theory: Advances and Applications* **29** (1988), pp. 21–44.

[2] A. Feintuch and B. Francis, "Uniformly optimal control of linear systems," *Automatica* **21** (1986), pp. 563-574.

[3] C. Foias and A. Tannenbaum, "On the four block problem, II: the singular system," *Integral Equations and Operator Theory* **11** (1988), pp. 726–767.

[4] C. Foias and A. Tannenbaum, "On the singular values of the four block operator and certain generalized interpolation problems," to appear in a volume dedicated to Professor Mischa Cotlar on his seventy-fifth birthday.

[5] C. Foias, A. Tannenbaum, and G. Zames, "Some explicit formula for the singular values of certain Hankel operators with factorizable symbol," *SIAM J. on Math. Analysis* **19** (1988), pp. 1081–1089.

[6] B. Francis, *A Course in $H^\infty$ Control Theory*, McGraw-Hill, New York, 1981.

[7] B. Sz.-Nagy and C. Foias, *Harmonic Analysis of Operators on Hilbert Space*, North-Holland Publishing Company, Amsterdam, 1970.

[8] G. Zames, "Feedback and optimal sensitivity: model reference transformations, multiplicative seminorms, and approximate inverses," *IEEE Trans. Auto. Control* **AC-26** (1981), pp. 301–320.

# ROBUST STABILIZATION OF DELAY SYSTEMS

Jonathan R. Partington and Keith Glover

**Abstract**

Given a delay system with transfer function $G(s) = h_2(s)/h_1(s)$, where $h_1(s) = \sum_0^{n_1} p_i(s)e^{-\gamma_i s}$, and $h_2(s) = \sum_0^{n_2} q_i(s)e^{-\beta_i s}$, with $0 = \gamma_o < \gamma_1 < \ldots < \gamma_{n_1}$, $0 \leq \beta_o < \ldots < \beta_{n_2}$, the $p_i$ being polynomials of degree $\delta_i$, and $\delta_i < \delta_o$ for $i \neq 0$, and the $q_i$ polynomials of degree $d_i < \delta_o$ for each $i$, the robust stabilization of a class of perturbed coprime factors of this system is considered. Asymptotic estimates are obtained based on recent explicit results on the approximation and stabilization of normalized coprime factors.

## 1. Introduction

Given a nominal system model with transfer function $G(s) = M^{-1}(s)N(s)$, where $M$, $N \in RH_\infty$ are left coprime, it has been demonstrated in [16] that a suitable class of perturbed models is given by $G_\Delta = (M + \Delta_1)^{-1}(N + \Delta_2)$, where $\Delta = [\Delta_1, \Delta_2] \in RH_\infty$ and $\|\Delta\|_\infty \leq \epsilon$. In the case when coprime factors are *normalized*, that is $MM^* + NN^* = I$, then [12, 13] have shown that the largest family of $\Delta$ that is stabilizable by a single controller is given by the formula

$$\epsilon_{\max} = \sqrt{1 - \|[N, M]\|_H^2}.$$

This robust stabilization result has in fact been used to generate an effective design scheme. The derivation of this result in [12] involved state-space manipulations and is not readily extended to infinite-dimensional systems. The derivation in [13] is an input/output argument and can be modified for a suitable class of infinite-dimensional systems; alternatively an operator theory derivation is given in [8] together with a number of results on the gap and graph metrics.

It is the purpose of the present paper to show how an effective approximation scheme can be derived for delay systems. Optimal convergence rates will be established, based on recent approximation results given in [9, 10, 11]. These are then applied to construct robust finite-dimensional controllers which approach optimality.

## 2. Retarded delay systems

We shall consider the class of retarded delay systems with scalar transfer function given by

$$G(s) = h_2(s)/h_1(s), \tag{2.1}$$

where

$$h_1(s) = \sum_{o}^{n_1} p_i(s) e^{-\gamma_i s}, \tag{2.2}$$

and

$$h_2(s) = \sum_{o}^{n_2} q_i(s) e^{-\beta_i s}, \tag{2.3}$$

with $0 = \gamma_o < \gamma_1 < \ldots < \gamma_{n_1}$, $0 \leq \beta_o < \ldots < \beta_{n_2}$, the $p_i$ being polynomials of degree $\delta_i$, and $\delta_i < \delta_o$ for $i \neq 0$, and the $q_i$ polynomials of degree $d_i < \delta_o$ for each $i$. These systems were first analysed by Bellman and Cooke [1] and have the property that they possess only finitely many poles in any right half plane. Asymptotic results on the Hankel singular values and errors in the approximation of $G$ were given in [10], [11], where it was shown that there exist constants $A, B > 0$ and an integer $r > 0$ such that the Hankel singular values $(\sigma_k)$ of $G$ decay as $k^{-r}$ and

$$Ak^{-r} \leq \inf\{||G_{\text{stable}} - \tilde{G}||_\infty : \deg \tilde{G} = k\} \leq Bk^{-r}. \tag{2.4}$$

In general we restrict our analysis to SISO systems as above, except that in section 3 we consider the class of matrix-valued transfer functions $G(s) = e^{-sT}R(s)$, where $R$ is rational and $T > 0$: for these an explicit solution to the normalized coprime factorization problem can be given.

We recall the following definition, due to Callier and Desoer [2] of a class of functions admitting normalized coprime factorizations (see also [3, 16]).

The algebra $\mathcal{A}$ comprises all distributions on $(0, \infty)$ of the form

$$f(t) = f_a(t) + \sum_{0}^{\infty} f_i \delta(t - t_i), \tag{2.5}$$

with $f_a \in L_1(0, \infty)$, $(f_i) \in l_1$ and $t_i \geq 0$ for each $i$. We write $\hat{\mathcal{A}} \subset H_\infty$ for the algebra of Laplace transforms of elements of $\mathcal{A}$. The ring $\mathcal{A}_-(\alpha)$ consists of all

$f(t)$ as in (2.5) with $e^{-\beta t}f_a(t) \in L_1$ and $(e^{-\beta t_i}f_i) \in l_1$ for some $\beta < \alpha$. Then its ring of Laplace transform is denoted $\hat{\mathcal{A}}_-(\alpha)$, and the subset of functions bounded away from zero in $\Re e\, s \geq \alpha$ will be written $\hat{\mathcal{A}}_-^\infty(\alpha)$. Finally $\hat{\mathcal{B}}(\alpha)$ is the ring $(\hat{\mathcal{A}}_-(\alpha))(\hat{\mathcal{A}}_-^\infty(\alpha))^{-1}$.

It is easily seen that transfer functions satisfying the conditions above are in the class $\hat{\mathcal{B}} = \hat{\mathcal{B}}(0)$ and hence possess coprime factorizations $G(s) = E(s)^{-1}F(s)$. One such is given by

$$E(s) = h_1(s)/(s+1)^{\delta_0} \qquad \text{and} \qquad F(s) = h_2(s)/(s+1)^{\delta_0},$$

since $E(s)$ is certainly bounded away from zero near infinity. More details may be found in [16, Chapter 8].

In general, however, ***normalized*** coprime factorizations cannot be calculated explicitly for infinite-dimensional systems, and rational approximations are required. In section 4 we show how this can be done, and estimate the convergence rates of such normalized coprime factors.

## 3. Explicit analytic solutions

For the special case $G(s) = e^{-sT}R(s)$, with $T > 0$ and $R(s)$ rational and matrix-valued, a normalized coprime factorization $R = M^{-1}N$ gives a normalized coprime factorization of $G$:

$$G(s) = M(s)^{-1}(e^{-sT}N(s)),$$

and the stability margin can be calculated explicitly.

Let us suppose that $G(\infty) = 0$ and $[N, M] = [0, I] + C(sI - A)^{-1}[B, H]$ has controllability and observability Gramians $P$ and $Q$ respectively. In order to calculate the singular values of the Hankel operator corresponding to $[e^{-sT}N, M]$ we can set up a two-point boundary-value problem as in [5, 6, 9, 10, 17]. Alternatively [10, Theorem 2.1] can be applied directly and the result then simplifies to give the following.

**Proposition 3.1.** *$\sigma$ is a Hankel singular value of $[e^{-sT}N(s), M(s)]$ if and only if*

$$\det\left\{[-\sigma^{-1}Q, I]\exp(KT)\begin{bmatrix}\sigma^{-1}P\\ I\end{bmatrix}\right\} = 0 \tag{3.1}$$

*where*

$$K := \begin{bmatrix} A & \sigma^{-1}BB' \\ -\sigma^{-1}C'C & -A' \end{bmatrix} \tag{3.2}$$

*Hence the maximum normalized coprime factor stability margin for such a system is given by* $\epsilon_{\max} = \sqrt{1-\sigma_1^2}$, *where* $\sigma_1$ *is the largest solution to (3.1).*

□

## 4. Convergence of coprime factors

Let $G(s) = h_2(s)/h_1(s)$, with $h_1$ and $h_2$ as given in (2.2) and (2.3), be the transfer function of a retarded delay system, and suppose that $G(s) = E(s)^{-1}F(s)$ is a coprime factorization. Then $G$ possesses a *normalized* coprime factorization $G(s) = M(s)^{-1}N(s)$ where $M(s) = E(s)/R(s)$, $N(s) = F(s)/R(s)$ and $R^*R = E^*E + F^*F$ (see [16, Chapter 7]). This cannot in general be calculated explicitly, but we can choose rational approximations $E_k(s)$ and $F_k(s)$ converging to $E, F$ respectively in the $H_\infty$ norm and normalize these. The following result guarantees that such a procedure converges.

**Theorem 4.1.** *Let* $G(s) = E(s)^{-1}F(s)$ *be a coprime factorization of a system* $G(s) \in \hat{\mathcal{B}}$, *and suppose that* $(E_k(s))$ *and* $(F_k(s))$ *are sequences of rational* $H_\infty$ *functions of degree at most* $k$ *such that* $E_k \to E$ *and* $F_k \to F$ *in* $H_\infty$. *Then* $E_k$ *and* $F_k$ *are coprime for* $k$ *sufficiently large. Now let* $G(s) = M(s)^{-1}N(s)$ *be a normalized coprime factorization of* $G$, *and* $E_k(s)^{-1}F_k(s) = M_k(s)^{-1}N_k(s)$ *be normalized coprime factorizations of the rational approximants. If* $r > 0$ *is such that* $\|E - E_k\|_\infty$ *and* $\|F - F_k\|_\infty$ *are both* $O(k^{-r})$, *then the distance from* $[N, M]$ *to* $[N_k, M_k]$ *in the graph and gap metrics is* $O(k^{-r})$ *and hence there exist invertible functions* $U_k$ *such that*

$$\|[N, M]U_k - [N_k, M_k]\|_\infty = O(k^{-r}) \tag{4.1}$$

*as* $k \to \infty$. *Moreover* $U_k^*U_k = 1 + O(k^{-r})$ *as* $k \to \infty$.

**Proof** To show that $E_k$ and $F_k$ are eventually coprime, we observe that since $E$ and $F$ are coprime they satisfy a Bezout identity

$$E(s)X(s) + F(s)Y(s) = 1 \tag{4.2}$$

over $H_\infty$. Hence

$$E_k(s)X(s)Z(s) + F_k(s)Y(s)Z(s) = 1,$$

where $Z(s) = (1 + (E_k - E)X + (F_k - F)Y)^{-1}$, which exists in $H_\infty$ provided that

$$\|E_k - E\|\,\|X\| + \|F_k - F\|\,\|Y\| < 1,$$

which is true for sufficiently large $k$. Thus $E_k$ and $F_k$ are coprime over $H_\infty$ and thus have no common zero in $\mathbf{C}_+ \cup \{\infty\}$. Hence they are also coprime over the ring of rational $H_\infty$ functions, by [16, Chapter 2].

Now there exist $\alpha, \beta > 0$ such that $0 < \alpha \leq |R(s)| \leq \beta$ for $s$ on the imaginary axis, so that

$$\|[N, M] - [F_k/R, E_k/R]\| \leq (1/\alpha)\|[F - F_k, E - E_k]\|$$
$$= \eta_k, \text{ say.}$$

It follows from [16, lemma 7.3.2], that the distance from $[N, M]$ to $[N_k, M_k]$ in the graph metric can be bounded, namely

$$d([N, M], [N_k, M_k]) \leq \frac{2\eta_k}{1 - \eta_k} = O(k^{-r}).$$

The existence of $(U_k)$ as required follows from section 7.3 of [16]. Similarly the gap metric estimate follows from [7]. Moreover

$$U_k^* U_k = (NU_k)^*(NU_k) + (MU_k)^*(MU_k)$$
$$= N_k^* N_k + M_k^* M_k + O(k^{-r}) = 1 + O(k^{-r}).$$

□

Thus a normalised approximant to the original system is close to being a normalisation of the system. We now identify $r$ as the optimal convergence index of *any* sequence of approximate coprime factorizations of $G$.

**Theorem 4.2.** *Let $G(s)$ be as in (2.1)–(2.3) and let $r$ be the index given in (2.4). Then for any coprime factorization $G = E^{-1}F$ and functions $E_k$, $F_k$ in $RH_\infty$ of degree at most $k$ there exists $C > 0$ such that*

$$\|[F, E] - [F_k, E_k]\| \geq Ck^{-r}.$$

*Moreover there exists a coprime factorization $G = E^{-1}F$ and rational functions $E_k$, $F_k$ of degree $k$ such that*

$$\|[F, E] - [F_k, E_k]\| = O(k^{-r}).$$

**Proof** If $G = E^{-1}F$ is any coprime factorization, then, since there exist $X$ and $Y$ such that (4.2) holds, we can choose $M > 0$ such that $G(s + M)$ is stable and hence

$$1 = |E(s)X(s) + F(s)Y(s)| \leq |E(s)|\,|X(s) + Y(s)F(s)/E(s)|,$$

implying that $|E(s)|$ is bounded away from zero, i.e. $|E(s)| > \delta$, say, in $\Re e\, s > M$. Thus if $\|F - F_k\| < \epsilon_k$ and $\|E - E_k\| < \epsilon_k$, then

$$\left\| \frac{F(s+M)}{E(s+M)} - \frac{F_k(s+M)}{E_k(s+M)} \right\| \leq \frac{\|F\|\epsilon_k + \|E\|\epsilon_k}{\delta(\delta - \epsilon_k)}$$

in $\Re e\, s \geq 0$. But, by [10], $\sigma_k(G(s+M))$ and $\sigma_k(G_{\text{stable}})$ are asymptotic (to within a constant independent of $k$), and hence $\exists C > 0$ such that $\epsilon_k \geq Ck^{-r}$ for all $k$.

Conversely, suppose that $G = E^{-1}F + G_{\text{stable}}$, where $E$, $F$ are rational, coprime and stable. Then $G = E^{-1}(F + EG_{\text{stable}})$ is a coprime factorization of $G$. Choosing suitable rational approximants $\hat{G}_k$ to $G_{\text{stable}}$ as in [11], we have $\|(F + EG_{\text{stable}}) - (F + E\hat{G}_k)\| \leq \|E\|\|G_{\text{stable}} - \hat{G}_k\|$, achieving the optimal convergence rate.

□

**Remark 4.3.** If a rational approximation $[N_k, M_k]$ of degree $k$ is found with

$$\|[N, M] - [N_k, M_k]\|_\infty < \delta,$$

then a controller of degree $k$ stabilizing a ball around $[N_k, M_k]$ will have maximum stability radius $\epsilon_k > \epsilon_{\max} - \delta$ and hence when applied to $[N, M]$ will have stability radius at least $\epsilon_{\max} - 2\delta$. A similar argument in the case that (4.1) holds shows that rational controllers can be derived for delay systems, whose performance approaches optimality as $k \to \infty$.

## 5. Examples

Consider the delay system $G(s) = e^{-sT}/s$. This can be analysed using the techniques of section 3, and a normalized coprime factorization is given by $G(s) = M(s)^{-1}N(s)$, with $M(s) = s/(s+1)$ and $N(s) = e^{-sT}/(s+1)$. The singular values of the Hankel operator with symbol $[N, M]$ are given by $\sigma_k = 1/\sqrt{1+\lambda_k^2}$, where $(\lambda_k)$ are the positive roots of $\tan \lambda T = (1-\lambda^2)/2\lambda$ and thus $\sigma_k$ is asymptotic to $T/k\pi$. The robust stability margin is given by $\epsilon = \sqrt{1-\sigma_1^2}$; this is asymptotic to $\dfrac{1}{\sqrt{2}}(1 - \dfrac{T}{2})$ as $T \to 0$, and to $\pi/2T$ as $T \to \infty$. Some typical values are as follows:

| $T$ | 0.5 | 1 | 2 | 5 | 10 |
|---|---|---|---|---|---|
| $\sigma_1$ | 0.819 | 0.874 | 0.928 | 0.976 | 0.992 |
| $\epsilon$ | 0.573 | 0.486 | 0.374 | 0.220 | 0.130 |

Approximation of the coprime factors may be achieved by the technique used in [11], which consists of replacing $e^{-sT}$ by its $(k,k)$ Padé approximant. This is guaranteed to converge at the optimal rate $(O(k^{-1}))$. In practice the stability margins $(\epsilon_k)$ converge much faster than the coprime factors themselves. For the hardest case to control above, $T = 10$, one finds that $\epsilon_{\max} = 0.130$, even for a 5th order approximation.

Recall that, for a stabilizing controller, we have

$$\epsilon^{-1} = \left\| \begin{bmatrix} K \\ I \end{bmatrix} (I+GK)^{-1}[G,I] \right\|_\infty .$$

The best ***constant*** controller for $T = 10$, namely $K = 0.065$, gives $\epsilon = 0.056$, whereas higher order controllers can be produced which approach optimality.

Similarly for $G(s) = \frac{2e^{-2s}}{1-s^2}$, considered in [4], $\epsilon_{\max} = 0.032$ (and the coprime factors converge with error $O(k^{-2})$); for $\frac{1}{s-e^{-s}}$ (not soluble by the methods of Section 3), $\epsilon_{\max} = 0.434$, again with a convergence rate of $O(k^{-2})$ – in this case the 'obvious' coprime factorization, $\left[\frac{1}{s+1}, \frac{s-e^{-s}}{s+1}\right]$, if unnormalized leads to a suboptimal convergence rate of $O(k^{-1})$.

As a final example we consider $G(s) = \frac{2e^{-s}}{2s^2+e^{-s}}$, an example which cannot be stabilized by a constant controller and for which an analytical normalised coprime factorization is not known. This time $\epsilon = 0.224$ and the convergence rate $O(k^{-2})$; by taking a seven state approximation to $G(s)$ and using the methods of [12] combined with model reduction on the controller as in [14] one can obtain an order six controller giving a stability margin of 0.212, over 95% of the optimal.

Thus in general satisfactory low-order controllers can be produced by combining the approximation techniques of [11] and the methods of [12, 13, 14].

## References


[1] R. Bellman and K.L. Cooke, *Differential-Difference Equations*, Academic Press, 1963.

[2] F.M. Callier and C.A. Desoer, "Simplifications and New Connections on an Algebra of Transfer Functions of Distributed Linear Time-Invariant Systems," *IEEE Trans. Circ. Sys.*, CAS-27 (1980), 320-323.

[3] R.F. Curtain, "Robust stabilizability of normalized coprime factors: the infinite dimensional case," Preprint.

[4] R.F. Curtain and K. Glover, "Robust Stabilization of infinite dimensional systems by finite dimensional controllers: derivations and examples," *Proc. MTNS-85*, Stockholm, 1985.

[5] C. Foias, A. Tannenbaum and G. Zames, "Weighted sensitivity minimization for delay systems," *Proc. IEEE Conference on Decision and Control*, (1985), 244-249.

[6] C. Foias, A. Tannenbaum and G. Zames, "Weighted sensitivity minimization for delay systems," *IEEE Trans. Auto. Control*, AC-31 (1986), 763-766.

[7] T.T. Georgiou, "On the computation of the gap metric," *Systems and Control Letters* 11 (1988), 253-257.

[8] T.T. Georgiou and M.C. Smith, "Optimal Robustness in the Gap Metric," Internal Report, 1989.

[9] K. Glover, J. Lam and J.R. Partington, "Balanced Realisation and Hankel-norm approximation of systems involving delays," *Proc. IEEE Conference on Decision and Control*, Athens (1986), 1810-1815.

[10] K. Glover, J. Lam and J.R. Partington, "Rational approximation of a class of infinite-dimensional systems I: Singular values of Hankel operators," *Math. Control Circ. Sys.*, to appear.

[11] K. Glover, J. Lam and J.R. Partington, "Rational approximation of a class of infinite-dimensional systems II: Optimal convergence rates of $L_\infty$ approximants," *Math. Control Circ. Sys.*, to appear.

[12] K. Glover and D.C. McFarlane, "Robust Stabilization of Normalized Coprime Factors: an explicit $H_\infty$ solution," *Proceedings 1988 A.C.C.*, pp. 842-847.

[13] K. Glover and D.C. McFarlane, "Robust Stabilization of Normalized Coprime Factor Plant Descriptions with $H_\infty$-bounded uncertainty," *IEEE Trans. Auto. Control*, 34 (1989), 821-830.

[14] D.C. McFarlane, K. Glover and M. Vidyasagar, "Reduced order controller design using coprime factor model reduction," *IEEE Trans. Auto. Control*, 1990, to appear.

[15] J.R. Partington, K. Glover, H.J. Zwart and R.F. Curtain, "$L_\infty$ approximation and nuclearity of delay systems," *Systems and Control Letters* 10 (1988), 59-65.

[16] M. Vidyasagar, *Control System Synthesis: a factorization approach*, MIT Press, 1985.

[17] K. Zhou and P.P. Khargonekar, "On the weighted sensitivity minimization problem for delay systems," *Systems and Control Letters* 8 (1987), 307-312.

# Topological Aspects of Transfer Matrices with Entries in the Quotient Field of $H_\infty$

M.L.J. Hautus & S.Q. Zhu

**Abstract**

Let $F^{n\times m}$ be the set of all n by m transfer matrices with entries in the quotient field **F** of $H_\infty$. This article investigates the properties of $F^{n\times m}$ with respect to the gap topology. First, we identify the subset $B^{n,m}$ of $F^{n\times m}$ consisting of all transfer matrices possessing right and left Bezout fractions, and we show that $B^{n,m}$ is an open subset of $F^{n\times m}$ in the gap topology. Moreover, a bound is given in terms of the gap metric which guarantees that if the distance of a transfer matrix $P_1 \in F^{n\times m}$ from $P_2 \in B^{n,m}$ is smaller than this bound, then $P_1$ is also in $B^{n,m}$. In addition, $P_1$ and $P_2$ can be stabilized by a same controller. Furthermore, a relation between the gap of two transfer matrices and the gap of their domains is given. Using this relation, we show that, if the gap of two scalar transfer functions $P_1$ and $P_2$ is smaller than a specified number, then $P_1$ and $P_2$ and must have the same number of poles in the open right half plane.

## 1 Introduction

Let $H_\infty$ be the Hardy space in the right half plane and **F** the quotient field of $H_\infty$. We denote by M(**F**) the set of all matrices with entries in **F**. It is known that M(**F**) provides a general framework for studying distributed linear time-invariant systems in the frequency domain. M(**F**) covers many cases of interest both in theory and in practice. For example, it includes: finite-dimensional linear time-invariant systems, the Pritchard-Salamon class (i.e. semigroup systems), the Callier-Desoer class, the Logemann class etc.

Let $F^{n\times m}$ be the subset of M(**F**) consisting of all $n \times m$ transfer matrices. This article investigates the properties of $F^{n\times m}$ with respect to the gap topology. First, we identify the subset $B^{n,m}$ of $F^{n\times m}$ consisting of all transfer matrices possessing right and left Bezout fractions, and show that $B^{n,m}$ is an open subset of $F^{n\times m}$ in the gap topology. Moreover, a bound $\alpha := \alpha(P_2)$ given in terms of the gap metric guaranteeing that if the distance of a transfer matrix $P_1 \in \mathbf{F}^{n \times m}$ from $P_2 \in \mathbf{B}^{n,m}$ is smaller than $\alpha$, then $P_1$ is also in $\mathbf{B}^{n,m}$. In addition, $P_1$ and $P_2$ can be stabilized by a same controller. Next, a relation between the gap of two transfer matrices and the gap of their domains is given. Using this relation, we show that if the gap of two scalar transfer functions $P_1$ and $P_2$ is smaller than a certain number $\gamma := \gamma(P_1,P_2)$, then $P_2$ and $P_2$ have the same number of poles and the same number of zeros in the open right half plane.

This paper is arranged as follows. In the preliminary Section 2, we introduce the definition of the gap metric and we give some basic lemmas. In Section 3, we show that $\mathbf{B}^{n,m}$ is an open subset of $\mathbf{F}^{n \times m}$ and give a guaranteed bound. In Section 4, we give a relation between the gap of two systems and the gap of their domains, and we present a consequence of this result.

## 2 Preliminaries

First we introduce the gap metric defined on the space of all closed subspaces of a Hilbert space **X**. Let M,N be two closed subspaces of **X**. Let $\Pi(M)$ denote the orthogonal projection from X onto M. In order to measure the distance between the subspaces, we introduce the *gap metric*, which is defined as

$$\delta(M,N) := \|\Pi(M) - \Pi(N)\|.$$

For each $P(.) \in \mathbf{F}^{n \times m}$, we can define a *linear operator* **P** mapping a subspace of $\mathbf{H}_2$ into $\mathbf{H}_2$. The *domain* of **P** is defined as the subset Dom(**P**) consisting of all $x(.) \in \mathbf{H}_2$ for which the product $P(s)x(s)$ is in $\mathbf{H}_2$. The action of **P** on $x(.) \in$ Dom(**P**) is defined by $(\mathbf{P}x)(s) := P(s)x(s)$. It easy to show that **P** is *bounded* iff $P(.) \in M(\mathbf{H}_\infty)$. The following lemma is quoted from [Zhu 1988].

LEMMA 2.1 The operator **P** induced by $P(.) \in \mathbf{F}^{n \times m}$ is *closed*, i.e., the graph $G(\mathbf{P}) := \{(x,\mathbf{P}x) \in \mathbf{H}_2^{m+n}: x \in \mathrm{Dom}(\mathbf{P})\}$ of **P** is a closed subspace.

We define the *gap* between two transfer matrices $P_1(.)$ and $P_2(.)$ as the gap between the graphs of the operators induced by $P_1(.)$ and $P_2(.)$. For simplicity we write $\Pi(P)$ for $\Pi(G(\mathbf{P}))$. Then the gap between $P_1(.)$ and $P_2(.)$ is

(2.1) $$\delta(P_1,P_2) = \|\Pi(P_1) - \Pi(P_2)\|.$$

Suppose $D(.), N(.) \in M(H_\infty)$. We say that $(D(.),N(.))$ is a *right Bezout fraction* (r.b.f.) of $P(.) \in \mathbf{F}^{n\times m}$ if

1) $D(.)^{-1}$ exists,
2) there exist matrices $X(.),Y(.) \in M(H_\infty)$ such that
   $X(.)D(.) + Y(.)N(.) = I$,
3) $P(s) = N(s)D(s)^{-1}$.

Moreover, an r.b.f. $(D(.),N(.))$ of $P(.) \in \mathbf{F}^{n\times m}$ is said to be *normalized* if

(2.2) $\quad \tilde{D}(.)D(.) + \tilde{N}(.)N(.) = I,$

where $D^{\sim}(s) := D(-s)^T$. *Left Bezout fractions* (l.b.f.) and *normalized left Bezout fractions* are defined similarly. It is well known that not every transfer matrix has an r.b.f. (or an l.b.f.). For example, according to [Smith 1989], $P(s) = se^{-S}$ does not have an r.b.f. Denote by $\mathbf{B}^{n,m}$ the subset of $\mathbf{F}^{n\times m}$ consisting of all elements which have both an r.b.f. and an l.b.f.*Notice that the fact that a matrix* $P \in \mathbf{F}^{n\times m}$ *has a right (or a left) Bezout fraction does not imply that each of its entries has one*, for instance, the matrix

$$P = \begin{bmatrix} se^{-S} & 0 \\ \frac{-1}{S+1} & 1 \end{bmatrix} = \begin{bmatrix} \frac{se^{-S}}{S+1} & 0 \\ 0 & 1 \end{bmatrix} \begin{bmatrix} \frac{1}{S+1} & 0 \\ 1 & 1 \end{bmatrix}^{-1} = ND^{-1}$$

has a right Bezout fraction (D,N) over $M(H_\infty)$, although $se^{-S}$ does not have a Bezout fraction. $B(\mathbf{H}_2^m,\mathbf{H}_2^n)$ denotes the set of all linear bounded operators mapping $\mathbf{H}_2$ into $\mathbf{H}_2$. If $m = n$, we write $B(\mathbf{H}_2^m)$ for $B(\mathbf{H}_2^m,\mathbf{H}_2^m)$. Let $\mathbf{D} \in B(\mathbf{H}_2^m)$ and $\mathbf{N} \in B(\mathbf{H}_2^m,\mathbf{H}_2^n)$. The pair $(\mathbf{D},\mathbf{N})$ is said to be a *generalized right Bezout fraction* (g.r.b.f.) of $P \in \mathbf{F}^{m\times n}$, if

1) $\mathbf{D}^{-1}$ exists;
2) there exist bounded operators $\mathbf{X},\mathbf{Y}$ such that $\mathbf{XD} + \mathbf{YN} = I$;
3) $\mathbf{P} = \mathbf{ND}^{-1}$.

Obviously, an r.b.f. of P is a g.r.b.f. but not conversely. The concept of g.r.b.f. is a useful tool of this paper. The following lemma is from [Zhu 1988]

LEMMA 2.2 Suppose $\mathbf{D} \in B(\mathbf{H}_2^m)$, $\mathbf{N} \in B(\mathbf{H}_2^m,\mathbf{H}_2^n)$ and $P \in \mathbf{B}^{n,m}$. Then $(\mathbf{D},\mathbf{N})$ is a g.r.b.f. of P iff $\begin{bmatrix} \mathbf{D} \\ \mathbf{N} \end{bmatrix}$ maps $\mathbf{H}_2^m$ onto G(P) bijectively.

The next lemma about the relation between the gap metric and the g.r.b.f.'s is quoted from [Zhu et al 1988].

LEMMA 2.3 Let $P_1 \in \mathbf{B}^{n,m}$, $P_2 \in \mathbf{F}^{n\times m}$ and let (D,N) be an r.b.f. of $P_1$. Then, $\delta(P_1,P_2) < 1$, iff

$$\begin{bmatrix} \mathbf{D} \\ \mathbf{N} \end{bmatrix} := \Pi(P_2)\begin{bmatrix} \mathbf{D}_1 \\ \mathbf{N}_2 \end{bmatrix}$$

is a g.r.b.f. of $P_2$.

## 3. $\mathbf{B}^{n,m}$ is open in $\mathbf{F}^{n\times m}$.

Consider the following *feedback system* shown in Figure 3.1, where $P \in M(\mathbf{F})$ represents the *plant* and $C \in M(\mathbf{F})$ the *compensator*.

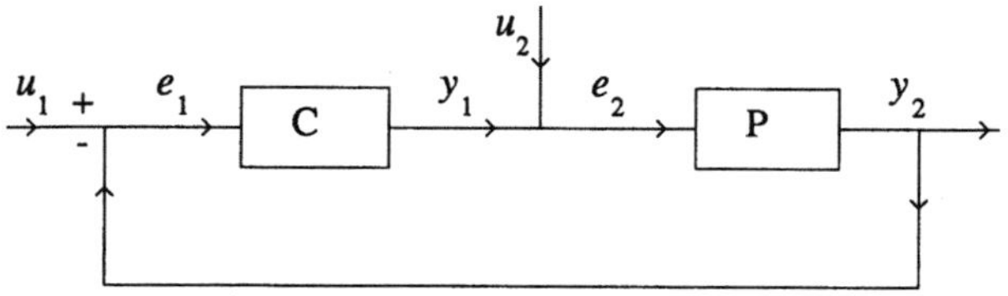


Figure 3.1 Feedback System

Suppose that $P,C \in M(\mathbf{F})$. The *transfer matrix* from $u := \begin{bmatrix} u_1 \\ u_2 \end{bmatrix}$ to $e := \begin{bmatrix} e_1 \\ e_2 \end{bmatrix}$ is

$$(3.1) \qquad H(P,C) := \begin{bmatrix} (I + PC)^{-1} & -P(I + CP)^{-1} \\ C(I + PC)^{-1} & (I + CP)^{-1} \end{bmatrix}$$

where we assumed that P and C have compatible dimensions, and that the well-posedness condition $|I + PC| \neq 0$ is satisfied, so that H(P,C) makes sense.

A transfer matrix is said to be stable iff it is in $M(\mathbf{H}_\infty)$, and the feedback system is said to be stable iff H(P,C) is stable. If for a system $P \in M(\mathbf{F})$ there is an element C in $M(\mathbf{F})$ such that H(P,C) is stable, we say that P is *stabilizable* and C is called a *stabilizing controller* of P.

Recall that a subspace M of $\mathbf{H}_\infty^m$ is said to be *shift invariant* if for any $\alpha > 0$, we have $e^{-\alpha s}M \subseteq M$. It is easy to prove that $G(\mathbf{P})$ is a shift invariant subspace of $\mathbf{H}_2^{m+n}$ for each $P(.) \in \mathbf{F}^{m\times n}$ We need following theorem to prove our main result of this section, which is an alternate version of Lax's theorem [Lax 1959].

THEOREM 3.1 [Lax 1959] Suppose that M is a closed shift invariant subspace of $\mathbf{H}_2^m$. Then, there is an integer $k > 0$ and an inner matrix $A \in \mathbf{H}_\infty^{m\times k}$, such that $\mathbf{A}$ maps $\mathbf{H}_2^k$ bijectively onto M.

The following statement is our main result in this section.

THEOREM 3.2 Let $(D_0,N_0)$ be an r.b.f. of $P_0 \in \mathbf{B}^{n,m}$. Suppose that $C \in \mathbf{B}^{n,m}$ is one of the stabilizing controllers of $P_0$ and $(X,Y)$ is an l.b.f. of C such that $XD_0 + YN_0 = I$.
Define

(3.2) $\quad K(P_0,\varepsilon) := \{ P \in \mathbf{F}^{n\times m} : \delta(P_0,P) < \varepsilon \},$

where

(3.3) $\quad \varepsilon := \left(\left\|\begin{bmatrix} D_0 \\ N_0 \end{bmatrix}\right\| \, \|(X,Y)\|\right)^{-1}.$

Then, $K(P_0,\varepsilon) \subseteq \mathbf{B}^{n,m}$ and each system $P(.) \in K(P_0,\varepsilon)$ can be stabilized by C.

PROOF According to Lax's theorem (Theorem 3.1), there are an integer k and an inner matrix $A \in \mathbf{H}_\infty^{(n+m)\times k}$ such that $\mathbf{A}$ mapping $\mathbf{H}_2^k$ bijectively onto $G(\mathbf{P})$.
If we define

$$\begin{bmatrix} \mathbf{D} \\ \mathbf{N} \end{bmatrix} := \Pi(P)\begin{bmatrix} \mathbf{D}_0 \\ \mathbf{N}_0 \end{bmatrix},$$

then, by Lemma 2.3, $\begin{bmatrix} \mathbf{D} \\ \mathbf{N} \end{bmatrix}$ maps $\mathbf{H}_2^m$ bijectively onto $G(\mathbf{P})$ (notice that the right-hand side of (3.3) is smaller than 1). Since,

$$\left\|(\mathbf{X},\mathbf{Y})\begin{bmatrix} \mathbf{D} \\ \mathbf{N} \end{bmatrix} - \mathbf{I}\right\| = \left\|(\mathbf{X},\mathbf{Y})\begin{bmatrix} \mathbf{D} \\ \mathbf{N} \end{bmatrix} - (\mathbf{X},\mathbf{Y})\begin{bmatrix} \mathbf{D}_0 \\ \mathbf{N}_0 \end{bmatrix}\right\| =$$

$$= \left\|(\mathbf{X},\mathbf{Y})\Pi(P)\begin{bmatrix} \mathbf{D}_0 \\ \mathbf{N}_0 \end{bmatrix} - (\mathbf{X},\mathbf{Y})\Pi(P_0)\begin{bmatrix} \mathbf{D}_0 \\ \mathbf{N}_0 \end{bmatrix}\right\| \le \|(\mathbf{X},\mathbf{Y})\| \, \|\Pi(P_0) - \Pi(P)\| \, \left\|\begin{bmatrix} \mathbf{D}_0 \\ \mathbf{N}_0 \end{bmatrix}\right\| =$$

$$= \|(X,Y)\| \, \|\Pi(P_0) - \Pi(P)\| \, \left\|\begin{bmatrix} D_0 \\ N_0 \end{bmatrix}\right\| < 1,$$

$(\mathbf{X},\mathbf{Y})\begin{bmatrix} \mathbf{D} \\ \mathbf{N} \end{bmatrix}$ is bijective. Thus, $(\mathbf{X},\mathbf{Y})$ maps $G(\mathbf{P})$ onto $\mathbf{H}_2^m$ bijectively. As a consequence, $(\mathbf{X},\mathbf{Y})\mathbf{A}$ is bijective. Hence, $(X,Y)A$ is unimodular. Consequently, $m = k$. If we partition A as $\begin{bmatrix} D \\ N \end{bmatrix}$, then $(D,N)$ is an r.b.f. of P and C stabilizes P. □

Let $(D_0,N_0)$ and $(\tilde{D}_0,\tilde{N}_0)$ be a normalized r.b.f. and a normalized l.b.f. of $P \in \mathbf{B}^{n,m}$, respectively (the existence of normalized Bezout fractions of transfer matrices with entries in the quotient field of $\mathbf{H}_\infty$ is proved in [Zhu 1988]). Assume that $X,Y,\tilde{X},\tilde{Y} \in M(\mathbf{H}_\infty)$ satisfy

$$(3.4)\qquad Y\tilde{X} = X\tilde{Y},\ \ XD_0 + YN_0 = I,\ \ \tilde{D}_0\tilde{X} + \tilde{N}_0\tilde{Y} = I.$$

Then it follows that $(Y + R\tilde{D}_0)(\tilde{X} - N_0R) = (X - R\tilde{N}_0)(\tilde{Y} + D_0R)$. In [Zhu 1988-1], it is shown that $|(X - R\tilde{N}_0)| \neq 0$ iff $|(\tilde{X} - N_0R)| \neq 0$. Consequently, $(X - R\tilde{N}_0)^{-1}(Y + R\tilde{D}_0) = (\tilde{Y} + D_0R)(\tilde{X} - N_0R)^{-1}$ for all $R \in M(\mathbf{H}_\infty)$ satisfying $|(\tilde{X} - N_0R)| \neq 0$. It is well known that the set of all stabilizing controllers of $P_0$ can parameterized as

$$(3.5)\qquad \{(X - R\tilde{N}_0)^{-1}(Y + R\tilde{D}_0):\ R \in M(\mathbf{H}_\infty),\ |(X - R\tilde{N}_0)| \neq 0\} =$$
$$= \{(\tilde{Y} + D_0R)(\tilde{X} - N_0R)^{-1}:\ R \in M(\mathbf{H}_\infty),\ |(\tilde{X} - N_0R)| \neq 0\}.$$

Furthermore, it was proved in [Georgiou and Smith, Theorem 3] and in [Zhu 1988-1]) that

$$(3.6)\qquad \|[X - R\tilde{N}_0, Y + R\tilde{D}_0]\| = \left\|\begin{bmatrix}\tilde{X} - N_0R \\ \tilde{Y} + D_0R\end{bmatrix}\right\|$$

for all $R \in M(\mathbf{H}_\infty)$.

THEOREM 3.3 Let $(D_0,N_0)$ and $(\tilde{D}_0,\tilde{N}_0)$ be a normalized r.b.f. and normalized l.b.f. of $P_0 \in \mathbf{B}^{n,m}$, respectively. Assume that $X,Y,\tilde{X},\tilde{Y} \in M(\mathbf{H}_\infty)$ satisfy (3.4). Let $C := (X - R_0\tilde{N}_0)^{-1}(Y + R_0\tilde{D}_0)$ satisfy

$$(3.7)\qquad \|(X - R_0\tilde{N}_0, Y + R_0\tilde{D}_0)\| = \inf_{R\in M(H_\infty)}\|(X - R\tilde{N}_0, Y + R\tilde{D}_0)\|.$$

Define

$$(3.8)\qquad \alpha := \alpha(P_0) := \left(\inf_{R\in M(H_\infty)}\|(X - R\tilde{N}_0, Y + R\tilde{D}_0)\|\right)^{-1}.$$

Then, $K(P_0,\alpha) \subseteq \mathbf{B}^{n,m}$ and all systems in $K(P_0,\alpha)$ can be stabilized by C.

Using (3.6), one can also express $\alpha$ in terms of $\tilde{X},\tilde{Y},N_0,D_0$. Notice that the existence of $R_0$ satisfying (3.7) is proved by [Glover et al. 1988]. We give an example to show that $\alpha$ can be reached. Let $P_0 = 0$. Then $C = 0$ is the controller obtained by solving (3.7). Consequently, $\alpha = \alpha(P_0) = 1$. According to [Kato *p.* 205, 1966], any system P with $\delta(0,P) = 1$ is unstable, and hence, can not be stabilized by $C = 0$. Therefore, in general, one can not give a larger number $\varepsilon$ than $\alpha$ such that $K(P_0,\varepsilon)$ is stabilized by one controller. A more general example illustrating that the bound cannot be improved was given in [Georgiou and Smith, Theorem 4].

**4 Relations of $\delta(P_1,P_2)$ with $\delta(\overline{Dom(P_1)},\overline{Dom(P_2)})$ and $\delta(\overline{Im(P_1)},\overline{Im(P_2)})$**

Let (D,N) be a normalized r.b.f. of $P \in \mathbf{B}^{n,m}$. For simplicity, we denote $\mathcal{D} := Dom(P)$ and $\mathcal{N} := Im(P)$. It is well known that $\mathcal{D} = Im(D)$ and $\mathcal{N} = Im(N)$.

LEMMA 4.1 Let $\alpha_1,\dots,\alpha_k$ be distinct numbers in $\mathbb{C}_+$ and let $j_1,\dots,j_k$ be nonnegative integers for i = 1,...,k. Define

$$\mathcal{E} := \{(s-\alpha_1)^{j_1}(s-\alpha_2)^{j_2}\dots(s-\alpha_k)^{j_k}x(s) \in \mathbf{H}_2^m : x(s) \in \mathbf{H}_2^m\}.$$

Then, $\mathcal{E}$ is closed and $\mathcal{E}^{\perp} = \operatorname{span}\{(s+\overline{\alpha}_p)^{-n}: n = 1,\dots,j_k;\ p = 1,\dots,k\}$.

PROOF It is obvious that $\mathcal{E}$ is closed. To prove the second claim, we notice that for each $y(.) \in \mathbf{H}_2^m$, the following formulas hold:

$$y(s) = \int_{-\infty}^{+\infty}(s - i\omega)^{-1}\, y(i\omega)d\omega,\ y^{(n)}(s) = \int_{-\infty}^{+\infty}(-1)^n n!(s - i\omega)^{-n-1}y(i\omega)d\omega.$$

As a consequence, we have

$$<(s+\bar{\alpha}_p)^{-n-1}, y(s)> = \int_{-\infty}^{+\infty}(\alpha_p - i\omega)^{-n-1}\, y(i\omega)d\omega = \frac{(-1)^n}{n\,!}\, y^{(n)}(\alpha_p).$$

for $n = 1,\dots,j_p$; $p = 1,\dots,k$. Hence,

$$y(.) \perp \operatorname{span}\{(s+\overline{\alpha}_p)^{-n-1}: n = 1,\dots,j_p;\ p = 1,\dots,k\}$$

iff $\alpha_p$ is a zero of $y(.)$ with order $j_p$ ($p = 1,\dots,k$) i.e. $y(.) \in \mathcal{E}$. □

THEOREM 4.2 Let $P_i \in \mathbf{B}^{n,m}$ and $(D_i,N_i)$ be a normalized r.b.f. of $P_i$ (i = 1,2). Assume that $D_i^{-1} \in M(\mathbf{L}_\infty)$ (i = 1,2). Then,

(4.1) $\qquad \delta(P_1,P_2) \geq d\delta(\bar{\mathcal{D}}_1,\bar{\mathcal{D}}_2),$

where

(4.2) $\qquad d := \min\{\|D_1^{-1}\|^{-1},\|D_2^{-1}\|^{-1}\}.$

PROOF The one-sided gap $\vec{\delta}(M,N)$ between two subspaces M and N is defined as

$$\vec{\delta}(M,N) := \sup_{x\in M,\ \|x\|\leq 1}\inf_{y\in N}\|x - y\|.$$

It is known that $\delta(M,N) = \max\{\vec{\delta}(M,N), \vec{\delta}(M,N)\}$. We have:

$$\vec{\delta}(P_1,P_2) = \vec{\delta}(G(P_1),G(P_2)) =$$
$$= \sup_{x\in G(P_1),\ \|x\|\leq 1}\inf_{y\in G(P_2)}\|x - y\| =$$
$$= \sup_{x\in \mathbf{H}_2^m,\ \|x\|\leq 1}\inf_{y\in \mathbf{H}_2^m}\left\|\begin{bmatrix}D_1\\N_1\end{bmatrix}x - \begin{bmatrix}D_2\\N_2\end{bmatrix}y\right\| \geq$$
$$\geq \sup_{x\in \mathbf{H}_2^m,\ \|x\|\leq 1}\inf_{y\in \mathbf{H}_2^m}\|D_1x - D_2y\| =$$

$$= \sup_{x\in \mathbf{H}^m_2,\ \|x\|\le\|D_1^{-1}\|} \inf_{y\in \mathbf{H}^m_2}\| \ \|D_1^{-1}\|^{-1}D_1x - D_2y\| =$$
$$= \|D^{-1}\|^{-1} \sup_{x\in \mathbf{H}^m_2,\ \|x\|\le\|D_1^{-1}\|} \inf_{y\in \mathbf{H}^m_2} \|D_1x - D_2y\|.$$

Since
$$\{D_1x\colon x \in \mathbf{H}^m_2, x \le \|D^{-1}\|\} \supseteq \{D_1x\colon x \in \mathbf{H}^m_2, \|D_1x\| \le 1\},$$
we have
$$\delta^{\rightarrow}(P_1,P_2) \ge \|D^{-1}\|^{-1} \sup_{x\in \mathbf{H}^m_2,\ \|D_1x\|\le 1} \inf_{y\in \mathbf{H}^m_2} \|D_1x - D_2y\| =$$

$$= \|D^{-1}\|^{-1} \sup_{x\in \mathcal{D}_1,\ \|x\|\le 1} \inf_{y\in \mathcal{D}_2} \|x - y\| =$$
$$= \|D_1^{-1}\|^{-1} \sup_{x\in \bar{\mathcal{D}}_1,\ \|x\|\le 1} \inf_{y\in \bar{\mathcal{D}}_2} \|x - y\| =$$

$$= \|D^{-1}\|^{-1} \delta^{\rightarrow}(\bar{\mathcal{D}}_1,\bar{\mathcal{D}}_2).$$

Hence,
$$\delta(P_1,P_2) \ge d\ \delta(\bar{\mathcal{D}}_1,\bar{\mathcal{D}}_2).$$ □

Now using Theorem 4.2, we show that

THEOREM 4.5 Let $P_1,P_2$ be biproper rational functions without poles on $i\mathbb{R}$ and $(D_i,N_i)$ be a normalized r.b.f. of $P_i \in \mathbf{B}$ $(i = 1,2)$. Let d be defined by (4.2). Then, if $\delta(P_1,P_2) < d$, then $P_1$ and $P_2$ have the same number of poles in the open right half plane.

PROOF By (4.1), we have $\delta(\bar{\mathcal{D}}_1,\bar{\mathcal{D}}_2) < 1$. We can easily check that $\delta(\bar{\mathcal{D}}_1^{\perp},\bar{\mathcal{D}}_2^{\perp}) = \delta(\bar{\mathcal{D}}_1,\bar{\mathcal{D}}_2) < 1$. According to [Kato p. 199, 1966], $\delta(\bar{\mathcal{D}}_1^{\perp},\bar{\mathcal{D}}_2^{\perp}) < 1$ implies $\dim \bar{\mathcal{D}}_2^{\perp} = \dim \bar{\mathcal{D}}_2^{\perp}$. Hence, it suffices to show that $\dim \overline{\mathcal{D}}_1^{\perp}$ is equal to the number of poles $P_1$ in open right half plane. This statement, however, follows from Lemma 4.1. □

**References**

[Georgiou et al. 1989]
Georgiou T.T. and Smith M.C. *Optimal robustness in the Gap Metric,* preprint 1989

[Glover et al. 1988]
Glover K. and D McFarlane *Robust stabilization of Normalized Coprime Factors: An explicit H∞ solution,* Proc. of the American Control Conference, *p.* 842-847, 1988.

[Kato 1966]
Kato T. *Perturbation Theory for Linear Operators,* Springer-Verlag, 1966.

[Lax 1959]
Lax P.D. *Translation Invariant Subspaces,* Acta Math., Vol. 101, *p.* 163-178, 1959.

[Smith 1989]
Smith M.C. *On Stabilization and existence of Coprime Factorizations,* IEEE Trans. on Automatic Contr.,*p.* 1005-1007, 1989.

[Zames et al. 1980]
Zames G. and A. El-Sakkary *Unstable Systems and Feedback : The Gap Metric,* in Proc. Allerton Conf. *p* 380-385, 1980.

[Zhu et al 1988]
Zhu S.Q., M.L.J. Hautus & C. Praagman *Sufficient Conditions for Robust BIBO Stabilization : Given by the Gap Metric,* Systems and Control Letters, Vol. 11, *p.* 53-59, 1988.

[Zhu 1988]
Zhu S.Q. *On normalized Bezout fractions of Distributed LTI systems,* Eindhoven 1988. (To appear in IEEE trans. Automat. Contr.).

[Zhu 1988-1]
Zhu S.Q. *Robustness of Feedback Stabilization : A Topological Approach.* Ph.D. Thesis, Eindhoven, 1988.

# ON THE NYQUIST CRITERION AND ROBUST STABILIZATION FOR INFINITE-DIMENSIONAL SYSTEMS

Hartmut Logemann

**Abstract**

A unified approach to the multivariable Nyquist stability criterion for various classes of nonrational transfer functions is presented. It is not asumed that the transfer matrix of the open-loop system can be extended meromorphically across the imaginary axis into the left-half plane. Applications to robust stabilization are discussed.

## 1. Introduction

It is worthwhile mentioning that Nyquist's original paper [17] on the stability of feedback amplifiers is not restricted to rational transfer functions, but includes certain infinite-dimensional systems as well. In the past 25 years there has been considerable interest in a *rigorous* treatment of Nyquist-type stability criteria for infinite-dimensional systems, cf. exempli gratia [10], [12], [9], [1] and [2] for single-loop systems and [11], [18] and [13] for multivariable systems (see the forthcoming survey paper [15] for a more complete bibliography). In Section 2 we give two statements of the Nyquist criterion: the first one is based on the determinant of the return difference matrix while the second one makes use of the eigenloci of the open-loop transfer matrix. The stability tests are presented in a unifying frequency-domain framework which covers various classes of nonrational transfer function including the Callier-Desoer class $\hat{\mathfrak{B}}(0)$ introduced in [3] and [4]. Their set-up has been sucessfully applied to a number of problems in infinite-dimensional linear systems theory (cf. [7] for an overview). However, there are situations, where classes of transfer functions, which do not require any knowledge of the system's impulse response, are easier to handle than $\hat{\mathfrak{B}}(0)$, which by definition is a set of Laplace-transformed impulse responses. Moreover there are quite a few infinite-dimensional systems (in particular systems described by Volterra integrodifferential equations or Volterra integral equations) whose transfer functions are meromorphic or holomorphic in the open right-half plane but don't belong to $\hat{\mathfrak{B}}(0)$. Take for example the convolution kernel $k(t) = (1+t^2)^{-1}$, $t \geq 0$, having a Laplace transform $\hat{k}$ which is holomorphic in $\mathrm{Re}(s) > 0$ and continuous in $\mathrm{Re}(s) \geq 0$. However, since $k(t)\exp(\varepsilon t)$ is not in $L^1(\mathbb{R}_+)$ for all $\varepsilon > 0$, it follows that $\hat{k}$ is not an element of $\hat{\mathfrak{B}}(0)$.

In Section 3 we indicate how the Nyquist criterion can be used in order to extend the results in [6] and [8] on robust stabilization of systems which belong to the Callier-

Desoer class to the more general set-up developed in Section 2

## 2. A generalized Nyquist criterion

In the following let $\mathbb{C}_+$ and $\bar{\mathbb{C}}_+$ denote the open and closed right-half plane, respectively. Denote the ring of functions which are holomorphic on $\mathbb{C}_+$ by $H(\mathbb{C}_+)$ and define $C(\bar{\mathbb{C}}_+)$ to be the ring of complex-valued functions which are continuous on $\bar{\mathbb{C}}_+$. Moreover, if $a \in \mathbb{C}$ and if $\varphi$ is a closed curve in the complex plane not passing through a, let $n(\varphi,a)$ denote winding number of $\varphi$ about a.

As we shall see later the following theorem can be interpreted as stability criterion for infinite-dimensional feedback systems.

### 2.1 Theorem

Let $N_1, D_1, N_2, D_2$ be matrices with entries in $H(\mathbb{C}_+) \cap C(\bar{\mathbb{C}}_+)$ of size $m \times p$, $m \times m$, $p \times m$ and $m \times m$ respectively, set $P_1 := D_1^{-1} N_1$, $P_2 := N_2 D_2^{-1}$ and assume

(A1) $\det(D_i(j\omega)) \neq 0 \quad \forall\, \omega \in \mathbb{R},\ i = 1,2.$

(A2) There exists a positive number $\rho_i$ such that
$$\inf_{|s| \geq \rho_i,\ s \in \bar{\mathbb{C}}_+} |\det(D_i(s))| > 0, \quad i = 1,2.$$

(A3) $\lim_{|s| \to \infty,\ s \in \bar{\mathbb{C}}_+} P_1 P_2(s) = A$, such that $-1$ is not an eigenvalue of A.

Then

$$\inf_{s \in \bar{\mathbb{C}}_+} |\det(N_1 N_2 + D_1 D_2)(s)| > 0 \tag{2.1}$$

if and only if

(i) $\det(I + P_1 P_2)(j\omega)) \neq 0 \quad \forall\, \omega \in \mathbb{R}$

and

(ii) $n(\det(I + P_1 P_2) \circ \gamma, 0) = -(\pi_1 + \pi_2)$,

where $\gamma$ is a parametrization of the $j\omega$-axis such that $\gamma(t)$ moves downwards from $j\infty$ to $-j\infty$ and $\pi_i$ is the number of zeros of $\det(D_i)$ in $\mathbb{C}_+$, $i = 1,2$.

### 2.2 Remark

Using the Principle of Isolated Zeros for holomorphic functions it follows from (A1) and (A2) that $\det(D_i)$ has at most finitely many zeros in $\mathbb{C}_+$, $i = 1,2$.

### Proof of Theorem 2.1:

<u>'If'</u>: First of all realize that

$$(2.2) \qquad \det(I + P_1P_2) = \frac{\det(N_1N_2+D_1D_2)}{\det(D_1D_2)}.$$

We have to show that (2.1) holds true. Using the identity (2.2) it follows from (A2) and (A3) that there exists $\rho > 0$ such that

$$(2.3) \qquad \inf_{|s|\geq\rho,\, s\in\bar{\mathbb{C}}_+} |\det(N_1N_2+D_1D_2)(s)| > 0.$$

Via (2.2) we obtain from (i) and (A1) that $\det(N_1N_2+D_1D_2)$ has no zeros on the $j\omega$-axis. Hence it remains to show that

$$(2.4) \qquad \det(N_1N_2+D_1D_2)(s) \neq 0 \quad \forall\, s \in B,$$

where $B := \{s \in \mathbb{C}_+ \,|\, |s| < \rho\}$.

We claim that $\det(N_1N_2+D_1D_2)$ has at most finitely many zeros in B. Assume the contrary, i.e there are infinitely many zeros $s_n \in B$. Since $\bar{B}$ is compact there exists a subsequence $(z_n)$ of $(s_n)$ converging in $\bar{B}$. Now it follows from the above that $\lim_{n\to\infty} z_n \notin \partial B$ and hence is in B. By the Principle of Isolated Zeros we have $\det(N_1N_2+D_1D_2) \equiv 0$ which can't be true because of (2.3). So far we have shown that $\det(N_1N_2+D_1D_2)$ has at most finitely many zeros in $\bar{\mathbb{C}}_+$ and all of them are in B. Hence there exists $\varepsilon_1 > 0$ such that all the zeros satisfy $\mathrm{Re}(s) > \varepsilon_1$. Moreover by (A1) and (A2) there exists $\varepsilon_2 > 0$ such that all zeros of $\det(D_1)$ and $\det(D_2)$ are in $\mathrm{Re}(s) > \varepsilon_2$. Now set $\varepsilon := \min(\varepsilon_1, \varepsilon_2)$ and define the curve $\gamma_\varepsilon$ by $\gamma_\varepsilon(t) := \gamma(t)+\varepsilon$, i.e. $\gamma_\varepsilon$ is a parametrization of the set $j\mathbb{R} + \varepsilon$. Using that the curves $\det(I + P_1P_2)\circ\gamma$ and $\det(I + P_1P_2)\circ\gamma_\varepsilon$ are homotopic in $\mathbb{C}\setminus\{0\}$ and that $\det(I+P_1P_2) \neq 0$ in $0 \leq \mathrm{Re}(s) \leq \varepsilon$ it follows from (ii)

$$(2.5) \qquad n(\det(I + P_1P_2)\circ\gamma_\varepsilon, 0) = -(\pi_1+\pi_2).$$

On the other hand it follows via (2.2) from the Argument Principle

$$(2.6) \qquad n(\det(I + P_1P_2)\circ\gamma_\varepsilon, 0) = Z - (\pi_1+\pi_2)$$

where $Z$ denotes the number of zeros of $\det(N_1N_2+D_1D_2)$ in B. By (2.5) and (2.6) we have $Z = 0$ which proves (2.4).

"Only if": The proof that conditions (i) and (ii) are necessary for (2.1) to hold is left to the reader.

In the following let $S$ be a subring of $H(\mathbb{C}_+) \cap C(\bar{\mathbb{C}}_+)$ satisfying the condition

(I) $\quad f \in S$ is invertible in $S$ if and only if $\inf_{s \in \mathbb{C}_+} |f(s)| > 0.$

Notice that (I) does not characterize the units of the ring $H(\mathbb{C}_+) \cap C(\bar{\mathbb{C}}_+)$ itself.

**2.3 Examples**

Let the superscript '^' denote the Laplace transformation and for $\alpha \in \mathbb{R}$ define $H_\alpha^\infty$ by $H_\alpha^\infty := \{f$ is a bounded holomorphic functions on $\mathrm{Re}(s) > \alpha\}$. The following subrings of $H(\mathbb{C}_+) \cap C(\bar{\mathbb{C}}_+)$ satisfy (I):

(a) $(\delta\mathbb{R} + L^1(\mathbb{R}_+))\hat{}$, where $\delta$ denotes the Dirac distribution with support in 0.

(b) $\hat{\mathcal{A}}$, where $\mathcal{A}$ is the ring of distributions f satisfying $f = f_a + \sum_{i=0}^{\infty} f_i\, \delta_{t_i}$, $t_0 = 0$, $t_i > 0$ $(i \geq 1)$, $\delta_{t_i}$ is the Dirac distribution with support in $t_i$, $f_a$ is an integrable real-valued function defined on $\mathbb{R}_+$ and the $f_i \in \mathbb{R}$ form a summable sequence.

(c) $\hat{\mathcal{A}}_-$, where $\mathcal{A}_- := \{f \in \mathcal{A} \mid \exists\, \varepsilon = \varepsilon(f) > 0\colon f e^{\varepsilon \cdot} \in \mathcal{A}\}$.

(d) $H_0^\infty \cap C(\bar{\mathbb{C}}_+)$.

(e) $H_-^\infty := \bigcup_{\alpha<0} H_\alpha^\infty$.

(f) $A(\mathbb{C}_+)$, where $A(\mathbb{C}_+)$ denotes the right-half plane analogue of the disc-algebra, i.e. $A(\mathbb{C}_+) := \{f \in H_0^\infty \cap C(\bar{\mathbb{C}}_+) \mid \lim_{|s| \to \infty,\, s \in \bar{\mathbb{C}}_+} f(s)$ exists$\}$.

The rings in (a) – (f) play an important role in the frequency-domain analysis of infinite-dimensional systems (cf. for example [3], [9], [13] and [16]). Moreover it should be noticed that the ring of stable proper rational functions is contained in each of the rings in (a) – (f).

Let $\mathcal{F}(S)$ denote the field of fractions of $S$. We can interpret $\mathcal{F}(S)$ as the set of all transfer functions of interest and $S$ as the subset of stable transfer functions.

**2.4 Definition**

Let $P \in \mathcal{F}(S)^{m \times p}$, a right-coprime factorization of P (over $S$) is a pair $(N,D) \in S^{m \times p} \times S^{p \times p}$ satisfying

(i) $\det(D) \not\equiv 0$

(ii) $P = ND^{-1}$

(iii) There exists $X \in S^{p\times m}$ and $Y \in S^{p\times p}$ such that $XN + YD \equiv I$.

Left-coprime factorizations are defined in an analogous way.

**2.5 Remark**

It is well-known that each $P \in \mathcal{T}(S)^{m\times p}$ will have a right-coprime factorization if and only if $S$ is a Bezout domain. Since the rings in Example 2.3 are not Bezout domains (cf. [13], [14] and [19]) coprime factorizations will fail to exist for certain irrational transfer matrices,. However, if $P \in \mathcal{T}(S)^{m\times p}$ is of the form $P = P_s + P_u$, where $P_s \in S^{m\times p}$ and $P_u$ is a proper rational transfer matrix having all its poles in $\bar{\mathbb{C}}_+$, then it is not difficult to show that P admits right-coprime and left-coprime factorizations.

Consider the feedback system in Fig. 1 which we will denote by $\mathcal{F}[P_1,P_2]$. Assume that $P_1 \in \mathcal{T}(S)^{m\times p}$, $P_2 \in \mathcal{T}(S)^{p\times m}$ and $\det(I + P_1P_2) \not\equiv 0$. Then we have

$$\begin{pmatrix} y_1 \\ y_2 \end{pmatrix} = \mathcal{H}(P_1,P_2)\begin{pmatrix} u_1 \\ u_2 \end{pmatrix},$$

where

$$\mathcal{H}(P_1,P_2) := \begin{pmatrix} P_1(I+P_2P_1)^{-1} & -P_1P_2(I+P_1P_2)^{-1} \\ P_2P_1(I+P_2P_1)^{-1} & P_2(I+P_1P_2)^{-1} \end{pmatrix} \in \mathcal{T}(S)^{(m+p)\times(m+p)}$$

The feedback system $\mathcal{F}[P_1,P_2]$ will be called $S$-stable if the matrix $\mathcal{H}[P_1,P_2]$ is in $S^{(m+p)\times(m+p)}$.

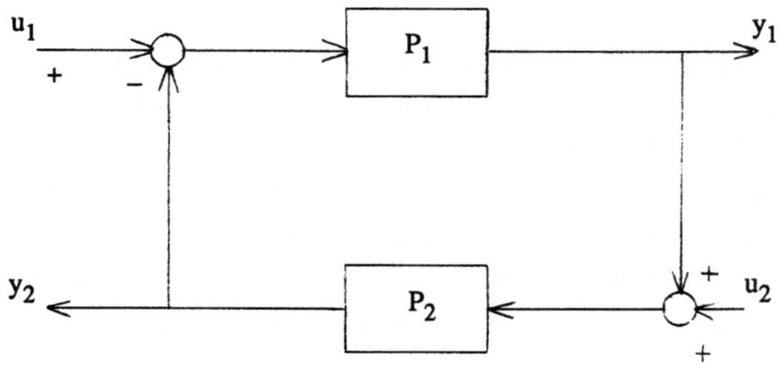


FIGURE 1

From Theorem 2.1 and Lemma 3.1 in [19] we obtain a Nyquist criterion for $\mathcal{S}$- stability.

**2.6 Theorem**

Let $P_1$ and $P_2$ be transfer matrices with entries in $\mathcal{F}(\mathcal{S})$, where $\mathcal{S}$ is a subring of $H(\mathbb{C}_+) \cap C(\overline{\mathbb{C}}_+)$ satisfying (I). Suppose that $P_1$ and $P_2$ have a left-coprime factorization $(D_1,N_1)$ and a right-coprime factorization $(N_2,D_2)$ respectively. If the assumptions (A1) - (A3) of Theorem 2.1 are satisfied the feedback system $\mathcal{F}[P_1,P_2]$ will be $\mathcal{S}$-stable if and only if (i) and (ii) of Theorem 2.1 hold.

In order to give an alternative formulation of Theorem 2.6 we introduce the notion of the eigencontour $E(P,\varphi)$ of a square matrix $P$ having entries in $H(\mathbb{C}_+) \cap C(\overline{\mathbb{C}}_+)$ with respect to a curve $\varphi$: $[a,b] \to \overline{\mathbb{C}}_+$: $E(P,\varphi)$ is formed by the path of the eigenvalues of $P(\varphi(t))$ as $t$ traverses the interval $[a,b]$.

**2.7 Theorem**

Suppose that the assumptions of Theorem 2.6 are satisfied. Using the notation of Theorem 2.1 we have: The feedback system $\mathcal{F}[P_1,P_2]$ is $\mathcal{S}$-stable if and only if

(i) $\qquad -1 \notin \text{image}\,(E(P_1P_2,\gamma),$

(ii) $\qquad n\,(E(P_1P_2,\gamma), -1) \;=\; -\,(\pi_1+\pi_2).$

**Proof:** Making use of some elementary algebraic function theory it is possible to show that $E(P_1P_2,\gamma)$ is a closed chain and $n(E(P_1P_2,\gamma), -1) = n(\det\,(I+P_1P_2)\circ\gamma, 0)$ (cf. [13] and [15] for details). Hence the result follows from Theorem 2.6.

**2.8 Remark**

(a) Theorem 2.7 retains the spirit of the Nyquist criterion for scalar systems, since plotting of the eigencontour of the open-loop transfer matrix allows one to check the closed-loop stability for a *family of gain parameters* by inspection.

(b) In Theorem 2.6 and Theorem 2.7 it is assumed that $P_1$ and $P_2$ have no singularities on the $j\omega$-axis. Under certain conditions the results remain true without making that assumption if we replace $\gamma$ by a curve $\gamma^*$ having indentations into the right-half plane whenever $P_1$ or $P_2$ have singularities on the $j\omega$-axis (cf. [15]).

Theorems 2.6 and 2.7 contain most Nyquist-type stability tests for infinite-dimensional systems which have been published in the literature (e.g. [9], [18], [11] and [5]). We mention however, that the graphical stability criterion, developed in [1] and [2] for scalar transfer functions $P$ of the form $P = P_s + P_u$, where $P_s \in \hat{\mathcal{A}}$ and $P_u$ is a proper rational functions having all its poles in $\overline{\mathbb{C}}_+$, does not require the assumption (A3) and hence is not contained in our approach (cf. [15] for a detailed discussion of the literature).

### 3. Applications to robust stabilization

In [6] a theorem was proven which provides a necessary and sufficient condition for robust stability of feedback systems belonging to the Callier-Desoer class. Having estabilished Theorem 2.6 it is not difficult to extend the criterion of [6] to the framework developed in Section 2. As a consequence it is possible to show that the results in [8] on robust *finite-dimensional* stabilization of systems with transfer functions in the Callier-Desoer class generalize to the set-up of Section 2. Although it should be clear, we remark that in both cases the plants and controllers being involved are required to have at most finitely many unstable poles.

**Acknowledgements**

Part of the work on this paper was done while I was visiting the University of Groningen. I would like to thank the members of the Systems and Control Group for their hospitality and for stimulating discussions.

**References**

[1] F.M. Callier, C.A. Desoer, "A graphical test for checking the stability of a linear time-invariant feedback systems", *IEEE Trans. Automat. Contr.*, vol. AC-17 (1972) 773-780.

[2] F.M. Callier, C.A. Desoer, "On simplifying a graphical stability criterion for linear distributed feedback systems", *IEEE Trans. Automat. Contr.*, vol. AC-21 (1976), 128-129.

[3] F.M. Callier, C.A. Desoer, "An algebra of transfer functions for distributed linear time-invariant systems", *IEEE Trans. Circuits Syst.*, vol. CAS-25 (1978), 651-663 (Correction CAS-25, 360).

[4] F.M. Callier, C.A. Desoer, "Simplifications and clarifications on the paper [3]", *IEEE Trans. Circuits Syst.* vol. CAS-27 (1980), 320-323.

[5] Y. Chait, C.R. MacCluer, C.J. Radcliffe, "A Nyquist stability criterion for distributed parameter systems", *IEEE Trans. Automat. Contr.*, vol. AC-34 (1989), 90-92.

[6] M.J. Chen, C.A. Desoer, "Necessary and sufficient conditions for robust stability of linear distributed feedback systems", *Int. J. Control,* vol. 35 (1982), 255-267.

[7] R.F. Curtain, "A synthesis of time and frequency-domain methods for the control of infinite-dimensional systems: a system-theoretic approach", Preprint, 1988.

[8] R.F. Curtain, K. Glover, "Robust stabilization of infinite-dimensional systems by finite-dimensional controllers", *Systems and Control Letters,* vol. 7 (1988), 41-47.

[9] J.H. Davis, "Fredholm operators, encirclements and stability criteria", *SIAM J. Control,* vol. 10 (1972), 608-622.

[10] C.A. Desoer, "A general formulation of the Nyquist criterion", *IEEE Trans. Circuit Theory,* vol. CT-12 (1965), 230-234.

[11] C.A. Desoer, Y.T. Wang, "On the generalized Nyquist stability criterion", *IEEE Trans. Automat. Contr.*, vol. AC-25 (1980), 187-196.

[12] B. Herz, "Das Nyquist-Kriterium für Regelkreise, deren Verhalten durch lineare Funktional-Differential-Gleichungen beschrieben wird", Doctoral Dissertation, TU Berlin, 1968.

[13] H. Logemann, "Funktionentheoretische Methoden in der Regelungstheorie unendlichdimensionaler Systeme", Doctoral Dissertation, Institut für Dynamische Systeme, Universität Bremen, 1986.

[14] H. Logemann, "Finitely generated ideals in certain algebras of transfer functions for infinite-dimensional systems, *Int. J. Control,* vol. 45 (1987), 247-250.

[15] H. Logemann, "The Nyquist criterion for infinite-dimensional systems - a survey", in preparation.

[16] H. Logemann, D.H. Owens, "Robust high-gain feedback control of infinite dimensional minimum-phase systems", *IMA J. Math. Contr. & Information,* vol. 4 (1987), 195-220.

[17] H. Nyquist,"Regeneration theory", *Bell. Syst. Techn. Journal,* vol 11 (1932), 126-147.

[18] J.M.E. Valenca, C.J. Harris, "Nyquist criterion for input-output stability of multivariable systems", *Int. J. Control,* vol. 31 (1980), 917-935.

[19] M. Vidyasagar, H. Schneider, B.A. Francis, "Algebraic and topological aspects of feedback stabilization" *IEEE Trans. Automat. Contr,* vol. AC-27 (1982), 880-894.

# Real Stability Radii for Infinite Dimensional Systems

*A. J. Pritchard, S. Townley*

**Abstract**

In a paper by Pritchard and Townley well-posedness and exponential stability of semi-groups under structured, unbounded, but *complex* perturbations is analysed, via contraction mappings on appropriate function spaces. The norm of a certain 'input-output' map yields an exact measure of the robustness of exponential stability of the perturbed semi-group.

In contrast, in this paper the class of allowable perturbations is restricted to unbounded linear maps defined on *real* Hilbert spaces. We impose conditions on the unperturbed semi-group and perturbation operators which allows us to pass between a time domain and a frequency domain analysis of the perturbed system. We define a *real* stability radius for well-posedness and exponential stability of the perturbed system. When applied to differential delay equations we obtain interesting and somewhat surprising qualitative results, as we vary the delay, (compared to the complex perturbations or zero delay).

## §0 Introduction

Let $(G,h,k) \in \mathbf{C}^{n\times n}\times\mathbf{C}^{n}\times\mathbf{C}^{n}$, with Re $\sigma(G) < 0$. Consider the uncertain linear system

$$\dot{w}(t) = Gw(t) + h\varepsilon k^{*}w(t-\alpha) \quad t > 0 \tag{1}$$

where $\alpha \geq 0$ is fixed and $\varepsilon \in \mathbf{C}$ is unknown. System (1) can be formulated as an abstract differential equation

$$\dot{z}(t) = Az(t) + DECz(t) \tag{2}$$

on a suitable complex Hilbert space $X$, (in this case $X = L^{2}(-\alpha,0;\mathbf{C}^{n})$ ). Under suitable regularity hypotheses (satisfied by (1)) it is possible to define a (complex) stability radius $r_{\mathbf{C}}(A;D,C)$, a measure of the maximal allowable $\|E\|$ for which stability of solutions of (2) is guaranteed. When applied to system (1) we find that

$$r_{\mathbf{C}}(A;D,C) = \inf_{\omega}|k^{*}(i\omega I-G)^{-1}h|^{-1}$$

(see [10]). Hence robustness with respect to uncertain $\varepsilon \in \mathbf{C}$ is independent of the value of $\alpha$ (which we can then take as uncertain, but fixed.) This demonstrates an obvious limitation in $r_{\mathbf{C}}(A;D,C)$ as a robustness measure. If G, h and k are real valued and $\varepsilon \in \mathbf{R}$ then we will show that the robustness is dependent on $\alpha$. Indeed, if $r_{\mathbf{R}}^{\alpha}(G;h,k^T)$ denotes this robustness measure we find, rather interestingly, that

$$\lim_{\alpha \downarrow 0} r_{\mathbf{R}}^{\alpha}(G;h,k^T) = r_{\mathbf{R}}(G;h,k^T)$$

and

$$\lim_{\alpha \uparrow \infty} r_{\mathbf{R}}^{\alpha}(G;h,k^T) = r_{\mathbf{C}}(G;h,k^T)^1.$$

The purpose of this paper is to develop a robustness analysis for system (2) when all the parameters are real. For differential-delay equations (more generally retarded or neutral functional equations) an obvious approach would be to exploit spectral characterizations of stability (see Logemann [8]) and follow Hinrichsen and Pritchard [7]. We do not persue this here since we can handle both functional and partial differential equations within a single framework. We adopt a semi-group approach, with Salamon type hypotheses to guarantee well-posedness and $L^p$-tests to establish exponential stability of the perturbed semi-groups. In section 1 we describe the well posedness of system (2) and review (and extend) the results for complex stability radii, in this general setting. In section 2 we consider the case of real stability radii. We comment on some questions raised by the characterisation of real stability radii and we illustrate the results with an example of the uncertain system (1) with a single delay. Throughout the paper $L_H^2$ denotes the space $L^2(0,\infty;H)$, of all square integrable functions on $[0,\infty)$ with values in a Hilbert space $H$.

## §1 Stability under unbounded, structured perturbations of strongly continuous exponentially stable semi-groups: Complex stability radii

Let $A$ be the generator of a strongly continuous, exponentially stable (SCES) semi-group $S(t)$ on a complex Hilbert space $X$. We suppose that $A$ is uncertain (neglected delay terms, uncertain boundary data, etc.) so that we must study stability of solutions to the abstract differential equation

$$\dot{z}(t) = Az(t) + DECz(t) \ , \ z(0) = z_0 .$$

To allow for the possible unboundedness in the perturbation term $DEC$ we assume that $D \in \mathcal{L}(V,X_{-1})$, $C \in \mathcal{L}(X_1,Y)$ are fixed and $E \in \mathcal{L}(Y,V)$ is arbitrary. $Y$ and $V$ are Hilbert spaces and the spaces $X_1$ and $X_{-1}$ are defined as follows: $X_1$ is $D(A)$ with $\|x\|_1 = \|Ax\|$. $X_{-1}$ is the completion of $X$ with respect to $\|x\|_{-1} = \|A^{-1}x\|$. It follows that $S(t)$ extends (restricts) to a semi-group on $X_{-1}$ ($X_1$) isomorphic to $S(t)$. In particular $S(t)$ defines a SCES semi-group on $X$, $X_1$, $X_{-1}$ with the same growth rate. A common approach (see Curtain, Pritchard [2]) is to consider the mild form of

---

1

For $\mathbf{K} = \mathbf{R}$ or $\mathbf{C}$ $r_{\mathbf{K}}(G;h,k^T) = \inf\{|\varepsilon| \,|\, \varepsilon \in \mathbf{K}, \sigma(G+h\varepsilon k^T) \cap \overline{\mathbf{C}_+} \neq \emptyset\}$, see Hinrichsen, Pritchard [6,7].

(2)

$$z(t) = S(t)z_0 + \int_0^t S(t-s)DECz(s)ds \tag{3}$$

Such a consideration leads to a formulation of hypotheses under which we can make sense of (2) and (3). The main difficulty we encounter is in handling $Cz(.)$. The natural space to handle $Cz(.)$ in is $L_Y^2$, since this solves the problem of point wise evaluation of $Cz(t)$ and also the stability of solutions. In [9], this is made possible by a fixed point solution of the auxiliary equation

$$y(t) = CS(t)z_0 + C\int_0^t S(t-s)DEy(s)ds$$

However this involves a rather unnatural assumption on the convolution of $S(.)D$ and $Ey(.)$ and also the restrictive assumption that $DEC$ is not as unbounded as $A$. (For arbitrary $DEC$ (like $A$ for example ) we cannot dispense with this condition, but for specially structured $D$ and $C$ we might do better). Exploiting techniques used in formulations for abstract control systems (see Salamon [12], Curtain, Weiss [3] and Weiss [13])

$$\dot{z}(t) = Az(t) + Dv(t), \quad y(t) = Cz(t), \quad z(0) = z_0$$

we consider the 'output' $Cz(.)$ defined by

$$y(t) = CS(t)z_0 + C\left[\int_0^t S(t-s)Dv(s)ds - (\beta I-A)^{-1}Dv(t)\right] + H(\beta)v(t) \tag{4}$$

for any $\beta \in \rho(A) := \{ \lambda \mid (\lambda I-A)^{-1} \in \mathcal{L}(X)\}$ with $v(.) = Ey(.)$ in mind. $H(\beta) \in \mathcal{L}(V,Y)$ and satisfies a compatability condition

$$H(\beta) - H(\alpha) = -(\beta-\alpha)C(\beta I-A)^{-1}(\alpha I-A)^{-1}D.$$

$H(\beta)$ is the transfer function for the triple $(A,D,C)$, see Salamon [12], Weiss [13]. The right hand side of (4) is valid for all $z_0 \in X_1$ and $v(.) \in W^{1,2}(0,t;V)$ with $Dv(0) \in X$. In order to make sense of (4) for all $z_0 \in X$ and $v(.) \in L_V^2$, and subsequently (3), we must impose certain regularity hypotheses on the triple $(A,D,C)$. These are identical to those which ensure well-posedness of the control system above.

(A1) The map $\varphi_\tau : L_V^2 \to X_{-1}$ defined by

$$\varphi_\tau\, v = \int_o^\tau S(\tau-s)Dv(s)ds$$

has its range in $X$. From this it follows that $\varphi_\tau \in \mathcal{L}(L_V^2,X)$ (see [13])

(A2) $\Lambda_\infty : X_1 \to L_Y^2$ defined by $(\Lambda_\infty x)(t) = CS(t)x, \quad t \in [0,\infty), \quad x \in X_1$
has a continuous extension to all of $X$.
For $v(.) \in W^{1,2}(0,\infty;V)$ with $Dv(0) \in X$ define $\mathcal{F}_\tau v \in L_V^2$ by

$$(\mathcal{F}_\tau v)(t) = C[\int_0^t S(t-s)Dv(s)ds - (\beta I - A)^{-1}Dv(t)] + H(\beta)v(t) \text{ for } t \in [0,\tau).$$

$$=0 \text{ for } t \geq \tau.$$

(A3) $\mathcal{F}_\tau$ extends to a bounded linear map from $L_V^2$ into $L_Y^2$.
It follows that $\mathcal{F}_\infty \in \mathcal{L}(L_V^2, L_Y^2)$ where

$$\mathcal{F}_\infty v = \lim_{\tau \to \infty} \mathcal{F}_\tau v$$

and $\mathcal{F}_\infty v$ is the output corresponding to input $v$ and $z_0 = 0$ (see Weiss [13]).
(A4) The triple $(A,D,C)$ is regular, i.e for all $u \in V$

$$\lim_{\tau \downarrow 0} \frac{1}{\tau} \int_0^\tau y_v(s) ds$$

exists, where $y_v = \mathcal{F}_\infty v$ and $v(t) = u$ for all $t \geq 0$.
We denote by $D_F \in \mathcal{L}(V,Y)$ the feedthrough operator guaranteed by assumption (A4), i.e

$$D_F u = \lim_{\tau \downarrow 0} \frac{1}{\tau} \int_0^\tau y_v(s) ds$$

and note that

$$H(s) = C_L(sI - A)^{-1}D + D_F$$

where $C_L$ is the Lebesgue extension of $C$.
Since we are interested in stability of solutions we also require $\int_o^\infty \|\varphi_t v\|_X^2 dt \leq k^2 \|v(.)\|_{L_V^2}^2$ but this follows applying proposition 2.1 (Weiss [13]). For a detailed discussion of the formulation invoked in (A1)-(A4) see Weiss [13], Salamon [12], Curtain and Weiss [3].

*Definition 1.1*

We say that system (3), (4) is well posed if there exists a strongly continuous semigroup $S_E(t)$ satisfying

$$S_E(t)z_0 = S(t)z_0 + \int_0^t S(t-s)DEy_E(s,z_0)\, ds \tag{5a}$$

$$y_E(t,z_0) = (\Lambda_\infty z_0)(t) + (\mathcal{F}_\infty E y_E(.,z_0))(t)\,. \tag{5b}$$

*Remark*

$S_E(t)z_0$ is then related to the formal equations (3),(4) through continuous extension of (4).

*Definition 1.2*

$$r_C(A;D,C) = \sup\{r \mid \|E\| < r \text{ implies there exists aSCES semi-group } S_E(t)\}.$$

*Remark*

Under hypotheses (A1)-(A4) we characterise $r_C(A;D,C)$ and also if $\|E\| < r_C(A;D,C)$ we gain information about the generator $A_E$ of $S_E$.

*Theorem 1.3*

Assume (A1)-(A4) hold, then

i) $r_C(A;D,C) = \|\mathcal{F}_\infty\|^{-1} = \inf_\omega\{\|H(i\omega)\|^{-1}\}$

ii) If $\|E\| < r_C(A;D,C)$ let $A_E\colon D(A_E) \to X$ denote the generator [2] of $S_E(t)$ then for each $x \in \{x \in X \mid Ax+Dv \in X,\ v \in V,\ v=EC(i\omega I-A)^{-1}(i\omega x-Ax-Dv)+EH(i\omega)v\}$ $x \in D(A_E)$ and $A_E x = Ax + Dv$.

*Proof*

If we assume part (i) then part (ii) follows using the guaranteed bounded invertibility of $(I-\mathcal{F}_\infty E)$ on $L^2_Y$ and Salamon [12] (p. 403 Theorem 4.3). To show part (i), sufficiency of the bound $\|\mathcal{F}_\infty\|^{-1}$ follows using Salamon [12] p. 402 and the fact that $(\mathcal{F}_\infty v)(\omega) = H(i\omega)\hat{v}(\omega)$ for any $v \in L^2_V$. $\hat{v}$ denotes the Fourier (Plancherel) transform of v. Exponential stability of $S_E(t)$ follows from the estimates

$$\|S_E(.)z_0\|_{L^2_X} \le \|S(.)z_0\|_{L^2_X} + k\|E\|\,\|y_E(.,z_0)\|_{L^2_Y}$$

$$\|y_E(.,z_0)\|_{L^2_Y} \le \|(I-\mathcal{F}_\infty E)^{-1}\|\,\|\Lambda_\infty z_0\|_{L^2_Y}$$

and the characterisation of exponential stability due to Datko [4]. To show necessity of the bound $\|\mathcal{F}_\infty\|^{-1}$ fix $\varepsilon > 0$. Choose $v \in V$ and $\omega \in \mathbf{R}$ such that

$$\|H(i\omega)v\|^{-1} \le \|\mathcal{F}_\infty\|^{-1} + \varepsilon$$

and define $E \in \mathcal{L}(Y,V)$ by

---

2

If $D_F \equiv 0$ then the generator of $S_E(t)$ is given as follows:-

Set $Z = D(A) + A^{-1}(Range D)$ then

$D(A_E) = \{z \in Z \mid (A+DEC_L)z \in X\}$ and for $z \in D(A_E), A_E z = (A+DEC_L)z$ .

$$\hat{Ey} = \frac{v\langle \hat{y}, y\rangle}{\|y\|^2} \qquad \text{where } y = H(i\omega)v.$$

It is easy to see that $\|E\| = \|H(i\omega)v\|^{-1}$. Now $x(t)$ and $y(t)$, defined by

$$y(t) = e^{i\omega t}y \qquad x(t) = e^{i\omega t}(i\omega I - A)^{-1}DEy$$

satisfy (5a),(5b) and hence $S_E(t)$ is not exponentially stable.
Since $\varepsilon > 0$ is arbitrary the result follows. □

*Remark*

Theorem 1.3 is an improvement on the results presented in [10]. Under more restrictive hypotheses we can derive a maximising sequence of feedbacks for the functional $F \to r_{\mathbf{C}}(A+BF;D,C)$ with $F = \frac{1}{\varepsilon^2}B^*P$ and $P$ a solution of the parameterised ARE

$$A^*Px + PAx - \rho^2C^*Cx - PDD^*Px + \frac{1}{\varepsilon^2}PBB^*Px = 0$$

for $x \in D(A)$. Here $F$ is a bounded feedback, $B$ is an unbounded input operator with $(A+BF,D,C)$ Pritchard Salamon well-posed. See Pritchard, Townley [11] for a detailed analysis.

*Example 1.4*

Consider the delay equation

$$\dot{w}(t) = Gw(t) + h\varepsilon k^T w(t-\alpha) \qquad t>0.$$

Posed as an abstract system (2) we have $X = \mathbf{C}^n \times L^2(-\alpha,0;\mathbf{C}^n)$, $D(A) = \{\, (f,f(.)) \mid f(.) \in H^1(-\alpha,0;\mathbf{C}^n), f=f(0) \,\}$,

$$Af = (Gf(0), \frac{\partial f}{\partial \theta})$$

and $D = \begin{bmatrix} h \\ 0 \end{bmatrix}$, $C = [0, k^TE_{-\alpha}]$ where $E_{-\alpha}f(.) = f(-\alpha)$. If Re $\sigma(G) < 0$ then (A1)-(A4) are readily verified. Computing $H(i\omega)$ we find that

$$r_{\mathbf{C}}(A;D,C) = \inf_{\omega} \|e^{-i\omega\alpha}k^*(i\omega I - G)^{-1}h\|^{-1} = \inf_{\omega} |k^*(i\omega I - G)^{-1}h|^{-1}.$$

*Remark*

If G, h, k and $\varepsilon$ are real then $r_{\mathbf{C}}$ is a poor conservative bound because variation with $\alpha$ is not captured. If, however, $\varepsilon$ is genuinely complex then this invariance with respect to $\alpha$ allows for $\alpha$ being fixed but uncertain, with $r_{\mathbf{C}}$ the exact bound. Moreover, G could be any stable matrix and hence using delay free feedback control we can optimise the robustness with respect to complex valued uncertainty (see [11]).

## §2 Real stability radii for infinite dimensional systems

Throughout this section we assume that (A1)-(A4) hold and $D_F \equiv 0$. It is clear that $r_{\mathbf{C}}$ is a sufficient bound, for guaranteed existence of a SCES semi-group $S_E(t)$, when we restrict to perturbed systems defined on real Hilbert spaces. However, as Example 1.4 clearly demonstrates, its applicability as a conservative bound on $\|E\|$ is rather unsatisfactory. In this section we develop an analysis for real stability radii. When applied to Example 1.4 we obtain an interesting qualitative behaviour in the robustness, as we vary $\alpha$. The results are somewhat surprising in comparison with the case $\alpha = 0$.

*Definition 2.1*

$$r_{\mathbf{R}}(A;D,C) = \sup \{ \, r \mid \|E\| < r \text{ implies a SCES semi-group } S_E(t) \text{ exists } \}$$

Again bounded invertibility of $(I-\mathcal{F}_\infty E)$ on $L_Y^2$ plays an important role.

*Definition 2.2*

$$r(\mathcal{F}_\infty) = \sup\{ r \mid \ \|E\| < r \text{ implies } (I-\mathcal{F}_\infty E) \text{ is boundedly invertible } \}$$

It is clear that $r(\mathcal{F}_\infty) \le r_{\mathbf{R}}(A;D,C)$. We show that $r_{\mathbf{R}}(A;D,C) = r(\mathcal{F}_\infty)$. In order to demonstrate this equality we convert loss of bounded invertibility for $(I-\mathcal{F}_\infty E)$ into a minimum norm problem involving the operator $H(s)$, $s \in i\mathbf{R}$.

*Notation*

For a real Hilbert space H, we denote by $\tilde{H}$ its (natural) complexification with

$$\|h_R + ih_I\|_{\tilde{H}}^2 = \|h_R\|_H^2 + \|h_I\|_H^2 .$$

If $E \in \mathcal{L}(H_1,H_2)$ then $\tilde{E} \in \mathcal{L}(\tilde{H}_1,\tilde{H}_2)$ is the natural extension of $E$ to a map on $\tilde{H}_1$ into $\tilde{H}_2$.

*Proposition 2.3*

a) $(I-\mathcal{F}_\infty E)$ (equivalently $(I-\mathcal{F}_\infty^* E^*)$) is boundedly invertible if and only if $\|(I-\mathcal{F}_\infty E)y\| > k\|y\|$ and $\|(I-\mathcal{F}_\infty^* E^*)v\| > k\|v\|$, for all $y \in L_Y^2$ and $v \in L_V^2$.

b) Given $\varepsilon > 0$ there exists $E \in \mathcal{L}(Y,V)$ with $\|E\| < r(\mathcal{F}_\infty)+\varepsilon$, such that $1 \in \sigma(EH(i\omega))$ for some $\omega \in \mathbf{R}$ where

$$\sigma(EH(i\omega)) = \mathbf{C} \backslash \{\lambda \mid (\lambda I - EH(i\omega))^{-1} \in \mathcal{L}(\tilde{V})\}.$$

*Proof*

See Appendix

*Notation*

Let $H_1$, $H_2$ be real Hilbert spaces. For given $h \in \tilde{H}_1$, $k \in \tilde{H}_2$ let $E(h,k)$ denote the set of minimum norm solutions $E \in \mathcal{L}(H_1, H_2)$ of $Eh=k$.

*Theorem 2.4*

Under the standing assumptions (A1)-(A4) and additionally that $H(i\omega)$ is compact for all $\omega \in \mathbf{R}$ then

$$r_{\mathbf{R}}(A;D,C) = \inf\{ \|E(y,v)\| \mid y = H(i\omega)v, \ \omega \in \mathbf{R}\}$$

$$= \inf\{\|E\| \mid \omega \in \mathbf{R}, \ 1 \text{ is an eigenvalue of } EH(i\omega)\}$$

$$= r(\mathcal{F}_\infty)$$

*Proof*

The equalities

$$\inf\{ \|E(y,v)\| \mid y = H(i\omega)v, \ \omega \in \mathbf{R}\} = \inf\{\|E\| \mid \omega \in \mathbf{R}, \ 1 \text{ is an eigenvalue of } EH(i\omega)\}$$

follow immediately. Moreover the inequalities

$$r(\mathcal{F}_\infty) \le r_{\mathbf{R}}(A;D,C) \le \inf\{\|E\| \mid \omega \in \mathbf{R}, \ 1 \text{ is an eigenvalue of } EH(i\omega)\}$$

follow arguing similarly to Theorem 1.3. To establish equality let $\varepsilon > 0$. By Proposition 2.3 there exists $E \in \mathcal{L}(Y,V)$ with

$$r(\mathcal{F}_\infty) \le \|E\| \le r(\mathcal{F}_\infty)+\varepsilon \text{ and } 1 \in \sigma(EH(i\omega)$$

for some $\omega \in \mathbf{R}$. By compactness of $H(i\omega)$, it follows that 1 is an eigenvalue of $EH(i\omega)$ and

$$r(\mathcal{F}_\infty) = \inf\{\|E\| \mid \omega \in \mathbf{R}, \ 1 \text{ is an eigenvalue of } EH(i\omega)\}.$$

*Remark*

$H(i\omega)$ (or $H^*(i\omega)$) is compact if, in particular, one of $Y$ or $V$ is finite dimensional. However this excludes some interesting examples. We can relax this constraint, without altering the formulas in Theorem 2.4 if we can replace the assumption that $H(i\omega)$ is compact by $EH(i\omega)$ has compact resolvent. These assumptions of compactness are extra to the conditions (A1)-(A4). We can still characterise $r_{\mathbf{R}}(A;D,C)$ without the restrictions of compactness if we strengthen the assumptions (A1)-(A4) but then

$$r_{\mathbf{R}}(A;D,C) = \inf\{ \|E\| \mid \omega \in \mathbf{R},\ 1 \in \sigma(EH(i\omega)) \}$$

It remains a conjecture as to whether this is true in the general setting of assumptions (A1)-(A4). [3]
However, $r(\mathcal{F}_\infty)$ is a lower bound for $r_{\mathbf{R}}(A;D,C)$ and

$$\inf\{ \|E\| \mid \omega \in \mathbf{R},\ 1 \text{ is an eigenvalue of } EH(i\omega)\}$$

an upper bound.
The characterisation of $r_{\mathbf{R}}$ can be converted into an eigenvalue type problem. We do not persue this here since computations are rather cumbersome. It is more interesting to consider the special case when one or both of $Y$ or $V$ is one dimensional. For equation (2), modelling a partial differential equation on a domain $\Omega$, $Y$ of dimension one might correspond to $C$ being an averaging sensor, over part of the domain.

*Corollary 2.5*

If $Y$ or $V$ is one dimensional then

$$r_{\mathbf{R}}(A;D,C) = \inf_{\omega}\{ d(H_R(i\omega),\mathbf{R}H_I(i\omega))\}^{-1}$$

where

$$d(u,\mathbf{R}v) = \inf_{\gamma\in\mathbf{R}} \|u - \gamma v\|. \qquad \square$$

*Proof*

Consider the case $V$=$\mathbf{R}$. We have by Theorem 2.4 that $r_{\mathbf{R}}$ is characterised as the $\inf_{w,v\in\mathbf{R}} \|E\|$ such that

$$v - EH(iw)v = 0 \text{ ( and } v - H^*(-iw)E^*v = 0\text{ )} \tag{6}$$

Let $E\in\mathcal{L}(Y,\mathbf{R})(\equiv Y^*)$ satisfy (6) for given $v$ and $\omega$ then we have immediately that

$$\langle E,H_R(i\omega)\rangle_Y = 1,\ \langle E,H_I(i\omega)\rangle_Y = 0.$$

Therefore $\langle E,(H_R(i\omega)-\alpha H_I(i\omega))\rangle_Y = 1$ for all $\alpha\in\mathbf{R}$ and hence

---

[3] The difficulty we encounter is converting the statement $1 \in \sigma(EH(i\omega))$ to $i\omega \in \sigma(A_E)$.

$$\|E\| \geq \sup_{\alpha} \|H_R(i\omega) - \alpha H_I(i\omega)\|^{-1}.$$

The result now follows and the case of $Y = \mathbf{R}$ is analogous.

*Remark*

If $X$, $V$ and $Y$ are finite dimensional then $r_{\mathbf{C}}(A;D,C)$ can be computed via tests on eigenvalues of the Hamiltonian matrix

$$\Sigma(\rho) = \begin{bmatrix} A & DD^* \\ -\rho^2 C^* C & -A^* \end{bmatrix}$$

since $\sigma(\Sigma(\rho)) \cap i\mathbf{R} \neq \emptyset$ if and only if $\rho < r_{\mathbf{C}}(A;D,C)$, (see Hinrichsen et al [5]). For $X$ infinite dimensional, one possibility for computing $r_{\mathbf{C}}(A;D,C)$ is to take a balanced realization of $H(s)$ (Glover, Curtain [1]). Truncation of this balanced realization, with *a priori* $L^\infty$ error bounds, could then be used for numerical approximation of $r_{\mathbf{C}}(A;D,C)$ via the corresponding Hamiltonian matrix. In computing $r_{\mathbf{R}}(A;D,C)$, when $Y$, $V$, are finite dimensional, approximation of $H(s)$ by a finite dimensional transfer function is inappropriate. This is because $\inf_{\omega} d(H_R(i\omega), \mathbf{R}H_I(i\omega))$ might occur when $H_I(i\omega)$ is small. One is then forced to consider approximations of the transfer function $H(i\omega)$, which take account of the zeros of $H_I(i\omega)$. This consideration is made even clearer when both $Y$ and $V$ are one dimensional.

*Corollary 2.6 (to Corollary 2.5)*

If $Y$ and $V$ are one dimensional then

$$r_{\mathbf{R}}(A;D,C) = \inf_{\omega}\{\ |H_R(i\omega)|^{-1}\ |\ H_I(i\omega) = 0\} \qquad \square$$

*Example 2.7*

Consider example 1.4 with

$$G = \begin{bmatrix} 0 & 1 \\ -1 & -2 \end{bmatrix},\ h = \begin{bmatrix} 0 \\ 1 \end{bmatrix},\ k = \begin{bmatrix} b \\ 1 \end{bmatrix}$$

where $b$ and $\varepsilon$ are real. If $|b| < \frac{1}{2}$ then

$$r_{\mathbf{C}}(G;h,k^T) < r_{\mathbf{R}}(G;h,k^T) < |g(0)|^{-1}$$

where $g(i\omega) = k^T(i\omega I - G)^{-1}h$. Now $H(i\omega) = e^{-i\omega\alpha}g(i\omega)$, and hence $H_I(i\omega) = 0$ if and only if $g_I(\omega)\cos\alpha\omega - g_R(\omega)\sin\alpha\omega = 0$,

where

$$g_I(\omega) = \frac{\omega(1-2b-w^2)}{(1-\omega^2)^2+4\omega^2} \text{ and } g_R(\omega) = \frac{(1-\omega^2)b+2\omega^2}{(1-\omega^2)^2+4\omega^2}.$$

Therefore

$$r_{\mathbf{R}}^{\alpha}(G;h,k^T) = \inf_{\omega}\{\,|g(i\omega)|^{-1} \mid \arg g(i\omega)=\alpha\omega \bmod 2\pi\}$$

Considering this characterisation for various $\alpha \in (0,\infty)$ it follows that

a) $\lim_{\alpha\downarrow 0} r_{\mathbf{R}}^{\alpha}(G;h,k^T) = r_{\mathbf{R}}(G;h,k^T)$

b) $\lim_{\alpha\uparrow\infty} r_{\mathbf{R}}^{\alpha}(G;h,k^T) = r_{\mathbf{C}}(G;h,k^T)$

c) Given any $\beta < 0$ there exists $\alpha \geq \beta$ such that $r_{\mathbf{R}}^{\alpha}(G;h,k^T) = r_{\mathbf{C}}(G;h,k^T)$

Hence for $\alpha$ small, we have robustness approximately equal to that when $\alpha = 0$. For *$\alpha$ large* the robustness deteriorates to that for *complex $\varepsilon$.* Also for certain critical *commeasurate* delays the robustness is *minimised.* Notice that an upper bound for $r_{\mathbf{R}}^{\alpha}(G;h,k^T)$ is always $|g(0)|^{-1}$ and in the case when $b = 0.49$,

$$r_{\mathbf{R}}^{\alpha}(G;h,k^T) = |g(0)|^{-1}$$

at least for $\alpha \in (0.4,1)$. Hence for an operating range $\alpha \in (0.4,1)$ the robustness is *maximised and bigger than* $r_{\mathbf{R}}(G;h,k^T)$. This is somewhat surprising since we can improve the robustness for

$$\dot{w}(t) = (G + h\varepsilon k^T)w(t)$$

by introducing delay as in system (1).

## Conclusions

Motivated by the limitations of $r_{\mathbf{C}}(A;D,C)$ as a robustness measure for differential delay equations, we have developed a robustness analysis for abstract uncertain linear systems defined on real Hilbert spaces. Under certain Salamon-Weiss type hypotheses we characterise this robustness as a stability radius $r_{\mathbf{R}}(A;D,C)$. These characterisations pose interesting numerical problems associated with calculation of $r_{\mathbf{R}}(A;D,C)$ via various approximation schemes. When applied to the simple motivating example, a qualitative analysis indicates both the importance of $r_{\mathbf{R}}(A;D,C)$ as a robustness measure and the difficulties raised by its computation.

## References

[1] R. F. Curtain and K. Glover, Balanced realizations for infinite dimensional systems, 87-104, Operator Theory and Systems, Proc. Amsterdam, 1985, Ed. Bart, Gohberg and Kaashoek, Birkhauser, Basel.

[2] R. F. Curtain and A. J. Pritchard, Infinite Dimensional Linear Systems Theory, LNCS8, Springer Verlag, Berlin, 1978.

[3] R. F. Curtain and G. Weiss, Well posedness of triples of operators (in the sense of linear systems theory), proc 4th Int. Conf on Contr. Dist. Parameter Systems, Vorau, Austria 1988.

[4] R. Datko, Uniform asymptotic stability of evolutionary processes in a Banach space, SIAM J. Math. Anal. 3, 1972, 428-445.

[5] D. Hinrichsen, B. Kelb and A. Linnemann, An algorithm for the computation of the structured stability radius with applications, Institut fur Dynamische Systeme, Universitat Bremen, Report No. 162.

[6] D. Hinrichsen and A. J. Pritchard, Stability radius for structured perturbations and the algebraic Riccati equation, Systems & Control Letters, vol 8, 1986, 105-113.

[7] D. Hinrichsen and A. J. Pritchard, New robustness results for linear systems under real perturbations, Proc. 27th CDC, Austin, 1988.

[8] H. Logemann, On existence of finite dimensional compensators for retarded and neutral systems, Int. J. Control, 1986, vol. 43, 1986, 109-121.

[9] A. J. Pritchard and S. Townley, Robustness of linear systems, J. Diff. Eqns. vol 77, 1989, 254-286.

[10] A. J. Pritchard and S. Townley, A stability radius for infinite dimensional systems, 272-291, Proc. 3rd Int. Conf. on Contr. Dist. Parameter Systems, Vorau, Austria, LNCIS 102, Springer Verlag, 1987, Ed F. Kappel.

[11] A. J. Pritchard and S. Townley, Robustness optimization for abstract uncertain control systems: Unbounded inputs and perturbations. 5th, IFAC Symposium on Cont. Distr. Parameter Systems, Perpignan, France, 1989.

[12] D. Salamon, Infinite dimensional linear systems with unbounded control and observation: a functional analytic approach, Trans. Amer. Math. Soc., 300, 1987, 383-431.

[13] G. Weiss, Representation of linear systems on Hilbert spaces, 4th Int. Conf. on Contr. Dist. Parameter Systems, Vorau, Austria, 1988.

*Appendix Proof of Proposition 2.3*

a) This is standard operator theory.

b) Let $\varepsilon > 0$ and choose $E \in \mathcal{L}(Y,V)$ such that $\|E\| < r(\mathcal{F}) + \varepsilon$ and $(I-\mathcal{F}_\infty E)$ is not boundedly invertible equivalent to $(I-E\mathcal{F}_\infty)$ not boundedly invertible. Either $\|(I-E\mathcal{F}_\infty)v\|>c\|v\|$ for all $v \in L_V^2$ or $\|(I-E^*\mathcal{F}_\infty^*)y\|>c\|y\|$ for all $y \in L_Y^2$ fails. Suppose without loss of generality that it is the former, then given $\delta > 0$ there exists $\omega \in \mathbf{R}$ and $v \in \tilde{V}$ such that $\|v\|=1$ and $\|(I-EH(i\omega))v\| < \delta$. (If not then there exists $\delta > 0$ such that for all $\omega \in \mathbf{R}, v \in V$, $\|(I-EH(i\omega))v\| \geq \delta\|v\|$ and hence $\|(I-E\mathcal{F}_\infty)v(.)\| > \delta\|v(.)\|$ for all $v \in L_V^2$.).

Hence $1 \in \sigma(EH(i\omega))$ for some $\omega \in \mathbf{R}$, since $H(i.)$ is continuous and $\lim_{\omega\to\infty} \|H(i\omega)\| = 0$.

# Robust Stability Condition for Infinite-Dimensional Systems with Internal Exponential Stability

**Yutaka Yamamoto and Shinji Hara**

**Abstract**

In the current study of robust stability of infinite-dimensional systems, internal exponential stability is not necessarily guaranteed. This paper introduces a new class of impulse responses called $\mathcal{R}$, in which usual notion of $L^2$-input/output stability guarantees not only external but also internal exponential stability. The result is then applied to derive a robust stability condition which also assures internal stability.

## 1 Introduction

In the current study of robust stability, especially that for finite-dimensional systems, the space $H^\infty(\mathbb{C}_+)$ plays a key role. This is crucially based on the fact that $H^\infty(\mathbb{C}_+)$ guarantees stability, i.e., bounded $L^2$ inputs–bounded $L^2$ outputs correspondence on one hand, and exponential stability of the internal minimal realization on the other. The former property remains intact for infinite-dimensional systems ([5]), and there are in fact a number of investigations of robust stability/stabilizability along this line ([2], [4], [7], to name a few).

On the other hand, it is well known that stability of infinite-dimensional systems may not be determined by location of spectrum ([19]), so that there is a question as to if the second property above remains valid. In fact, Logemann [8] recently gave an example of a transfer function of a neutral delay-differential system whose transfer function belongs to $H^\infty(\mathbb{C}_+)$, and yet *its canonical (irredundant) realization is not exponentially stable.*

This gap between external stability (e.g., transfer functions in $H^\infty(\mathbb{C}_+)$), and internal stability (e.g., exponential stability of the canonical realization) has attracted the recent research interest. There are now several attempts to establish the equivalence between the two for a suitably defined class of transfer functions or impulse responses. Jacobson and Nett [9] and Callier and Winkin [1] have worked with the algebra $\mathcal{B}$ of transfer functions which are expressible as a ratio of functions in $\mathcal{A}$ (cf. [5]) whose denominator is invertible on a right half complex plane. They proved this equivalence under the hypotheses of i) bounded input/output operators, and ii) the system is stabilizable and detectable. Since the first assumption is restrictive in dealing with delay or boundary-control systems, Curtain [3] generalized their results to those with unbounded input/output operators. These results, however, do not apply to Logemann's example, The system there is canonical but neither stabilizable nor detectable. This situation is entirely different from the finite-dimensional case. Assuming irredundancy in realization is *not* enough to guarantee stabilizability, and it may be often difficult to assure stabilizability to begin with.

Yamamoto and Hara [17] gave a necessary and sufficient condition (with possibly unbounded output operators) for internal exponential stability for a different class of systems called *pseudorational.* While this class does not require a priori stabilizability/detectability, the condition given there requires a higher-order condition on transfer matrices, and hence is not fully adequate for the study of robust stability.

In this paper we present a different approach. We restrict the class of transfer functions further, but give a stronger result on stability. We do not require a priori stabilizability/detectability, nor do we restrict the system to have bounded observation maps. Therefore, the system will not be presumed to have finitely many unstable poles. On the other hand, we do require that the impulse response satisfy some mildness condition. This class is called class $\mathcal{R}$. Typically, *retarded* delay-differential systems belong to this class. Although somewhat restrictive in that it excludes neutral delay systems or some partial differential equations, this class is often large enough to cover important applications such as repetitive control ([6]). The advantage here is that we do not need any characterization involving state space realizations, so that suitable for studying closed-loop properties and robust stability from the external viewpoint. We prove that *the canonical realization of an impulse response in $\mathcal{R}$ is exponentially stable if and only if the poles of the transfer matrix belong to the strict left-half complex plane*. Under a mild assumption, this also implies that the so-called small-gain theorem ([5]) guarantees not only $L^2$ input/output stability, but also exponential stability of the internal realization. This result will then be used to prove a sufficient condition for robust stability in Section 4. The result resembles to the known counterparts in finite-dimensional systems or those using algebra $\mathcal{A}$ (and fractions derived from it) (e.g., [2], [7]). The difference here is that exponential stability is guaranteed.

## 2 Pseudorational Impulse Responses

Let us start by specifying the class of impulse responses we deal with. Because of the limitation of space, we only indicate the general idea; the details can be found in [15], [16], [17].

Let $A$ be an $p \times m$ impulse response matrix. We allow $A$ to be as singular as a measure on $[0, \infty)$, but no higher singularities such as differentiation are allowed [15]. We assume that $A$ can be decomposed as

$$A = A_0 \cdot \delta + A_1, \tag{1}$$

where $A_0$ is a constant matrix and $A_1$ is a regular distribution (i.e., a function type) in a neighborhood of the origin. When $A_0$ is zero, $A$ is said to be *strictly causal.* We say that an impulse response matrix $A$ is *pseudorational* if $A$ can be written as

$A = Q^{-1} * P$ for some distributions $Q, P$ with compact support contained in $(-\infty, 0]$ (There is an additional requirement, which does not concern us here; see [15]).

The next question is realization. This can be done by the standard shift realization procedure. Let $\Gamma := L^2_{loc}[0,\infty)$. This is the space of all output functions. For a $p \times m$ impulse response $A$, define

$$X_A := \overline{\{\pi(A * \omega) \in \Gamma^p; \omega \in \Omega^m\}}, \tag{2}$$

where $\pi$ is the truncation mapping $\pi\varphi := \varphi\,|_{[0,\infty)}$, and the closure is taken in $\Gamma^p$. In particular, if $A$ is of the form $A = Q^{-1}$, the space $X_A$ is denoted by $X^Q$.

Taking $X_A$ as a state space, the following state equations give a canonical realization of a pseudorational impulse response $A = A_0\delta + A_1$ ([15]):

$$\frac{d}{dt}x_t(\cdot) = Fx_t(\cdot) + A_1(\cdot)u(t) \tag{3}$$

$$y(t) = x_t(0) + A_0u(t) \tag{4}$$

$$Fx(\tau) := \frac{dx}{d\tau}, \qquad D(F) = W^1_{2,loc}[0,\infty) \cap X_A. \tag{5}$$

That is, one takes all free output functions on $[0, \infty)$, take their closure in the space $L^2_{loc}[0,\infty)$ of locally square integrable functions, and then regard it the state space ([14], [15]). Although somewhat not widely appreciated, this procedure always gives rise to the canonical realization, and it often agrees with very familiar models such as the $M_2$-model for delay-systems. Furthermore, it turns out that because of the bounded support property of $Q$, this state space is isomorphic to a Hilbert space. Indeed, one needs only to take the bounded-time function pieces on $[0, T]$, say, for any $T$ greater than the length of the support of $Q$, thereby isomorphic to a closed subspace of $L^2[0,T]$ ([14]). This is one of the consequences of pseudorationality. We denote this canonical realization of $A$ by $\Sigma_A$. Its free state transition is induced by the left shift semigroup $\sigma_t$ in $L^2_{loc}[0,\infty)$. (Note, however, that if we restrict ourselves to such a subspace of $L^2[0,T]$, the semigroup hardly looks like left shifts since this procedure involves computation of equivalence classes, and the representation of $\sigma_t$ is changed accordingly.)

We may thus unambiguously speak about internal stability of $A$:

**Definition 2.1** Let $\Sigma_A$ be as above, and let $\sigma_t$ be the semigroup of its free state-transition. We say that $\Sigma_A$ is *exponentially stable* if there exist $M, \beta > 0$ such that $\|\sigma_t\| \le Me^{-\beta t}$ for all $t \ge 0$.

The proof of the following lemma is quite involved, so will not be given here.

**Lemma 2.2** *Let $A = (a_{ij})$ be a pseudorational impulse response. Then its canonical realization $\Sigma_A$ is exponentially stable if and only if each $a_{ij}$ has the same property.*

# 3 Stability in the Class $\mathcal{R}$

Lemma 2.2 in the previous section states that for a pseudorational impulse response $A$, stability can be discussed separately on each entry of $A$. Furthermore, it is easily shown that $A$ is pseudorational if and only if each entry is pseudorational. Thus, without loss of generality, we now confine ourselves to the single-input/single-output case.

**Definition 3.1** A pseudorational impulse response $A$ belongs to the class $\mathcal{R}$ if there exists a factorization $A = q^{-1} * p$ such that

$$\text{ord } q^{-1} \mid_{(T,\infty)} \ < \ \text{ord } q^{-1} \quad \text{for some } T > 0. \tag{6}$$

where ord $\alpha$ denotes the *order* of a distribution $\alpha$. A multivariable impulse response matrix $A(t)$ belongs to class $\mathcal{R}$ if and only if each entry of $A$ does.

This means that the regularity of $q^{-1}$ becomes higher after $T > 0$. Finite-dimensional systems satisfy this property. For example, consider the unit step Heaviside function $H(t)$. Its global order is $-1$, since it has jump at the origin and differentiation of $H(t)$ yields the Dirac distribution $\delta$ which is of order zero. However, for any $T > 0$, its restriction $H(t) \mid_{(T,\infty)}$ is a $C^\infty$-function, so (6) is satisfied. Another example that satisfies this condition is given by the impulse responses of retarded delay-differential systems. For, as is well known, impulse responses in this class becomes smoother after some finite duration of time. (Consider, for example, the impulse response $A(t) = (\delta'_{-1} - \delta)^{-1} * \delta$.) On the other hand, impulse responses of neutral delay-differential systems do *not* satisfy this condition. This is because, in general, neutral systems exhibit perpetual jump behavior in their impulse responses, thereby maintaining its irregularity as high as that around the origin. One of the importance in the class $\mathcal{R}$ lies in the fact that it shares a mildness property as retarded systems have but is also characterized in terms of the external behavior and not in terms of the state equation. This is particularly fruitful in discussing robust stability, which we will witness in subsequent sections.

The following theorem has been obtained in [18]:

**Theorem 3.2** *Suppose that $A \in \mathcal{R}$. Then the canonical realization $\Sigma_A$ of $A$ is exponentially stable if and only if either one of the following conditions holds:*

1. *The poles of $\hat{A}(s)$ belong to the strict left-half plane $\{s \in \mathbb{C} : Re\ s < -c\}$ for some $c > 0$.*

2. *$\hat{A}$ belongs to $H^\infty(\mathbb{C}_+)$.*

We here give only an outline of the proof. As noted in Section 2, the state space $X_A$ is a closed subspace of $L^2[0, a]$ for some $a > 0$. The value $T > 0$ appearing in the definition of class $\mathcal{R}$ can be taken larger than $a$. The crux of the proof relies on showing that the semigroup operator $\sigma_t$ is a *compact* operator for any $t > T$. Roughly speaking, the proof goes as follows: Invoking the definition of class $\mathcal{R}$, one can show that for any $x \in X_A$ the shifted state $\sigma_t x$ has higher regularity than that of $x$. This implies that $\sigma_t$ is compact. For details, see [18].

## 4 Closed-Loop Stability

**Theorem 4.1**

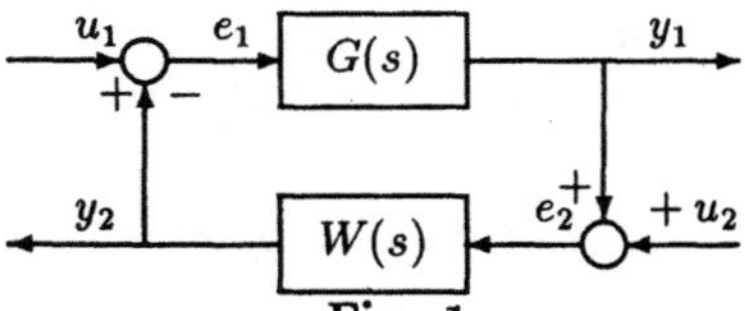

**Fig. 1**

*Consider the closed-loop system given by Fig.1, where $G(s)$ and $W(s)$ denotes the canonical realizations of these transfer matrices, respectively. Suppose that their inverse Laplace transforms $\check{G}$ and $\check{W}$ belong to the class $\mathcal{R}$.*

*Decompose $\check{G}$ and $\check{W}$ as*

$$\check{G} = \check{G}_0 \cdot \delta + \check{G}_1, \; \check{W} = \check{W}_0 \cdot \delta + \check{W}_1, \tag{7}$$

*as in (1). Suppose that*

1. *the matrix $I + \check{W}_0\check{G}_0$ is nonsingular, and*
2. *$\check{G}_1$ and $\check{W}_1$ are functions in $\Gamma$.*

*Then the closed-loop system is internally exponentially stable if and only if the matrix*

$$\begin{pmatrix} (I+GW)^{-1} & -G(I+WG)^{-1} \\ W(I+GW)^{-1} & (I+WG)^{-1} \end{pmatrix} \tag{8}$$

*belongs to $H^\infty(\mathbb{C}_+)$.*

**Proof** In view of the assumptions, the closed-loop matrix (8) is well defined. Since $\check{G}_1$ and $\check{W}_1$ are functions, an easy argument using the standard Neumann type expansion shows that matrix (8) is indeed of class $\mathcal{R}$.

Since the matrix (8) gives the correspondence $(u_1, u_2) \mapsto (e_1, e_2)$, and since $(e_1, e_2)$ is related to $(y_1, y_2)$ via $e_1 = -y_2 + u_1, e_2 = y_1 + u_2$, matrix (8) belongs to $H^\infty(\mathbb{C}_+)$ if and only if the closed-loop system is $L^2$-input/output stable. In view of Theorem 3.2, this is enough to guarantee internal exponential stability if the closed-loop system is a canonical realization. But this can be shown by the following easy argument: First note that each pair $(x_1, x_2)$ of reachable states in two boxes is clearly reachable, so that the reachable subspace is dense. Secondly, each subsystem is topologically observable ([15]), so that the dual of each is exactly reachable ([13]). The dual of the total closed-loop system is obtained by reversing the signal arrows and taking the dual of each subsystem. Repeating the same argument for reachability, we see that the dual closed-loop system is exactly reachable and the closed-loop system is topologically observable by duality (in the sense of [13]). This completes the proof. □

**Theorem 4.2** (Small-Gain Theorem with Internal Stability) *Consider the same closed-loop system Fig.1 as above, under the same hypotheses as in Proposition 4.1. Assume the following conditions:*

1. *$G, W \in H^\infty(\mathbb{C}_+)$,*

*2.* $\|W(s)G(s)\|_\infty < 1$.

*Then the closed-loop system is internally exponentially stable.*

**Proof** By Theorem 4.1, it suffices to show that (8) belongs to $H^\infty(\mathbb{C}_+)$. Since $H^\infty(\mathbb{C}_+)$ is a Banach algebra, condition 2 makes the Neumann series

$$(I + WG)^{-1} = \sum_{i=0}^{\infty}(WG)^i$$

convergent. Thus $(I + WG)^{-1} \in H^\infty(\mathbb{C}_+)$, and by

$$(I + GW)^{-1} = I - G(I + WG)^{-1}W,$$

$(I + GW)^{-1} \in H^\infty(\mathbb{C}_+)$. Thus the four blocks in (8) all belong to $H^\infty(\mathbb{C}_+)$. This completes the proof. □

## 5 Robust Stability Condition in the Class $\mathcal{R}$

**Theorem 5.1**
*Consider the feedback system shown in Fig. 2, where the inverse Laplace transforms of $P, C, \Delta P$ belong to $\mathcal{R}$, and each block denotes the canonical realization. Assume the following:*

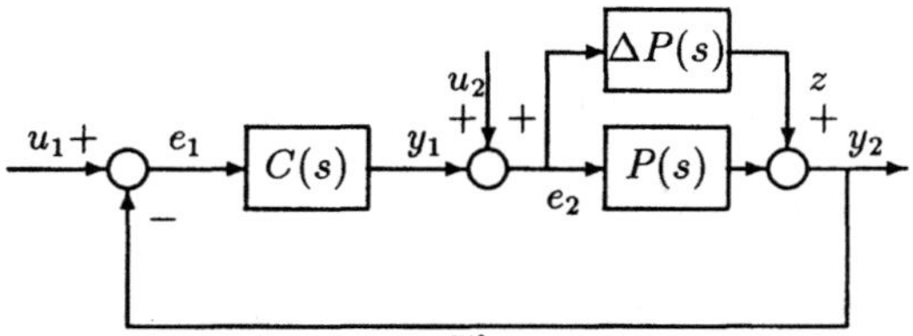


**Fig. 2**

1. *$C(s)$ stabilizes the nominal plant $P(s)$, i.e., the nominal closed-loop system without the perturbation $\Delta P$ is internally stable.*

2. *The unknown perturbation $\Delta P$ is stable, i.e., $\Delta P \in H^\infty(\mathbb{C}_+)$.*

3. *This perturbation $\Delta P$ admits the frequency-domain bound:*

$$| \Delta P(j\omega) | < r(j\omega) \text{ for all } \omega \in \mathbb{R} \tag{9}$$

*for some $H^\infty(\mathbb{C}_+)$ function $r(s)$.*

*Now suppose that*

$$\|r(s)C(s)(I + P(s)C(s))^{-1}\|_\infty < 1. \tag{10}$$

*Then the perturbed closed-loop system remains internally exponentially stable.*

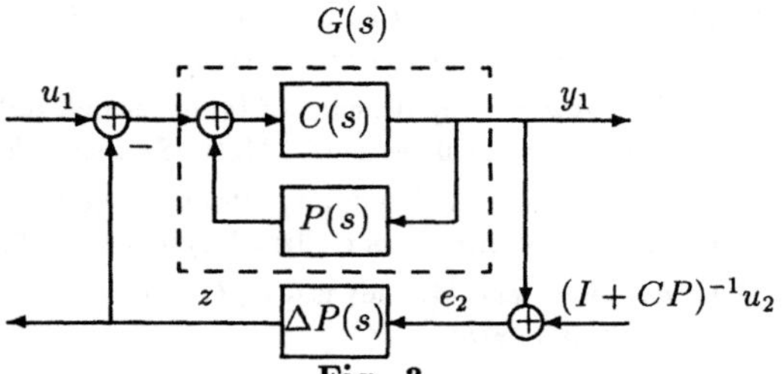


**Fig. 3**

**Proof** Let us first show that the internal stability of the closed-loop system Fig. 2 is equivalent to that of the closed-loop system given by the diagram Fig. 3. In fact, from Fig. 2, we have

$$e_2 = y_1 + u_2;\ y_1 = Ce_1 = Cu_1 - Cy_2;\ y_2 = z + Pe_2, \tag{11}$$

so that

$$e_2 = Cu_1 - Cy_2 + u_2 = Cu_1 - Cz - CPe_2 + u_2. \tag{12}$$

This implies

$$e_2 = (I + CP)^{-1}[-Cz + Cu_1 + u_2] = G(s)(u_1 - z) + (I + CP)^{-1}u_2, \tag{13}$$

because $G(s) = (I + CP)^{-1}C$. This means that we can take out the input $u_2$ outside the loop of $G(s)$. Correspondence of the rest of the variables is easy to check, so that we have Fig. 3 as an equivalent diagram.

Now suppose that the inputs $u_1$ and $u_2$ are in $L^2$. Then by stability of the nominal feedback system and by Theorem 4.1, $(I + CP)^{-1}$ is in $H^\infty(\mathbb{C}_+)$. Hence $(I + CP)^{-1}u_2 \in L^2$. Now $G(s)$ and $\Delta P(s)$ belong to $H^\infty(\mathbb{C}_+)$ by our hypotheses. Also, the condition $\|r(s)C(s)(I + P(s)C(s))^{-1}\|_\infty < 1$ implies that the loop gain of the system Fig. 3 is less than 1. Therefore, we can apply the small-gain theorem 4.2 to show $e_2, z \in L^2$. Hence $y_1 = e_2 - u_2 \in L^2$, and

$$\begin{aligned} y_2 &= z + Pe_2 \\ &= z + P(I + CP)^{-1}C(z - u_1) + P(I + CP)^{-1}u_2 \\ &= z + (I - (I + PC)^{-1})(z - u_1) + P(I + CP)^{-1}u_2 \in L^2 \end{aligned} \tag{14}$$

where the second equality follows from (13). This shows that the perturbed feedback system Fig. 2 maps any $L^2$ inputs $u_1, u_2$ to $L^2$ outputs $y_1$ and $y_2$. Therefore, by Theorem 4.1, the closed-loop system is internally exponentially stable, and this concludes the proof. □

**Acknowledgments** This work was supported in part by the Inamori Foundation, Kyoto, JAPAN. The author would like to thank the Ministry of Education, Japan, for the travel grant.

## References

[1] F. M. Callier and J. Winkin, "Distributed system transfer functions of exponential order," *Int. J. Control*, **43**: 1353-1373, 1986.

[2] M. J. Chen and C. A. Desoer, "Necessary and sufficient conditions for robust stability of linear distributed feedback systems," *Int. J. Control*, **35**: 255-267, 1982.

[3] R. F. Curtain, "Equivalence of input-output stability and exponential stability for infinite dimensional systems," *Math. Systems Theory*, **21**: 19-48, 1988.

[4] R. F. Curtain and K. Glover, "Robust stabilization of infinite dimensional systems by finite dimensional controllers," *Syst. Contr. Letters*, **7**: 41-47, 1986.

[5] C. A. Desoer and M. Vidyasagar, *Feedback Systems: Input/Output Properties*, Academic Press, 1975.

[6] S. Hara, Y. Yamamoto, T. Omata and M. Nakano, "Repetitive control system - a new-type servo system," *IEEE Trans. Autom. Control*, **AC-33**: 659-668, 1988.

[7] P. P. Khargonekar and K. Poola, "Robust stabilization of distributed systems," *Automatica*, **22**: 77-84, 1986.

[8] H. Logemann, "On the transfer matrix of a neutral system: Characterizations of exponential stability in input-output terms," *Syst. Contr. Letters*, **9**: 393-400, 1987.

[9] C. A. Jacobson and C. N. Nett, "Linear state-space systems in infinite-dimensional space: the role and characterization of joint stabilizability/detectability," *IEEE Trans. Autom. Control*, **33**: 541-549, 1988

[10] A. Pazy, *Semigroups of Linear Operators and Applications to Partial Differential Equations*, Springer, 1983.

[11] A. J. Prichard and J. Zabczyk, "Stability and stabilizability of infinite dimensional systems," *SIAM Review*, **23**: 25-52, 1983.

[12] L. Schwartz, *Théorie des Distributions*, 2me Edition, Hermann, 1966.

[13] Y. Yamamoto, "A note on duality: reconstructibility and controllability", *Proc. 8th IFAC World Congress*, 277-282, 1981.

[14] Y. Yamamoto, "A note on linear input/output maps of bounded type," *IEEE Trans. Autom. Control*, **AC-29**: 733-734, 1984

[15] Y. Yamamoto, "Pseudo-rational input/output maps and their realizations: a fractional representation approach to infinite-dimensional systems," *SIAM J. Control & Optimiz.*, **26**: 1415-1430, 1988.

[16] Y. Yamamoto, "Reachability of a class of infinite-dimensional linear systems: an external approach with applications to general neutral systems," *SIAM J. Control & Optimiz.*, **27**: 217-234, 1989.

[17] Y. Yamamoto and S. Hara, "Relationships between internal and external stability with applications to a servo problem," *IEEE Trans. Autom. Control*, **AC-33**: 1044-1052, 1988.

[18] Y. Yamamoto, "Equivalence of external and internal stability for a class of infinite-dimensional systems," to appear in *Proc. 4th Int. Conf. on Control of Distributed Parameter Systems* at Vorau, Austria, Birkhäuser.

[19] J. Zabczyk, "A note on $C_0$-semigroups," *Bull l'Acad. Pol. de Sc. Serie Math.*, **23**: 895-898, 1975.

# Progress in Systems and Control Theory

*Series Editor*

Christopher I. Byrnes
Department of Systems Science and Mathematics
Washington University
Campus P.O. 1040
One Brookings Drive
St. Louis, MO 63130-4899
U.S.A.

*Progress in Systems and Control Theory* is designed for the publication of workshops and conference proceedings, sponsored by various research centers in all areas of systems and control theory, and lecture notes arising from ongoing research in theory and applications of control.

We encourage preparation of manuscripts in such forms as LATEX or AMS TEX for delivery in camera ready copy which leads to rapid publication, or in electronic form for interfacing with laser printers.

Proposals should be sent directly to the editor or to: Birkhäuser Boston, 675 Massachusetts Avenue, Suite 601, Cambridge, MA 02139, U.S.A.

PSCT1 Computation and Control: Proceedings of the Bozeman Conference, Bozeman, Montana, 1988
*K. Bowers and J. Lund*

PSCT2 Perspectives in Control Theory: Proceedings of the Sielpia Conference, Sielpia, Poland, 1988
*B. Jakubczyk, K. Malanowski, and W. Respondek*

PSCT3 Realization and Modelling in System Theory: Proceedings of the International Symposium MTNS-89, Volume I
*M.A. Kaashoek, J.H. van Schuppen, and A.C.M. Ran*

PSCT4 Robust Control of Linear Systems and Nonlinear Control: Proceedings of the International Symposium MTNS-89, Volume II
*M.A. Kaashoek, J.H. van Schuppen, and A.C.M. Ran*

PSCT5 Signal Processing, Scattering and Operator Theory, and Numerical Methods: Proceedings of the International Symposium MTNS-89, Volume III
*M.A. Kaashoek, J.H. van Schuppen, and A.C.M. Ran*

GPSR Compliance

*The European Union's (EU) General Product Safety Regulation (GPSR) is a set of rules that requires consumer products to be safe and our obligations to ensure this.*

*If you have any concerns about our products, you can contact us on ProductSafety@springernature.com*

In case Publisher is established outside the EU, the EU authorized representative is:

Springer Nature Customer Service Center GmbH
Europaplatz 3
69115 Heidelberg, Germany

**Batch number: 10725962**

Printed by Printforce, the Netherlands